(17-6) $I_p = 1.414 I_{rms}$

(17-7) $V_{p-p} = 2.828 V_{rms}$

(17-8) $I_{p-p} = 2.828 I_{rms}$

(17-9) $V_{avg} = 0.637 V_p$

(17-10) $I_{avg} = 0.637 I_p$

(17-11) $I_p = \dfrac{V_p}{R}$

(17-12) $I_{p-p} = \dfrac{V_{p-p}}{R}$

(17-13) $I_{rms} = \dfrac{V_{rms}}{R}$

(17-14) $V_{avg} = I_{avg} \times R$

CHAPTER 18

(18-1) $V_{instantaneous} = V_{maximum} \times \sin\phi$

(18-2) $f(Hz) = \dfrac{\text{number of magnetic poles} \times \text{rotational speed (rpm)}}{120}$

(18-3) $V_L = V_p$

(18-4) $I_L = I_p \times 1.73$

(18-5) $V_L = V_p \times 1.73$

(18-6) $I_L = I_p$

CHAPTER 19

(19-1) $k = \dfrac{\text{flux lines between } L_1 \text{ and } L_2}{L_1 \text{ flux}}$

(19-2) $L_m = k\sqrt{L_1 L_2}$

(19-3) $\text{turns ratio} = \dfrac{N_P}{N_S}$

(19-4) $\dfrac{V_P}{V_S} = \dfrac{N_P}{N_S}$

(19-5) $V_S = \dfrac{N_S}{N_P} \times V_P$

(19-6) $\dfrac{I_P}{I_S} = \dfrac{N_S}{N_P}$

(19-7) $P_P = I_P \times V_P$ and $P_S = I_S \times V_S$

(19-8) $I_S = \dfrac{N_P}{N_S} \times I_P$ and $V_S = \dfrac{N_S}{N_P} \times V_P$

(19-9) $P_S = V_P I_P = P_P$

(19-10) $\text{efficiency} = \dfrac{\text{output}}{\text{input}} \times 100$

CHAPTER 20

(20-1) $R_T = R_L + R_{load}$

(20-2) $I_T = \dfrac{V_T}{Z_T}$

(20-3) $X_L = \dfrac{V}{I}$

(20-4) $X_L = 2\pi f L$

(20-5) $L = \dfrac{X_L}{2\pi f}$

(20-6) $X_{LT} = X_{L1} + X_{L2} + X_{L3} + \ldots + X_{Ln}$

(20-7) $\dfrac{1}{X_{LT}} = \dfrac{1}{X_{L1}} + \dfrac{1}{X_{L2}} + \dfrac{1}{X_{L3}} + \ldots + \dfrac{1}{X_{Ln}}$

(20-8) $V = I \times X_L \qquad I = \dfrac{V}{X_L} \qquad X_L = \dfrac{V}{I}$

(20-9) $I_{avg} = \dfrac{V_{avg}}{R}$

(20-10) $I_{rms} = \dfrac{V_{rms}}{R}$

(20-11) $I_{pk} = \dfrac{V_{pk}}{R}$

CHAPTER 21

(21-1) $Z^2 = R^2 + X_L^2$

(21-2) $I_T = \sqrt{I_R^2 + I_{XL}^2}$

(21-3) $Z_T = \dfrac{V_T}{I_T}$

(21-4) $Q = \dfrac{X_L}{r_i} = \dfrac{2\pi f L}{r_i}$

(21-5) $Q = \dfrac{P_L}{P_{ri}} = \dfrac{I^2 X_L}{I^2 r_i} = \dfrac{X_L}{r_i} = \dfrac{2\pi f L}{r_i}$

(21-6) $Q = \dfrac{X_L}{R_e}$

(21-7) $V_L = \dfrac{L \Delta i}{\Delta t}$

CHAPTER 22

(22-1) $C = \dfrac{Q}{V}$

(22-2) $C = 8.85 \times 10^{-12} K\left(\dfrac{A}{D}\right)$

(22-3) $C_T = C_1 + C_2 + C_3 + \ldots + C_n$

(22-4) $V_T = V_{c1} = V_{c2} = V_{c3} = \ldots = V_{cn}$

(22-5) $\dfrac{1}{C_T} = \dfrac{1}{C_1} = \dfrac{1}{C_2} + \dfrac{1}{C_3} + \ldots + \dfrac{1}{C_n}$

(22-6) $C_T = \dfrac{C}{n}$

(22-7) $V_T = V_{c1} = V_{c2} = V_{c3} = \ldots = V_{cn}$

(22-8) $V_C = \left(\dfrac{C_T}{C_x}\right) V_T$

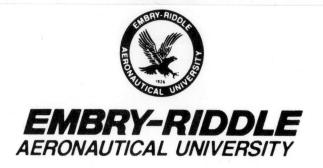

EMBRY-RIDDLE
AERONAUTICAL UNIVERSITY

3200 NORTH WILLOW CREEK ROAD • PRESCOTT. ARIZONA 86301

DATE DUE

AUG 0 3 1995			
AUG 2 3 1995			
OCT 1 9 1995			
MAR 0 6 1996			
DEC 0 1 1996			

ELECTRONICS FUNDAMENTALS
Circuits and Devices

ELECTRONICS FUNDAMENTALS
Circuits and Devices

Joel Goldberg

Macomb Community College, Warren, MI

Prentice-Hall, Englewood Cliffs, NJ 07632

Library of Congress Cataloging-in-Publication Data

Goldberg, Joel
  Electronic fundamentals.

  Includes index.
  1. Electronics. 2. Electronic apparatus and
appliances. 3. Electronic circuits. I. Title.
TK7816.G65 1988   621.381   87-7172
ISBN 0-13-251323-4

Interior design: Jayne Conte
Cover design: Jayne Conte
Manufacturing buyer: Peter Havens
Cover photograph: Steven Hunt, The Image Bank

 © 1988 by Prentice-Hall, Inc.
A Division of Simon & Schuster
Englewood Cliffs, New Jersey 07632

Printed in the United States of America

10 9 8 7 6 5 4 3 2

ISBN 0-13-251323-4

PRENTICE-HALL INTERNATIONAL (UK) LIMITED, *London*
PRENTICE-HALL OF AUSTRALIA PTY. LIMITED, *Sydney*
PRENTICE-HALL CANADA INC., *Toronto*
PRENTICE-HALL HISPANOAMERICANA, S.A., *Mexico*
PRENTICE-HALL OF INDIA PRIVATE LIMITED, *New Delhi*
PRENTICE-HALL OF JAPAN, INC., *Tokyo*
SIMON & SCHUSTER ASIA PTE. LTD., *Singapore*
EDITORA PRENTICE-HALL DO BRASIL, LTDA., *Rio de Janeiro*

This book is dedicated to my parents and to my parents-in-law. The influence and examples set by these people have been a force that has encouraged me to exert the effort required to write several text books. Without their support the course of my life would not have been in education and the potential talents they recognized would not have been developed.

Contents

7

Parallel Circuits 176

8

Series–Parallel Circuits 204

9

Voltage and Current Dividers 246

10

Kirchhoff's Laws 272

11

Network Theorems 290

12

DC Meters 324

13

Chemical Energy Sources 360

22

Capacitance 560

23

Capacitive Reactance 592

24

Capacitive Circuits 612

25

RC and L/R Time Constants 630

26

AC Circuits 652

27

The J Operator and Complex Numbers 674

Appendices 858

Index 897

Preface

Electronics Fundamentals: Circuits and Devices is an entry-level text which covers the essential topics in DC and AC circuits and electronic devices. The text is composed of 33 chapters, each developing a particular concept as it relates to the basic theories. Mathematics is limited to the use of algebra and basic trigonometry. Each chapter is short in length (approximately 20–30 pages) so that the student is not overwhelmed by the amount of material presented. This also allows the instructor flexibility in the sequence of topics covered.

Helpful Learning Aids for the Student

- Functional use of 2-color throughout
- Frequently used equations are printed on the endpapers for quick reference.
- Chapter opening objectives
- 'Real-world' chapter opening vignettes
- Troubleshooting applications are integrated throughout and are highlighted by color tints
- Review Questions appear within each chapter
- Section Reviews are presented after several chapters
- "Helpful Hints" are included in each chapter and are designated by this symbol ☼. It indicates to the student that this is additional helpful information
- Over 1,000 illustrations
- Over 300 examples are boxed and highlighted
- Key terms are *OUTLINED* at the beginning of each chapter and are *REVIEWED* in the margins of the text pages
- End-of-chapter material includes a Summary, True/False Questions, Multiple-Choice Problems, Essay Questions, Practice Problems, and occasionally Troubleshooting Problems
- A Math Review Supplement which reviews basic math procedures, including the use of the calculator, is shrink-wrapped to every text. *Free*
- End of text Glossary
- Computerized Student Study Guide

Organization

The author chose to cover the introduction of resistors and conductors prior to Ohm's Law and circuit analysis because the relationships of resistances is included in many of the initial laboratory assignments as well as the use of resistance as a load in many electrical circuit examples. This choice is based on the idea that a student must first understand the tool he/she is working with before the theory related to its use can be learned.

Chapter 10, Kirchoff's Laws, received excellent reviews because the author's presentation of the laws and his approach to the mathematical analysis of these circuits is clear and concise. The logical sequence in which the mathematical information is presented helps to break down the barriers normally experienced in understanding the common circuit analysis laws. This chapter is devoted to the use of DC circuits for analysis, while in Chapter 28, the same laws are explained using AC circuit values. This material follows the introduction of AC circuits and terminology.

The meter is presented in Chapter 12. In laboratory exercises, students often have problems understanding why the current meter is placed in series with the circuit rather than parallel to it. It is also difficult, sometimes, for students to understand circuit loading and the concepts related to the basic meter. The information in this chapter resolves these concerns and will probably go a long way to save accidental destruction of meters during laboratory activities.

Chapter 13 explains DC sources. The material in this chapter explains cell construction, including the newer nickle-cadmium and lithium cells. The use of series and parallel cell connections is also discussed as well as combination series-parallel cell connections. The concept of internal resistance is introduced and explained at this time.

Inductive Coupling and Transformer Action, Chapter 19, has been noted by the reviewers as an outstanding chapter. It takes a simplistic approach to teaching transformer relationships and transformer loading

The Appendices cover many relevant topics such as schematic symbols, a glossary of terms, metric conversion rules, and basic formulas for series and parallel circuit analysis. Also included is material on how to identify and replace chip components which are currently used in industry.

A separate section is devoted to troubleshooting electrical and electronic circuits. The final section contains answers to selected chapter questions.

Development Process

Over 35 reviewers were commissioned in the development of this text. These reviewers aided in the development of refining the necessary topics and confirming the accuracy of the text.

Extensive Teaching/Learning Package

■ *Lab Manual:* Written by Capon/Warner of Suburban Technical School; Includes over 30 class-tested labs and incorporates information on drawing schematics and writing lab reports. (25136-3)

- *Instructor's Manual to Lab Manual:* Contains suggested lab results. Free upon adoption (25139-7)
- *Instructor's Manual to the Text:* Provides fully worked-out solutions to the problems within the text. Free upon adoption (25135-5)
- *Instructor's Transparency Masters:* Over 150 transparency masters to aid in class lectures. Free upon adoption (25140-5)
- *Math Review Supplement:* Provides a review of the necessary math functions for a beginning DC/AC circuits course. Includes step-by-step calculator examples throughout. Free upon adoption of the text.
- *Instructor's Test Item File:* Provides the instructor with an additional 500 problems for class/test use. Free upon adoption (25141-3)
- *Computerized Student Study Guide:* Authored by Henderson of Computer Processing Institute. Allows the student to obtain an extensive review of each chapter along with 20 additional problems per chapter. These problems will be available on a disk so your students can become familiar with the use of the computer while reviewing pertinent material in DC/AC circuits. This Study Guide will also contain an appendix on booting up your computer and an introduction to BASIC programming. Saleable (25142-1)
- *Video Tape:* One free video to the school, upon adoption of *Electronics Fundamentals: Circuits and Devices*. Prentice-Hall will give you a video tape entitled "Technical BASIC for DC/AC Circuits." This ½" VHS video tape (approximately 120 min.) will include the following: Booting Up Your Computer, An Overview of Basic Programming, Basic Programming Applications as it applies to DC/AC Circuits, Debugging Your Program, and Troubleshooting Your Program.

About the Author

The final analysis of any textbook is the author's ability to present information in a clear, concise, and interesting fashion. The author has over 25 years of teaching experience in electrical and electronics theory and is active in the evaluation process of electronic educational programs used at a variety of schools throughout the United States. He has been active in the field of electronics communications, as an advanced class amateur radio licensee and as the owner and active operator of a wholesale electronics parts business for nine years. These combined experiences have served to hone the author's expertise in the basic concepts and theories related to electricity and electronics as well as his ability to communicate this material effectively. As a teacher, the author understands that students wishing to enter the field of Electronics are sometimes turned away because of the intense level of math and the high concentration of theory without practical application as it is presented in traditional electronics curricula. It is with this in mind that *Electronics Fundamentals: Circuits and Devices* and its accompanying supplements has been developed.

Joel Goldberg has written several other texts with Prentice-Hall which are directly related to electrical and electronics education. They are as follow:

- *Radio, Television, and Sound System Repair: An Introduction,* 1978
- *Fundamentals of Electricity,* 1981

- *Fundamentals of Television Servicing, 1982*
- *Fundamentals of Stereo Servicing, 1983*
- *Satellite Television Reception: A Personal User's Guide, 1984*
- *Electronic Servicing of Robotic Equipment, 1985*

Credits

A book cannot be written without the aid of many other people. This is very true in the development of this book. The author wishes to acknowledge the help and support of many of his colleagues during the book's preparation. He is particularly grateful for the assistance and suggestions from the editorial support staff of Prentice-Hall. Finally, this project, or any other of its magnitude, could not begin or be completed without the support of his family. The author wishes to acknowledge the support of his wife, Alice, for her aid in the preparation of the manuscript, and her dedication to the use of the word processor for inputting the written material into the computer and her understanding and forgiveness of the additional hours required to make this undertaking develop into a finished document.

Joel Goldberg
and
Prentice-Hall

Acknowledgments

My appreciation and thanks are extended to the instructors who reviewed and contributed greatly to the text and lab manual. They are as follows;

George Andrews, Lee County Area Vocational Technical Institute
Don Barrett, Jr., DeVry Institute of Technology
Richard Bridgeman, DeVry Institute of Technology
Anthony R. Canniff, Pennco Technical School
Robert Capon, Suburban Technical School
Gerard Colgan, Bowling Green State University
Vincent Coupal, Eastern Maine Vocational Technical Institute
Matthew Deady, Mt. Holyoke College
Robert W. Derby, DeVry Institute of Technology
Howard Duhon, Lee College
Stanley L. Eisenberg, DeVry Institute of Technology
Albert J. Gabrysh, St. Phillips College
John Hart, DeVry Institute of Technology
Thomas Koryu Ishii, Marquette University
Kenneth L Jauch, William Rainey Harper College
Fred S. Kerr, DeVry Institute of Technology
Leonard Laabs, Walla Walla College
Wallace A. McIntyre, DeVry Institute of Technology
William E. Mullins, Middle Tennessee State University
Terrence D. Nelson, Lincoln Technical Institute
Richard Parrett, ITT Technical Institute
L. Kirk Ray, DeVry Institute of Technology
Andrea Rutherford, DeVry Institute of Technology
Warren B. Thiers, Brevard Community College
Frank Voeffray, Macomb County Community College
Kenneth Whitehead, Chipola Jr. College
James D. Wilkes, Heart of Georgia Area Vocational Technical Institute
Kenneth C. Williams, ITT Technical Institute
Edmund D. Wilson, St. Petersburg Vocational Technical Institute

Joel Goldberg

ELECTRONICS FUNDAMENTALS
Circuits and Devices

CHAPTER 1

Background and History

Perhaps the greatest influence that any human finding has had on our daily lives began with the discovery of electricity. Actually, the greatest influence was not its discovery but the ways that were developed to use electrical energy to make our lives more comfortable. In both home and workplace, electrical energy is used instead of human muscle power in the performance of daily work. A brief review of the background and history of electrical energy development is essential to all of us who use this energy to function in today's world.

One example of the importance of electrical energy is the use of satellites orbiting the earth for communications purposes. We now have the capability of almost instant communication from and to any section of the earth's surface using communication satellites as relay stations for the broadcasting of television programs, telephone conversations, and all forms of data transmission. Operation of these satellites is totally dependent on the use of electrical energy. This is only one major application of electrical energy as it affects our daily lives.

KEY TERMS

Ampere   **Ohm's Law**
Capacitance   **Robotics**
Henry   **Volt**

OBJECTIVES

Upon completion of the material in this chapter, you will

1. Know the history of the development of electrical and electronic concepts.
2. Understand the application of electrical and electronic discoveries.
3. Realize the impact of the discoveries on our present life-style.

The discovery of electricity was one of the most significant events in the history of the human race. Consider, if you will, that this discovery occurred less than 400 years ago. Compare this relatively short span of time with the thousands of years that earth has been occupied by human beings. Also compare the evolution of technology over the past 400 years. When we begin to relate the approximate date and time of the various discoveries in the field of electricity to our present level of technological sophistication, the realization of the speed at which we are improving our life-style becomes apparent.

Consider the type of life that you would have if we did not have electrical energy. Knowledge of what is happening in various part of the world would not be known until days or weeks after it occurred. We would not have the use of the telephone or telegraph. Television would not exist. Travel would be limited to carts or wagons pulled by animals. We could possibly have some steam-powered vehicles, but even then, travel would be limited by the need to stop often for fuel and water. Air travel would be limited to the use of lighter-than-air craft. Certainly, your imagination could add a great deal to this list.

Fortunately for us, we have discovered the applications of electrical energy. We use it to perform heavy work such as manufacturing and to provide motor power for machines. We also use electrical energy to perform smaller amounts of work. We have upgraded and developed one branch of the science of electricity and have called this branch *electronics*. As we know and use the term "electronics," it often relates to the controlling portion of an electrical circuit. Electronics plays a significant role in our ability to communicate with each other. The development of vacuum tubes, and later transistors and integrated circuits, led to the production of some very sophisticated electronic products.

Since we are fortunate to have the advantages of electrical energy, let us look at the short span of time from its discovery to the present. Let us also identify some of the people involved in the discovery of electrical concepts. These people are important because their names are often related to the concepts they discovered and presented to the world. Some of these names should be familiar to you, whereas others are obscure and hardly mentioned in our textbooks; but they are all important to us. As you read this section, consider the number of years that passed from the date of the original concept's discovery to the date when the next significant discovery occurred.

An English physician, William Gilbert, discovered the properties of magnetism in approximately the year 1600. His discoveries related to the fact that substances amber and lodestone would attract different types of materials. Gilbert used the Latin word *electron* to represent "amber" and the word *electrical* for other materials that acted like amber.

A second Englishman Stephen Gray, discovered that some substances were able to conduct electricity and other substances did not conduct electricity. Gray, who lived from 1696 to 1736, felt that there were two types of electricity. One was called *resinous* and the other, *vitreous*. This led to the discovery of the two types of electrical charges that we know today as *positive* and *negative*.

Actually, the terms "positive" and "negative" were first used by Benjamin Franklin. His studies were conducted during the eighteenth century. Franklin felt that electricity was a fluid. Among other things, he is noted for his experiments flying a kite during an electrical storm. Fortunately for all of us, he did not get electrocuted while conducting these tests. He did discover that lightning is a form of electrical energy, which was a great surprise to the scientific world.

Experimentation with electricity was not limited to the United States and England. It was also being conducted on the European continent. Charles Coulomb, a Frenchman, suggested the laws that explain the relationship between positive- and negative-charged bodies. His work was published in 1785. The unit of electrical charge is called the *coulomb* in his honor.

During the same period, an Italian named Luigi Galvani was experimenting with current electricity. His work was published in 1786 and is the basis of our present understanding of electrical current movement. Some types of electrical measuring meters in present-day use are called *galvanometers*, a name that is a carryover from early days, when electrical current was called "galvanism."

Another Italian physicist, Alexander Volta, discovered that the chemical action between moisture and two dissimilar metals would produce an electrical potential. His work was done in the latter half of the eighteenth century. Volta constructed the first battery. It was made with layers of copper and zinc plates. The plates were separated with layers of paper that had been moistened with salt water. This battery, called a *voltaic pile,* provided the first constant source of electrical current. Today we identify the unit of electrical pressure as the **volt** in honor of Volta.

Volt *Difference in electrical pressure between two points in a circuit.*

During the early nineteenth century a Danish scientist discovered the properties of electromagnetism. This man, Hans Oersted, found that an electrical current that moved through a wire would make a compass needle move from its normal position. This work established that magnetic fields existed around a wire in which a current flowed. Units of magnetism are called *oersteds.*

In 1800 another significant discovery was made. André Ampère, a Frenchman, measured the magnetic effect of the electrical current. He found that wires that carry an electrical current can either repel or attract each other in a manner that is similar to the effects of permanent magnets. Units of electrical current are discussed in terms of the **ampere.**

Ampere *Unit of electrical current flow.*

The relationship of voltage, current, and resistance was proposed by a German, Georg Simon Ohm, in 1826. His work is the basis for one of the best known and widely used rules in electrical theory, called **Ohm's law.**

Ohm's law *Relationship of voltage, current and resistance in an electrical circuit.*

Also having a significant impact on the world of electricity was the work of Joseph Henry, Heinrich Hertz, James Maxwell, and Michael Faraday. In 1831, Henry and Faraday, working independently, discovered the laws of electromagnetic induction, a principle based on the work of Oersted. Both Henry and Faraday found that electricity could produce magnetism. This discovery, the exact opposite of the results of Oersted's work, is the basis for the rule which states that a moving magnet will induce a voltage in a wire that is placed in the field of the magnet. Units of electromagnetic inductance are called **henrys**. Work related to the effect of the electrical fields that exist between two charged bodies is called **capacitance**. Units of capacitance are termed *farads*, in honor of the work done by Faraday.

Radio and television wave laws were proposed by Maxwell and by Hertz in the late nineteenth century. Maxwell suggested that electromagnetic waves that traveled at the speed of light (186,000 miles per second) could be produced. These waves were, then, actually produced by Hertz. This work is the foundation for all radio and television communication. Electromagnetic waves used for communications are known as *Hertzian waves*. The unit of measurement for the frequency of these waves is the *hertz*, or "cycle per second." We use the phrase "60 hertz" instead of "60 cycles per second" to describe one common frequency or repetition rate of an electrical wave.

Science moved from the development of electrical theories to the identification of electronic devices starting around the year 1900. The electron as we know it today was discovered by Jean Perrin in the 1890s. Edison patented a vacuum-tube diode but never put the tube to work. Fleming used the vacuum-tube diode to detect electromagnetic waves. We could say that this heralded the age of vacuum-tube electronics.

Application of the vacuum tube was limited until another discovery took place—the addition of a third active element to the tube. This work, done by Lee deForest, produced a tube known as a *triode*, that was used to amplify an electrical signal. This work led to the development of a transcontinental telephone system in the United States. Further work by deForest and Edwin Armstrong gave us the FM radio and many other sophisticated electrical devices. Additonal developmental work by others improved the vacuum tube's efficiency. In 1920 the first television picture tube was developed. This tube was called a *kinescope*. Further developments in the 1930s gave us the *magnetron tube*, used for microwave applications, including microwave cooking, satellite communications, and radar.

Recently it has been shown that some of the discoveries attributed to Edison and Hertz were really those of Nikola Tesla. Tesla is now credited with development of the alternating-current distribution system we use today for virtually all of our electrical power. He opposed Edison's concept of using direct-current electricity for powering appliances and industrial equipment. The radio as we know it was also developed by Tesla. He used radio-controlled ships to demonstrate this new concept before Marconi made his first radio transmission. In addition to these developments, Tesla invented the radar system we use today for navigation and location of vessels. Despite all of these inventions, Tesla's work failed to receive recognition until late in 1984.

The development in 1947 of a solid-state device, the *transistor*, opened the door to an entirely new world of electronics. Transistor applications reduced the size of electronic devices considerably. These applications also reduced the power requirements for electronic devices—an outstanding example of this being the computer. The first electronic digital computer, which used vacuum tubes for operation, occupied a large room. It also required a large air-conditioning unit to keep it cool enough to operate. Today we have personal computers that one can carry and hold on one's lap to operate. These sophisticated computers cost far less, require about one one-hundredth of the operating power of the first computers, operate at much greater speed, and have huge memory systems compared to that of the first computers.

The production of small, high-speed computers was aided by still another electronic device, the *integrated circuit* (IC) developed in the 1960s. This device now plays an enormously significant role in our lives.

Before we review some applications of electronic devices, let us take a look at some of the time lines involved in the history of the development of electrical and electronic devices. Development started in the year 1600 with the discovery of electricity, which opened the door for further work. Applications of electricity were not as rapid as its discovery. Faraday built an experimental electric motor in 1829, over 200 years after the discovery of electricity. Another 59 years passed before the first practical electric motor was produced and put to work. Compare this with the time involved in the development of the transistor: Less than 5 years passed from its invention until it found a practical use in electronic devices.

The following list outlines some of the "who, what, and when" behind important electrical discoveries. Some of the later discoveries are not identified with a single person, but rather, represent team efforts, and in some cases the discoverer is unknown.

Date	Individual	Discovery/Application
1600	Gilbert	Electromagnetism
Late 1600s	Gray	Electrical charges
1700s	Franklin	Fluid electricity
1785	Coulomb	Charged bodies
1786	Galvani	Current electricity
Late 1700s	Volta	Voltaic pile battery
Early 1800s	Oersted	Electromagnetism in wires
1800	Ampere	Current in wires
1826	Ohm	Relationship of voltage, current, and resistance
1829	Faraday	Capacitance
1831	Henry	Electromagnetic induction (motor)
Late 1800s	Maxwell	Electromagnetic propagation
Late 1880s	Hertz	Radio waves
Late 1800s	Tesla	Radio transmission, ac power
Early 1900s	Edison	Vacuum-tube diode
Early 1900s	deForest	Vacuum-tube triode
Early 1900s	Armstrong	FM radio
1920		Television tube
1930s		Magnetron tube
1947		Transistor
1960		Integrated circuits

This rapid technological advance has a direct effect on our lives. It is significant for those who wish to enter the work force. The rapid development and application of new technology forces us to learn basic concepts and their applications. The basic concepts really do not change in the field of electricity and electronics; only the applications of these concepts have changed.

Rather than continuing to compare data, let us look at several current developments in the field of electronics. The world of electronics has crept into almost every aspect of our lives. There are very few areas left in which electronics does not appear in one form or another. Probably the most familiar area is the home. In the line of consumer products we can identify television, radio, tape decks, and personal computers. Consider also digital watches and clocks and electronic controls for conventional and microwave ovens, washing machines, dryers, and heating systems. Continue with a burglar alarm, the telephone, and the electronic games we work through our television receivers. Enter the garage, which may be equipped with an automatic door opener, and look at the electronics used in late-model cars. Radio, tape deck, two-way citizens band radio, and electronic ignition are just a few of the devices now in common use. Some cars use electronic digital readout dashboard instruments to replace the conventional electromechanical devices used in earlier models. Many cars also have a microprocessor-controlled emission system, trip mileage indicators, and control systems that speak to the operator when things are out of their normal operating order.

Another area that affects our lives directly is related to electronic communication. Telephones, radio, and television are accepted as normal in most of our homes. The technology used to get this service to the home has changed dramatically in the past few years. Television signals, a form of electromagnetic energy, are transmitted from earth to a stationary satellite. The satellite is 22,300 miles above the earth and centered over the equator. It rebroadcasts the television signal back to the earth. Further processing is carried out by a local TV broadcast station or a cable TV company. Finally, the signal is received and processed by your TV receiver. Direct broadcasts from satellite to individual homes have begun as well.

Personal communications has also taken a giant step forward. The availability of citizens' band radios provided thousands of U.S. citizens with another form of communications. Cordless telephones opened still another communications channel for the average person. Now we also have cellular radio-telephones to use in our cars. Boaters welcomed the availability of two-way radios, depth finders, fume detectors, and navigational aids.

Data processing as used by business and industry permits accurate and rapid record keeping. Billings are done faster and money collected more quickly since computers have been available for data processing. The success of many businesses today is directly dependent on the computer.

Industry has extensively incorporated the use of electronic control devices and computers. One of the latest forms of such computer control is **robotics**. In this system a microprocessor computer controls the manufacturing processes of a machine. Complete systems using computer control are called *process control systems*. In this format the computer is used to coordinate and control an entire process in the manufacturing field. This could be the process of producing gasoline from oil, production of an assembled automobile or engine, or one of many other applications.

Robotics *Use of computer to control manufacturing processes.*

Another area that has benefited greatly from developments related to electronics is the field of medicine. Electronic units display electrical waveforms of the heart and brain, computer scanning devices are able to evaluate the state of each organ of the body, and heart pacemakers supplement the normal impulses required for life-sustaining heartbeats. A visit to any hospital emergency room, intensive care unit, or cardiac care unit will disclose an unbelievable quantity of electronic equipment in use or available for use.

The applications of electrical and electronic devices is endless and their influence is profound. Uses for these devices will continue to expand as long as we are able to produce electrical energy. Those who are interested in learning about this fascinating subject will find that the basic rules discovered in the seventeenth and eighteenth centures are still valid. The development of the transistor is based on the concepts established by Ohm, Ampère, and Volta. These rules do not change with technological growth; it is the applications of these rules that have changed. All who wish to learn about this subject ragardless of occupational level or emphasis, need to have a clear understanding of each of the concepts that make up the field of electronics. Engineers who design sophisticated electronic devices require and use these basic rules constantly. Technicians working as part of an engineering design or application team use the same body of knowledge as do field service engineers or technicians. The hobbyist–home experimenter uses the same concepts and rules when building circuits.

The body of facts that we identify as electricity and electronics is requisite knowledge for all who have any interest in this field, either for work or for pleasure. The concepts have remained the same for several centuries, but the applications have changed radically. However, the basic fact that an applied voltage causes a current to flow in a circuit has not been altered. These factors still produce some amount of work, or make something happen.

Electrical concepts must be learned thoroughly, and once learned, we must be able to apply them to any electrical or electronic circuit or system. This fact is basic for all who wish to be successful in the field. The only way to be successful is to be able to apply the basic rules and concepts to evolving technological changes. This knowledge helps the practitioner stay up to date despite increasingly rapid changes. It is almost impossible to know and remember everything about electricity and electronics. The only way to stay current technically is to be able to apply the basics to the latest technological applications. With this statement in mind, the author wishes success to all who enter this exciting field. Whether it is for employment or for hobby purposes, electronics is a fascinating field. You should be able to enjoy the subject no matter what your learning goals might be. Do not be afraid to question and seek answers. Learning is a two-way street: Traffic, in both directions—from learner to teacher and from teacher to learner—is necessary for success. Many years ago a man named Elbert Hubbard stated: "The best service a book can render you is, not to impart truth, but to make you think it out for yourself." With this quotation in mind, you should be ready to start your journey to learn more about the fascinating field of electricity and electronics. As you progress, keep in mind that the person who is able to apply this knowledge is the one most likely to be successful.

CHAPTER 2

Nature of Electricity

Electrical energy is a force that we cannot see or feel. We do, however, have the ability to create this energy from some other form and to utilize it in the performance of some type of work. Our normal daily life-style requires an almost constant use of electrical energy. We normally do not even consider how it is produced or how it is made available at the electrical outlets in the walls of our buildings.

Try to imagine your daily life without electrical energy. It would be similar to life-styles before the early nineteenth century. There would be no telephones, radio, television, automobiles, or airplanes. Communications between cities, or even from one side of a city to the other, would depend on the speed of a horse, or a person walking, to deliver a message. Today, we use the telephone or a two-way radio to accomplish the same thing in fractions of a second.

KEY TERMS

Alternating Current	Law of Charges
Ammeter	Load
Ampere	Negative
Conductance	Nonconductor
Conductor	Ohm
Conventional Current	Open Circuit
Coulomb	Photoelectric
Direct Current	Positive
Electric Circuit	Protons
Electric Field	Resistance
Electromotive Force	Semiconductor
Electron Current	Short Circuit
Electron Flow	Sine Wave
Electrons	Static Electricity
	Voltage

OBJECTIVES

Upon completion of the material in this chapter, you will

1. Understand the nature of electricity.
2. Be able to discuss insulators, conductors, and semiconductors.
3. Be able to use basic electrical terms: voltage, current, resistance, ac and dc.
4. Understand how electrical energy is created.

We all seem to accept the fact that we have electricity available for use. We usually do this without thought about how or where this electrical energy is developed. We also expect to have it available in the proper quantities and types for our use. When a wall switch in a home is turned on, we expect the light to be turned on also. Rarely do we think of how, why, or where the electrical energy is produced. The material presented in this chapter covers the fundamentals of electricity and the terms, formulas, and values used in this field.

2-1 Polarity and Charge

Few of us have not had the experience of replacing some type of battery, either in a portable electronic unit or in an automobile. Each terminal of the replacement battery is marked with a polarity sign, either positive ($+$) or negative ($-$). The unit in which the battery is placed has similar signs. To operate correctly, the battery must be installed with its positive terminal connected to the positive terminal of the unit.

Batteries used in the electrical/electronics industry are illustrated in Figure 2-1(a). Each battery has polarity markings at or near its terminals. When only one polarity symbol is provided, the user is to assume that the unmarked terminal is of opposite polarity from the marked one. The graphic symbols used to represent batteries on an electrical schematic or diagram are shown in Figure 2-1(b). The basic symbol is often used to represent any power source that provides power in the same form as a battery. Often, numbers that represent an operating voltage value are also provided. The alternate symbol is similar to the basic one except that there are multiple sets of lines. The longer of the two lines is used to represent the positive terminal in both sets of symbols. The shorter line indicates the negative terminal.

Polarity signs have significance to those working with electricity. All known materials have electrical charges, based on the chemical makeup of the material. There are two types of charges, **protons** and **electrons**. The proton, a part of the atom of a material, has a positive charge. The electron, also a part of the atomic structure of a material, has a negative charge.

You may be wondering how a material can be classified as either positive or negative if it contains both protons and electrons. The answer is that the polarity of a given substance depends on the number of electrons contained in the electron orbital rings that surround its nucleus. When an atom of copper, for instance, loses the one electron in its outer, or valence, ring it becomes a posi-

Protons *Positively charged particles found in nucleus of natural elements.*

Electrons *Negatively charged particles found in all elements.*

(a)

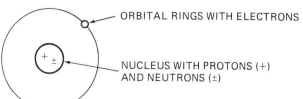

(b)

FIGURE 2-1 Typical chemical cells, or batteries, used in electrical/electronic work. All use a similar graphic symbol. Often a multicell battery is shown by indicating its voltage level. (Photo courtesy of Eveready Battery Company.)

tively charged ion. In this state it will attract an electron from another atom of copper. It can also attract free electrons from any other atom of material. Copper becomes an excellent conductor of electrons when an external force is applied to it.

When the number of electrons in the outer ring of an atom of matter is less than half of the number required to completely fill its outer ring it is considered to be a conductor. This occurs because the atom will give up these electrons easily.

If the outer orbital ring of the atom is over half filled with electrons it does not readily give these up. In this condition the atom or material acts as a nonconductor, or insulator.

2-2 Atomic Structure

All of the elements known to us have an atomic structure. This structure consists of a nucleus and orbital rings (Figure 2-2). The nucleus contains protons and neutrons, and the orbital ring(s) contain the electrons. Figure 2-2 shows the element we know as hydrogen. Hydrogen has an atomic weight of 1: It has one proton and one electron. Other elements have different numbers of protons

ORBITAL RINGS WITH ELECTRONS

NUCLEUS WITH PROTONS (+)
AND NEUTRONS (±)

FIGURE 2-2 The hydrogen atom has a nucleus and two electrons in its orbital ring.

and electrons. A second example is the element copper (Figure 2-3). Copper, one of the elements used most commonly in electrical systems, has 29 protons and 29 electrons, in its atomic structure.

The electrons are arranged in a very specific manner in every element. A series of electron orbital rings surround the nucleus of the atom. There is a maximum number of electrons that each orbital ring can contain. Each ring has to be filled before the next ring is formed. The copper element is typical of this arrangement. In an atom of copper the first orbital ring can hold 2 electrons, the second ring can hold 8, the third ring has a maximum of 18 electrons, and the outer ring has 1 electron, although this ring could hold up to 8 electrons. The number of electrons in the rings differs for each chemical element. When the outer ring of the atom of any element has half or less than the maximum possible number of electrons, the electrons can easily drift or move to another atom of any material. These electrons are known as *free electrons* because of their ability to move in a random pattern in the material. Copper is a good conductor of electrical energy because of the free electron in its outer ring. Materials that have free electrons are known as electrical **conductors.** They are considered to be *negative* in nature because of their ability to *give up their electrons*.

When the outer orbital ring of a material has over half of its orbital ring filled with electrons it is not considered to be a conductor. The difference between this type of material and others is that it has a tendency to *accept electrons* rather than give them up. This type of material is thus considered to be an insulator.

A material that has its outer orbital ring completely filled will neither accept nor give up electrons. This type of material is considered as a **nonconductor** of electricity. It is also known as an *insulator*. It will not be assigned any type of polarity sign. In the electrical field we find materials that are positive, negative, and neutral in their electrical properties. In the following section we look at some of these materials and see how they function.

Conductors *Material having ability to easily give up its electrons.*

Nonconductor *Material not able to accept or give up its electrons freely.*

FIGURE 2-3 The copper atom has several electron orbital rings. Each is complete except the outer one, which has one electron.

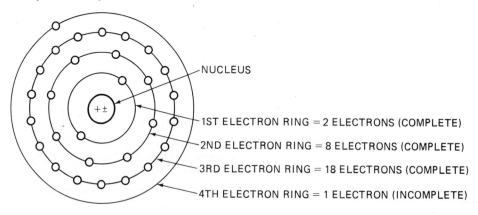

NUCLEUS

1ST ELECTRON RING = 2 ELECTRONS (COMPLETE)

2ND ELECTRON RING = 8 ELECTRONS (COMPLETE)

3RD ELECTRON RING = 18 ELECTRONS (COMPLETE)

4TH ELECTRON RING = 1 ELECTRON (INCOMPLETE)

2-3 Electrical Charge

Under normal conditions both paper and plastic are considered to be electrically neutral. If you comb your hair with a plastic comb and then hold the comb near a piece of paper, the paper will be attracted to the comb. The process of passing the comb through the hair has given the comb a polarity. The protons and electrons have separated, yielding a negative polarity. The result of this action has provided a force of attraction between the comb and the paper. This action is the basis for all electrical processes.

The charges are an example of *static electricity*. The creation of static electrical charges occurs in a variety of forms. These include the action of charging your body when you walk across a carpet containing wool or plastic material. When you touch a light switch plate or other metallic surface, you may receive an electrical *shock*, or discharge. While you were walking, your body has been charged with a quantity of positive charges. The metallic surface has a quantity of electrons. The difference between the amount of charges on the two surfaces produces the electrical discharge as the surfaces attempt to balance the number of charges. This type of electrical discharge is called **static electricity** because the electrical charges are not considered to be in motion. Although they are moving around the individual nucleus of their atoms, they do not have a specific direction to their movement.

Static electricity *Type of electricity considered to be stationary and not able to give up its electrons.*

The unit of electrical charge is called the **coulomb** (C). This is a basic measurement for electrical charge. The coulomb contains a rather large quantity of electrons. It is equal to a charge of 6.25×10^{18} electrons stored in an electrical material.

Coulomb *Unit of electrical charge.*

The symbol used to represent the quantity of electrical charge is the letter Q. In a practical example a charge of 6.25×10^{18} electrons is written as $Q = 1$ C. The charge contained in one electron is 1.6×10^{-19} C.

Rules for Electrical Charges Electrical charges can be identified as either positive or negative in nature. A positive electrical charge is written as $+Q$ and a negative electrical charge is written as $-Q$. The positive charge has an excess of protons. Another way of stating this is to say that this material has a shortage, or lack, of electrons—it is capable of accepting electrons. The negative charge has an excess of electrons—it will give these up to another atom under normal conditions. These facts are used to state two rules concerning the properties of charged electrical bodies, known collectively as the *laws of charges*.

Law of charges *Law stating like charges repel each other and opposite charges attract each other.*

Laws of Charges These laws identify the relationship between two bodies each of which has an electrical charge. We consider first the relationship between the bodies when they have the same charge and then their relationship when they have opposite charges.

Opposite Charges Attract Consider the situation where two lightweight balls are suspended from a string. The balls are close to each other but do not touch. One ball is given a positive charge and the second ball, a negative charge. Both charges are provided by an outside source. The two balls will be attracted to each other and will actually attempt to touch. This exemplifies the rule that two bodies that have opposite electrical charges have a tendency to be attracted to each other. Simply stated: Electrical charges of opposite polarity attract. Should the two charged balls touch each other, the charges will transfer from one to the other, resulting in a balance of charges. The charged balls become electrically neutral and are no longer attracted to each other.

Like Charges Repel Now, imagine the same two balls are both given a positive charge. Each ball contains the same quantity of charges. The result is that the balls will move away from each other. When both have an equal positive charge, they repel each other.

The two balls are caused to touch each other in order to neutralize their charges. They are then both charged negatively. The result is that, again, they move away from each other. When both balls have an equal negative charge, they also repel each other.

The results of these experiments display a fundamental fact regarding electrical energy. This is the *law of attraction and repulsion*. The law can be simply stated: *Opposite charges attract each other and like charges repel each other*.

Electric field *Area around a charged body.*

Electric Fields The examples describing the action and relationship of two charged bodies can be used to explain another electrical rule. The ability of two charged bodies to interact is due to the fact that each body creates an **electric field** around itself. The field is found all around the body. This electric field is equal in strength at each point that is equidistant from the body. If we were to measure the strength of the electric field $\frac{1}{4}$ in. from the surface of the body, the strength of the field will be the same at all places measured. When another measurement is taken at a point $\frac{5}{8}$ in. away from the body, the strength of the field will have diminished. The amount of this reduction will be great, as it is equal to the inverse of the square of the distance from the body.

Increasing the amount of the original charge on the body will produce a larger field around it. Conversely, reducing the amount of the original charge will decrease the measured values. The relationship of electrical charges on two bodies was identified by Coulomb and is his *law of electrostatic principles*. The various principles of electrostatic charges and their relationships are all used in electrical work, one of the principal uses being to store electrical charges on metallic plates. A detailed description of this process and its application is provided in Chapter 22.

REVIEW PROBLEMS

2-6. What is the name of the unit of electrical charge?
2-7. How many electrons are in the electrical charge?
2-8. State the laws of charges.
2-9. What is an electric field?
2-10. What happens to the strength of the electric field as it is measured at different distances from a charged body?
2-11. What is the relationship between a charged body and the strength of its electric field?

2-4 Electrical Voltage

In electrical terms we have both static electricity (described in Section 2-3) and current electricity. Whereas static electricity is a nonmoving electricity, current electricity (described in Section 2-5) is electricity in motion. There are times when the conditions exist for current electricity except that the system's switch has not been turned on. Under these conditions the *potential* exists for making current electricity. In other words, we have the ability to perform some type of useful function but that function is not occurring at this time.

A body is charged by performing some amount of work. The work is done when the body's electrons and protons are separated. Under normal conditions the protons and electrons are in balance and there is no positive or negative charge on the body. The opposite charges have a great desire to combine. The greater the number of charges that are separated, the greater their "desire" to recombine and return to their neutral state. Thus we have the ability to change the amount of electrical potential by varying the strength of the charges. This produces the potential for performing some work. The amount of work that can be accomplished is equal to the amount of work used to separate the charges initially.

The effect of differently charged bodies creates the potential difference between those bodies. Earlier it was stated that the charge on a body could have either a $+Q$ or a $-Q$ value. If we have one body that has a charge of $+10$ C and it is paired with a second body that has a charge of -4 C, we have a difference in potential of 14 C. When the first body has a charge of $+4$ C and the second body's charge is $+2$ C, the difference between them is only 2 C.

The quantity of charges that can be moved is directly related to the strength of the potential difference between two bodies. A greater difference is described as a larger value of potential difference between the two bodies. This can also occur between two wires or two terminals in any electrical system.

The repulsion or attraction of charged bodies is described as an electrical force. This force develops movement in the electrical charges. It provides the charges with a specific direction as well as movement. The potential electrical difference between two charged bodies is called the *electromotive force*, or EMF. The EMF in an electrical system is described in units of the *volt* (V).

Voltage is the amount of energy required to move an amount of electrical charge. This is expressed by the formula

$$V = \frac{W}{Q} \qquad (2\text{-}1)$$

where V = voltage value

W = amount of energy

Q = quantity of charges

Energy is expressed in units of the joule (J) and charge in coulombs (C). By definition: One volt is the amount of potential difference between two points when 1 joule of energy is used to move 1 coulomb of charge from one point to another.

Use of this definition requires some clarification and further identification of values. When any quantity of charges is moved, some amount of work is performed. The joule is the same as 0.7376 foot-pound of work. The foot-pound (ft-lb) is the scientific measure for performance of some work. In this definition the charges are moved from one point to another. The electrical force, or voltage, has accomplished this work. It is measured by comparing its value at two different points in the system. Voltage is the potential difference measured between two points in an electrical system. A battery is a source for voltage. Its terminals have a difference of potential, which is a measurable value. The potential difference, or voltage, is usually marked on the case of the battery. The potential difference of the battery terminals is the **electromotive force** available from the battery. The EMF applied to an electrical system is one factor in determining the quantity of charges, or electrical current flow, in the system.

Electromotive force *Quantity of voltage available to move electrons in a circuit.*

EXAMPLE 2-1

If 48 J of energy is used to move 12 C of charge through a wire, what is the voltage across the wire?

Solution

$$V = \frac{W}{Q} = \frac{48J}{12C} = 4 \text{ V}$$

EXAMPLE 2-2

Energy of 400 J produces 80 V measured across a conductor. How many coulombs of charge is used?

Solution

$$Q = \frac{W}{V} = \frac{400J}{80V} = 5 \text{ C}$$

Measurement of Voltage **Voltage** is defined as the *difference* in electrical potential between two points. When voltage is measured, the measuring device is used to compare the difference in potential between two points. A diagram showing an electric meter called a voltmeter is shown in Figure 2-4. This figure has several significant points. First, we should define an **electric circuit**. A minimum of three components are required for all electric circuits: a *source of energy*, a *load*, and connecting *wires*. The source of energy can be a generator, a solar cell unit, a chemical battery, or any other device used to produce electricity. The load is a unit that uses electrical energy to perform some type of work. Loads are devices that require electrical energy. Any device that fits this description is considered to be a load. Computers, radio broadcast transmitters, stereo receivers, machine tools, and electric razors are all examples of such devices. The connecting wires can be printed etched circuit boards, multiple-conductor cables, or the frame and body of an automobile. In other words, any conducting surface can be used to replace discrete wiring. The load for this circuit is shown as a box. Often the load is illustrated using the schematic symbol or symbols for specific electrical components. This type of symbolic drawing is discussed later in the book.

Often the schematic symbol for a meter, or measuring device, is a circle. Inside the circle is a letter. The letter represents the type of measuring meter used for the particular application. In Figure 2-4 the letter V is used to represent a voltage-measuring device. Electric meters have two wires or test probes. One is connected to each of the meter's two terminals. When any value is measured, both leads are required. Since voltage is defined as the difference in electrical pressure between two points in a circuit, one of the leads from the meter is placed at each of the two points. In a sense, a reading of 10 V is actually a measurement of a difference of 10 V between the two specified points in the circuit. The meter in Figure 2-4 shows this method of measuring voltage.

Voltage *Difference in electrical potential measured between two points in a circuit.*

Electric circuit *A complete path for electron flow, consisting of a source, a load, and connecting wires.*

2-5 Electric Current

The application of an EMF to a circuit causes the electrical charges in that circuit to move in a specific way. The movement of the electrical charges is

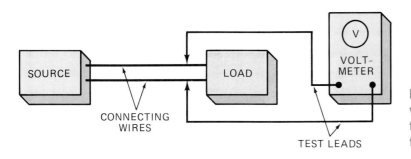

FIGURE 2-4 A device called a voltmeter is used to measure the potential difference between two points in a circuit.

called the *electric current*. A voltage must be applied to two different points in an electric circuit to produce a flow of electric current. There is a movement of free electrons in all materials. Until a difference of potential is applied, the electron movement is limited to a random drift. The application of voltage gives the electrons a very specific movement. When the voltage is applied to the conductor, electrons from the negative terminal of the voltage source are forced onto the conductor. The free electrons that were originally at the end of the conductor have to move. These electrons are forced away from the negative terminal and are attracted to the positive terminal of the voltage source. This fact is related directly to the laws of charges discussed in Section 2-3.

The movement of electrons through a conductor such as a copper wire occurs in a manner that is similar to a chain reaction. An electron does not enter at one end of the conductor and move through it to its other end. The first electron bumps into an atom of copper. The bumping action forces the valence electron in that atom of copper away from its spot in the orbiting ring. The first electron replaces the one that was bumped away. This second free electron is forced by the applied voltage to bump into a second atom of copper. It forces that atom's electron out of orbit and replaces it in orbit around the second atom. This bumping and replacing action occurs continuously until a free electron from the last atom of copper reaches the positive terminal of the power source. Here, the free electron combines with a positive charge. In a working electrical circuit this occurs for great quantities of electrons, not for just one or two electrons.

We can state that an electric current is a continuous flow of electrons. We can also say that this electron flow has a controlled direction. Further, it can be stated that the controlled flow of electrons in a given direction is produced when a source of EMF, or voltage, is applied to two different points in an electrical circuit. It is the application of the voltage to the circuit that produces the electron flow.

Keep in mind that the voltage that is applied to a conductor or circuit is actually a difference in potential that is applied to two different points in the circuit. If the same potential is applied to the two points, there would be no difference and the electron current would not flow. If only one wire from a voltage source were applied to the circuit, the same results would occur. There would be no current flow. Reversing the polarity of the voltage source wires will reverse the direction of electric current flow in the conductor.

Another influence on current flow is the amount of EMF that is applied to the circuit. An increase in the pressure, or voltage level, will make more electrons travel in the circuit. It can be stated that the quantity of the electric current is directly proportional to the amount of applied voltage.

The physical size of the wire or conductive path also has an influence on the amount of current. This can be compared to the amount of water one can get from a garden hose. The diameter of the hose is a determining factor in this situation. When the hose size is increased, more water will flow, and when it is reduced, there will be less water flow for a given period of time.

The final point in this discussion is directed to electron movement in the conductor. The action that develops when the voltage source is applied to the circuit is equal throughout the circuit. The forces that cause the current flow are not equal, but the results they produce are equal. The negative terminal of the source has a large electron-repelling factor and is affected only minimally by the electron-attracting force of the positive terminal. At the positive terminal just the opposite is true. In the center of the conductor there are equal attracting and repelling forces. This produces the effect of a constant current flow throughout the circuit.

Electron current is defined as the quantity of electric charges that move past a given point during a period of 1 second(s). The charges moving at a rate of 6.25×10^{18} electrons during 1 second are considered to constitute one *ampere* (A) of current. This is equal to a rate of 1 C of charge per second.

Electron current *Movement of electrified charges.*

The SI symbol for electric current is the letter I. This represents the *intensity* of the charge movement. Electric current is described in units of the **ampere**. The quantity of amperage is often large for industrial users, and the prefix "kilo," meaning 1000, is added to "amperes" to form a more useful unit, the kiloampere (kA). On the other hand, in electronic work the quantities of current are often small. The reference is then usually in values of milliamperes (mA) or microamperes (μA)—1/1000 and 1/1,000,000 of an ampere, respectively.

Ampere *Descriptive term used to identify quantity of electrons moving past a given point in one second.*

One of the significant factors in electrical studies is related to movement of the electrical charge. References in this section so far have been to the electron current and its movement. There are, however, other types of electrical charges that are considered important in the field of electricity. These include the ion and the hole. *Ions* are usually found in liquids or gases and are formed when the gases or liquids are ionized. These ions are also conductors of electrical current. They are used in chemical cells.

A *hole* is unique. It is formed when an electron leaves its orbit, thus creating a vacancy for an electron. When an electron moves from left to right through a conductive surface, the hole it creates moves from right to left (Figure 2-5). Holes are considered to be positive charges. They are usually discussed

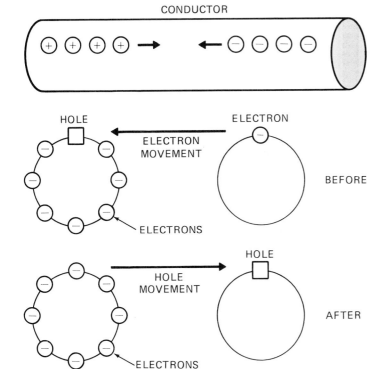

FIGURE 2-5 Electron flow is from an atom with an excess of electrons to an area with a shortage of electrons. The hole represents a place for an electron. Hole movement is in the opposite direction to that of the electron.

in relation to conduction in the types of semiconductive materials used for transistors, diodes, and integrated electronic circuits. Some semiconductor materials are identified as *P type* because they are characterized by the movement of the *positive* charges.

In all situations the relationship of charges, time, and current is the same. The formula

$$I = Q/T \qquad\qquad (2\text{-}2)$$

is used to determine the intensity of charge flow. I is the intensity of current measured in amperes, Q is the quantity of coulombs, and T refers to the amount of time measured in seconds. The direction of charge movement does not make a difference. The polarity of the charges being measured also does not matter. The important factors are the quantity of charges that are moving and the rate at which the charges are moving.

EXAMPLE 2-3

A charge of 38 C is moving past a point every second. What is the intensity of the current flow?

Solution

$$I = \frac{Q}{T}$$

$$= \frac{38C}{1s}$$

$$= 38 \text{ A}$$

EXAMPLE 2-4

A charge of 8 C is moving past a point in 1 s. What is the rate of flow?

Solution

$$I = \frac{Q}{T}$$

$$= \frac{8C}{1s}$$

$$= 8 \text{ A}$$

Another way of expressing this formula is $Q = I \times T$. Using this arrangement, one can determine the quantity of the charge.

EXAMPLE 2-5

What is the amount of charge that develops when a current of 1.6 A can charge the material for a period of 0.4 s?

Solution

$$Q = I \times T$$
$$= 1.6A \times 0.4s$$
$$= 0.64 \text{ C}$$

Figure 2-6 compares the types of current that can move in a circuit. It also shows the polarity of the current, the type of current, and some practical applications for the current. In each example the rate of movement is comparable.

The charge that is used more than others for reference is the electron. Electrons are forced to move through metallic conductive paths when an external voltage force is applied to the circuit. The direction of electron movement is always from the negative terminal of the source, through the external load, to the positive terminal of the voltage source. When the polarities of the voltage source are reversed, the electron movement through the external circuit still moves from the negative terminal, through the load, to the positive terminal (Figure 2-7).

Other methods of producing electron flow still depend on the application of a voltage source. Electron movement occurs in metallic devices such as

FIGURE 2-6 Electrical charges for current and some typical applications.

TYPE	POLARITY	CURRENT	AMOUNT	TYPICAL APPLICATIONS
ELECTRON	NEGATIVE	ELECTRON FLOW	$Q = 0.16 \times 10^{-18}$ C	WIRES, SEMICONDUCTORS, AND VACUUM TUBES
HOLE	POSITIVE	HOLE CURRENT	$Q = 0.16 \times 10^{-18}$ C	SEMICONDUCTORS
ION	EITHER POSITIVE OR NEGATIVE	ION CURRENT	Q OR MULTIPLES OF Q	GASES AND LIQUIDS

FIGURE 2-7 Electron flow (a) is the movement of charges from negative, through the load, and to the positive terminal of the source. When source polarity is reversed (b) electron flow is still from negative to positive.

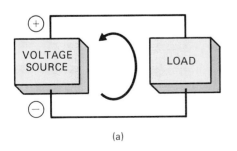
(a)

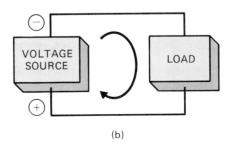
(b)

copper wire because there are free electrons in the metal. A vacuum tube is a device that can also be used for conduction of electrons. Vacuum tubes are used in video display terminals, computer monitors, television receivers, and for signal processing in all types of communications equipment. In the vacuum tube the electrons are released from a metallic surface by a process called *thermionic emission*. This process uses heat to literally boil the electrons out of the surface of the heated element of the tube. Electrons are then attracted to a positive terminal inside the tube. This terminal is then connected to the load and voltage source.

Semiconductor *Type of material permitting electron flow when voltage is applied.*

Electron movement can also occur in special materials called **semiconductors**. These are materials such as germanium and silicon that normally are nonconductors. The addition to molten germanium or silicon of a controlled quantity of impurities changes their electrical properties. This addition will produce either N-type material or P-type material. The N type has an excess of electrons and the P type has a lack of electrons. Another way of stating this is to say it has an excess of holes. The electrons in N-type semiconductor material flow from negative to positive. P-type semiconductors use positive *holes* to create a movement of charges. Hole flow is from positive to negative.

Magnetic Properties of Electric Currents The flow of an electric current of any type will produce a magnetic field around a conductor. This is the same concept that was found to be true for a charged body. The magnetic field surrounds the conductor and is formed at right angles to the conductor. The strength of this magnetic field is dependent on the intensity of the electric current. This concept is applied to electrical transformers and electromagnetic devices and also in many diversified electronic applications. These are discussed in detail in Chapter 16.

2-6 Resistance

All mechanical devices have a factor called *friction*. Friction has to be overcome in order to make a device operate correctly. Often, friction will produce heat as a by-product. There are times when the heat produced by friction is useful. In most instances one wishes to reduce friction to a minimum in order to increase the efficiency of the system.

Resistance *Opposition to movement of electrons in a circuit. Measured in units of the Ohm.*

Resistance is a factor in electrical circuits that is similar to mechanical friction. The purpose of resistance is to limit or control the amount of electric current in a system or circuit. Resistance is also used correctly in order to produce heat. The heat then is used for a second purpose. An example of this is the heat created from electricity in a toaster or electric oven.

The resistance component of any electrical conductor depends on its atomic structure. Copper has many free electrons and is said to have a very low resistance factor. Carbon, on the other hand, has very few free electrons and has a higher resistance factor than copper for a similar size or length of material.

The application of a specific amount of voltage to a piece of copper wire and a comparable piece of carbon wire will show that electron flow in the copper wire is much greater than that in the carbon wire. It is possible to increase the amount of current flow in any material by increasing the amount of applied voltage, but this changes the values used for comparison and makes the comparison invalid.

The Ohm A practical unit of measurement of electrical resistance is the **ohm**. This unit is represented by the Greek capital letter omega (Ω). The value for the ohm was developed by George Simon Ohm. He built an experimental circuit that applied 1 V to a resistance value. The current flow in the circuit was held to 1 A. The resistance value was adjusted to limit the current to this value. When 1 A of current flowed for 1 s, the resistance created some heat. This heat could raise the temperature of water 1 degree Celsius (°C). The amount of resistance in the circuit was considered to be a basic unit of 1 ohm. Resistance in a circuit is represented by the letter R and is stated in units of the ohm. There are varoius types of resistor devices used in electrical circuits. Two of these are shown in Figure 2-8(a): a high-power resistor that is formed by winding a high-resistance wire on a ceramic form and a low-power resistor that is made of a carbon and clay mixture that is held in a plastic form. The universal symbol for all resistance is shown in Figure 2-8(b).

Ohm Term used to describe units of resistance.

2-7 Conductance

When we discuss the opposition to current, the term "resistance" is used. The description of the ability or ease with which a material conducts current is called its **conductance** factor. Conductance is the reciprocal of resistance. The unit of conductance in present use is the siemens (S), which has largely replaced the mho, an earlier unit for conductance. The letter G is used to represent conductance in a circuit. Almost all work on circuits and circuit analysis is discussed in units of resistance, but there are times when conductance is a more convenient

Conductance Term describing ability of material to permit flow of electrons.

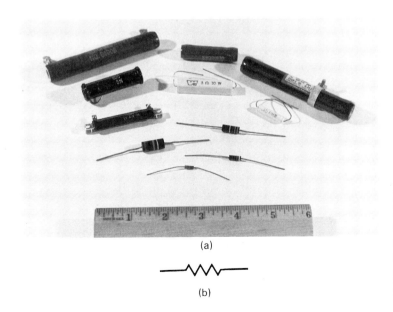

(a)

(b)

FIGURE 2-8 Resistors used in electrical circuits are produced in a variety of sizes and shapes. A general rule is that the power rating is proportional to the size of the resistor.

method of looking at circuits. One such time is during the analysis of parallel circuits (see Chapter 7).

EXAMPLE 2-6

For a resistance of 1000 Ω

$$\text{conductance} = \frac{1}{1000\ \Omega} = 0.001\ \text{S}$$

2-8 Electric Circuits

A complete electrical circuit has a minimum of three components: a voltage source, wires, and a load. These components are illustrated in Figure 2-9. The three parts of the figure each show the same basic circuit. Figure 2-9(a) shows the actual set of components used in the circuit. These consist of a chemical battery, a pair of wires, and a lamp bulb. When all of these are connected as illustrated, the bulb will glow. The reason for this is that the free electrons at the negative terminal of the battery are able to flow through one wire, the bulb, and the second wire until they reach the positive terminal of the battery.

Figure 2-9(b) shows the same set of components. They are redrawn using a standardized set of graphic symbols. The battery is replaced with the symbol used to represent it. The bulb is replaced by a symbol representing it, also. This drawing is similar in appearance to those used in the electrical and electronics industry. Persons working with these components and others have to be able to make the transition from the actual components to their equivalent schematic symbols.

Figure 2-9(c) illustrates how the components in the circuit act. The bulb is now represented as a resistance. In other words, the bulb has specific resistive qualities. Often the circuit power requirements are determined by the resistive properties of the load. This is true for this circuit. The bulb in this circuit has a resistance of 500 Ω. The dashed lines represent the direction of electron current flow in the circuit. Figure 2-9(c) is similar to many of the schematics used in this book for clarification of circuit action. The use of a resistance value instead of an actual load is a common occurrence. Remember that the resistance is used to represent the working load of the circuit. It is much easier to draw a resistor symbol than it is to draw the actual radio, computer, or machine that the resistor represents.

There are three characteristics that are fundamental to all electric circuits.

1. A source of voltage must be connected to the circuit. If there is no voltage source, there is no force available to make the current flow in the circuit.
2. The circuit must have a complete path for current flow. This path starts at the negative terminal of the source when one considers electron flow. The current then flows through a wire or conductive path to the load. It travels through the load and a second wire or conductive path to the positive terminal of the voltage source.

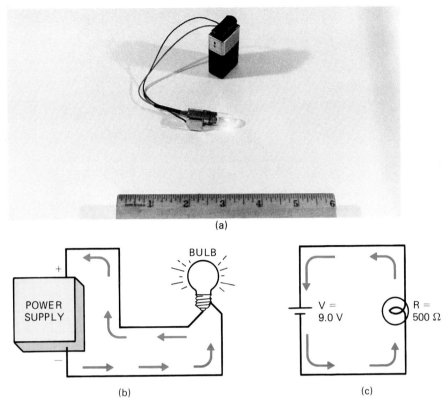

FIGURE 2-9 A basic electrical circuit has three components: the source, load, and the connecting wiring. These may be illustrated in three different formats: (a) photograph; (b) drawing; (c) schematic drawing.

3. The current path must have some resistive qualities. The resistance is used to represent a device where some form of work is being accomplished. The resistance may be used to limit current, create heat, or operate an electrical device.

The Difference Between Voltage and Current One important factor in this discussion is that the voltage is applied to a circuit. Voltage *never flows*, it only *exists*. The application of a specific quantity of voltage to a circuit will create a flow of *electron current* in the circuit. Current is the movement of electrical charges past a given point in a period of 1 s.

Figure 2-10 shows how the current flows in a complete circuit. Note how the voltage and current are illustrated. Current flow, being constant throughout the circuit, can be measured at any point from the terminals of the source to the terminals of the load. Usually, this is accomplished by breaking, or opening, a connection in the circuit and inserting a current flow meter (Figure 2-11). The wire is disconnected, thus breaking open the circuit. This device, called an **ammeter** is inserted between the two open ends of the wire and reconnecting the circuit and current path. Since current flow is equal throughout the system, the ammeter will indicate the quantity of electron flow. The ammeter can be added at any point, or more than one ammeter can be inserted. All should show the same quantity of electron flow.

Ammeter *Measuring device having ability to quantify electron flow.*

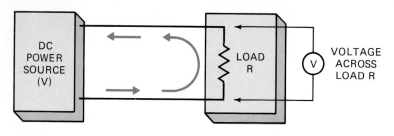

FIGURE 2-10 Current flow through the basic circuit is from the source, through the load, and back to the source.

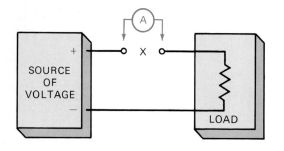

BREAK WIRES AT POINT X—INSERT LEADS FROM AMMETER, ONE ON EACH SIDE OF BREAK

FIGURE 2-11 A current flow, or ammeter, is added to the circuit by separating the wires and using the meter to complete the circuit.

The difference between voltage and current is more apparent when one sees how voltage is measured. Voltage is, by definition, a pressure difference between two points. Voltage exists in order to produce a current flow. Voltage, being a pressure difference between two given points, is measured as illustrated in Figure 2-12. Each measuring meter has two wires or test leads. One of these is connected to the most negative side of the points to be measured. The other lead is connected to the most positive point to be measured. The difference in potential that is created by current flow is then displayed on the meter as a voltage value.

There is one exception to this example. In the simple circuit shown, we would be able to measure a pressure difference when the load device is removed. The reason for this is that voltage is still present at the source. When the circuit is complete with the load, the meter leads are actually connected to both the source and the load. When the load is removed, the meter is still connected to the source. There is still current flow in the circuit because the meter has completed the current path. The amount of current is far less than the amount that flows when the load is connected. The voltage difference developed by the source is still present and is measured as the same value as that found when the load is connected.

FIGURE 2-12 Voltage is measured by placing one lead from the voltmeter on each side of the load, or desired circuit.

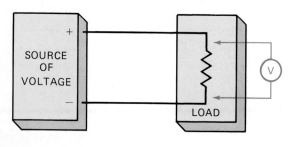

PLACE ⊕ LEAD OF VOLTMETER ON + SIDE OF LOAD; PLACE ⊖ LEAD ON − SIDE OF LOAD

When one considers the three components of a basic electrical circuit it becomes apparent that the source and the load control the quantity of electron current. The circuit used for the example has a 6.0-V source and the load is equal to a resistance of 500 Ω. Under these conditions the current flow is determined by the formula $I = V/R$, or $I = 6$ V/500 Ω. The actual current flow in this circuit is 0.012 A, or 12 mA. This value is called the load current, or just **load**.

Changing the value of the load resistance will also change the quantity of current. When the load resistance is doubled to 1000 Ω, the current is reduced. Using the same formula yields

$$I = \frac{V}{R} = \frac{6 \text{ V}}{1000 \text{ Ω}} = 0.006 \text{ A} \quad \text{or} \quad 6 \text{ mA}$$

When the voltage is increased to 12 V and the resistance left at 500 Ω, the current is increased. Using the same formula, we have $I = V/R = 0.024$ A or 24 mA. This variation in load conditions can be related to any value of load resistance and current. It is normal to discuss the size of the load in relation to the amount of current that flows in the circuit. For example, a *heavy load* is used to describe a load that requires large amounts of current. These terms also represent the amount of work that is being done by the load. The energy for this work, of course, is supplied by the source in the circuit.

Current Flow Direction The discussion so far has concentrated on electron flow in the system. **Electron flow**, or electron current, is described as the movement of negatively charged electrons. They are found at the negative terminal of the source. The movement is from an area that has a high concentration of electrons to an area that has a lack of, or low, quantity of electrons. This second area is the positive terminal of the source. The electrons cannot move through the source. They must seek an outside path for their movement. This outside path is the load and the wires that connect it to the source. We therefore say that electron flow is from negative to positive in the circuit. When this action occurs, the result is a quantity of work that is performed by the load.

There is an alternative method of discussing current flow in an electric circuit. This method describes the movement of positive charges rather than that of the negatively charged electrons. This description is called **conventional current** flow. The concept used to describe this type of current is that the positive charges have a higher value than do negatively charged bodies. If this is true, the positive charges will *fall* down from their place to the lower-ranked negative terminal in the system.

Two examples of this type of charge movement are the electrolyte battery and the charge movement in P-type semiconductor material. In an electrolyte battery, conduction occurs inside the battery because of ionization of the liquids in the electrolyte. Ionized charge movement is from positive to negative. This is therefore a conventional current that is produced by an ionization process. P-type semiconductors have a majority of positively charged carriers. Their current flow is also conventional.

It really does not make a lot of difference which method one uses to describe or determine current flow in an electrical circuit. The major consideration is that when there is a movement of charged bodies through a load, some amount of work will be performed. The presentation in this book is limited to the movement of the electron. This is done in an effort to minimize confusion in describing the action of an electrical circuit.

It is important to understand that the electromotive force (V) is used to force electrons (I) through a path (wires) through a load (R). When these conditions are met, one has a complete electrical circuit, current flow occurs, and work is performed by the load. For this to occur, all three components of the circuit are required.

Circuit Terms There are two terms that are used in the field to describe circuit conditions that are not considered to meet the conditions of a working circuit. These terms are *open* and *short*.

Open circuit *An incomplete circuit in which electrons normally do not flow.*

Short circuit *Abnormal conductive path permitting excess current flow.*

Open Circuit An open circuit is one that does not have a path for current flow. The open is considered as a break in the normal current path (Figure 2-13). If the circuit should be completed, the *potential* for work still exists. The characteristics of an open circuit are:

1. There is no current flow due to the broken path for current.
2. The source voltage is present and can be measured at the point of the break, or open. This is true even when there are other resistive components still in the circuit.

Short Circuit A short circuit is described as a circuit that has an abnormally low resistance path for current flow. Two types of short circuits are shown in Figure 2-14. One of these is a total, or complete short circuit [part (a)] and the other is a partial short circuit [part (b)]. The circuit is still complete. Usually, the resistance is close to 0 Ω when a total short circuit is present. With a partial short, the resistance is less than the normal value. This unusually low resistance value will provide an alternative path for current. When the short-circuit path is across the load, there is not an adequate amount of current going through the load. The load does not perform any work under these conditions. The short circuit will create an unusually large amount of current. The result of this can generate heat in the conductors, which can fail under these conditions. The insulation used around the conductors can also start to burn if heated in this manner. Circuit protection devices are used to protect against these over-current conditions. These devices include fuses, circuit breakers, and electronic shutdown circuits. The characteristics of the short circuit are:

FIGURE 2-13 The open circuit stops current flow. The source voltage can be measured across the point of the open.

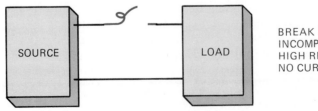

BREAK IN CONDUCTIVE PATH:
INCOMPLETE CIRCUIT
HIGH RESISTANCE
NO CURRENT FLOW

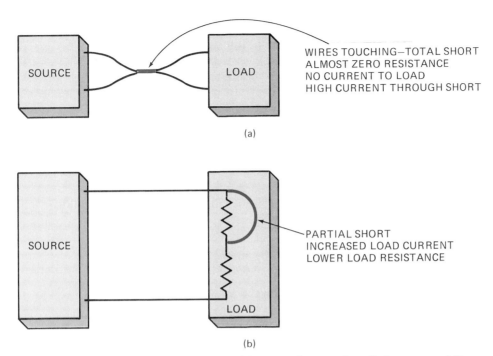

WIRES TOUCHING—TOTAL SHORT
ALMOST ZERO RESISTANCE
NO CURRENT TO LOAD
HIGH CURRENT THROUGH SHORT

(a)

PARTIAL SHORT
INCREASED LOAD CURRENT
LOWER LOAD RESISTANCE

(b)

FIGURE 2-14 The short circuit provides an abnormal path for current flow. A complete short (a) will bypass the normal path. A partial short (b) will bypass one or more components.

1. Lower-than-normal circuit resistance
2. Much-higher-than-normal current flow
3. Often, a burned or charred component in the circuit
4. An open fuse or circuit breaker (as a result of the above)

2-9 Voltage Sources and Types

Static and current electricity have been identified and described. Two types of sources have been also identified: the chemical cell and the electrolyte cell. There are some other types of voltage sources that are equally important to us. Let us look briefly at several of these and describe how they are used to create electrical energy.

The first type of electrical energy discovered is called static electricity. This is created by rubbing two dissimilar materials together. The process can be identified as the creation of electrical energy using friction.

A system that is in common use today is the creation of electrical energy by means of chemical or electrolyte batteries. Chemical reactions are utilized in these processes to create two opposing polarity terminals in the battery.

Gases and some liquids can create electrical energy by utilizing the ions of a material. The result will create both positive and negative charges.

Thermal emission is the process of heating a material to a point where the electrons in the material are "boiled off" its surface. This process works best in a vacuum. An example of this process is the heating of a cathode element in a vacuum tube. The electrons are then attracted to a positively charged anode. This provides a conductive path in the tube for electrons.

Some materials used for electrical work are sensitive to light. These materials are said to be **photoelectric** in nature. Electrons and positive charges are rearranged in the materials when they are exposed to light rays. The atomic structure of the material determines whether it is sensitive to light or to dark. These photosensitive materials are used in cathode ray tubes, photocells, photovoltaic cells, and phototransistors. Two types of activity can occur. One of these creates a voltage when light shines on the surface of the material. These are called *photovoltaic* substances. The second changes the electrical field in the substance. These materials are called *photoresistive*.

Possibly the greatest source of electrical energy is electromagnetism. The fields created by magnetic energy are used to create electricity. When a wire is moved through a magnetic field, a voltage is created on the wire. Moving the field past the wire also creates a voltage on the wire. This concept is used in all power generators. Electric power companies use this principle to create all of the power we use in home and industry. Electric power to keep the battery charged for cars, airplanes, and small engine devices is created using electromagnetic principles. You may question these statements and offer the use of a waterfall generator or atomic energy generator as an example. These generators use water or atomic energy in order to operate a turbine. The turbine is then used to operate the generator. Once the power for the turbine is created, the principle of creating electrical energy does not change.

Direct current *System permitting current flow from source in only one direction.*

Our discussion so far has described electron movement flow as occurring from negative to positive in the circuit. This fact does not change. We have shown a power source that has a constant polarity. This type of power source is called a **direct-current** (dc) source. The current flow through the circuit is in one direction because the terminals of the power source always have a constant polarity. A typical source for dc is a battery.

Another source for electrical energy is a generator (Figure 2-15). A generator rotates a wire in a magnetic field in order to create its output voltage. The polarity of the voltage is maintained by a device called a *segmented commutator*. Details on this are presented in another chapter. It is possible to use a pair of second devices called *slip rings* to remove the voltage from the generator. When slip rings are used the polarity and amplitude of the voltage change periodically. The shape of the dc voltage created by the two sets of output devices is shown in Figure 2-16. Dc voltage has a constant amplitude value. It normally does not vary from this level. The symbol for a dc source and the graph showing the level of voltage are both shown.

Sine wave *Form used to describe an alternating current or voltage source.*

Alternating current *System permitting a cyclical reversal of voltage and current.*

The level of the voltage from a generator that uses slip rings is always varying. The specific amount depends on the relationship of the wires in the generator and magnetic field. The "signature" of the output voltage is the **sine wave** (Figure 2-17). This wave alternates from positive to negative during one rotation of the generator. It is called an **alternating-current** voltage because of this. Figure 2-17 also illustrates the form and polarity of the wave. Alternating current (ac) is the type of voltage that is created by almost all electric power companies. The rate at which it alternates from positive to negative is called the *frequency* of the wave. The standard ac frequency on the North American continent is 60 Hz, or 60 times a second.

Electron flow in an ac circuit changes direction 120 times a second. This

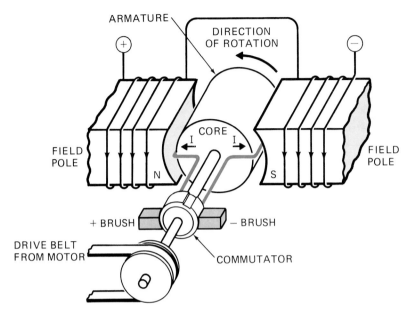

ARMATURE

DIRECTION OF ROTATION

CORE

FIELD POLE

FIELD POLE

N

S

I ← → I

+ BRUSH

− BRUSH

DRIVE BELT FROM MOTOR

COMMUTATOR

FIGURE 2-15 This electric generator consists of stationary magnets and a rotating armature. The rotation of the wire through the magnetic field will produce a voltage on the wire. (From Joel Goldberg, *Fundamentals of Electricity*, 1981, p. 234. Reprinted by permission of Prentice-Hall, Englewood Cliffs, New Jersey.)

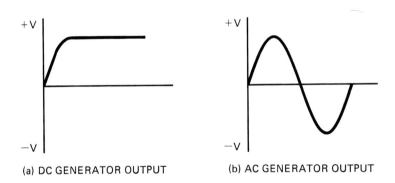

(a) DC GENERATOR OUTPUT

(b) AC GENERATOR OUTPUT

FIGURE 2-16 Direct current (dc) voltage is created as a constant value by the generator (a) Alternating-current (ac) voltage generators produce a varying polarity voltage (b).

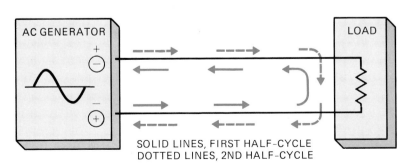

AC GENERATOR

LOAD

SOLID LINES, FIRST HALF-CYCLE
DOTTED LINES, 2ND HALF-CYCLE

FIGURE 2-17 Alternating-current (ac) generators have a polarity reversal each half-cycle. Current flow through the load is also reversed at these times.

is also shown in Figure 2-17. During the first half-cycle the upper terminal is positive. During the second half-cycle the polarities are reversed. The pattern repeats 60 times a second. Electron flow, shown by the dashed lines, is now in the opposite direction in the circuit. Details of dc and ac circuits are presented in subsequent chapters.

CHAPTER SUMMARY

1. The terminals of all batteries have a specific polarity assigned to them.
2. All substances have an atomic structure consisting of a nucleus and electrons.
3. The number of free electrons used to complete the outermost ring of the atom determines if it is a conductor or an insulator.
4. Materials that give up their outer ring electrons are said to be conductors. Materials that accept electrons in order to fill their outer ring are said to be nonconductors, or insulators.
5. Electrical charges that are not in motion are called static electricity.
6. The coulomb is the unit of electrical charge. It is equal to 6.25×10^{18} electrons.
7. Opposite charges attract each other, whereas like charges repel each other.
8. A charged body creates an electric field around itself. The strength of this field is inversely proportional to the square of the distance away from the body.
9. The size of the electric field is directly proportional to the amount of charge used to create it.
10. Potential is a difference in electrical charge that exists between two points in a system.
11. The quantity of charge that can be moved by an electrical potential is directly related to the strength of this potential difference, or voltage.
12. Electromotive force (EMF) is measured in units of the volt.
13. A voltage-measuring meter is connected across the points of potential difference in a circuit.
14. The quantity of electric charge that moves past a given point in 1 s is called the electric current. It is measured in amperes.
15. Electron current flows from negative to positive in a circuit.
16. The amount of voltage that is applied to an electric circuit determines the current flow. Increased voltage produces increased current flow.
17. Electric current flow creates an electromagnetic field around a conductor.
18. The opposition to electron flow is called resistance. It is measured in ohms.
19. A basic electric circuit contains a source of power, a load, and connecting wiring.
20. Electron current flow is from negative to positive, and conventional current flow is from positive to negative.
21. An open circuit is a break in the electron flow path. It is a very high resistance.
22. A short circuit is an undesirable low-resistance path. It permits unusually high levels of current flow.

23. Electrical energy is created by chemical action, ionization action, heat, light, and electromagnetic induction.

24. Direct current (dc) flows constantly in one direction in the system.

25. Alternating current (ac) reverses its polarity periodically in the system. Current flow is constantly being reversed by this process.

Answer true or false; change all false statements into true statements.

2-1. Copper atoms do not have free electrons.

2-2. Electron flow is from negative to positive in the circuit.

2-3. Similar electrical charges attract each other.

2-4. The opposition to electron flow is called resistance.

2-5. The unit of electromotive force is the ampere.

2-6. The unit of electron flow is the ampere.

2-7. The unit of electrical opposition is the ohm.

2-8. There are five charge units between $+8$ and a -3 charge.

2-9. Voltage is the force used to move electrons in a circuit.

2-10. A hole is a positive charge.

2-11. The positive charge in an atom of a material is the

   (a) neutron
   (b) proton
   (c) electron
   (d) nucleus

2-12. Electrons are contained in the
   (a) outer rings of the element
   (b) nucleus of the element
   (c) free air around the rings
   (d) none of the above

2-13. Electricity at rest is called _____ electricity.
   (a) current
   (b) EMF
   (c) static
   (d) ready

2-14. Electromotive force is measured in
   (a) ohms
   (b) amperes
   (c) watts
   (d) volts

2-15. The difference in electrical potential between two points is called
   (a) voltage
   (b) current

(c) resistance
(d) pressure

2–16. Current flow is determined by
(a) voltage and resistance
(b) amperage and voltage
(c) resistance and amperage
(d) resistance only

2–17. Current flow is measured in
(a) ohms
(b) volts
(c) watts
(d) amperes

2–18. The positive charge current is measured in
(a) electrons
(b) holes
(c) amperes
(d) volts

2–19. Opposition to charge movement is measured in
(a) ohms
(b) volts
(c) watts
(d) amperes

2–20. A circuit has a source of 100 V and a load of 200 Ω. Its current flow is
(a) 10.0 A
(b) 1.0 A
(c) 0.5 A
(d) 5.0 A

ESSAY QUESTIONS

2–21. Provide the name of the units used to describe values of each of the following and the letter used for these values.
(a) electromotive force
(b) current
(c) opposition to current
(d) conduction of current

2–22. State the two types of polarities used in electrical work.

2–23. Draw the schematic symbol for a 6-V battery. Include polarity markings with the symbol.

2–24. Draw a picture of the atomic structure of the copper atom.

2–25. Describe the relationship of a negatively charged atom and the number of electrons in the outer ring of the atom.

2–26. How many electrons are in one charge unit?

2–27. State the two laws of charges.

2–28. What is an electric field?

2–29. How does the strength of the electric field relate to the amount of current flowing in a conductor?

2–30. State how the size of an electric field around a conductor can be changed.

2–31. Define the term *voltage*.

2–32. What force in electrical circuits is used to move electrons?

2–33. How is the force in question 2-32 measured?

2–34. Draw a schematic diagram for a basic electric circuit consisting of a resistance load and a battery. (*Sec. 2-8*)   **PROBLEMS**

2–35. Show the schematic symbols for the following devices. (*Sec. 2-8*)
  (a) battery (indicate polarity)
  (b) resistor
  (c) voltmeter
  (d) ammeter

2–36. Draw another electric circuit in which you include a voltmeter. The voltmeter should be measuring the voltage across the load resistance. (*Sec. 2-8*)

2–37. Draw a third electric circuit in which you include an ammeter. The ammeter should be measuring the current flow between the load and the positive terminal of the battery.(*Sec. 2-8*)

2–38. State three factors that influence the quantity of electron flow in a circuit. (*Sec. 2-8*)

2–39. Define the term *ampere*. (*Sec. 2-5*)

2–40. How large (or small) is the milliampere? (*Sec. 2-5*)

2–41. How large (or small is the microampere? (*Sec. 2-5*)

2–42. Define the term *hole* as used in electrical terms. (*Sec. 2-3*)

2–43. How much amperage is there in a circuit that has 140 charges moving past a reference point in 3 s?(*Sec. 2-8*)

2–44. What is the quantity of electrical current that is present in a circuit when 10 charges flow during 1 s? (*Sec. 2-3*)

2–45. How many charges are developed when 3.8 A flow in a circuit for 1.6 s? (*Sec. 2-3*)

2–46. What is the difference between electron current and conventional current? (*Sec. 2-5*)

2–47. What term is used to describe electrical friction in a circuit? (*Sec. 2-6*)

2–48. State three factors that are present in every electrical circuit. (*Sec. 2-8*)

2–49. How does alternating current differ from direct current in a circuit? (*Sec. 2-9*)

2–50. What is meant by the term *hertz*? (*Sec. 2-9*)

2–51. Draw the shape of the ac waveform and state its name. (*Sec. 2-9*)

2–52. What is the difference between electrical current and voltage? (*Sec. 2-5*)

2–53. Determine the amount of difference in charge between the following charged bodies. (*Sec. 2-3*)
(a) +3 C and + 8 C
(b) −2 C and −4 C
(c) −2 C and +4 C
(d) +6 C and − 24 C
(e) − 400 C and − 180 C

2–54. Describe the following units in decimal terms. (*Sec. 2-5*)
(a) 10 mA
(b) 1.6 kA
(c) 2.8 μA
(d) 1800 mA

2–55. In a given circuit a source voltage of 100 V is applied to a load whose resistance is 50 Ω. What is the current flow in this circuit? (*Sec. 2-8*)

2–56. A circuit has a resistance of 16 kΩ. How much current flow occurs when 48 V is applied to it? (*Sec. 2-8*)

2–57. A circuit is composed of a 5-kΩ load and a 200-V source. What is the current flow? (*Sec. 2-8*)

2–58. A 0.6-Ω short is applied across the 5-kΩ resistance in problem 2-57. What is the current flow under these conditions? (*Sec. 2-8*)

2–59. The resistance value of a 15-kΩ resistance increases due to heating to a value of 1.5 megohms (MΩ). The source voltage is 100 V. Determine the current flow in this circuit for the following resistances. (*Sec. 2-8*)
(a) 15 kΩ
(b) 1.5 MΩ

2–60. The resistance value of the resistor used in problem 2–59 increases to a new value of 150 gigaohms (GΩ). What is the current flow under these conditions? (*Sec. 2-8*)

ANSWERS TO REVIEW PROBLEMS

2–1. Positive and negative

2–2. (+) and (−)

2–3. ⊣⊢⁺ or ⊣⊨⁺

2–4. Positive protons and negative electrons

2–5. Electrons

2–6. Coulomb

2–7. 6.25×10^{18} electrons

2–8. Opposites attract, likes repel.

2–9. An electric field is a force field that surrounds a charged body.

2–10. It is reduced in strength with distance away from the body.

2–11. It diminishes by the square of the distance away from the body.

CHAPTER 3

Conductors, Insulators, and Basic Components

A person's ability to design or repair devices using electrical and electronic components depends on the ability to recognize how a circuit works, the use of the correct types of components, and the actual shape of the components. Circuit rules presented by early scientists have not changed since they were discovered. The only changes that have occured involve application of the rules. The integrated circuits in use today use the same principles of current flow and voltage differences as those of light bulbs or motors.

Persons involved successfully in electrical and electronic occupations rely on their ability to analyze circuits and systems. It would be very difficult for anyone to attempt to test a circuit without knowing what or where to test. Only after a knowledge of circuits, components, and the application of circuit rules is acquired can this be accomplished successfully.

KEY TERMS

Circuit Breaker   Relay

Circular Mil   Schematic Diagram

Common   Signal

Dielectric Strength   Solid Wire

Fuse   Stranded Wire

Ground   Switch

Ohmmeter   Voltage Drop

Printed Circuit

OBJECTIVES

Upon completion of the material in this chapter, you will

1. Understand conductor sizes and ratings.
2. Be able to "read" a schematic diagram.
3. Be able to explain the purposes of fuses and circuit breakers.
4. Understand electric switch concepts and applications.

In Chapters 1 and 2 we have covered some of the basics of electrical theory. We have explored the concepts of the discovery of the various properties of electricity. In addition, details of the concepts of electrical charge movement and the properties of an electrical pressure called voltage have been presented. This foundation of electrical knowledge needs to be expanded to a greater degree. Some of the electrical properties covered on an introductory level up to this point include resistance and conductance. These two topics are presented in detail in this chapter.

There is some additonal information that is relevant to an understanding of how electrical circuits operate. Included in this information is how we "read" an electrical schematic diagram. Some of the more common components found in electrical systems are also described. These include fuses, circuit breakers, lamps, and switches. Other components are described in later chapters as they become pertinent to the discussion. All of these are relative to the fundamental understanding of electrical and electronic circuits and systems. As stated previously, once the theories are learned, the knowledge is easily transferred to a variety of circuits and applications.

3-1 Conductors

Any description of a basic electrical circuit includes a pair of wires. These wires are used to connect the source and the load. Their purpose is to carry electrons from the negative terminal of the source to one end of the load and from the other end of the load to the positive terminal of the source. Every effort is made to allow all possible electrons to travel this route.

Perhaps you are aware of the fact that most of the wire used for this purpose is made of copper. Copper has an abundance of free electrons. This permits an easy movement of electrons between the source and the load. The question one should ask at this point is: Can wires made of other materials also be used? The answer is: "Yes, but. . . ." The reason for the "but" is that certain materials can be used as conductors and others cannot be used. Metallic materials can be arranged on a continuum (Figure 3-1), that groups to one side materials that will allow ease of electron movement. These are known as *conductors*. At the opposite end of the chart is a group of materials that do not allow any electron movement under normal conditions. These are classified as *insulators*. Between these two extremes one finds a third group of materials that we call *semiconductors*. The term "semiconductors" as it is used here is different from its use in solid-state or transistor theory. The conditions used here mean that the material can conduct electrons but it does so with difficulty.

CONDUCTORS	SEMICONDUCTORS	NONCONDUCTORS
GOLD	NICKEL	PURE SILICON
SILVER	NICHROME	PURE GERMANIUM
COPPER		MICA
ALUMINUM		CERAMICS

FIGURE 3-1 Typical types of conductors, semiconductors, and insulators.

Almost any material can conduct electrical energy under certain conditions. When we remember that the quantity of electrons that move through a given material is influenced by the strength of the EMF that is applied to the system, it is clear that it is possible to make any material a conductor. Normally, the materials that are classified as insulators are rated by their **dielectric strength** (Figure 3-2). This term is used to indicate the insulating qualities of the material. Higher numbers in Figure 3-2 indicate a higher dielectric-strength rating. These materials, which include air, mica, oil, and paper, are typically used as insulating material in electrical and electronic components. Capacitors, for example, use these four materials in their construction to provide insulation. Should this dielectric material fail, the insulating quality is destroyed and the capacitor fails to perform its function of storing an electric charge. Further discussion of this action is presented in Chapter 22.

Dielectric strength *Describes the insulating qualities of a material.*

Copper is the electron conductor in the most common use. The reasons for this are simple. There is a great quantity of copper available from the earth at reasonably low cost, and copper has free electrons. These factors have made copper conductors a standard. Other materials, including silver and aluminum, have also been used successfully as conductors. Silver works very well as a conductor. The fact that it is considered a precious metal and is relatively costly compared to copper has limited its use. Aluminum is also a good conductor. Aluminum alloys are used for conductors for the high-voltage power lines that connect power generators to local power distribution centers. Its lighter weight makes it ideal for the long nonsupported spans of wire between the towers. Aluminum has been used in home wiring in a limited way, but problems occur when aluminum wire and copper wire are joined. If not done properly, the connection can become resistive and create an undesired amount of heat. Therefore, aluminum wiring is no longer used in homes.

REVIEW PROBLEMS

3-1. Name three types of wire conductor materials.

3-2. State the difference between a conductor and an insulator.

3-3. What is meant by the term *dielectric strength*?

MATERIAL	DIELECTRIC STRENGTH (V/0.001 IN.)	DIELECTRIC CONSTANT
AIR	20–100	1.0
CERAMIC	600–1200	75–1000
GLASS	200–250	7.6–8
MICA	500–1500	3–7
OIL	275	2–6
PAPER	1250	2–4

FIGURE 3-2 Dielectric strength ratings for insulating materials used in electrical and electronic work. The specific values will depend upon the exact composition of the material used. These values are the broad range covered by the material.

Basis of Conductors The major purpose of any conductor is to present the source EMF to the load. When these two circuit components are close together, the theory behind this arrangement works very well. When a great distance separates the source from the load, an additional factor is introduced. This factor, called **voltage drop**, can reduce the amount of EMF that is present at the load. In the real world almost everything has some friction, and this is also true in electrical circuits. Electrical friction is known as resistance. All electrical conductors have some resistance factor. The theory we use is to try to keep this factor at a very low level. Let us describe how resistance influences load performance before we describe the ratings and standards for copper wire.

Voltage drop Term describing difference in voltage at two different points in circuit.

In Section 2-8 an application of Ohm's law was used to determine current flow in a circuit. Ohm's law can also be used to calculate the voltage drop that occurs due to conductor resistance. Figure 3-3 shows a typical electrical circuit. The source voltage is 9 V and the radio load is 500 Ω. Using Ohm's law, $I = V/R = 9/500$, we find that the current under ideal conditions is 0.018 A or 18 mA. This value of load current is calculated with zero-resistance conductors in the circuit.

Now look at Figure 3-4. The resistance of the wires connecting the load and the source is 40 Ω for each wire. This thus adds an additional 80 Ω of resistance to the circuit. The addition of 80 Ω increases the total resistance in the circuit from 500 Ω to 580 Ω. The total current flow is now reduced from 18 mA to 15 mA.

Using this new value of current the load current has been reduced to 83% of its ideal value. This reduces the efficiency of the work performance of the load from 100% to 83%. What has happened is that the wires have become low-value resistances. These resistances are a part of the total circuit and their effect on the total circuit has to be considered. We can consider this factor as a voltage drop due to current flow through a resistance. When Ohm's law is applied, the exact amount of voltage that is developed across the resistances can be determined. Using the formula $V = I \times R$ will provide this value. In this example,

$$V = I \times R$$
$$= 0.015 \text{ A} \times 40 \text{ }\Omega \qquad (3\text{-}1)$$
$$= 0.6 \text{ V}$$

FIGURE 3-3 Correct selection of conductor sizes will minimize any voltage drop except at the source.

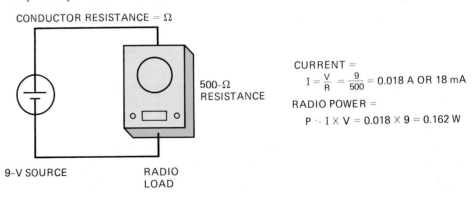

CONDUCTOR RESISTANCE = Ω

500-Ω RESISTANCE

9-V SOURCE

RADIO LOAD

CURRENT =
$$I = \frac{V}{R} = \frac{9}{500} = 0.018 \text{ A OR } 18 \text{ mA}$$

RADIO POWER =
$$P = I \times V = 0.018 \times 9 = 0.162 \text{ W}$$

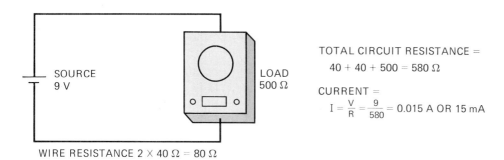

TOTAL CIRCUIT RESISTANCE =
40 + 40 + 500 = 580 Ω

CURRENT =
$$I = \frac{V}{R} = \frac{9}{580} = 0.015 \text{ A OR } 15 \text{ mA}$$

SOURCE 9 V

LOAD 500 Ω

WIRE RESISTANCE 2 × 40 Ω = 80 Ω

FIGURE 3-4 A voltage drop due to conductor resistance will reduce voltage available at the load.

The voltage drop in each conductor is equal to 0.6 V. The total voltage drop due to $I \times R$ loss is 1.2 V. The voltage that is actually delivered to the load is not 9.0 V but 9.0 − 1.2 or 7.8 V. This loss of desired operating voltage often affects adversely the operation of a radio.

The values used in the preceding example are relatively small. The listening pleasure of the radio owner may not be affected by this voltage drop. Let us look at another example, perhaps an extreme one, but still a practical example. An electric generator (Figure 3-5) produces an operating EMF of 4800 V. It is located 100 miles from one of its customers. The load resistance is equal to 100 Ω. The wires that are used to carry the generated voltage to the load, or consumer's equipment, have a factor that produces a voltage drop of 1.2 Ω for each 1000 ft of wire. This is equal to 528 pieces of wire that are each 1000 ft long. The total resistance in one of the two wires required to complete the circuit is 633.6 Ω (528 × 1.2 = 633.6). The total resistance in both wires is twice 633.6, or 1267.2 Ω. When the resistance of the wires is added to the load resistance, the total resistance is equal to 1367.2 Ω. In this example the wire resistance is over 13 times the resistance of the load! The amount of voltage that is available at the load in this circuit is determined as follows:

$$\text{ideal} = \frac{4800 \text{ V}}{100 \text{ Ω}}$$

$$= 48 \text{ A}$$

$$\text{total resistance} = \text{wires} + \text{load}$$

$$= 1267.2 \text{ Ω} + 100 \text{ Ω}$$

$$= 1367.2 \text{ Ω}$$

FIGURE 3-5 The voltage drop, or loss, that occurs in long-distance electric power lines can reduce the operating voltage at the load to a very low value.

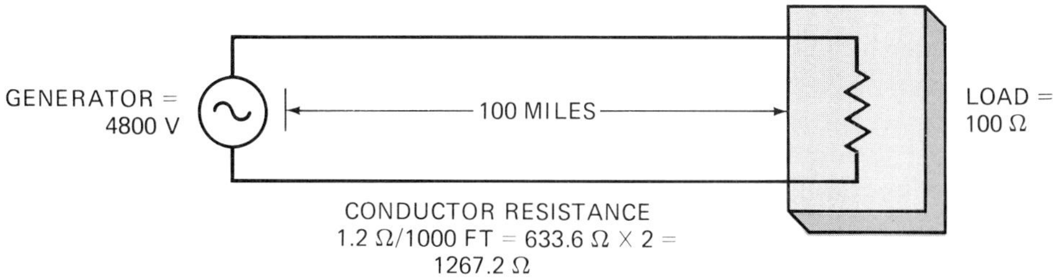

GENERATOR = 4800 V

100 MILES

LOAD = 100 Ω

CONDUCTOR RESISTANCE
1.2 Ω/1000 FT = 633.6 Ω × 2 =
1267.2 Ω

$$\text{actual current} = \frac{4800 \text{ V}}{1367.2 \text{ }\Omega}$$

$$= 3.5 \text{ A}$$

$$\text{wire voltage drop} = I \times R$$

$$= 3.5 \text{ A} \times 1267.2 \text{ }\Omega$$

$$= 4435.2 \text{ V}$$

$$\text{efficency} = \frac{3.5}{48}$$

$$= 7.29\%$$

In this example almost all of the EMF generated would be dropped by the resistance of the wires; less than 8% would be available to operate the load. The type of voltage drop described in the foregoing two examples occurs in all wiring. The purpose of the design engineer is to select a wire of the proper size, capacity, and resistance value to minimize this factor. Even when this is done there is still the possiblity of developing undesired voltage drops. We notice these in our home when the size of the television picture is reduced during periods of high energy demand. We also notice it when the motor speed of an air-conditioning unit seems to slow down or when the wires used to deliver the operating EMF develop heat due to electrical resistance, or friction.

REVIEW PROBLEMS

3-4. What is the major purpose of any conductor?
3-5. What does the term *resistance* mean?
3-6. What is meant by the term *load*?
3-7. How does wire resistance affect load operation?

Conductor Ratings and Size The common language we are developing also carries into the methods of classifying the size and ratings of wire. Copper wire has become the industry standard during the past years. As a result, a reference chart has been developed by the industry. This chart is called a *copper wire table* (Figure 3-6). The wires are given a gage size, as shown in the left-hand column. The number of the gage is inversely proportional to the size of the wire. A number 00 gage wire is much larger than a number 30 gage wire.

The next column on the chart is titled **circular mils.** This is the measurement of the cross-sectional area of the wire. Our experience with decimals identified the mil as one one-thousandth of an inch (0.001). This is also true for wire measurement. One mil is equal to 0.001 in. A circular mil is equal to the square of the diameter of the wire in units of the mil.

Circular mil *Describes cross sectional area of a conductor or wire.*

STANDARD ANNEALED COPPER WIRE SOLID AMERICAN WIRE GAGE							
1	2	3	4	5	6	7	8
					ALLOWABLE CURRENT CAPACITY, A		
GAGE NUMBER	DIAMETER, MILS	AREA CIR MILS	RESISTANCE Ω/1000 FT 25°C (77°F)	WEIGHT LB/1,000 FT	RUBBER INSULATION	VARNISHED CAMBRIC INSULATION	OTHER INSULATIONS
0000	460.0	211,600.0	0.0500	641.0	225	270	325
000	410.0	167,800.0	0.0630	508.0	175	210	275
00	365.0	133,100.0	0.795	403.0	150	180	225
0	325.0	105,500.0	0.100	319.0	125	150	200
1	289.0	83,690.0	0.126	253.0	100	120	150
2	258.0	66,370.0	0.159	201.0	90	110	125
3	229.0	52,640.0	0.201	159.0	80	95	100
4	204.0	41,740.0	0.253	126.0	70	85	90
5	182.0	33,100.0	0.319	100.0	55	65	80
6	162.0	26,250.0	0.403	79.5	50	60	70
7	144.0	20,820.0	0.508	63.0			
8	128.0	16,510.0	0.641	50.0	35	40	50
9	114.0	13,090.0	0.808	39.6			
10	102.0	10,380.0	1.02	31.4	25	30	30
11	91.0	8,234.0	1.28	24.9			
12	81.0	6,530.0	1.62	19.8	20	25	25
13	72.0	5,178.0	2.04	15.7			
14	64.0	4,107.0	2.58	12.4	15	18	20
15	57.0	3,257.0	3.25	9.86			
16	51.0	2,583.0	4.09	7.82	6		
17	45.0	2,048.0	5.16	6.20			
18	40.0	1,624.0	6.51	4.92	3		
19	36.0	1,288.0	8.21	3.90			
20	32.0	1,022.0	10.4	3.09			
21	28.5	810.0	13.1	2.45			
22	25.3	642.0	16.5	1.95			
23	22.6	509.0	20.8	1.54			
24	20.1	404.0	26.2	1.22			
25	17.9	320.0	33.0	0.970			
26	15.9	254.0	41.6	0.769			
27	14.2	202.0	52.5	0.610			
28	12.6	160.0	66.2	0.484			
29	11.3	127.0	83.4	0.384			
30	10.0	100.0	105.0	0.304			
31	8.9	79.7	133.0	0.241			
32	8.0	63.2	167.0	0.191			
33	7.1	50.1	211.0	0.152			
34	6.3	39.8	266.0	0.120			
35	5.6	31.5	335.0	0.0954			
36	5.0	25.0	423.0	0.0757			
37	4.5	19.8	533.0	0.0600			
38	4.0	15.7	673.0	0.0476			
39	3.5	12.5	848.0	0.0377			
40	3.1	9.9	1070.0	0.0299			

FIGURE 3-6 Copper wire table, showing wire gage number, size, resistance, and current-carrying capacities.

> **EXAMPLE 3-1**
>
> Find the cross-sectional area of a wire whose diameter is 0.012 in.
>
> *Solution*
>
> The first step is to convert the measurement into mils. In this example 0.012 in. is equal to 12 mils.
>
> $$\text{Circular-mil area} = (12 \text{ mil})^2$$
> $$= 144 \text{ circular mils}$$

The conversion from the linear measurement in decimal units of the inch into circular-mil area is shown in the third column in Figure 3-6. This is a handy reference guide.

The next column gives the amount of resistance for each 1000 ft of copper wire. Here, the resistance value increases as the wire diameter decreases. The use of a temperature measurement of 25°C is common. This temperature is considered to be equal to a normal room temperature of 77°F.

The final column provides a current rating for the wire. There is a direct relation between the size of the conductor and the quantity of current in amperage that the conductor can safely handle. A general rule is that as the size of the diameter increases, the conductor is able to safely carry more current. In the home, number 14 (14 gage) wire is used for circuits that normally carry a maximum of 15 A of current. A number 12 wire is used for 20-A circuits. Many electronic circuits use wire that ranges between number 20 and number 24 for connecting components. The normal term used is to *hook up* the components with pieces of wire. As a result of this terminology, this type of wire is often called *hook up wire*.

REVIEW PROBLEMS

3-8. What information is provided by a copper wire table?
3-9. How do wire gage numbers and wire size compare?
3-10. What is the resistance of 2500 ft of number 20 wire?
3-11. What is the maximum safe current rating for a number 14 copper wire?

Conductor Types There are a great many different types of wires that are in common use in electrical circuits. These range from a single bare wire to a plastic-covered group of wires called a *cable*. Some of these are shown in Figure 3-7. There are many ways of describing electrical wire. One of these is a description of the conductor itself. For example, some wires are made of one solid piece of copper. This type of wire is identified as **solid wire.** Other wires are made from several smaller wires twisted together to form a larger-diameter

Solid wire *Conductor produced from one piece of wire.*

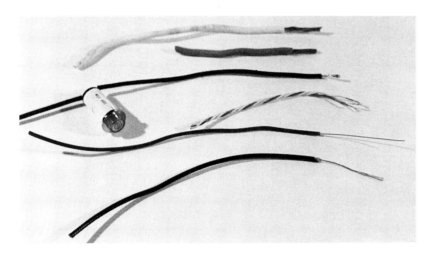

FIGURE 3-7 Wires used in electrical and electronic circuits, including bare wires, insulated wires, and multiple-wire cables.

wire. This type of wire is identified as **stranded wire** because it is composed of several individual strands of wire. In use the stranded wire is more flexible and less subject to breakage under conditions of vibration. Both the solid and stranded types of wires are usually covered with a plastic insulation material. This is true for wires used to connect circuits or components. Wire used for motor windings is also insulated. The insulation is a clear plastic, and such wire often gives the appearance of not having any insulation. Clear plastic insulation is much thinner than the colored insulation used on some wires. The insulation is required to keep two adjacent wires from accidentally touching and developing a short circuit. The insulation is available in several different colors. The colors used in electrical and electronic systems have a numerical value (Figure 3-8). The recommended color coding assigns number 1 to the color brown, two to red, and so on. Black, being zero, is used for ground in electronic circuits. This is an industry standard format. With numbers above 9 a two-color system is used. The unit color–number relation is the body of the wire, and traces of a second color are added to this basic color. Often, the second color is called a *tracer* because of this marking. For example, one could state that the color of the wire is green with a red tracer. This means that the basic color is green and there are red markings in addition to the green color of the insulation on the wire.

Often, two or more individual wires are manufactured as a unit. These multiple conductor wires are called *cables*. An electric cable may be produced in one of several forms, including flat ribbon cables, woven flat cables, and round cables (Figure 3-9). The round cable will often be covered with a rubber or plastic jacket. These are shown in the figure.

Stranded wire *Conductor produced from several small pieces of wire wrapped together to form a larger conductor.*

COLOR	NUMBER	COLOR	NUMBER
BLACK	0	GREEN	5
BROWN	1	BLUE	6
RED	2	VIOLET	7
ORANGE	3	GRAY	8
YELLOW	4	WHITE	9

FIGURE 3-8 Standard color code used by the electronic industries in the United States.

FIGURE 3-9 Assortment of multiple-wire assemblies, called cables by the industry.

Another type of conductor is shown in Figure 3-10. A single conductor cable is woven from many strands of thin conductors. This is the type of cable that is used to connect the negative terminal of an auto battery to the body and frame of the vehicle. The flat woven cable has a very low value of resistance and can carry large amounts of current.

The work done by Faraday showed us that an external magnetic field could have an influence on a conductor. In many instances this effect is undesirable, particularly when one uses wires for microphones and radio or TV receiver circuits. A method of keeping this from occuring is to cover the wire with a shielding material. The shield is made of a woven wire braid wrapped around the conductors. The shield is connected to a common point or a negative point in the circuit. Any electric fields that are present on the outside of the shield will not be able to influence the wires on the inside of the shield. The opposite is true, also. Any electric fields that are present and desired on the wires inside the shield cannot get out to affect other wiring. In many circuits the shield braid is used as one of the two required conductors, normally the common or ground conductor.

A further refinement of the shield cable is also shown in Figure 3-10. This is called a *coaxial cable*. It is used to carry radio-frequency signals from antennas to receivers in television installations. The wire used for transmission of cable

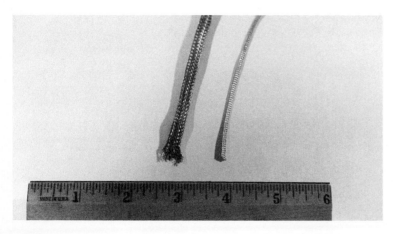

FIGURE 3-10 A braided or woven wire is used for grounds and connecting the commons of various units to each other.

TV signals is also coaxial cable. Antenna wires used for amateur radio and citizens' band radio as well as the wires used to carry signals from a computer to a video display monitor are also coaxial cables. Coax, as it is called, has specific power, frequency, and impedence characteristics. These must be considered when selecting a specific type of coax cable for use in a given installation. Refer to cable manufacturers' specifications for this information.

REVIEW PROBLEMS

3-12. What is the purpose of insulation on a wire?

3-13. What type of wire—solid or stranded—should be used for wiring in an automobile?

3-14. What is meant by the term *cable*?

3-15. Why is a shield used around a wire or cable?

3-2 Printed Circuits

Technological advances in the electronics industry have given us the **printed circuit** (PC) board assembly. Automation processes used to produce electronic controls and other products require that components be assembled and mounted in specific places in all of hundreds of the same unit. This is accomplished by using a prefabricated board that contains all the wiring for one or more circuits. A typical PC board is shown in Figure 3-11. The board is made of a nonconducting base material. Almost all production boards use a glass–epoxy base material. The raw PC board is produced with a coating of copper foil on either one or both of its sides. The foil is usually photo-processed and then etched in a chemical solution. The result is a base material that has copper conductive paths left on it. Holes for components to be mounted on the board are drilled or punched through the copper and the base. Next, the components are attached and soldered into place. After testing, the board is ready to be used.

    Proper care has to be taken when working with PC boards. The glass–epoxy base is reasonably strong but can fracture in extreme cases, such as

Printed circuit Electric circuit formed on insulated base material. Copper paths form conductive pattern.

FIGURE 3-11 The printed wiring or circuit board is now used in many systems. It replaces individual wiring in the circuit.

FIGURE 3-12 A small braided wire is used as an arch to remove solder from a connection.

dropping or extreme bending. Even if the base should not break, it is possible to develop a break in one or more of the conductive paths on the board.

When it is necessary to replace a defective or damaged component on a board, one has to be careful that the conductive foil paths are not damaged. The copper foil is attached to the base of the board with a strong adhesive. Under normal conditions the adhesive is able to withstand the effects of heat. When a component is replaced it usually has to be unsoldered. This process may possible melt the adhesive material. The result is that the foil will lift off the base material. It may even break off. Selection of a proper heat soldering iron will minimize this problem. Also, proper soldering techniques are required to keep the foil on the board, where it belongs.

Removal of components from PC boards is done carefully. One has to be certain that the wire lead of the component can be pulled through its mounting hole. The solder that is originally used to make the electrical connection between the lead and the PC board has to be removed. This can be accomplished by using a vacuum desoldering tool. It can also be done by placing a piece of braiding over the soldered connection and heating the braid with a solder iron (Figure 3-12). The braid will act as a wick and the solder will be absorbed from the board and component onto the wick. This usually leaves a clean, bare connection and the component can easily be removed from the board.

If the foil should break on the PC board, it, too, can be repaired. If the base board is cracked, it must be repaired first. Use of an instant-contact type of glue will repair the board. Soldering a small-diameter wire across the break in the foil (Figure 3-13) will repair it. Often, the process of tack soldering a piece of hookup wire across the break or breaks in the foil will make a temporary repair until the board or module can be replaced. There are some more so-

JUMPER WIRE USE
TO CROSS BREAK

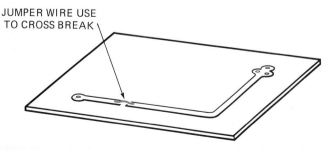

FIGURE 3-13 Breaks in circuit-board conductive paths may be repaired by soldering a piece of bare wire over the break.

phisticated techniques for board repair. These are usually performed in the repair shop. Some of these include soldering an overlay of copper foil to the original conductive pattern.

REVIEW PROBLEMS

3-16. What is a *printed circuit*?
3-17. Why should minimum heat be used when soldering on a PC board?
3-18. How can a cracked foil be repaired?

3-3 Schematic Diagrams

Persons employed in the electrical and electronics field have to be able to do much more than learn the theories and their applications. They must also be able to read and understand the blueprints or **schematic diagrams** of circuits. This is true for design, construction, repair, and maintenance purposes. In addition to the ability to use a common set of symbols to represent the components, they must be able to identify the actual components, that is, be able to locate a specific component on the actual circuit from the schematic circuit diagram. This ability requires a knowledge of typical component shapes as well as their general schematic symbol form. A person may also have to locate a specific point on the schematic diagram and relate it to the actual component in the circuit.

A further requirement for students, hobbyists, and designers is to be able to construct a circuit from its schematic diagram. A typical electronic schematic diagram is shown in Figure 3-14. This diagram is for a small dc power supply. To understand how to read this schematic, you first have to know how the system operates. This is best explained by using a group of building blocks to help explain everything. A diagram made up of blocks is shown in Figure 3-15. The process used to develop a dc voltage from the ac power lines requires several stages of operation. The block diagram reads from left to right, just like the lines of print in this book. The first block is called the *input block*. The power supply input is connected to a 120-V ac source, probably at a wall power outlet. The ac voltage has to be changed in value in order to develop

Schematic diagram *Electrical blueprint showing circuit configurations.*

FIGURE 3-14 A schematic diagram of an electronic power supply and voltage regulator circuit.

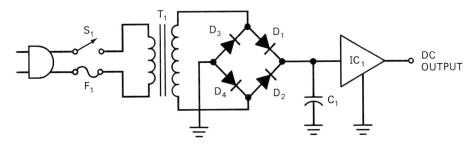

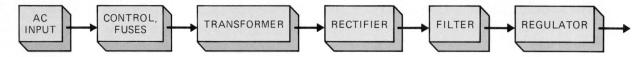

FIGURE 3-15 Block diagram of the regulated power supply circuit shown in Figure 3-14.

the correct operating voltage at the output of the power supply. Usually, a device called a *transformer* is used to change one value of ac voltage to another value of ac voltage. This is accomplished in the third block. The fourth block contains some electronic devices called *rectifiers*. Rectifiers are one-way switches. They are used to change the ac wave's changing polarity into a constant polarity, or dc. After the rectification process the dc voltage requires further processing. At this point the dc wave looks like one half of the original ac sine wave. The desired output voltage should not have any variation in its amplitude. The process of reducing or removing any voltage variations developed by rectification is called *filtering*. The output of the dc power supply system is also shown in this block.

The block diagram provides us with a diagram of the functional systems used in the power supply. We have not been shown any actual components or how the blocks relate to schematic symbols. This is illustrated in Figure 3-16, where the blocks are overlaid on the schematic diagram. This will provide us with a better understanding of the components located in specific blocks. Here, as in the block diagram, the process flow is from the left to the right. If a schematic diagram needed additional space on the page it would still show the process flow from left to right on one line and then be continued on additional lines. The process flow would continue to be similar to the lines in a book, using as many as required to complete the diagram.

Figure 3-17 shows the schematic diagram and photo of the actual unit. Letters are used to show the relation of a specific component to its schematic symbol. At this point you will notice that the actual layout of the components is not an "in-line" type of drawing as we saw on the schematic. The art of "packaging" electronic components is a separate subject and too extensive to be presented in this book. Several excellent publications relating to electronic construction are available to readers.

FIGURE 3-16 Blocks over each section of the power supply showing how the two match.

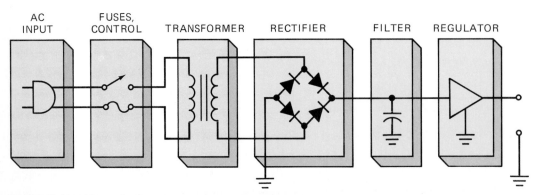

(a)

FIGURE 3-17 A commercial regulated power supply and its schematic diagram.

There are two other types of formats that are present on most electronic schematic diagrams. One of these is related to signal processing. The other is the general layout format for schematics. A layout for a traditional schematic is shown in Figure 3-18. Almost all electronic devices are used to control or condition electrical waves. The electrical wave inside the unit is actually a desired variation in the operating voltage of the unit. These controlled variations are called **signals**. Signal processing flow diagrams are laid out in the manner described previously. Signal processing flow is from left to right and usually one line at a time. If necessary, the process flow path is continued on the second, third, and additional lines, as required to show the complete diagram. Traditionally, signal input to the system is shown in the upper left-hand section of the schematic. The signal output is shown on the right-hand side of the last line in the schematic. A power source is usually shown in the lower left-hand section of the diagram.

Signals *Term used to describe information-carrying components of voltage in system.*

REVIEW PROBLEMS

3-19. What area of a schematic diagram shows the input?
3-20. What area of a schematic diagram shows the output?
3-21. How does a block diagram differ from a schematic diagram?
3-22. How is signal flow "read" on a schematic?

A problem that is present in any schematic diagram is that the printed schematic cannot show depth conveniently. This may cause some confusion when one sees that the connection wiring is shown as lines drawn on the schematic. A problem arises when one drawn line has to cross another line. A method is used to show that the lines, which represent wiring, cross each other but do not touch. Another method has to be used to show that the lines connect at the point of their intersection. Both of these methods are illustrated in Figure 3-19. One method uses a dot to indicate an electrical connection. When a dot

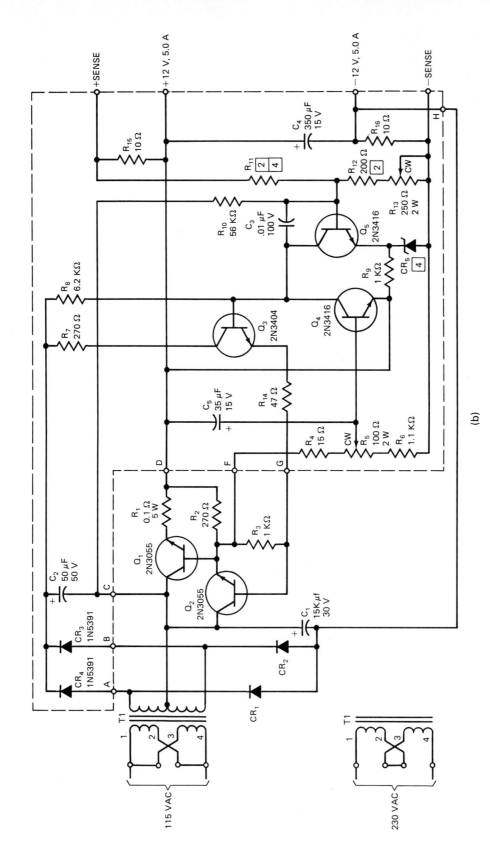

FIGURE 3-17 *(continued)*

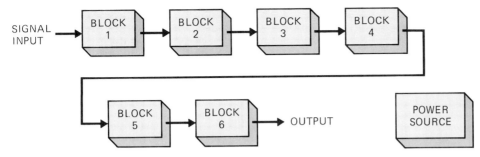

FIGURE 3-18 The method of layout for a schematic diagram showing signal flow and placement of the power source.

is shown covering an intersection, there is a connection at that point. In this format if there is no dot, the lines indicate that the wires cross but are not connected electrically. The second method uses a half-circle to indicate that the lines cross and are not connected. In this system when the half-circle is not used, there is a connection at that point. When reading a schematic it is wise first to determine which of the two systems is being used.

Almost all of manufacturers and authors in the electronics field have a set of graphic symbol standards. The Institute of Electrical and Electronics Engineers (IEEE) has standards for graphics as well as for other electronic products. A complete set of these is available from their offices. These are recommended and one will find that there are deviations from these by different companies. Overall, the recommended standards help to provide us with a common means of communicating. A partial set of these standards is included in Appendix B.

Included in the format for a schematic diagram is an order to the voltage values shown on the schematic. Each circuit has a format that seems to be consistent for all manufacturers. In this arrangement the lowest voltage is at the bottom, or lowest point, of the circuit. As the voltages increase in value their position in the schematic also rises. Usually, the highest level of operating voltage is at the upper part of the circuit containing the circuit schematic (Figure 3-20).

Connections and Symbols Often, the draftsperson will attempt to minimize the number of lines and connections on a diagram. When diagrams are very complex the number of lines used to connect components from one section to another can produce a drawing that looks like a solid mass of lines. It could

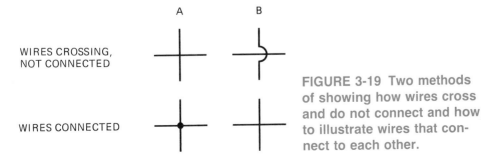

FIGURE 3-19 Two methods of showing how wires cross and do not connect and how to illustrate wires that connect to each other.

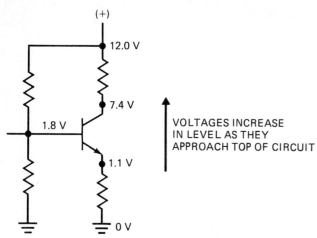

VOLTAGES INCREASE
IN LEVEL AS THEY
APPROACH TOP OF CIRCUIT

**FIGURE 3-20 The arrange-
ment of voltages in a circuit.
Usually the lowest level of
voltage is at the bottom of
the circuit.**

be difficult to follow the path of one line. Often the draftsperson will use alternative methods of showing these connections. Some of these are illustrated in Figure 3-21. One of the most often used symbols is called *ground* or *common*. In a great many circuits the negative teminal of the power source is connected to many different places in the circuit. Do not forget that this is the source of all electrons for the system. They flow from the negative terminal through the load circuits and then to the positive terminal of the power source. The use of a graphic symbol instead of lines on paper is an accepted method of showing these connections. This is the symbol labeled **ground** in Figure 3-21. Sometimes the term **common** is used to indicate these connections. Keep in mind that the actual circuit has a connection from the power source to each of these places. The schematic symbol indicates this fact.

The positive connection to the power source can be drawn using another symbol. This symbol is the plus sign (+) shown at the top of the drawing. Other types of line-limiting art include the use of pairs of letters or numbers

Ground *Term used to identify
a common connection, usually
for the negative terminal of the
power source.*

Common *Term used to
represent a reference point for
voltage measurments and
ground.*

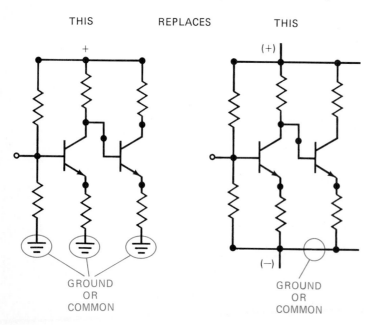

THIS   REPLACES   THIS

GROUND
OR
COMMON

GROUND
OR
COMMON

**FIGURE 3-21 Use of the
"common" or "ground" sym-
bol to replace a continuous
line on the page.**

to indicate two connecting points on a schematic diagram. This format also reduces the quantity of black ink on the drawing.

REVIEW PROBLEMS

3-23. What two methods are used on the schematic to show that a connection is made at a point?

3-24. What is the normal order of voltage values that is used on a schematic?

3-25. What does the term *common* mean?

3-26. Why is a symbol used to show common connections?

Other connections that are required in almost all equipment and assemblies include what are called input and output jacks. There are a great many different types of connecting jacks. Catalogs are available that show a good many of these. For the purposes of this book we will show some of the more common types. As you become familiar with specific equipment, you will recognize other types of termination. Some of the more common jacks and their mating plugs are shown in Figure 3-22 (a). The names used to identify these terminators are the ones in common use in the industry. Schematic symbols [Figure 3-22(b)] are shown where applicable. Some of them are single-contact units. Others, like the RCA type and the phone jack, have two or more connections on the individual item. The header is a multiple-contact connector. It is used in integrated-circuit construction and cabling.

FIGURE 3-22 Typical electrical connectors and some of the more common schematic symbols used to represent them.

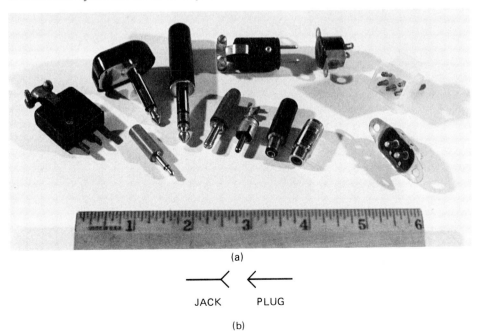

(a)

JACK   PLUG

(b)

3-4 Switches

A convenient method of turning a circuit on or off is to utilize a **switch** in addition to the source and the load. A switch could be described as a two-condition resistance. When the switch is *closed*, or *on*, its resistance is close to 0 Ω. Current will easily flow through it. When the switch is *open*, or *off*, its resistance is infinitely high and no current will flow in the circuit. Switches are identified by their physical shape and electrical connection capability. Some typical switch types are shown in Figure 3-23(a). The knife-type switch is probably one of the early types of switches used. It has a blade and a spring connector. When the blade is inserted into the spring connector, the switch is closed. This is illustrated by its graphic symbol [Figure 3-23(b)]. The circle indicates a movable hinge, the straight line with the arrow is used to show the blade, and the second circle is the spring connector. Wires are connected to each circle in order to use the switch. Each blade or circuit on the switch is called a *pole*. A knife switch that has two insulated sets of blades is called a *double-pole* switch. The number of different positions that can be connected to one pole is an additional way of describing the switch. When there is only one connection the switch is called a *single-throw* type. Each additional connection for the blade adds an additional throw number. A twin-blade knife switch can be used to connect either one of two spring contactors. It is called a *double-pole double-throw* switch. Often the first letters of the functional

(a)

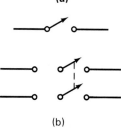

(b)

FIGURE 3-23 Typical switches used in equipment and their schematic symbols.

description are used instead of the whole name. Using this system, this switch is identified as a DPDT type.

The knife switch has been refined over the years. Almost all switches of this type are now built into a plastic housing. A handle is used to change the internal switch contacts. The handle *toggles* the contacts from one setting to the other. These are called *toggle-type* switches because of this. They, too, are described using SPST, DPST, and similar letter identifiers. Toggle switches have become miniaturized, as have many of the other components we find in circuits. One of the smallest types of toggle switches is shown in the same illustration. It is called a *DIP* switch. It is the same size and shape as the sockets used for integrated electronic circuits and fits into these sockets.

Still another method of describing a switch is the reference to the position of the contact and blade when the switch is in its usual position. The terms *normally open* (NO) and *normally closed* (NC) are used to show how the contacts are positioned. The NO and NC terms are usually used with switches that have spring-loaded handles or contact assemblies. These terms are not normally used with toggle or knife switches.

The **relay** shown in Figure 3-24 is an electromagnet type of switch. Relays have two distinct sections. One of these is a coil assembly. Coil assemblies are wound around a steel core. Current flow through the coil makes it an electromagnet. Coil specifications include ac or dc voltage levels as well as, in some types, a current or resistance rating. This, of course, depends on the application. The second section of the relay consists of one or more sets of contacts. The contacts are connected to a metal plate. The contact and plate assembly rotates on a wire pivot. It is also spring loaded so that one set of contacts is always

Relay *An electromechanical device used to control a circuit.*

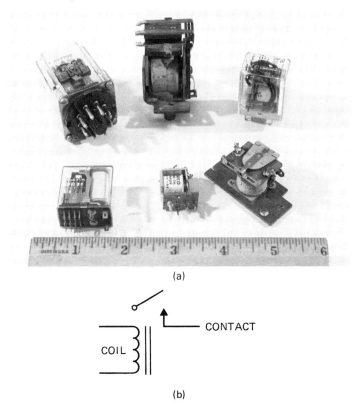

(a)

(b)

FIGURE 3-24 Electromagnetic switches, or relays, and their schematic symbol.

closed. When the coil is energized the plate is pulled against the coil's core by magnetic attraction. This breaks the first set of contacts and makes a second set. The relay is used to control secondary circuits in this manner. Relay assemblies often have four or more contact sets. One relay is able to control several circuits when it has multiple contact sets.

REVIEW PROBLEMS

3-27. What is a *jack* in electrical terminology?
3-28. What is meant by the term *DPST switch*?
3-29. Explain in electrical terms the word *open*.
3-30. What do *NO* and *NC* stand for?
3-31. What is a relay?

3-5 Circuit Protectors

A short circuit develops an undesired overcurrent condition in a system. If it is not stopped, this condition could result in damage to other units or to additional components in the circuit. There are certain types of devices used in electrical and electronic systems that are specifically designed to fail. These are known as fuses, circuit breakers, and fusible wire links. In each of the devices the extra heat that is generated by an overcurrent condition produces the failure. The simplest of these devices is the *wire link*, used extensively in automobiles and occasionally in television receivers. The link is simply a short piece of wire held in place in the circuit by mounting it on a terminal strip. The wire will melt under overcurrent conditions. This produces an open-circuit condition and results in the shutting down of the circuit.

Fuse *Circuit-protective device designed to fail under excess current conditions.*

The **fuse** is similar to the wire link operational theory. The fuse is built into a standard-size body to hold its wire link. Most fuses used in the electronics industry are similar to those shown in Figure 3-25. Fuses are rated in four ways. One of these is used to indicate its physical dimensions. The fuse will have the letters *AG* in its type number if it is similar to the glass type shown in the illustration. One manufacturer uses numbers in front of the "AG" to indicate specific size. The numbers used include 1AG, 3AG, 4AG, and 8AG. Another manufacturer uses a third letter. Types using AGA, AGC, and AGX are common. A catalog that provides specific size information is necessary to determine the exact characteristics. Both styles of identification also have a current rating. This is the second rating of a fuse. The glass-type fuses used for electronic systems will range from a low value of $\frac{1}{100}$A to 30A, with many intermediate values.

The third fuse rating is its operating voltage. Glass cartridge fuses are rated at either 32 or 250 v. This rating provides a safety factor when the fuse blows. Voltage can jump across an open when it is high enough to break down the insulating qualities of the air inside the fuse. If this should occur, the circuit will still be completed, current will flow, and the normal protection quality of the fuse is gone.

The final rating for a fuse is related to the amount of time the overcurrent

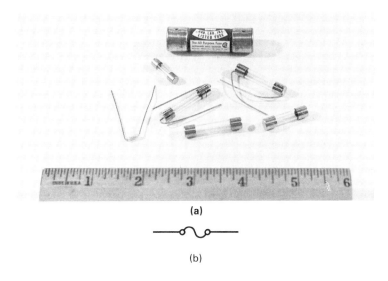

(a)

FIGURE 3-25 Cartridge fuses used in the electronics industry look like these. (Courtesy of Bussman Division, Cooper Industries.)

(b)

is allowed to occur before the link melts. A standard fuse link will melt immediately when a 10% overcurrent occurs. This fuse is called a *fast-blow* type. Sometimes it is necessary to allow an overcurrent for a short period. Devices that use electromagnetic principles, such as motors or relays, have a relatively high starting or turn-on current factor. Once the device is energized, its current is reduced to a lower level. If it were rated at its normal operating current, the fuse would blow each time it was turned on. On the other hand, if the system was fused at the starting current level, the circuit would not have adequate protection. A time-lag or *slow-blow*, fuse is designed to accommodate an overcurrent for a few seconds. This type of fuse provides adequate protection under both types of operating conditions.

Fuses are not normally repairable. Some large industrial types have a replaceable link. The link is held in place in a tubular body with removable end caps. The end caps are taken off in order to replace the destroyed link. Most glass cartridge types are throwaway items. A more sophisticated circuit-protection device is the **circuit breaker**. It uses a bimetallic strip contact arm. The heat produced by overcurrents will cause the bimetallic arm to bend away from the normally closed contact. A spring inside the breaker then keeps the bimetallic strip from restoring the contact to its original closed position. A reset button is incorporated in the breaker for resetting.

Circuit breaker *Resettable circuit-protective device using a heating element and contact points for control of excess current.*

Testing Protective Devices When they have not failed, the various circuit-protection devices provide a continuous path for current in the circuit. A device that can be called a *continuity tester* is used to test the circuit. One such device is shown in Figure 3-26. It consists of a battery, a lamp, and a set of test leads. This tester should only be used when the power source of the system is turned off or disconnected. If the device to be tested is good, it will complete the circuit and the lamp will light. If the device is open, the circuit is not completed and the lamp remains dark. This simple continuity tester can be replaced with a more sophisticated unit called an **ohmmeter**. This unit measures resistance in the circuit. A good fuse or breaker will have little or no measurable resistance. An open fuse will measure an infinitely high value of resistance.

Ohmmeter *Instrument used to measure resistance quantities in circuit.*

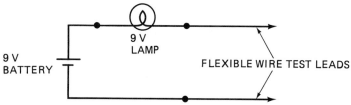

9 V
BATTERY

9 V
LAMP

FLEXIBLE WIRE TEST LEADS

FIGURE 3-26 Schematic circuit diagram for a simple continuity tester, and a tester built with readily available components.

Other methods of testing these devices require that the power to the circuit be connected. One way of testing the device is to measure the voltage drop that develops across it as current flows through it. Under normal working conditions the zero resistance at that point in the circuit will not indicate any voltage drop (Figure 3-27). If the device is open, the full value of source voltage will be measured across its terminals. The final way to test is to place the leads of a current-reading meter, or ammeter, across the device. If the device is good, the current, following the path of least resistance, will bypass the meter. There will not be a reading on it. If the device is open, the ammeter will display the amount of current actually flowing in the circuit. This type of test is both good and bad. The good point is that there are times when fuses or circuit breakers fail for no apparent reason. Some of the less expensive breakers seem to "get tired," or fail prematurely, after a few failures. Their point of failure is then less than their rated value. The use of a meter to measure actual circuit current will indicate if this is the situation. The bad part of the test is that the circuit-protection device is now bypassed. A rapid momentary test of the circuit current is permissible. Leaving the current on for any other period of time could damage other components if a short existed in the system.

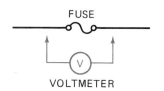

FUSE

VOLTMETER

0 V = FUSE GOOD

SOURCE V = OPEN FUSE

FIGURE 3-27 The voltmeter can be used to test a fuse while it is in the circuit.

REVIEW PROBLEMS

3-32. Name three circuit-protective devices.
3-33. What is the purpose of a voltage rating on a fuse?
3-34. What is a "slow-blow" type of fuse?
3-35. How is a circuit-protective device tested?

3-6 Circuit Construction

One of the many talents a beginner in electronics needs to develop is the ability to construct a successful circuit. This is normally done using a schematic diagram as a reference. Circuits are built in several different ways. These range from a highly sophisticated final design that is ready for production to a test design that is built on a universal layout board. The beginner in the electrical and electronics fields usually constructs test circuits on a temporary layout board. One type of board is shown in Figure 3-28(a). This board has several sets of contacts. The partial layout of these contacts is shown in Figure 3-28(b). Components are pushed into holes on the board. Up to five components can be connected in this manner in the groups of holes that are vertical in the illustration. Two groups of holes can be connected together by using a short jumper wire between sets of holes. On this board there are multiple sets of five holes on each half of the board. The spacing of the holes and the distance between the upper and lower halves of the board is designed to accommodate integrated-circuit chips. The horizontal holes in the long row along the top of the board are all connected together. This group of connections, in turn, is usually connected to the positive terminal of the power source. The holes in the lower horizontal row are also all connected to each other. This row is usually connected to the negative terminal of the source. It becomes the common connection for the circuit.

The easiest way to build a test circuit on a prototype board is to build it exactly like the schematic diagram. If possible, try to transfer the printed schematic form to the parts layout on the board. A sample of this type of

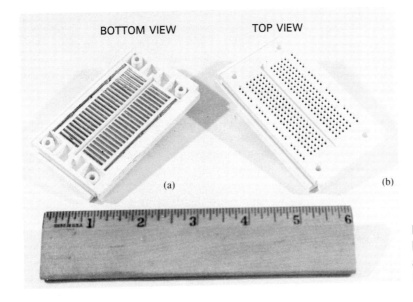

BOTTOM VIEW   TOP VIEW

(a)   (b)

FIGURE 3-28 A method of building a test circuit utilizes a prefabricated prototyping board such as this one.

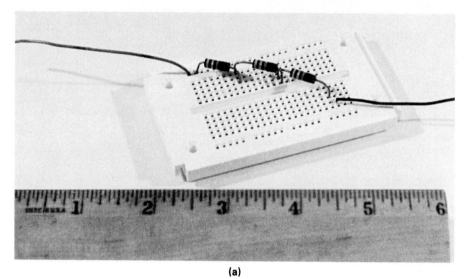

(a)

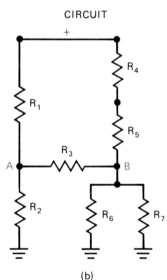

CIRCUIT

(b)

FIGURE 3-29 An electrical circuit and one method of constructing it on the prototyping board.

transfer is shown in Figure 3-29. The circuit shown is a rather simple resistive network. It is not designed to be a functional circuit. The major purpose is to provide an example of how to do this type of work.

Before starting with the actual development, let us examine the schematic diagram. Each resistor in the circuit has an identifying number. These are the subscripts given after each R letter. They range from R_1 to R_7. Each dot on this schematic represents an electrical connection between components and possibly the power source. This fact should help when constructing the circuit. For example, R_1 has one of its leads connected to the positive terminal of the power source. Resistor R_4 also has one of its leads connected to the positive power source. When the circuit is complete it should have two resistor leads and one wire to the power source in one set of connectors on the board.

Resistors R_1, R_2, and R_3 all have one of their leads connected together. There are three lines on the schematic which all converge at the common dot. This dot is identified by the letter A. The other lead on resistor R_2 is connected to circuit common. Both ends of R_1 and R_2 should be connected and one end of R_3 should also be connected. The loose end of R_3 can now be inserted into any convenient vacant set of holes.

Resistors R_4 and R_5 are connected in a line between the positive rail and the right-hand side of R_3. Any vacant set of holes is used for the middle connection of R_4 and R_5. There are four components to connect at point B on the schematic. These include one end of R_3, one end of R_5, R_6, and R_7. When the leads of these components are all inserted into one set of holes, four of the five holes in that set should be filled. The other ends of R_6 and R_7 are both connected to the negative, or common, rail in the bottom of the board. The final result should look similar to the one shown in Figure 3-29. Any creative variations are always acceptable in this type of prototype layout. Circuits similar to this are used in laboratory classes to help students understand the theories of electricity and electronics. You may wish to procure a board of this type in order to build and test some of the circuits used in this book. These boards are available from several sources of electronic components. The sources are retail outlets and electronic parts wholesale stores. In addition, check the mail-order advertisers in magazines related to electronics or in your community's telephone directory.

3-7 Basic Troubleshooting Procedures

TROUBLESHOOTING APPLICATIONS

When one constructs electrical or electronic circuits there often is a need to test the circuit. There are times in the life of most of us when we may make an error in the construction of a circuit. If the circuit is simple, a check of the actual circuit against the circuit shown in the schematic diagram is reasonably easy to do. When the circuit has several sections or blocks, the process of troubleshooting is more complex. A block diagram of a multiple-section system is shown in Figure 3-30. This diagram is a display of the functional blocks of an eight-section system. The signal processing in this system is from left to right. There is only one path for the signal in this system. It is called a *linear*, or straight-in-line system. The linear system is one of the most common types of flow-path systems used today. Linear systems are also used in the basic electrical circuit and in many simple current-flow-path circuits. The method of troubleshooting a system of this type is relatively simple. One should learn to establish a regular procedure for doing this. The best technique to use for this type of circuit is called the *half-split method*. This method is used successfully for both signal flow systems and current flow systems. All that one has to do is to check for the proper signal, current, or voltage at the midpoint of the

FIGURE 3-30 A basic multiple section block diagram for a linear system.

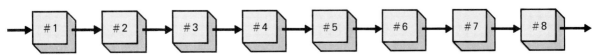

system. This is the third test to be made. It is done after checking the input and the output. A flowchart for this half-split system is shown in Figure 3-31. When one has a good input and there is no output, the next logical test is at, or near, the midpoint of the system. This test eliminates one-half of the system. It is determined using the rationale that if the proper operating conditions are found at the midpoint, everything from the input to the test point is working properly and the system problem is in the area between the point of the test and the output point. If, on the other hand, the test at the midpoint of the system shows an improper operating condition, the problem is located between the input and the test point. One test has thus eliminated 50% of the system. This type of test is efficient and very productive to anyone performing either service or general system testing.

The fourth step in this process is to make a test at the midpoint of the section of the system that was determined to be nonfunctional by the first test. The results of this test are analyzed in the same manner as before. This process is continued until only one section remains. The trouble is then determined to be in that section.

If one were making tests on a linear signal processing system and reduced the area of trouble to one section, the next step would require use of the half-split method to analyze the current flow in the suspected system. Use Figure 3-32 as a reference for this description. It shows a linear system consisting of two resistors, R_1 and R_2, connected to elements of a transitor. This is a *series* type of circuit. It is actually a linear circuit. Electron flow is from common, through R_2, the transistor, and R_1, then to the positive terminal of the power

FIGURE 3-31 A flow chart showing how to approach and test the system shown in Figure 3-30.

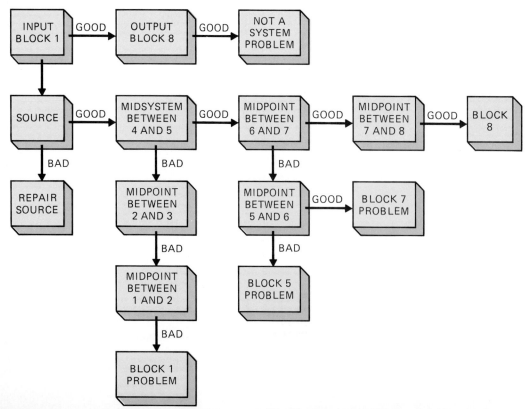

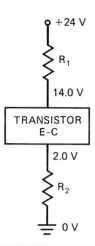

+24 V

R_1

14.0 V

TRANSISTOR
E–C

2.0 V

R_2

0 V

FIGURE 3-32 Another form of linear circuit is shown here in this transistor circuit.

source. The test procedure for this system is the same type as that used for the signal flow path system. First test the input electron flow from the power source. In this circuit the test is to measure the source voltage. Never, never overlook the fact that the problem could be originating at the power source. Once it is determined that the source power is correct, the next test is made at the midpoint of the circuit. In this circuit there is no specific midpoint. The technician has to make a choice as to the specific test point. It should be done at the center of the components that make up the circuit. It does not make a lot of difference whether the test is made at the lower end of R_1 or at the upper end of R_2.

A test of the voltage at the lower end of R_1 should indicate 14 V. If the voltage measured at this point is equal to the 24 V at the positive terminal of the power source, the circuit is open between the measured points. In this case it would be between the upper terminal, or collector element, of the transistor and circuit common. The explanation for this is that we expect to measure less than the source voltage here. A measurement equal to source voltage indicates that there is a no-current fault in the system. Normally, when current flows we expect to have a voltage drop across each component in the system. The test result shows no voltage drop across R_1.

The fact that there is no voltage drop across a component does not always mean that the component is defective. Another test has to be made before a final decision is made. This test is made at the upper part of R_2. If there is 0 V at this point, the problem area is between the two test points. It indicates that the fault is the transistor. If the test shows source voltage (24 V) at this point, the fault is not the transistor. It is between the test point and common. We would be correct to state that the fault is associated with resistor R_2. Either the resistor is open or one of its connections to the circuit is open. In either case the problem is located quickly using a common test procedure. This test procedure can be used successfully for any type of linear circuit. The student of electronics or electricity should learn to think the problem through before actually making any tests. The time spent in thinking and planning a troubleshooting approach may seem like nonproductive time, but in reality it is a part of the proper approach to troubleshooting. In practice this time will actually contribute to a more efficient process.

1. Electrical conductors should have free electrons and low-resistance factors. Typical conductors are copper, silver, gold, and aluminum.
2. Copper is the major conductor in use because of its availability and relatively low cost.
3. An insulator is a material that normally does not have any free electrons. Typical insulators are oil, mica, air, and ceramics.
4. The insulating quality of a material is called its dielectric strength.
5. The purpose of any conductor is to permit electron flow from the source to the load and back to the source. This forms a basic electrical circuit.
6. Wire resistance affects the current-carrying capability of a conductor.
7. Wire gage number is the system used to describe wire size and resistance. It is inversely proportional to the physical size of the wire.
8. Wire conductors are available as single conductors of multiple-conductor cables and may also have a shielding over the conductor.
9. Printed circuit boards are used instead of discrete wiring in automated production processes.
10. A schematic diagram is used to show the electrical wiring of a circuit or system.
11. A block diagram shows a functional section diagram of a circuit or system.
12. Schematic diagrams are "read" like the printed page of a book. Voltage values rise to higher levels as the wires on the circuit rise on the schematic. (The lowest line has the lowest voltage value.)
13. Components are shown on a schematic diagram using standard graphic symbols.
14. A switch is a mechanical device used to make an electrical connection. A relay is an electromechanical device used for the same purpose.
15. Circuit-protective devices include wire links, fuses, and circuit breakers.
16. Fuses are rated by size, current capacity, operating voltage, and failure-time values.
17. A circuit can be constructed on a prototype board. One of the easiest methods of doing this is to follow the schematic diagram.

SELF TEST

Answer true or false; change all false statements into true statements.

3–1. Pure silicon is a good electrical conductor.
3–2. A fuse is designed to fail.
3–3. The major metal used for electrical conductors is steel.
3–4. Current flow in a conductor is affected by the resistance of the conductor.
3–5. Schematic symbols are used to represent actual components.
3–6. The term *common* is used to represent a reference point in an electrical circuit.
3–7. A relay is an electromechanical switch.

3-8. A short is an abnormally high resistance.

3-9. Dielectric strength is related to a voltage insulating factor.

3-10. Electronic schematic diagrams are read from left to right and line by line.

3-11. An electrical conductor's purpose is to

   (a) carry voltage to the source

   (b) create circuit resistance

   (c) insulate the source from the load

   (d) carry current between the load and the source

3-12. The most commonly used metal in a conductor is

   (a) aluminum

   (b) copper

   (c) gold

   (d) silver

3-13. Voltage loss in a conductor is caused by

   (a) source current

   (b) load resistance

   (c) conductor resistance

   (d) use of ac on the wires

3-14. The largest-diameter wire of those listed here is

   (a) number 28

   (b) number 14

   (c) number 32

   (d) number 8

3-15. The term *circular mils* refers to the

   (a) length of a wire

   (b) diameter of a wire

   (c) cross-sectional area of a wire

   (d) current capacity of a wire

3-16. Electrical cables are

   (a) multiple-conductor wire sets

   (b) often shielded

   (c) sometimes made like a flat ribbon

   (d) all of the above

3-17. Wire used to carry radio-frequency signals is

   (a) called coaxial cable

   (b) never shielded

   (c) flat, like a piece of ribbon

   (d) none of the above

3-18. The diameter of a conductor determines its

   (a) length

   (b) current capacity

   (c) voltage rating

   (d) resistance per mil

MULTIPLE CHOICE

3–19. On a schematic diagram, the power source is usually located in the
  (a) lower left
  (b) upper left
  (c) lower right
  (d) upper right

3–20. On a schematic diagram, signal processing is
  (a) from right to left, line by line
  (b) from left to right, line by line
  (c) from bottom to top, line by line
  (d) independent of any lines

ESSAY QUESTIONS

3–22. Explain the term *conductor* using resistance values.

3–23. What happens when the dielectric strength of a device is exceeded?

3–24. Describe the terms *stranded wire* and *solid wire*.

3–25. What is a tracer, and how it is used?

3–26. Describe a printed circuit.

3–27. How is the signal path "read" on a schematic diagram?

3–28. What is meant by the term *wires crossing* on a schematic diagram?

3–29. Explain the difference between a schematic diagram and a block diagram.

3–30. Why is the symbol for "common" used on a schematic diagram?

3–31. A switch has the description 3PDT. What does this mean? (*Sec. 3-4*)

PROBLEMS

3–32. Find the circular-mil area for the following wires. (*Sec. 3-1*)
  (a) number 16
  (b) number 24
  (c) number 40
  (d) number 2

3–33. A circuit normally has 100 Ω of resistance. What current flows when each of the following source voltages is applied to it? (*Sec. 3-1*)
  (a) 24 V
  (b) 100 V
  (c) 400 V
  (d) 9 V

3–34. A circuit normally has 9 V applied to a 2.5-kΩ load. What current flows when a 0.2-Ω short occurs? (*Sec. 3-7*)

3–35. Draw the schematic symbols for the following components. (*Sec. 3-3*)
(a) common
(b) resistor
(c) dc source
(d) fuse

3–36. Draw the schematic symbols for two wires crossing and connected and two wires crossing but not connected. (*Sec. 3-3*)

3–37. A voltmeter is used to measure the voltage across a circuit breaker's terminals. The source voltage is 120 V and the voltage measured across the terminals is 120 V. Is the breaker open or closed? (*Sec. 3-7*)

3–38. An ohmmeter is used to measure a fuse's element. The measurement is 0.1 Ω. Is the fuse good? (*Sec. 3-7*)

3–39. What size wire can safely carry 20 A of current? (*Sec. 3-1*)

3–40. Draw a schematic diagram for a basic electrical circuit with a source and a load. Show how you would add a SPST switch and a fuse to the circuit. (*Sec. 3-3*)

3–41. What is the total resistance of 1500 ft of number 18 wire? (*Sec. 3-1*)

3–42. What would be the total resistance of the wire of problem 3-41 if it were cut in half? (*Sec. 3-1*)

3–43. The coil of a relay is wound with number 28 wire. There are 4000 turns of this wire on the coil. Each turn is approximately 1.5 in. What is the total resistance of the coil? (*Sec. 3-1*)

3–44. What is the space required to separate a pair of plates in free air when a voltage of 15 kV is applied to the circuit? (*Sec. 3-1*)

3–45. What is the voltage breakdown value for a piece of porcelain 3.1 cm thick? (*Sec. 3-1*)

TROUBLESHOOTING PROBLEMS

3–46. The resistance between the terminals of a plug and its mating jack is 230 Ω. Is this an acceptable value?

3–47. A circuit has a 100-V source, a 3-A fuse, and a load of 50 Ω. The voltage measured across the fuse is 100 V. Will the source operate the load under these conditions?

3–48. A total of 5000 ft of wire is required for an outdoor application. How much voltage drop occurs when the source voltage is 480 V, the wire size is number 6, and the load is 100 Ω?

3–49. Draw a schematic diagram of a circuit that uses a 12-V source to control a relay coil. The control device is a SPST switch and the relay contacts are SPST NC type.

3–50. A linear type of electronic unit has 14 blocks, or sections, in it. The input section works properly but there is no output from the last block. Using block numbers as a reference, show how to set up a method for rapid testing of the unit.

3–51. Using the block diagram shown in Figure 3-30, write out a plan for locating a problem of "no signal output" when the problem is in block 5.

3–1. Copper, aluminum, and silver

3–2. A conductor allows electron flow; an insulator does not allow electron flow.

3–3. Dielectric strength is the insulating quality of a material. It is measured in units of kV/cm.

3–4. It permits electron flow to and from the source.

3–5. Resistance is the opposition to electron flow or charge movement.

3–6. *Load* is a term used to represent the device that performs work.

3–7. As wire resistance increases the efficiency of the load decreases, due to voltage drops in the wire.

3–8. Gage number, area diameter, resistance per 1000 ft, and some current capacities

3–9. They are inverse to each other.

3–10. $2.5 \times 10.75 = 23.375 \ \Omega$

3–11. 15 A

3–12. To protect the wire from touching other metallic surfaces

3–13. Stranded; it is flexible and has less tendency to break due to vibration.

3–14. A cable is a multiple-conductor assembly.

3–15. To keep undesirable voltages from being induced onto the wire or to keep voltages on the wire inside the shield from affecting other wiring.

3–16. A printed circuit is a conductive copper path made on a base material.

3–17. Excessive heat may melt the glue that holds the foil to the board's base material.

3–18. Use a conductor and solder it across the break.

3–19. Usually, the upper left-hand area.

3–20. Usually, the lower right-hand area.

3–21. Blocks represent functional areas; schematic shows wiring of the circuit.

3–22. The same way that one reads a book: left to right and line by line from top to bottom.

3–23. (1) Wires crossing with a dot, or (2) wires crossing without a dot.

3–24. A point of reference in the circuit, usually the negative terminal of the power source.

3–25. A reference point in the circuit used for measurements.

3–26. It makes less clutter on the diagram.

3–27. A socket for an external connector.

3–28. Two poles and one way of throwing the switch blade.

3–29. A high-resistance situation, usually with a no-current-flow condition in the circuit.

3–30. Normally open and normally closed.

3–31. A electromagnetic switch.

3–32. Fuses, fusible wire, and circuit breakers.

3–33. Shows the safe operating voltage for the circuit.

3–34. A time-lag or time-delay fuse.

3–35. A continuity tester or ohmmeter will show if the device is good.

CHAPTER 4

Resistors

One of the most commonly used components in electronic circuits is the resistor. Its purposes include development of a voltage drop and the limitation of current flow in the circuit. Resistors are available in a great many different sizes, shapes, and values. The ability to recognize and select the proper resistance unit is critical knowledge for persons working with electrical and electronic circuits.

An interesting story is told about the scientist Georg Simon Ohm. This story relates to his development of the relationship of voltage, current, and resistance in an electrical circuit. The story states that Ohm's first attempts to show this relationship were proven wrong. He had to revise his theories and try again. We now use the second published version of his law as the correct version.

KEY TERMS

Axial	Radial
Coefficient	Resistance
EIA Color Coding System	Resistor
Fusible Resistor	Rheostat Circuit
Heat Sinking	Series
Multimeter	Thermistor
Multiplier	Tolerance
Ohmmeter	Variable Resistor
Parallel	Varistor
Photoresistor	Watt
Potentiometer Circuit	Zero-Ohms Resistor
Power Resistor	

OBJECTIVES

Upon completion of the material in this chapter, you will
1. Be able to explain the concept of resistance.
2. Be able to identify resistance values by color coding.
3. Understand resistor power and voltage ratings.
4. Be able to identify a basic series or parallel circuit.
5. Understand the concept of temperature coefficients.
6. Understand the operation of special types of resistors.
7. Know why resistors fail.

Resistance *Term used to describe quantity of opposition to current flow in circuit.*

The term **resistance** is defined as an opposition to the flow of electrons in a circuit. It is also described as being similar to friction in an electrical circuit. Both of these descriptions may make one feel that circuit resistance is an undesirable quality. Quite the opposite is true. Probably the most used component in any electrical or electronic circuit is the resistor. Resistors are used in circuits to limit current flow. They are also used to develop a desired voltage drop in the circuit. In many instances the operating voltage at the elements of a transistor or an integrated circuit is less than the source voltage. Currently flow through a resistance produces a voltage drop. This is used to provide the proper level of voltage for the device.

There are many factors that influence the size, shape, and rating of resistors. There are also some standards for size, shape, and ratings. In addition to this, some resistors are marked using a standard color coding system to indicate their numeric value in ohms. All of these topics and others related to resistance are presented in this chapter. Since resistors are used to control electric power it is necessary to discuss the concepts related to electric power and work. The first section of this chapter covers this topic. Once you have an understanding of electric power, the role of the resistor in a circuit will have more significance.

4-1 Resistor Ratings

Resistors are normally rated in two major ways: in terms of power and in terms of resistance. Power ratings are in units of the watt (W). Resistance ratings are in units of the ohm. Resistors' ohmic values are totally independent of their power ratings. For example, a 10-Ω resistor can be purchased with one of several different power ratings. The specific requirement for a power rating is a factor that must be understood by those working in electrical and electronic fields.

Watt *Unit of electric power.*

Electric Power The unit of electric power is the **watt**. It is defined as the amount of work done in 1 s by 1 V moving 1 C of charge. The formula for this is

$$P = I \times V$$

where P = power, watts
I = current, amperes
V = potential, volts

The letter V is used here to represent the amount of voltage that is required to move a charge. There are three ways of expressing the power formula:

$$P = I \times V \qquad (4\text{-}1)$$

$$I = \frac{P}{V} \qquad (4\text{-}2)$$

$$V = \frac{P}{I} \qquad (4\text{-}3)$$

EXAMPLE 4-1

A television receiver is rated at 660 W. It requires 120 V for operation. What is its current?

Solution

$$I = \frac{P}{V}$$

$$= \frac{660 \text{ W}}{120 \text{ V}}$$

$$= 5.5 \text{ A}$$

EXAMPLE 4-2

A toaster uses 12 A of current when operated from a 120-V source. What is its wattage rating?

Solution

$$P = I \times V$$

$$= 12 \text{ A} \times 120 \text{ V}$$

$$= 1440 \text{ W}$$

EXAMPLE 4-3

A light bulb is rated at 200 W. What is its current rating when operated from a 120-V source?

Solution

$$I = \frac{P}{V}$$

$$= \frac{200 \text{ W}}{120 \text{ V}}$$

$$= 1.67 \text{ A}$$

Electric power is more simply defined as the time rate of performing work. In commercial and home applications the electric energy used to operate machines or appliances is rated in units of the watthour. The quantities of electric power are so large that the more common method of kilowatthours (kWh) is used. Electric power is equal to a value of mechanical power. Mechanical power is rated in units of *horsepower* (hp). One unit of mechanical horsepower is equal to 746 W of electrical power. The formula for this is

$$1 \text{ hp} = 746 \text{ W} \qquad (4\text{-}4)$$

Electric power sold by power companies is rated in kilowatthours. The average rate for this power is presently about 6 cents per kWh. An electric appliance rated at 1800 W that is operated for 1 h will use 1.8 kWh of energy. The cost to the user is 1.8 × 0.06, or 10.8 cents.

EXAMPLE 4-4

What is the watthour rating for a 120-V 200 W lamp that is used for 3 h?

Solution

$$kWh = W \times \frac{h}{1000}$$

$$= 200 \text{ W} \times \frac{3 \text{ h}}{1000}$$

$$= \frac{600 \text{ Wh}}{1000}$$

$$= 0.6 \text{ kWh}$$

EXAMPLE 4-5

An electric device is operated from a 120-V power line for 5 h. Its current rating is 10 A. What is the kWh rating?

Solution

This is a two-part problem. First find the wattage rating and then find the kWh rating.

$$P = I \times V$$

$$= 10 \text{ A} \times 120 \text{ V}$$

$$= 120 \text{ W} \quad \text{or} \quad 1.2 \text{ kW}$$

$$kWh = kW \times h$$

$$= 1.2 \text{ kW} \times 5 \text{ h}$$

$$= 6.0 \text{ kWh}$$

Power in Resistance Earlier it was stated that one of the products of resistance is heat. The heat develops because of the friction of movement of the electrons in the conductor. The heat generated is a result of electric power application to the system. The application of voltage caused the movement of current in the circuit. These factors can be used in the power formula to determine the amount of energy required to do work. Evidence of this action is observed when the link of a fuse melts due to an overcurrent condition. The link of a fuse has a low-temperature melting point. The heat created by the current flow in excess of the design value makes the fuse link heat and melt.

In any circuit the quantity of power used by the circuit is equal to the amount of power created by the source. This is true whether the circuit is designed to generate heat or to perform another function. There is a direct relationship between electric power and heat. One watt of power that is used for 1 s is equal to 0.24 calories (cal) of heat energy. Power that is used during the operation of a system is often expressed in terms other than the watt. A common method of doing this is to use the formula I^2R. This formula develops from a combination of Watt's and Ohm's laws. It is accomplished in this manner:

$$V = I \times R \qquad \text{(Ohm's law)}$$

$$P = I \times V \qquad \text{(Watt's law)}$$

In any equation one equality may be substituted for another:

$$P = I \times (I \times R)$$

Combining terms gives us

$$P = I^2 \times R \qquad (4\text{-}5)$$

This is a commonly used formula for expressing electric power. The reason for this is that the heat produced in the system is the result of the resistance in the circuit. Another substitution in the formula is when V/R is substituted for I.

$$P = V \times I$$

$$= V \times \frac{V}{R} \qquad (4\text{-}6)$$

$$= \frac{V^2}{R}$$

Another example of how this works can be illustrated by the circuit shown in Figure 4-1. In this circuit the power dissipated by the resistance is 72 W. This is calculated as follows:

$$I = \frac{V}{R}$$

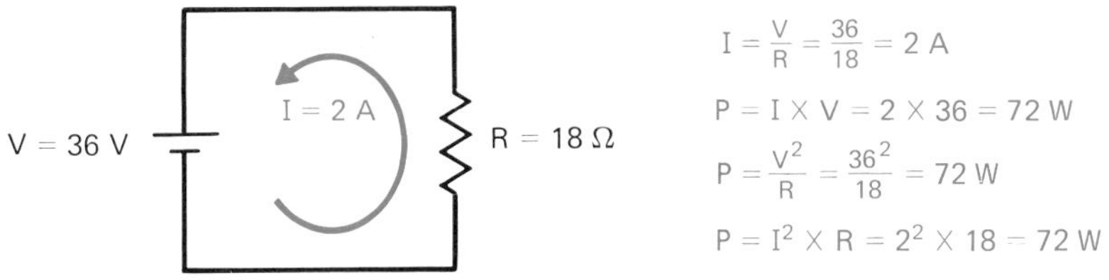

$$I = \frac{V}{R} = \frac{36}{18} = 2 \text{ A}$$

$$P = I \times V = 2 \times 36 = 72 \text{ W}$$

$$P = \frac{V^2}{R} = \frac{36^2}{18} = 72 \text{ W}$$

$$P = I^2 \times R = 2^2 \times 18 = 72 \text{ W}$$

FIGURE 4-1 The power provided by the source is used to operate the load. Three power formulas are illustrated here.

$$= \frac{36 \text{ V}}{18 \text{ }\Omega}$$

$$= 2 \text{ A}$$

Then

$$(1) \ P = I \times V$$

$$= 2 \text{ A} \times 36 \text{ V}$$

$$= 72 \text{ W}$$

or

$$(2) \ P = I^2R$$

$$= 2 \text{ A} \times 2 \text{ A} \times 18 \text{ }\Omega$$

$$= 72 \text{ W}$$

or

$$(3) \ P = \frac{V^2}{R}$$

$$= \frac{36 \text{ V} \times 36 \text{ V}}{18 \text{ }\Omega}$$

$$= 72 \text{ W}$$

Each of the three methods provides the same answer. In this circuit the heat dissipated by the resistance is 72 W. The power rating of the resistance has to be equal to or greater than this value if the resistance is to operate properly.

EXAMPLE 4-6

Find the power in a circuit that has a source voltage of 120 V and a resistance of 1.8 kΩ.

Solution

$$P = \frac{V^2}{R}$$

$$= \frac{120 \text{ V} \times 120 \text{ V}}{1800 \text{ }\Omega}$$

$$= 8 \text{ W}$$

EXAMPLE 4-7

Find the power in a circuit that uses a source of 9.0 V to produce a current of 100 mA through a resistance of 90 Ω.

Solution

$$P = \frac{V^2}{R}$$

$$= \frac{9 \text{ V} \times 9 \text{ V}}{90 \text{ } \Omega}$$

$$= 0.9 \text{ W}$$

or

$$P = I^2 \times R$$

$$= 0.1 \text{ A} \times 0.1 \text{ A} \times 90 \text{ } \Omega$$

$$= 0.9 \text{ W}$$

There is a very definite relationship between voltage and current in a power circuit. When one of them increases the other will decrease in order to maintain the same level of power. Another way of showing this is to use an example of a portable battery-operated lamp. Under normal operating conditions the lamp will consume about 2.2 A of current in order to create 20 W of heat. The heat is then converted into light energy by the filament of the lamp. If the operating voltage should drop 1.0 V, the current flow in the circuit would have to be increased from its original level of 2.2 A to a higher level of 2.5 A if the original circuit power values are to be maintained. On the other hand, if the voltage source is changed from 9.0 V to 12.0 V, the current required to develop the same amount of power drops to a value of about 1.67 A. In summary, a lower voltage will require higher levels of current in order to maintain a constant power level. Also, a higher voltage requires a reduced current level for the same amount of power.

4-2 Types of Resistors

The **resistor** is a device used to oppose the flow of current in an electric circuit. It is produced as a device containing some type of resistive material and a non-conductive housing. Two wires are used to connect the resistance element with the electrical circuit.

Resistor *Device made of resistive materials used to control circuit current.*

Resistors are usually categorized as either being low power or high power. The breakover point for these ratings is around 3 W. Low-power resistors have some general physical values. Their size is related to specific wattage rating (Figure 4-2). The standard power ratings for low-power resistors are $\frac{1}{8}$, $\frac{1}{4}$, $\frac{1}{2}$, 1, and 2 W. These resistors and their dimensions are shown in Figure 4-2. Note that the physical size increases with the wattage rating. The reason for this is that the larger physical size is able to dissipate more heat.

The low-power resistor is produced in one of three major forms. An early method of producing a resistor used a mixture of clay and carbon. These resistors are called *carbon-composition* types. The ratio of clay to carbon de-

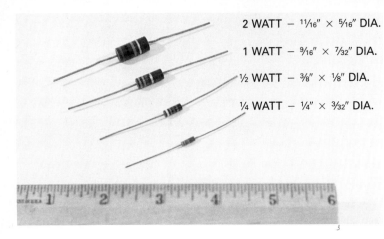

2 WATT — 11/16" × 5/16" DIA.

1 WATT — 9/16" × 7/32" DIA.

1/2 WATT — 3/8" × 1/8" DIA.

1/4 WATT — 1/4" × 3/32" DIA.

FIGURE 4-2 Low-power resistors. Usually, the physical size of the resistor body is represented by these standard dimensions.

termined the specific value of resistance. The mixture is held in a formed plastic housing (Figure 4-3). Two wire leads are utilized to allow current flow through the resistor.

As technology improved, the methods of producing resistors also improved. One current method used in the manufacture of resistors results in a film type of resistor, shown in Figure 4-4. It has a ceramic form on which is deposited a film of resistive material. The film is deposited in a spiral-like form to make its length as long as possible. An epoxy coating covers the resistive material. Wires are also used to permit current flow through the resistor.

The third form of low-power-resistor manufacturing results in the device shown in Figure 4-5. A thin resistance type of wire is wound onto a ceramic form. The length of the wire and its diameter determine the amount of resistance. The resistor is coated with an epoxy cover in a manner similar to that used for the other types of resistors. The use of a wire-wound resistor permits the production of ohmic values that are very precise.

Power resistor *Resistor having a power rating in excess of 3 W.*

High-power resistors are often called **power resistors** and low-power resistors are called simply resistors. Power resistors are also constructed on a ceramic form. A resistance wire is wound around the form. Its ends are connected to either wires or terminals. A select group of power resistors is shown in Figure 4-6. The shape of a power resistor may vary. This is because of differences in manufacturing methods. In general, the larger the power rating of the resistor, the larger its physical size will be. Power resistors are available in many wattage ratings. These will range from a low value of about 5 W to

FIGURE 4-4 Construction of the film-type low-power resistor. (From Joel Goldberg, *Fundamentals of Electricity*, 1981, p. 33. Reprinted by permission of Prentice-Hall, Englewood Cliffs, New Jersey.)

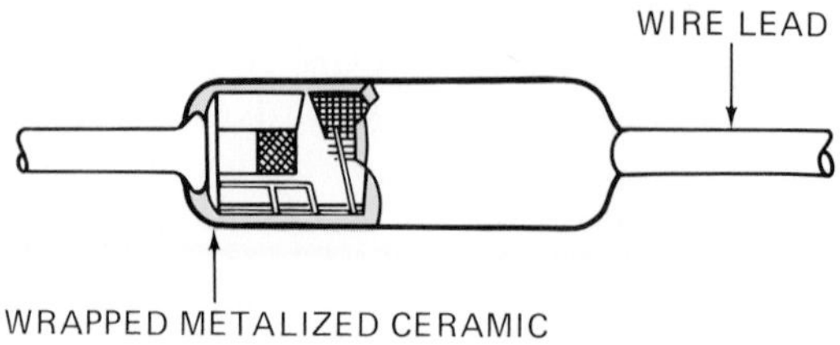

FIGURE 4-3 Construction of the carbon-composition resistor.

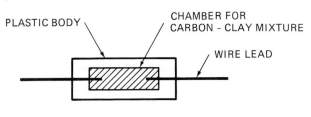

PLASTIC BODY

CHAMBER FOR CARBON - CLAY MIXTURE

WIRE LEAD

WIRE LEAD

SPIRAL WRAPPED METALIZED CERAMIC

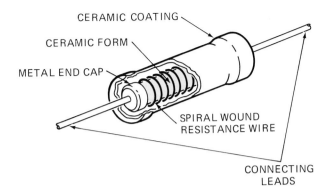

CERAMIC COATING

CERAMIC FORM

METAL END CAP

SPIRAL WOUND
RESISTANCE WIRE

CONNECTING
LEADS

FIGURE 4-5 Construction details of the low-power wire-wound resistor.

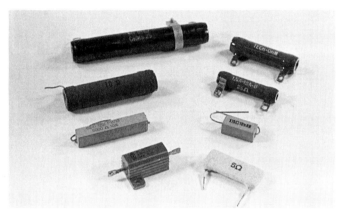

FIGURE 4-6 Typical power resistors and their representative forms.

well over 200 W. In every application the power rating of the resistor is determined by the power requirement of the circuit in which it is used.

The surface area of the body of the resistor is a major factor in its ability to dissipate the heat that is generated when it is functioning properly. Resistors that have a power rating of 10 W or more often are formed on a hollow ceramic core. The open center of the core allows additional airflow and aids in the dissipation of the heat. In situations of extreme heat it is often necessary to add a small fan for cooling purposes.

The power requirements for any resistor are determined by the application of Watt's law. The information that is necessary for this is usually the ohmic resistance value and the voltage drop that will occur.

EXAMPLE 4-8

What power rating is required for a resistor with a value of 2.5 kΩ and a required voltage drop of 100 V?

Solution

$$P = \frac{V^2}{R}$$

$$= \frac{100 \text{ V} \times 100 \text{ V}}{2500 \text{ }\Omega}$$

$$= 4 \text{ W minimum}$$

EXAMPLE 4-9

What is the required rating for a resistor in a circuit with a resistance of 1 MΩ and a voltage drop of 450 V?

Solution

$$P = \frac{V^2}{R}$$

$$= \frac{450 \text{ V} \times 450 \text{ V}}{1{,}000{,}000 \text{ } \Omega}$$

$$= 0.2025 \text{ W minimum}$$

The calculated rating for a resistor in a specific application is usually a minimum value. This means that in many applications a resistor with a higher power rating will operate at a cooler surface temperature because the large resistor has a greater surface area and will dissipate its heat over this larger area. One limiting factor in this is if the physical size of the larger resistor will permit its installation in a given space. A second limiting factor is related to the use of the resistor as a fuse or current-limiting device. More on this topic is discussed later in this chapter.

Heat sinking *Method of mounting component on a metallic base to dissipate any heat generated during normal operation.*

Heat Sinking The purpose of the larger resistor body is to dissipate heat. The larger the surface area, the higher the amount of heat that is dissipated. It is not possible in every installation to install the higher-power-rated resistor, due to space limitations. Another method of dissipating additional heat is shown in Figure 4-7. This method uses the surface area of the metal case or frame of the device as an area from which to dissipate the heat. This is called **heat sinking**. Two types of heat sinks are illustrated. One of these has several fins. The power resistor is mounted tightly against the heat sink. The fins provide additional surface area. The flat bottom of this type of heat sink is mounted

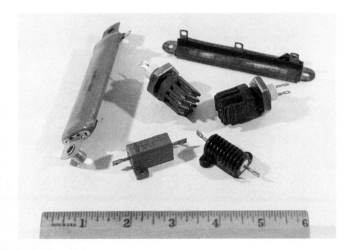

FIGURE 4-7 The heat sink is a device that is used to permit larger-than-normal power levels of operation.

on a metal surface. It will also dissipate the heat created by the resistor. The other type of heat sink looks like a cover for the resistor. The body of the resistor is built into a hole in the body of the heat sink. The unit is mounted on a metallic surface just as the finned heat sink is mounted. This second method will permit 25 W of heat power to be dissipated by a 2-W-rated resistor.

REVIEW QUESTIONS

4-1. What is the purpose of the power rating of a resistor?

4-2. A 100-W-rated resistor is installed in a circuit that has 0.85 A flowing through a resistance of 500 Ω. Is the size of the resistor adequate for this application?

4-3. What wattage rating is required for a 10-kΩ resistor that has a 100-V drop across its terminals?

Resistor Values Resistors are available in values of less than 1.0 Ω to over 10 MΩ. The quantity of values between these two extremes led to a standardization of resistor values that are commercially available. The reason for this is that resistors are used by all product manufacturers but they are made by only a few companies.

In many applications there is little need for a resistor that has an exact value. Almost all applications permit some variance from the calculated value for the circuit. This is a **tolerance** factor. The electronic industry has an association of its manufacturing members. This is called the Electronic Industries Association (EIA). The EIA has developed several recommended standards for components. These include size, marking systems for values, and a standard number set for values. The number set that is recommended is shown in Figure 4-8. The number set is established by use of a percentage factor. In the left-hand column a set that has a tolerance limit of 20% is shown. There is a difference of 20% between each of the numbers shown. The middle column has a 10% spread between its numbers. The right-hand column has a spread of 5% between each number. What this indicates to persons working in the field is that the *exact* measured value of the component is permitted to be in

Tolerance Permissible deviation from the marked value of a component.

EIA PREFERRED NUMBER SERIES						
±5% TOLERANCE	10	15	22	33	47	68
	11	16	24	36	51	75
	12	18	27	39	56	82
	13	20	30	43	62	91
±10% TOLERANCE	10	15	22	33	47	68
	12	18	27	39	56	82
±20% TOLERANCE	10	15	22	33	47	68

FIGURE 4-8 Electronic Industries Association (EIA) preferred number set. (From Joel Goldberg, *Fundamentals of Electricity*, 1981, p. 36. Reprinted by permission of Prentice-Hall, Englewood Cliffs, New Jersey.)

the range of the numeric value and its tolerance. A component that is related with a 10% tolerance is permitted to measure anywhere from less than 10% to up to 10% above its marked value.

EXAMPLE 4-10

A resistor is marked as being 1 kΩ and has a tolerance value of 10%. Find its ohmic limitation.

Solution

$$\text{Tolerance} = 1000 \times 0.1 \ (10\%)$$
$$= 100$$
$$\text{upper tolerance range} = +100 + 1000 = 1100 \ \Omega$$
$$\text{lower tolerance range} = -100 + 1000 = 900 \ \Omega$$
$$\text{limits of acceptable resistor} = 900 \text{ to } 1100 \ \Omega$$

EXAMPLE 4-11

A 15-kΩ 5% resistor is measured as a value of 9 kΩ. Is it in tolerance?

Solution

$$15 \text{ k}\Omega - 750 \ \Omega$$
$$15,000 \ \Omega - 750 \ \Omega = 14,250 \ \Omega$$

The resistor is out of tolerance.

Tolerance values are provided as a percentage figure. The implied factor here is that the tolerance figure should be read as ±5%, ±10%, and so on. The plus or minus signs are not always provided. You are expected to know this as a standard representation.

REVIEW QUESTIONS

Find the permissible ohmic value range for the following resistors.

4-4. 56 kΩ ± 10%
4-5. 100 kΩ ± 20%
4-6. 120 Ω ± 5%
4-7. 1.8 MΩ ± 10%

Resistor Markings Resistors are normally marked with an ohmic value. The methods of doing this are standardized for low-power resistors and are reasonably standardized for power types. Two different methods are used, one

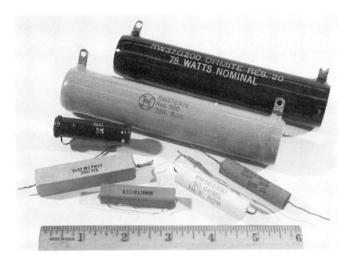

FIGURE 4-9 Ratings for power resistors are usually printed on the body of the resistor, as shown here.

for each type of resistor. We discuss power resistors first because their markings are easier to understand. Several different power resistors are shown in Figure 4-9. Each of these is marked with its ohmic value. The value of resistance is printed in digital form on the body of the resistor.

For some resistors, the power rating is also provided. For example, a resistor that is marked "10K, PW7" is a 10-kΩ resistor with a power rating of 7 W. Some power resistors are not marked in this manner. Sometimes power resistors that are purchased for installation by the manufacturer of the unit are marked only with the part number that the manufacturer uses. Under these circumstances one would have to refer to a parts list to find the design ratings.

Often the manufacturer will use a catalog number to identify the physical specifications for a power resistor. The marking "PW7" is one manufacturer's catalog number for a 7-W-rated resistor. Another manufacturer will use a catalog number of "$1\frac{3}{4}$ A" for a specific power rating. There is little standardization for marking power ratings. Probably the best way to handle this is to obtain a catalog describing power resistors from the manufacturers or from an independent electronic parts distributor.

The second method of marking ohmic value on resistors is standard for almost all low-power resistors. It uses the EIA color coding system to show the ohmic values. The **EIA color coding system** is a system used by the electronics industry. It uses a standard set of colors to represent a group of numbers between 0 and 9. Groups of colors are used to indicate the specific numeric value of components. The standard color-coding number set is shown in Figure 4-10. This color and number set is standardized for all components. The method of displaying it has become standard for low-power resistors. This standard

EIA color coding system *Method of relating numbers to colors for component-value identification.*

FIGURE 4-10 EIA color code chart is used to relate colors with specific numbers. (From Joel Goldberg, *Fundamentals of Electricity*, 1981, p. 35. Reprinted by permission of Prentice-Hall, Englewood Cliffs, New Jersey.)

BLACK	0	GREEN	5
BROWN	1	BLUE	6
RED	2	VIOLET	7
ORANGE	3	GRAY	8
YELLOW	4	WHITE	9

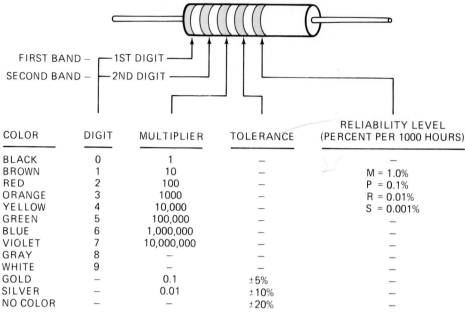

COLOR	DIGIT	MULTIPLIER	TOLERANCE	RELIABILITY LEVEL (PERCENT PER 1000 HOURS)
BLACK	0	1	–	–
BROWN	1	10	–	M = 1.0%
RED	2	100	–	P = 0.1%
ORANGE	3	1000	–	R = 0.01%
YELLOW	4	10,000	–	S = 0.001%
GREEN	5	100,000	–	–
BLUE	6	1,000,000	–	–
VIOLET	7	10,000,000	–	–
GRAY	8	–	–	–
WHITE	9	–	–	–
GOLD	–	0.1	±5%	–
SILVER	–	0.01	±10%	–
NO COLOR	–	–	±20%	–

EIA COLOR CODING SYSTEM FOR 5% , 10%, 20%
TOLERANCE LOW POWER RESISTORS

FIGURE 4-11 EIA color code is used to mark low-power resistors, as illustrated here. (From Joel Goldberg, *Fundamentals of Electricity*, 1981, p. 35. Reprinted by permission of Prentice-Hall, Englewood Cliffs, New Jersey.)

resistor color coding is shown in Figure 4-11. Resistors in use today employ a minimum of three color bands to represent their ohmic value. Some low-power resistors have as many as five color bands. Let us start the explanation by first learning how to "read" the first three color bands. Examination of Figure 4-11 will show that the resistor has one color band that is close to one end of the resistor. This is the starting point. Position a resistor with the banded end on the left. The colors are now interpreted in this manner:

Band 1: first significant figure (a number)
Band 2: second significant figure (a number)
Band 3: multiplication factor

Multiplier *Describes third color band value of resistor*

Note that the first two color bands represent whole numbers. This is always true. There are no variances from this rule. The numbers represented must be one of the EIA standard numbers. We are not discussing tolerance at this point. The third color is *never* a whole number. It is always a **multiplier** expressed in powers or multipliers of 10 of the value of resistance. Note that the colors gold and silver are also used for a multiplier when located in the third column. The gold color in this column represents multiples of 0.1 (or divide by 10), and silver is a multiple of 0.01 (or divide by 100). Let us see how this system functions.

EXAMPLE 4-12

A resistor has color bands of red, red, and orange. What is its ohmic value?

Solution

$$\text{First-band value} = 2$$
$$\text{second-band value} = 2$$
$$\text{third band value} = \times 1000 \text{ or } 1 \times 10^3$$

When combined: 22,000 Ω or 22 kΩ.

EXAMPLE 4-13

Determine the ohmic values for the following resistors.
(a) Brown, green, yellow
(b) Orange, white, gold
(c) Yellow, violet, green

Solution

1   5   × 10,000 = 150,000 Ω or 150 kΩ or 15 × 10⁴ Ω

Solution

3   9   × 0.1 = 39 × 0.1 or 3.9 Ω or 39 × 10⁻¹

Solution

4   7   × 1,000,000 = 4,700,000 Ω or 4700 kΩ or 4.7 MΩ or 47 × 10⁵ Ω or 4.7 × 10⁶ Ω

Almost all of the resistors found in electronic circuits have a fourth color band. The information used on this band is independent of the ohmic value coding for the resistor. This band is used to indicate the tolerance factor of the resistor. The standard set of tolerance levels for resistive components has been 5, 10, and 20%. The color gold on the fourth band indicates a tolerance of ±5% and silver is used to represent a ±10% tolerance. When the tolerance value is ±20%, no color band is used. Recently, an additional set of tolerances has been added to this group. They are 1.0, 2.0, 0.5, 0.25, and 0.1%. These tolerance values are used for film-type resistors. They are represented by the colors brown, red, green, blue, and violet, respectively. Film-type resistors do not have tolerances greater than ±5%. The methods of producing them permit the manufacture of these close tolerance values. Keep in mind that the fourth color band on any low-power resistor is always used to indicate the tolerance value of the resistor. It never contains any portion of the ohmic value.

The demands of our technology have added one additional color band to the body of the resistor. This is a fifth band. Its function is to identify a reliability level for one batch of resistors. The reliability factor describes the number of resistors that will fail during 1000 h under some exacting test conditions. The need for this type of testing is apparent when one considers the difficulty and cost of making repairs on a satellite located somewhere in space.

The complete set of color bands is read in a very simple manner. Usually, just the first four color bands are read when one converts the colors into a numeric value. The reliability band is one that is needed by manufacturers of equipment. This band identifies the projected reliability factor, or failure rate, of the resistor.

Use of the fourth band permits the user some leeway in the selection or testing of a specific resistor. If the tolerance is listed as ±10%, the actual resistance value of a resistor is accepted if the resistor value falls between a figure that is up to 10% less than the rated value or up to 10% higher than the value.

EXAMPLE 4-14

Determine the tolerance range for the following resistors:
(a) Red, red, orange, silver
    = 2   2   × 1000   ±10%
    = 22,000 Ω − 2200 Ω = 19,800 Ω (low)
    = 22,000 Ω + 2200 Ω = 24,200 Ω (high)
    acceptable range = 19,800 to 24,200 Ω
(2) Brown, black, red, gold
    = 1   0   × 100   ±5%
    = 1000 − 50 = 950 Ω (low)
    = 1000 + 50 = 1050 Ω (high)
    acceptable range = 950 to 1050 Ω

The reason for a tolerance value will become apparent when one considers how a carbon-composition resistor is constructed. In the beginning of the electronics industry we did not have methods of producing accurate and close tolerance resistors. A compound of clay and carbon was mixed and inserted into the body of the resistor. The finished resistor was tested for its ohmic value. Those resistors that were within the 20% tolerance range were marked as such. Additional testing was done to select the specific resistors whose ohmic value fell into either the 10% or 5% range. These resistors always were more expensive because of the extra handling involved. Also, these close tolerance values were not usually required for use in older vacuum-tube circuits. As the manufacturing process improved and the demand for close-tolerance resistor values increased, the production of 5% resistors increased to a point where perhaps the greatest percentage of resistors sold today are of this tolerance value.

Tolerance values for components are determined by the circuit designer. Those who are engaged in design and repair should always attempt to use replacement components that are equal to, or better than, the original value. A circuit that requires a resistance of 1.5 kΩ ± 10% requires a resistance value

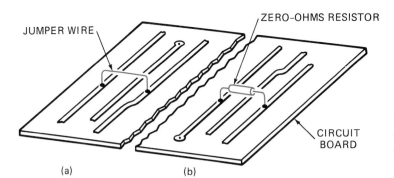

ZERO-OHMS RESISTOR

CIRCUIT
BOARD

(a) (b)

FIGURE 4-12 The 0-Ω resistor is actually a jumper wire that is inserted by a machine on a printed circuit board.

that falls between the values 1350 to 1650 Ω. Any resistor that meets this requirement can be used. For example, a 1.5-kΩ resistor with a tolerance of ±20% can be used successfully and correctly as long as its measured value falls between 1350 and 1650 Ω. In this example the measured value of the resistor has a better tolerance value than that of its rated value. The tolerance value makes no difference to the circuit as long as the actual value is within the specified range.

Recently, a new value of resistance is being used in the production of electronic modules. This is the **zero-ohms resistor**. It may sound foolish to produce a resistor that has no resistance, but there is a definite need for this value in the industry. Many of the printed circuit boards in use today require that a wire be used to "jump" across a foil path (Figure 4-12). This is done to connect two paths together even though there is another path between them. A piece of hookup wire is used to connect the two paths in Figure 4-12(a).

Zero-ohms resistor A jumper wire housed in a standard resistor body used when components are machine inserted in PC boards.

Figure 4-12(b) shows a zero-ohms resistor. The body of the resistor looks like any other resistor. The difference is that the inside of the resistor is a solid piece of wire. The need for this type of resistor came about as machines were utilized instead of people in order to insert components on the board. The machine is not able to hold, form, and insert a piece of bare wire when it is set up for resistor insertion. This resulted in the development of the zero-ohms resistor because it can be handled by the machine.

Special Color Coding As the electronics industry continues to grow and tends to become more specialized, the use of color coding for resistance values also changed. There are two new types of color coding systems now in use. This is in addition to the "standard" color code described previously. One of these is used on low-power metal-film resistors. This color coding applies to resistors falling between 0.1 and up to 5% tolerance value. This family of resistances uses a five-band color coding system. This system is shown in Table 4-1 which contains a chart showing the standard values of resistance for metal-film resistors. It also shows how the values of these resistors are read. In this system there are three significant numbers before the tolerance band. The fourth band is the multiplier and the fifth band is used for the tolerance range. Often, the tolerance band is slightly wider than the other color bands on the body of the resistor.

Another system is used for a family of resistances called "chip" types. This is in addition to the metal-film system for color coding resistances. In some applications the chip resistor is used in place of the standard-bodied types

TABLE 4-1 Metal Film Resistor Color Code

STANDARD RESISTANCE VALUES FOR THE 10-TO-100) DECADE

RESISTANCE TOLERANCE (± %)

0.1% 0.25% 0.5%	1%	2% 5%	0.1% 0.25% 0.5%	1%	2% 5%	0.1% 0.25% 0.5%	1%	2% 5%	0.1% 0.25% 0.5%	1%	2% 5%	0.1% 0.25% 0.5%	1%	2% 5%	0.1% 0.25% 0.5%	1%	2% 5%
10.0	10.0	10	14.7	14.7	—	21.5	21.5	—	31.6	31.6	—	46.4	46.4	—	68.1	68.1	68
10.1	—	—	14.9	—	—	21.8	—	—	32.0	—	—	47.0	—	47	69.0	—	—
10.2	10.2	—	15.0	15.0	15	22.1	22.1	22	32.4	32.4	—	47.5	47.5	—	69.8	69.8	—
10.4	—	—	15.2	—	—	22.3	—	—	32.8	—	—	48.1	—	—	70.6	—	—
10.5	10.5	—	15.4	15.4	—	22.6	22.6	—	33.2	33.2	33	48.7	48.7	—	71.5	71.5	—
10.6	—	—	15.6	—	—	22.9	—	—	33.6	—	—	49.3	—	—	72.3	—	—
10.7	10.7	—	15.8	15.8	—	23.2	23.2	—	34.0	34.0	—	49.9	49.9	—	73.2	73.2	—
10.9	—	—	16.0	—	16	23.4	—	—	34.4	—	—	50.5	—	—	74.1	—	—
11.0	11.0	11	16.2	16.2	—	23.7	23.7	—	34.8	34.8	—	51.1	51.1	51	75.0	75.0	75
11.1	—	—	16.4	—	—	24.0	—	24	35.2	—	—	51.7	—	—	75.9	—	—
11.3	11.3	—	16.5	16.5	—	24.3	24.3	—	35.7	35.7	—	52.3	52.3	—	76.8	76.8	—
11.4	—	—	16.7	—	—	24.6	—	—	36.1	—	36	53.0	—	—	77.7	—	—
11.5	11.5	—	16.9	16.9	—	24.9	24.9	—	36.5	36.5	—	53.6	53.6	—	78.7	78.7	—
11.7	—	—	17.2	—	—	25.2	—	—	37.0	—	—	54.2	—	—	79.6	—	—
11.8	11.8	—	17.4	17.4	—	25.5	25.5	—	37.4	37.4	—	54.9	54.9	—	80.6	80.6	—
12.0	—	12	17.6	—	—	25.8	—	—	37.9	—	—	56.6	—	—	81.6	—	—
12.1	12.1	—	17.8	17.8	—	26.1	26.1	—	38.3	38.3	—	56.2	56.2	56	82.5	82.5	82
12.3	—	—	18.0	—	18	26.4	—	—	38.8	—	—	56.9	—	—	83.5	—	—
12.4	12.4	—	18.2	18.2	—	26.7	26.7	—	39.2	39.2	39	57.6	57.6	—	84.5	84.5	—
12.6	—	—	18.4	—	—	27.1	—	27	39.7	—	—	58.3	—	—	85.6	—	—
12.7	12.7	—	18.7	18.7	–-	27.4	27.4	—	40.2	40.2	—	59.0	59.0	—	86.6	86.6	—
12.9	—	—	18.9	—	—	27.7	—	—	40.7	—	—	59.7	—	—	87.6	—	—
13.0	13.0	13	19.1	19.1	—	28.0	28.0	—	41.2	41.2	—	60.4	60.4	—	88.7	88.7	—
13.2	—	—	19.3	—	—	28.4	—	—	41.7	—	—	61.2	—	—	89.8	—	—
13.3	13.3	—	19.6	19.6	—	28.7	28.7	—	42.2	42.2	—	61.9	61.9	62	90.9	90.9	91
13.5	—	—	19.8	—	—	29.1	—	—	42.7	—	—	62.6	—	—	92.0	—	—
13.7	13.7	—	20.0	20.0	20	29.4	29.4	—	43.2	43.2	43	63.4	63.4	—	93.1	93.1	—
13.8	—	—	20.3	—	—	29.8	—	—	43.7	—	—	64.2	—	—	94.2	—	—
14.0	14.0	20.5	20.5	20.5	—	30.1	30.1	30	44.2	44.2	—	64.9	64.9	—	95.3	95.3	—
14.2	—	—	20.8	—	—	30.5	—	—	44.8	—	—	65.7	—	—	96.5	—	—
14.3	14.3	—	21.0	21.0	—	30.9	39.9	—	45.3	45.3	—	66.5	66.5	—	97.6	97.6	—
14.5	—	—	21.3	—	—	31.2	—	—	45.9	—	—	67.3	—	—	98.8	—	—

	NOMINAL	MULTIPLIER	TOLERANCE (%)
BLACK	0	1	
BROWN	1	10	1
RED	2	100	2
ORANGE	3	1 k	
YELLOW	4	10 k	
GREEN	5	100 k	0.5
BLUE	6		0.25
VIOLET	7		0.1
GRAY	8		
WHITE	9		
SILVER		0.01	
GOLD		0.1	5

	PRECISION	GENERAL PURPOSE
1ST BAND	NOMINAL	NOMINAL
2ND BAND	NOMINAL	NOMINAL
3RD BAND	NOMINAL	MULTIPLIER
4TH BAND	MULTIPLIER	TOLERANCE
5TH BAND	TOLERANCE	SOLDERABILITY (OPTIONAL)

usually used for low-power applications. When used, the chip resistor has a three-digit number stamped on the bottom of its body. The system used for this is:

First number: first significant digit

Second number: second significant digit

Third number: multiplier

For example, a chip resistor having the numbers 561 on the underside of its body has a value of 56×10^1, or 560 Ω. In this system a zero-ohms resistor has a body color of either red or green. These resistors are small, on the order of about $\frac{1}{4}$ in. in width and length.

Lead Description There are two forms of attaching leads to resistors. One of these is called **axial** and the other is known as **radial**. This is also true for other components. The two methods are illustrated in Figure 4-13(a) and (b). Almost all low-power resistors are made with axial leads. This term indicates that the leads are connected on the axis of the body of the resistor. To save space, some miniaturized electronic devices use radial lead resistor components. Both leads are brought out from one end of the resistor when this method is used. If one needs to replace a radial lead component and one is not available, the method shown in Figure 4-13(c) may be used. One end of an axial lead resistor is folded over the body of the resistor. It now looks like a radial lead type. A word of caution: This alternative method exposes one of the leads of the resistor. It is wise to try to have the exposed lead at zero volts potential, or as close to zero as possible. This minimizes any problem related to electric shock if one should accidentally touch this exposed lead. This type of installation should only be done when the original type of component is not available.

Axial *Term describing component lead mounting on axis of component.*

Radial *Term describing lead mounting when both leads are on same end of component.*

REVIEW PROBLEMS

Identify the following resistor values.

4-8. Yellow, violet, orange, gold
4-9. Brown, red, red, silver
4-10. Orange, orange, orange, gold, red
4-11. Green, blue, gold, gold

Identify the colors for the following resistors.

4-12. 2200 Ω ± 10%
4-13. 1 MΩ ± 20%
4-14. 1.0 Ω ± 5%
4-15. 180 Ω ± 5%

Resistor Failures Fixed-value resistors in a circuit often fail. The failure is usually due to excess heat produced in an overcurrent situation, often caused

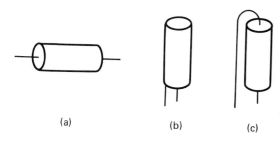

(a)   (b)   (c)

FIGURE 4-13 Many electronic components are available with either axial (a) or radial (b) levels. Selection depends on space requirements. (c) An axial lead component can be used in a radial installation.

by the failure of some other component in the circuit. An example of this is illustrated in Figure 4-14. The original circuit has a source of 100 V and a load that consists of two 500-Ω resistances. Ohm's law, $I = V/R = 100/1000$, tells us that under normal conditions the current in this circuit is 0.1 A. The power dissipated by each resistor is determined: $P = I^2 \times R = 5$ W. A short circuit in the upper component places the 100-V source directly across the second resistance. The current now has increased to 0.2 A. Ohm's law, again, gives us this value; $I = V/R = 100$ V/500 $\Omega = 0.2$ A. The smaller-value resistor's power dissipation now has increased to four times its original value: $P = I^2R = 0.2^2$ A $\times$ 500 $\Omega = 20$ W. The resistor will now fail due to the excessive heat. The cause of the resistor failure is the other component. Changing the resistor will not repair the circuit. Further testing is required to determine what caused this failure.

Resistor failures will cause a change in the ohmic value of the resistor. When the resistor is a carbon-composition type, the failure will usually be an increase in the ohmic value. The method of failure in carbon-composition resistors is shown in Figure 4-15. When the carbon compound is first over-heated, the carbon tends to fuse. This results in a below-normal resistance value. The reduced resistance will permit a larger-than-normal current flow through the resistor and produce an increase in its power dissipation. This is based on the formula $I = V^2/R$. The increased heat will now make the resistance increase until the resistor becomes so hot that it will split into two pieces. Its resistance at this point is infinitely higher and the resistor acts like an open circuit. The ohmic values in other types of low-power resistors will normally increase only as they fail. This is also true for the failure conditions of power resistors. When a power resistor fails it usually results in an open in the resist-

FIGURE 4-14 A partial short circuit across R_1 will result in higher-than-normal circuit flow and power consumption in the circuit.

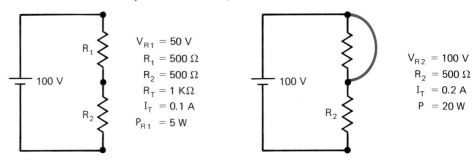

$V_{R1} = 50$ V
$R_1 = 500$ Ω
$R_2 = 500$ Ω
$R_T = 1$ KΩ
$I_T = 0.1$ A
$P_{R1} = 5$ W

$V_{R2} = 100$ V
$R_2 = 500$ Ω
$I_T = 0.2$ A
$P = 20$ W

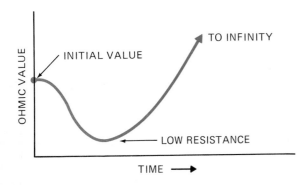

FIGURE 4-15 Composition resistor failure curve. One may find defective resistors with either higher- or lower-than-normal values.

ance wire wound on the ceramic form. Often these are not visible. Visual clues to a failure of this type are a broken or cracked resistor body or discoloration on the resistor's body due to excess heat.

4-3 Resistor Variations

The resistors described thus far in the chapter are known as fixed-value types. There are several types of resistors that do not have a fixed value. These are in use in many circuits. They need to be described together with some typical applications for each. The resistors can generally be categorized as mechanically variable or adjustable and electrically variable or adjustable. There are also some current- or voltage-limiting types that can be included in the second group. In this section we describe several of the more common types.

Mechanically Variable Types Mechanically variable resistors may be divided into two categories. There are those that are adjustable and those that are variable. This may seem like a rather fine dividing line. It is the method used by the industry to describe the two types of resistors.

Adjustable Resistors This type of resistor is shown in Figure 4-16. This resistor is basically a power resistor. It has one area on its body that is not covered by the ceramic coating. The resistance wires that are wound on the core are exposed in this area. One or more circular clamps that have a contact on them are placed around the body of this resistor. The clamps include a fastener that can be tightened to hold them firmly at a predetermined place on the body of the resistor.

This assembly could be considered as a fixed value of resistance when one measures its ohmic value from one end to the other end of the body. This is illustrated by the schematic symbol shown for the adjustable resistor. The total resistive value is measured between these points. The value of resistance from one side of the adjustable clamp to one end of the resistor is less than the total value. The same is true for the amount of resistance from the clamp to the

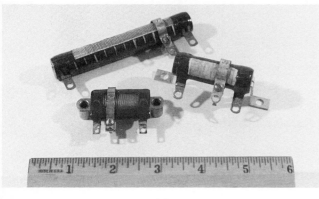

(a)

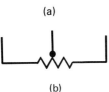

(b)

FIGURE 4-16 Typical adjustable resistors used in the industry. These are called "fixed adjustable" because the adjustment is fixed at one specific point by the position of the clamp when the resistor is installed in the circuit.

other end of the resistor. These two values will equal the total resistance value when they are added. Once the adjustment is made, the clamp is normally not changed. This is considered to be a "fixed" type of adjustable resistor.

Variable resistor *A mechanically adjustable resistor.*

Variable Resistors The second type of mechanically adjustable resistor is called a **variable resistor**. The reason for this is that the adjustment is one that can readily be changed. Some of the more common variable types of resistors are shown in Figure 4-17(a). The specific mechanical configuration used depends on the requirements of the circuit. The variable resistor has a fixed-value resistive element that is usually made in the form of a circle. Terminals are connected to each end of the element. This provides a fixed value of resistance.

(a)

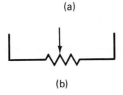

(b)

FIGURE 4-17 Typical variable resistors. Most of these can be varied, or adjusted, by either a technician or the customer after being installed in the circuit.

A slider that can be rotated from the center of the resistor is connected to a terminal that is located between the other two terminals. The rotating slider often is connected mechanically to a rotatable shaft. Some variable resistors use a short slotted piece instead of the shaft. The slotted piece is adjusted by use of an adjusting tool or screwdriver. The resistors that have shafts are usually used as panel controls. They are mounted to the control panel of the unit. Knobs are fastened to the shaft of the variable resistor to aid in the rotation of the shaft. These installations have been given the name *controls*.

Some controls have more than one section. These multiple-section controls are called *concentric* types. They are illustrated in Figure 4-17(a). One type has multiple resistive elements and uses a common shaft to rotate them both at the same time. The other type has two separate sections and uses two shafts. One of the shafts is larger than the other. It is hollow and is connected to the front control. The rear control has a narrower diameter shaft. This shaft rotates the rear control only. It fits into the hollow center of the front shaft. In some cases a mechanical switch is also connected to the rear control. The switch is electrically isolated from the control. Its function is to turn on or off a specific circuit. An example of this application is the volume-and-switch combination used in many consumer products. This combination of switch and control may be used on either single- or multiple-section controls.

The resistive element for a variable resistor is made of a carbon compound or with a resistive wire. Carbon resistors, like the fixed low-power types, are limited to low-wattage applications. Higher-power variable resistors use resistive wire for the element. The upper power limitation is not an electrical factor, but rather, a physical factor. The wattage rating is limited only by the physical size of the resistor and its placement.

Some of the many types of variable resistors that are commonly used in the industry are shown in Figure 4-17(a). These range from the miniature to large high-power types. The ohmic value of any type of variable resistor will range from a low value of 5 or 10 Ω to a high value of over 1 MΩ. Here, as before, the specific value is determined by the requirements of the circuit.

Applications of Mechanically Variable Resistors The purpose of any resistor in a circuit is either to limit current flow or to develop a voltage drop as current flows in the circuit, or both. This is true for variable and adjustable resistors. There are two general ways in which these devices are connected. One is known as a **rheostat circuit**. The other is called a **potentiometer circuit**. The same variable resistor may be used for either circuit. The difference is in the method by which they are wired into a circuit. The *rheostat circuit* (Figure 4-18) has a source, a lamp for a load, and a variable resistance. For this circuit only two of the three leads of the variable resistor are used. Changing the value of resistance between the movable arm and one end of the resistor will change

Rheostat circuit *Used to control circuit current by adjusting amount of resistance.*

Potentiometer circuit *Used to control circuit voltage.*

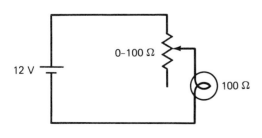

FIGURE 4-18 The rheostat circuit is used to control current in the circuit.

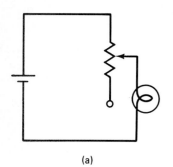 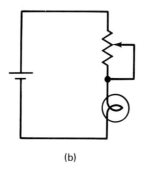

(a)   (b)

FIGURE 4-19 Rheostats may be connected as shown here. The preferred method (b) has a built-in safety factor.

the total amount of resistance in the circuit. Using Ohm's law shows us that the current will be changed as the resistance value is changed. A reduction in the resistance will permit a larger current flow. When the full 100 Ω of the rheostat's resistance is in the circuit, the current is $I = V/R = 12\ \text{V}/200\ \Omega = 0.06$ A. When the rheostat's arm is touching the source side of the circuit, the resistance of this device is zero and the only resistance in the circuit is now the lamp, or 100 Ω. The current has now increased to 0.12 A.

Often the rheostat is wired as shown in Figure 4-19. This is a fail-safe type of circuit. The weakest point in the circuit is usually at the slider arm of the variable resistor. Should this fail the circuit will open and be shut off. A preferred method is shown in the illustration. The arm is connected to the terminal of the resistor that is connected to the load. In this circuit, if the arm should not make its contact, the full value of resistance is still in the circuit and the lamp, or load, will still function.

The second use for a variable resistor is called the *potentiometer circuit*. The term "potentiometer" is a development of the use of this circuit. It is probably taken from early attempts to describe electricity as a fluid. The control of the amount of fluid in a system is called *metering*. In the potentiometer circuit shown in Figure 4-20, the quantity of potential, or voltage, is being controlled. This then develops into a circuit that is designed to meter, or control, the amplitude of the potential that is applied to the load. The same variable resistor is used for the potentiometer (pot) circuit as is used for a rheostat circuit. The significant difference is in the way the circuit is wired. The total value of resistance is connected across the source. The load is connected between the variable arm and the circuit common end of the variable resistor. This type of circuit is usually limited to low-current applications. Its major limiting factor is the current that is required by the load. A typical application for the pot circuit is the audio volume control for consumer entertainment devices. An electric signal is used as the source of energy. The volume control is connected to this source, as shown in Figure 4-21. The arm of this control is connected to an audio amplifier's input [Figure 4-21(a)]. The common connection is also connected to the common in the audio amplifier. The position of the arm of the control determines the amplitude of the signal voltage that is applied to the amplifier. As the arm is rotated away from the common connection, the signal voltage that is measured from the arm to the circuit common increases. It will continue to increase until the arm and the upper connection as seen on the circuit schematic act as one connection. This is the maximum amount of signal that is available to the amplifier. Figure 4-2(b) shows a similar connection using the lamp instead of a audio signal. An increase in voltage applied to the lamp causes more current flow through its filament.

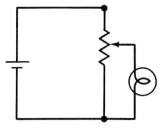

FIGURE 4-20 The potentiometer circuit is used to vary the potential, or voltage, in a circuit.

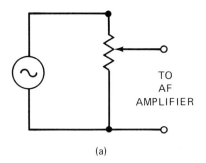
(a)

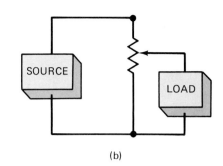
(b)

FIGURE 4-21 Applications of the potentiometer include (a) signal level control (volume control) and (b) dc operating voltage controls.

The result is a brighter light emitted from it. This circuit is the potentiometer, or voltage control circuit application for the variable resistor.

Failures in Mechanically Variable Resistors The failure of a mechanically variable resistor is usually limited to the rotating arm or to the element of the resistor. The contact assembly may become oxidized or dirty due to the environment. This effect increases the normal amount of resistance in the circuit. The resistance may become so high in value that it acts like an open circuit. Another condition that occurs is that the element of the variable resistor may arc and develop a higher-than-normal resistance at the place where the arc occurs. The third problem that may occur is that the element may overheat between one end and the arm of the resistor. The result is either an increase in the resistance of the element or an open occurring at that point. Another problem that may occur is that the element becomes dirty and the arm does not make good contact with it. In any of these conditions the variable resistor has to be replaced.

TROUBLESHOOTING APPLICATIONS

Electrically Variable Resistors There are several different types of resistors whose value is varied by an application of light, heat, voltage, or current. These include photoresistors, thermistors, voltage-dependent resistors (VDRs), and current-sensitive resistors. The application of each of these types is a major factor in the operation of the circuit in which they are used. This section describes each type and shows some of the applications of each type. The schematic symbols for each type of resistor is shown in Figure 4-22. Each of these uses the same basic schematic symbol as a base. The difference is shown by the addition of a symbol or a letter inside the circle surrounding the resistor symbol. Most of these are readily identified. Arrows pointing toward the resistor indicate that its value is affected by light. The letters T and V indicate that these devices are affected by temperature and voltage, respectively. Variations of the symbols may be used by companies that have their own graphic standards. There are other types of electrically controlled resistors, including some in the semiconductor family. These will provide a general introduction to these types.

FIGURE 4-22 The schematic symbol for special resistors will provide information as to type of resistor.

Coefficient *Ratio of change in value due to temperature, etc.*

Positive and Negative Temperature Coefficients The term "**coefficient**" means that this factor also affects the outcome of an operation. The description of the components used in electrical work includes three types of coefficients: positive, negative, and zero. When a coefficient is positive it adds to the basic value of a component. When one has a negative coefficient the factor is subtracted from the basic value. A zero coefficient means that there is no effect on the basic value of a component.

Examples of the use of this term will provide a better understanding of how it is used. Certain resistors have a positive temperature coefficient. This means that under normal use the addition of heat will produce an increase in the basic value of the resistor. If the resistor has a negative temperature coefficient, the addition of heat would cause a reduction in the ohmic value of the resistor. A resistor that has a zero temperature coefficient does not change its value when its temperature is varied. These terms are used to describe the effect of various outside influences on electrically variable resistances.

Photoresistor *Resistor having ability to change resistance due to exposure to light.*

Photoresistors These are devices that are light sensitive. They have a negative coefficient of resistance; that is, the resistance of the photoresistor is decreased as light strikes its surface. The amount of resistance decreases in proportion to an increase in the intensity of the light. An application for a photoresistor is shown in Figure 4-23. A photoresistor is placed across the wires for a lamp. When the resistance of the photoresistor is less than that of the lamp, the circuit current will flow around the lamp through the photoresistor. The lamp will not light. As the environment darkens the photoresistor's value increases. This permits a current flow through the lamp and it is illuminated.

Thermistor *Resistor having ability to change value due to variations in heat.*

Thermistors The **thermistor** may be produced with either a negative or a positive temperature coefficient. The specific coefficient depends on the materials used to make it. A thermistor with a negative temperature coefficient is used to limit surges of current that occur in some circuits. Often an electronic or electric load will have a positive temperature coefficient. After it is turned on its resistance will increase, due to the effects of heat. This is true for the filament of a light bulb or of a vacuum tube. Both of these devices have a "cold" resistance and a "hot" resistance. A low value of cold resistance permits a reasonably large amount of current flow during the initial moments after turn-on occurs. The filaments of these devices could be damaged by repeated

FIGURE 4-23 Photoresistor circuit. In this circuit the intensity of the light will vary the value of resistance and control circuit current flow.

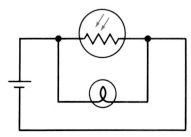

FIGURE 4-24 The thermal resistor will change in value as it is heated by normal current flow.

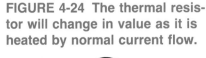
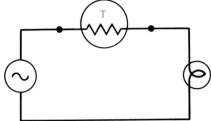

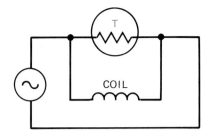

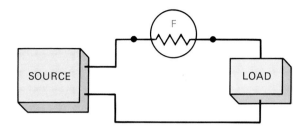

FIGURE 4-25 The thermistor is used to control current flow through the coil.

FIGURE 4-26 The fusible resistor is used to limit current flow in this circuit.

cycling of their on and off times. A method of protecting the filaments is shown in Figure 4-24. A negative temperature coefficient thermistor is connected between the lamp and the source as shown. The initial temperature of the thermistor keeps its resistance high. The thermistor's resistance is reduced by the effects of the heat produced in its body as current flows in the circuit. After a controlled period of time, usually about 10 s, the resistance value of the thermistor is so low that it becomes ineffective in the circuit. When power is removed from the circuit the thermistor's resistance value increases as it cools and it is ready to repeat its performance the next time the set is turned on.

Another circuit using a thermistor is shown in Figure 4-25. This time a different type of thermistor is used. This thermistor has a negative temperature coefficient. Its function in this circuit is to provide an alternative path for current flow. This circuit is similar to those used in color television receivers. The purpose of this circuit is to permit current flow through the coil during the first few seconds that the receiver is turned on. The coil shown on the diagram is used to demagnetize the face of the color picture tube. If this is not done, the colors presented there might not be as true as expected. Once the demagnetization is done, the circuit is turned off by the action of the negative temperature coefficient thermistor. Current will flow around the coil because of the reduced resistance of the thermistor.

Fusible Resistors Another type of resistor is one that is designed to fail. This occurs only under overcurrent conditions in the circuit. An example of this is shown in Figure 4-26. The fusible resistor normally has a very low resistance value. It is often less than 10 Ω. The design of the resistor is similar to that of a fuse except that the fusible resistor's value is never 0 Ω. The fusible resistor will limit surges of current in the circuit when it is first energized. When an overcurrent condition occurs, the resistive element with its positive temperature coefficient increases in resistance until a point where the element melts. The circuit is then shut down until the problem is located and repaired.

Fusible resistor *Low-value resistance used as a fuse in a circuit.*

REVIEW PROBLEMS

Complete each statement using the terms "positive," "negative," or "zero."

4-22. The value of a resistor increases as it heats. Its temperature coefficient is _____ .

Varistor *Resistor having ability to change value due to variations in voltage across its terminals.*

Varistors Another type of variable resistor is the **varistor**, or voltage-dependent resistor. This device's resistance has the ability to vary, or change, due to variations in voltage. The change is limited to a specific range. The change occurs as the voltage applied to its terminals changes. The amount of change is inversely proportional to the applied voltage. A rise in voltage reduces the resistance value of the varistor. This action reduces the voltage drop across the varistor to its original value before the rise in voltage occurred. A drop in the level of the applied voltage will increase the resistance of the varistor, thus raising the voltage back to its original level. The result of this action is a regulation of the amplitude of the voltage applied to a specific device in the circuit.

Resistor Combinations It is possible and permissible to combine individual resistors. The need for this occurs when one requires a specific resistance value and finds that the value is not readily available. This usually occurs either late at night after the supply houses are closed, on a weekend or holiday, or when the supply house is in a distant location. There are two ways that individual resistors can be combined. The specific way depends entirely on the requirements of the circuit. The two ways for connecting multiple resistor values, called **series** and **parallel**, are shown in Figure 4-27. The series connection is shown in part (a). The two resistors are connected end to end. In this type of connection the total resistance is the sum of the values of the individual resistors. This type of connection is not limited to two resistors. Any number of resistors may be wired in this manner to achieve a specific resistance value for the total string of resistors.

Series *Circuit connection having only one path for current flow.*

Parallel *Circuit connection having multiple paths for current flow.*

When selecting resistors for a circuit, the wattage rating of the circuit must not be overlooked. Each resistor must be able to dissipate the required amount of heat. The wattage rating for each resistance is determined by the formula $P = I^2R$. You must know the rules for resistors connected in series to be able to select the proper ohmic and wattage values. These rules are presented in detail in Chapters 5 and 6.

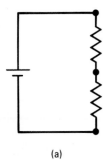

(a)

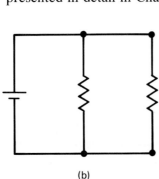

(b)

FIGURE 4-27 Resistors are often combined in (a) series connections or (b) in parallel connections.

Resistors may also be connected in parallel with each other. When this is done the total resistance of the combination is reduced to less than the value of the single lowest resistor. The rules for this are presented in Chapters 5 and 7. Using a simple example [Figure 4-27(b)], if two 20-Ω resistors are parallel wired, the total resistance presented by the combination is 10 Ω. The wattage values for the combination will add in this type of circuit arrangement.

Resistors may also be connected in combinations that have both series and parallel components in one total circuit. This type of circuit is complex. It requires a thorough understanding of the basic rules for series and parallel wiring. Since almost all electric and electronic circuits have some combination series and parallel wired circuits, it is necessary to have this knowledge. You should be able to identify a series circuit and a parallel circuit. That is why a full chapter is dedicated to each of these topics.

4-4 Testing Resistors

During the process of troubleshooting you will probably want to test the ohmic value of a resistor. The test instrument designed to do this is called an *ohmmeter*. Usually, the ohmmeter is a part of a more universal tester called a **multimeter**. A switching arrangement on the multimeter's panel permits the selection of voltage, current, or resistance measurements. One multimeter is shown in Figure 4-28.

Multimeter *Instrument capable of measuring voltage, current, and resistance.*

Inspection of the circuit used to measure resistance will show that in reality, one is measuring the current flow through the resistor. The manufacturer of the meter has calibrated the dial, or a numerical digital readout, to display a resistance value for a given voltage. Ohm's law, $R = V/I$, is used to convert the voltage and current values into a resistance value.

There are some very important things to keep in mind when measuring the resistance of a circuit or a component.

FIGURE 4-28 Typical multimeter used to measure voltage, current, and resistance. (Courtesy of Simpson Electric Company.)

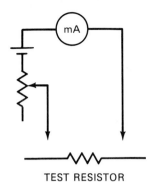

FIGURE 4-29 The basic ohmmeter circuit has a battery, a meter, and a rheostat to control initial current flow.

TEST RESISTOR

1. The circuit power must be turned off. The ohmmeter has its own power source (Figure 4-29). If the power in the circuit is present, it will provide an invalid reading or it may damage the internal components of the meter.

2. One of the leads of the meter is placed on each side of the resistor to measure its ohmic value.

3. The best way to measure the value of a resistor is to isolate it from the circuit. At least one end should be free to ensure an accurate reading. This is not always practical and it is a not good practice to remove anything from a circuit until you are reasonably certain that you have located the problem.

4. The reading for a good resistor will probably be close to the ohmic value or less than this value. If it is greater, then in all probability the resistor has failed.

5. Be aware that the resistance of your own body can affect a reading. The source voltage for an ohmmeter is usually around 1.0 to 1.4 V. There is little danger of an electric shock at this level of voltage. The resistance of the human body can easily give a false reading on a high value of resistance if your hands are holding the metallic portion of the probes.

CHAPTER SUMMARY

1. The most common discrete component used in electronic circuits is the resistor.

2. Resistors are used to limit current and to provide a voltage drop in the circuit.

3. Resistors are generally grouped in either "power" (high power) or "resistor" (low power) types. Low-power resistors are rated at $\frac{1}{8}$, $\frac{1}{4}$, $\frac{1}{2}$, 1, and 2 W. High-power resistors are rated at 3 W or higher.

4. Two ratings are used for resistors. These are power and ohmic value. They are independent of each other.

5. Electric power is a product of voltage and current in the circuit. It is also described as an I^2R factor. Power is rated in watts or kilowatts.

6. Low-power resistors are made of a carbon and clay composition, metallized film, or wire-wound elements inside a plastic housing. Power resistors are made of a resistance wire wound on a ceramic form.

7. Heat buildup on the body of the resistor is removed by use of a heat sink or a fan.

8. Standard numeric values for electronic components have been established by the Electronic Industries Association.

9. All resistors have a tolerance rating. If the resistor is good, the measured value of the resistor must be within the extremes of the tolerance.

10. An industry standard color code is used to represent numbers, multipliers, and reliability ratings for components.

11. Resistors are produced as either fixed or adjustable ohmic values.

12. Variable resistors may be wired as either rheostats or potentiometers. The rheostat is used to control current, and the potentiometer is used in voltage-control circuits.

13. A group of electrically variable resistors include varistors, thermistors, photoresistors, and fusible resistors.

14. Temperature coefficients describe the change in original value that occurs when a device is heated during normal operation.

15. Resistor values are measured with an ohmmeter.

Answer true or false; change all false statements into true statements.   **SELF TEST**

4–1. Resistance is the opposition to current flow in a circuit.

4–2. Electric power is rated in the units volt-hour.

4–3. One product of resistance in a circuit is heat.

4–4. The quantity of power consumed by a load is greater than that provided by the source.

4–5. Resistors are rated for their power and their ohmic values.

4–6. A device used to increase the power rating of a component is called a heat sink.

4–7. The color code for the value 180 is brown, gray, brown.

4–8. The fifth color band on a resistor's body is used for tolerance values.

4–9. Resistors are produced as either fixed or variable types.

4–10. A rheostat is used to control voltage in the circuit.

4–11. The color code for a 6.8-kΩ ± 5% resistor is   **MULTIPLE CHOICE**
    (a) blue, gray, orange, gold
    (b) green, blue, orange, gold
    (c) blue gray, red, gold
    (d) yellow, red, red, silver

4–12. A component with a negative temperature coefficient will
    _____ when heated.
    (a) reduce its value
    (b) raise its value

(c) remain the same

(d) none of the above

4-13. The device used to measure resistance is the

(a) ohmmeter

(b) wattmeter

(c) voltmeter

(d) ammeter

4-14. When two resistors with values of 1 and 1.8 kΩ are wired in series, the total value of resistance is

(a) 1 kΩ

(b) 1.8 kΩ

(c) 800 Ω

(d) 2.8 kΩ

4-15. The varistor's value is related to _____ changes.

(a) current

(b) voltage

(c) power

(d) heat

4-16. A resistor that is made of a solid piece of wire is technically called a _____ resistor.

(a) zero-ohms

(b) no-ohms

(c) super-low-ohms

(d) jumper

4-17. Current flow in a circuit that develops a partial short will

(a) stop flowing

(b) decrease

(c) not change

(d) increase

4-18. A power resistor that fails will

(a) decrease its ohmic value

(b) keep a constant ohmic value

(c) burn to a crisp

(d) increase its ohmic value

4-19. The total resistance value for a 10-kΩ linear taper variable resistor when its arm is midway between its ends is

(a) 10 kΩ

(b) 7.5 kΩ

(c) 5 kΩ

(d) 2.4 kΩ

4-20. The total resistance in a potentiometer circuit

(a) increases with arm rotation

(b) decreases with arm rotation

(c) remains constant with arm rotation

(d) is not affected by arm rotation

4–21. Why is it necessary to turn off the circuit power when making a resistance measurement?

4–22. Draw a schematic diagram showing how to connect a rheostat between source and load.

4–23. Draw a schematic diagram showing how to connect a potentiometer between source and load.

4–24. Draw a schematic diagram showing how you would measure the resistance of a resistor before inserting it into the circuit.

4–25. Two 10-kΩ 5-W resistors are connected in series. What is the total resistance? What is the total wattage rating? (*Sec. 4-2*)

4–26. What reading on an ohmmeter would you obtain when the resistor is open? When the resistor is shorted? (*Sec. 4-4*)

4–27. A current of 25 mA flows through a 4-kΩ resistor. What is the wattage rating required for the resistor? (*Sec. 4-2*)

4–28. State two differences between rheostats and potentiometers. (*Sec. 4-2*)

4–29. In Figure 4-30 the values for the resistors are $R_1 = 10$ kΩ $\pm 10\%$, $R_2 = 470$ Ω $\pm 5\%$, and $R_3 = 680$ Ω $\pm 5\%$. State the color codes for each. (*Sec. 4-2*)

4–30. Write the color codes for the following resistors. (*Sec. 4-2*)

	Band 1	Band 2	Band 3	Band 4
(a) 1.0 Ω $\pm$ 5%	_____	_____	_____	_____
(b) 0.47 Ω $\pm$ 5%	_____	_____	_____	_____
(c) 180 kΩ $\pm$ 10%	_____	_____	_____	_____
(d) 74 Ω $\pm$ 10%	_____	_____	_____	_____
(e) 2.2 MΩ $\pm$ 20%	_____	_____	_____	_____

4–31. Write the resistance values for the following resistors. (*Sec. 4-2*)

	Band 1	Band 2	Band 3	Band 4	Resistance
(a)	orange	orange	orange	silver	_____
(b)	brown	black	yellow	none	_____
(c)	red	gray	black	gold	_____
(d)	white	black	red	gold	_____
(e)	green	brown	brown	gold	_____

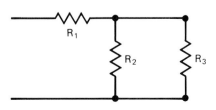

FIGURE 4-30 Circuit for problem 4-29.

4–32. Write the tolerance range for the following resistors. (*Sec. 4-2*)

Value	Low	Nominal	Upper
(a) 100 Ω ± 5%	——	——	——
(b) 820 kΩ ± 10%	——	——	——
(c) 47 kΩ ± 5%	——	——	——
(d) 10 kΩ ± 20%	——	——	——
(e) 0.47 Ω ± 5%	——	——	——

4–33. Determine the power dissipation for the following resistors under the conditions shown. (*Sec. 4-2*)

Resistance	Voltage	Current	Wattage
(a) 10 kΩ	50 V		——
(b) 10 kΩ		1.2 A	——
(c) 120 kΩ		0.1 A	——
(d) 680 Ω	12 V		——
(e) 47 Ω		0.01 A	——

4–34. A precision resistor with a value of 31 Ω is to be constructed using number 20 wire. How many feet of wire are needed? (*Sec. 4-1*)

4–35. A circuit has a current flow of 500 mA and a source voltage of 500 V.
(a) What is the value of the resistance in the load? (*Sec. 4-2*)
(b) What is the power dissipation of the load?

4–36. Listed below are several resistance values. Select two from the list to produce the indicated resistance. Tell how they are wired to achieve the resistance value. (*Sec. 4-2*)

Two each: 100 Ω, 470 Ω, 1.2 kΩ, 4.7 kΩ, 10 kΩ, 15 kΩ

Required Resistance	Resistor 1	Resistor 2	Wiring (Series or Parallel)
(a) 1.67 kΩ	——	——	——
(b) 600 Ω	——	——	——
(c) 7.5 kΩ	——	——	——
(d) 5.9 kΩ	——	——	——
(e) 10.1 kΩ	——	——	——

TROUBLESHOOTING PROBLEMS

4–37. A circuit has three 500-Ω resistances connected in a manner that provides a total of 1500 Ω. When the circuit is connected to a 15 V source, the total current flow measured is 0.015 A. What, if anything, is the problem in this circuit?

4–38. The same circuit described in the previous problem has a current flow of 0.010 A. What, if anything, is the problem in this circuit?

4–39. A resistor having the color code bands of red, green, red, and gold measures a value of 2700 Ω. Is it in tolerance?

4–40. Three resistances having the color codes of brown, black, orange, gold on the first one, yellow, violet, brown, gold on the second one, and brown, black, brown, gold on the third are connected together in series. Their total resistance measures 8500 Ω. Is this acceptable?

4–41. When the three resistors in the previous problem are parallel connected, their total resistance measures 817.8 Ω. Is this acceptable? If not, what should their total resistance measurement be?

4–42. A circuit diagram states that the total resistance for a series circuit containing four 10-kΩ resistances measures 45000 Ω. If the tolerance of each resistance is $\pm 5\%$, is this an acceptable value for the resistances?

4–43. A resistance measures 0 Ω in the circuit. It is connected in parallel with a 5-Ω resistance. Is this an acceptable measurement?

4–1. Identifies its heat dissipation rating

4–2. No; a rating of over 360 W minimum is required.

4–3. 1.0 W

4–4. 50.4 to 61.6 kΩ

4–5. 80 to 120 kΩ

4–6. 114 to 126 Ω

4–7. 1.62 to 1.98 MΩ

4–8. 47 k$\Omega \pm 5\%$

4–9. 1.2 k$\Omega \pm 10\%$

4–10. 33 k$\Omega \pm 5\%$

4–11. 5.6 $\Omega \pm 5\%$

4–12. Red, red, red, silver

4–13. Brown, black, green

4–14. Brown, black, gold, gold

4–15. Brown, gray, brown, gold

4–16. Yes (tolerance range is 2565 Ω to 2835 Ω)

4–17. No (tolerance range is 162 Ω to 198 Ω)

4–18. Yes (tolerance range is 376 kΩ to 564 kΩ)

4–19. No (tolerance range is 2250 Ω to 2750 Ω)

ANSWERS TO REVIEW PROBLEMS

4–20. No (tolerance range is 5.7 Ω to 6.3 Ω)

4–21. No (tolerance range is 95 Ω to 105 Ω)

4–22. Positive

4–23. Zero

4–24. Negative

REVIEW

Chapters 1–4

The fundamental concepts related to electronic theories are applied throughout this book. They are also applied almost daily in the world of work. The material presented in this section is an introduction to these concepts. The major points for the first four chapters are reviewed here.

1. The terminals to all batteries have a specific polarity assigned to them.

2. All substances have an atomic structure consisting of a nucleus and electrons.

3. The number of electrons in the outermost ring of the atom determines the polarity of the material.

4. Materials that give up electrons are said to be negative, materials that accept electrons are positive, and those that do neither are neutral in polarity.

5. Electrical charges that are not in motion are called static electricity.

6. The coulomb is the unit of electrical charge. It is equal to 6.25×10^{18} electrons.

7. Opposite charges attract each other, whereas like charges repel each other.

8. A charged body creates an electric field around itself. The strength of this field is inversely proportional to the square of the distance away from the body.

9. The size of the electric field is directly proportional to the amount of charge used to create it.

10. Potential is a difference in electrical charge that exists between two points in a system.

11. The quantity of charge that can be moved by an electrical potential is directly related to the strength of this potential difference, or voltage.

12. Electromotive force (EMF) is measured in units of the volt.

13. A voltage-measuring meter is connected across the points of potential difference in a circuit.

14. The quantity of electric charge that moves past a given point in 1 s is called the electric current. It is measured in amperes.

15. Electron current flows from negative to positive in a circuit.

16. The amount of voltage that is applied to an electric circuit determines the current flow. Increased voltage produces increased current flow.

17. Electric current flow creates an electromagnetic field around a conductor.

18. The opposition to electron flow is called resistance. It is measured in ohms.

19. A basic electric circuit contains a source of power, a load, and connecting wiring.

20. Electric current flow is from negative to positive, and conventional current flow is from positive to negative.

21. An open circuit is a break in the electron flow path. It is a very high resistance.

22. A short circuit is an undesirable low-resistance path. It permits unusually high levels of current flow.

23. Electrical energy is created by chemical action, ionization action, heat, light, and electromagnetic induction.

24. Direct current (dc) flows constantly in one direction in the system.

25. Alternating current (ac) reverses its polarity periodically in the system. Current flow is constantly being reversed by this process.

26. Electrical conductors should have free electrons and low-resistance factors. Typical conductors are copper, silver, gold, and aluminum.

27. Copper is the major conductor in use because of its availability and relatively low cost.

28. An insulator is a material that normally does not have any free electrons. Typical insulators are oil, mica, air, and ceramics.

29. The insulating quality of a material is called its dielectric strength.

30. The purpose of any conductor is to permit electron flow from the source to the load and back to the source. This forms a basic electron circuit.

31. Wire resistance affects current-carrying capability of a conductor.

32. Wire gage number is the system used to describe wire size and resistance. It is inversely proportional to the physical size of the wire.

33. Wire conductors are available as single conductors or multiple-conductor cables and may also have a shielding over the conductor.

34. Printed circuit boards are used instead of discrete wiring in automated production processes.

35. A schematic diagram is used to show the electrical wiring of a circuit or system.

36. A block diagram shows a functional section diagram of a circuit or system.

37. Schematic diagrams are "read" like the printed page of a book. Voltage values rise to higher levels as the wires on the circuit rise on the schematic. (The lowest line has the lowest voltage value.)

38. Components are shown on a schematic diagram using standard graphic symbols.

39. A switch is a mechanical device used to make an electrical connection. A relay is an electromechanical device used for the same purpose.

40. Circuit-protective devices include wire links, fuses, and circuit breakers.

41. Fuses are rated by size, current capacity, operating voltage, and failure-time values.

42. A circuit can be constructed on a prototype board. One of the easiest methods of doing this is to follow the schematic diagram.

43. The most common discrete component used in electronic circuits is the resistor.

44. Resistors are used to limit current and to provide a voltage drop in the circuit.

45. Resistors are generally grouped in either "power" (high power) or "resistor" (low power) types. Low-power resistors are rated $\frac{1}{8}$, $\frac{1}{4}$, $\frac{1}{2}$, 1, and 2 W. High-power resistors are rated at 3 W or higher.

46. Two ratings are used for resistors. These are power and ohmic value. They are independent of each other.

47. Electric power is a product of voltage and current in the circuit. It is also described as an I^2R factor. Power is rated in watts or kilowatts.

48. Low-power resistors are made of a carbon and clay composition, metallized-film, or wire-wound elements inside a plastic housing. Power resistors are made of a resistance wire wound on a ceramic form.

49. Heat buildup on the body of the resistor is removed by use of a heat sink or a fan.

50. Standard numeric values for electronic components have been established by the Electronic Industries Association.

51. All resistors have a tolerance rating. If the resistor is good, the measured value of the resistor must be within the extremes of the tolerance.

52. An industry standard color code is used to represent numbers, mulltipliers, and reliability ratings for components.

53. Resistors are produced as either fixed or adjustable ohmic values.

54. Variable resistors may be wired as either rheostats or potentiometers. The rheostat is used to control current, and the potentiometer is used in voltage-control circuits.

55. A group of electrically variable resistors include varistors, thermistors, photoresistors, and fusible resistors.

56. Temperature coefficients describe the change in original value that occurs when a device is heated during normal operation.

57. Resistor values are measured with an ohmmeter.

CHAPTER 5

Circuit Laws

Now that you know what resistors are, the next major step is to be able to utilize this knowledge in a practical manner. Since the purposes of resistors include voltage-drop development and current limiting, it is desirable to know how to select the proper resistance value for use in a circuit.

Many electronic components have very specific operating voltage and current limitations. Selection of the proper type of component and its successful use are directly related to the ability to "build" a circuit for proper operation. In addition, should the component fail after the circuit is designed and operational, you must know how to test the circuit and system to determine which component caused the failure and how to select the proper values of components to repair the circuit.

KEY TERMS

Closed Loop	**Ohm's Law**
Junction	**Voltage Drop**
Kirchhoff's Law	

OBJECTIVES

Upon completion of the material in this chapter, you will

1. Be able to apply Ohm's law to basic circuits.
2. Recognize the need for safe operating practices.
3. Understand the concept of circuit common, or ground.
4. Understand the basic concept of Kirchhoff's laws.
5. Understand the concept of electrical power.
6. Be able to apply Ohm's and Watt's laws to basic circuits.

Every person who works in the fields of electricity and electronics uses the basic circuit laws developed by Ohm, Watt, and Kirchhoff. Often the use of these laws is so natural that we do not realize that they are being used. Each of these circuit laws is required knowledge. Almost every known certification test, licensing test, or aptitude test related to employment in the electrical and electronic fields examines for knowledge of how these three laws are used. In addition, how well the person knows how to apply them to perform circuit design or analysis is also examined. The importance of learning these laws and how to apply them cannot be overemphasized. The material in this chapter will help in overcoming any lack of knowledge about the laws presented by Ohm, Watt, and Kirchhoff.

5-1 Ohm's Law

Ohm's law *Relationship of voltage, current, and resistance in a circuit.*

Ohm's ideas about the relationships among voltage, current, and resistance are explained by his law. He found that the voltage drop that develops across a resistance is directly proportional to the amount of current that flows through the resistance. When the voltage that is applied to the resistance is increased, the current flow through the resistance is also increased. The opposite is also true. When the voltage across a resistance is decreased, the current flow through the resistance also decreases. This is valid only when the value of the *resistance* is held at a constant value. This is illustrated in Figure 5-1. Part (a) has a 10-V source and a 10-Ω resistance. The current flow is 1 A under these conditions. When the voltage is increased to 20 V and the resistance held at 10 Ω [Figure 5-1(b)], the current has now increased to 2 A. If the voltage is reduced to 5 V [Figure 5-1(c)], the amount of current drops to 0.5 A. This proves the statement originally made about the relationship of voltage, current, and resistance in a simple circuit.

Another way of stating Ohm's law is used to describe conditions that occur when the *voltage* is held at a constant value and the amount of resistance in the system is varied. Under these conditions a different set of actions occur. When the resistance is increased, the current flow will decrease. When the resistance is decreased, the current flow will increase in the circuit (Figure 5-2). When the resistance is increased from 10 Ω to 20 Ω, the current decreases to 0.5 A. When the resistance is reduced to 5 Ω, the current will increase to 2 A.

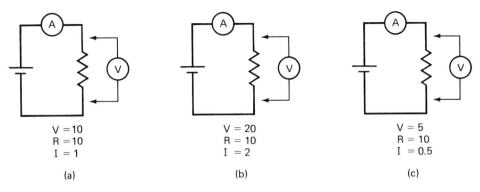

V = 10
R = 10
I = 1

(a)

V = 20
R = 10
I = 2

(b)

V = 5
R = 10
I = 0.5

(c)

FIGURE 5-1 Comparison of voltage, resistance, and current in a common circuit. When the value of resistance is kept constant, the voltage and current values are inversely proportional.

These relationships can be presented in formula form. The formula for current in an electric circuit is

$$I = \frac{V}{R}$$

This formula describes the action shown and described in Figure 5-1 and 5-2. When a constant value of resistance is used in a circuit any change in the amount of voltage will directly affect the quantity of current flow. As the voltage rises, the quantity of current also rises. Any reduction in the amount of the applied voltage also reduces the quantity of current flow in the circuit. When the voltage is maintained at a constant value there is an inverse relationship to resistance and current flow. Any increase in resistance will reduce the amount of current flow. A reduction in the amount of resistance will produce an increase in current flow in the circuit. We can state that current and resistance are inversely proportional when the voltage is a constant value in the circuit. The formula for this relationship is

$$V = I \times R$$

FIGURE 5-2 When the value of voltage is kept constant, the current and resistance are inversely proportional.

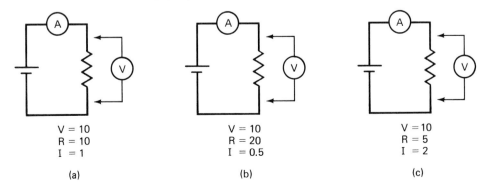

V = 10
R = 10
I = 1

(a)

V = 10
R = 20
I = 0.5

(b)

V = 10
R = 5
I = 2

(c)

This formula is derived from the previous relationship in the following manner:

$$I = \frac{V}{R} \qquad \text{(original formula)}$$

$$I \times R = V \times \frac{R}{R} \qquad \text{(multiply both sides by } R\text{)}$$

$$I \times R = V \qquad \text{(cancel } R/R\text{)}$$

Thus, by multiplying both sides of the equation by the same value (R) we are able to develop an equivalent form of Ohm's law. This modification of the formula permits us to determine a voltage value when both the resistance and current values are known.

A third formulation can be developed from the original $I = V/R$ relationship. This is done in a manner that is similar to the one used to find voltage values. We can also determine resistance values from the original formula:

$$I = \frac{V}{R} \qquad \text{(original formula)}$$

$$I \times \frac{R}{I} = \frac{V}{R} \times \frac{R}{I} \qquad \text{(multiply both sides by } R/I\text{)}$$

$$R = \frac{V}{I} \qquad \text{(cancel } I/I \text{ and } R/R\text{)}$$

This provides us with the third relationship in Ohm's law. We can now determine the amount of resistance in an electrical circuit if we know both the voltage and the current values.

These three formulas are all developed from the relationship Ohm identified in his research. The three formulas are all required knowledge for persons working in either electrical or electronic occupations.

Figure 5-3 shows a method that is often used to aid in the memory of these relationships. The aid to your memory is often drawn as either a circle or a triangle. The specific form is of little importance. The really important factor here is to visualize the positions of voltage, current, and resistance. Using

FIGURE 5-3 The three forms of Ohm's law can be identified easily through the circle or triangle format.

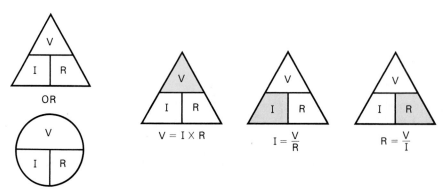

this visual or mental aid one can easily derive the correct formula. This, of course, is using the assumption that for any use of the formula, two of the three factors are known. Often, people will mentally place a finger over the one factor they wish to find. When this is done the formula relationship is quickly identified. For example, if one wants to find resistance (R) and knows the values of voltage and current, then by covering the letter R the formula $R = V/I$ is readily apparent. Replacing the letters V and I with numbers representing the values of voltage and current will provide a mathematical equation that can be solved for the value of resistance.

EXAMPLE 5-1

Find the value of the missing component in each of the following.

	V	I	R
(a)	10	2	___
(b)	___	50	1000
(c)	75	___	25

Solution

(a) $R = V/I = 10 \text{ V}/2 \text{ A} = 5 \text{ }\Omega$
(b) $V = I \times R = 50 \text{ A} \times 1000 \text{ }\Omega = 50{,}000 \text{ V or } 50 \text{ kV}$
(c) $I = V/R = 75 \text{ V}/25 \text{ }\Omega = 3 \text{ A}$

Let us use some practical examples to further clarify our understanding of how Ohm's law works. The electrical circuit shown in Figure 5-4 consists of six 1.5-V cells, an ammeter, a lamp, and a selector switch. The purpose of the switch is to select a voltage. This voltage is 1.5 V for one cell of the system. Each time one additional cell is added to the system the voltage applied to the lamp is increased by 1.5 V. The six cells in this system develop a total voltage of 9 V (6 V × 1.5 V = 9 V). Figure 5-5 will help your understanding of how this is accomplished. In the figure, one cell switched into the lamp circuit produces a current flow of 0.15 A. Three cells produce a current of 0.45 A while the resistance is held at a constant value. As the applied voltage increases,

FIGURE 5-4 Switching additional cells into the source will increase the current through the lamp load.

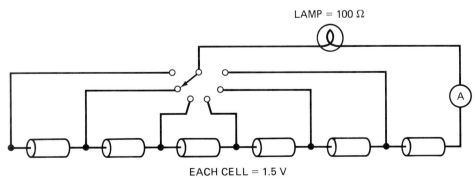

LAMP = 100 Ω

EACH CELL = 1.5 V

R	V	I
10	1.5	0.15
10	3.0	0.30
10	4.5	0.45
10	6.0	0.60
10	7.5	0.75
10	9.0	0.90

FIGURE 5-5 The resistance shown in Figure 5-4 is held at a constant value (10 Ω). The current flow depends on the value of applied voltage.

the current also increases. This continues until the highest value, 9.0 V is applied. This produces a current flow of 0.9 A in the circuit.

This example illustrates the necessity of matching the operating voltage of a lamp to the applied voltage from the cells. When the correct voltage is applied, the lamp will be bright. As the applied voltage is decreased, the lamp brilliance will also be diminished. A voltage that is higher than the design value for the lamp will produce a larger-than-normal current flow. The result is the failure of the filament in the lamp. The observed result of this is that the lamp will no longer produce light.

REVIEW PROBLEMS

Calculate the voltage for each of the following.

5-1. 0.016 A through 120 Ω
5-2. 10.01 A through 16 Ω
5-3. 0.15 A through 2400 Ω

Current (*I*) In many applications in electronic circuits the current is limited to a very low value. The size of the conductors used to carry the current often influences this factor. This application is used often in the values required to operate the cathode ray tube in a video display terminal. The value of voltage required for the tube ranges between 6000 V (6 KV) and 20,000V (20KV). The actual amount depends on the characteristics of the specific tube used in the system. The resistance in these circuits is usually in the range of 3 MΩ (3,000,000Ω). When we calculate the current flow using a value of 10,000 V and the resistance value of 3,000,000 Ω, we find

$$I = \frac{V}{R}$$

$$= \frac{10,000}{3,000,000}$$

$$= 0.0033 \text{ A}$$

Thus the high level of operating voltage for this circuit requires a very low value of current. This is an example of a circuit that requires a high voltage and a low current for its operation.

Often, the resistance value for a specific circuit or system is very low. It can be under 1 Ω with successful operation of the circuit. If a high voltage of 10 kV were applied to this circuit, a current of several thousand amperes would flow. If the exact value of resistance were 1.0 Ω, the current would be determined as follows:

$$I = \frac{V}{R}$$

$$= \frac{10,000}{1.0}$$

$$= 10,000 \text{ A } (10 \text{ kA})$$

The conductors required to carry this amount of current are too large even to comprehend. This is not a practical system. Usually, circuits that have a very low operating resistance require a low operating voltage. This voltage is normally anywhere from 3 to 12 V. One common value for operating voltage is 5.0 V. Let us determine the current flow with a resistance of 0.5 Ω.

$$I = \frac{V}{R}$$

$$= \frac{5}{0.5}$$

$$= 10 \text{ A}$$

This value, although large, can be developed by a commercial power source. The current values can be handled by readily available conductors. In this example the low level of voltage permits a larger value of current.

EXAMPLE 5-2

How much current is required to operate a 14-Ω load from a 120-V power source

Solution

$$I = \frac{V}{R}$$

$$= \frac{120 \text{ V}}{14 \text{ Ω}}$$

$$= 8.57 \text{ A}$$

EXAMPLE 5-3

A radio with a resistance of 800 Ω is operated from a 12.0-V battery. What is the current flow?

Solution

$$I = \frac{V}{R}$$

$$= \frac{12\text{ V}}{800} \, \Omega$$

$$= 0.015\text{ A}$$

EXAMPLE 5-4

The radio of Example 5-3 is operated from a 120-V source. What is the current flow now?

Solution

$$I = \frac{V}{R}$$

$$= \frac{120\text{ V}}{800\ \Omega}$$

$$= 0.15\text{ A}$$

Examples 5-2 through 5-4 show how the operating voltage affects current flow. They also illustrate the effect of different resistances with the same operating voltage.

Voltage drop *Difference in voltage between two points in circuit created by current flow through a resistance.*

Voltage (*V*) When attempting to analyze an electrical circuit, certain rules must be kept in mind. One of these is that the circuit must contain a voltage source, a load, and a set of connecting wires or conductors. Another extremely important rule states that the only way current will flow in a circuit is when a voltage is applied to the circuit. The circuit shown in Figure 5-6 contains all the components required for current flow. In this simple circuit, the voltage applied to the load is equal to the voltage measured across the load.

Voltage is determined by the formula $V = I \times R$. The voltage that is developed across the load is determined by the use of the product of the values of I and R. This is often called the *IR drop* in technical literature. The *IR* drop is equivalent to the **voltage drop**. In this or any other circuit, there would be

FIGURE 5-6 Use of Ohm's law to determine circuit values in an electrical circuit.

SOURCE 9 V   WIRES   LOAD 12 Ω   9 V

$$V = IR = 0.75 \times 12 = 9\text{ V}$$

$$R = \frac{V}{I} = \frac{9}{0.75} = 12\ \Omega$$

$$R = \frac{V}{R} = \frac{9}{12} = 0.75\text{ A}$$

no voltage drop unless current flows through the load; that is, there is resistance. This factor is used in any electrical circuit or in any portion of a circuit. When we attempt to determine the amount of voltage that *drops*, or develops, across a resistance, we use the amount of current in amperes and the amount of resistance in ohms to find the value.

REVIEW QUESTIONS

Calculate the current for each of the following.

5-4. 24 V applied to 1200Ω
5-5. 120 V applied to 500 Ω
5-6. 9 V applied to 45 Ω

Resistance (R) The third factor that is used in Ohm's law is resistance. It is measured in units of the ohm. The relationship of voltages and current to resistance is shown in the formula.

$$R = \frac{V}{I}$$

In Figure 5-6 the amount of resistance is found when the applied voltage is divided by the current flow in the load. What is being stated here is that the resistance is the relationship of voltage and current. The exact values of voltage and current may change in a given system. The amount of resistance may stay at a fixed value as long as the ratio of the voltage and current remain the same. For example, if the voltage in our sample circuit should double and become 18 V, the current will now be 1.5 A. If the value for the voltage is reduced to 6 V and the load resistance of 12 Ω is maintained, the current will drop to a value of 0.5 A. The ratio between the values of voltage and current stays at a constant value of 12 Ω in this example regardless of the specific values of the voltage and current in the system.

The makeup of the resistance in a given circuit is not always necessary. The graphic symbol used to represent a resistance can represent almost any device as long as the device has some resistive qualities. The load resistance can represent a computer, a radio, a complex industrial machine, or any other device. In many applications the only information that is required is to show the symbol. The specific load is often shown as a rectangular or square box that has a resistance value (Figure 5-7). The resistance of this unknown unit,

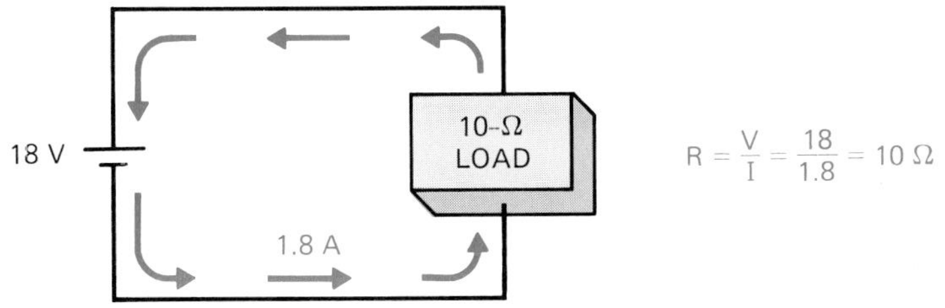

18 V   10-Ω LOAD   1.8 A

$R = \frac{V}{I} = \frac{18}{1.8} = 10\ \Omega$

FIGURE 5-7 The load is normally represented by a value of resistance, since the current is dependent on the quantity of applied voltage and the load resistance.

or box, is determined by Ohm's law, $R = V/I = 10\ \Omega$. The contents of the box are not important at this time. The major concern for the engineer or technician relates to the specific circuit values that develop from this resistance. The material or components that produce this value of resistance are not of great concern at this time. Further analysis is required for their identification. This analysis is not required at this time.

Linear Qualities of Voltage and Current The study of the relationship between voltage and current as they were explained by Ohm shows that they have some linear, or in-line, qualities. When the circuit resistance is maintained as a fixed value, this relationship is clear. In Figure 5-8 the circuit resistance is maintained as 10 Ω. The voltage applied to the circuit will start at 10 V and rise by 10-V steps until it reads 100 V. Using Ohm's law, we find that the current for a 10-V source is 1 A. When the voltage is increased to 20 V the current increases to 2 A. Each 10-V rise in the applied source voltage produces an increase in the circuit current by 1 A. This is an example of a *linear*, or in-line, increase in the system. The diagonal line on the graph shows the Ohm's law value that is held at a constant value of 10 Ω.

A second illustration of this effect is shown in Figure 5-9. Using Ohm's law, an applied voltage of 100 V will produce a current of 0.01 A through a load resistance of 10 kΩ. Increasing the applied voltage to 200 V will produce a current of 0.02 A through the same load. When the voltage is increased to 800 V the current increases to 0.08 A. This also supports the linear relationship between voltage and current with a constant value of resistances in the circuit.

FIGURE 5-8 The relationship of voltage and current may be represented as a linear expression when the resistance in maintained at a constant value in the circuit.

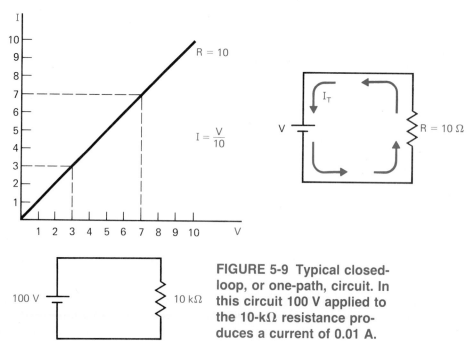

FIGURE 5-9 Typical closed-loop, or one-path, circuit. In this circuit 100 V applied to the 10-kΩ resistance produces a current of 0.01 A.

REVIEW PROBLEMS

Calculate the resistance for each of the following.

5-7. 105 V producing 6 A
5-8. 80 V producing 1.2 A
5-9. 5 V producing 24 A
5-10. 120 V producing 4 A

Practical Units for Ohm's Law When we use electrical energy we often find the need to express the values of voltage, current, and resistance in rather large values. When expressing these values in electronic circuits there is a need to use some very small values. Some of these values are applied to both electrical and electronic systems. As an example, resistances are often expressed in units of thousands or millions of ohms. Voltage is expressed in thousands of volts and also in thousandths of a volt. Current values are often given in values of thousandths or millionths of the ampere.

The confusion that can result when one attempts to read or use a numeric value that contains five, six, or seven or more zeros and perhaps a decimal point is minimized by the use of metric equivalents instead of all the zeros. Some of the more common metric equivalents are shown in Figure 5-10. These units can be considered as conversion factors for common electrical values. One should learn these conversion factors. They are in common use in almost all electrical and electronic circuit diagrams and descriptions. This is a portion of the complete set of prefixes used to describe the values.

FIGURE 5-10 Typical common values used to express units, multiples, and subunits of electrical values.

PREFIX	SYMBOL	BASIC UNIT RELATION	EXAMPLE
mega	M	1,000,000 OR 1×10^6	8 MΩ OR 8 MEGOHMS = 8,000,000 = 8×10^6 OHMS
kilo	k	1000 OR 1×10^3	3 kV = 3000 V = 3×10^3 V
milli	m	0.001 OR 1×10^{-3}	6 mA = 0.006 A = 6×10^{-3} A
micro	μ	0.000 001 OR 1×10^{-6}	24 μA = 0.000 024 A = 24×10^{-6} A
pico	p	0.000 000 000 001 OR 1×10^{-12}	18 pA = 0.000 000 000 018 A = 18×10^{-12} A

In actual operational devices it is not unusual to find values of 28 mA or 47 kΩ. The value of the voltage required for tube operation of the video display terminal is described in units of the kilovolt. Be sure that you can differentiate between capital letters and lowercase letters as they have different meanings. For example, the capital letter M is used to indicate millions, or 1×10^6, and the lowercase letter m is used to indicate one-thousandths, or 1×10^{-3}.

EXAMPLE 5-5

How much current flows through a resistance of 18 kΩ when 12 V is applied

Solution

$$I = \frac{V}{R} = \frac{12 \text{ V}}{18 \times 10^3 \text{ } \Omega}$$

$$= 0.000667 \text{ A} \quad \text{or} \quad 667 \times 10^{-6} \text{ A} \quad \text{or} \quad 667 \text{ } \mu\text{A}$$

EXAMPLE 5-6

What is the voltage (*IR* value) when an *I* of 62 mA flows through a 13-kΩ *R*?

Solution

$$V = I \times R$$

$$= 0.062 \text{ A} \times 13{,}000 \text{ } \Omega \quad \text{or} \quad (62 \times 10^{-3}) \text{ A} \times (13 \times 10^3) \text{ } \Omega$$

$$= 806 \text{ V}$$

In Example 5-6 we see that when we multiply *kilo* values by *milli* values the answer is in whole units. The reason for this is that the factors of 10^{-3} and 10^3 cancel each other out. Other combinations can also be found. For example, when we have one value in *volts* and another value in *kilohms*, the answer is in units of *milliamperes* of current when solving with Ohm's law. In the same way, when working with voltage and *megohms*, the result is reported in units of the *microamperes*.

Some of the more common combinations of these factors are shown in Figure 5-11. These conversion factors are commonly used for values in electronic circuits. The reason for this is because of the small quantities normally used by the components currently in use in these circuits.

mA × kΩ = V $\quad$ ($10^{-3} \times 10^3$) $\qquad$ $\dfrac{V}{k\Omega}$ = mA $\quad$ $\left(\dfrac{V}{10^3} = 10^{-3}\right)$

μA × MΩ = V $\quad$ ($10^{-6} \times 10^6$) $\qquad$ $\dfrac{V}{M\Omega}$ = μA $\quad$ $\left(\dfrac{V}{10^6} = 10^{-6}\right)$

FIGURE 5-11 Conversion factors frequently used in describing electrical values.

5-2 Electric Power

Electric power is defined as the time rate of performing some work using electrical energy. The unit of electrical power is the watt (W). A practical unit for the measurement of electric power consumption is the kilowatthour (kWh). This is explained as the number of kilowatts used in 1h in the performance of work. An illustration of this action and its description can be seen when one considers the amount of power required to operate a medium-sized computer for a period of 24 h. The current rating of the computer is 41.1 A. When operated for 1h from an 120-V source the power consumption is 4932 W, or 4.93 kW. This is the quantity of power required for any given moment of time. When the computer is operated for 24 h, this power rating is multiplied by the total period of time. In this illustration the watthour rating is 118.368 kWh.

REVIEW PROBLEMS

5-19. What is the power required to operate a 120-V toaster rated at 11 A?
5-20. What is the current required to operate a 1.2-kW-rated appliance from a 240-V source?
5-21. What is the current required to operate a pocket calculator from a 3-V source? The calculator consumes 3.0 mW of power.

5-3 Interrelationship of Ohm's Law and Watt's Law

One of the basic rules for mathematics is that one value can be substituted for another as long as they are equal. This is also applicable in electrical work. Because of this rule, we can show an interrelationship between Ohm's law and Watt's law. Start by using the formula $P = V \times I$. We have found that Ohm used $V = I \times R$ in his formulation. He stated that the value of V is equal to the product of I and R. It is possible to use the equality $I \times R$ as a substitute for the value of V, if desired. Let us try it out in the following example:

$$P = I \times V \quad \text{and} \quad V = I \times R$$

$$P = I \times (I \times R) \qquad \text{by substitution}$$

$$= I \times R \quad \text{or} \quad I^2R \quad \text{by combination of like terms}$$

Another relationship is shown in this example:

$$P = I \times V \quad \text{and} \quad I = \frac{V}{R}$$

$$P = V \times \frac{V}{R} \qquad \text{by substitution}$$

$$= \frac{V^2}{R} \qquad \text{by combination of like terms}$$

This procedure may be followed for each of the six interrelationships between these two laws. Figure 5-12(a) shows these relationships as well as the basic formulas of Watt and Ohm. The wheel displayed in Figure 5-12(b) shows how we can use any one of three formulas to find the value shown in the center of

$$V = \frac{P}{I} \qquad \text{OR} \qquad V = \sqrt{PR} \qquad \text{OR} \qquad V = IR$$

$$P = \frac{V^2}{R} \qquad \text{OR} \qquad P = I^2\,R \qquad \text{OR} \qquad P = IV$$

$$I = \frac{V}{R} \qquad \text{OR} \qquad I = \frac{P}{V} \qquad \text{OR} \qquad I = \sqrt{\frac{P}{R}}$$

$$R = \frac{V}{I} \qquad \text{OR} \qquad R = \frac{V^2}{P} \qquad \text{OR} \qquad R = \frac{P}{I^2}$$

(a)

OHM'S LAW & WATT'S LAW RELATIONSHIPS

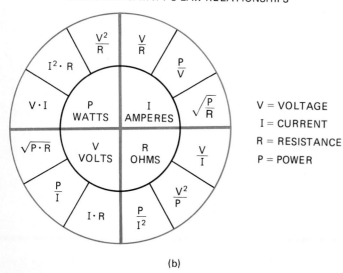

V = VOLTAGE
I = CURRENT
R = RESISTANCE
P = POWER

FIGURE 5-12 Interrelationships of the laws of Ohm and Watt are shown on this wheel.

(b)

the wheel. All the variations are based on these mathematical interrelationships and substitutions.

REVIEW PROBLEMS

5-22. What quantity of power is consumed when 600 mA of current flows through a 3.3-kΩ resistor?

5-23. What power is consumed when 600 V is connected to a 60-Ω resistor?

5-24. A calculator uses 5 mW of power. How much power is consumed during a 24-h period?

5-25. A 2-W power supply creates an operating voltage of 50 V. What current does it supply?

5-4 Kirchhoff's Laws

Another scientist, Gustav Kirchhoff, developed two laws that help make circuit analysis easy. One of these laws is related to voltages and the other, to current flow in a circuit. Both laws are explained in detail in Chapter 10. They need to be introduced at this point to help you understand the basics of circuit analysis. The circuit shown in Figure 5-13 will be used to help explain one of the laws. This circuit is more complex than any of the others used so far in this book. Resistors R_1 and R_2 are connected together at point B. Any current flow from point C to point A will also pass through point B. A third resistor, R_3, is connected from point C to point A. This provides a second path for current from the source V. This type of circuit, one that has more than one current path, has a special name. It is called a *parallel* type of circuit. The two paths for current are (1) through the combination of R_1 and R_2, and (2) through R_3. Parallel current path circuits are discussed in detail in Chapter 7.

Kirchhoff's law of currents states that a current entering a **junction** is equal to the current leaving the same junction. A junction is defined as a point in an electric circuit where two or more components or wires are connected to each other. There are several junctions in Figure 5-13. One of these is at point B. The current flowing through R_2 enters that junction and also leaves it to go through R_1. In addition to this, there is a junction at point C. Current from the source enters the junction and then flows into the branches. Part of the

Kirchhoff's laws *Describes relationship of voltage drops in circuit and current flow at specific junctions in circuit.*

Junction *Connection of two or more components in circuit.*

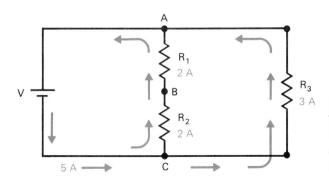

FIGURE 5-13 Parallel, or multipath, circuit. Current entering each of junctions *A*, *B*, and *C* is equal to the current leaving the same junction.

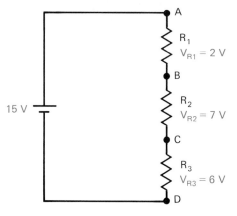

FIGURE 5-14 **The series circuit, or closed loop. The sum of the voltage drops developed across each of the loads is equal to the value of the applied voltage.**

current flow is to R_2 and R_1 and the rest is to R_3. In any case, the sum of the currents leaving the junction at C is the exact value of the current entering that junction. The sum of the currents of 3 A and 2 A is 5 A. Thus we have 5 A entering junction C and 5 A leaving the same junction. The same rule applies to point A. Here 3 A from R_3 and 2 A from R_1 enter the junction. The result is a current of 5 A that flows from the junction back to the power source.

When we analyze the current at point B the same rule is applied. In this part of the circuit there is only one path for current flow. All of the 2-A current flow in R_2 enters point B. It all leaves this point and then flows through R_1. This supports the current law when there is only one path for circuit flow.

The second of Kirchhoff's laws is illustrated in Figure 5-14. In this circuit there is only one path for current flow. This type of circuit is called a **closed-loop**, or series, circuit. The currents entering each of junctions A, B, C, and D also leave these junctions. In this circuit we are looking at the various voltage drops that develop as current flows through the circuit. Kirchhoff stated that the sum of the voltage drop in a closed-loop circuit will equal the voltage applied to the total circuit. In this illustration a source voltage of 15 V is applied to the string of three resistors. The sum of IR drops or voltage drop for these three resistors is equal to the applied voltage from the source. This rule is also always true. One cannot have any values *left over* using this method. The two rules that relate to current flow and to voltage drops will work in any similar situation. Keep these in mind as you learn more details about electrical circuit analysis in subsequent chapters.

Closed loop *Term describing a single path circuit, or series circuit.*

REVIEW PROBLEMS

Use Figure 5-15 to find the following values.

5-26. Current through R_2.
5-27. Voltage measured across $R_2 + R_3$.
5-28. Current through R_4.
5-29. Source voltage.

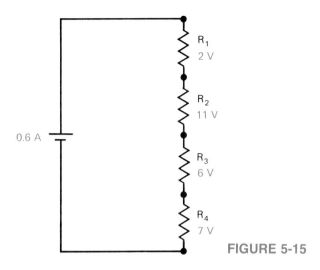

FIGURE 5-15

5-5 Electrical Safety

One of the major factors involved with the use of electrical energy is that under normal conditions one does not see voltage or current flow. Under the proper set of conditions one can *feel* the effects of an applied voltage. The statement that almost all materials are able to conduct electrical current is to be respected. The human body is made of about 90% water or liquid. When the skin is dry, its resistance is in the hundreds of thousands of ohms. For wet skin this resistance drops to levels between 25 Ω and 2 kΩ or less. Let us relate these resistive values to current flow through a resistance.

If you were to hold a bare wire in each hand and have someone connect the bare wires to a voltage source, a current will flow through the resistive path of your body. Consider what current will flow when the hands and skin are dry. (Figure 5-16). It is not unusual to find voltages of 120 V or higher in electrical and electronic equipment. If you should accidentally touch the sources of these voltages, a current will flow through the completed circuit of your body. If one wire is touched by each hand at the same time, the current path is through the hand, the arm, the central trunk of the body, and the other arm, to the second hand. A current of 0.02 A is sufficient to interrupt the normal heartbeat rhythm. This value will differ for men and women, but it is an excellent point of reference for this discussion. When the normal heartbeat is

VOLTAGE	CURRENT (A)		
	DRY SKIN	WET SKIN	BROKEN SKIN
120 V	0.00002	0.0048	1.2
240 V	0.0005	0.0096	2.4
480 V	0.001	0.0192	4.8
600 V	0.0012	0.024	6.0
10 kV	0.02	0.4	100

FIGURE 5-16 Values of current flow through the human body for a variety of applied voltages.

interrupted, the body is no longer able to exchange the gas wastes from the blood for fresh oxygen. The result is the death of the victim of this severe electrical shock.

The severity of electrical shock will produce different results when the current value is less than the lethal amount. Less current will cause an involuntary contraction of the muscles. This reaction could cause a person to jump back from the circuit. This involuntary reaction, if it occurred in a confined work area, could make the victim hit another device. The result could be bleeding from an open wound or even the fracture of a bone.

Under normal conditions the resistance of the skin will minimize current flow through the body. Unfortunately, the skin has a tendency to blister due to the current flow at the specific point of contact. The blisters will break and the current path will have much less resistance when this occurs. Using Ohm's law for a reference, the reduction of the value of resistance in a circuit (and we certainly have a complete circuit under these conditions) will result in an increase of current flow. The end result of this action is usually death by electrocution. The best way of preventing this from happening is to plan ahead how you are going to make the specific test. Electrical tests are conducted daily in a safe manner by many thousands of persons employed or involved in electrical and electronic circuit analysis. They must be following a safe set of standards or they certainly would not be present to report the results of their efforts. You, too, can be successful and have a reasonably long life if you will follow safe operating procedures.

How to Avoid Electrical Shock One of the best ways of avoiding the dangers of electric shock is to develop a set of personal safe work habits. One example of this is to work with only one hand when making circuit measurements. Learn to keep the second hand away from the wires or any of the metal parts of the system. Many electronic devices use the metal chassis as one side of the conductive path. This is usually the negative, or ground, part of the circuit. Insulating yourself from this path is very wise. A second important safety rule is to try to make connections to the circuit for measuring voltage when the power is turned off. The preferred method to use is to turn off the circuit power and then connect the leads from the meter. After the leads are in place the power is restored to the circuit and the measurement is read and noted. Power is then again turned off in order to move the leads to another position or to remove them from the system altogether.

Perhaps the best advice that can be offered at this time is to take your time when involved in circuit analysis and troubleshooting. Plan what you intend to do and plan how you are going to accomplish what you intend to do. Often, a person who is in a hurry to complete work is prone to make mistakes. Even if the mistakes are not fatal to the person, they may be fatal to the measuring equipment or to the circuits being measured. In any event, the final result is not the solution of the problem but the creation of another problem.

If, by chance, another person should come in contact with an electrical circuit and you are present, the most important thing for you to do is to attempt to remove the person from the circuit. This must be done carefully. If you are not careful, you may also become a part of the circuit. If possible, turn off the source of electricity. If this is not convenient, attempt to remove the person from the source. Use insulating materials to do this. After the person is separated, arrange for professional help. If you know how, artificial resuscitation may help restore the normal body functions to the victim. Again, be extremely careful—keep yourself from becoming a second victim.

1. The rules established by Ohm, Watt, and Kirchhoff have universal use in the electrical and electronic fields.
2. Ohm's law shows the relationships among voltage, current, and resistance.
3. Three ways of using Ohm's law are: $V = I \times R$, $I = V/R$, and $R = V/I$.
4. When resistance R is held at a constant value, current I increases directly with increases in voltage V.
5. When voltage V is held at a constant value, current I decreases as resistance R increases.
6. The ohm may be expressed in units of 1000 by using the metric prefix "kilo" and in units of 1,000,000 by the metric prefix "mega."
7. There is a linear relationship between voltage and current.
8. Submultiples of electrical values for 1/1000, 1/1,000,000, and 1/1,000,000,000,000 are "milli," "micro," and "pico."
9. Electric power is the time rate of performing work. It is expressed in units of the watt.
10. There are interrelationships between the laws presented by Ohm and Watt. These are accomplished by substituting equal quantities in the formulas.
11. Kirchhoff's law for voltage drop in a closed loop states that the individual drops must equal the voltage applied by the source.
12. Kirchhoff's law for current at a given junction states that the current(s) leaving the junction must equal the current(s) entering the junction.
13. When not handled properly, electric energy can give a person a strong shock. Under certain conditions, current flow through the human body can kill.
14. The wise technician or engineer will turn off circuit power before touching any of the components in a circuit.
15. Planning ahead before getting involved with electrical circuits will prevent shock in most situations.

Answer true or false; change all false statements into true statements.

5–1. Ohm's law shows the relationships of voltage, power, and resistance.
5–2. Watt used the equation $P = I \times R$ in order to show his work.
5–3. In an electric circuit, when the voltage is held at a constant value, the current will decrease as the resistance increases.
5–4. When resistance is held at a constant value, an increase in voltage will produce a decrease in current flow.
5–5. When current is held at a constant value, an increase in resistance will require an increase in the value of applied voltage.

5-6. In any mathematical equation one value may be replaced by any other value if they are equal to each other.

5-7. In a closed-loop circuit the voltages are always equal in any part of the circuit.

5-8. In a closed-loop circuit the source voltage and total resistance determine the amount of current flow.

5-9. In any circuit the current flow that enters a junction is equal to all the currents leaving that junction.

5-10. Electric current is able to flow through all conductive paths, including the human body.

MULTIPLE CHOICE

5-11. A value of 1.8 kV is equal to
(a) 1.8 V
(b) 1800 V
(c) 18,000 V
(d) 180 V

5-12. A resistance value of 120 kΩ is equal to
(a) 1.20 Ω
(b) 1200 Ω
(c) 12,000 Ω
(d) 0.12 MΩ

Select the best answer for each expression.

	(a)	(b)	(c)	(d)
5-13. 0.012 A	12 A	12 kA	1.2 mA	12 mA
5-14. 1.2 kΩ	1200 Ω	1,2000,000 Ω	1.2 Ω	12 kΩ
5-15. 0.0003700 A	370 A	3.7 μA	370 μA	37 μA
5-16. 16,000,000 W	16 kW	0.016 kW	16 MW	1.6×10^3 W
5-17. 15,000 V	1500 V	15 kV	1.50 kV	0.15 mV
5-18. 120,000 Ω	120 kΩ	12×10^4 Ω	0.012 kΩ	120×10^6 Ω

5-19. A circuit contains a 18-V source and a 10-kΩ resistor. The total current in the circuit is
(a) 18 mA
(b) 1.8 mA
(c) 0.18 mA
(d) 0.018 mA

5-20. A voltage is applied to a resistance of 5.6 kΩ, causing a current of 3.8 mA. The value of voltage is
(a) 43.6 V
(b) 31.65 V
(c) 21.28 V
(d) 18.34 V

ESSAY QUESTIONS

5-21. State the relationships among voltage, current, and resistance used in Ohm's law.

5–22. What circuit value appears in Watt's law that does not appear in Ohm's law?

5–23. What is meant by the term *linear*?

5–24. What is meant by the terms *kilo*, *mega*, *milli*, *micro*, and *pico*?

5–25. What is meant by the term *closed loop*?

5–26. Describe an electrical junction.

5–27. State the basic rules as they relate to electrical safe operating practices.

5–28. Why do we advise the reader to plan ahead before attempting to work on any electrical device or circuit?

5–29. Is it possible to combine the work of Ohm and Watt? Give an example of how this is done.

5–30. What is meant by the term *voltage drop*?

5–31. How many volts are in 60 kV? (*Sec. 5-2*)

PROBLEMS

5–32. How many watthours are used when a television set that uses 600 W is operated for 3 h? (*Sec. 5-2*)

5–33. How much power is used when 800 mA of current flows through 1.2 kΩ? (*Sec. 5-2*)

5–34. What power is used when 60 V produces a current of 200 A? (*Sec. 5-2*)

Use Figure 5-17 to answer problems 5–35 through 5–37.

5–35. Calculate the current for each value of resistance. (*Sec. 5-1*)
   (a) $R = 10\ \Omega$
   (b) $R = 25\ \Omega$
   (c) $R = 60\ \Omega$
   (d) $R = 180\ \Omega$

5–36. Calculate the currents for each resistance when the supply voltage is tripled. (*Sec. 5-1*)
   (a) $R = 10\ \Omega$
   (b) $R = 25\ \Omega$
   (c) $R = 60\ \Omega$
   (d) $R = 180\ \Omega$

5–37. What resistances are required to develop each current when the voltage is held at a constant value of 25? (*Sec. 5-1*)
   (a) $I = 5A$
   (b) $I = 0.5\ A$
   (c) $I = 1.0\ A$
   (d) $I = 0.0004\ A$

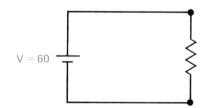

FIGURE 5-17

5–38. A piece of test equipment has a resistance value of 5.1 MΩ. What current flows when it is used to measure 6 V? (*Sec. 5-1*)

5–39. An 18-kΩ resistor has 50 mA flowing through it. What voltage is applied to the resistor to make this current flow? (*Sec. 5-1*)

5–40. What power is produced in a wire with a resistance of 600 Ω when 120 V is applied to it? (*Sec. 5-2*)

5–41. Convert the following to kilowatts. (*Sec. 5-2*)
(a) 15×10^4 W
(b) 0.5 MW
(c) 8700 W
(d) 180 W

5–42. Convert the following to milliwatts. (*Sec. 5-2*)
(a) 0.35 W
(b) 2 W
(c) 0.0025 W
(d) 0.384 W

5–43. Convert the following to volts. (*Sec. 5-2*)
(a) 6.3 kV
(b) 490 mV
(c) 1.683 mV
(d) 0.18 kV

5–44. Convert the following to ohms. (*Sec. 5-2*)
(a) 18 mΩ
(b) 61 kΩ
(c) 8.2 kΩ
(d) 0.19 kΩ

5–45. How much current flows when a 100-Ω resistor is connected to the terminals of a 9.0-V battery? (*Sec. 5-1*)

5–46. What is the operating voltage for a lamp rated at 2 A and a power rating of 100 W?

5–47. What is the current flow through a lamp rated at 60 W and 120 V? (*Sec. 5-2*)

5–48. What is the resistance of a toaster rated at 8 A and operated from a 120-V source? (*Sec. 5-1*)

5–49. What is the power consumption for the toaster described in problem 5–48? (*Sec. 5-2*)

5–50. A circuit is protected with a 20-A fuse. Several devices are connected to the same circuit. Each one adds to the current flow.

Order of Connection	Device	Voltage	Current or Power
1	Toaster	120 V	10 A
2	Coffee pot	120 V	140 W
3	Microwave oven	120 V	600 W
4	TV	120 V	700 W
5	Lamp	120 V	60 W
6	Juicer	120 V	2 A

At what point does the fuse "blow," or fail? (*Sec. 5-2*)

For problems 5-51 through 5-60, draw a circuit diagram and label all components where applicable. Show each formula used. Start with the basic formula, add numeric values, and show all of your work in solving the problem.

5–51. A video display terminal consumes 4 A when 12 V is applied to it. What is the resistance of the unit? (*Sec. 5-1*)

5–52. The unit described in problem 5–51 has a high-voltage section that requires 12 kV and consumes 9 mW of power. What is the current flow for the high-voltage section? (*Sec. 5-2*)

5–53. A current of 14 A flows through a load. What is the resistance of the load when 240 V is applied to it? (*Sec. 5-1*)

5–54. A transistor circuit has a resistance of 0.56 Ω as a part of the circuit. What is the current flow when a voltage drop of 1.68 V is measured across the resistance? (*Sec. 5-1*)

5–55. In a similar circuit, there is a resistance of 390 Ω. The measured current flow is 100 mA. What is the amount of increase in applied voltage that will produce a current of 180 mA? (*Sec. 5-1*)

5–56. A loudspeaker is rated at 8 Ω and 40 W. What is the maximum current flow for this speaker? (*Sec. 5-3*)

5–57. A television receiver's antenna system has a design factor of 300 Ω. What current flows when 10,000 μV of signal is received at the antenna? (*Sec. 5-2*)

5–58. A piece of wire is 4800 ft long. Its resistance is 240 Ω per 1000 ft. How much current will flow through the wire when 120 V is applied to each end? (*Sec. 5-1*)

5–59. A variable power source is able to produce an output voltage ranging from 0 to 100 V. Draw a graph showing the current flow that develops through a 100-Ω resistor when the power source is connected to the resistor. Use the graph shown in Figure 5-18 as a guide for the answer. (*Sec. 5-1*)

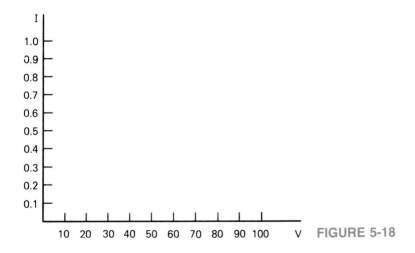

FIGURE 5-18

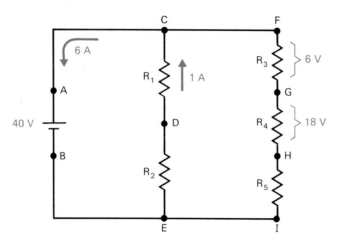

FIGURE 5-19

5-60. Use the two laws presented by Kirchhoff to determine the following values (all questions refer to Figure 5-19). (*Sec. 5-4*)
 (a) current from *B* to *E*
 (b) current from *E* to *D*
 (c) current from *B* to *E* and *I*
 (d) voltage from *H* to *G*
 (e) voltage from *I* to *F*
 (f) voltage from *I* to *H*
 (g) current from *F* and *C* to *A*

5-61. Why should you use only one hand to make circuit measurements? (*Sec. 5-5*)

5-62. Why should circuit power be turned off when attaching test equipment to the circuit? (*Sec. 5-5*)

5-63. How much current will cause death (with a 120-V source)? (*Sec. 5-5*)

5-64. Why does body resistance change when the skin surface is wet? (*Sec. 5-5*)

5-65. What is the correct procedure for you to follow if someone becomes the victim of electrical shock? (*Sec. 5-5*)

TROUBLESHOOTING PROBLEMS

Use the terms **open circuit** *and* **short circuit** *to complete the following statements.*

5-66. A circuit normally has 2 A flowing through the load when 120 V is applied. The current rises to 100 A due to a malfunction. The problem is a(n) _____.

5-67. A circuit uses 150 W of power when 120 V is applied. Power consumption drops to 0 W due to a malfunction. The problem is a(n) _____.

5-68. The total voltage drop across a string of two resistors is 40 V. The voltage drop across one of the resistors measures 40 V due to a malfunction. The problem is a(n) _____.

5-69. The total voltage drop across a string of two resistors is 40 V. The voltage drop across one of the resistors measures 0 V due to a malfunction. The problem is a(n) —————.

5–70. A circuit contains a power source of 48 V and a load resistance of 240 Ω. Circuit current, when measured, is 2 A due to a malfunction. The problem is a(n) —————. What areas of the system could be at fault?

5–1. 1.92 V

5–2. 160.16 V

5–3. 360 V

5–4. 0.02 A

5–5. 0.24 A

5–6. 0.2 A

5–7. 17.5 Ω

5–8. 66.67 Ω

5–9. 0.21 Ω

5–10. 30 Ω

5–11. 28×10^{-3} A

5–12. 6×10^3 Ω

5–13 12×10^{-2} A

5–14. 3×10^6 Ω

5–15. 3,000,000 V or 3 MV

5–16. 0.219 A or 219 mA

5–17. 0.000061 A or 61 μA

5–18. 12 kΩ

5–19. 1.32 kW

5–20. 5 A

5–21. 0.001 A

5–22. 1.188 kW

5–23. 6 kW

5–24. 0.12 Wh or 120 mWh

5–25. 0.04 A or 40 mA

5–26. 0.6 A

5–27. 17 V

5–28. 0.6 A

5–29. 26 V

CHAPTER 6

Series Circuits

One of the two basic types of electrical circuits is known as the series circuit. Recognition of this type of circuit and the rules related to current flow, resistance, and voltage drops are required knowledge for all who design or service any type of electrical product. This knowledge can be used to analyze or create electrical and electronic circuits using any present component or any device developed as a result of technological advances in the future.

KEY TERMS

Chassis Ground	Partial Short Circuit
Circuit Common	Series Aiding
High Side	Series Opposing
Low Side	Total Resistance
Kirchhoff's Voltage Law	Voltage Divider
Pad	Voltage-Drop Polarities

OBJECTIVES

Upon completion of the material in this chapter, you will

1. Be able to discuss the principles of the series circuit.
2. Understand the concept of voltage drop in a circuit.
3. Recognize the concept of circuit common.
4. Understand the concepts of open and short circuits.
5. Be able to explain the concept of the voltage divider.
6. Understand how to troubleshoot a series circuit.

One of the two types of basic circuits is called the *series circuit*. The series circuit has certain characteristics that are important for circuit analysis. These characteristics include the factors of resistance, voltage, current, and power as identified by Ohm and Watt. The ability to design any type of electrical or electronic circuit demands a thorough knowledge of how these factors are affected by operation in a series circuit. Those who are able to analyze, troubleshoot, and repair either electrical or electronic equipment also need to understand and apply the factors mentioned above. The ability to plan what to test, how to test it, and what equipment to use to make the test is based on knowledge of how circuits operate. In addition, we must be able to recognize the type of circuit and its basic characteristics. After the proper tests are concluded, we must apply the concepts for successful analysis and repair procedures.

The series circuit has several significant characteristics. Probably the most important is the fact that there is only one path for current flow in a series circuit. One type of series circuit is illustrated in Figure 6-1(a). This circuit has a source, two resistances, and the necessary connecting wires that identify it as a complete system. Electron flow in any circuit is from the negative terminal of the source through the wire to the resistor R_2. The flow then continues to resistor R_1 and then through another wire to the positive terminal of the source. It should be fairly obvious to you that there is only one path for electron flow in this circuit. This is the characteristic that identifies it as a series circuit.

The same circuit is illustrated in Figure 6-1(b) in a different manner. In this drawing the typical symbol for a power source is not shown. The plus (+) sign and the value 40 V is used to represent the positive connection of the

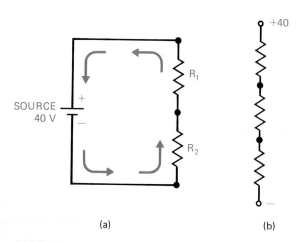

(a)   (b)

FIGURE 6-1 The closed loop, or series circuit, may be shown in either of these types of schematic drawings.

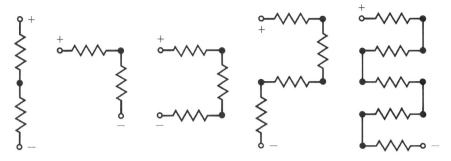

FIGURE 6-2 Different methods of showing the series circuit.

power source. The negative connection is shown as a minus (−) sign. This indicates the negative power source terminal and the source of electrons. Many schematic diagrams use this format instead of showing a complete circuit with the power source. The second portion of this illustration may make it easier for one to visualize a series circuit. In many practical systems there are several series circuits that are a portion of the total system. It is important to be able to identify these as series components, or sections, of a complex system. Their analysis can be accomplished as if they were a stand-alone type of circuit.

One of the more difficult tasks facing beginners in electrical or electronic studies is the identification of a series circuit. The concept of only one path for current flow is perhaps the best recognition method. The problem that occurs is that very few series circuits in practical use appear as simple as those illustrated in Figure 6-1. To identify them, we have to be able to *read* a circuit, or schematic, diagram. In Figure 6-2 several circuits are shown. Each is a series circuit. Some have a minimum of two components; others have more than two components. In all the circuits the components are joined together in an end-to-end arrangement. The key factor in each of these circuits is that there is only one path for current flow through each of them. We could state that the resistors are arranged in the fashion of a chain. They are linked together end to end.

When components are mounted in a printed circuit board they can function in the same manner. The difference between this type of circuit and one that is hand wired is illustrated in Figure 6-3. Part (a) shows one version of a hand-wired series circuit. It contains three resistors and a terminal strip. The resistors are mechanically fastened to the individual terminals and are then soldered to the terminal. This makes a valid electrical connection. Part (b) shows the same circuit on a printed circuit board. The board has conductive copper paths to

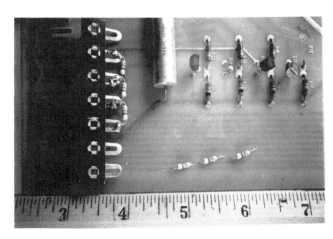

FIGURE 6-3 Two practical methods of physically connecting series resistances used in electrical equipment: (a) resistances mounted on a printed circuit board; (b) hand-wired resistances mounted on a terminal strip.

Pad *Term describing place on a printed circuit board where component is connected to board.*

carry electrons from one component to another. Each connection to the copper conductive path requires a **pad**. The pad is considered to be a mounting point as well as a connecting point. The wire lead from one end of the component is soldered to the pad in order to make a valid electrical current connection. As you can see, the identification of the series circuit on the circuit board is not as easy as when one observes the components wired together. Since the state of the art in electronics has moved to circuit boards the ability to trace the circuit becomes another requirement for success.

One more factor to remember is that the components found in a series circuit are not always resistances. Any electrical or electronic device that can handle charges or permit current flow can be used. Components can also be mixed. One may find resistors, inductors, and capacitors connected in series with each other or with both solid-state or vacuum-tube devices. The use of resistors will help make your understanding of how these circuits function easier than if circuits using mixed components were illustrated. Let us now examine each factor that is significant to the series circuit.

6-1 Current Flow in the Series Circuit

In the process called current flow, electrons move from one atom of material to another atom. The effect is somewhat in the nature of a chain reaction rather than one or more electrons starting at the negative terminal of the power source and flowing through the circuit outside the source toward the positive source terminal. Since the electron movement is like a chain reaction it can also be compared to a string of freight cars in a train. In both of these examples the movement is at a constant speed. In both examples the number of electrons or train cars that move past a given point determines the rate of movement. When discussing trains, we call this the speed of the train; in electrical terms it is called the electrical current. By definition the quantity of electrons that move past a given point in 1 s is the electron current.

A device called an *ammeter* is used to measure the flow of electrons in a circuit. The ammeter is actually an electron flow meter, inserted into the circuit as shown in Figure 6-4. When an ammeter is added to a circuit the electrons have to flow through just as they flow through the components in the circuit. Using four ammeters as we do here shows that the amount of electron flow is of the same value at every part of the circuit. Each black dot in this illustration

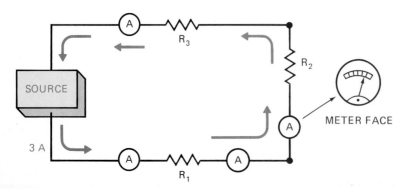

FIGURE 6-4 The measurement of electrical current is done by inserting the ammeter into the wiring of the circuit. In the series circuit this may be done at any point, since the current flow is equal in the circuit.

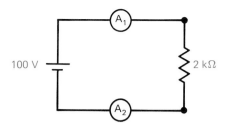

FIGURE 6-5

represents a junction, which allows us to use Kirchhoff's current law to explain the action. The current entering the junction at each black dot has to be equal to the current leaving that junction. If there is a measurement of 3 A at the negative terminal of the power source, the same value would enter and leave the first meter in the circuit. The same quantity of 3 A would also leave the meter and enter the resistance, R_1. The action will continue at each junction in the system. This will help prove that the current is equal in every part of a series circuit.

REVIEW PROBLEMS

Use Figure 6-4 to answer problems 6-1 and 6-2.

6-1. What current flows through R_3?
6-2. If we were to add another ammeter at the positive terminal of the power source, what amount of current will it indicate?

Use Figure 6-5 to answer problems 6-3 and 6-4.

6-3. What is the current flow through the resistor?
6-4. What current flow will be measured by meters A_1 and A_2?
6-5. What can be stated about the amount of current at any point in a series circuit?

6-2 Resistors in the Series Circuit

As indicated earlier, resistors or any other components can be series wired when used in a series string. There is a very simple and important rule that is used to determine the total amount of resistance in the circuit. In the series circuit the **total resistance** is equal to the sum of each resistor value. This can be shown by the formula

$$R_T = R_1 + R_2 + R_3 + \cdots + R_n \qquad (6\text{-}1)$$

Total resistance *Value of all resistances in circuit when measured from point of reference.*

where R_T is equal to the total ohmic value and R_n represents the highest-numbered resistor in the circuit. The series of dots reminds you to include every resistor between the third and the last. When we wish to find the total resistance in a series circuit, then, all we need do is add together the value of each individual resistance.

When calculating the total resistance value in a circuit, the order of the resistances is not important. A circuit may use its resistance values in various orders depending on the needs of the circuit. The critical item in this situation is the total. It also makes little difference if you start at the top of the string and add resistance values until you reach the last resistance, or start at the bottom and add values until you read the top of the string. Just be sure to include every resistance in the string.

One method of finding the total resistance when several of the resistors have the same value is to count the number of equal-value resistors and multiply that ohmic value by the quantity of equal resistors. In electrical and electronic work it is convenient to use subscripts to identify specific resistances. For example, the symbol R_T is equal to the total of all resistance values. The designations R_1, R_2, and R_5 refer to specific resistors in the system. In many large schematic diagrams there may not be adequate space in the drawing to show specific values for the components. The alternative is to identify the component by a letter and a numeric subscript, the letter R if it is a resistor. These identifiers are then listed in a separate parts listing for all components. You will find this system used throughout this book.

It is possible to build a specific resistance value by using two or more resistors, which total close to the value needed, in a series connection. For example, if you needed a 47-kΩ resistor, you could use a 22-kΩ and a 24-kΩ in series. This does not total 47 kΩ, but it does total 46 kΩ. Selection of resistors

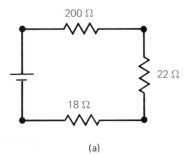

(a)

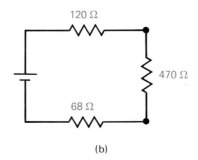

(b)   **FIGURE 6-6**

that actually measure slightly higher than their marked values can create a 47-kΩ resistance value. Remember that almost all resistors have a tolerance rating. This rating permits a variance from the ideal marked value. A 5% tolerance rating for a 22-kΩ resistor permits an actual value of between 23.1 and 20.9 kΩ. The same tolerance gives us a range of 25.2 to 22.8 kΩ for the 24-kΩ resistance value. If resistors that measure on the high side of the tolerance range are selected, these can be used to replace the original 47-kΩ value.

Another way of looking at this type of situation is that a great many electrical and electronic circuits have a built-in tolerance factor. This is usually an "understood" factor. It is seldom stated on any of the design or service literature. Using the earlier example of a 47-kΩ resistance, we could use in a series combination any resistors whose ohmic value came within 5% of the value specified for the circuit. The need to measure and select specific resistors is not always necessary. One of the best methods of determining if this will work in a specific application is to try the system temporarily using a value close to the one desired. If the system functions properly, that resistance may be installed in the circuit permanently. As you become more familiar with electrical systems and circuits, you will find this judgment easier to make. Always check the manufacturer's literature first to see if the replacement resistor has a critical value. If it does, you *must* replace it with a resistor close to the original ohmic value. This suggestion does not imply that you should redesign the circuit. Selected values should be reasonably close to the original ohmic rating of the resistor to be replaced.

REVIEW PROBLEMS

6-11. A circuit requires a resistance value of 82 kΩ. You have several resistors available, but not one of value 82 kΩ. The values you have include one each of 10, 22, 39, 1, 47, and 12 kΩ. Which of these values can be used to make a value of 82 kΩ?

6-12. Using the chart of EIA preferred values illustrated in Chapter 5, find resistors that can be series connected to provide the required value of resistance.

	Required	Used (Two or More)			
(a)	1.7 kΩ	____	____	____	____
(b)	26 kΩ	____	____	____	____
(c)	130 kΩ	____	____	____	____
(d)	99 kΩ	____	____	____	____
(e)	0.8 kΩ	____	____	____	____

6-3 Voltage Drops in the Series Circuit

Kirchhoff found that a flow of current in a closed loop, or series circuit, did more than provide equal current. He measured the voltage drops that developed across the terminals of each resistance and found that their sum was equal to the applied voltage. Before we explore this concept, let us discuss why the voltage drop occurs.

When a current flows through a resistance a difference in electrical potential develops across the terminals of that resistance. The exact amount of

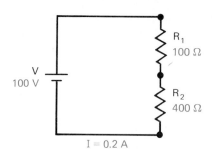

R_1
$100\ \Omega$

V
$100\ V$

R_2
$400\ \Omega$

$I = 0.2\ A$

FIGURE 6-7 Voltage drops in the series circuit are directly proportional to the values of resistance in the circuit.

potential difference depends on several factors. These include the value of the applied voltage, the quantity of resistances in the circuit, and the amount of current flow. Returning to one of the formulas presented by Ohm, we find that $V = I \times R$, or voltage is equal to the current times the resistance in a circuit or a component of the circuit. In Figure 6-7 a current of 0.2 A flows through the total resistance of 500 Ω when a source of 100 V is applied to the circuit. These values are determined in the following manner:

$$I_T = \frac{V_T}{R_T}$$

$$= \frac{100\ V}{500\ \Omega}$$

$$= 0.2\ A$$

Note that all of the subscripts used in this formula consist of the letter T. To use Ohm's law we must always work with the same units. In this example the common element was the total value. The preliminary work of finding current requires finding R_T before it can be used to find total current, I_T. A more proper method of showing how to find the current is as follows:

$$I_T = \frac{V_T}{R_T} = \frac{100\ V}{R_T}$$

$$R_T = R_1 + R_2 = 500\ \Omega$$

$$I_T = \frac{V_T}{R_T} = \frac{100\ V}{500\ \Omega} = 0.2\ A$$

Once this is accomplished we can find individual IR, or voltage, drops for each component in the circuit. Since this is a series circuit, the current is equal in all parts of the circuit. This being true, we can use the value 0.2 A for I_{R1}, I_{R2}, and I_T. Stating this as a formula, we have

$$I_T = I_{R1} = I_{R2}$$

When we have equalities, one may be substituted for any other in the formula without changing its values. This is used to find the IR drop across each of the two resistors.

$$V_{R1} = I_{R1} \times R_1 \qquad V_{R2} = I_{R2} \times R_2$$

$$= 0.2\ A \times 100\ \Omega \qquad = 0.2\ A \times 400\ \Omega$$

$$= 20\ V \qquad\qquad = 80\ V$$

In this example the value of 0.2 A for I is used consistently. The values for V and R are identified as we attempt to find the voltage drop (IR drop) for R_1 and R_2. Note that in each formula the subscript is the same for that formula. This is a *must* if you wish to find these values successfully.

Returning now to the statement made earlier about Kirchhoff's voltage law, we can see that the sum of the voltage drops in the circuit are equal to the voltage applied from the source. This is always true in a series circuit. It is possible to have a small voltage difference between actual and calculated values. This is usually due to the rounding off of the mathematical answers. This does not make the mathematical analysis invalid. Usually, there are tolerances in the system that give us measured values that differ slightly from their printed counterparts. In general, if the difference is less than 5%, the analysis is probably correct.

The rule by Kirchhoff can be stated as a formula:

$$V_T = V_{R1} + V_{R2} + V_{R3} + \cdots + V_n. \tag{6-2}$$

This formula is the third one that shows the relationships of the total voltage, current, and resistance to the values that develop across each component in the series circuit. They are expressed as follows:

Voltage in the series circuit: $V_T = V_{R1} + V_{R2} + \cdots + V_{Rn}$
Current in the series circuit: $I_T = I_{R1} = I_{R2} = \cdots = I_{Rn}$
Resistance in the series circuit: $R_T = R_1 + R_2 + \cdots + R_n$

These rules apply only to the series circuit. There are other rules and relationships that are used for a parallel circuit. These are presented and explained in Chapter 7.

Another concept that is important is shown in the example. That is the relationship of voltage drop to specific resistance values. When a circuit has resistances of different values, such as the one in Figure 6-7, the voltage drops that develop as current flows are not equal. The largest voltage drop will always develop across the largest resistance value. In this example the 80-V drop developed across the 400-Ω resistor, and the 20-V drop developed across the 100-Ω resistor. This concept is true for two or more resistances in a series circuit.

Another way of looking at this concept is to consider the proportions of the two resistances. In this example there is a 1 : 4 ratio between the resistances (100 : 400). When these are considered as individual parts of the system, we have five parts (1 + 4). The 400-Ω value is $\frac{4}{5}$ of the total resistance of 500 Ω and the 100-Ω resistor is $\frac{1}{5}$ of the total resistance value. This ratio or proportion may be used to determine the voltage drop across each component:

$$V_{R1} = \tfrac{1}{5} \times V_T = 20 \text{ V}$$

$$V_{R2} = \tfrac{4}{5} \times V_T = 80 \text{ V}$$

This method provides a rapid way of approximating the voltage drops in any circuit. This holds true as long as a ratio or proportion of each value to the total can be established. This process works well with two or more resistors in a series circuit.

6-4 Applying Ohm's Law to the Series Circuit

Ohm's law is often applied to individual components in any circuit to find the operating condition for the component. The method of doing this is not difficult when one follows some very easy but specific rules.

1. Always work with the same set of subscripts. Any attempt to mix subscripts will usually result in total confusion and wrong answers.
2. Apply the rules for total circuit values when attempting to find current in the circuit.
3. Attempt to approximate your answer before working it out. This will provide a "ballpark" figure toward which you can work. If your answers do not come close to agreeing, one of the processes is incorrect. Start again in an effort to obtain correct results.

Use the circuit shown in Figure 6-8 to find all the values for each component in this circuit. Let us start by first *thinking* about what we should know about these values. Your thinking should follow these lines:

1. This is a series type of circuit.
2. Current is equal in all parts of the circuit.

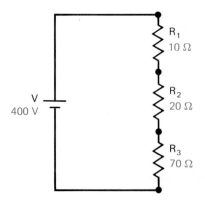

FIGURE 6-8

 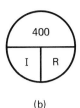 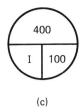 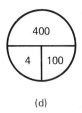

(a)　　　　　(b)　　　　　(c)　　　　　(d)

FIGURE 6-9 Use of the Ohm's law circle to aid in the determination of missing circuit values.

3. The sum of the voltage drops across each component is equal to the source voltage.
4. The total resistance is the sum of each resistor's value.

The first step in finding circuit values for this circuit is to fill in the blanks in an Ohm's law circle [Figure 6-9(a)]. In this circuit the only total value that is immediately seen is the source voltage. This value of 400 V should be placed in the upper portion of the circle, replacing the letter V [Figure 6-9(b)].

It is still not possible to solve for the other *total* values. Both the I and the R sections are still unknown values. The easiest way of adding one more value is to add the resistances values and thus provide a total figure. In this circuit the total resistance is 100 Ω. This figure is derived by adding the ohmic value of each resistor. This *total* value is added to the circle [Figure 6-9(c)].

We now have two of the three components necessary to show the relationships among voltage, resistance, and current. Using Ohm's law, $I = V/R$, the value of the current is found. In this circuit V/R is equal to 400 V/100 Ω or 4 A. This value is added to the circle [Figure 6-9(d)].

Once *total* circuit values are obtained, it is possible to find the values for each component in the circuit. This is accomplished using Ohm's law circles as illustrated in Figure 6-10. This is the same circuit as that described previously. An Ohm's law circle has been added by each component. In addition, the known facts are added to each of the wheels. Part (a) provides us with the *total*

FIGURE 6-10 Use of the Ohm's law circles (1) aids in determination of values that are present and (2) establishes the formulas for finding those values that are missing.

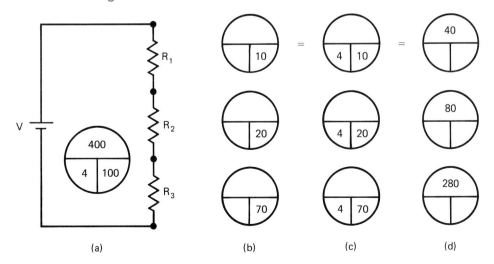

(a)　　　　　(b)　　　　　(c)　　　　　(d)

circuit values that we obtained earlier. It is possible to replace the letter R in each individual circle with the ohmic value for that resistor [part (b)].

The next step is to attempt to identify any other known values. In this circuit the current is equal in all parts. Therefore, the total current value may be substituted for the letter I in each of the individual circles [part (c)]. In case you are wondering how this may be possible, be reminded that one equality may be substituted for another without changing the equation. In this situation, I_T and I_{R1}, I_{R2}, and I_{R3} are all equal. The value of I_T can be substituted successfully for any of the others.

The information now known will permit us to find the individual voltage-drop values for this circuit.

$$V_{R1} = I_{R1} \times R_1 = 4 \text{ A} \times 10 \text{ }\Omega = 40 \text{ V}$$

$$V_{R2} = I_{R2} \times R_2 = 4 \text{ A} \times 20 \text{ }\Omega = 80 \text{ V}$$

$$V_{R3} = I_{R3} \times R_3 = 4 \text{ A} \times 70 \text{ }\Omega = 280 \text{ V}$$

These values are added to the circles in part (d). This completes the information for each component and the total circuit.

Another way of finding only the voltage drop is to use the ratio-proportion method described earlier in this chapter. Using this method, first add the values of resistance. This provides an answer of 100 Ω. Next find the ratio of each resistor to the entire set: $10 : 100 = 1 : 10$, $20 : 100 = 2 : 10$, $70 : 100 = 7 : 10$. Once it is seen that there are 10 parts to the whole ($1 + 2 + 7 = 10$), the relationship is easily established. The voltage applied to the total circuit is 400 V. Resistor R_1 has a ratio of $1 : 10$, or $\frac{1}{10}$ of the 400 V, or 40 V. Resistor R_2 has a ratio of $2 : 10$, or $\frac{2}{10}$ of the total. This is equal to $\frac{2}{10} \times 400$ V, or 80 V. The third resistor has a ratio of $7 : 10$, or $\frac{7}{10}$ of 400 V, or 280 V. These values are easily proven when they are added: $40 \text{ V} + 80 \text{ V} + 280 \text{ V} = 400$ V. This is a short but correct method of finding voltage drops when there is no need to find the current in the circuit.

6-5 Polarities of Voltage Drops

Voltage drop polarities *Term describing charge on each end of component in circuit.*

Any time a difference in potential develops across a component, each of the leads, ends, or wires has a specific polarity. This means that one of the ends of the device has a positive value and the other end has a negative value. The specific polarity of a component is determined by the relation of that end to the power source. In Figure 6-11(a) one resistor is connected to the power

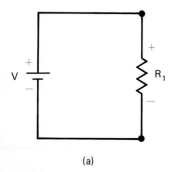

(a)

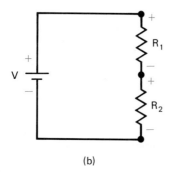

(b)

FIGURE 6-11 How polarities for each component are established. The end of each component closest to the power source terminal has the same polarity as the power source terminal.

source. The polarities of the ends of this resistor can be identified easily. The upper end has a positive polarity because it is connected directly to the positive terminal of the power source. The lower end of the resistor is negative because of its direct connection to the negative terminal of the power source.

When more than one resistance is connected in series to the load another set of rules applies [Figure 6-11(b)]. When multiple components are connected, each has its own polarity. The polarity for each is identified by its relation to the power source. Resistor R_1 has its upper lead connected directly to the positive terminal of the source. This lead is positive. The lower lead of R_1 is closer to the negative power source terminal. It is therefore identified as the negative lead *of that resistor*. The leads of R_2 are polarized in the same manner. Its upper lead is closer to the positive terminal of the source and it is classified as being positive. Similarly, its lower lead is connected directly to the negative terminal and it has negative polarity. You may be confused by the fact that the junction of R_1 and R_2 has both a positive and a negative polarity. Do not let this disturb you. Just keep in mind that the polarities at this point have to be referenced to a component or a source of power. When this fact is applied, each negative sign is related to R_1 and the positive sign to R_2. Under these conditions it should make sense to you.

6-6 Kirchhoff's Voltage Law

Kirchhoff's voltage law states that in a closed loop the sum of all the voltage drops that are present will equal zero. This rule is explained in a very simple manner using Figure 6-12. Each of the three components in this circuit has an assigned polarity. This includes the power source, V. The application of this law is as follows: We can start at any point in the circuit and add all voltage drops. The direction, clockwise or counterclockwise, makes no difference. Just add each voltage drop and be sure to include its polarity.

Kirchhoff's voltage law Sum of voltage drops in a closed loop or series circuit is equal to zero.

In this example, start at A and add each voltage in a counterclockwise direction, as the current would flow. The formulation for this is:

$$0 = V_{R2} + V_{R1} + V \qquad \text{(basic formula)}$$
$$= + (-V_{R2}) + (-V_{R1}) + V \qquad \text{(add signs)}$$
$$= -V_{R2} - V_{R1} + V \qquad \text{(simplify)} \qquad (6\text{-}3)$$
$$= -34 \text{ V} - 16 \text{ V} + 50 \text{ V} \qquad \text{(and insert values)}$$
$$= 0 \qquad \text{(add)}$$

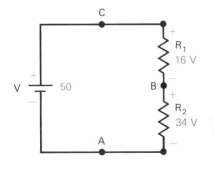

FIGURE 6-12 Use of the polarity assignment rule as it applies to Kirchhoff's law for voltage drops in the closed-loop circuit.

The same results are proven by starting at any of the points *A*, *B*, or *C* and adding in either direction. This was Kirchhoff's method of proving his law. We have been using it all along in a modified practice. This modification is the statement that the voltage drops in the circuit are equal to the applied voltage from the source.

6-7 Circuit Common and Ground

Circuit common *Reference point for measurements, usually the negative connection to a circuit.*

Chassis ground *Term of reference for circuit common when connected to metal chassis of unit.*

In reality, all voltage measurements are taken from a point of reference. When we discuss the voltage drop that develops as current flows through a resistance, one end of the resistor is the reference point. When working with circuit measurements the typical way of discussing a difference in potential, or voltage drop, across a component is simply to state: "There is 24 V across that resistor." When making a measurement to a specific point, then, we state: "There is 24 V at that point." In the first example the reference point for the measurement is included in the statement. The second example uses an inferred point of reference. In this example the inference is to a common point of measurement in the circuit.

Almost all circuits have a common reference point. Usually, this is the negative terminal of the power source. This point is often called *ground* even though there may not be any connection from it to the earth or ground. This term is a carryover from earlier days when one terminal of the source was actually connected to the earth. Even today we refer to the negative terminal of the automobile battery as being connected to ground. Actually, the tires of the auto help to insulate it from the earth. The terms *ground* and *circuit common* have tended to become merged over the years. Often, the circuits we use today have no connection to the ground, even through a power source that may be connected to the house wiring. Many circuit boards use a border of conductive material as the common point of the circuits. The negative terminal of the power source is connected to it. All electron flow is from this common, through the components, to the positive terminal of the source. Often, the positive terminal of the source is called the **high side** of the source and the negative terminal is considered the **low side**. This reference probably comes from the relationship of the wiring on a schematic diagram, as described in Chapter 4. In the early days of radio when batteries were used as a power source, the "B" battery was used to provide the required high voltage for tube operation. The positive terminal was connected to the tube circuit and the negative terminal was connected to the chassis or metal base of the radio. This connection arrangement is carried over to today's power sources. Usually, the negative terminal is called *chassis ground* or *chassis common* and the positive terminal is called *B +*. Another battery was used to heat the filaments of the tubes. This was the "A" supply. The term *A +* has the same reference connotation as the term *B +* in circuit descriptions.

High side *Describes end of component in circuit closest to positive terminal of power source.*

Low side *Describes end of component in circuit closest to negative terminal of power source.*

Be aware that every circuit does not use a negative-to-common connection. There are circuits in use today that have their *positive* source terminal connected to circuit common. There is little problem with this type of arrangement as long as you recognize this situation. Electron flow is still the same. It is just that the circuit requirements make this arrangement better. Two similar circuits

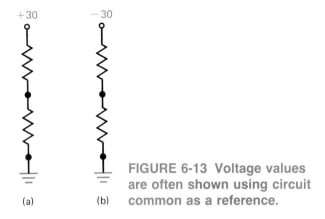

FIGURE 6-13 Voltage values are often shown using circuit common as a reference.

(a)   (b)

are shown in Figure 6-13. The circuit in part (a) has a negative common connection. This is understood when one sees the plus (+) sign at the top of the diagram. The circuit in part (b) has a positive common connection. This, too, can be observed when one sees the negative (−) polarity sign at the top of the circuit drawing.

One other situation to be aware of is when the common of the power source is not the same connection as the common for the circuit. One example of this is shown in Figure 6-14. In this circuit the power source has one common connection. This means that all electron flow originates at one common point in the power source. A second common connection is shown for the load circuit. The common connection for the load circuit is at the junction of the load-balancing resistors R_1 and R_2. Electron flow is still the same, as shown in the drawing, but the two circuit commons are not connected to each other.

Another type of system that uses two different common points is found in some recent-production video terminals or video monitors. These units have two separate power sources. There is no direct connection between the two circuits. They are electrically isolated from each other. The video signal is transferred from one circuit to the second circuit by the use of optical coupling devices, thus eliminating the need for a signal common path between the two units.

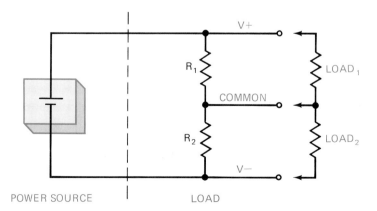

FIGURE 6-14 Some circuits have a common point that is not the same as the negative terminal of the power source.

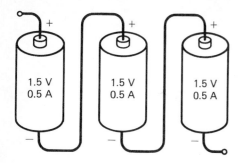

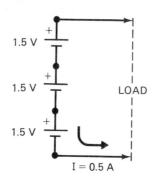

FIGURE 6-15 Series-connected power sources have the same current, and the individual voltages of each source add to provide a total value.

6-8 Power Sources in the Series Circuit

It is possible to require some power sources that are series connected to each other. When this is necessary the total voltage in the system becomes the sum of each individual source. This procedure may be used with any type of voltage source. It can be accomplished in two different ways. One of these is called *series aiding* and the other is called *series opposing*.

Series Aiding The most commonly used of the two types of series power sources is the **series-aiding** type, illustrated in Figure 6-15. Each of the three cells is capable of producing a voltage of about 1.5 V. When the three cells are connected as illustrated, the voltages of each cell add. This is an application of Kirchhoff's voltage law. The characteristics of the three-cell combination are similar to those for any series circuit. The voltage drops across the components add to give us a total voltage of 4.5 V. The circuit has the same characteristics as those of a series resistor circuit. Each cell has the capability of permitting 0.5 A of current to flow. Since they are wired in series the total current flow through the cells and any load will be 0.5 A. In the series power source the voltages add while the current is the same throughout the system.

Current in the series-aiding power source is limited by the current rating of the lowest-value source. If, in this example, one of the cells had a current capability of 0.01 A, the maximum current available to the load would also be limited to 0.01 A. Imagine that we have three sections of garden hose. Two sections have an inside diameter of 1 in. while the third has an inside diameter of $\frac{1}{2}$ in. The amount of water that flows in the system is limited to the water-holding capabilities of the smaller hose.

Series Opposing When power sources are wired in a **series-opposing** manner (Figure 6-16), the total available voltage from the system is also the algebraic sum of each of the sources. Using the rules presented by Kirchhoff, the total voltage for this system is 1.5 V instead of the 4.5 V available from the previous example. This is because one of the cells is connected in a reverse manner. The total voltage is actually the value of one of the cells, since the upper one cancels the effect of the middle cell.

The need for series-opposing sources will best be understood by studying the effects of a signal on a device such as a transistor, integrated circuit, or

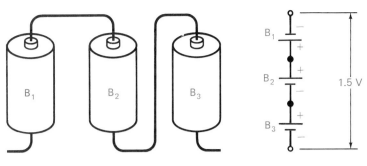

FIGURE 6-16 When the terminals of one source in a series circuit are reversed, the source is added as a negative value to provide the total voltage.

vacuum tube. These devices have a constant value of voltage applied to their control elements. A varying voltage is then introduced to the system. The varying voltage adds to the value of the constant voltage. The varying voltage usually has both a positive and a negative value. This is added to the constant voltage, either increasing or decreasing the total control voltage level. This changes the operational level of the device, which can thus perform some work.

REVIEW PROBLEMS

6.16. A circuit has four power sources connected in series. Their voltages are $+4$, $+8$, -3, and $+2$ V. What is the total voltage?

6-17. A power source uses three batteries to achieve the proper operating voltage. These cells are rated as follows: 3 V at 4 A, 6 V at 1 A, and 18 V at 1.5 A. What is the maximum current that is available from the set of cells when they are series connected?

6-18. Two batteries are series connected. They are rated as follows: 12 V at 4 A and -3 V at 1.8 A. What is the total voltage available from the series? What is the maximum current?

6-19. Two power sources are series connected. One has an output voltage of 130 V. The voltage that is measured across the load is 68 V. What is the output voltage and polarity of the second source?

6-9 Power in the Series Circuit

Total power in a series circuit is equal to the sum of each component's power. The formula for this is

$$P_T = P_1 + P_2 + P_3 + \cdots + P_n \qquad (6\text{-}4)$$

where P_1 is the power consumed by the first unit and P_n is the power consumed in the final unit.

The circuit shown in Figure 6-17 illustrates this role. Total power in the circuit is determined by multiplying total current by the total voltage, $P = I \times V$. Substituting values, we have

$$P = I \times V$$
$$= 4 \text{ A} \times 60 \text{ V}$$
$$= 240 \text{ W}$$

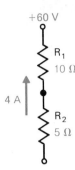

+60 V

R_1
10 Ω

4 A

R_2
5 Ω

FIGURE 6-17 Circuit used to demonstrate the power formula.

The individual power rating for each component is determined in a similar manner:

$$P_{R1} = I^2 \times R_1 \qquad\qquad P_{R2} = I^2 \times R_2$$
$$\qquad\quad = 4A \times 4\,A \times 10\,\Omega \qquad\qquad = 4\,A \times 4\,A \times 5\,\Omega$$
$$\qquad\quad = 160\ W \qquad\qquad\qquad\qquad\quad = 80\ W$$
$$P_{R1} + P_{R2} = P_T$$
$$160 + 80 = 240\ W$$

In this example there are two methods of determining the power. Any of Ohm's or Watt's laws or combinations of them can be utilized to calculate the power level of the circuit or the components in the circuit. The reference to total power here is considered the same as the amount of power available from the power source. This is the same value as the amount of power consumed by the load.

When one of a group of power sources is connected in a series-opposing circuit, the power created by it also has to be considered. The circuits shown in Figure 6-18 illustrate this point. The series-aiding sources produce a total of 36 W of power for use by the load. When one of the sources is connected as to oppose the others, the total power available for a load is 12 W. Each source is still creating 12 W of power, but since one of them is reverse connected, its power is subtracted from the total power available for use by the load. This condition will be true even when sources of different voltage and current ratings are connected in a series-opposing circuit.

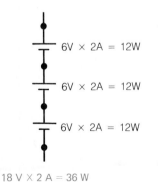

6V × 2A = 12W

6V × 2A = 12W

6V × 2A = 12W

6V × 2A = 12W

6V × 2A = 12W

6V × 2A = 12W

FIGURE 6-18 In a series-opposing circuit total power is still the sum of individual power values, even though one is negative when compared to the other polarities.

18 V × 2 A = 36 W

6 V × 2 A = 12 W

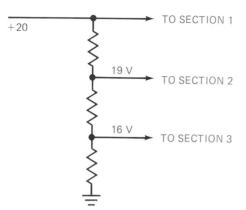

TO SECTION 1

19 V   TO SECTION 2

16 V   TO SECTION 3

+20

FIGURE 6-19 Voltage-divider circuit. Each of the voltage drops provides the proper operating voltage for one circuit or section.

6-10 Voltage Dividers

The series circuit that contains two or more resistances is also a **voltage-divider** network. This term is used to indicate that the applied voltage from the source is proportional, or divided, across each of the resistances in the circuit. The use of a voltage-divider network is very important in electrical and electronic circuits. Its use establishes the correct operational voltage for each component in the circuit. Without the voltage divider only those components that operated with the full amount of source voltage could be used.

    An example of a voltage-dividing network is shown in Figure 6-19. Each section of the device operated from the 20-V power source requires a different level of operating voltage. The use of a voltage-divider network permits operation of all the sections of the device from one common 20-V power source. In addition, when the voltage divider is connected across the terminals of the power source, as in this example, it will discharge any of the components holding a charge that are in the source. This acts as a safety factor and eliminates electrical shock from a power source that is turned off.

    Another application of a voltage divider is the potentiometer circuit. The potentiometer circuit is used to meter, or control, the amount of voltage for a circuit. A basic potentiometer circuit is shown in Figure 6-20. The potentiometer (pot) is actually a variable resistor. It is constructed as shown in part (a). It has a resistance element. This element has a constant resistance value that was established during the manufacturing process. This constant, or fixed, resistance

Voltage divider *Series-resistive circuit designed to provide voltage levels less than applied voltage.*

FIGURE 6-20 The potentiometer is actually a voltage-dividing device.

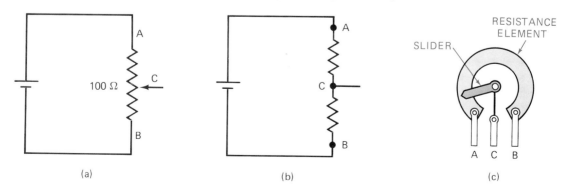

(a)   (b)   (c)

AUDIO SIGNAL INPUT

AUDIO SIGNAL OUT

HIGH LEVEL

MEDIUM LEVEL

LOW LEVEL

FIGURE 6-21 The potentiometer as used in a signal voltage-dividing circuit.

value is shown as the semicircular component in the illustration. For the purposes of discussion, let us establish this value as 100 Ω. This resistance value is measured between points A and B in the drawings. A third connection of the variable resistor is at point C. This component is a slider. It rotates around a center pivot point. The slider has a connection to the outside, as do the ends of the fixed-resistance element.

The purpose of the variable resistor in this application is to change the ratio of resistances for R_{1A} and R_{1B}. The sum of these two resistances is equal to the total resistance in the circuit. Rotating the slider changes the ohmic values of R_{1A} and R_{1B}. When the slider is close to part A the resistance of R_{1A} is minimal and the resistance of R_{1B} is close to 100 Ω. Moving the slider close to part B reverses the ohmic values for R_{1A} and R_{1B}.

One application for the voltage-divider potentiometer circuit is its use as a volume control in a radio, tape player, or TV set (Figure 6-21). The audio signal in any of these devices is a varying voltage. It is processed by the unit and amplified to a predetermined value. This acts as the source for this circuit and the full value of the pot is the load. When the slider of the control is near the input connection of the circuit the audio signal level at the output is high. This corresponds to the relationship of voltage drop that develops across two resistances in series. The larger drop develops across the largest component. When the slider is close to the lower end of the control, the output is at minimum. Setting the slider near the midpoint provides a medium level of audio.

Changing the names and functions for the circuit provides another type of voltage-divider control using the same components. In Figure 6-22 the system is used to control the operating point of an IC or transistor. The control is adjusted to set the point of operation at a predetermined value. This is another one of many applications for the voltage-divider circuit. Either fixed-value resistances or the adjustable potentiometer circuit can be used as a voltage divider. The specific use depends on the application and requirements of the circuit.

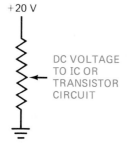

+20 V

DC VOLTAGE TO IC OR TRANSISTOR CIRCUIT

FIGURE 6-22 The potentiometer as used in an operating voltage-dividing circuit.

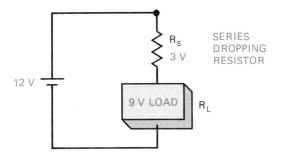

SERIES DROPPING RESISTOR

FIGURE 6-23 Some circuits require the use of a series voltage-dividing, or voltage-dropping, resistance in order to provide the proper level of operating voltage for the load.

Voltage Dropping Resistors Often, we find a resistance in series with another component in a working circuit. The reason for this is that proper operation of the other component requires a value of voltage that is less than that of the source voltage. One example of this type of circuit is shown in Figure 6-23. The load in this circuit has a maximum operating voltage of 9.0 V. The source voltage is 12.0 V. If the source and the load were directly connected to each other, the higher voltage and its resulting above-normal current flow would destroy the load. A series voltage-dropping resistor is required to operate the load properly. In this example the requirements of the load are 9.0 V and 0.050 A, or 50 mA. Since this is a series circuit the current through the source and the series dropping resistor is also 0.05 A. Once the current in the circuit is identified, we need to determine the amount of voltage drop that occurs across the series resistor. When the current and voltage drop are known, it is easy to use Ohm's law to calculate the ohmic value of the series resistor. The process for doing this is accomplished by determining the voltage drop across the series resistor:

$$V_s = V_T - V_L$$

where V_s is the drop across the series resistor and V_L is the drop across the load.

$$V_s = 12.0 \text{ V} - 9.0 \text{ V} = 3.0 \text{ V}$$

The next step is to use Ohm's law to find the ohmic value for R_s:

$$R_s = \frac{V_s}{I_s} = \frac{3 \text{ V}}{0.05 \text{ A}} = 60 \text{ } \Omega$$

Thus the value is 60 Ω for this resistor.

REVIEW PROBLEMS

Use Figure 6-24 to find the following values.

6-20. Voltage drop across load
6-21. Current through R_s
6-22. Source current
6-23. Source power
6-24. Voltage drop across R_s
6-25. Ohmic value of R_s
6-26. Power rating of R_s

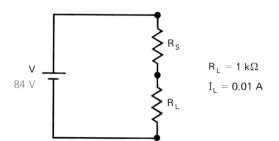

FIGURE 6-24

6-11 Opens, Shorts, and Partial Shorts in the Series Circuit

Three faults that occur in the series circuit are the open, the short, and the partial short. Each of these has its own set of characteristics. When they are applied to the troubleshooting of circuits, the location of the fault area or device is easy. Since each fault has its own unique characteristics, we examine them individually.

Open Circuit The open circuit is a physical break in the normal current path circuit. Let us use the series circuit of Figure 6-25 to show how the open affects the normal circuit characteristics. When a current flows through the circuit, each component develops a voltage drop. The sum of these drops, we now know, equals the applied source voltage. The drops are proportional to the relative size of the series resistances. All values for current and individual voltage drops are shown in the drawing.

An open, by definition, is a break in the normal current path. This break has the characteristics of an infinitely large value of resistance. For purposes of our discussion, we will assign a value of 1×10^{24} Ω (1,000,000,000,000,000,000,000,000 Ω). This huge amount of resistance is added to the series circuit at the point of the break. The total resistance has now increased from 1×10^2 Ω to 1×10^{24} Ω. The added resistance will reduce any current flow from 0.1 A to an unmeasurable trickle of electrons, if one desires to calculate it.

When we consider that the size of the voltage drops in the series circuit are proportionate to the individual resistance values, the size of the "resistor"

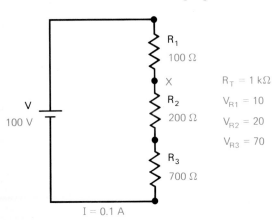

$R_T = 1$ kΩ

$V_{R1} = 10$

$V_{R2} = 20$

$V_{R3} = 70$

FIGURE 6-25 Closed-loop, or series, circuit. A break at point X will create an infinitely high value of resistance at that point in the circuit.

at the break will provide a voltage drop across the break that is almost equal to the source voltage value. For all practical purposes it can be stated that the source voltage at the point of the break is an excellent indication of an open circuit at that specific point.

Refer again to Figure 6-25. Assume that a break in the circuit occurred at the point marked with an "X." If one wished to troubleshoot this circuit to locate the break, there are certain procedures that can be applied. When these are used, the analysis of the problem can be limited to a few basic tests. Start out the analysis by expecting the circuit to function properly. Let us consider the device labeled R_2 to be an electronic device that controls a motor. When power is applied to the circuit, the motor is expected to rotate. In this example, when power is applied the motor does *not* rotate, indicating a system fault. The most appropriate test instrument to use for analysis of circuit problems in this circuit is a voltmeter. The negative lead from the voltmeter is connected to circuit common and the positive level is first connected to the upper end of R_1 [Figure 6-26(a)]. This initial test tells us that power is applied to the circuit and that the source is operating properly. An expert technician will always test the power source as an early indication of the location of the problem. At this point in the circuit the voltage measurement should be 100 V.

The second place to make a test is somewhere near the midpoint of the motor, R_2. This is not a practical place to make a test because it requires disassembly of the motor, and this is not a wise step early in troubleshooting. A better place for the second test is the upper end of the motor connection. A test at this point [Figure 6-26(b)] should show a voltage value of 90 V if the circuit is functional. Since we have indicated that a break has occurred at point X the resulting voltage at this point is zero. The reason for this is that there is

FIGURE 6-26 Measurement of voltage values in an open circuit. When the measurement is obtained from circuit common the source voltage may be measured up to the point of the break because there is no current flow in the circuit.

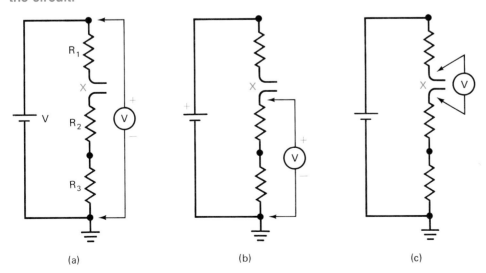

(a)   (b)   (c)

no possible way that electrons could get to the + terminal of the power source. The zero-voltage indication at this point tells us that the break is between the points of the first and second tests. The use of an ohmmeter with the circuit power turned off will confirm this. A resistance measurement from the upper end of R_2 to the upper end of R_1 should show about 100 Ω if the circuit is functional. When the break is between these points the ohmmeter will indicate an infinitely high resistance. In most cases the meter would probably indicate an unreadably high value, or no indication of resistance on the meter.

The next step to take to locate the faulty component can be made with either a voltmeter or an ohmmeter. Let us look at the results that will be obtained using each instrument.

When using a voltmeter the next step is to move both of the meter leads. The negative lead is moved to the upper part of R_2 [Figure 6-26(c)]. The positive meter lead is connected to the lower end of R_1. If the break is in the wire connecting these two components, the voltmeter will give a reading of close to source voltage, 100 V. This reading shows that the break is between the two meter leads. A replacement of the conductor between the two points should repair the problem.

When one uses an ohmmeter a similar procedure is used. The meter leads are moved from their original measuring points to new points that are closer to each other. The ohmic reading will be infinitely high until one of the leads crosses the break. Then the resistance reading will be the value of one of the components in the circuit, or it will be close to 0 Ω is there is continuity in the circuit between the test points. Once continuity is established, the break is between the point of continuity and the previous, infinite resistance measurement.

Short Circuit The short circuit is described as an unusually low resistance path in the circuit. It can be either a total short or a partial short. Let us discuss how the partial short affects the operation of the circuit. A circuit with a **partial short circuit** is illustrated in Figure 6-27. Often the cause for a short circuit of this type is the failure, or breakdown, of one of the components in the series circuit. The motor identified as R_2 will develop a short circuit in this example. Let us assume that the short-circuit resistance for the motor is 0 Ω instead of its normal 200-Ω value.

Partial short circuit Condition where part of normal current path has unusually low resistance.

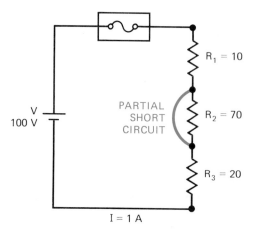

FIGURE 6-27 A partial short circuit creates a close-to-zero-ohm value across the shorted component, but not across any other component in the circuit.

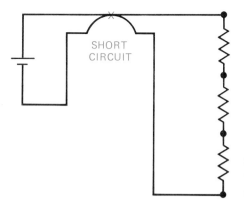

FIGURE 6-28 A total short circuit creates a 0-Ω condition in parallel with all the components in the circuit.

The total circuit resistance will decrease from its original value of 100 Ω to a value of 30 Ω, due to the motor short circuit. Circuit current will increase due to the reduction in resistance. The value of current will increase from 1 A to 3.3 A, about 3⅓ times its original value. Probably the first action that occurs is that the fuse in the positive level will melt, and this will turn off the voltage connected to the load and the motor will stop. A problem of this type is best approached by looking first at the circuit diagram. The open fuse is not a cause for the failure; it is a *result* of the failure of the motor. Replacing the fuse will not repair the circuit at this time, although it will be done before the circuit will be operable.

TROUBLESHOOTING APPLICATIONS

When we study the circuit diagram we note the total resistance of 100 Ω. A measurement of the total circuit resistance with an ohmmeter should give a reading of 100 Ω. When the measurement is taken the reading is found to be 30 Ω. The only way the circuit can have that resistance is for the motor, R_2 to be shorted. A second resistance measurement, this time across the motor's terminals, will verify the location of the problem area.

Two types of circuit problems, the open and the partial short circuit, are relatively easy to locate. This is particularly true when one applies a basic, logical approach to an analysis of the circuit. The ability to apply the knowledge one has as it relates to how the circuit is supposed to work will make the job of troubleshooting and repair much easier. The third type of problem, the total short (Figure 6-28), is discussed in Chapter 7. The total short affects the circuit in a different manner. Since it most closely resembles the characteristics of a parallel circuit, it is presented in the chapter that deals with the parallel circuit.

REVIEW PROBLEMS

Use Figure 6-29 to answer problems 6-27 through 6-29.

6-27. Voltages are radio = 0 V, fuse = 7.0 V. Which component is open?
6-28. What is the current flow for the radio when the circuit is working properly?

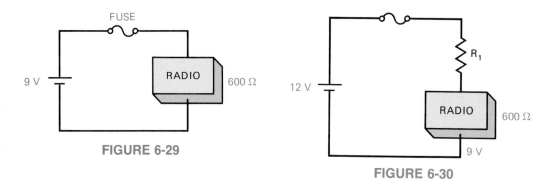

FIGURE 6-29

FIGURE 6-30

6-29. What is the fuse resistance when the circuit is working properly?

6-30. How is a fuse tested with (a) a voltmeter and (b) an ohmmeter?

Use Figure 6-30 to answer problems 6-31 and 6-32.

6-31. What is the voltage drop across R_1.

6-32. If R_1 is open, what is the current flow through it?

CHAPTER SUMMARY

1. Resistors in a series string are connected end to end.
2. The total resistance in a series circuit is the sum of the value of each resistance.
3. There is only one path for current flow; therefore, the current is equal throughout the circuit.
4. The voltage drops in a series circuit add to equal the source voltage.
5. In a series circuit, each voltage drop, including the source, has a polarity.
6. The sum of all voltages in a series circuit, including the source, is equal to zero (Kirchhoff's voltage law).
7. A voltage drop is the difference in electrical potential measured across a component due to current flow in the circuit.
8. A voltage divider consists of two or more series-connected resistors.
9. The amount of voltage drop that occurs across a component is equal to the ratio of the component's resistance to the total circuit resistance.
10. A potentiometer is a variable resistance that is used as a variable voltage divider.
11. Total power in a series circuit is equal to the sum of the power developed by each component.
12. An open circuit acts as an extremely high resistance in the circuit.
13. When all resistors in a series circuit are equal in ohmic value, the total circuit resistance is determined by multiplying the value of one resistor by the quantity of resistors in the circuit.

6–1. The resistance of a short circuit is very high.

6–2. The total resistance in a circuit containing four 400-Ω resistors is 1200 Ω.

6–3. A closed-loop circuit has more than one current path.

6–4. A series circuit contains a 100-Ω and a 200-Ω resistor. The current through the 100-Ω resistor is equal to the current through the 200-Ω resistor.

6–5. Two 100-W 120-V light bulbs are series connected to a 240-V source. The total circuit power is 200 W.

6–6. A voltage-dropping resistor is series wired between the source and the load to reduce the operating voltage for the load.

6–7. The voltage available from two 6-V batteries wired in series aiding is 12 V.

6–8. The current available from three 6-V 1.5-A batteries wired in series aiding is 4.5 A.

6–9. Kirchhoff's current law does not apply to series-wired circuits.

6–10. An open in a series circuit can be detected if source voltage can be measured across the open area.

6–11. Four series-wired resistors each develop 0.41 W of power. The total **MULTIPLE CHOICE** power is
(a) 0.41 W
(b) 0.82 W
(c) 1.64 W
(d) 8.2 W

6–12. Six 18-Ω resistors are series connected. The total resistance is
(a) 360 Ω
(b) 108 Ω
(c) 36 Ω
(d) 180 Ω

6–13. A 36-V source is connected to three 78-Ω series-connected resistors. The current through the system is
(a) $\frac{1}{2}$ A
(b) 1.2 A
(c) $\frac{1}{4}$ A
(d) $\frac{1}{6}$ A

6–14. A 100-V source is connected to a 100-Ω R_1, a 400-Ω R_2, and a 500-Ω R_3. The current through R_2 is
(a) 111 mA
(b) 0.5 A
(c) 250 mA
(d) 100 mA

6–15. When a 2-, an 18-, and a 40-kΩ resistor are series connected, the total resistance is
(a) 15 kΩ
(b) 10 kΩ
(c) over 60 kΩ
(d) 60 kΩ

6–16. A 100-, a 1000-, and a 5-kΩ resistor are series wired. Current through the 5-kΩ resistor is limited to 0.05 A. The power dissipated by the 1-kΩ resistor is
(a) 25 W
(b) 12.5 W
(c) 7500 W
(d) 2.5 W

6–17. The total current flow in the circuit shown for Figure 6-31 is
(a) 0.15 A
(b) 0.10 A
(c) 0.05 A
(d) 0.01 A

6–18. The total resistance in Figure 6-31 is
(a) 160 Ω
(b) 200 Ω
(c) 40 Ω
(d) 240 Ω

6–19. Five 100-Ω resistors are series wired to a source. The current through the second resistor is 1.5 A. The total current
(a) is 1.5 A
(b) is 7.5 A
(c) is 15 A
(d) cannot be determined

6–20. If the total current for problem 6–19 is 3.5 A, the voltage source is
(a) 175 V
(b) 1750 V
(c) 350 V
(d) unknown

ESSAY QUESTIONS   **6–21.** Name the significant factors relating to current flow, total resistance, and voltage drops as they relate to the series circuit.

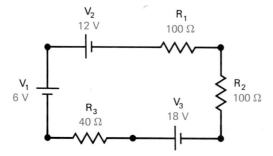

FIGURE 6-31

6–22. State how current-measuring meters are used to monitor the current in a series circuit.

6–23. Explain why current-measuring meters are placed in the series circuit.

6–24. State Kirchhoff's law as it relates to the series circuit.

6–25. Why is the statement "the applied voltage is equal to the sum of the voltage drops across each load in a series circuit" applicable?

6–26. Explain why it is necessary to use total current in the series circuit when attempting to determine individual voltage drops for each load.

6–27. Explain why it is necessary to use total resistance when solving for the current through individual resistances in the series circuit.

6–28. Explain what is meant by the term *circuit common*.

6–29. How is circuit common used for the measurement of voltages in the series circuit?

6–30. Explain what is meant by the terms *series opposing* and *series aiding*.

Use Figure 6-32 to answer problems 6–31 through 6–33.

6–31. Find the source voltage. (*Sec. 6-4*)

6–32. Find current through R_2. (*Sec. 6-4*)

6–33. Find voltage across R_2 and R_3. (*Sec. 6-4*)

6–34. Three 6-V 5-A sources are series connected to a 5-kΩ resistor. What is the current through the middle power source? (*Sec. 6-10*)

6–35. How much power is available from a 6-V 4-A, a 3-V 2-A, and a 24-V 2-A power source that are series connected? (*Sec. 6-8*)

Use Figure 6-33 to answer problems 6–36 through 6–42.

6–36. What is the current from the source? (*Sec. 6-4*)

6–37. What is the voltage drop across R_1? (*Sec. 6-4*)

6–38. What is the voltage drop across R_2? (*Sec. 6-4*)

6–39. What is the voltage drop across R_3? (*Sec. 6-4*)

6–40. What is the voltage drop across R_4? (*Sec. 6-4*)

FIGURE 6-32

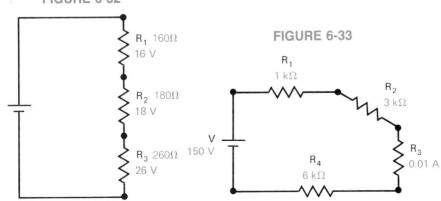

FIGURE 6-33

6-41. What is the total circuit resistance? (*Sec. 6-4*)

6-42. What is the total circuit power? (*Sec. 6-9*)

6-43. In a series circuit, what is the source voltage if the voltage drop across $R_3 = 2.37$ V, $R_1 = 6.8$ kΩ, $R_2 = 2.2$ kΩ, and $R_3 = 10$ kΩ? (*Sec. 6-4*)

Use Figure 6-34 to answer problems 6–44 and 6–45.

6-44. The current through R_2 is 50 mA. The voltage across R_1, 50 V. What is the resistance of R_1? (*Sec. 6-4*)

6-45. What is the current through R_1? (*Sec. 6-4*)

Use Figure 6-35 to answer problems 6–46 through 6–51.

6-46. What is the current through the transistor? (*Sec. 6-4*)

6-47. What is the voltage drop across R_2? (*Sec. 6-4*)

6-48. What is the voltage drop across R_1? (*Sec. 6-4*)

6-49. What is the voltage drop across the transistor? (*Sec. 6-4*)

6-50. What is the current through R_1? (*Sec. 6-4*)

6-51. What percentage of voltage is developed across the transistor connection? (*Sec. 6-4*)

Use Figure 6-36 to answer problems 6–52 through 6–60.

6-52. What is the total resistance, R_T? (*Sec. 6-4*)

6-53. Find the total current, I_T. (*Sec. 6-4*)

6-54. What is the voltage drop that develops across each resistor? (*Sec. 6-4*)

6-55. What is the voltage drop that is measured from D to B? (*Sec. 6-4*)

6-56. What current flow occurs between C and A? (*Sec. 6-4*)

6-57. What amount of power is delivered by the source? (*Sec. 6-4*)

6-58. If the polarity of the source is reversed, what is the voltage drop from D to B? (*Sec. 6-4*)

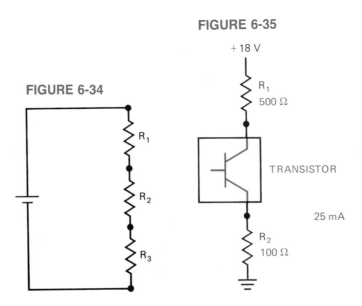

FIGURE 6-35

FIGURE 6-34

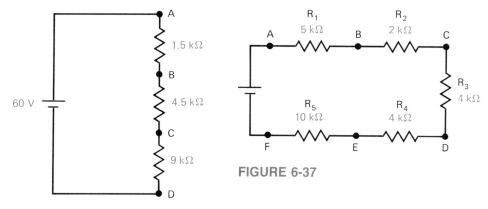

FIGURE 6-36

FIGURE 6-37

6–59. What is the total current flow when the source polarity is reversed? (*Sec. 6-4*)

6–60. What percentage of voltage drop develops across each resistor in the circuit? (*Sec. 6-4*)

Use Figure 6-37 to answer problems 6–61 through 6–68.

6–61. What is the total resistance in this circuit? (*Sec. 6-4*)

6–62. If the voltage drop across R_2 is 20 V, what is the total current flow in the circuit? (*Sec. 6-4*)

6–63. What is the source voltage in the circuit when a 20-V drop is measured across R_2? (*Sec. 6-4*)

6–64. What is the power dissipated by R_5? (*Sec. 6-9*)

6–65. What is the voltage drop across each component when the current through R_4 is 0.015 mA? (*Sec. 6-9*)

6–66. What is the total circuit power when the current through R_3 is 0.02 A? (*Sec. 6-9*)

6–67. How much current will flow through R_5 when the source voltage is 200 V? (*Sec. 6-4*)

6–68. Write a statement comparing voltage, current, and power for R_3 and R_4. (*Sec. 6-9*)

TROUBLESHOOTING PROBLEMS

Answer problems 6–69 through 6–73 based on the condition that a short circuit developed across R_2 and R_3 (from B to D).

6–69. What is the total resistance under these conditions?

6–70. What is current through R_5?

6–71. What are the voltage drops across each of the five resistors?

6–72. What is the current flow through R_3?

6–73. What power is provided by the source?

Answer problems 6–74 through 6–76 based on the condition that an open develops in R_1.

6–74. What voltage is measured across each component?

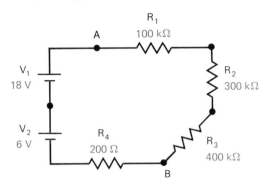

FIGURE 6-38

6–75. What current flow occurs through each component?

6–76. What power is provided by the source?

Use Figure 6-38 to answer problems 6–77 through 6–80. Answer in terms of opens or shorts and be specific. The answer should be related to specific components.

6–77. The current through R_2 is measured and found to be 0.04 A. What is wrong in the circuit?

6–78. The voltage across R_2 measures 0 V. What is wrong?

6–79. The voltage drop across R_4 is 24 V. What is wrong?

6–80. The voltage measured between *A* and *B* is zero. Each individual part measures the proper resistance when removed from the circuit. What could be wrong? (There is more than one answer to this question. List all you can think of for a solution.)

ANSWERS TO REVIEW PROBLEMS

6–1. 3 A

6–2. 3 A

6–3. 0.05 A

6–4. 0.05 A

6–5. It is equal.

6–6. 6727 Ω

6–7. 1.075 MΩ

6–8. 50 Ω

6–9. (a) 240 Ω; (b) 658 Ω

6–10. 48,600 Ω

6–11. 22, 1, 47, and 12 kΩ

6–12. (a) 1.6 kΩ + 100 Ω; (b) 24 kΩ + 2 kΩ; (c) 120 kΩ + 10 kΩ; (d) 62 kΩ + 36 kΩ + 1 kΩ; (e) 680 Ω + 120 Ω

6–13. $V_{R1} = R_1/(R_T + R) \times 50 = 4700/9400 \times 50 = 25$ V

6–14. $V_{R1} = 10$ kΩ/84 kΩ $\times 100 = 11.9$ V; $V_{R2} = 14$ kΩ/84 kΩ $\times 100 = 16.67$ V; $V_{R3} = 60$ kΩ/84 kΩ $\times 100 = 71.43$ V

6–15. $V(R_1 + R_2) = 200$ kΩ/500 kΩ $\times 1000 = 400$ V

6–16. +11 V

6–17. 1.0 A

6–18. 9 V and 1.8 A

6–19. −62 V

6–20. 10 V

6–21. 0.01 A

6–22. 0.01 A

6–23. 840 mW

6–24. 74 V

6–25. 7400 Ω

6–26. 0.74 W

6–27. The fuse

6–28. 0.015 A

6–29. 0 Ω

6–30. (a) Measure *V* across fuse, should be 0 V; (b) resistance across fuse, should be 0 Ω

6–31. 3 V

6–32. 0 A

Chapters 5–6

In this section we cover the basic rules for electrical circuits. Also covered is material on the fundamental circuit laws presented by Ohm and Kirchhoff. These rules are then used to explain the action of the series electrical circuit.

1. The rules established by Ohm, Watt, and Kirchhoff have universal use in the electrical and electronic fields.

2. Ohm's law shows the relationships among voltage, current, and resistance.

3. Three ways of using Ohm's law are: $V = I \times R$, $I = V/R$, and $R = V/I$.

4. When resistance R is held at a constant value, current I increases directly with increases in voltage V.

5. When voltage V is held at a constant value, current I decreases as resistance R increases.

6. The ohm may be expressed in units of 1000 by using the metric prefix "kilo" and in units of 1,000,000 by the metric prefix "mega."

7. There is a linear relationship between voltage and current.

8. Submultiples of electrical values for 1/1000, 1/1,000,000, and 1/1,000,000,000,000 are "milli," "micro," and "pico."

9. Electric power is the time rate of performing work. It is expressed in units of the watt.

10. There are interrelationships between the laws presented by Ohm and Watt. These are useful in solving problems related to either of the laws, since equal values of one formula may be substituted for equal values in any other formula.

11. Kirchhoff's law for voltage drop in a closed loop states that the individual drops must equal the voltage applied by the source.

12. Kirchhoff's law for current at a given junction states that the current(s) leaving the junction must equal the current(s) entering the junction.

13. When not handled properly, electric energy can give a person a strong shock. Under certain conditions, current flow through the human body can kill.

14. The wise technician or engineer will turn off circuit power before touching any of the components in a circuit.

15. Planning ahead before getting involved with electrical circuits will prevent shock in most situations.

16. Resistors in a series string are connected end to end.

17. The total resistance in a series circuit is the sum of the value of each resistance.

18. There is only one path for current flow in a series circuit; therefore, the current is equal throughout the circuit.

19. The voltage drops in a series circuit add to equal the source voltage.

20. In a series circuit, each voltage drop, including the source, has a polarity.

21. The sum of all voltages in a series circuit, including the source, is equal to zero (Kirchhoff's voltage law).

22. A voltage drop is the difference in electrical potential measured across a component due to current flow in the circuit.

23. A voltage divider consists of two or more series-connected resistors.

24. The amount of voltage drop that occurs across a component is equal to the ratio of the component's resistance to the total circuit resistance.

25. A potentiometer is a variable resistance that is used as a variable voltage divider.

26. Total power in a series circuit is equal to the sum of the power developed by each component.

27. An open circuit acts as an extremely high resistance in the circuit.

28. When all resistors in a series circuit are equal in ohmic value, the total circuit resistance is determined by multiplying the value of one resistor by the quantity of resistors in the circuit.

CHAPTER 7

Parallel Circuits

The second basic circuit configuration is the parallel circuit. This circuit, too, has basic rules that apply to all components connected in a parallel configuration. This circuit, like its series counterpart, has specific rules and applications. Recognition of its action and how current flows is essential knowledge for those who wish to use parallel-circuit applications.

KEY TERMS

Branch Circuit   **Open Branch**
Conductance   **Parallel Circuit**
Equivalent Resistance   **Short-Circuit Branch**

OBJECTIVES

Upon completion of the material in this chapter, you will

1. Understand the concepts of voltage, current, and resistance in the parallel circuit.
2. Have the ability to solve problems related to parallel circuit analysis.
3. Be able to explain Kirchhoff's current law.
4. Be able to explain the concept of an open and a short in a parallel circuit.
5. Be able to troubleshoot a parallel circuit.

In Chapter 6 you learned about the series circuit. It has one continuous set of connections for all the components in the circuit, thus establishing only one path for current flow. Each component in the series circuit developed its own voltage drop. The sum of these voltage drops was equal to the applied source voltage. The parallel circuit described in this chapter differs greatly from the series circuit. A comparison of these two circuits is shown in Figure 7-1. The difference between these two types of circuits becomes obvious as the comparison is made.

The parallel circuit consists of two or more loads. Each load is connected to the same source. Each load has its own set of characteristics. Each load has its own current path, although a common path for all the loads may be observed in part of the circuit. The basic difference, then, between the series and the parallel circuit is the method in which the loads are connected to the power source. A second difference is that current in this circuit has more than one path for its flow from source negative to source positive. Stated still another way, the **parallel circuit** provides at least two paths for current flow. It can also be stated that any two (or more) components that are connected as shown in Figure 7-1(b) are considered to be wired as a parallel circuit. There are some components that do not normally permit electron flow. These may be wired in parallel with others that do normally permit current flow. The main point here is that any two (or more) components may be connected as a parallel circuit. For the purposes of this chapter, we limit the presentation to parallel-wired conductive devices. These will be resistors, since they best illustrate the concepts necessary to understand this circuit. A common term often used to describe the parallel path is the **branch circuit**.

Parallel circuit *Circuit having more than one current path connected to a source.*

Branch circuit *Name for the parallel portion of a complex circuit.*

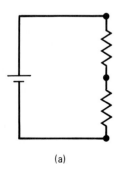

(a)

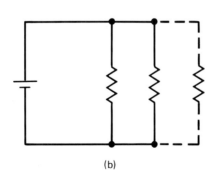

(b)

FIGURE 7-1 (a) Series and (b) parallel circuits. The parallel circuit has multiple current paths.

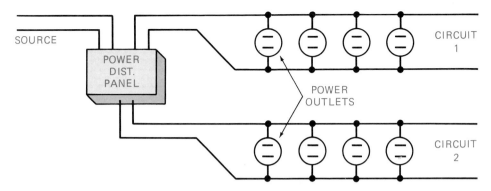

FIGURE 7-2 Most house wiring configurations use the parallel circuit arrangement.

7-1 Voltage in the Parallel Circuit

An example of a parallel-wired circuit is shown in Figure 7-2. In our homes we have several parallel-wired circuits. Each of these starts in the electrical control center. This center is often called the fuse box or circuit breaker box. One set of wires from the source, the electric company, is connected as input in the center. There are several *branches* that are then wired into the center. Each branch is a parallel-wired circuit. It is connected to several different power outlets in the house. Each outlet has the same amount of voltage present at its terminals. This is one of the facts related to a parallel-wired circuit. The voltage at each connection in the circuit is equal to the source voltage. The formula for this is

$$V_T = V_1 = V_2 = V_3 = \cdots = V_n \qquad (7\text{-}1)$$

If you consider the operating voltage for the plug-in appliances in your home you will realize that they all operate with the same value of voltage. In the United States this value is 120 V.

REVIEW PROBLEMS

7-1. Compare the voltage drops in a parallel circuit to the applied voltage of the source.
7-2. In terms of current paths and voltage drops, what is the basic difference between a parallel-wired circuit and a series-wired circuit?

7-2 Current in the Parallel Circuit

While the current in a series circuit is equal throughout the circuit, this is not the situation for a parallel circuit. The currents in a parallel circuit may be different for each branch of the circuit. Let us consider the appliances found in a kitchen (Figure 7-3). Each of these operates from a 120-V source. Each has its own current rating. If this is true, each appliance also has its own power

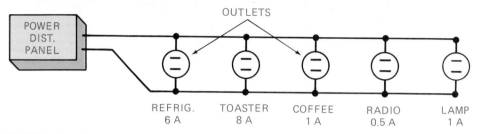

FIGURE 7-3 Current through each path of the branch adds to equal total branch current.

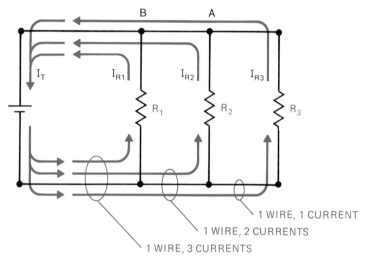

1 WIRE, 1 CURRENT

1 WIRE, 2 CURRENTS

1 WIRE, 3 CURRENTS

FIGURE 7-4 Current flow in the parallel circuit.

rating and its own value of resistance. In the parallel circuit the current increases each time an additional load is added to the circuit. The formula for this action is

$$I_T = I_1 + I_2 + I_3 + \cdots + I_n \qquad (7\text{-}2)$$

Simply stated, the formula shows that the total current in the circuit is equal to the sum of each branch current. This rule may be classified when Kirchhoff's law for current at a junction is considered. Using Figure 7-4 as an example, the current at the upper junction of R_3 and R_2 is the sum of the individual currents for R_3 and R_2. These enter junction B. The result is that the sum of these currents leaves the junction and is attracted to the positive terminal of the power source. When the currents $(R_3 + R_2)$ enter junction A, the current from R_1 also enters this junction. The result is that the sum of all of the currents $(R_3 + R_2 + R_1)$ travel from junction A and flow together to the power source. When analyzing a parallel-wired circuit, one must be aware of this concept.

7-3 Resistors in The Parallel Circuit

The resistance of each component in the parallel-wired circuit determines the amount of current that will flow through the same component. The two resistors shown in Figure 7-5 will illustrate this point. Ohm's law is used to calculate the resistance values.

FIGURE 7-5 Current flow in each of the parallel paths is determined by Ohm's law.

The circuit shows voltage source V = 100, with resistor R_1 rated 2 A and resistor R_2 rated 10 A in parallel.

$$R_1 = \frac{V_1}{I_1} = \frac{100\ \text{V}}{2\ \text{A}} = 50\ \Omega$$

$$R_2 = \frac{V_2}{I_2} = \frac{100\ \text{V}}{10\ \text{A}} = 10\ \Omega$$

This illustrates that the resistance values for individual resistances, or loads, in the parallel circuit determine the amount of current that flows through each resistor, or load. Keep in mind that the purpose of the applied voltage from the source is to establish a movement for the electrons in the circuit. The voltage stays at a fixed value and forces electron flow through the circuit. The quantity of electron current that flows is determined by the amount of resistance in the circuit.

REVIEW PROBLEMS

Use Figure 7-5 to find the following values.

7-3. Voltage across R_1
7-4. Current through R_1
7-5. Voltage across R_2
7-6. Current through R_2

7-4 Total Current and Branch Current

When current flows through the circuit (Figure 7-5), each of the branch currents is independent of any other current. They do, however, add to provide a value of total current for the circuit. In this example, the current through R_1 is 2 A and the current through R_2 is 10 A. The sum of these two currents is 2 A + 10 A, or 12 A. Schematic diagrams for parallel-wired circuits are normally drawn in the manner shown in this illustration. In an application for this circuit the actual wiring is done in a similar manner. There is a very good reason for this. Less wire is used for this type of construction than when individual pairs of wires are run from the source to each load.

A rule that applies to this type of circuit is that the sum of each of the branch currents is equal to the total current supplied by the source. The wires from the source may be considered as the major source, or main line, for the source power. Each of the loads may be considered as branches, or branch lines that are connected to the main-line system. This concept may help you understand the difference in the terminology used to describe the components in the system.

EXAMPLE 7-1

Three branch circuits are connected to a 240-V source. The source current is 28 A when the current through one branch is 8.2 A and the current through the second is 3.8 A. What is the current through the third branch?

Solution

The sum of the current through each branch equals total circuit current. Therefore, it is easy to find the missing current value:

$$I_T = I_1 + I_2 + I_3$$

$$I_3 = I_T - I_2 - I_1$$

$$16 = 28 \text{ A} - 8.2 \text{ A} - 3.8 \text{ A}$$

Combining like terms, the result is

$$I_3 = 28 \text{ A} - 8.2 \text{ A} - 3.8 \text{ A} = 16 \text{ A}$$

EXAMPLE 7-2

Four branch currents are 1.3, 2.4, 10, and 0.5 A. What is total current?

Solution

The total current is the sum of each branch current.

$$1.3 \text{ A} + 2.4 \text{ A} + 10 \text{ A} + 0.5 \text{ A} = 14.2 \text{ A}$$

EXAMPLE 7-3

Two resistances of 240 and 4800 Ω are connected in parallel to a 24-V main line. What is the total current flow?

Solution

The total current is found by first using Ohm's law to find each branch current value and then adding these figures.

$$I_1 = \frac{24 \text{ V}}{240 \text{ } \Omega} = 0.1 \text{ A}$$

$$I_2 = \frac{24 \text{ V}}{4800 \text{ } \Omega} = 0.005 \text{ A}$$

$$I_T = 0.005 \text{ A} + 0.1 \text{ A} = 0.105 \text{ A}$$

EXAMPLE 7-4

Branch currents are found to be 20 mA, 208 mA, and 1600 μA. Find the total current.

Solution

These values cannot be added unless they are all in the same format. We will convert them all to milliamperes. They are 20 mA, 208 mA, and 1.6 mA when converted. The figures may now be added. The total current is 229.6 mA.

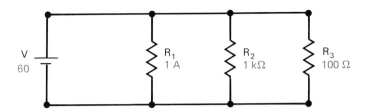

FIGURE 7-6

When attempting to use Ohm's law to determine the individual circuit values, the basic rules for a parallel circuit must be applied. For example, all voltage values are equal in the parallel circuit. This fact needs to be used in order to determine resistance and current for each component in the system. The rule stating that the total current is the sum of the individual branch currents must also be remembered and used. Let us apply these rules to the circuit shown in Figure 7-6.

First, let us recognize this circuit as a parallel-wired resistance circuit. The first concept that we need to remember is that the voltage across each resistor is equal to the source voltage. Therefore, we can make this statement:

$$V_{R1} = 60 \text{ V} \qquad V_{R2} = 60 \text{ V} \qquad V_{R3} = 60 \text{ V}$$

Using this set of facts, we can now apply Ohm's law to find either the current or the resistance values for each resistor.

The ohmic value of resistor R_1 is unknown. We use the formula $R = V/I$:

$$R_1 = \frac{V}{I} = \frac{60 \text{ V}}{1 \text{ A}} = 60 \text{ } \Omega$$

The current through R_2 and R_3 is determined in a similar manner. This time the formula $I = V/R$ is used:

$$I_{R2} = \frac{V}{R_2} = \frac{60 \text{ V}}{1 \text{ k}\Omega} = 0.06 \text{ A}$$

$$I_{R3} = \frac{V}{R_3} = \frac{60 \text{ V}}{100 \text{ } \Omega} = 0.6 \text{ A}$$

The total current is determined by adding each branch circuit's current:

$$I_T = I_1 + I_2 + I_3 = 1 \text{ A} + 0.6 \text{ A} + 0.06 \text{ A} = 1.66 \text{ A}$$

The format used in this illustration may be applied to any parallel-wired resistive circuit.

REVIEW PROBLEMS

7-7. Parallel circuit branch currents are 4, 3, and 8 A. Calculate the total current, I_T.

7-8. If $I_T = 16$ A, I_1 is 3 A, I_3 is 2.5 A, and I_4 is 7 A, find I_2.

7-5 Resistances in the Parallel Circuit

There is a significant difference between consideration of the factors concerning individual resistances in the parallel circuit and the *effect* of individual resistors on the total operational values of the parallel circuit. When two or more resistances are connected in parallel, they both have an effect on the total current in the circuit. We know that the currents for each resistance can be added in order to identify the total current. The circuit shown in Figure 7-7 illustrates this point. The total current is 0.75 A, the sum of 0.5 A and 0.25 A. We can use the value of 0.75 A to determine the *total* circuit resistance. In this circuit $R_T = \dfrac{V_T}{I_T}$, or 100 V/0.75 A. The answer is 133.3 Ω. This value of 133.3 Ω is considered to be what the source "sees" as it is applied to the circuit. A source cannot identify the number of resistances in the circuit. It can only react to the total value of all the resistances. In any parallel circuit the total resistance is always less than the ohmic value for the lowest single value of resistance. This is certainly true for the example used in Figure 7-7. The lowest individual value of resistance is 200 Ω and the total circuit resistance is 133.3 Ω.

The reason for this low resistance value should become obvious when we consider that the parallel circuit provides multiple paths for current flow. Each time an additional path is provided, the source current increases. The only way that current can increase in any circuit when the source voltage is held at a constant value is when the resistance in the circuit is reduced. If this does not make sense to you, try it in an Ohm's law problem. Use a value of 10 V and a resistance of 10 Ω. The current is going to be 1 A in this example. When the resistance is reduced to 5 Ω, the current rises to 2 A. This is the same effect as adding an additional resistance to the circuit. Two 10-Ω resistors wired in parallel present a 5-Ω resistance to the source (Figure 7-8). When one 10-Ω resistor is used, the Ohm's law current is 1 A. Adding a second 10-Ω resistance increases the total current to 2 A. The Ohm's law calculation for total resistance is now 5 Ω. When a third 10-Ω resistor is added to the circuit, the total current is now 3 A and the total resistance is reduced to 3.3 Ω.

Finding the Total Resistance in a Parallel Circuit One basic formula is used to calculate the total value of parallel resistances. The formula states that the reciprocal of the total resistance is equal to the sum of the reciprocal of each resistance. It looks like this:

$$\frac{1}{R_T} = \frac{1}{R_1} + \frac{1}{R_2} + \frac{1}{R_3} + \frac{1}{R_n} \qquad (7\text{-}3)$$

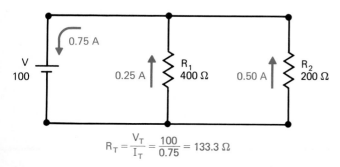

FIGURE 7-7 In the parallel circuit the sum of the currents through each branch will determine total current and total resistance. Total resistance is *always* less than the lowest single resistance value in the circuit.

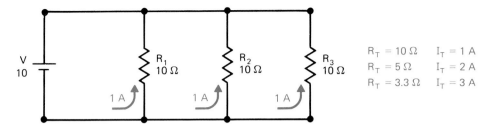

FIGURE 7-8 Illustration of the statement that total resistance is less than the lowest single resistor value in the parallel circuit.

When we use this formula to find the total resistance for our latest example, we have

$$\frac{1}{R_T} = \frac{1}{R_1} + \frac{1}{R_2} + \frac{1}{R_3}$$

$$= \frac{1}{10\ \Omega} + \frac{1}{10\ \Omega} + \frac{1}{10\ \Omega}$$

$$= \frac{3}{10\ \Omega}$$

$$R_T = \frac{10\ \Omega}{3} = 3.3\ \Omega$$

The value of R_T is inverted in the last two lines of this process. This is because we wish to find the actual value and we have been working its reciprocal. The reciprocal of any number is that number divided into 1. Using this conversion method, the reciprocal of $1/R_T$ is equal to R_T.

This method may be used for all values of resistance. It works very well with either mixed resistance values or equal resistance values. Our first example used equal values of resistance. The next sample uses mixed resistor values.

EXAMPLE 7-5

Find the total resistance when resistors of 400, 800, 500 and 1200 Ω are wired in parallel.

Solution

$$\frac{1}{R_T} = \frac{1}{R_1} + \frac{1}{R_2} + \frac{1}{R_3} + \frac{1}{R_4}$$

Use your hand calculator to find the reciprocal of each ohmic value: $1/400 = 0.0025$, $1/800 = 0.0013$, $1/500 = 0.002$, and $1/1200 = 0.0008$. These figures have been rounded slightly to simplify the answers. The sum of these reciprocals is 0.0066. The next step is to find the reciprocal of 0.0066, for this will equal the total resistance of the circuit. The answer is 151.9 Ω. This value is not exact because of our rounding off. It probably will not matter in a circuit. This is due to the tolerance values for the

components. An ohmmeter used to measure the total resistance would probably not be able to be as specific as the answer derived mathematically.

When all resistances in the system are equal, a shortcut method may be used to find the total circuit resistance. This method uses the ohmic value for one of the resistors and the quantity of equal resistors in the circuit. Simply divide this single resistor value by the number of equal-value resistors. For example, in a circuit having six 1800-Ω resistors in parallel, divide 1800 by 6. The total resistance is 300 Ω. The equal resistance method works with two or more equal resistors. This is actually an adaptation of the basic reciprocal rule for parallel resistors.

Another adaptation for the reciprocal rule may be used when two unequal-value resistors are parallel connected. The formula for this is:

$$R_T = \frac{R_1 \times R_2}{R_1 + R_2} \qquad (7\text{-}4)$$

The total resistance for a circuit using a 600-Ω and a 30-Ω resistor is calculated in this way:

$$R_T = \frac{600\ \Omega \times 30\ \Omega}{600\ \Omega + 30\ \Omega}$$

$$= \frac{18,000\ \Omega}{630\ \Omega}$$

$$= 28.57\ \Omega$$

This is a special-situation formula and will function only when there are two resistances in the circuit. When more than two resistances are wired in parallel, the reciprocal or equal-value formulas must be used.

Equivalent Resistance   Since the source cannot determine the specific quantity of resistors in a circuit, we can state that the source "sees" a value of resistance that is equivalent to the total resistance. It is possible to use the methods described earlier in this chapter in order to simplify a circuit. The circuit shown in Figure 7-9(a) will illustrate this type of problem solving. We may use two equal-value resistors to derive an equivalent value for them. This circuit has two sets of equal-value resistors. Each of these may be combined and an equivalent value for each pair established. The result of the combining is two resistances with values of 25 and 20 Ω [Figure 7-9(b)]. These two resistance values are now reduced to a single value by using the formula $R_T = (R_1 \times R_2)/(R_1 + R_2)$. The values are 500 Ω/45 Ω and the total equivalent resistance for this four-resistor circuit is 11.1 Ω [Figure 7-9(c)].

Equivalent resistance   Single value of resistance representing all components in one section of a complicated circuit.

Total Resistance Is Less Than the Lowest Single Value   In each of the examples the total circuit resistance is less than the ohmic value of the smallest single resistor. This fact provides you with a reasonably good method of checking your calculations or measurements. If you are making some resistance tests in a parallel circuit and the measured value is less than the value you expected,

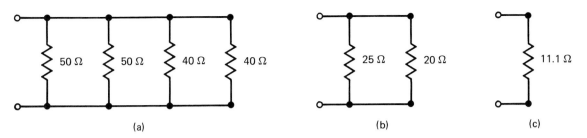

(a)   (b)   (c)

FIGURE 7-9 The resistance meter "sees" a value of resistance that is equivalent to the total resistance in the circuit.

it is possible that a resistance is in parallel with the component you are measuring. Before removing the tested component, be sure to check the schematic
diagram for a parallel current path. This will explain the lower resistance value.

Unknown Branch Resistance There are times when it is necessary to parallel a resistor with a second resistor. This is done when the required value of
resistance is not available. The desired resistance is identified in R_T and the
unknown value is called R_x. The original formula for this is

$$R_T = \frac{R \times R_x}{R + R_x}$$

This is transposed to obtain

$$R_x = \frac{R \times R_T}{R - R_T}$$

EXAMPLE 7-6

What value of resistance (R_T) is required to provide a 120-Ω resistance
(R_T) when it is parallel wired to a 500-Ω resistor?

Solution

$$R_x = \frac{R \times R_T}{R - R_T}$$

$$= \frac{500\ \Omega \times 120\ \Omega}{500\ \Omega - 120\ \Omega}$$

$$= \frac{60,000\ \Omega}{380\ \Omega}$$

$$= 157.9\ \Omega$$

This formula is successful when used with two resistance values, or two
parallel branches.

REVIEW PROBLEMS

7-9. Calculate the total resistance (R_T) for four 680-kΩ resistors wired in parallel.

7-10. Calculate R_T for a 1.2-MΩ and a 470-kΩ resistor wired in parallel.

7-11. Calculate R_T for three 1.2-kΩ resistors wired in parallel with one 400-Ω resistor.

7-6 Conductances in the Parallel Circuit

Conductance *The ease in which a wire permits flow of electrons.*

Conductance is the reciprocal of resistance. It is expressed $1/R$. The letter G is used to represent conductance and the unit of conductance is the siemens (S). The formula for conductance is

$$G_T = G_1 + G_2 + G_3 + \cdots + G_n \tag{7-5}$$

The method of expressing conductance and expressing the same values as resistances is shown in Figure 7-10. In this example G_1 is 1/100 Ω and G_2 is 1/50 Ω. The total conductance is calculated

$$G_T = G_1 + G_2$$

$$= 0.01 \text{ S} + 0.02 \text{ S}$$

$$= 0.03 \text{ S}$$

One does not require the use of reciprocals when working with conductances. This method may be used to find the total resistance of a parallel circuit. First convert each resistance value into a conductance value. This is done by finding the reciprocal of each individual resistance in the circuit. Use the for-

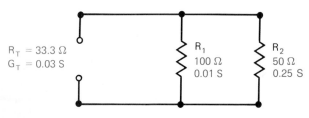

$R_T = 33.3\ \Omega$
$G_T = 0.03\ S$

R_1
100 Ω
0.01 S

R_2
50 Ω
0.25 S

FIGURE 7-10 Use of conductance values to calculate total resistance in the parallel circuit.

mula to find the total amount of conductance. Then use this formula to convert conductance into resistance:

$$R = \frac{1}{G} \qquad (7\text{-}6)$$

The answer should be the same using this method as when the resistance formulas are utilized.

Another concept used to solve circuit value problems is that the current in each branch of a parallel circuit is directly proportional to its conductance value. This concept is related to the proportional voltage drop that develops across series-connected resistances. When the branch current is directly proportional to the branch conductance value, the total current in the circuit is directly proportional to the total conductance in the circuit.

REVIEW PROBLEMS

7-12. Find the total conductance for a circuit containing $G_1 = 35$ S, $G_2 = 55$ S, and $G_3 = 125$ S.

7-13. In a circuit with $G_1 = 0.5$ mS, $G_2 = 0.01$ mS, $G_3 = 1.01$ mS, and $G_4 = 0.7$ mS, find G_T and R_T.

7-7 Power in the Parallel Circuit

All power used in any circuit originates from the power source. The power source provides a specific voltage value, while current flow depends on the size of the load or loads in the system. Since all of the power is produced by the source, the total power consumed by the individual loads is equal to the source power. Actually, the total circuit power is the sum of each load's power consumption. The formula for total power in the circuit is

$$P_T = P_1 + P_2 + \cdots + P_n \qquad (7\text{-}7)$$

The parallel circuit shown in Figure 7-11 is used to show how the total power is determined. The source voltage of 40 V is applied to a three-branch parallel circuit. The circuit loads are 10, 40, and 80 Ω. The current through each branch is 4, 1, and 0.5 A, respectively. These values are determined by applying Ohm's law to each resistance: $I_{R1} = 40$ V/10 Ω = 4 A, $I_{R2} = 40$ V/40 Ω = 1 A, and $I_{R3} = 40$ V/80 Ω = 0.5 A. The total current is 4 A + 1 A + 0.5 A = 5.5 A. The total resistance is then $R_T = V_T/I_T = 40$ V/5.5 A = 7.27 Ω.

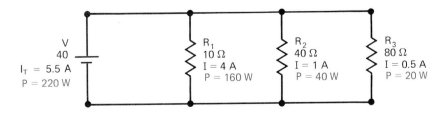

FIGURE 7-11 Total power in the parallel circuit is equal to the sum of the individual power values.

The power consumed by each branch is calculated using Watt's law, $P = V \times I$: $P_1 = 40\text{ V} \times 4\text{ A} = 160\text{ W}$, $P_2 = 40\text{ V} \times 1\text{ A} = 40\text{ W}$, and $P_3 = 40\text{ V} \times 0.5\text{ A} = 20\text{ W}$. When these values are added, the total is 220 W. This value is the total power used by the branches. It is also equal to the total power supplied by the source. Another way of stating this is to say that the power consumed by the load(s) is equal to the power created (or supplied) by the source. This fact is an important one to remember when considering the transfer of power occurring from one circuit to another circuit or device. These values must be equal.

The fact that the total power in a circuit is equal to the power supplied by the source is the same for both series and parallel wired circuits. In both of these types of circuits the total power supplied by the source is equal to the sum of the power used by each of the loads in the circuit.

REVIEW PROBLEMS

7-14. A parallel circuit has branch powers of 3, 4, and 10 W. Calculate the total circuit power.

7-15. A parallel-wired system has branch resistances of 10, 20, and 40 Ω. Calculate the total power, P_T, when a source of 100 V is used.

7-16. Use the ohmic values of problem 7-15 and a source of 50 V. Calculate the individual power consumption and the total power consumption.

7-8 Analysis of the Parallel Circuit

When analyzing a parallel-wired circuit, there are certain factors that you can use. These will make the determination of circuit values easier.

1. In the parallel circuit the voltage drop that develops across each load is equal to the voltage applied to one of the loads. There is only one voltage value that is measured across the branches under these conditions.

2. When the current through one branch of a two-branch parallel circuit is known, the current through the other branch can be found when the known current is subtracted from the total current.

Use the circuit in Figure 7-12 to illustrate the points that we have made. In this problem you are asked to find the applied voltage V_A and the ohmic value of R_1. The first thing to do is to look at the circuit to determine the facts that you have available for your use. One cannot determine the Ohm's law values unless any two of the three components in the law are known. In this circuit R_3 has two of the facts provided. We are able to find the voltage drop across R_2 by using the formula $V = I \times R = 4\text{ V} \times 100\ \Omega = 400\text{ V}$. Since this is a parallel circuit, the value of 400 V is equal everywhere in the circuit; this means that V_A as well as V_{R1} and V_{R2} are equal to 400 V.

To find the value of R_1, the current through R_3 must first be determined. We now know the voltage and resistance value for R_3. Ohm's law is used again in order to determine the current. $I_{R1} = V/R_1 = 400\text{ V}/50\ \Omega = 8\text{ A}$. This is

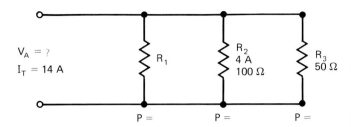

FIGURE 7-12 Total power values, when found, can be used to determine applied voltage in the circuit.

easily accomplished when the currents through R_2 and R_3 are subtracted from the source current.

$$I_{R1} = I_T - I_{R2} = I_{R3}$$
$$= 14 \text{ A} - 4 \text{ A} - 8 \text{ A} = 2 \text{ A}$$

Once the current through R_1 is determined, its ohmic value is easily determined by again using Ohm's law.

$$R_1 = \frac{V}{I_{R1}} = \frac{400 \text{ V}}{2 \text{ A}} = 200 \text{ }\Omega$$

Using the facts that are available and applying Ohm's law makes circuit analysis much simpler. The major factor is to remember to think out the method you will use to find the unknown values. A little bit of preplanning goes a very long way in circuit problem analysis. After all, you usually do not take a trip without first planning the route you are going to use to reach your destination.

REVIEW PROBLEMS

Use Figure 7-12 to find the following values.

7-17. Voltage drop across R_2
7-18. Current through R_3
7-19. Power consumption for each resistance
7-20. Total R and total P

7-9 The Open Branch in a Parallel Circuit

The open is defined as an infinitely high resistance. When a circuit is considered to be open there is no current flow and no work is performed by the load in the circuit. The circuit shown in Figure 7-13 consists of two parallel-wired resistance loads. One of these has a value of 800 Ω and the other a value of 1.2 kΩ. Together they represent a load of 480 Ω to the source. When both of the loads are functioning, the source provides a current of 0.25 A. This amount of current divides between the two loads. R_1 uses 0.15 A and R_2 uses 0.1 A in its part of the circuit. A failure in the R_1 circuit of either the wiring or the actual component could result in an open. The open current path interrupts the normal current flow. Current in that path is reduced to zero. The load

Open branch *An abnormally high resistance path in a parallel circuit.*

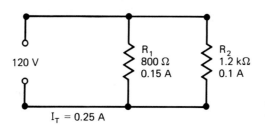

FIGURE 7-13 A failure such as an open in R_1 will reduce total circuit current.

ceases to function under these conditions. If the load was a device that produced a visible output, such as a lamp, we would know immediately that a problem existed.

Measurement of total circuit current would show show that the current has dropped from its normal 0.25 A to a lower value, 0.1 A. This would be an indication of an open, nonfunctioning portion of the total system.

The open branch in a parallel-wired circuit interrupts only the components in the specific branch. All other branches continue to function in a normal manner. This is one advantage of parallel-wired circuits. It is the major reason why our homes and most appliances are wired in parallel.

REVIEW PROBLEMS

7-21. Four lamps are parallel wired and the filament of lamp 1 opens. How can you tell by measurement that it has failed?

7-22. In problem 7-21, will total circuit resistance increase or decrease due to the lamp failure?

7-23. A parallel-wired circuit has six loads. Each is rated at 0.5 A and 100 V. Two of the lamps burn out (open). What is the total voltage and current for the circuit under these conditions?

7-10 The Short-Circuit Branch in a Parallel Circuit

Short-circuit branch *One path of a branch circuit having an unusually low resistance path.*

For all practical purposes the short-circuit represents a resistance close to 0 Ω. Let us use a value of 0.1 Ω to represent the short circuit. This is added to the parallel circuit shown in Figure 7-14. This circuit normally measures 90 Ω and draws 1.33 A of current. When a short circuit occurs it acts just like another resistance added to the other components in the parallel circuit. The major

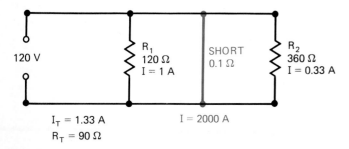

FIGURE 7-14 A short circuit in a parallel branch reduces total circuit resistance to less than 0.1. Ω

difference, of course, is the value of this added resistance. Its value of 0.1 Ω reduces the total circuit resistance to less than 0.1 Ω. Remember, the total resistance in a parallel-wired circuit is always less than the lowest single resistance value in the circuit. Resistances of 120, 360 and 0.1 Ω produce an equivalent resistance value of 0.0998 Ω. Under these conditions the total current flow from the 120-V source will increase from the normal 1.33 A to over 1200 A. This, of course, is a mathematical analysis of what *could* happen when a short occurred across the parallel circuit wiring.

Effects of Short-Circuit Current The conditions produced by a short circuit do not result in the amounts of current flow indicated. There are some limiting factors to consider. These include the power source rating, current capabilities of the wiring in the system, and the use of any circuit protection devices. Let us look at each of these to see their effect on an overcurrent condition.

7-11 Power Source Limitations

Almost all power sources have both a voltage and a current rating. This rating tells us the values we can expect to have available for load operation. If the power source is unregulated, the output voltage will diminish as the load current is increased. This is due to the reduction of load resistance values. In theory the voltage could drop to close to zero as the load resistance is reduced to 1 Ω or less.

A great number of power sources used in electronic equipment are designed to include a voltage and/or current regulation system. These systems maintain the design voltage and current until the load demand exceeds the power source rating. The power source has a safety shutdown circuit. When an overdemand occurs, the shutdown circuit turns the power source off. This keeps the power source from burning out and may also help keep the load from being totally destroyed.

A second type of safety device used in almost all wired circuits is a type of overcurrent protector. These devices are known as fuses and circuit breakers. Their function is unique—they are designed to fail! A fuse contains a metallic link that will melt if the current through the fuse goes beyond its current rating. The heat produced by excess current flow melts the link. The fuse is wired in series between the source and the load. When it opens it interrupts the circuit current and all load activity ceases. A circuit breaker operates in a similar manner. One major difference between the fuse and the circuit breaker is that the circuit breaker may be reset. The open condition of a circuit breaker is described by stating that it has *tripped*. This term indicates an open-circuit condition for the breaker. Once the source of the overcurrent is eliminated or corrected the circuit breaker may be reset. The circuit is then restored to normal operating conditions.

The final limiting factor for any circuit is the wiring. The size, or gage, of the wire used determines the quantity of current. When an overcurrent condition occurs the wires start to heat. The heat is produced when current larger than the capability of the wire attempts to flow through the wire. Two effects of overcurrent in a wire are heat or rupture. The excess current heats the wire. It may heat to the point where it either melts the insulation around the wire or causes the insulation to burn. When it burns it can also start a fire in components near the wire. The other effect is the rupture of the wire. Excess current can melt the wire at its weakest point. This produces an open circuit. It will interrupt circuit current and shut off load current.

REVIEW PROBLEMS

7-24. What is the resistance of a short circuit?

7-25. What type of circuit protection devices are used to protect the power source from overloads or short circuits?

7-12 Troubleshooting Parallel Circuits

TROUBLESHOOTING APPLICATIONS

The methods used to troubleshoot a parallel-wired circuit differ from those used for series-wired circuits. One of the best methods of determining a circuit fault in a parallel circuit is to start with some resistance measurements. This method may seem to take a bit longer than some others. Its advantage is the elimination of the initial need to remove any components from the circuit. Later you may be required to do this, but it should be considered as a last resort in troubleshooting. *Remember:* Resistance measurements are made only when the source power is turned off! The circuit in Figure 7-15 is used for the test circuit in this example. The first step is to look at the schematic diagram for the circuit. There are two loads in this circuit. Their total effective resistance should be less than the value of the lowest single resistance, or less than 25 Ω. A measurement of resistance between points *A* and *B* should provide a reading of about 17 Ω if all is working properly. A reading of 25 Ω indicates an open in the R_1 circuit. A resistance reading of 50 Ω indicates an open in the R_2 branch circuit. A reading of infinitely high resistance indicates an open between the loads and the place where the resistance is measured. It may also indicate that both loads are open, but this is not the usual result of a component failure.

A reading of close to 0 Ω indicates a short in the system; the short can be anywhere. A different technique is required for location of the problem. When there is a system short, each branch has to be isolated in order to locate the problem. Either a wire has to be cut or a printed circuit path has to be opened in order to isolate each branch. The branch that has an unusually low resistance value will indicate the problem area.

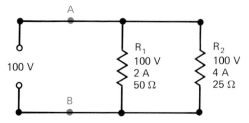

FIGURE 7-15

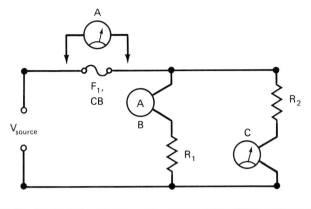

FIGURE 7-16 Use of an ammeter to determine actual circuit current values. The meter is connected across the open fuse or circuit breaker terminals.

It is possible to use an ammeter to measure currents through each branch. The method of measuring current requires insertion of the ammeter into the circuit. The ammeter is placed in series with the load in order to measure current through the load. (Remember that current in the series circuit is equal in all parts of the circuit.) In Figure 7-16, ammeter A is placed in series between the source and all of the loads. It will indicate total circuit current. A good place to insert this ammeter is to connect it in parallel with the terminals of the fuse or circuit breaker. These devices must be open in order to have current flow through the ammeter. The value of current measured at this point, when power is momentarily applied, often indicates whether the problem is a total system short or a short through one branch.

The other places for ammeters are at points B and C in Figure 7-16. Since the ammeter is series connected between the wires and the load, it may be placed on either side of the load. For safety reasons ammeter C is preferred. An abnormally high current flow through either meter will help locate the problem branch circuit.

REVIEW PROBLEMS

7-26. What happens to total circuit resistance when one branch circuit opens?

7-27. What would be the resistance of the open branch circuit?

7-28. What would be the resistance of a short-circuited branch?

1. Parallel-wired resistances are connected to the same points in the circuit.
2. The parallel circuit has two or more current paths.
3. Total resistance in the parallel circuit is always less than the lowest single resistor value.
4. Voltage across each branch is equal to the source voltage in the parallel circuit.

CHAPTER
SUMMARY

5. Currents through each branch add to equal source current.

6. Power in the parallel circuit is equal to the sum of each branch's power use.

7. When one branch of a parallel circuit opens, the total resistance left in the circuit increases. The result is less current flow.

8. When one branch circuit opens, the current still flows through the remaining branches.

9. A short circuit in a parallel branch affects all the branches.

10. An ohmmeter may be used to locate an open or a short in a parallel circuit.

11. A short circuit in one branch will result in no current flow in all the branches in the system.

12. Safety devices are used to protect the system when an overcurrent occurs.

SELF TEST

Answer true or false; change all false statements into true statements.

7–1. A parallel-wired circuit has only one path for current flow.

7–2. Loads in the parallel circuit may be other than resistors.

7–3. Each branch in the parallel circuit has its own current flow value.

7–4. Adding branches to the parallel circuit will decrease total current flow.

7–5. Total power in the parallel circuit is the sum of the power used by each branch.

7–6. Total resistance in the parallel circuit increases as loads are added to the system.

7–7. When one branch in the parallel circuit opens, total current flow stops.

7–8. Conductance is the same as resistance.

7–9. An open fuse or circuit breaker is indicative of an overcurrent problem in the circuit.

7–10. The resistance value of an open circuit breaker is close to 0 Ω.

MULTIPLE CHOICE

7–11. A parallel circuit has a minimum of _____ loads.
(a) four
(b) three
(c) two
(d) one

7–12. The voltage drop across any component in a parallel system is _____ the source voltage.
(a) greater than
(b) less than
(c) equal to
(d) independent of

7–13. Total resistance in a parallel circuit is
(a) equal to each resistance
(b) greater than each resistance value
(c) less then the sum of all resistance

7–14. When the number of loads is increased in the parallel circuit:
    (a) total resistance increases
    (b) total resistance decreases
    (c) applied voltage is reduced
    (d) there is no effect on total resistance

7–15. Total parallel circuit power is
    (a) greater than the sum of all load power values
    (b) equal to each load's consumption
    (c) independent of load values
    (d) equal to the sum of all load consumption

7–16. The reciprocal of 160 is
    (a) 0.0625
    (b) 0.00625
    (c) 0.000625
    (d) 0.625

7–17. When the resistance of one load in the parallel circuit is increased:
    (a) total load current increases
    (b) total load current remains the same
    (c) total load current decreases
    (d) current in that load will increase

7–18. When source voltage is decreased:
    (a) total resistance decreases
    (b) load current decreases
    (c) load current increases
    (d) load current remains the same

7–19. Increasing source voltage will
    (a) increased load current
    (b) increase current through the highest load only
    (c) decrease total resistance
    (d) have no effect on current

7–20. Adding an additional load to a parallel-wired circuit will
    (a) increase total resistance
    (b) decrease total resistance
    (c) decrease total current
    (d) not affect circuit values

ESSAY QUESTIONS

7–21. Is it possible to utilize Ohm's law to measure current through a resistance? Explain your answer.

7–22. Is it possible to convert a voltage measurement into a current reading? Explain your answer.

7–23. Explain why total circuit resistance decreases with the addition of an additional load.

7–24. Explain why current in a parallel circuit is greatest through the smallest branch resistance.

7–25. Explain why Kirchhoff's law for current applies to the parallel circuit.

7–26. Two 100-Ω resistors are parallel connected to a 500-V source. What are the total resistance, total current, and current through each resistance? (*Sec. 7-8*)

7–27. Two resistors are parallel connected to a 100-V source. One resistor value is 100 Ω and the current through the other one is 5 A. What is the resistance of the second resistor? (*Sec. 7-8*)

7–28. A 5.6- and a 6.8-kΩ resistor are connected in parallel. What is the total resistance of this pair? (*Sec. 7-3*)

Use Figure 7-17 to answer problems 7–29 through 7–31.

7–29. Each branch current is 0.15 A and the source is 40 V. Find the total current and total resistance. (*Sec. 7-8*)

7–30. The current through R_1 is 150 A and the current through R_2 is 300 A. When a 24-V source is applied, what are the total resistance and total current? (*Sec. 7-8*)

7–31. When the source voltage is 30 V, $R_1 = 27$ Ω, and $R_2 = 68$ Ω, what are the total current and power? (*Sec. 7-8*)

Use Figure 7-18 to answer problems 7–32 and 7–33.

7–32. What are the source voltage and power when $R_1 = 100$ Ω, $R_2 = 200$ Ω, and $R_3 = 400$ Ω? The current through R_1 is 400 mA. (*Sec. 7-8*)

7–33. What is the total power? (*Sec. 7-7*)

Use Figure 7-19 to answer problems 7–34 and 7–35.

7–34. Find the value of R_4. (*Sec. 7-8*)

7–35. What are the total resistance and total current? (*Sec. 7-8*)

Use Figure 7-20 to answer problems 7–36 and 7–37.

7–36. Find the value of R_2. (*Sec. 7-8*)

7–37. Find R_T and I_T. (*Sec. 7-8*)

7–38. You have a circuit consisting of a 180-V source and a series 9-A fuse. How many 90-Ω loads can be parallel connected to the source before the fuse will blow and turn off the load current? (*Sec. 7-8*)

Use Figure 7-21 to answer problems 7–39 through 7–42.

7–39. What is the total current flow? (*Sec. 7-8*)

7–40. What is the current flow through R_2? (*Sec. 7-8*)

7–41. What is the total load resistance? (*Sec. 7-8*)

7–42. What is the resistance value of R_2? (*Sec. 7-8*)

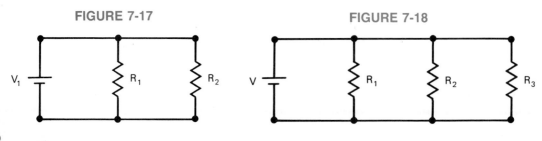

FIGURE 7-17   **FIGURE 7-18**

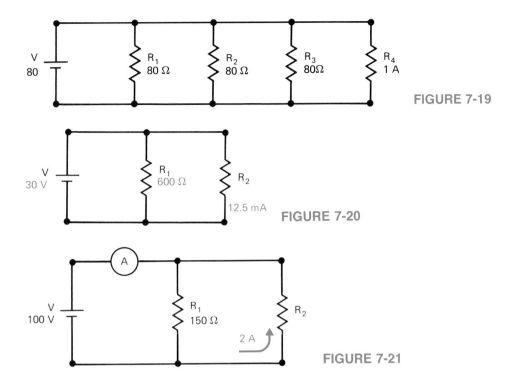

FIGURE 7-19

FIGURE 7-20

FIGURE 7-21

Use Figure 7-22 to answer problems 7–43 through 7–47.

7–43. What is the total circuit current? (*Sec. 7-8*)
7–44. What is the voltage across R_1? (*Sec. 7-8*)
7–45. What is the total circuit resistance? (*Sec. 7-8*)
7–46. What is the value of V? (*Sec. 7-8*)
7–47. What is the resistance value of R_1? (*Sec. 7-8*)

Use Figure 7-23 to answer problems 7–48 through 7–52.

7–48. What is the resistance of R_3? (*Sec. 7-8*)
7–49. What is the total circuit current flow? (*Sec. 7-8*)
7–50. What is the total circuit power? (*Sec. 7-8*)
7–51. What is the current through R_3? (*Sec. 7-8*)
7–52. What is the ohmic value of R_1? (*Sec. 7-8*)

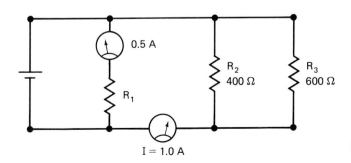

FIGURE 7-22

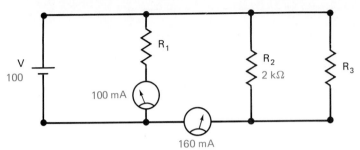

FIGURE 7-23

Use Figure 7-24 to answer problems 7–53 through 7–55.

7–53. What are the currents through each resistor? (*Sec. 7-8*)

7–54. What is the source voltage if R_3 is equal to 1.2 kΩ? (*Sec. 7-8*)

7–55. Find the ohmic values for each resistor when the source voltage is 150 V? (*Sec. 7-8*)

Use Figure 7-25 to answer problems 7–56 through 7–65.

7–56. One resistor is faulty in this circuit. Identify it and explain what has happened to it (open or shorted?) (*Sec. 7-8*)

7–57. What is the total circuit resistance? (*Sec. 7-8*)

7–58. What is the resistance value of R_3? (*Sec. 7-8*)

7–59. If the value of R_2 is 100 Ω, what is the correct current through R_2 and R_3? (*Sec. 7-8*)

7–60. What is the correct total current flow when all resistors are in working order? (*Sec. 7-8*)

7–61. What voltage should you expect to measure across R_1? (*Sec. 7-8*)

7–62. What total resistance should you expect to measure when all components are working correctly? (*Sec. 7-8*)

7–63. What resistance value would you measure across R_3 when all resistors are working correctly? (*Sec. 7-8*)

7–64. If a short should develop across R_3, what voltage will be measured across R_1? (*Sec. 7-12*)

7–65. If meter M_2 should open, what current will be measured by meter M_1? (*Sec. 7-12*)

FIGURE 7-24

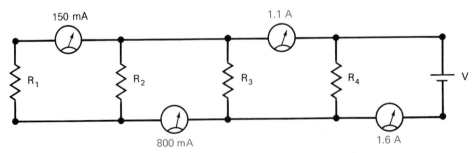

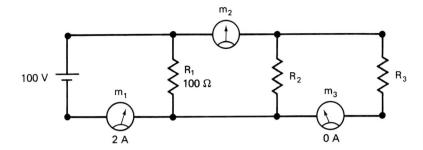

FIGURE 7-25

Use Figure 7-26 to answer problems 7–66 through 7–68.

7–66. If a source voltage of 400 V is used, what is the voltage measured across each resistance? (Three answers) (*Sec. 7-8*)

7–67. R_1, R_2, and R_3 are equal in resistance value and total 200 Ω and a total current of 10 A is flowing from the source. What is the ohmic value for each resistor? (*Sec. 7-8*)

7–68. What is the source voltage under the conditions described in problem 7-67? (*Sec. 7-8*)

7–69. Three resistances are connected in parallel across a 120-V source. The source current is 12 A. Current through R_1 is 4 A and the ohmic value of R_3 is 60 Ω. Draw a schematic diagram of the circuit and solve it to find I_{R2} and R_T. (*Sec. 7-8*)

7–70. What is the total resistance for each of the following parallel-wired combinations? (*Sec. 7-8*)
(a) 10-kΩ and 40-kΩ
(b) eight 6-kΩ
(c) three 6-kΩ and two 2-kΩ
(d) ten 100-kΩ and three 9-kΩ

7–71. How much resistance must be connected in parallel to a 10-kΩ resistor in order to have a current of 100 mA when 10 V is applied to the circuit? (*Sec. 7-8*)

7–72. A power source is delivering 1.2 kW at 120 V to a load. The load consists of two equal resistances. What is the power consumption for each resistance? (*Sec. 7-3*)

7–73. What are the values of each resistance and its current flow for the circuit in problem 7-72? (*Sec. 7-7*)

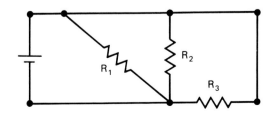

FIGURE 7-26

Use Figure 7-26 and values of V = 100V, R = 100 Ω each for problems 7-74, 7-75 and 7-77.

TROUBLESHOOTING PROBLEMS

7–74. If R_2 should open, what are the resulting voltage, current, and resistance for the circuit?

7–75. If R_1 should open, what are the resulting voltage, current, and resistance for R_2 and R_3?

7–76. If R_1 = 100 Ω, R_2 = 400 Ω, and R_3 = 1 kΩ, what is the total resistance in the circuit?

7–77. If R_2 should increase in value (due to a circuit malfunction) to 4 kΩ, what are the values of the voltage and current through R_2 under these conditions?

7–78. What is the total current consumption for problem 7-29 when R_3 is open?

7–79. What would the voltage be in problem 7-27 when R_1 is shorted?

7–80. If one of the loads in problem 7-26 opened, what are the resulting current, voltage, and power used by the load?

7–81. A failure in a load reduces circuit current to two-thirds its normal amount. What happens to the missing one-third of the current?

7–82. Three resistors are connected in parallel. R_1 = 10 kΩ, R_2 = 20 kΩ, and R_3 = 5 kΩ. Their combined measured resistance is actually 4 kΩ. What has occurred in the circuit?

7–83. Four resistances are parallel wired to a 100-V source. Their individual values are R_1 = 6 kΩ, R_2 = 12 kΩ, R_3 = 600 Ω, and R_4 = 1.2 kΩ. Current flow from the source is 0.267 A. What, if anything, has occurred in the circuit?

7–84. If measured current in problem 7-78 is 0.0083 A, what has happened in the circuit?

ANSWERS TO REVIEW PROBLEMS

7–1. They are equal.

7–2. Parallel circuit has multiple current path; series circuit has one path.

7–3. 100 V

7–4. 2 A

7–5. 100 V

7–6. 10 A

7–7. 15 A

7–8. 3.5 A

7–9. 170 kΩ

7–10. 337,725 Ω (approximate)

7–11. 200 Ω

7–12. 215 S

7–13. 2.22 mS, 450 Ω

7–14. 17 W

7–15. 1.75 kW

7–16. P_1 = 250 W, P_2 = 125 W, P_3 = 62.5 W, P_T = 437.5 W

7–17. 400 V

7–18. 8 A

7-19. $P_1 = 800$ W, $P_2 = 1.6$ kW, $P_3 = 3.2$ kW

7-20. $R_T = 28.57$ Ω, $P_T = 5.6$ kW

7-21. Total resistance is higher.

7-22. Increase

7-23. 100 V and 2.0 A

7-24. Almost 0 Ω

7-25. Fuses, circuit breakers, and shutdown circuits

7-26. Increases

7-27. Infinitely high

7-28. Close to 0 Ω

CHAPTER 8

Series–Parallel Circuits

Almost all electrical systems use a combination of series and parallel circuits. The ability to recognize the complex circuit and to reduce it to a basic configuration is important. The ability to recognize circuit faults and to correct these faults is essential. Only after this is accomplished are we able to locate defective components and replace them.

KEY TERMS

Balanced Bridge Circuit   **Resistor Bank**
Branch Current   **Parallel Resistance Strings**
IR **Drop**   **Series–Parallel**

OBJECTIVES

Upon completion of the material in this chapter, you will

1. Be able to explain voltage drops and current flow in the combination circuit.
2. Understand the concept of the bridge circuit.
3. Understand the effects of open and shorts in the combination circuit.
4. Be able to troubleshoot the combination circuit.

A small percentage of the circuits we use are wired either strictly parallel or strictly series. Almost all of the circuits are made up of a combination of series- and parallel-wired components. An example of one type of combination circuit is shown in Figure 8-1. This circuit is typical of those used in industry, the home, or business. Its function is to provide a constant value of operating voltage to each of the three outlets. A fuse is connected between the 120-V source and the three outlets. The fuse is wired in series between the source and the loads. Each of the three outlets is wired in parallel across the 120-V wires connected to the source. This is a prime example of a combination series–parallel circuit. The series-parallel circuit has both a series section and a parallel section. The two sections are wired in series with each other and then connected to the power source, forming one complete circuit. It is rather simple and easy to understand. Not all combination circuits are this simple. Some, in fact, are very complicated. One needs to be able to reduce the complex circuit to its simpler components if it is desirable to calculate values of voltage, current, and resistance for each branch and component in the system.

8-1 Identification of the Series-Parallel Circuit

The ability to identify the combination series–parallel circuit is made easier when one examines the configuration of the components in the circuit. The material presented in Chapter 6 described the series circuit configuration. One characteristic of that circuit is the fact that there is only one path for current flow. Any portion of the series–parallel that is limited to one current path can be identified as a *series* section of the circuit. The circuits shown in Figure 8-2 are combination series–parallel circuits. They are the same circuit drawn in different ways. Resistor R_1 is connected betwen the positive terminal of the source and the junction of resistors R_2 and R_3. Resistor R_1 allows only one

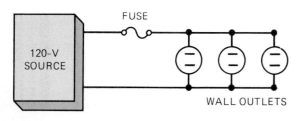

FIGURE 8-1 A combination series and parallel circuit. The fuse is connected in series with the parallel-wired wall outlets.

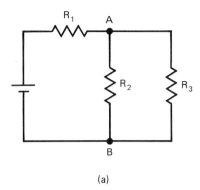

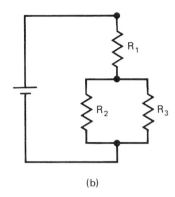

FIGURE 8-2 Alternative diagrams for representing a series–parallel circuit.

path for current between these two points of the circuit. It, therefore, is a series-wired component.

Resistors R_2 and R_3 have a different configuration. They are connected to each other in parallel. This part of the circuit is similar to those described in Chapter 7. The two resistances, wired in parallel, provide a dual path for current flow between points A and B in the circuit. This is the parallel-wired component in the circuit. In this example resistors R_2 and R_3 are connected in parallel to each other. The parallel combination of R_2 and R_3 is connected in series with R_1 to the power source.

The identification of the series component in a combination series–parallel circuit is done when we apply this rule: *If all the current flow in that section of the circuit is through one component or limited to one path, that section of the circuit is wired in series.* The laws regarding series circuits then apply to that section of the circuit.

The identification of the parallel-wired component in a combination series–parallel circuit is accomplished by applying this rule: *If the total current between two given points divides and flows through more than one path, or branch, those paths are parallel to each other.* This is true only when the voltage drops across each branch are equal. The laws regarding a parallel-wired circuit then apply to that portion of the combination circuit

The example in Figure 8-3 will help you clarify the identification of the series and parallel portions of the series–parallel circuit. As you review these circuits, please remember that any electrical circuit can be composed of components other than resistors. The circuits may also be composed of a mixed combination of resistors and other electrical components. To keep them relatively simple, resistors are used in the circuits illustrated in these examples.

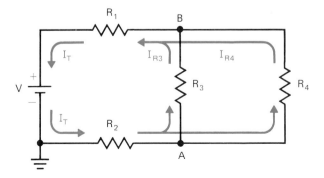

FIGURE 8-3 Current flow quantities in the series–parallel circuit.

EXAMPLE 8-1

Identify the series–parallel relationships in Figure 8-3.

Solution

The best method of identifying the type of circuit is to follow the current path in the circuit. Electron flow in this circuit starts at the negative terminal of the source. It then flows through R_2 until it reaches point A. At this point the current separates into two paths. Part of the current flows through R_3 and the balance through R_4. Since the same voltage is found across these two resistors, they are in parallel with each other. The two currents come together at point B. The total current flow now is through R_1 and to the positive terminal of the source.

Since the total current flows through R_1 and R_2, they are connected in series between the source and the combination of R_3 and R_4. Since the current separates, or divides, as it flows through R_3 and R_4, they are considered as a parallel combination in this circuit.

EXAMPLE 8-2

Identify the series–parallel relationships in Figure 8-4.

Solution

In this circuit there is one current path through R_1, R_2, and R_3. They are wired in series. The current flow separates through the combination of R_4 and R_5. These two resistors are connected in a parallel configuration.

EXAMPLE 8-3

Describe the series–parallel combination in the circuit shown in Figure 8-5.

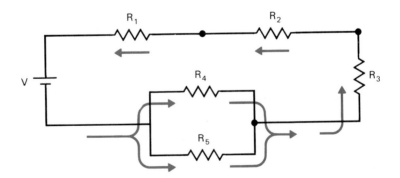

FIGURE 8-4 A series–parallel circuit combination showing the division and rejoining of currents.

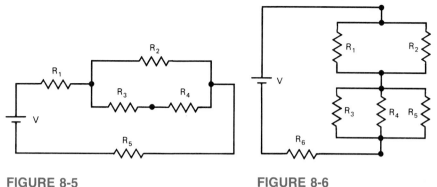

FIGURE 8-5   **FIGURE 8-6**

Solution

This circuit introduces a subcircuit, resistors R_3 and R_4. They are connected in series with each other. The combinaion of R_3 and R_4 is connected in parallel with R_2. Resistors R_1 and R_5 are connected in series with the source and the parallel portion of the total circuit.

EXAMPLE 8-4

Describe the series–parallel combination in Figure 8-6.

Solution

There are two sets of parallel-connected resistances in this circuit. One set consists of R_1 and R_2. The second set has three parallel-wired resistances. These are R_3, R_4, and R_5. Resistor R_6 is wired in series wih the two sets of parallel-wired resistances.

In Examples 8-1 through 8-4 it was necessary to determine the current path as an initial step. Once the current path is found, the subcircuit can be identified as either a series- or a parallel-wired section of the total circuit. After the identification is done, one can calculate values for voltage, current, and resistance. These calculations are done by using Ohm's law.

REVIEW PROBLEMS

Use Figure 8-7 to answer problems 8-1 and 8-2.

8-1. Which resistors are series connected?
8-2. Which resistors are parallel connected?

Use Figure 8-8 to answer problems 8-3 and 8-4.

8-3. Which resistor combinations are series wired?
8-4. Which resistor combinations are parallel wired?

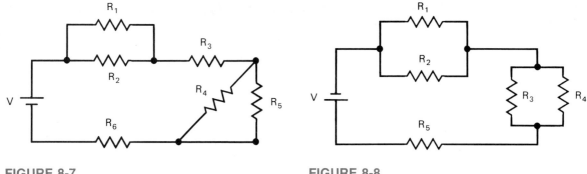

FIGURE 8-7

FIGURE 8-8

8-2 Analysis of the Series–Parallel Circuit

The first step in the mathematical analysis of a combination series–parallel circuit is to identify its basic sections. When these are known, the rest of the steps involved are relatively easy. We will start this process by looking at two different circuits and deciding whether each is basically a series circuit or a parallel circuit configuration.

The circuit shown in Figure 8-9 has both series and parallel sections. The combination of R_1 and R_2 forms a parallel circuit. This parallel circuit is in series with R_3. This circuit will be considered as a series type of circuit with one parallel subsystem in it. The circuit shown in Figure 8-10 also has three resistances. In this circuit resistors R_2 and R_3 are series connected to each other. The combination of R_2 and R_3 is then wired in parallel to resistor R_1. This circuit will be classified as a parallel circuit with a series subsystem in it. This type of identification is necessary if one is to attempt to find the exact values of voltage, current, and resistance in the circuit as well as the total voltage, current, and resistance values in the system.

Finding Total Resistance You have learned how to determine total resistance in a series circuit in Chapter 6. You also learned how to find total resistance in the parallel resistive circuit in Chapter 7. When you want to determine total resistance in the series–parallel circuit, all that is required is to identify the

FIGURE 8-9 Two parallel-wired resistances connected in series with a third resistance and the load.

FIGURE 8-10 Two series-connected resistances connected in parallel with a third resistance and the load.

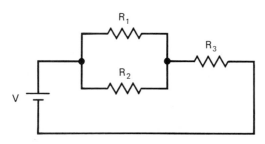

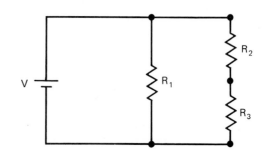

series and parallel sections of the total circuit and apply the correct formula for total resistance to that section of the circuit. When you have completed this activity the next step is to combine the figures into a total resistive value. The examples that follow will clarify how this is accomplished.

EXAMPLE 8-5

Find the total resistance for the circuit shown in Figure 8-11.

Solution

This circuit is basically a series circuit with a parallel component. Resistors R_1 and R_2 are connected in parallel. This combination is then connected in series to R_3. First reduce the parallel resistor combination to its equivalent resistance R_{EQ}. The sum of the reciprocals method is used to do this. This value is then added to the value of R_3 for a total resistance figure.

$$\frac{1}{R_{EQ}} = \frac{1}{R_1} + \frac{1}{R_2}$$

$$= \frac{1}{100 \ \Omega} + \frac{1}{400 \ \Omega}$$

$$= 0.01 \text{ S} + 0.0025 \text{ S} = 0.0125 \text{ S}$$

$$R_{EQ} = \frac{1}{0.0125 \text{ S}}$$

$$= 80 \ \Omega$$

The resistance that is measured across the parallel combination of R_1 and R_2 is 80 Ω. This value is equivalent to the two individual resistor values when they are parallel connected.

The total circuit resistance is determined by applying the law for series resistors:

$$R_T = R_{EQ} + R_3$$

$$= 80 \ \Omega + 120 \ \Omega$$

$$= 200 \ \Omega$$

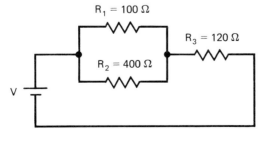

FIGURE 8-11

EXAMPLE 8-6

Determine total resistance in the circuit shown in Figure 8-12.

Solution

This circuit is basically a parallel circuit with a series subcomponent. Resistors R_2 and R_3 are connected in series with each other. This combination is then connected in parallel with resistor R_1. The first thing to do is to find the equivalent resistance value of R_2 and R_3. Since these resistors are connected in series, we use the formula

$$R_{EQ} = R_2 + R_3$$
$$= 40\ \Omega + 60\ \Omega$$
$$= 100\ \Omega$$

This equivalent value is then used in the formula for parallel resi:

$$\frac{1}{R_T} = \frac{1}{R_{EQ}} + \frac{1}{R_1}$$
$$= \frac{1}{100\ \Omega} + \frac{1}{100\ \Omega}$$
$$= 0.01\ \text{S} + 0.01\ \text{S}$$
$$= 0.02\ \text{S}$$
$$= 50\ \Omega$$

Another method of finding this value, since the two resistors have equal resistance, is to use the formula

$$R_T = \frac{R}{n} \qquad (n = \text{number of equal resistors})$$
$$= \frac{100\ \Omega}{2} = 50\ \Omega$$

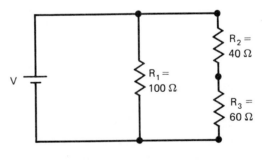

FIGURE 8-12

Finding Total Current Total current flow in any resistive circuit depends on the value of the applied voltage and the amount of resistance in the entire circuit. Normally, one has to know these two values before total current can be found. One exception to this rule is the fact that the current through any component that is series connected to the source also has all of the current flowing through it. If there is any component through which all the current flows in the circuit and if it is possible to calculate the current through it, the answer is equal to total current flow from the source.

Ohm's law is used to find total circuit current. Figures 8-11 and 8-12 are used to show how this is done.

EXAMPLE 8-7

Find the total circuit current for the circuit shown in Figure 8-11. Use a value of 100 V for a source.

Solution

$$I_T = \frac{V_T}{R_T}$$

$$= \frac{100 \text{ V}}{200 \ \Omega}$$

$$= 0.5 \text{ A}$$

EXAMPLE 8-8

Find the total circuit current for the circuit shown in Figure 8-12. Use a value of 100 V for the source.

Solution

$$I_T = \frac{V_T}{R_T}$$

$$= \frac{100 \text{ V}}{50 \ \Omega}$$

$$= 2 \text{ A}$$

REVIEW PROBLEMS

8-8. Find the total circuit current when the voltage in Figure 8-11 is 400 V.

8-9. Find the total circuit current for Figure 8-11 when the source voltage is 10 V.

8-10. Find the total circuit current for the circuit in Figure 8-12 when the source voltage is 24 V.

8-3 Series–Parallel Circuit Terminology

The study and understanding of these combination series–parallel circuits requires the use of some additional terminology. Some of the terms commonly used to describe portions of the total circuit include *branch circuit, resistor strings, branch current*, and *resistor bank*. These terms are defined below.

A *branch circuit* is one current path for current in a parallel circuit or parallel portion of a series–parallel circuit. The resistance identified as R_3 in Figure 8-13 is one branch in the circuit. Resistor R_4 is the other branch in that circuit. A portion of the total current flow occurs through each of the branches. These currents are identified as the **branch currents** for that circuit.

Branch currents *Current flow in one branch of a circuit.*

Resistors R_1 and R_2 are series connected in this circuit. They are described as being a *series string*. Any time that two or more resistors are series connected, they are described in this manner.

When two or more resistors are connected in parallel with the source, they are described as a *resistance bank*. The three resistances wired as shown in Figure 8-14 are called a **resistor bank**. It is also possible to have a resistor bank as a portion of a total circuit. Resistors R_3 and R_4 form such a bank in Figure 8-13. A general statement describing the resistor bank is that it is used to identify a group of parallel-wired resistances in a circuit.

Resistor bank *Group of resistors connected as a composite part of total circuit.*

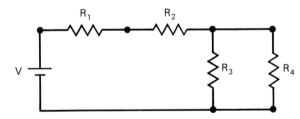

FIGURE 8-13 R_3 and R_4 are branch paths in this circuit.

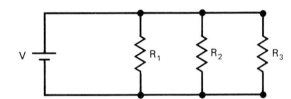

FIGURE 8-14 All three resistances form resistance bank in this circuit.

REVIEW PROBLEMS

8-11. Identify the type of resistance network in each of these figures.
   (a) Figure 8-2: R_2 and R_3
   (b) Figure 8-4: R_1, R_2, and R_3
   (c) Figure 8-4: R_4 and R_5
   (d) Figure 8-6: R_3, R_4, and R_5

8-12. Calculate the total resistance in Figure 8-13 when each resistor has a value of 100 Ω.

8-13. Calculate the total resistance in Figure 8-14 when each resistor has a value of 300 Ω.

8-4 Parallel Resistance Strings

Many of the circuits used in electrical or electronic systems are made up of several series strings. Each string normally establishes its own current path. The string may be used to control a specific circuit or active component. Several series strings are connected in parallel to a common power source. The circuit shown in Figure 8-15 will illustrate this concept. One string consists of three resistances: R_1, R_2, and R_3. The second string contains R_4 and R_5. Each string has series-connected resistors. Both strings are connected to the same power source. The total circuit can be considered as a resistive bank. The bank is made up of two branch circuits. The branch circuits are the two strings described previously.

Parallel resistance strings Two or more series circuits connected in parallel with each other.

Finding Branch Currents The current flow through each branch depends on the amount of applied voltage and the total resistance in that branch. When voltage and resistance values for the circuit shown in Figure 8-16 are established, one can calculate the current in each of the branches, the total resistance, and the total current in the circuit.

The first step in this analysis is to reduce the resistance in each of the strings, or branches, to an equivalent value. These are series strings. The rule for resistances in series is used:

$$R_{T1} = R_1 + R_2 + R_3 \qquad\qquad R_{T2} = R_4 + R_5$$

$$= 100\ \Omega + 50\ \Omega + 150\ \Omega \qquad = 125\ \Omega + 475\ \Omega$$

$$= 300\ \Omega \qquad\qquad\qquad\qquad = 600\ \Omega$$

FIGURE 8-15 Two strings of series resistances connected in parallel with each other.

FIGURE 8-16 The addition of voltage and resistance values offers the opportunity to solve for circuit current values.

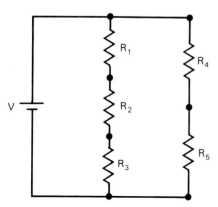

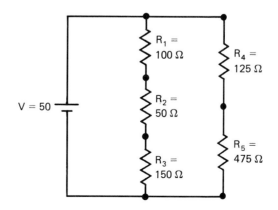

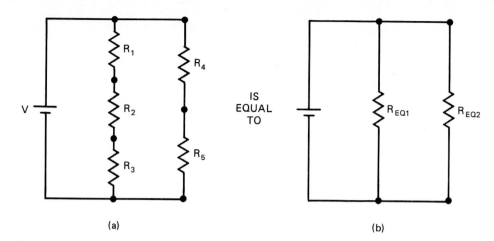

FIGURE 8-17 Circuit (a) is equivalent to circuit (b).

Now there are two equivalent resistances wired in parallel to each other (Figure 8-17). The values of equivalent resistance are 300 and 600 Ω, respectively. These two values are used to find the current in each of the branches in the circuit. The values of the applied voltage from the source and the equivalent resistance are used in Ohm's law. The answer is the current in each of the branches.

$$I_{EQ1} = \frac{V}{R_{EQ1}}$$

$$= \frac{50 \text{ V}}{300 \text{ }\Omega}$$

$$= 0.167 \text{ A}$$

$$I_{EQ2} = \frac{V}{R_{EQ2}}$$

$$= \frac{50 \text{ V}}{600 \text{ }\Omega}$$

$$= 0.083 \text{ A}$$

Finding Voltage Drops in the Branch Circuit. When the resistance value of each component and the current through the component in a branch circuit is known, one is able to calculate the **IR drop**, or voltage drop, for the individual components. The rule for voltages in a series circuit still applies and the sum of each *IR* drop is equal to the applied source voltage.

IR drop *Voltage drop in a specific portion of circuit (I * R = V)*

EXAMPLE 8-9

Using the values for Figure 8-16, calculate the voltage drop across each component.

Solution

$$V_{R1} = I_{R1} \times R_1 = 0.167 \text{ A} \times 100 \text{ } \Omega = 16.7 \text{ V}$$

$$V_{R2} = I_{R2} \times R_2 = 0.167 \text{ A} \times 50 \text{ } \Omega = 8.35 \text{ V}$$

$$V_{R3} = I_{R3} \times R_2 = 0.167 \text{ A} \times 150 \text{ } \Omega = 25.05 \text{ V}$$

$$V_{R4} = I_{R4} \times R_4 = 0.083 \text{ A} \times 125 \text{ } \Omega = 10.375 \text{ V}$$

$$V_{R5} = I_{R5} \times R_5 = 0.083 \text{ A} \times 475 \text{ } \Omega = 39.425 \text{ V}$$

When the *IR* drops that develop across R_1, R_2, and R_3 are added, they equal 50.1 V. The applied voltage is 50 V. The reason for the 0.1-V difference is due to rounding off of the answers done during the calculations. Usually, this type of answer is close enough for normal mathematical analysis. This is also true for the *IR* drop calculated for R_4 and R_5. Their mathematical total is 49.8 V. This is reasonably close to the 50 V applied to the string to be acceptable. The reason for the derived answers is that the circuit current values were rounded when they were determined. In a working circuit these values would be acceptable. One does have to allow a little bit for meter calibration error or some undesirable meter lead resistance. Answers that are within 1 or 2% of the required value are usually acceptable when comparing actual values in a working circuit to the design value in the circuit diagram.

Finding Total Circuit Resistance, R_T The total circuit resistance is found by taking the values of total resistance for each branch and using these values in the parallel resistance formula.

$$\frac{1}{R_T} = \frac{1}{R_{EQ1}} + \frac{1}{R_{EQ2}}$$

$$= \frac{1}{300 \text{ } \Omega} + \frac{1}{600 \text{ } \Omega}$$

$$= 0.033 \text{ S} + 0.00167 \text{ S}$$

$$= 0.005 \text{ S}$$

$$R_T = 200 \text{ } \Omega$$

Finding Total System Current I_T The total system current is found either by adding the currents through each of the branches or by first finding the total circuit resistance, the applied voltage, and then using Ohm's law. The formulas for this are

$$I_T = I_1 + I_2 = 0.167 \text{ A} + 0.083 \text{ A} = 0.250 \text{ A}$$

or

$$I_T = \frac{V_T}{R_T} = \frac{50 \text{ V}}{200 \text{ } \Omega} = 0.250 \text{ A}$$

8-5 Series Resistance Banks

The circuit shown in Figure 8-18 has both a series and a parallel section. In this circuit the parallel-connected resistors R_1 and R_2 are series connected to R_3. R_3 is identified as a series component because all of the circuit current must flow through it. Resistors R_1 and R_2 are parallel wired and the voltage will be equal across their terminals.

Finding Total Circuit Resistance The first step in this process is to reduce the parallel bank of resistors to one equivalent value:

$$\frac{1}{R_{EQ}} = \frac{1}{R_1} + \frac{1}{R_2}$$

$$= \frac{1}{10\ \Omega} + \frac{1}{90\ \Omega}$$

$$= 0.1\ S + 0.011\ S$$

$$= 0.111\ S$$

$$R_{EQ} = \frac{1}{0.111\ S}$$

$$= 9\ \Omega$$

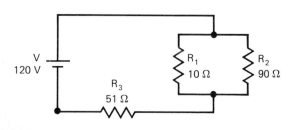

FIGURE 8-18 The combination of R_1 and R_2 form a resistance bank in series with R_3.

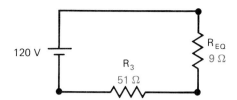

FIGURE 8-19 Reduction of the parallel bank of Figure 8-18 into one equivalent resistance value.

This value is now added to the series resistor R_2. The answer is the total resistance in the circuit (Figure 8-19).

$$R_T = R_{EQ} + R_3$$

$$= 9\ \Omega + 51\ \Omega$$

$$= 60\ \Omega$$

Finding Total Current When both the source voltage and the total resistance are known, Ohm's law is used to determine the circuit current.

$$I_T = \frac{V_T}{R_T}$$

$$= \frac{120\ V}{60\ \Omega}$$

$$= 2\ A$$

Finding *IR* Drop The first step toward finding the *IR* drop is to recognize the series resistances in this circuit. R_3 is series connected to the parallel bank of R_1 and R_2. The *IR* drops across the two resistances in the bank are the same, or equal. Since these two resistors are parallel wired, the voltage that is measured across the components has the same value. This value and that of R_3 are added together to determine the total *IR* drop.

$$V_T = V_{REQ} + V_{R3}$$

This value is calculated by using Ohm's law:

$$V_{REQ} = I_{REQ} \times R_{EQ}$$

$$= 2\ A \times 9\ \Omega$$

$$= 18\ V$$

and

$$V_{R3} = I_{R3} \times R_3$$

$$= 2\ A \times 51\ \Omega$$

$$= 102\ V$$

The values of 18 and 102 add to equal the applied source voltage of 120.

Finding Current Through Each Component All of the current in this circuit flows through R_3, so its current value is the same as the total circuit current. Calculating the current through resistors R_1 and R_2 is accomplished by use of Ohm's law. The voltage across each of these resistors is 18 V. The ohmic values for each is also known.

$$I_{R1} = \frac{V_{R1}}{R_1}$$

$$= \frac{18 \text{ V}}{10 \text{ }\Omega}$$

$$= 1.8 \text{ A}$$

$$I_{R2} = \frac{V_{R2}}{R_2}$$

$$= \frac{18 \text{ V}}{90 \text{ }\Omega}$$

$$= 0.2 \text{ A}$$

The sum of these two currents is 2.0 A. The current through R_3 is also 2.0 A. Thus we have determined the branch currents for the components in the resistor-bank portion of the circuit and seen that they are equal to the current flow in the balance of the circuit.

The method used to determine the actual IR drop and current flow through individual components in the series–parallel circuit can be used for any combination of resistances. It is possible to have a circuit with more than one parallel resistance bank and be able to calculate individual IR drops and currents. This is also true for circuits with more than one series resistance string. It is important to remember to reduce the circuit to its equivalent resistance values and to apply Ohm's law properly when doing this.

REVIEW PROBLEMS

Use Figure 8-18 to find the following values.

8-17. Total circuit resistance when $R_1 = 500 \text{ }\Omega$, $R_2 = 500 \text{ }\Omega$, and $R_3 = 750 \text{ }\Omega$.

8-18. Total circuit current and branch current

8-19. Individual IR drops

8-6 Combination Series–Parallel Resistance Banks and Strings

Circuit values are determined by reducing the circuit to its equivalent values. Once this is done, Ohm's law is used to solve for specific values. Figure 8-20 shows how this is done. Several steps are required in order to find the total

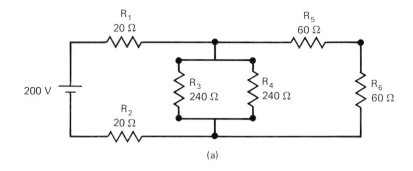

(a)

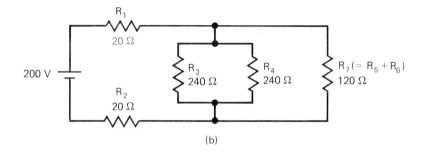

(b)

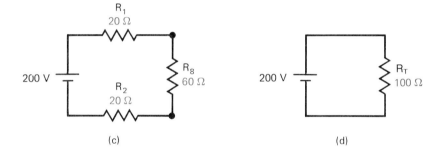

(c)

(d)

FIGURE 8-20 The reduction of a complex series–parallel circuit, step by step, into one equivalent resistance value.

circuit resistance and current flow. Start at the right-hand side of the circuit. This side is the farthest away from the 200-V source. Resistors R_5 and R_6 form a series string whose equivalent value is 120 Ω. Identify these as R_7. The combination is shown in Figure 8-20(b). Resistors R_3, R_4, and R_7 form a parallel bank. Their total resistance is

$$\frac{1}{R_T} = \frac{1}{R_3} + \frac{1}{R_4} + \frac{1}{R_7}$$

$$= \frac{1}{240 \ \Omega} + \frac{1}{240 \ \Omega} + \frac{1}{120 \ \Omega}$$

$$= 0.01667 \ \text{S}$$

$$R_T = 60 \ \Omega \ \text{(approximately)}$$

The balance of the circuit is shown in Figure 8-20(c). The result of reducing the R_3, R_4, R_5, and R_6 to one equivalent value is identified as R_8. These three resistances are identified as a series string in the figure. The total resistance of this string is determined by adding the three resistances:

$$R_T = R_1 + R_8 + R_2$$

$$= 20\ \Omega + 60\ \Omega + 20\ \Omega$$

$$= 100\ \Omega$$

When the total resistance of the circuit is calculated, it can be used with the applied source voltage to find the total circuit current. Ohm's law is used for this purpose:

$$I_T = \frac{V_T}{R_T}$$

$$= \frac{200\ \text{V}}{100\ \Omega}$$

$$= 2\ \text{A}$$

Once total current is determined, it is possible to return to the original circuit to solve for individual branch currents and *IR* drop (Figure 8-21). The analysis of these facts is done as follows:

All of the 2-A current flows through both R_1 and R_2. Their values are 20 Ω each. The *IR* drop across each of these resistances is 2 A × 20 Ω or 40 V. The sum of these two *IR* drops is subtracted from the source voltage. The remainder is 120 V (200 V − 40 V − 40 V = 120 V). This value is the voltage that is measured across the resistance bank of R_3, R_4 and the subseries string of R_5 and R_6.

The total current of 2 A divides as it flows through these three paths. The current through R_3 is calculated by using the formula

$$I_{R3} = \frac{V_{R3}}{R_3}$$

$$= \frac{120\ \text{V}}{240\ \Omega}$$

$$= 0.5\ \text{A}$$

FIGURE 8-21 When voltage values are known, Ohm's law is used to solve for unknown circuit values.

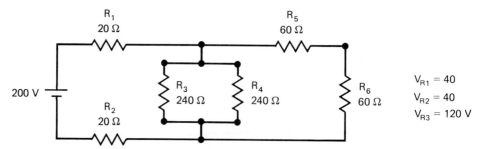

Since R_3 and R_4 are of the same value, the current through R_4 is also 0.5 A. The balance of the total 2-A current flows through the series string of R_5 and R_6. This may be proven using either of two methods. The first is:

$$I_T = I_{R3} + I_{R4} + I(R_5 + R_6)$$

$$I_T - I_{R3} - I_{R4} = I(R_5 + R_6)$$

$$2 - 0.5 - 0.5 = I(R_5 + R_6)$$

$$1 = I(R_5 + R_6)$$

The second method is:

$$I(R_5 + R_6) = \frac{V(R_5 + R_6)}{R_5 + R_6}$$

$$= \frac{120 \text{ V}}{120 \text{ }\Omega}$$

$$= 1 \text{ A}$$

There are many methods of solving for individual circuit component values. Each is dependent on the recognition of the type of circuit and the application of the correct circuit rules. The following problems illustrate some of these points and methods of solving for the individual circuit values.

EXAMPLE 8-10

Find the value of R_3 and I_{R2} in Figure 8-22.

Solution

Two basic rules for circuit solution are shown in this problem. They are:

1. The sum of the currents in each branch equals the total circuit current.
2. The rule for *IR* drop in a parallel bank states that the *IR* drops across each of the components in that bank are equal and the same.

The branch current I_{R2} is found when one subtracts the branch current I_{R2} from the total current I_T. Branch current I_{R3} is then determined by subtracting 4 A from the total current of 10 A. The difference between these two currents is the current through R_2, or 6 A.

The *IR* drop across R_2 is found by use of Ohm's law as it relates to the component R_3. First we must find the *IR* drop in the parallel resistance bank. We know the resistance and the current values for R_2. Using these values we can find the *IR* drop. $V = IR = 20 \text{ A} \times 6 \text{ }\Omega = 120 \text{ V}$. Since R_2 and R_3 are in parallel, the *IR* drop across R_2 equals the *IR* drop across R_3, or 120 V. When this value is known we can use the figure to calculate the resistance value of R_3. $R_3 = V_{R3}/I_{R3} = 120 \text{ V}/4 \text{ A} = 30 \text{ }\Omega$.

If we desire to find the source voltage in this circuit, we first must

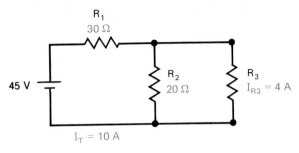

FIGURE 8-22

find the *IR* drop across R_1. This resistance of 30 Ω has all the current in the circuit flowing through it since it is in series between the source and the resistor bank of R_2 and R_3. We use the 10-A current value and the resistance value of 30 Ω for this. $V = I \times R$, or $V = 10 \text{ A} \times 30 \text{ Ω}$ = 300 V. This value represents the *IR* drop across resistance R_1. It is added to the *IR* drop of 120 V that develops across the parallel resistance bank, resulting in a total source voltage value of 420 V. This value is the voltage that is applied to the circuit in order to create a flow of electrons and the *IR* drops.

EXAMPLE 8-11

Find all circuit values for Figure 8-23.

Solution

This problem is best resolved by determining the branch currents as an initial step. The branches are connected as a parallel bank across the voltage source of 45 V. Branch ($R_1 - R_2$) has two series-connected resistances. They must be added together and an equivalent value determined if we wish to find the branch current for this portion of the circuit. The basic rules for this circuit are:

1. When one has parallel strings across a main line, it is possible to find the currents in the branches without finding circuit resistance.
2. Each series-connected resistance has its own *IR* drop. This *IR* drop is less than the applied voltage from the source.
3. Any branch current must be less than the total circuit current.

When we look at this circuit we see R_1 and R_2 in series in one branch and R_3 as a stand-alone component in the other branch. We also see that the source voltage of 45 V is provided. Resistors R_1 and R_2 form an equivalent value of 90 Ω. The current through these resistances is 45 V/90 Ω, or 0.5 A, found by using Ohm's law, $I = V/R$. This current flows through both of the resistances because they are series connected. The *IR* drop across R_1 is equal to 0.5 A × 40 Ω, or 20 V, and the *IR* drop across R_2 is equal to 0.5 A × 50 Ω or 25 V.

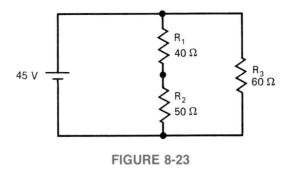

FIGURE 8-23

Current flow through R_3 is found by using the formula $I = V/R$, or 45 V/60 Ω, giving a value of 0.75 A. The total current, therefore, is the sum of the branch currents, 0.5 A + 0.75 A, or 1.25 A. The total circuit resistance can now be determined. $R_T = V_T/I_T$, or 45 V/1.25 A. The answer is 36 Ω.

EXAMPLE 8-12

Find all voltages and current for the circuit in Figure 8-24.

Solution

The best way to find the values of voltage and current in this circuit is first to find R_T. This can then be used to find I_T. The rules given for the preceding examples are required to solve this problem.

To find R_T, the circuit must first be simplified. Start at the right-hand side with resistances R_6 and R_7. These two are parallel connected. Their equivalent resistance is 40/2, or 20 Ω. This bank is series connected to R_3. The total resistance in this series string is 20 Ω + 20 Ω, or 40 Ω. The 40-Ω string is parallel connected to R_5. When we calculate the equivalent for this parallel bank, we find that 40/2 equals 20 Ω.

The equivalent bank of R_5, R_3, R_6, and R_7 is series connected to R_2. The sum of the resistances in this string is 40 Ω + 20 Ω, or 60 Ω. This equivalent bank of 60 Ω is parallel connected to resistance R_4. The 60-Ω values of R_4 and the equivalent value of 60 Ω provide an equivalent value of 30 Ω. This value is series connected to R_1. The ohmic value of R_1 (120) and the ohmic value of all of the other resistances (30) are added. The total circuit resistance is 150 Ω.

Total current I_T is now calculated. $I_T = V_T/R_T$, or 450 V/150 Ω = 3 A. This current flows through the entire circuit. All of it flows through the series resistance of R_1. The IR drop for R_1 is $V = I \times R$, or 3 A × 120 Ω, or **360** V. Once this is found we can work back from the source to determine the voltage and current values for each of the other components in the circuit. If there is a 360-V drop across R_1, then 450 V − 360 V, or 90 V, is applied across the balance of the circuit. The IR drop across R_4 is 90 V and the current through R_4 is 90 V/60 Ω, or 1.5 A. Using the rule about current dividing at a parallel junction, we find the current through R_2 to be 3 A − 1.5 A, or 1.5 A. The IR drop across

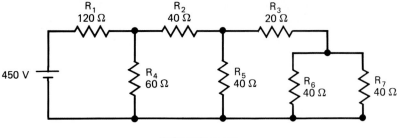

FIGURE 8-24

R_2 is 1.5 A × 40 Ω, or 60 V. The IR drop across R_5 is 90 V − 60 V, or 30 V. Current through R_5 is 30 V/40 Ω, or 0.75 A. This leaves a current of 0.75 A for the balance of the circuit. The IR drop across R_3 is 0.75 A × 20 Ω or 15 V. Since R_6 and R_7 are in parallel, they have the same IR drop of 30 V − 15 V, or 15 V. Both resistances have the same value; therefore, the current divides equally through them; 0.75/2 = 0.375 A for each branch resistance. The IR drop across this bank is 0.375 × 40, or 15 V.

The answers for each component in this circuit are provided in the summary shown below.

$$R_T = 150\,\Omega \qquad I_{R2} = 1.5\,A \qquad I_{R6} = 0.375\,A$$

$$I_T = 3\,A \qquad V_{R2} = 60\,V \qquad V_{R6} = 15\,V$$

$$I_{R1} = 3\,A \qquad I_{R5} = 0.75\,A \qquad I_{R7} = 0.375\,A$$

$$V_{R1} = 360\,V \qquad V_{R3} = 30\,V \qquad V_{R7} = 15\,V$$

$$I_{R4} = 1.5\ A \qquad I_{R3} = 0.75\ A$$

$$V_{R4} = 90\ V \qquad V_{R3} = 15\ V$$

REVIEW PROBLEMS

Use Figure 8-23 to answer problems 8-20 and 8-21.

8-20. What current flows through R_1?
8-21. What is the current through R_2?

Use Figure 8-22 to answer problems 8-22 and 8-23.

8-22. Which resistance is in series with R_2?
8-23. Which resistance is in parallel with R_2?

Power in the Series–Parallel Circuit When it is necessary to calculate the power requirements in a series–parallel circuit, it is helpful to recall that the rules for power in each type of circuit are identical. Power in a series–parallel circuit is determined by use of Watt's law for power. Once the power values for each component are found they can be added together for a total power consumption figure. The power consumption in the circuit shown in Figure 8-23 is calculated in this manner:

$$P_T = V_T I_T = 45 \text{ V} \times 1.25 \text{ A} = 56.25 \text{ W}$$

The power dissipation for R_3 is

$$P_{R3} = \frac{V^2}{R} = \frac{2025 \text{ V}}{60 \text{ }\Omega} = 33.75 \text{ W}$$

The power dissipation for R_1 is

$$P_{R1} = I^2 \times R_1 = 0.25 \text{ A} \times 40 \text{ }\Omega = 10 \text{ W}$$

The power dissipation for R_2 is

$$P_{R2} = I^2 \times R_2 = 0.25 \text{ A} \times 50 \text{ }\Omega = 12.5 \text{ W}$$

These three values add to equal the total circuit power of 56.25 W.

The Balanced Bridge Circuit One application of the series–parallel circuit configuration is used in measuring devices. It is called the balanced bridge. An application of this circuit is known by the name of Wheatstone's bridge. A schematic diagram for the bridge circuit is shown in Figure 8-25. It consists of two series strings. The two series strings are connected in parallel with each other. A current flow through each branch establishes an IR drop across the two resistances in the branch. This provides a specific value of voltage at points A and B. Any device connected to these points completes the circuit between them. If point A has a higher voltage potential than point B, the current flow will be from B to A. Conversely, if B has a higher voltage potential than A, the current will flow through the component from A to B. When these two points have the same voltage potential, there is no current flow and the circuit is said to be *balanced*.

In a practical application the value of R_1 would be unknown. Resistance R_2 would be a variable resistance. It would also have a convenient means of reading the value of its resistance, such as a calibrated dial marker. The purpose of this circuit is to find the value of the unknown resistance, R_1.

The circuit is completed when an unknown resistance value is inserted at points R_4 on the circuit shown in Figure 8-26. R_2 is adjusted until the meter (G) shows that there is no current flow in any direction through it. Normally, a very sensitive meter called a *galvonometer* is used for this purpose. This meter's indicating needle rests at the center of its scale with no circuit con-

Balanced bridge circuit Parallel circuit having two series strings with the center points of each string equal or balanced.

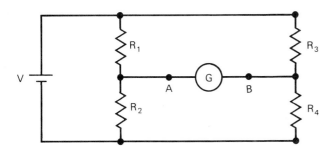

FIGURE 8-25 A bridge circuit has two parallel branches, each having two series-connected resistances. The galvonometer (G) is used to indicate a balance of voltage at each midpoint.

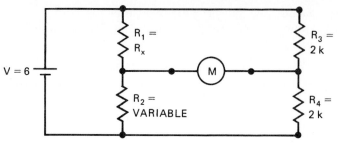

FIGURE 8-26

nections. When current flows, the deflection of the indicator needle shows the direction of current flow. The circuit is balanced by adjusting R_2. When balance occurs, the reading of the dial on R_2 will provide the value of resistance for R_1. The circuit is balanced when the ratio of R_4 and R_2 is equal to the ratio of R_3 and R_4. The values of the two branches do not have to be equal as long as the ratio of their resistances is equal. A formula for this is

$$\frac{R_1}{R_2} = \frac{R_3}{R_4}$$

Another way of showing this is

$$R_1 = R_2\frac{R_3}{R_4}$$

EXAMPLE 8-13

Find the value of the unknown resistance, R_1, when the value of $R_2 = 68\ \Omega$, $R_3 = 100\ \Omega$, and $R_4 = 1000\ \Omega$.

Solution

$$R_1 = R_2\frac{R_3}{R_4}$$

$$= 68\ \Omega \times \frac{100\ \Omega}{1000\ \Omega}$$

$$= \frac{6800\ \Omega}{1000\ \Omega}$$

$$= 6.8\ \Omega$$

A second method of determining the answer is to use the ratio of R_3 to R_4. In this circuit it is 1:10. The value of R_1 is 68 Ω × 1/10, or 6.8 Ω.

REVIEW PROBLEMS

8-24. What is the voltage measured across R_4 in Figure 8-26?

8-7 Voltage Measurements to Common or Ground

The electrical term *ground* refers to a point in the circuit used as a common reference for measurements. This term is also called *circuit common* or just *common* by many persons working with electrical and electronic circuits. All voltage measurements are actually a comparison of the electrical potential between a point of reference and a second point in the circuit. In many electrical circuits the point of reference is called circuit *ground* or circuit *common*. One of the two leads from the voltmeter is connected to this point in the circuit. The values of voltage measured in this manner may be described as either positive or negative in value. The level of pressure is described in units of the *volt*. Thus we may have either a positive or a negative voltage value at any test point in the circuit. The symbol for electrical ground and common is shown in Figure 8-27(a). This figure also illustrates the two basic types of voltage-referenced grounds we use. Part (b) shows the negative terminal of the power source connected to ground. We call this a negative ground system. It is probably the most often used type of circuit. The positive lead is usually called the *hot* or *high* side of the circuit and the negative lead is the *cold* or *low* side of the system. Part (c) shows a positive ground system. This type of circuit is used but not as often as its negative ground counterpart. Both of the circuits can be connected to the same load. The difference in operation is the direction of current flow and the polarity of the voltages in the system.

Many electrical devices use the metal chassis as the ground for the system. An example of this is the automobile. The negative lead from the battery is connected to the motor block, the chassis, and the body sheet metal. This forms an electron path that has a very low resistance for electron flow. In electronic applications the metal chassis of the device is usually the circuit ground or common. Printed circuit boards usually have a portion of the metal foil connected to the negative power source. This is the common or ground for the circuit board. A major reason for using the common or ground is that its use reduces the quantity of wires that are used as a common electron source from the power supply. When one large surface is considered the common source of electrons, there is minimal resistance from that source to the circuits requiring these electrons. The result is a more efficient system that uses fewer wires to make it function properly.

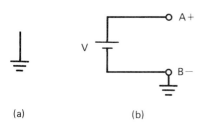

(a)   (b)

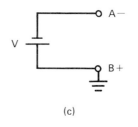

(c)

FIGURE 8-27 (a) Symbol used to indicate circuit common on a schematic diagram. Circuit common may be either negative (b) or positive (c).

Identifying Circuit Ground There are several different ways of drawing a circuit with a common. Three of the more commonly used methods are shown in Figure 8-28. The three circuits are identical in values. The differences occur in showing how the power source is connected. Part (a) shows a complete circuit. Part (b) is the same circuit with the power source voltage value given and the power source symbols not shown. Part (c) uses the ground symbols to indicate that these points are connected to a common reference point. The latter method is most often used on large schematic diagrams.

It is possible to move the circuit reference point to any point in the circuit (Figure 8-29). Part (a) shows a typical system with a negative ground. In part (b) the reference point is the positive terminal of the power source. All voltages in this system are negative when measured from the ground reference point. A third reference point is shown in part (c). The common for this circuit is located between R_1 and R_2. Using this system, we have both positive and negative voltage values. The voltage polarities are also shown here. The voltage drop across R_1 is positive with respect to circuit ground. The voltage drops that develop across the remaining resistors are all negative with respect to circuit ground. This method provides the dual voltage source required for operation of a circuit.

It is easy to become confused when attempting to identify circuit component polarities. The easiest method of keeping these organized is to find the power source reference and to check its polarity. Once this is accomplished it is simple to analyze the circuit. The three circuits shown in Figure 8-29 use the same components. The only difference in the circuits is the point for reference. Each circuit has a source voltage of 20 V and a total of 20 Ω of resistance. The current flow is 20/20, or 1 A, for the circuits. Circuit (a) has a negative ground. All the voltage drops are positive when measured from ground. Circuit (b) has its positive point connected to circuit ground. All voltages are negative when they are measured from this reference point. Circuit (c) has the midpoint of the two resistances connected to circuit ground. Neither of the power supply leads is connected to the common reference point. The two resistances have

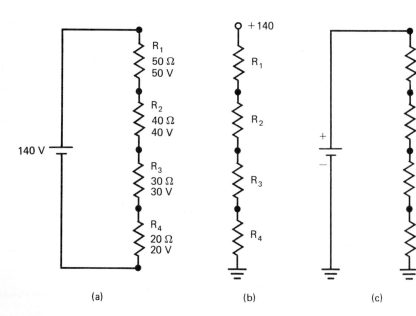

FIGURE 8-28 Three methods of illustrating the same electrical circuit.

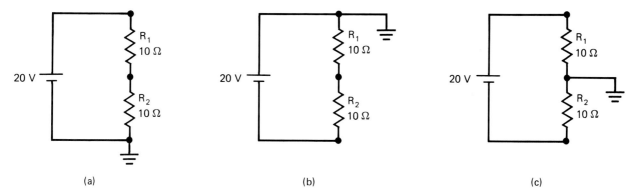

FIGURE 8-29 Circuit common may be at the negative (a), at the positive (b), or in the middle (c) of a group of loads in a circuit.

opposite polarities when their voltage drops are measured from the ground reference point. You may be confused by the interchange of the terms *common*, *common ground*, *common reference*, and *ground reference*. In the world of electrical and electronics work these terms are all used interchangeably to indicate the circuit's reference point for voltage measurements.

REVIEW PROBLEMS

8-27. Calculate the voltage drops for Figure 8-29 (c). Include the polarities of the voltage drops.
8-28. Calculate the current for Figure 8-29 (b) when $R_1 = 100\ \Omega$ and $R_2 = 400\ \Omega$.
8-29. Calculate the voltage drops under the conditions described in problem 8-28. Include polarities of the voltages.

8-8 Troubleshooting Series–Parallel Circuits

The most common problems in series or parallel circuits relate to either an open or a short circuit. Methods of finding these for both the series and the parallel circuit were discussed in Chapters 6 and 7. Some techniques for troubleshooting the combination series–parallel circuit are described in this section.

In any troubleshooting process the first step is to attempt to determine if the system is working properly. Certain tests must be conducted and the results of the tests analyzed. In addition, one can usually identify excess heat problems because of the telltale odor or signs of smoke. In total short-circuit conditions the protective fuse or circuit breaker will shut the circuit down. If we can assume that the circuit is part of a system, the visual signs of indicator lights glowing will tell us that the power is turned on. These, together with data obtained from the schematic diagram and service literature, help in the analysis and location of faults.

There are certain facts related to proper circuit operation that should aid

TROUBLESHOOTING APPLICATIONS

in the diagnosis of faults and their ultimate correction. These all relate to a circuit that is operating correctly.

1. The flow of current through a resistance produces a measurable voltage drop across its leads.
2. The only place one should measure zero volts is at circuit common.
3. The schematic diagram is used as a reference for comparing measured values to operational values.
4. A total short circuit creates an extremely high current flow.
5. A partial short circuit produces an abnormally high current flow.
6. A partial open circuit reduces the normal amount of current flow.

The following examples illustrate these points.

EXAMPLE 8-14

The voltage measured across R_3 is 18 V. Using the circuit in Figure 8-30, indicate if this is a correct value. If not, what is the problem?

Solution

Circuit analysis must be done first to see if this value is correct. The circuit has R_1 in series with the resistive bank of R_2 and R_3. The equivalent value for the R_2–R_3 combination is

$$\frac{1}{R_{EQ}} = \frac{1}{R_2} + \frac{1}{R_3}$$

$$= \frac{1}{5000 \ \Omega} + \frac{1}{10,000 \ \Omega}$$

$$= 0.0002 \ S + 0.0001 \ S$$

$$= 0.0003 \ S$$

$$R_{EQ} = \frac{1}{0.0003 \ S}$$

$$= 3333.3 \ \Omega$$

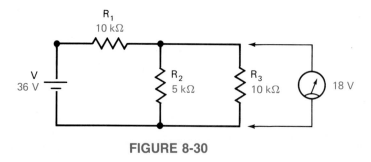

FIGURE 8-30

The total resistance in this circuit is

$$R_T = R_1 + R_{EQ}$$

$$= 10,000 \ \Omega + 3333 \ \Omega$$

$$= 13,333 \ \Omega$$

The current flow is

$$I_T = \frac{V_T}{R_T}$$

$$= \frac{36 \ V}{13,333 \ \Omega}$$

$$= 0.0027 \ A$$

The *IR* drop across R_2 and R_3 *should* be

$$V_{REQ} = 0.0027 \ A \times 3333 \ \Omega$$

$$= 9 \ V \ (approximately)$$

The answer to the question is that the circuit is malfunctioning. The voltage measured across R_3 should be about 9.0 V. A higher than normal voltage at this point indicates that a resistance is open. There is current flow because a voltage drop is measured. This indicates that either R_2 or R_3 is open and the remaining resistance is functioning properly. When one looks at the circuit in order to analyze it, certain facts start to become obvious.

We have determined that either R_2 or R_3 is open. If either was shorted, the total resistance across the voltmeter's leads would be zero and there would be no reading. There are two conditions that can happen: either R_2 or R_3 is open. If R_3 is open, we have a 10-kΩ and a 5-kΩ resistance in series. Any voltage drop across the two resistances has the same ratio as their values. The voltage at this point should be one-third of the 36-V source value, or 12 V. The fact that we have a measurement of 18 V at this point tells us that the probability of R_3 being open is very low.

If R_2 is open, we have two 10-kΩ resistors in series across a 36-V source. The applied source voltage divides equally across the two equal-value resistors. The voltage measured across resistance R_3 is one-half of the 36-V source, or 18 V. This is the value of the measured voltage. The analysis of this circuit indicates that R_2 is open. The next check for the technician is to determine why it is open. Further investigation is needed to determine if the resistance is bad or if the connections to the circuit are bad.

EXAMPLE 8-15

In Figure 8-31 a voltage value of 48 V is measured with the voltmeter. Is the circuit open or shorted? What component, if any, is at fault?

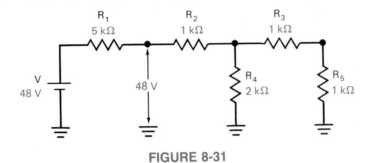

FIGURE 8-31

Solution

There is no voltage drop across resistance R_1. This indicates an open circuit between the point of measure and circuit ground. The suspect components could be any of the other resistances. Before the meter lead is moved to a new test point, let us review the circuit. If R_4 should open, a current path is still left through R_3 and R_5. The voltage at the test point would be less than the 48-V source value. If either R_3 or R_5 would open, there still is a current path through R_4 and, again, the voltage at the test point would be less than the 48-V source. If any of the resistances were to short, there would still be a current flow and some value of voltage drop across the remaining resistances. The only exception to this is if R_1 would short. Then the voltage as measured would be 48 V. Under most operating conditions, resistances do not develop a 100-Ω short. Usually the value of resistance is close to zero ohms.

The only possibility left for a circuit fault is an open resistance. The specific component that failed has to be resistance R_2. It is the only component that, with an open, would produce the voltage measurements shown in this example.

EXAMPLE 8-16

Check the voltmeter readings in Figure 8-32 to see if they are correct. Determine if there are any opens or shorts in the circuit and identify them.

Solution

The first thing to do is to check to see if the two voltmeters are producing correct readings. Meter M_1 is connected between the 5-kΩ R_3 and circuit common. It is in a series string consisting of R_1, R_2, and R_3. The total resistance in this string is 24 kΩ. The total current is calculated as

$$I_T = \frac{V_T}{R_T}$$

$$= \frac{48 \text{ V}}{24,000 \text{ } \Omega}$$

$$= 0.002 \text{ A}$$

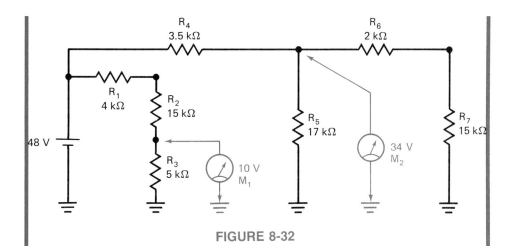

FIGURE 8-32

The *IR* drop across R_3 is

$$V_{R3} = I_{R3} \times R_3$$

$$= 0.002 \text{ A} \times 5000 \text{ }\Omega$$

$$= 10 \text{ V}$$

This value is correct as measured. There is no problem with this part of the circuit. If either R_1 or R_2 were open, the meter would indicate a 0-V reading. If R_3 were open, there would be no current flow and the meter reading would be 48 V.

Another way to check the voltage at the test point is to attempt to approximate the expected answer. Resistance R_3 is $\frac{5}{24}$ of the total resistance value. This is approximately 20% of the total resistance (actually close to 21%). Twenty percent of the total resistance should establish an *IR* drop of 20% of applied voltage. In this circuit 20% of the 48-V source is 9.6 V. The measured value of 10 V, as shown in the circuit, is acceptable. This type of calculation can often be a mental exercise. It will provide the person analyzing the circuit with a rapid and reasonably accurate check for proper circuit voltage values.

The second part of this circuit is a series–parallel configuration. The meter in this circuit indicates a value of 35 V. Let us determine if this value is correct. One portion of the total circuit is a series–parallel circuit. In addition, one of the branches in the parallel section contains a series string consisting of R_6 and R_7. When these resistances are combined, they produce an equivalent value of 17 kΩ. This value is parallel with resistance R_5. It, too, has a value of 17 kΩ. The equivalent value of the two equal resistance parallel branches is

$$R_T = \frac{R}{n}$$

$$= \frac{17,000 \text{ }\Omega}{2}$$

$$= 8500 \text{ }\Omega$$

Total resistance in this section of the circuit is determined when the value of R_4 is added to the equivalent value of the branch resistances.

$$R_T = R_4 + R_{EQ}$$
$$= 3500 \ \Omega + 8500 \ \Omega$$
$$= 12,000 \ \Omega$$

The total current in this section of the system is

$$I_T = \frac{V_T}{R_T}$$
$$= \frac{48 \ V}{12,000 \ \Omega}$$
$$= 0.004 \ A$$

The *IR* drop that should be at this point is

$$V_{REQ} = I_T \times R_{EQ}$$
$$= 0.004 \ A \times 8500 \ \Omega$$
$$= 34 \ V$$

There is a problem in this circuit because the measured voltage is higher than it is supposed to be. A higher-than-normal voltage measurement indicates that the resistance in the circuit is higher than normal. When the resistance of the two branches is equal, the fault could be in either of the branches. One additional voltage measurement will tell us which of the branches has the wrong value of resistance. The junction of R_6 and R_7 indicates a voltage of 35 V. This, by calculation, is incorrect. The proper voltage value for this point is

$$V_{R7} = I_{R7} \times R_7$$
$$= 0.002 \ A \times 15,000 \ \Omega$$
$$= 30 \ V$$

The problem, then, is determined when one finds the *IR* drop across R_7 if R_5 is not included in the circuit.

$$R_T = R_4 + R_6 + R_7$$
$$= 3.5 \ k\Omega + 2 \ k\Omega + 15 \ k\Omega$$
$$= 20.5 \ k\Omega$$

$$I_T = \frac{V_T}{R_T}$$
$$= \frac{48 \ V}{20.5 \ k\Omega}$$
$$= 0.00234 \ A$$

$$V_{R7} = I_{R7} \times R_7$$

$$= 0.00234 \text{ A} \times 15,000 \text{ }\Omega$$

$$= 35 \text{ V}$$

Using this method, the problem is identified as an open resistance at R_7.

---

1. A series–parallel circuit combines both series paths and parallel paths.
2. Determination of total resistance requires the application of both series and parallel circuit laws, then adding the answers.
3. Total current is found when total voltage is divided by total resistance.
4. Circuit ground is a point for common reference in the circuit.
5. Voltage measurements are usually made between circuit ground and a point in the circuit.
6. Circuit ground is considered as a zero voltage point.
7. Voltage values may be either positive or negative with respect to circuit ground.
8. Circuit faults are either opens or shorts.
9. Opens decrease the normal current flow and are a higher-than-normal resistance.
10. Shorts increase the normal current flow and are a lower-than-normal resistance.
11. Resistances usually increase in value when they become defective.

Answer true or false; change all false statements into true statements.   **SELF TEST**

8–1. Total resistance in a parallel circuit is always greater than any single value of resistance in that circuit.
8–2. Total resistance in a parallel branch is greater than any one resistor's value.
8–3. Current flow in the parallel portion of a series parallel circuit is equal to the current flow in the series portion of the circuit.
8–4. The sum of all the *IR* drops in a series–parallel circuit will equal the source voltage.
8–5. Current through the series resistances in a series–parallel circuit is equal to the current flow in one branch of the circuit.
8–6. An open circuit in the series portion of a series–parallel circuit stops all current flow.
8–7. An open circuit in one branch of a series–parallel circuit will produce an increase in the level of voltage across the parallel bank.

8–8. A short circuit in one branch of a series–parallel circuit increases the *IR* drop across the series components in the circuit.

8–9. Defective resistive circuits can be identified by mathematical analysis of the circuit.

8–10. The terms *circuit ground* and *circuit common* refer to a similar point for making measurements.

MULTIPLE CHOICE

8–11. In Figure 8-4:
  (a) R_4 and R_5 are wired in series
  (b) R_2 and R_5 are wired in parallel
  (c) R_4 and R_5 are wired in parallel
  (d) R_1 and R_4 are connected in series

8–12. In Figure 8-4, if all resistances are 100 Ω, the total resistance is
  (a) 100 Ω
  (b) 200 Ω
  (c) 350 Ω
  (d) 500 Ω

8–13. In Figure 8-4, if the voltage across R_4 is 100 V:
  (a) the voltage across R_3 is 100 V
  (b) the voltage across R_5 is 100 V
  (c) all voltages are 100 V
  (d) the source voltage is 400 V

8–14. In a series string with three resistances, the
  (a) current is highest in the largest resistance
  (b) the voltage is equal throughout the circuit
  (c) biggest *IR* drop occurs across the smallest value
  (d) biggest *IR* drop occurs across the largest value

8–15. In Figure 8-11, the current flow is
  (a) largest through R_1
  (b) largest through R_2
  (c) largest through R_3
  (d) equal throughout the circuit

8–16. In Figure 8-11, the *IR* drop is
  (a) equal across R_2 and R_3
  (b) equal across both branches
  (c) larger across R_3 than R_1
  (d) larger across R_2 than R_3

8–17. In Figure 8-12, the current
  (a) through R_1 equals the current through R_2
  (b) is greatest through R_3
  (c) is less at the source
  (d) is greatest through R_1

8–18. In Figure 8-16 the
  (a) resistance is equal in both branches
  (b) current through string 2 is less than current through string 1
  (c) current through string 1 is less than current through string 2
  (d) current is the same in both strings

8–19. In Figure 8-18 the
    (a) current through R_1 is higher than current through R_2
    (b) current through R_2 is higher than current through R_1
    (c) current through R_3 is higher than current through R_1
    (d) current through R_3 is less than current through R_1

8–20. In the balanced bridge circuit (Figure 8-25):
    (a) the current is equal in both branches
    (b) the voltages applied to both branches are equal
    (c) the voltage at the middle of the strings is not critical
    (d) when V_A is greater than V_B, the bridge is balanced

Sketch circuit diagrams for questions 8-21 through 8-25.

8–21. R_1 in series with the parallel combination of R_2, R_3 and R_4.

8–22. R_1 in parallel with the series combination of R_2, R_3, and R_4.

8–23. R_1 in parallel with a branch containing R_2 in series with a parallel combination of R_3, R_4, and R_5.

8–24. A parallel combination of two branches, one containing two series resistances and the other containing three series resistances.

8–25. A series combination of four parallel circuits, each circuit containing three resistances.

8–26. Explain why the voltage drops across all loads in a parallel section of a combination circuit are equal.

8–27. Explain why the sum of each voltage drop in the series section of a combination circuit is equal to the total voltage applied to that portion of the circuit.

8–28. Explain why one may have different values of current through the branches of the parallel section of a combination circuit.

8–29. Explain why the currents through the branch section loads are equal to the current through the series section of a combination circuit.

8–30. Explain why the addition of an additional resistance in the branch section of a combination circuit will increase the total circuit current.

8–31. A parallel resistive circuit has two resistances; one is 2 kΩ and the total resistance is 666.7Ω. What is the ohmic value of the second resistance? (*Sec. 8–4*)

Use Figure 8-3 to answer problems 8-32 and 8-33.

8–32. How could you measure the *IR* drop across R_4 without connecting a meter to circuit common? (*Sec. 8–7*)

8–33. How could you determine the *IR* drop across R_4 when one meter lead is connected to circuit ground? (*Sec. 8–7*)

8–34. In the bridge circuit shown in Figure 8-26, the voltage at $A = 3.8$ V and at $B = 4.2$ V. What is the *IR* drop across the meter? (*Sec. 8–6*)

8–35. In problem 8-34, how much current will flow through the meter M when its resistance is 500 Ω? (*Sec. 8–6*)

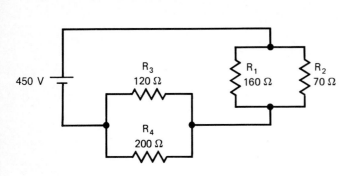

FIGURE 8-33

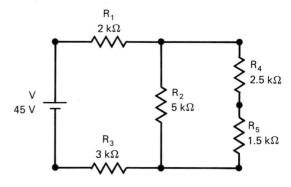

FIGURE 8-34

Use Figure 8-33 to answer problems 8-36 through 8-38.

8–36. Find total I and R. (*Sec. 8–2*)

8–37. Find the IR drop across each resistor. (*Sec. 8–2*)

8–38. Find the current through R_1 and R_3. (*Sec. 8–2*)

Use Figure 8–34 to answer problems 8-39 through 8-41.

8–39. Find total I and R. (*Sec. 8–2*).

8–40. Find the current through each component. (*Sec. 8-2*)

8–41. Find the IR drop across each component. (*Sec. 8-2*)

Use Figure 8-35 to answer problems 8-42 and 8-43.

8–42. Determine R_T. (*Sec. 8-2*)

8–43. Calculate the current through each component. (*Sec. 8-6*)

8–44. In Figure 8-36 what size resistance is required in order to have 16 mA flow in the circuit? (*Sec. 8-6*)

8–45. In Figure 8-37 find values of V, I, and R for all components, including the power source. (*Sec. 8-6*)

FIGURE 8-35

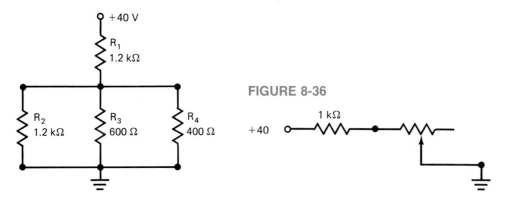

FIGURE 8-36

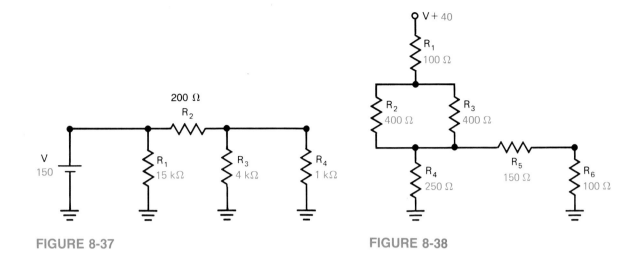

FIGURE 8-37   FIGURE 8-38

Use Figure 8-38 to answer problems 8-46 through 8-49.

8–46. Find R_T. (*Sec. 8-6*)

8–47. Find the *IR* drops across each resistance. (*Sec. 8-6*)

8–48. Find I_T if R_4 is open. (*Sec. 8-6*)

8–49. Find I_T if R_1 is shorted. (*Sec. 8-6*)

Use Figure 8-39 to answer problems 8-50 through 8-54.

8–50. $R_1 = 150\ \Omega$, $R_2 = 200\ \Omega$, $R_3 = 600\ \Omega$, $R_4 = 100\ \Omega$, $R_5 = 200\ \Omega$, and $I_T = 3$ A. Find all *IR* drops and the source voltage. (*Sec. 8-6*)

8-51. Find the power ratings for each resistance in problem 8-50. (*Sec. 8-6*)

8-52. $V = 600$ V, $V_{R4} = 250$ V, $V_{R3} = 200$ V, $I_{R1} = 1$ A, $R_5 = 250\ \Omega$, and $I_T = 1.5$ A. Find R_T and P_T. (*Sec. 8-6*)

8-53. Find R_1, V_{R2}, I_{R3}, I_{R4}, R_4, and I_{R5} for problem 8-52. (*Sec. 8-6*)

8–54. Find V_{R4} when the branch of R_1, R_2, and R_3 is shorted. (*Sec. 8-6*)

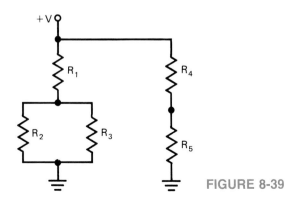

FIGURE 8-39

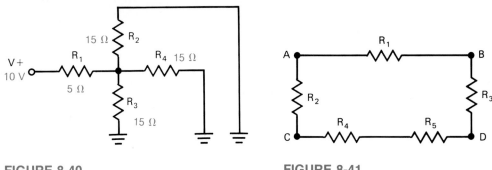

FIGURE 8-40   FIGURE 8-41

Use Figure 8-40 to answer problems 8-55 through 8-59.

8–55. Calculate all *IR* drops in this circuit. (*Sec. 8–6*)

8–56. Calculate I_{R2} with R_4 open. (*Sec. 8–6*)

8–57. Calculate current through R_2 with R_4 shorted. (*Sec. 8–6*)

8–58. Calculate I_{R4} with R_2 open. (*Sec. 8–6*)

8–59. Calculate I_{R4} with R_3 open (*Sec. 8–6*)

Use Figure 8-41 to answer problems 8-60 through 8-62.

8–60. Calculate R_T across terminals *A–B* when all resistances equal 20 Ω. (*Sec. 8–6*)

8–61. Calculate R_T across *C–D* when all resistances equal 10 Ω. (*Sec. 8–6*)

8–62. Show how to connect five 470-Ω and one 390-Ω resistors to obtain an equivalent resistance of about 75 Ω. (*Sec. 8–4*)

Use Figure 8-42 to answer problems 8-63 and 8-64.

8–63. Show the voltages and their polarities when (a) *X* is grounded; (b) *Y* is grounded; (c) *Z* is grounded. (*Sec. 8–7*)

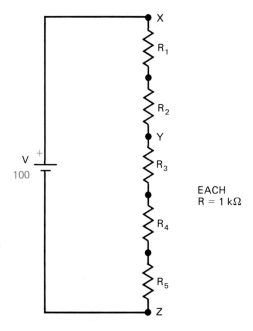

FIGURE 8-42

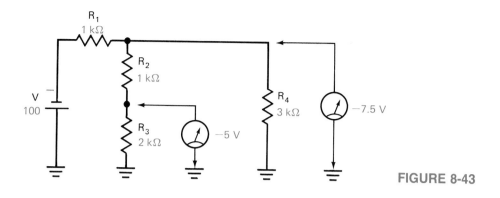

FIGURE 8-43

8–64. Repeat problem 8-63 when each resistance value is changed to 2.5 kΩ. (*Sec. 8–7*)

8–65. Locate the fault(s) in the circuit shown in Figure 8-43.

Use Figure 8-44 to answer problems 8-66 through 8-72.

8–66. If R_1 opens, what is the voltage at A?

8–67. If R_5 opens, what is the voltage at C?

8–68. If R_3 shorts, what is the voltage at B?

8–69. If R_4 opens, what is the voltage at B?

8–70. If R_3 increases to 2.5 kΩ, what is the voltage at C?

8–71. A reading of 0 V at point B indicates what kind of trouble?

8–72. A reading of 15 V at point C indicates what kind of trouble?

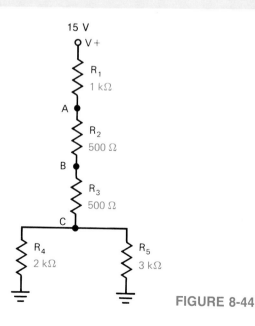

FIGURE 8-44

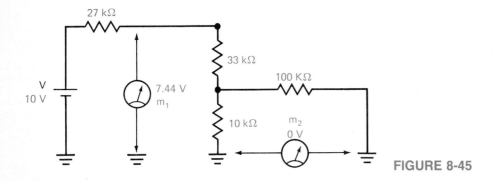

FIGURE 8-45

Use Figure 8-45 to answer problems 8-73 and 8-74.

8–73. Is the voltage reading for M_2 correct?

8–74. Is the voltage reading for M_1 correct?

ANSWERS TO REVIEW PROBLEMS

8–1. R_3 and R_6

8–2. R_1 and R_2, R_4 and R_5

8–3. R_5

8–4. R_1 and R_2, R_3 and R_4

8–5. 1 kΩ

8–6. 500 Ω

8–7. 480 Ω

8–8. 2.0 A

8–9. 0.05 A

8–10. 0.48 A

8–11. (a) parallel; (b) series; (c) parallel; (d) parallel;

8–12. 250 Ω

8–13. 100 Ω

8–14. String 1 = 1.3 A, string 2 = 0.67 A

8–15. String 1 = 0.03 A, string 2 = 0.01 A

8–16. $V_{R1} = 15$ V, $V_{R2} = 6.0$ V, $V_{R3} = 27$ V, $V_{R4} = 36$ V, $V_{R5} = 12$ V

8–17. 1 kΩ

8–18. $I_T = 120$ mA, $I_{R1} = 60$ mA, $I_{R2} = 60$ mA

8–19. $V_{R3} = 90$ V, $V_{R1} = 30$ V, $V_{R2} = 30$ V

8–20. 0.5 A

8–21. 0.5 A

8–22. R_1

8–23. R_3

8–24. 3.0 V

8–25. 1 kΩ

8–26. None

8–27. $V_{R1} = +10$ V, $V_{R2} = -10$ V

8–28. 0.04 A

8–29. $V_{R1} = -4.0$ V, $V_{R2} = -16$ V

CHAPTER 9

Voltage and Current Dividers

Division of voltage and current values in a circuit is commonly accomplished. The purpose of doing this is to present the proper values to the specific sub-circuit in the system. In addition, it is important to recognize the effects of adding additional components to an existing circuit. This often changes the original circuit values and modifies the operation of the circuit. This may be accomplished by the process of attempting to measure a value of voltage in a circuit. Design of the proper types of divider networks is essential for correct operation of all electrical and electronic circuits.

Bleeder Resistor **Output Voltage**
Circuit Loading **Potentiometer**
Current Divider **Voltage Divider**
Loaded Voltage Divider

OBJECTIVES

Upon completion of the material in this chapter, you will
1. Understand the concept of a series voltage-divider network.
2. Be able to analyze a circuit for voltage drops.
3. Be able to analyze a circuit for current division.
4. Understand the relationship of voltage and current in divider circuits.
5. Be able to explain the loaded voltage divider.
6. Be able to troubleshoot divider networks.

Voltage divider *Voltage values from source are divided and fractions of the total are used in individual sections.*

Current divider *Circuit in which current divides or takes separate paths.*

As we continue our study of electron flow, an understanding of some additional circuit concepts is required. Two of these concepts are **voltage dividers** and **current dividers**. The material related to series circuits presented the concept of a voltage drop, or *IR* drop, across components in the circuit. Each series circuit is in essence a voltage-dividing network. The current flow through the circuit creates a difference in potential across each of the components in the circuit. The value of the voltage drop is directly proportional to the ohmic value of each resistance in the circuit. The characteristic is typical for every series-wired circuit. The voltage drops will add to equal the value of the source voltage.

It is also possible to have circuits with current dividing subsystems. This factor is typical for a parallel-wired bank of resistances. The total circuit current divides through each of the branches in the bank. The current quantity for each bank is inversely proportional to the ohmic value for that bank. In other words, the largest resistance value has the least amount of current. The sum of the currents in each of the branches is equal to the current flow from the source in this type of circuit.

Voltage-divider and current-divider circuits are used extensively in electrical and electronic equipment. Quite often it is necessary to provide one or more values of operating voltage from a common source of power. This is accomplished by use of a voltage-dividing network. These networks have certain characteristics that permit them to be designed on paper before they are built. Current-division circuits are also necessary in a wide variety of circuits. These two concepts are discussed in this chapter. In addition to learning how they work, you will learn how to design them and how to troubleshoot some basic circuits.

9-1 Series Voltage Dividers

The current is the same in all parts of a series circuit. It becomes very simple to calculate the voltage drop for each of the components in these circuits. The voltage drop across each component is directly proportional to the ohmic value of the component. The highest resistance value has the largest voltage drop. The smallest resistance value has the smallest voltage drop. When two or more resistances have equal values, the voltage drop across them is also equal. One way of showing this is to consider the voltage drop as being proportional to the total resistance. Using the formula

$$V = \frac{R}{R_T} \times V_T \qquad (9\text{-}1)$$

will determine the voltage drop across each component in the system. This formula is called the *general voltage-divider formula*. It is stated: The voltage drop across any resistor or combination of resistors in a series circuit is equal to the ratio of that amount of resistance to the total resistance, multiplied by the applied source voltage value.

Another method of determining voltage drops is to use Ohm's law:

$$V = I \times R$$

When the total resistance is known, the total current can be determined. The total current I_T and the current through each component are the same value in a series circuit. This value can be used to find the voltage drop, or *IR* drop, for each component:

$$V_{R1} = I_T \times R_1$$

$$V_{R2} = I_T \times R_2, \qquad \text{etc.}$$

EXAMPLE 9-1

Four 100-Ω resistances are series connected to a 40-V source. Find the *IR* drop across each component.

Solution

The voltage drop is 10 V for each of the resistances. This is true because the 40 V is divided equally across the four equal-value resistances. This is proven by use of the formula

$$V = \frac{R}{R_T} \times V_T$$

$$= \frac{100\Omega}{400\ \Omega} \times 40\ \text{V}$$

$$= \frac{1}{4} \times 40\ \text{V} = 10\ \text{V}$$

The circuit shown in Figure 9-1 is typical for a voltage-dividing network. It consists of three resistances of different value. Using the voltage proportional method, it is relatively easy to find the voltage drop across R_2.

$$V_{R2} = \frac{R_2}{R_T} V_T$$

$$= \frac{100\ \Omega}{200\ \Omega} \times 400\ \text{V}$$

$$= \frac{1}{2} \times 400\ \text{V} = 200\ \text{V}$$

We can use this same procedure to prove that the voltage across R_1 is 120 V.

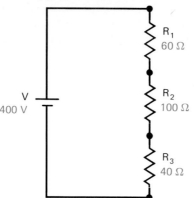

FIGURE 9-1 A typical voltage-divider network consists
of two or more series-connected resistances.

$$V_{R1} = \frac{R_1}{R_T}V_T$$

$$= \frac{60\ \Omega}{200\ \Omega} \times 400\ \text{V}$$

$$= \frac{3\ \Omega}{10\ \Omega} \times 400\ \text{V} = 120\ \text{V}$$

V_{R3} is calculated as being 80 V.

$$V_{R3} = \frac{R_1}{R_T}V_T$$

$$= \frac{40\ \Omega}{200\ \Omega} \times 400\ \text{V}$$

$$= \frac{1}{5} \times 400\ \text{V} = 80\ \text{V}$$

As proof, the sum of the voltage drops across R_1, R_2, and R_3 is equal to 400 V. This value is the same as the source voltage.

Another method of doing this is to find total resistance and total current.

$$R_T = R_1 + R_2 + R_3 = 200\ \Omega$$

$$I_T = \frac{V_T}{R_T} = \frac{400}{200} = 2\ \text{A}$$

The 2-A value of I_T is the same as the values for I_{R1}, I_{R2}, and I_{R3}. The IR drops are calculated:

$$V_1 = I \times R_1 = 2\ \text{A} \times 60\ \Omega = 120\ \text{V}$$

$$V_2 = I \times R_2 = 2\ \text{A} \times 100\ \Omega = 200\ \text{V}$$

$$V_3 = I \times R_3 = 2\ \text{A} \times 40\ \Omega = 80\ \text{V}$$

These are identical to the values calculated by the proportional method. They also add to equal the applied source voltage of 400 V.

Before moving on to other concepts, let us review the facts about the series voltage-divider network. First, when current flows through a circuit, a voltage drop develops across each of the series components. These voltage drops add to equal the value of the applied source voltage. Two methods are used to determine the values of individual voltage drops. One of these methods is the general voltage-divider formula. The second method requires knowing the circuit current. Once this value is known, it is possible to use Ohm's law to find the value of each resistance's voltage drop.

When two resistances are in series it is possible to determine the voltage drop across one of them using either of the methods just described. When the value is found, the voltage drop for the second component may be identified by subtracting the found value from the source voltage value.

An example of this is a series-wired circuit shown in Figure 9-2. It contains a 10-kΩ and a 15-kΩ resistance. The resistances are wired to a 25-V source. The voltage drop across the 10-kΩ resistance is 10 V. The voltage drop across the 15-kΩ resistance may be determined by subtracting 10 V from the 25-V source: 25 − 10 = 15 V. The voltage drop across the 15-kΩ resistance is 15 V in this circuit.

It is also possible to approximate values as an initial step in problem solving. The circuit shown in Figure 9-2 has a ratio of 10 : 15, or 2 : 3. This ratio identifies five parts, two for the 10-kΩ and three for the 15-kΩ resistance. The value of resistance for the 10-kΩ resistance is two-fifths of the total resistance. Since voltage drops are directly proportional to the value of the resistances, the voltage drop across the 10-kΩ resistance is two-fifths of 25 V. This value happens to be 10 V. This method is offered as a means of helping you find out if the calculated answer is close to being correct. When the approximate answer and the calculated answer agree, this helps prove the correctness of your work. This method also helps us find voltage values when they are not provided on the schematic diagram or in the service literature.

REVIEW PROBLEMS

Use Figure 9-1 to find the following values.

9-1. What is total R?
9-2. What percentage of total resistance is R_2?
9-3. What percentage of applied voltage is V_{R2}?

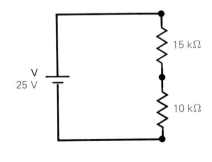

V
25 V

15 kΩ

10 kΩ

FIGURE 9-2 Voltage division in this network is directly proportional to the ratio of the two resistances to the total resistance in the circuit.

The examples that follow illustrate the methods for solving voltage-divider problems.

EXAMPLE 9-2

Find the voltage drop across R_1 and R_2 in Figure 9-3.

Solution

Use the voltage-divider formula:

$$V_{R1} = \frac{R_1}{R_T} V_T$$

$$= \frac{50 \ \Omega}{150 \ \Omega} \times 20 \ \text{V}$$

$$= 0.33 \times 20 \ \text{V} = 6.67 \ \text{V}$$

The voltage drop across R_2 is found by subtraction:

$$V_{R2} = V_T - V_{R1}$$

$$= 20 - 6.67 = 13.33 \ \text{V}$$

We could also use the voltage-divider formula:

$$V_{R2} = \frac{R_2}{R_T} V_T$$

$$= \frac{100 \ \Omega}{150 \ \Omega} \times 20 \ \text{V}$$

$$= 0.67 \times 20 \ \text{V} = 13.33 \ \text{V}$$

In either way, the results are the same, $V_{R2} = 13.33$ V.

EXAMPLE 9-3

Find the voltage drops between the following points in Figure 9-4.
(a) *A* to *B*
(b) *B* to *C*
(c) *A* to *C*
(d) *B* to *D*
(e) *C* to *E*

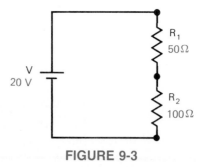

FIGURE 9-3

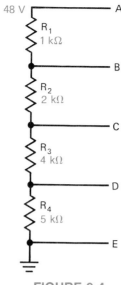

FIGURE 9-4

Solution

The first step is to find R_T.

$$R_T = 1 \text{ k}\Omega + 2 \text{ k}\Omega + 4 \text{ k}\Omega + 5 \text{ k}\Omega = 12 \text{ k}\Omega$$

Next use the voltage-divider formula to determine each voltage. In each problem, the letters A, B, and so on, indicate the two points in the circuit for measurement. The value V_{BC} means the voltage measured between points B and C in the circuit. In this problem this value is 2 kΩ.

(a) $V_{AB} = \dfrac{1 \text{ k}\Omega}{12 \text{ k}\Omega} \times 48 \text{ V} = 4.0 \text{ V}$

(b) $V_{BC} = \dfrac{2 \text{ k}\Omega}{12 \text{ k}\Omega} \times 48 \text{ V} = 8.0 \text{ V}$

(c) $V_{AC} = \dfrac{3 \text{ k}\Omega}{12 \text{ k}\Omega} \times 48 \text{ V} = 12.0 \text{ V}$

(d) $V_{BD} = \dfrac{6 \text{ k}\Omega}{12 \text{ k}\Omega} \times 48 \text{ V} = 24 \text{ V}$

(e) $V_{CE} = \dfrac{9 \text{ k}\Omega}{12 \text{ k}\Omega} \times 48 \text{ V} = 36 \text{ V}$

9-2 The Potentiometer Circuit

One of the devices described in Chapter 4 is the variable resistor. One method of using this device is in a *potentiometer* circuit. This term is derived from early ideas that electricity was a fluid. The terminology describing the use of the

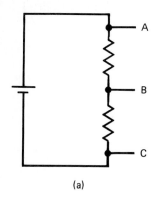

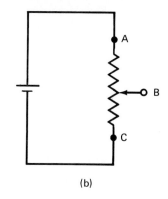

(a)   (b)

FIGURE 9-5 A potentiometer may be used as a variable voltage-divider control in a circuit.

variable resistance comes from the control of fluids. Variable fluid flow control is described as *metering* the fluids. In this application electrical potential is being controlled. Hence the term **potentiometer**, or control of the potential.

The variable resistor [Figure 9-5(a)] is, in essence, two resistors in a series circuit. Each of the resistances are variable in their ohmic value. The specific value for the two resistances depends on the position of the variable arm connected to point B in Figure 9-5(b). In other words, $R_{AB} + R_{BC} = R_T$. Again, since the voltage drops in a series circuit are directly proportional to the resistance value, we can state that $V_{AB} + V_{BC} = V_T$.

The voltage drop between the arm and circuit ground is used to control another circuit (Figure 9-6). Usually, the source voltage is applied to the total resistance value of the potentiometer, or *pot*, as it is called. The position of the arm in relation to circuit ground determines the **output voltage** value. When the arm is close to the top of resistance, the output voltage is equal to the source voltage. Moving the arm to the center of the resistance gives an output voltage of one-half of the source, or 5 V in this example. When the arm is close to the ground terminal, the output voltage (between the arm and ground) is zero. This provides a variable-voltage output.

Applications for Potentiometers

One very common use for the potentiometer is as a volume control in an audio system. The source voltage is changed from dc to the amplified audio signal. This signal is connected to the fixed terminals of the pot. The output is connected between the arm and circuit ground. The movement of the arm between the ground end and the upper or "hot" end of the control determines the amount of signal that is amplified by the system. Another use for the pot is to set a level of operating voltage. Often a fixed value of operating voltage is required at some point in a circuit. This

FIGURE 9-6 The output of a potentiometer circuit is dependent on the position of the moving arm of the control.

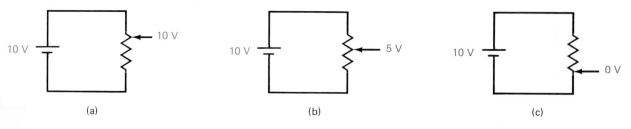

(a)   (b)   (c)

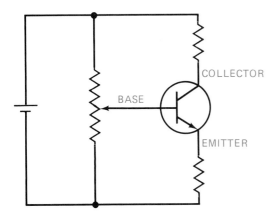

FIGURE 9-7 Use of a potentiometer to "set" or adjust the base-to-common voltage on a transistor.

value is less than the source value and needs to be adjusted. Such a circuit is shown in Figure 9-7, which shows a transistor and a pot. The level of voltage at the base determines the operational characteristics for the transistor. This voltage needs to be adjusted on occasion to correct for changing component values in the circuit. A potentiometer is used to set this voltage level for proper circuit operation.

REVIEW PROBLEMS

9-4. What is meant by the term *voltage divider*?
9-5. What is the difference between *IR drop* and *voltage drop*?
9-6. What is the general formula for *voltage dividers*?

9-3 Current Dividers

A parallel circuit is a current-divider circuit. The circuit shown in Figure 9-8 is an example of this. The total resistance in this circuit is 3.3 Ω. When a source voltage of 20 V is applied, a total current of 6.0 A flows. This current divides and a part of the total flows through each branch in the circuit. The current flow is inversely proportional to the value of resistance. Current through R_1, the larger resistance, is 2.0 A, and the current through R_2 is 4.0 A. When added, these two currents equal the 6.0-A total current flow.

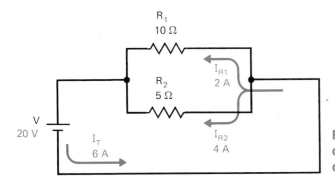

FIGURE 9-8 The combination of R_1 and R_2 form a current-divider network in this circuit.

When attempting to find the currents through one branch of a two-branch system, the current-divider formula is used. This formula is similar to the one used for finding the voltage drop across a series resistance. It compares the resistance of one of the branches to the sum of the resistances in the circuit. This ratio is then multiplied by the total current flow. The formula is

$$I_1 = \frac{R_2}{R_1 + R_2} I_T \qquad (9\text{-}2)$$

The circuit in Figure 9-8 is used to find I_{R1}:

$$I_{R1} = \frac{5\ \Omega}{15\ \Omega} \times 6\ \text{A}$$

$$= 0.33\ \Omega \times 6\ \text{A} = 2\ \text{A}$$

The formula is also used to find the current through the second resistance:

$$I_{R2} = \frac{R_1}{R_1 + R_2} I_T$$

$$= \frac{10\ \Omega}{15\ \Omega} \times 6\ \text{A} = 4\ \text{A}$$

It is not necessary to calculate both of the currents in a two-branch system. The second current is found when I_1 is subtracted from the total current:

$$I_1 = I_T - I_2$$

$$= 6\ \text{A} - 4\ \text{A} = 2\ \text{A}$$

The general current-divider formula is stated by saying that the current through one branch of a parallel circuit equals the total parallel resistance divided by the resistance of that branch and multiplied by the total current with the parallel circuit. This formula will work for any number of parallel resistances in one bank.

EXAMPLE 9-4

Find the current through each branch in Figure 9-9.

Solution

The first step is to calculate total resistance:

$$\frac{1}{R_T} = \frac{1}{R_1} + \frac{1}{R_2} + \frac{1}{R_3}$$

$$= \frac{1}{100\ \Omega} + \frac{1}{150\ \Omega} + \frac{1}{300\ \Omega}$$

$$= 0.01\ \text{S} + 0.0067\ \text{S} + 0.0033\ \text{S}$$

$$= 0.02\ \text{S}$$

$$R_T = \frac{1}{0.02\ \text{S}} = 50\ \Omega$$

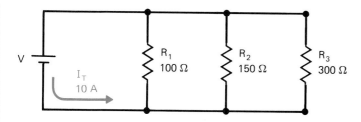

FIGURE 9-9 A parallel branch circuit is a current-dividing network.

The total current is 10 A and the current through each branch is

$$I_1 = \frac{R_T}{R_1} \times 10 \text{ A} = \frac{50 \text{ }\Omega}{100 \text{ }\Omega} \times 10 \text{ A} = 5 \text{ A}$$

$$I_2 = \frac{R_T}{R_2} \times 10 \text{ A} = \frac{50 \text{ }\Omega}{150 \text{ }\Omega} \times 10 \text{ A} = 3.33 \text{ A}$$

$$I_3 = \frac{R_T}{R_3} \times 10 \text{ A} = \frac{50 \text{ }\Omega}{300 \text{ }\Omega} \times 10 \text{ A} = 1.6 \text{ A}$$

If the method of using the reciprocal of the resistance value bothers you, remember that conductance (G), in siemens, is the reciprocal of resistance. Current and resistance are indirectly proportional, which current and conductance are directly proportional to each other. The formula for current using the conductance method is

$$I_1 = \frac{G}{G_T} \times I_T \qquad (9\text{-}3)$$

where G is the conductance of one branch and G_T is the total circuit conductance. Using this formula is simple, as shown in the following example.

EXAMPLE 9-5

Find the current through each branch of the circuit in Figure 9-10 using the conductance formula.

Solution

First find the conductance for each resistance. If you have a calculator with a reciprocal key ($1/x$), all that is necessary is to enter the ohmic

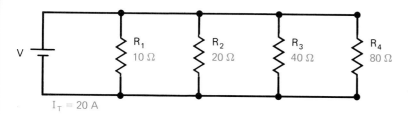

FIGURE 9-10 Use this circuit and the conductance formula to determine individual current values.

value and push this key. If your calculator does not have this feature, divide the ohmic value *into* 1.

$$G_1 = \frac{1}{10 \ \Omega} = 0.1 \qquad G_2 = \frac{1}{20 \ \Omega} = 0.05$$

$$G_3 = \frac{1}{40 \ \Omega} = 0.025 \qquad G_4 = \frac{1}{80 \ \Omega} = 0.0125 \ S$$

When these are added, the total conductance is

$$G_T = 0.1875 \ S$$

The conductance formula may now be used to find individual currents, since we know that the total current flow is 20 A.

$$I_1 = \frac{G_1}{G_T} \times 20 \ A = \frac{0.1}{0.1875 \ S} \times 20 \ A = 10.67 \ A$$

$$I_2 = \frac{G_2}{G_T} \times 20 \ A = \frac{0.05 \ S}{0.1875 \ S} \times 20 \ A = 5.33 \ A$$

$$I_3 = \frac{G_3}{G_T} \times 20 \ A = \frac{0.025 \ S}{0.1875 \ S} \times 20 \ A = 2.67 \ A$$

$$I_4 = \frac{G_4}{G_T} \times 20 \ A = \frac{0.0125 \ S}{0.1875 \ S} \times 20 \ A = 1.33 \ A$$

The sum of these, when rounded, equals 20 A. This formula may be used for circuits with any number of parallel branches. It is usually easier to use than any of the other methods illustrated for finding branch current. The major reason for its ease of use is the direct proportional relationship between conductance and current flow in a parallel bank.

REVIEW PROBLEMS

Use Figure 9-10 to find the following values.

9-7. Ratio of G_2 to G_4
9-8. Ratio of I_3 to I_T

9-4 Loaded Voltage Dividers

The purpose of a voltage-divider network is to provide an operating voltage to a load. This operating voltage is always less than the source voltage. The load is defined as some electrical device. This device requires an operating voltage. The operating voltage produces a current flow through the load. One characteristic of the load is its resistance. When a voltage-divider network is

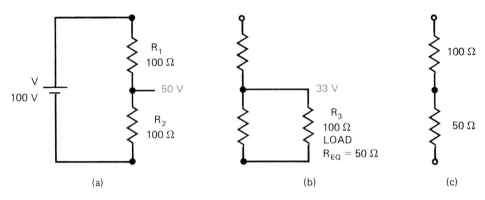

FIGURE 9-11 (a) The voltage divider is unloaded. (b) A load is placed in parallel with the voltage divider. (c) The effective circuit with the load connected in parallel.

developed without a load, the values of resistances determine the voltage drops across each resistance in the series circuit [Figure 9-11(a)]. Two equal resistances of 100 Ω are series connected to a 100-V source. The current flow in this circuit creates a voltage drop of 50 V across each of the two resistances.

In Figure 9-11(b) a load of 100 Ω is connected in parallel with resistance R_2. The result of adding this third resistance creates an equivalent resistance of 50 Ω in the parallel bank consisting of R_2 and R_3. This changes the ratio of resistances in the series portion of the circuit from 1 : 1 to 2 : 1. It also creates a different voltage drop across the two sections of the circuit. The new values are $V_{R1} = 67$ V and $V_{R2} = 33$ V.

TROUBLESHOOTING APPLICATIONS

Loaded voltage divider *Characteristic of voltage-divider circuit with load connected.*

As far as the circuit is concerned, it acts as a two-resistance circuit [Figure 9-11(c)]. The voltage drop across the parallel bank has dropped to 33 V from its original value of 50 V. This lower value may not be adequate for proper circuit operation. It illustrates why one must take voltage measurements in a circuit with all parts connected. To do otherwise usually gives the wrong value—a value that is not correct for operating conditions in the circuit. A situation like this fails to meet the design criteria for the circuit. A method of developing a circuit that will compensate for the load resistance is available. When the voltage divider and the load need to be considered, the combination is identified as a **loaded voltage divider**.

Any resistive device that is added to a circuit can change the values of operating voltage and current at that point. Circuit loading occurs when a working load, such as an electronic amplifier, is connected to its power source. The resistance of the load is placed in parallel with one resistance in the voltage-divider network. This action creates a parallel bank circuit. The total bank resistance and current change due to this addition. The voltage drop across the branch circuit components drops due to the added resistance.

This is also true when testing equipment is connected to any circuit. It is possible to obtain a false reading when the resistance of the test equipment creates an abnormal voltage drop across the components in the circuit being tested. When we use test and measuring equipment having a high internal resistance (on the order of 1 MΩ or higher), this loading effect is minimized. Keep this in mind as you make voltage measurements in the circuits. This factor of **circuit loading** is undesirable and should be kept to a minimum value at all times.

*Circuit loading Operating
conditions of circuit when load
is applied.*

9-5 Loaded Voltage-Divider Design

The loaded voltage divider is a series–parallel circuit. An example of a simple loaded voltage divider is shown in Figure 9-12. In this circuit the source voltage is 150 V. The load requires 60 V at 0.003 A. The ohmic value for the load is based on its voltage and current requirements:

$$R_L = \frac{V_L}{I_L} = \frac{60 \text{ V}}{0.003 \text{ A}} = 20 \text{ k}\Omega$$

The circuit shows R_1 with a value of 10 kΩ. The voltage drop across R_1 is found when the load voltage of 60 V is subtracted from the source voltage.

$$V_{R1} = V_T - V_L = 150 \text{ V} - 60 \text{ V} = 90 \text{ V}$$

This value is used to calculate current through R_1.

$$I_{R1} = V_{R1}/R_1 = 90 \text{ V} \times 10 \text{ k}\Omega = 0.009 \text{ A}$$

When the current through R_1 is known, the current through R_2 is found by subtracting the load current from the total current:

$$I_{R2} = I_T - I_L = 9 \text{ mA} - 3 \text{ mA} = 6 \text{ mA}$$

The value of current through R_2 is used to calculate its resistance value:

$$R_2 = \frac{V_{R2}}{I_{R2}} = \frac{60 \text{ V}}{0.006 \text{ A}} = 10 \text{ k}\Omega$$

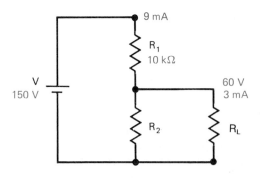

FIGURE 9-12 Use of the vol-tage-divider network to de-sign the correct values for load requirements.

The value of R_2 is 10 kΩ for this circuit. This is proven when the equivalent resistance for R_2 and R_L is found:

$$\frac{1}{R_{EQ}} = \frac{1}{R_2} + \frac{1}{R_L}$$

$$= \frac{1}{10 \text{ kΩ}} + \frac{1}{20 \text{ kΩ}}$$

$$= 0.0001 \text{ S} + 0.00005 \text{ S}$$

$$R_{EQ} = \frac{1}{0.00015 \text{ S}}$$

$$= 6667 \text{ Ω}$$

The equivalent value is then used to prove that the voltage is 60 V at this point:

$$V_{REQ} = I_T \times R_{EQ}$$

$$= 0.009 \text{ A} \times 6666 \text{ Ω} = 60 \text{ V}$$

The Bleeder Resistor Most of the voltages obtained for use in electronic circuits are produced by a power source. The requirements of the source include a voltage-divider network. The network has to form a complete circuit in order to permit current flow and to develop the appropriate levels of operating voltages. A series resistance is needed between the parallel bank formed by the load and the source. This resistance has all the circuit current flowing through it. The resistor, R_1, and the resistor identified as R_L in Figure 9-12 are called **bleeder resistors**. Bleeder resistors are a portion of the loaded voltage divider. The current, voltage, and resistance values for the series resistance must be considered when calculating the values for a loaded voltage-divider network.

Bleeder resistors *Resistors used to stabilize the operating conditions of the circuit.*

Finding Voltage-Divider Resistances The values of both voltage and current must be known as a start on the design of a loaded voltage divider. The circuit shown in Figure 9-13 is used as a reference for this design. The next value to identify is total circuit current. This is found by use of the rule stating that all the circuit current has to flow through any series component. In this circuit R_1 is in series with the parallel resistance bank.

$$I_T = I_{R1} = I_{R3} + I_{R4} + I_{R5}$$

$$= I_{T1} = 0.006 \text{ A} + 0.012 \text{ A} + 0.010 \text{ A}$$

$$= 0.028 \text{ A}$$

When total circuit current is known, the value of the series resistance can be found. The voltage drop across this resistance is equal to the difference in

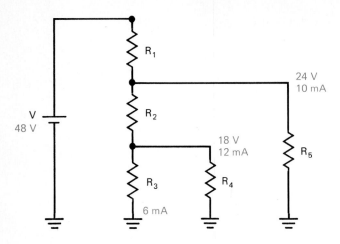

24 V
10 mA

18 V
12 mA

R₅

FIGURE 9-13 Voltage-divider
network resistance values
may be calculated when individual current values and
voltage values are known.

voltages across its terminals. This value is found when 24 V is subtracted from the source voltage.

$$V_{R1} = V_T - V_{R5}$$

$$= 48\ V - 24\ V = 24\ V$$

The ohmic value for R_1 is

$$R_1 = \frac{V_{R1}}{I_{R1}}$$

$$= \frac{24\ V}{0.028\ A} = 857\ \Omega$$

Next find the values for R_3 and R_4.

$$R_3 = \frac{V_{R3}}{I_{R3}}$$

$$= \frac{18\ V}{0.006\ A} = 3\ k\Omega$$

$$R_4 = \frac{V_{R4}}{I_{R4}}$$

$$= \frac{18\ V}{0.012\ A} = 1.5\ k\Omega$$

The value for R_5 is the next to be determined.

$$R_5 = \frac{V_{R5}}{I_{R5}}$$

$$= \frac{24\ V}{0.010\ A} = 2.4\ k\Omega$$

To find the value of R_2 we must first find the voltage drop across its terminals:

$$V_{R2} = 24 - 18 = 6\ V$$

Now we can find the ohmic value for this resistance.

$$R_2 = \frac{V_2}{I_{R2}}$$

$$= \frac{6 \text{ V}}{0.018 \text{ A}} = 333 \ \Omega$$

The values for each resistance in this loaded voltage-divider network are: R_1 = 857 Ω, R_2 = 333 Ω, R_3 = 3 kΩ, R_4 = 1.5 kΩ, and R_5 = 2.4 kΩ.

EXAMPLE 9-6

Find the values for all resistances in the voltage divider in Figure 9-14.

Solution

Find the total circuit current:

$$I_T = I_{R1} = I_{R2} + I_{R3}$$

$$= I_{R1} = 0.25 \text{ A} + 0.5 \text{ A} = 0.75 \text{ A}$$

Find the value for R_1:

$$R_1 = \frac{V_{R1}}{I_{R1}}$$

$$= \frac{35 \text{ V}}{0.75 \text{ A}} = 46.67 \ \Omega$$

Find R_2:

$$R_2 = \frac{V_{R2}}{I_{R2}}$$

$$= \frac{65 \text{ V}}{0.25 \text{ A}} = 260 \ \Omega$$

Find R_3:

$$R_3 = \frac{V_{R3}}{I_{R3}} = \frac{65 \text{ V}}{0.5 \text{ A}} = 130 \ \Omega$$

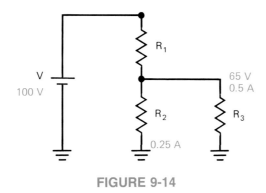

FIGURE 9-14

REVIEW PROBLEMS

Use Figure 9-13 to find the following values.

9-9. How much current flows through R_3 when R_4 is disconnected from the circuit?

9-10. What current flows through R_1 and R_2 when R_4 and R_5 are not connected?

9-11. What voltage is measured across R_3 with all components connected?

9-6 Troubleshooting the Divider Circuit

TROUBLESHOOTING APPLICATIONS

Methods of diagnosing troubles in the voltage and current divider are the same as one uses for combination series–parallel circuits. The initial step is to make some voltage measurements. If these measurements agree with the values provided on the schematic diagram or in the service literature, the circuit is operating properly and there is no need to do any further troubleshooting.

When a voltage is higher than normal, one can expect to find an open component. If the component is not open, the alternative is a higher-than-normal resistance in one of the circuits. Either of these do not load the voltage divider properly. The result is the higher-than-normal value of voltage drop.

If the measured voltage is lower than expected, the cause is a lower-than-normal resistance in the circuit. This could be either a total short or a partial short circuit. In either condition the reduced resistance value produces a low voltage at that point.

Once it is determined if the problem is a short or an open, the troubleshooting can continue as described in the chapters covering series, parallel, and combination series–parallel circuits.

CHAPTER SUMMARY

1. The voltage drops in a series circuit are directly proportional to resistance values.

2. In the series circuit the largest-value resistor has the biggest voltage drop.

3. The current divides in a parallel circuit.

4. The amount of current in each branch is inversely proportional to the ohmic value of the components. The largest current flow occurs through the lowest-value resistance.

5. Current in a parallel circuit's branches is directly proportional to the conductance values in the branches.

6. A loaded voltage divider is used to provide proper levels of operating voltage in a circuit or an electronic device.

7. The value for individual resistances in a loaded voltage divider is found by

calculating (1) total I and (2) individual V values and using these data to find values for each R.

8. Analysis for the voltage-divider network is accomplished by use of Ohm's law and the rules for series and parallel circuit.

Answer true or false; change all false statements into true statements.

9–1. The general voltage-divider rule is used for circuit current analysis.

9–2. The conductance in each branch circuit is directly proportional to the current flow in that branch.

9–3. The current in a parallel branch is directly proportional to total circuit current.

9–4. Current in a parallel branch circuit is inversely proportional to the value of resistance in that branch.

9–5. The potentiometer is a variable resistance.

9–6. The percentage of total formula is used to find the voltage drop in a divider network.

9–7. A resistor used in a voltage divider to maintain current flow is called a bleeder resistor.

9–8. A load represents a resistive device.

9–9. A load does not reduce the divider voltage.

9–10. A short circuit produces a higher-than-normal voltage.

Use Figure 9–15 to answer questions 9–11 through 9–14.

9–11. The current flow through R_2 is
(a) 0.001 A
(b) 0.002 A
(c) 0.003 A
(d) 0.004 A

9–12. The voltage drop across R_2 is
(a) 30 V
(b) 45 V

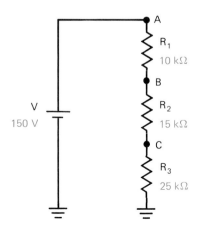

FIGURE 9-15

(c) 60 V
(d) 150 V

9–13. The voltage drop from B to ground is
   (a) 120 V
   (b) 100 V
   (c) 130 V
   (d) 90 V

9–14. The voltage drop from A to C is
   (a) 90 V
   (b) 70 V
   (c) 75 V
   (d) 80 V

Use Figure 9–16 to answer questions 9–15 through 9–20.

9–15. The current flow in the circuit without R_4 is
   (a) 0.001 A
   (b) 0.0015 A
   (c) 0.002 A
   (d) 0.0133 A

9–16. If R_4 is 10 kΩ, the approximate total circuit current is
   (a) 0.010 A
   (b) 0.014 A
   (c) 0.016 A
   (d) 0.008 A

9–17. The voltage drop across R_3 with R_4 in the circuit is
   (a) 5 V
   (b) 48 V
   (c) 69 V
   (d) 54.5 V

9–18. The voltage drop across R_2 with R_4 in the circuit is
   (a) 40 V
   (b) 20 V
   (c) 53 V
   (d) 35 V

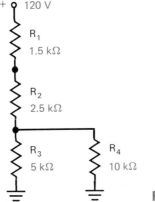

FIGURE 9-16

9–19. The current through R_4 is approximately
   (a) 0.016 A
   (b) 0.010 A
   (c) 0.075 A
   (d) 0.005 A

9–20. Changing the value of R_3 to 10 kΩ will
   (a) maintain original circuit current
   (b) increase circuit current
   (c) decrease circuit current
   (d) decrease VR_3

ESSAY QUESTIONS

9–21. Explain why there is a difference in source current between an unloaded and a loaded voltage divider.

9–22. Explain why there is a difference in source voltage when a loaded voltage divider is used.

9–23. Explain why the term *voltage drop* is used to describe the voltage in a circuit.

9–24. Explain why it is easier to calculate total current using the conductance value method.

9–25. What is the relationship between resistance values and voltage drops in the series circuit?

9–26. What is the relationship between resistance values and circuit branch currents?

9–27. Explain why circuit current total will increase when an additional load is placed in parallel with the voltage-divider network.

9–28. Explain why the voltage drop across the load may decrease as an additional resistance is added in parallel with the load.

Use Figure 9–17 to answer problems 9–29 through 9–32.

PROBLEMS

9–29. Determine the unloaded output voltage and current. (*Sec. 9–1, Sec. 9–2*)

9–30. Determine the output voltage and R_1 current when a load of 6 kΩ is added. (*Sec. 9–4*)

9–31. Determine load current when a 12-kΩ load is added. (*Sec. 9–4*)

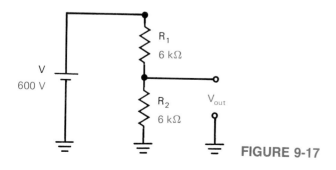

FIGURE 9-17

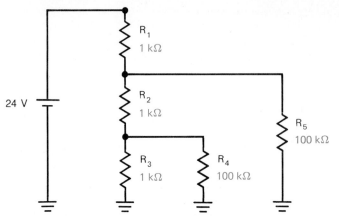

FIGURE 9-18

9–32. Determine load voltage when a bank of three 6-kΩ resistors is added to the output. (*Sec. 9–4*)

Use Figure 9–18 to answer problems 9–33 through 9–38.

9–33. Calculate the total unloaded circuit current. (*Sec. 9–4*)

9–34. Calculate the loaded circuit current. (*Sec. 9–4*)

9–35. Calculate the load voltages with loads connected. (*Sec. 9–4*)

9–36. Calculate the load currents (IR_4 and IR_5). (*Sec. 9–4*)

9–37. What change in voltage occurs when R_5 is removed from the circuit? (*Sec. 9–5*)

9–38. What change in R_5 voltage and current occurs when R_4 is removed from circuit? (*Sec. 9–5*)

9–39. Design a loaded voltage divider that supplies +5 V at 30 mA, +4 V at 15 mA, + 20 V at 10 mA from a 24-V source. The bleeder current is 0.01 A. (*Sec. 9–5*)

9–40. Design a loaded voltage divider that supplies + 10 V at 0.1 A from a 24-V source. The bleeder current is 0.1 A. (*Sec. 9–5*)

9–41. Design an unloaded voltage divider that supplies + 16 V, + 24 V, and + 36 V from a 120-V source. (*Sec. 9–1*)

9–42. Design a loaded voltage divider that supplies + 16 V at 0.02 A, + 10 V at 1.0 A, and + 5 V at 10 A from a 24-V source. The bleeder current is 100 mA. (*Sec. 9–5*)

9–43. A current-divider network has a total of 140 mA. Branch R_1 is 10 kΩ and branch R_2 is 50 kΩ. Find IR_1 and IR_2 in this circuit. (*Sec. 9–5*)

9–44. A parallel bank has a total of 0.08 A, G_1 = 105 S, G_2 = 205 S, and G_3 = 1005 S. Find I_1, I_2, and I_3. (*Sec. 9–3*)

Use Figure 9–19 to answer problems 9–45 through 9–48.

9–45. Find I_1, I_2, and I_3. (*Sec. 9–3*)

9–46. Find V_{R1}, V_{R2}, and V_{R3}. (*Sec. 9–3*)

9–47. Find R_T when R_2 is short circuited to ground.

9–48. What is I_T when R_1 is open?

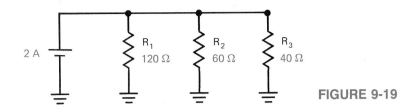

FIGURE 9-19

9–49. Current flow measured in the circuit shown in Figure 9–9 is 6.67 A. What is the problem?

9–50. Current flow through the combination of R_1 and R_2 shown in problem 9–8 is 6.0 A. What, if anything, is the problem?

9–51. Resistance measured in the circuit of Figure 9–10 is about 6.15 Ω. What is the problem?

9–52. What would you expect the voltage across R_3 of Figure 9–10 to be when R_1 is open? Explain your answer.

9–53. What would you expect the voltage across R_3 of Figure 9–10 to be when R_1 is shorted? Explain your answer.

9–54. Voltage measured across R_4 in Figure 9–16 is 85.7 V. What is the problem?

9–55. Voltage at the 'output' terminals of the circuit in Figure 9–17 measures 200V from common to the V_{out} terminal. What is the problem?

9–1. 200 Ω

9–2. 50%

9–3. 50%

9–4. A voltage divider is a circuit in which the applied voltage is proportional to the ohmic values of the resistors.

9–5. None

9–6. $R_1/R_T \times V_T$

9–7. 4 : 1

9–8. 2.6 : 20

9–9. 0.0054 A or 5.4 mA

9–10. 11.5 mA

9–11. 18 V

ANSWERS TO REVIEW PROBLEMS

Chapters 7–9

The major points in this section are reviewed here. This section deals with the basic rules for circuit analysis. This information is essential for all persons using electrical and electronic principles. Learn how to recognize a basic electrical circuit. Also learn how to analyze the circuit for its components of resistance, current flow, and voltage drops.

1. Parallel-wired resistances are connected to the same points in the circuit.
2. The parallel circuit has two or more current paths.
3. Total resistance in the parallel circuit is always less than the lowest single resistor value.
4. Voltage across each branch is equal to the source voltage in the parallel circuit.
5. Currents through each branch add to equal source current.
6. Power in the parallel circuit is equal to the sum of each branch's power use.
7. When one branch of a parallel circuit opens, the total resistance left in the circuit increases. The result is less current flow.
8. When one branch circuit opens, the current still flows through the remaining branches.
9. A short circuit in a parallel branch affects all the branches.
10. An ohmmeter may be used to locate an open or a short in a parallel circuit.
11. A short circuit in one branch will result in no current flow in all the branches in the system.
12. Safety devices are used to protect the system when an overcurrent occurs.

13. A series–parallel circuit combines both series paths and parallel paths.

14. Determination of total resistance requires the application of both series and parallel circuit laws, then adding the answers.

15. Total current is found when total voltage is divided by total resistance.

16. Circuit ground is a point for common reference in the circuit.

17. Voltage measurements are usually made between circuit ground and a point in the circuit.

18. Circuit ground is considered as a zero voltage point.

19. Voltage values may be either positive or negative with respect to circuit ground.

20. Circuit faults are either opens or shorts.

21. Opens decrease the normal current flow and are a higher-than-normal resistance.

22. Shorts increase the normal current flow and are a lower-than-normal resistance.

23. Resistances usually increase in value when they become defective.

24. The voltage drops in a series circuit are directly proportional to resistance values.

25. In the series circuit the largest-value resistor has the biggest voltage drop.

26. The current divides in a parallel circuit.

27. The amount of current in each branch is inversely proportional to the ohmic value of the components. The largest current flow occurs through the lowest-value resistance.

28. Current in a parallel circuit's branches is directly proportional to the conductance values in the branches.

29. A loaded voltage divider is used to provide proper levels of operating voltage in a circuit or an electronic device.

30. The value for individual resistances in a loaded voltage divider is found by calculating (1) total I and (2) individual V values and using these data to find values for each R.

31. Analysis for the voltage-divider network is accomplished by use of Ohm's law and the rules for series and parallel circuits.

CHAPTER 10

Kirchhoff's Laws

Additional analysis of the electrical circuits is often simplified by the use of Kirchhoff's two laws. Each offers a direct relation to the values of current and voltage in a circuit. Application of these laws for circuit analysis and design makes the work much easier. The complexity of a circuit containing more than one power source requires the use of Kirchhoff's rules to make identification of circuit values convenient.

KEY TERMS

Branch Current   **Mesh Current**

Closed Loop   **Node Voltage**

Kirchhoff's Current Law   **Polarity**

Kirchhoff's Voltage Law

OBJECTIVES

Upon completion of the material in this chapter, you will

1. Be able to apply the closed-loop rules.
2. Be able to assign polarities to components in the closed-loop circuit.
3. Be able to apply branch current rules.
4. Be able to analyze branch circuits.
5. Be able to apply node and mesh analysis.

Our studies to this point have concentrated on the use of Ohm's law for circuit analysis. There are many situations where the use of Ohm's law is not practical or even possible. In some of the illustrations used in this book in previous chapters, problems could have been solved using methods that are easier to use than those required by Ohm's law. These additional methods will help you in the ability to solve circuit problems and develop new circuits or values for existing circuits.

One method of analyzing electrical circuits was developed by a German scientist, Gustav Kirchhoff. This method is related to circuit voltages and currents. In fact, Kirchhoff developed rules for both voltage and current. These rules, or laws, are named in his honor. They are known as *Kirchhoff's voltage law* and *Kirchhoff's current law*.

These laws are based on a common theory. This theory assigns polarities to each component in an electrical circuit. It also assigns polarities to each junction in any electrical circuit. **Kirchhoff's current law** uses this method of polarity assignment. The law states that the *current entering a junction is equal to the current leaving a junction*. This is true for all electrical junctions. We must first assign a polarity to each current at the junction. Kirchhoff identified currents entering the junction as *negative*. He identified all currents leaving the junction as *positive* for their polarity (Figure 10-1). The illustration has two currents flowing into the junction. These currents are negative. Their values are −5 A and −3 A, for a total of −8 A. There are three currents leaving this junction. Their values are +1 A, +5 A, and 2 A. Their total is also +8 A. When all of these currents are added, their total value is zero:

$$I_1 + I_2 + I_3 + I_4 + I_5 = 0 \qquad (10\text{-}1)$$

$$-3\,A - 5\,A + 2\,A + 5\,A + 1\,A = 0\,A$$

Figure 10-2 has one current entering the junction and three currents leaving the same junction. When these currents are added, their sum is also zero:

$$I_1 + I_2 + I_3 + I_4 = 0$$

$$-6\,A + 2\,A + 1\,A + 3\,A = 0$$

Kirchhoff's current law *Rule related to conditions of current flow in a circuit.*

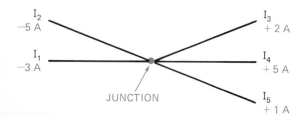

FIGURE 10-1 Currents entering a junction are designated as negative, and currents leaving the same junction are designated as positive.

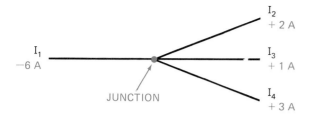

FIGURE 10-2 Some circuits have a single current entering the junction and multiple current paths leaving the same junction.

This law does not indicate that the current flow is zero. It does illustrate a simple but necessary rule. This rule states that the current entering a junction is equal to the current leaving the same junction. This is not a new rule. It is a variation of the statement of Kirchhoff's current law. One cannot have current "left over" in an electrical circuit. All current flow in any circuit is based on the quantity of applied voltage and the resistance in the circuit.

EXAMPLE 10-1

A third way of looking at this rule for circuit current is illustrated in Figure 10-3. This is a combination series–parallel circuit. There are four junctions identified in this circuit. They are points A, B, C, and D. Current from the power source first enters junction D. It has a value of -6 A. This current leaves the same junction as $+6$ A. The current at junction D is -6 A $+ 6$ A $= 0$ A. At junction A the current is -6 A. It leaves the junction and has two paths. One path is through R_2 and the other path is through R_3. The analysis of this junction is $I_1 + I_2 + I_3 = 0$, or -6 A $+ 4$ A $+ 2$ A $= 0$ A. The two currents meet at junction B. Here $I_2 + I_3 + I_4 = 0$, or $+4$ A $+ 2$ A $- 6$ A $= 0$ A. The current flows through R_1 and at junction C the same condition exists. $I_4 + I_5 = 0$. Here, too, -6 A $+ 6$ A $= 0$. Each junction has a total of 6 A and the total circuit current is also 6 A.

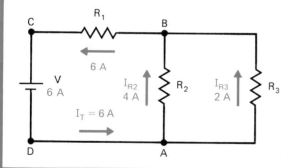

FIGURE 10-3 Each junction must be considered when a combination series–parallel circuit is analyzed.

REVIEW PROBLEMS

10-1. What is the output current from a junction when the input currents are $I_1 = 2$ A, $I_2 = 0.5$ A, and $I_3 = 4.2$ A?

10-2. A circuit has 4 A flowing into a junction. One output from the junction is 1.5 A; what is the other output current?

Use Figure 10-4 to answer the following questions.

10-3. What quantity of current enters junction *C*?
10-4. What quantity of current enters junction *B*?
10-5. What quantity of current enters junction *A*?
10-6. What quantity of current enters the + terminal of the power source?

Kirchhoff's voltage law *Sum of voltage drops in a closed loop, or series circuit equal zero.*

Polarity *Positive and negative charges on each end of a component, related to connections to power source.*

Closed loop *Term describing a series or single path circuit.*

Kirchhoff's voltage law also uses a **polarity** method. This law is based on the *IR* voltage drops that develop in a series circuit. The law states that *the algebraic sum of the voltage drops in a closed loop equal zero*. The term **closed loop** is another way of identifying a series circuit. Our purposes here may offer more than one closed loop in a total circuit. The number of closed loops in a total circuit is not important. The purpose of this law is to prove that the voltage drops in a closed loop, or series circuit, equal the applied voltage for that loop.

The basic loop circuit shown in Figure 10-5 illustrates how this system works. Each component in the circuit is assigned a polarity. This is done on the basis of the relationship of the end of the component to the power source. Resistance R_1 is typical. Its left side is connected directly to the positive terminal of the source. This end has a (+) polarity. The right side of R_1 is closer to the negative (−) terminal. It has a (−) polarity. Resistance R_2 has its polarities assigned in the same manner. One of its two ends is closer to the positive source than the other end. This end has the (+) sign and the other end receives the (−) sign. The polarities for resistance R_3 are assigned in the same number as those for R_1 and R_2.

Each resistance in this circuit has an identified voltage drop. It is possible to start at any point in this circuit in order to add these voltage drops. It does not matter where you start or whether you go clockwise or counterclockwise around the circuit. When you add all of the − voltage drops and the source voltage the answer will be zero. Start at point *A* and add all voltages. Add the voltages as you go in a clockwise manner.

$$V_{R3} + V_{\text{SOURCE}} + V_{R1} + V_{R2} = 0$$
$$+4 \text{ V} - 25 \text{ V} + 6 \text{ V} + 15 \text{ V} = 0$$

(10-2)

FIGURE 10-4

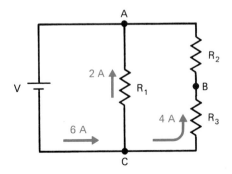

FIGURE 10-5 Current analysis for the closed-loop circuit, showing polarity designations for each component.

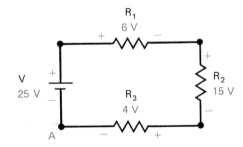

When the direction for adding the voltages is reversed the answer is the same:

$$-15\ \text{V} - 6\ \text{V} + 25\ \text{V} - 4\ \text{V} = 0$$

EXAMPLE 10-2

Find the source voltage in Figure 10-6.

Solution

The source voltage is equal to the sum of the voltage drop in the closed loop. This may be solved using two methods.

Method 1:

$$V_{R1} + V_{R2} + V_{R3} - V_s = 0$$

$$18\ \text{V} + 30\ \text{V} + 24\ \text{V} - V_s = 0$$

$$72\ \text{V} - V_s = 0 \quad \text{or} \quad V_s = 72\ \text{V}$$

Method 2:

$$V_s = V_{R1} + V_{R2} + V_{R3}$$

$$= 18\ \text{V} + 30\ \text{V} + 24\ \text{V} = 72\ \text{V}$$

The circuit illustrated in Figure 10-7 is a combination series–parallel circuit. It has three loops. These are identified as the outside loop, an inside loop with a power source, and an inside loop without a power source. Specifically, the outside loop consists of V_s, R_1, R_2, R_4, and R_5. One inside loop is constructed from V_s, R_1, R_3, and R_5, while the other inside loop contains R_2, R_4, and R_3. The loop equation for each of these is:

Loop 1:

$$V_S + V_{R1} + V_{R2} + V_{R4} + V_{R5} = 0\ \text{V}$$

$$-200\ \text{V} + 40\ \text{V} + 40\ \text{V} + 60\ \text{V} + 60\ \text{V} = 0\ \text{V}$$

FIGURE 10-7 Analysis of a series–parallel circuit requires the identification of each loop in the system.

FIGURE 10-6

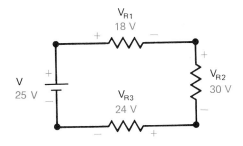

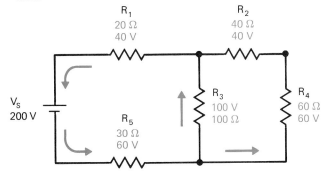

Loop 2:

$$V_S + V_{R1} + V_{R3} + V_{R5} = 0 \text{ V}$$

$$-200 \text{ V} + 40 \text{ V} + 100 \text{ V} + 60 \text{ V} = 0 \text{ V}$$

Loop 3:

$$V_{R2} + V_{R4} + V_{R3} = 0 \text{ V}$$

$$40 \text{ V} + 60 \text{ V} - 100 \text{ V} = 0 \text{ V}$$

It is possible and permissible in any equation to apply the same process to both sides of the equation. When this is done the equation is still correct. This can be done for each of the three equations described above. When the source voltage is added to both sides of the equation representing loop 1, it looks like this:

$$V_S = V_S + V_{R1} + V_{R2} + V_{R4} + V_{R5} + V_S$$

When values are subtracted we have

$$+200 \text{ V} = -200 \text{ V} + 40 \text{ V} + 40 \text{ V} + 60 \text{ V} + 60 \text{ V} + 200 \text{ V}$$

When the equation is simplified and like terms are combined we have

$$+200 \text{ V} = -200 \text{ V} + 400 \text{ V} = 200 \text{ V}$$

Therefore, the sum of the voltage drops across each component in the loop equals the applied voltage from the source.

When the loop does not have a voltage source, the sum of the voltage drop in that loop must equal zero. This is the situation for loop 3. There is no voltage source in that loop. The sum of the voltage drops across each resistance does equal zero.

REVIEW PROBLEMS

10-7. Using the circuit shown in Figure 10-8, establish the voltage polarity for each end of every component: R_{1A}, R_{1B}, R_{2C}, R_{2D}, R_{3E}, R_{3F}, V_G, and V_H.

Use Figure 10-9 to answer problems 10-8 and 10-9.

10-8. If the values of voltage for loop 2 are changed to $V_{R4} = 100$ V and $V_{R5} = 160$ V, what is the voltage drop across R_2?

10-9. Changing the voltages in loop 2 to $V_{R1} = 80$ V, $V_{R5} = 120$ V, and $V_{R6} = 240$ V, what is the source voltage for this loop?

10-1 Solving for Branch Currents

Branch currents *Currents in each path of a branch circuit.*

We often use Kirchhoff's current and voltage laws to solve for currents in each branch of a complex circuit. When the current in each branch is known, it is then possible to find the voltages in the circuit.

Figure 10-10 shows a circuit containing two voltage sources. This circuit

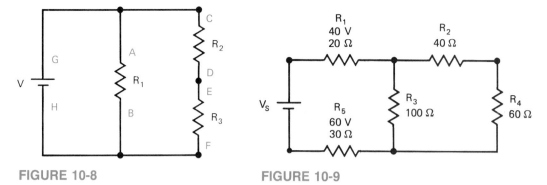

FIGURE 10-8   FIGURE 10-9

has two current loops. One of these consists of V_1, R_1, and R_3. The other consists of V_2, R_2, and R_3. This circuit becomes complex because both currents flow through resistance R_3. There are two closed loops in this circuit. One is identified as loop V_1 and the other as loop V_2. In addition to these two loops there are four junctions in this circuit. These junctions are also called *nodes*. They are identified by the letters A, B, C, and D. The node is simply a junction where two or more current paths join together in a circuit.

There are some general steps that are used to solve for circuit values using the *branch current method*. These are:

1. Assign a current in each loop. The current direction is not important. What is important is recognition of the current loop. Try to follow the electron current path if you are not sure of which direction to go.
2. Use the method described earlier and assign a polarity to each component in the loop. The polarities are shown in Figure 10-10.
3. Use Kirchhoff's voltage law for each loop in the system:

Loop 1: $-V_1 + V_{R1} + V_{R3} = 0$ or
$+ V_1 = V_{R1} + V_{R3}$

The voltage drops for any of the resistances in either of these loops are not known. Substitute IR for each voltage drop. This is possible since $V = IR$.

$$V_{R1} = I_1 R_1 = I_1 \times 100 = 100 I_1$$

$$V_{R2} = I_2 R_2 = I_2 \times 80 = 80 I_2$$

$$V_{R3} = I_3 (R_1 + R_2) R_3 = I(I_1 + I_2) \times 40$$

$$= 40(I_1 + I_2) = 40 I_1 + 40 I_2$$

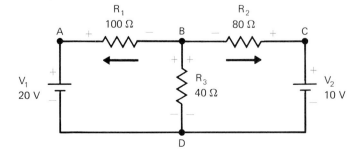

FIGURE 10-10 A complex circuit having two voltage sources requires the identification of each basic loop as a start in the analysis process.

When the values are substituted in each loop we have

$$20 - 100I_1 - 40(I_1 + I_2) = 0$$

$$10 - 80I_2 - 40(I_1 + I_2) = 0$$

I_1 and I_2 are multiplied by 40 in each equation to remove the parentheses:

$$20 - 100I_1 - 40I_1 - 40I_2 = 0$$

$$10 - 80I_2 - 40I_1 - 40I_2 = 0$$

Next, combine like terms:

$$-140I_1 - 40I_2 = 0 - 20$$

$$-40I_1 - 120I_2 = 0 - 10$$

Divide the loop 1 equation by -20 and loop 2 equation by -10. This will make them smaller and change all numbers to positive values:

Loop 1: $7I_1 + 2I_2 = 1$
Loop 2: $4I_1 + 12I_2 = 1$

The currents for each loop can now be solved by applying the method for simultaneous equations. To do this, one of the currents has to be equal in both equations. Multiply the loop 1 equation by 6. This will make I_2 equal in both of the equations.

$$42I_1 + 12I_2 = 6$$

$$4I_1 + 12I_2 = 1$$

Subtract the bottom equation from the upper one. This process eliminates I_2.

$$38I_1 = 5$$

$$I_1 = \frac{5}{38} = 0.132 \text{ A}$$

The value for I_1 can be substituted for I_1 in the other loop:

$$4I_1 + 12I_2 = 1$$

$$4 \times 0.132 \text{ A} + 12I_2 = 1$$

$$12I_2 = -0.528 \text{ A} + 1 \text{ A}$$

$$12I_2 = 0.472 \text{ A}$$

$$I_2 = \frac{0.472 \text{ A}}{12 \text{ A}} = 0.039 \text{ A}$$

When the currents for each loop are identified, they can be combined in order to find the current through R_3.

$$I_{R3} = I(R_1 + R_2) = I_{R1} + I_{R2}$$

$$= 0.135 \text{ A} + 0.128 \text{ A} = 0.263 \text{ A}$$

The final step is to substitute the current values in the equation to determine the IR drop across each component.

$$V_{R1} = I_1 \times R_1 = 0.132 \text{ A} \times 100 \text{ } \Omega = 13.2 \text{ V}$$

$$V_{R2} = I_2 \times R_2 = 0.039 \text{ A} \times 80 \text{ } \Omega = 3.12 \text{ V}$$

$$V_{R3} = I_{R3} \times R_3 = 0.171 \text{ A} \times 40 \text{ } \Omega = 6.84 \text{ V}$$

This calculation can be proven by adding the voltage drops in each of the loops:

$$V_1 = V_{R1} + V_{R3} = 13.5 \text{ V} + 6.84 \text{ V} = 20.34 \text{ V}$$

$$V_2 = V_{R2} + V_{R3} = 3.12 \text{ V} + 6.84 \text{ V} = 9.96 \text{ V}$$

These are correct values. The reason for the slight difference is the rounding of numbers when doing the mathematical solution. This is an accepted procedure.

This circuit's values have been found by using only the two laws presented by Kirchhoff. Neither the rules for parallel circuits nor the rules for series circuits were utilized in this process.

EXAMPLE 10-3

Solve for branch current in Figure 10-11 using the branch current method.

Solution

Right loop: $\quad 40I_2 + 80I_3 = 20$
Left loop: $\quad 200I_1 + 80I_3 = 40$
Node X: $\quad I_1 - I_3 = I_2 \quad$ or $\quad I_1 = I_3 + I_2$

$$200(I_3 - I_2) + 80I_3 = 40$$

$$200I_3 - 200I_2 + 80I_3 = 40$$

$$280I_3 - 200I_2 = 40$$

$$70I_3 - 8I_2 = 1 \text{ (divide by 40)}$$

$$70I_3 - 1 = 8I_2$$

$$I_2 = \frac{70I_3 - 1}{8 \text{ A}}$$

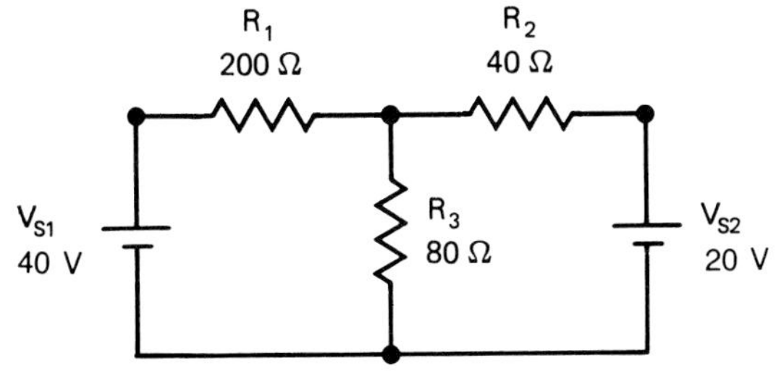

R₁ 200 Ω   R₂ 40 Ω   V_S1 40 V   R₃ 80 Ω   V_S2 20 V

FIGURE 10-11

$$40 \frac{70I_3 - 1}{8} + 80I_3 = 20$$

$$320I_3 - 8 \text{ A} + 80I_3 = 20 \text{ A}$$

$$400I_3 = 28 \text{ A}$$

$$I_3 = \frac{28}{400} = 0.07 \text{ A}$$

$$200I_1 + 80 \, (0.07) = 40$$

$$200I_1 + 56 = 40$$

$$200I_1 = -16$$

$$I_1 = 0.08 \text{ A}$$

$$0.08 - I_2 + 0.07 = 0$$

$$0.15 = I_2$$

10-2 Using Node-Voltage Analysis

Node voltage *Junction point of a circuit.*

When solving for branch currents the current values are used for identifying the voltage drops in the loops. These equations are then expressed in the format required to meet the rules for Kirchhoff's voltage law. When the equation for each loop is identified and solved, the final step is to calculate the current value for each branch.

A second method of solving multiple loop equations is to use the voltage drop at each *node* in the circuit. This is done by use of Kirchhoff's current law. Nodes as used in this example have two or more current paths. In general, the steps involved in this method are as follows:

1. Identify the number of nodes. The circuit shown in Figure 10-12 is the example for the solution of this problem. There are four nodes in this circuit. They are identified as points A, B, C, and D.

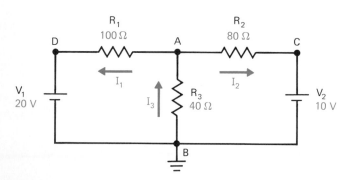

FIGURE 10-12 Node analysis may be used for circuits having more than one power source.

2. Use one node as a reference point. Node B is connected to circuit ground. It will be the reference point for this example. Assign all known voltages to each node:

$$\text{Node } A = \text{unknown}$$

$$\text{Node } B = \text{reference}$$

$$\text{Node } C = 10 \text{ V}$$

$$\text{Node } D = 20 \text{ V}$$

3. At each node where the voltage is not known a voltage is assigned. In this example this unknown voltage is assigned a value of V_A.

4. Identify and assign currents at each node where the voltage is unknown. In our example this current is identified as I_3.

5. Use Kirchhoff's current law at each node with an assigned current value. The method for doing this is:

$$I_1 - I_3 + I_2 = 0$$

6. Express each current equation in terms of voltages. Use Ohm's law to do this. When the values are assigned the problem can be solved for the unknown node voltages.

$$I_1 = \frac{V_1}{R_1} = \frac{V_{S1} - V_A}{R_1}$$

$$I_2 = \frac{V_2}{R_2} = \frac{V_{S2} - V_A}{R_2}$$

$$I_3 = \frac{V_3}{R_3} = \frac{V_A}{R_3}$$

When these are substituted with the current equation identified in step 4, the result is

$$\frac{V_{S1} - V_A}{R_1} - \frac{V_A}{R_3} + \frac{V_{S2} - V_A}{R_2} = 0$$

The only unknown value in this formula is V_A. It is found by substituting known values and combining terms. When the voltage V_A is identified, all branch currents can be calculated.

$$\frac{V_{S1} - V_A}{R_1} - \frac{V_A}{R_3} + \frac{V_{S2} - V_A}{R_2} = 0$$

$$\frac{20 \text{ V} - V_A}{100 \text{ }\Omega} - \frac{V_A}{40 \text{ }\Omega} + \frac{10 \text{ V} - V_A}{80 \text{ }\Omega} = 0$$

$$\frac{20 \text{ V}}{100 \text{ }\Omega} - \frac{V_A}{100 \text{ }\Omega} - \frac{V_A}{40 \text{ }\Omega} + \frac{10 \text{ V}}{80 \text{ }\Omega} - \frac{V_A}{80 \text{ }\Omega} = 0$$

When the lowest common denominator is found, the equation looks like this:

$$\frac{80 \text{ V}}{400 \text{ }\Omega} - \frac{4 V_A}{400 \text{ }\Omega} - \frac{10 V_A}{400 \text{ }\Omega} + \frac{50 \text{ V}}{400 \text{ }\Omega} - \frac{5 V_A}{400 \text{ }\Omega} = 0$$

Combining like terms gives us

$$\frac{130 \text{ V}}{400 \text{ } \Omega} - \frac{19V_A}{400 \text{ } \Omega} = 0$$

$$\frac{130 \text{ V}}{400 \text{ } \Omega} = \frac{19V_A}{400 \text{ } \Omega}$$

$$19V_A = 130 \text{ V}$$

$$V_A = 6.84 \text{ V}$$

When V_A is found and identified as 6.84 V, the voltages for R_1 and R_2 can be found.

$$V_{R1} = V_{S1} - V_A = V_A = 20 \text{ V} - 6.84 \text{ V} = 13.16 \text{ V}$$

$$V_{R2} = V_{S2} - V_A = 10 \text{ V} - 6.84 \text{ V} = 3.16 \text{ V}$$

The current through each resistance is

$$I_{R1} = \frac{V_{R1}}{R_1} = \frac{13.16 \text{ V}}{100 \text{ } \Omega} = 0.132 \text{ A}$$

$$I_{R2} = \frac{V_{R2}}{R_2} = \frac{3.16 \text{ V}}{80 \text{ } \Omega} = 0.040 \text{ A}$$

$$I_{R3} = \frac{V_{R3}}{R_3} = \frac{6.84 \text{ V}}{40 \text{ } \Omega} = 0.171 \text{ A}$$

Here, too, the values have been rounded in order to make the numbers a little simpler.

EXAMPLE 10-4

Find node voltages for the circuit in Figure 10-13.

Solution

Establish a reference node (B). The unknown voltage is V_A.

$$I_1 - I_3 + I_2 = 0$$

$$\frac{40 - V_A}{200 \text{ } \Omega} - \frac{V_A}{80 \text{ } \Omega} + \frac{20 \text{ V} - V_A}{40 \text{ } \Omega} = 0$$

$$\frac{40 \text{ V}}{200 \text{ } \Omega} - \frac{V_A}{200 \text{ } \Omega} - \frac{V_A}{80 \text{ } \Omega} + \frac{20 \text{ V}}{40 \text{ } \Omega} - \frac{V_A}{40 \text{ } \Omega} = 0$$

$$- \frac{V_A}{200 \text{ } \Omega} - \frac{V_A}{80 \text{ } \Omega} - \frac{V_A}{40 \text{ } \Omega} = \frac{-40 \text{ V}}{200 \text{ } \Omega} - \frac{20 \text{ V}}{40 \text{ } \Omega}$$

$$\frac{(-2V_A - 5V_A - 10V_A)}{400 \text{ } \Omega} = \frac{-40 \text{ V} + 100 \text{ V}}{200 \text{ } \Omega}$$

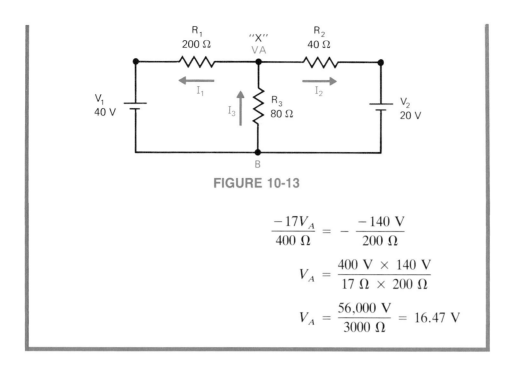

FIGURE 10-13

$$\frac{-17V_A}{400\ \Omega} = -\frac{-140\ \text{V}}{200\ \Omega}$$

$$V_A = \frac{400\ \text{V} \times 140\ \text{V}}{17\ \Omega \times 200\ \Omega}$$

$$V_A = \frac{56{,}000\ \text{V}}{3000\ \Omega} = 16.47\ \text{V}$$

10-3 Using Mesh Current Analysis

This method of identifying circuit values works with loop currents instead of branch currents. Our previous discussions identified a branch current as the actual current through a branch. This is the value of current that an ammeter will measure when it is inserted into the circuit. The term *loop current* is an abstract quantity used to make circuit analysis easier. A system for using the mesh analysis process is described using Figure 10-14. The process is as follows:

Mesh current *Point in circuit where two or more currents meet and join resulting in the sum of the individual values.*

1. Assign a current in a clockwise direction around each closed loop in the circuit. The current flow's actual direction is not important at this stage of the process.
2. Mark the polarities of the voltage drops in each loop. The polarities should be based on the assigned direction of current flow.

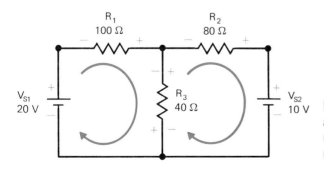

FIGURE 10-14 Mesh current analysis is another method used to solve for circuit values.

3. Apply Kirchhoff's voltage law around each closed loop. A component that has more than one circuit flowing through it may have more than one voltage drop assigned to it.

The loop currents show in Figure 10-14 are both drawn in a clockwise direction. It would be possible to draw a third loop. This loop would go around the perimeter of the circuit. This loop is not necessary because the current already flows through every component. Another current path would be a duplication, and this is undesirable.

Currents I_1 and I_2 flow in opposite directions through R_3. This is because R_3 is common to both loops. We now have two voltage polarities shown for R_3. These are imaginery values because if we were actually to try to measure them, we would only be able to measure one value. It would be based on the influence of the larger current source.

The next step is to apply Kirchhoff's voltage law for each loop:

Loop 1: $-I_1R_1 - I_1R_3 + I_2R_3 - V_S = 0$

Loop 2: $-I_2R_3 + I_1R_3 - I_2R_2 + V_{S2} = 0$

The equations are rearranged and like terms are combined. When this is completed, all the branch currents can be calculated.

Loop 1: $-I_1(R_1 + R_3) + I_2R_3 = V_{S1}$

Loop 2: $-I_2R_3 + I_1R_3 - I_2R_2 = -V_{S2}$

When known values are substituted we have

Loop 1: $-I(100\ \Omega - 40\ \Omega) + 40I_2 = 20\text{ V}$

$-100I_1 - 40I_1 + 40I_2 = 20\text{ V}$

$-140I_1 + 40I_2 = 20\text{ V}$

Loop 2: $40I_1 - I_2(80\ \Omega + 40\ \Omega) = -10\text{ V}$

$40I_1 - 80I_2 - 40I_2 = -10\text{ V}$

$40I_1 - 120I_2 = -10\text{ V}$

The answers for this problem are found by use of determinants.

$$I_1 = \frac{(40\ \times\ -10) - (-120 \times 20)}{(-140 \times\ -120) - (40 \times 40)}$$

$$= 0.1315 \quad \text{or} \quad 0.132\text{ mA}$$

$$I_2 = \frac{(-140\ \times\ -10) - (20 \times 40)}{(-140\ \times\ -120) - (40 \times 40)}$$

$$= 0.0394 \quad \text{or} \quad 40\text{ mA}$$

After these currents are identified the voltage drops across each of the components can be determined:

$$V_{R1} = I_{R1} \times R_1 = 0.132 \text{ A} \times 100 \ \Omega = 13.2 \text{ V}$$

$$V_{R2} = I_{R2} \times R_2 = 0.040 \text{ A} \times 80 \ \Omega = 3.2 \text{ V}$$

$$V_{R3} = (I_{R1} + I_{R2}) \times R_3 = (0.132 \text{ A} + 0.04 \ \Omega) \times 40 \ \Omega$$

$$= 0.172 \text{ A} \times 40 \ \Omega = 6.8 \text{ V}$$

The three methods of determining the voltage and current values in a complex circuit are now identified. Use any one or more of them you feel comfortable with. The choice is up to the individual.

CHAPTER SUMMARY

1. *Kirchhoff's voltage law.* The sum of the voltage drops in a closed loop are equal to zero. When a source voltage is present, the sum of the voltage drops equal the source voltage.
2. *Kirchhoff's current law.* The sum of the currents entering a junction are equal to the currents leaving that junction.
3. Polarity values can be assigned to each component in a circuit depending on the polarity of the source voltage.
4. "Closed loop" is another term for a series circuit.
5. Complex circuit analysis may be accomplished by using loop, node, or mesh analysis.

SELF TEST

Answer true or false; change all false statements into true statements.

10–1. The sum of all *IR* voltage drops in a parallel circuit equal zero.
10–2. A node is a circuit point where two or more currents merge.
10–3. When a loop does not have a source voltage the sum of all *IR* drops is equal to zero.
10–4. Series voltage drop rules are based on Kirchhoff's laws.
10–5. Parallel current rules are based on Ohm's law.

MULTIPLE CHOICE

10–6. The positive end of a component in a circuit is
   (a) in the middle of the voltage-divider string
   (b) closest to the + terminal of the source
   (c) closest to the − terminal of the source
   (d) anywhere as assigned by the problem solver
10–7. In the closed loop:
   (a) currents through each component add to equal source current
   (b) resistances add to equal source resistance
   (c) voltage drops add to equal source voltage

(d) current is independent of applied voltage and total resistance in the circuit

10-8. In the closed loop, the largest individual voltage drop develops across the
   (a) largest value of resistance
   (b) smallest value of resistance
   (c) total circuit
   (d) two smallest resistance values

10-9. Electron current flow in the closed loop is
   (a) from source negative, through the load, and then to source positive
   (b) from source positive, through the load, and then to source negative
   (c) independent of applied source voltage
   (d) bidirectional, depending on type of dc source

10-10. In a parallel branch containing two 10-kΩ resistances, the Kirchhoff current is
   (a) larger through R_1
   (b) larger through R_2
   (c) not applied
   (d) equal through both

ESSAY QUESTIONS

10-11. Explain how the current directions are assigned in the closed loop.

10-12. Explain how the polarities are assigned to each component.

10-13. What effect will a source voltage polarity reversal have on total current flow in a closed loop?

10-14. Explain why, when two opposite polarity sources are series connected, their total voltage is smaller than the sum of their individual voltages.

10-15. Explain why current in each branch of a parallel circuit must flow in the same direction.

PROBLEMS

10-16. In Figure 10-15, if source current is 10 A, what is the current through R_2? (*Sec. 10–3*)

Use Figure 10-16 to answer problems 10-17 and 10-18.

10-17. Use the branch current method to find the currents. (*Sec. 10–1*)

10-18. Use the mesh current method to find the current through R_3 when V_{S1} = 24 V and V_{S2} = 20 V. (*Sec. 10–3*)

Use Figure 10-17 to answer problems 10-19 through 10-23.

10-19. Identify the polarity of each resistance in this figure. (*Sec. 10–1*)

10-20. Solve for currents through R_1, R_2, and R_3. (*Sec. 10–3*)

10-21. Solve for the current through R_4. (*Sec. 10–3*)

10-22. Determine V_{R4}. (*Sec. 10–3*)

10-23. Solve for currents when V_{S2} has its polarity reversed. (*Sec. 10–3*)

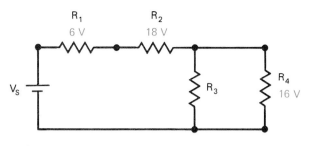

FIGURE 10-15

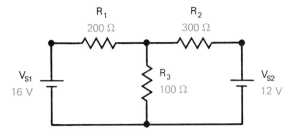

FIGURE 10-16

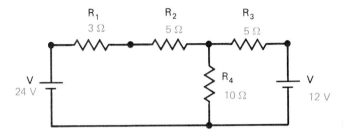

FIGURE 10-17

10–1. 6.7 A
10–2. 2.5 A
10–3. 6 A
10–4. 6 A
10–5. 6 A
10–6. 6 A
10–7. $R_{1A} = +$, $R_{1B} = -$, $R_{2C} = +$, $R_{2D} = -$, $R_{3E} = +$, $R_{3F} = -$, $V_G = +$, $V_H = -$
10–8. 60 V
10–9. 440 V

CHAPTER 11

Network Theorems

The analysis of complex electrical circuits is made simpler when we are able to reduce the complex circuit to a simple equivalent circuit. Several common rules have been developed to accomplish this. In many instances we do not need to know the specific components in a circuit. All we require is the equivalent value of the voltage, current, or resistance of the total circuit. The theorems presented in this chapter offer methods of accomplishing this and thus being able to use this information for further circuit analysis.

KEY TERMS

Current Source	**Superposition Theorem**
Delta Network	**Thévenin Equivalent Circuit**
Linear Manner	**Thévenin Equivalent Resistance (R_{TH})**
Millman's Theorem	**Thévenin Voltage (V_{TH})**
Norton Current (R_N)	**Thévenin's Theorem**
Norton Equivalent Circuit	**Wye Network**
Norton Equivalent Resistance (R_N)	**Voltage Source**
Norton's Theorem	

OBJECTIVES

Upon completion of the material in this chapter, you will

1. Be able to explain and apply the following basic network theorems:
   (a) Superposition
   (b) Thévenin's
   (c) Norton's
   (d) Millman's
   (e) Delta or wye
2. Be able to convert between delta and wye circuits.

Some circuit analysis can be very complicated. In many situations there is little need to know the voltage and current as it relates to every component in the circuit. When this situation occurs, there are some simple ways of representing the circuit. These ways can also be used to represent an *equivalent* value for the entire circuit. The equivalent value will tell us how the circuit looks and functions when connected to a load.

A general group of network theorems exists. This group includes theorems presented by Thévenin, Norton, and Millman. The superposition theorem is still another theorem that is often used in resolving values in circuits. Thévenin's theorem is the most often used of this group. Most of the material in this chapter is presented in an attempt to further an understanding of this theorem and how it is used. The other theorems are also discussed, but with less emphasis. Methods of converting data from one theorem to another are also presented. The final section covers delta and wye circuits and their conversion factors.

11-1 Thévenin's Theorem

Thévenin's theorem
Replacement of complex circuit by equivalent circuit of one dc source in series with one resistance.

The basic statement for **Thévenin's theorem** is: Any linear dc circuit, no matter how complicated, can be replaced by an equivalent circuit containing one dc voltage source in series with one resistance. This concept is illustrated in Figure 11-1. The actual circuit may contain one or more voltage sources. It may also contain one or more current sources. In addition to these sources the original circuit may contain a quantity of fixed components. These include resistors or any other component with resistance quantities as long as these devices act in a linear manner. The term **linear manner** means that any change in the action of the device occurs on a linear curve. When voltage is increased the current also increases, and these changes, when plotted on a graph, would produce a straight diagonal line.

Linear manner *Equal change in value for each step of change in circuit.*

The advantage of using Thévenin's theorem becomes apparent when we consider the amount of circuit problem solving that occurs using the Kirchhoff-related formulas presented in Chapter 10. When the Thévenin equivalent value for a circuit has been found, it can be used as often as required with a variety of loads. This application speeds circuit analysis and is a great aid to those involved in electrical and electronic design work.

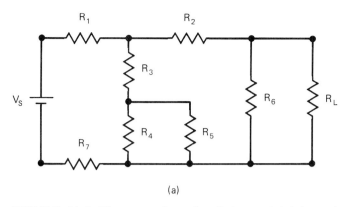

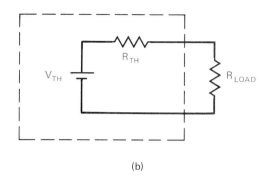

(a)   (b)

FIGURE 11-1 The complex circuit in part (a) is reduced in part (b) to an equivalent single source and a series-connected load using Thévenin's theorem.

The use of the Thévenin theorem is fairly simple. Certain steps are required when we use this process.

1. Disconnect or remove the load from the circuit. The load is normally represented by a value of resistance.

2. When the load has been removed, the next step is to calculate the open-circuit voltage that would be found at the load terminals. This may require the use of one of the systems analysis procedures described in Chapter 10. This **voltage** is identified as V_{TH}.

    Thévenin voltage (V_{TH}) Value of voltage with no load connected in Thévenin circuit analysis.

3. All source voltages are removed. The terminal connections for each source must be short circuited. If you are working with an actual circuit, the source connection can be short circuited by replacing the source with a piece of wire. Any current sources are also removed. Their terminal connections are open circuited.

4. When all voltage and current sources are eliminated, the next step is to determine the circuit resistance. This is the amount of resistance one will "see" looking at the power source from the terminals of the load. This value is identified as R and identified as R_{TH}, the **Thévenin resistance** value. The determination of this resistance value may require the use of the circuit analysis processes described in Chapter 10.

    Thévenin resistance (R_{TH}) Value of circuit resistance.

5. The value of V_{TH} is connected in series with the value of R_{TH} as indicated in Figure 11-1(b).

6. When the values of R_{TH} and V_{TH} are known, the original load may be reconnected to its terminals. This may be an actual connection or it may be done on paper as a design problem. When the load is reconnected, values of voltage, current and power for the load may now be calculated.

Determination of the Thévenin equivalent values for a circuit requires finding both the Thévenin equivalent voltage, V_{TH}, and the Thévenin equivalent resistance, R_{TH}. The first section of the presentation reveals how to establish the Thévenin equivalent voltage. Following the examples is a second section illustrating how the Thévenin equivalent resistance R is identified.

Finding Thévenin Voltage

EXAMPLE 11-1

Determine V_{TH} for the circuit shown in Figure 11-2.

Solution

This is a simple circuit. Removal of the load results in a series circuit containing R_1 and R_2. Use the voltage-divider formula in order to find V_{R2}.

$$V_{R2} = \frac{R_2}{R_1 + R_2} V_s$$

$$= \frac{100\ \Omega}{300\ \Omega} \times 50\ V = 16.68\ V$$

The Thévenin equivalent voltage is equal to the IR voltage drop across R_2 since the load is connected in parallel with R_2.

$$V_{TH} = V_{R2} = 16.68\ V$$

Let us now look at a more complex circuit and analyze it for its Thévenin equivalents.

EXAMPLE 11-2

Find the Thévenin equivalent voltage V_{TH} for the circuit shown in Figure 11-3.

Solution

In this circuit the Thévenin voltage V_{TH} appears across terminals of the load resistor R_L. When the load is removed from the circuit, there is no current flow through R_4. The basic system is a series voltage-divider

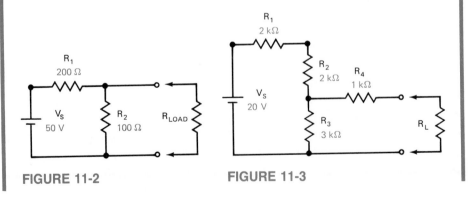

FIGURE 11-2   FIGURE 11-3

---

network consisting of R_1, R_2, and R_3. The voltage-divider formula is used to determine V_{TH}.

$$V_{TH} = V_{R3} = \frac{R_3}{R_1 + R_2 + R_3} V_S$$

$$= \frac{3000\ \Omega}{7000\ \Omega} \times 20\ V = 8.57\ V$$

The open terminal voltage, V_{TH}, is 8.57 V.

The term *seen* is used to describe the resistance that is across the terminals for load connection in a circuit. This is the value of resistance one would measure with an ohmmeter using the condition described to establish a **Thévenin equivalent circuit**. The meter does not have the ability to analyze a circuit and identify the actual components in the circuit. It is limited to measuring a total quantity of resistance in the circuit at the point of measurement.

Thévenin equivalent circuit *Ohmic value measured using Thévenin analysis.*

EXAMPLE 11-3

Find the Thévenin equivalent voltage V_{TH} in Figure 11-4.

Solution

Analysis of this circuit shows that resistances R_2 and R_3 affect the load voltage. The load is connected in parallel with R_3. Resistance R_1 is in the circuit and it will draw current from the source. It is not a part of the load circuit and is not used in determining V_{TH}.

$$V_{TH} = V_{R3} = \frac{R_3}{R_2 + R_3} V_s$$

$$= \frac{80\ \Omega}{120\ \Omega} \times 10\ V = 6.7\ V$$

The Thévenin voltage V_{TH} is 6.7 V.

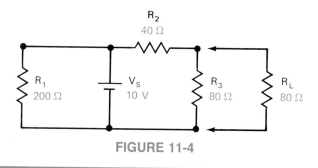

FIGURE 11-4

Finding Thévenin Equivalent Resistance This section establishes the method used to find **Thévenin equivalent resistance R_{TH}**. The value of R_{TH} is identified as the total resistance appearing across the terminals for the load when all

Thévenin equivalent resistance (R_{TH}) *Total resistance across load in Thévenin circuit analysis.*

voltage and current sources are eliminated. The process replaces a voltage source with a short circuit and a current source with an open circuit. The components remaining in the circuit form a resistive network. The value for the resistive network is determined by use of any of the circuit analysis processes presented in Chapter 10.

EXAMPLE 11-4

Find the Thévenin equivalent resistance in Figure 11-5.

Solution

This circuit has two of its components removed. The power source is replaced by a short circuit and the load is not connected. This results in a circuit with two parallel connected components. R_1 and R_2 form a parallel bank under these conditions. The Thévenin equivalent resistance is

$$\frac{1}{R_{TH}} = \frac{1}{R_1} + \frac{1}{R_2}$$

$$= \frac{1}{500 \ \Omega} + \frac{1}{1500 \ \Omega}$$

$$= 0.002 \ S + 0.0007 \ S = 0.0027 \ S$$

$$R_{TH} = \frac{1}{0.0027 \ S} = 370 \ \Omega$$

The Thévenin equivalent resistance $R_{TH} = 370 \ \Omega$.

EXAMPLE 11-5

Find R_{TH} for the circuit in Figure 11-6.

Solution

Analysis of this circuit with the power source short circuited shows R_3 in parallel with the series string R_1 and R_2. Resistance R_4 is in series with the parallel bank. The solution for R_{TH} is

$$R_{TH} = R_4 + \frac{R_3(R_1 + R_2)}{R_1 + R_2 + R_3}$$

$$= 500 \ \Omega \ \frac{2000 \ \Omega(500 \ \Omega + 1500 \ \Omega)}{500 \ \Omega + 1500 \ \Omega + 2000 \ \Omega}$$

$$= 500 \ \Omega + \frac{4,000,000 \ \Omega}{4000 \ \Omega}$$

$$= 500 \ \Omega + 1000 \ \Omega$$

$$= 1500 \ \Omega$$

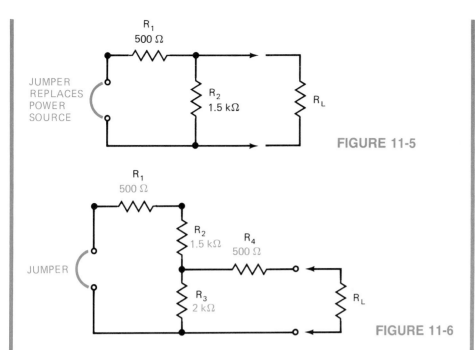

FIGURE 11-5

FIGURE 11-6

EXAMPLE 11-6

Find the Thévenin equivalent resistance R_{TH} as seen by the load resistance in Figure 11-7.

Solution

Resistances R_2 and R_3 form a parallel bank when the power source is short circuited. A short circuit also negates any resistance for R_1 since the total resistance in a parallel circuit is always less than the lowest single resistance value. A short circuit is considered to be 0 Ω. There is no lesser ohmic value than zero, so this component can be ignored when calculating the value of R_{TH}.

Both R_2 and R_3 have the same value. They have formed a parallel bank. Their resistance is the same as R_{TH}.

$$R_{TH} = \frac{R_1}{R_n} = \frac{800 \ \Omega}{2} = 400 \ \Omega$$

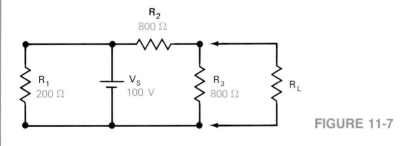

FIGURE 11-7

The Thévenin equivalent resistance R_{TH} is 400 Ω.

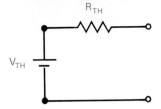

FIGURE 11-8 Thévenin equivalent circuits are drawn in this manner.

Any Thévenin circuit may be redrawn and shown in its simplest form. This form consists of a voltage source V_{TH} and a resistance value R_{TH} (Figure 11-8). The method of obtaining these values is illustrated in the following examples.

EXAMPLE 11-7

Draw a circuit showing the Thévenin equivalent voltage and resistance for Example 11-6.

Solution

The Thévenin equivalent voltage V_{TH} is found by applying the voltage-divider formula. The source voltage, V_S, is 100 V.

$$V_{TH} = \frac{R_3}{R_3 + R_2} \times V_S$$

$$= \frac{800 \ \Omega}{1000 \ \Omega} \times 200 \ \text{V}$$

$$= 50 \ \text{V}$$

The Thévenin equivalent resistance is calculated:

$$R_{TH} = \frac{R_1}{R_n} = \frac{800 \ \Omega}{2}$$

$$= 400 \ \Omega$$

The complete circuit, including both of the Thévenin equivalents, is shown in Figure 11-9.

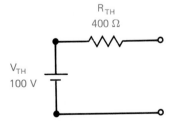

FIGURE 11-9 Thévenin equivalent circuit with its values.

Summary of Thévenin's Theorem Process The circuit resulting from this representation is always shown as a series circuit. It contains only two components. These are the resistance R_{TH} and the voltage V_{TH}. When any load is connected to this circuit, it will have the same voltage and current characteristics as those available from the original circuit. The steps involved in solving for these equivalents are summarized in this manner:

1. Open the terminals and remove the load.
2. Determine the voltage across the terminals (V_{TH}).
3. Determine the resistance across the same terminals with voltage sources shorted and current sources open circuited.
4. Connect V_{TH} in series with R_{TH} to create the Thévenin equivalent of the original circuit.

Finding R_{TH} and V_{TH} by Measurement The use of Thévenin's theorem is normally limited to a paper-and-pencil activity. Its purpose is to create the circuit in theoretical form before it is actually constructed. In this form it is easy to use a calculator or a computer to see how a variety of loads will affect the operation of the system. There are moments, however, when it may be more practical actually to build the test circuit and measure its values. These values, of course, should agree with the calculated values. The procedure for measuring the values is as follows:

1. Remove the load from the output terminals (Figure 11-10).
2. Use a voltmeter to measure the voltage at the terminals for the load. This value is V_{TH}.
3. Connect a variable resistor across the load terminals.
4. Adjust the variable resistor while measuring V_{TH}. When the value of V_{TH} is equal to one-half of the original V_{TH} value, remove the variable resistor from the circuit without any more adjustment.
5. Measure the ohmic value of the variable resistance. This value is equal to the value of R_{TH}.

The reasons for using this procedure is that the original circuit drawing or schematic diagram may not be available or it may not be practical to disconnect power sources for the purposes of measurement. The original power source must be designed to support the additional load current required for step 4 in this process.

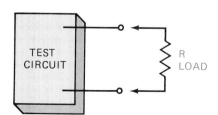

FIGURE 11-10 When measuring circuit values for Thévenizing, the circuit the load is removed.

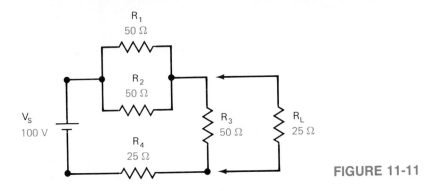

FIGURE 11-11

REVIEW PROBLEMS

11-1. State Thévenin's theorem in your own words.
11-2. What two components make up the Thévenin equivalent circuit?
11-3. How is R_{TH} determined?
11-4. How is V_{TH} determined?
11-5. Find the Thévenin equivalent circuit as seen by R_L for the circuit in Figure 11-11.

11-2 Norton's Theorem

Norton's theorem Reduction
*of complex circuit to a current
source and a parallel
resistance value.*

Norton's theorem is also used to reduce a circuit from a complex form to a simple form. The use of Thévenin's theorem is restricted to the identification of a voltage source in series with a resistance value. **Norton's theorem** is related to the reduction of a circuit to *a current source and a parallel resistance value*. The format for showing this is illustrated in Figure 11-12. The current source, I_s, is shown as an arrow enclosed in a circle. The Norton resistance is shown as a resistor with the identifier of R_N.

When applying Norton's theorem, there are two quantities that must be found. These are identified as I_N and R_N. When these values are identified they are connected in parallel to form the Norton equivalent circuit.

Norton current (R_N) Short
*circuit current between two
points in circuit.*

Finding Norton Equivalent Current The **Norton current** is defined as the *short-circuit current* between two points in a circuit. A component connected to the two points "sees" a current source. The current source, I_N, is in parallel with a Norton resistance value of R_N.

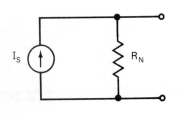

**FIGURE 11-12 A Norton
equivalent circuit consists of
a current source in parallel
with a load.**

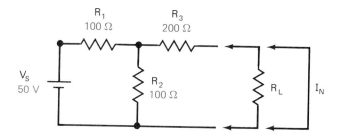

FIGURE 11-13 Circuit used to clarify the Nortonizing of a circuit.

The circuit illustrated in Figure 11-13 will help to clarify the development of Norton values. The original circuit consists of the source V_s, resistances R_1, R_2, R_3, and the load R_L. The purpose of this is to develop a Norton circuit that is equivalent to the circuit "seen" by the load resistance R_L. The first step in the process is to short circuit the load. This is indicated by the dashed lines and the arrow bracket marked I_N in the drawing.

When the load terminals are short circuited, the Norton equivalent resistance value is calculated. The circuit now appears to have R_2 in parallel with R_3. This parallel bank is series connected to R_1. The total resistance in the circuit is

$$R_T = R_1 + \frac{R_2 \times R_3}{R_2 + R_3}$$

$$= 100 \ \Omega + \frac{200 \ \Omega \times 100 \ \Omega}{200 \ \Omega + 100 \ \Omega}$$

$$= 100 \ \Omega + \frac{20{,}000 \ \Omega}{300 \ \Omega} = 167 \ \Omega$$

Next the total source current is found by using the values of V_s and R_T.

$$I_T = \frac{V_s}{R_T}$$

$$= \frac{50 \ V}{167 \ \Omega} = 0.3 \ A$$

The source current is 0.3 A.

The last step in this process is to use the current-divider formula to find the short-circuit current I_N.

$$I_N = \frac{R_2}{R_2 + R_3} I_T$$

$$= \frac{100 \ \Omega}{100 \ \Omega + 200 \ \Omega} \times 0.3 \ A = \frac{100 \ \Omega}{300 \ \Omega} \times 0.3 \ A$$

$$= 0.1 \ A$$

The Norton equivalent current is found to be equal to 0.1 A.

EXAMPLE 11-8

Determine the Norton equivalent current in Figure 11-14.

Solution

Short circuit the output terminals. This places R_2 and R_3 in parallel. This combination is in series with R_1. The first step in the process requires determining R_T.

$$R_T = R_1 + \frac{R_2 \times R_3}{R_2 + R_3}$$

$$= 400 \ \Omega + \frac{600 \ \Omega \times 1200 \ \Omega}{600 \ \Omega + 1200 \ \Omega}$$

$$= 400 \ \Omega + \frac{720 \ \text{k}\Omega}{1.8 \ \text{k}\Omega}$$

$$= 800 \ \Omega$$

Next, find I_T for this circuit:

$$I_T = \frac{V_S}{R_T} = \frac{160 \ \text{V}}{800 \ \Omega}$$

$$= 0.2 \ \text{A}$$

Now, apply the current divider formula:

$$I_N = \frac{R_3}{(R_2 + R_3)} \times I_T$$

$$= \frac{1.2 \ \text{k}\Omega}{(600 + 1.2 \ \text{k}\Omega)} \times 0.2$$

$$= 0.133 \ \text{A}$$

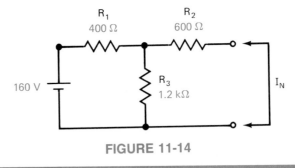

FIGURE 11-14

Norton equivalent resistance (R_N) *Resistance value of circuit when voltage source terminals are short circuited.*

Finding Norton Equivalent Resistance

When we wish to identify the **Norton equivalent resistance** the process involved is similar to that used for determining current. When finding current the load terminals of the circuit are shorted. To find the Norton equivalent resistance (R_N), the terminals of the voltage source are short circuited. After doing this the total resistance in the system is calculated using any of the several circuit solution processes.

EXAMPLE 11-9

Find the Norton equivalent resistance, R_N, for the circuit shown in Figure 11-14.

Solution

Short circuit V_s. This places R_1 and R_2 in parallel. R_2 is series connected to the parallel bank of R_1 and R_2.

$$R_N = R_2 + \frac{R_1 \times R_3}{R_1 + R_3}$$

$$= 600\ \Omega + \frac{400\ \Omega \times 1200\ \Omega}{400\ \Omega + 1200\ \Omega}$$

$$= 600\ \Omega + 300\ \Omega = 900\ \Omega$$

The Norton equivalent resistance of this circuit is 900 Ω.

When the Norton equivalent current and resistance values are identified, the circuit may be redrawn. This results in the circuit shown in Figure 11-15. This is the usual form, with its "N" subscripts. These subscripts indicate Norton values for the circuit.

Summary of Norton's Theorem A load resistor connected between the terminals of a **Norton equivalent circuit** has the same voltage across it and the same current through it as if it were connected to the original circuit. The steps involved in identifying these values through the use of Norton's theorem are as follows:

Norton equivalent circuit Values of I and R determined using Norton circuit analysis.

1. Short circuit the two terminals where you wish to find the Norton equivalent circuit.
2. Calculate current through the short-circuited terminals (I_N).
3. Calculate the resistance in the circuit with the terminals open and the power source short circuited (R_N).
4. Connect I_N in parallel with R_N.

This is the Norton equivalent of the original circuit.

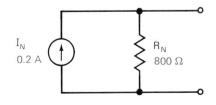

I_N
0.2 A

R_N
800 Ω

FIGURE 11-15 A Norton equivalent circuit is drawn as this schematic illustrates.

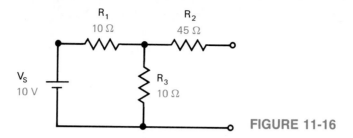

R₁
10 Ω

R₂
45 Ω

Vₛ
10 V

R₃
10 Ω

FIGURE 11-16

REVIEW PROBLEMS

11-6. State Norton's theorem in your own words.

11-7. What are the two components that make up the Norton equivalent circuit?

11-8. How is I_N determined?

11-9. How is R_N determined?

11-10. Find the Norton equivalent for the circuit shown in Figure 11-16.

11-3 The Superposition Theorem

Superposition theorem
Determination of circuit current values using individual sources in a multisource circuit.

The **superposition theorem** is very handy when one needs to analyze a circuit containing more than one voltage source. This theorem permits us to find the current in any one branch of a multiple-source circuit. The theorem is stated as follows: Currents in any branch of a multisource circuit are identified by determining the current in each source individually when all other sources are replaced by their internal resistance values. The current through the specific branch is in the algebraic sum of the current from each source in the specific branch.

In a practical sense this statement says that one needs to find the current in a branch of a multisource circuit by calculating the currents one at a time. This is done by short circuiting all sources except one and calculating currents. This process is repeated for each power source. The currents identified in each step are algebraically added to find the total current through the specific current in that branch or component. The following steps are involved in this process:

1. Replace all voltage sources *except* one with a short circuit. Any current sources are removed or left as open circuits.

2. Calculate the current produced by the remaining power source.

3. Repeat steps 1 and 2 for each power source in the system. This will result in the same number of currents as power sources.

4. Identify each current's polarity as it flows through the components in the circuit.

5. Add each current value algebraically.

Use the following examples to clarify the process.

EXAMPLE 11-10

Calculate the current through R_3 in Figure 11-17.

Solution

Short circuit V_{S2} and calculate I_{R3}. Use the current-divider formula. First find R_T.

$$R_T = R_1 + \frac{R_2 \times R_3}{R_2 + R_3}$$

$$= 600 \ \Omega + \frac{400 \ \Omega \times 400 \ \Omega}{400 \ \Omega + 400 \ \Omega}$$

$$= 600 \ \Omega + 200 \ \Omega$$

$$= 800 \ \Omega$$

Next, find I_T.

$$I_T = \frac{V_{S1}}{R_T}$$

$$= \frac{20 \ \text{V}}{800 \ \Omega} = 0.025 \ \text{A}$$

Use the current-divider formula to find I_{R3}.

$$I_{R3} = \frac{R_2}{R_3 + R_2} I_T$$

$$= \frac{400 \ \Omega}{800 \ \Omega} \times 0.025 \ \text{A} = 0.0125 \ \text{A}$$

The direction of current flow through R_3 is from the bottom of the resistance to the top. This is indicated by the arrow on the left side of the circuit.

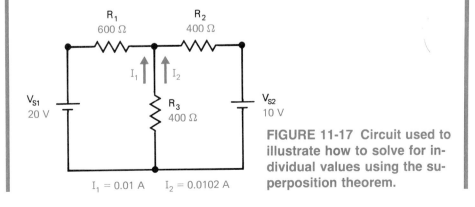

$I_1 = 0.01 \ \text{A}$ $I_2 = 0.0102 \ \text{A}$

FIGURE 11-17 Circuit used to illustrate how to solve for individual values using the superposition theorem.

The process is now repeated for the current flow from V_{s2}. The first step, again, is to short circuit V_{s1} and find R_T.

$$R_T = R_3 + \frac{R_1 \times R_2}{R_1 + R_2}$$

$$= 400 \ \Omega + \frac{600 \ \Omega \times 400 \ \Omega}{600 \ \Omega + 400 \ \Omega}$$

$$= 400 \ \Omega + 240 \ \Omega = 640 \ \Omega$$

Now find I_T.

$$I_T = \frac{V_{S2}}{R_T}$$

$$= \frac{10 \ V}{640 \ \Omega} = 0.0156 \ A$$

The final step for this source is to find I_{R3}.

$$I_{R3} = \frac{R_1}{R_1 + R_3} \times I_T$$

$$= \frac{600 \ \Omega}{1000 \ \Omega} \times 0.0156 \ A = 0.00936 \ A$$

The current due to this source is from the bottom to the top of the resistance. This is shown by the arrow on the right side of R_3. Total current through R_3 equals the algebraic sum of all currents through R_3.

$$I_{R3} = I_3 \ (\text{from } V_{S1}) + I_3 \ (\text{from } V_{S2})$$
$$= 0.0125 \ A + 0.00936 \ A = 0.02186 \ A$$

The superposition theorem may be used with circuits having more than two power sources. The next example has three power sources.

EXAMPLE 11-11

Calculate currents through each resistance for the circuit shown in Figure 11-18.

Solution

To find I_{VS1}, replace V_{S2} and V_{S3} with short circuits. Calculate R_T and I_T for V_{S1}:

$$R_T = 20 \ \Omega + \frac{30 \ \Omega \times 40 \ \Omega}{30 \ \Omega + 40 \ \Omega}$$

$$= 37.14 \ \Omega$$

$$I_T = \frac{V_{S1}}{R_T} = \frac{60 \ V}{37.14 \ \Omega} = 1.615 \ A$$

Next calculate I_{R2} and I_{R3}:

$$I_{R2} = \frac{R_3}{R_2 + R_3} \times I_T = \frac{40\ \Omega}{70\ \Omega} \times 1.615\ A = 0.923\ A$$

$$I_{R3} = I_T - I_{R2} = 1.615\ A - 0.923\ A = 0.692\ A$$

Current produced by $V_{S1} = I_{R1} = 1.615\ A; I_{R2} = 0.923\ A; I_{R3} = 0.692\ A$.
To find I_{VS2}, replace V_{S1} and V_{S3} with short circuits. Calculate R_T and I_T for V_{S2}:

$$R_T = R_3 + \frac{R_1 \times R_2}{R_1 + R_2} = 40\ \Omega + \frac{20\ \Omega \times 30\ \Omega}{20\ \Omega + 30\ \Omega}$$

$$= 40\ \Omega + \frac{600\ \Omega}{50\ \Omega} = 52\ \Omega$$

$$I_T = \frac{V_{S2}}{R_T} = \frac{40\ V}{52\ \Omega} = 0.769\ A$$

Next, calculate I_{R1} and I_{R2}:

$$I_{R1} = \frac{R_2}{R_1 + R_2} \times I_T = \frac{30\ \Omega}{50\ \Omega} \times 0.769\ A = 0.461\ A$$

$$I_{R2} = I_T - I_{R1} = 0.769\ A - 0.461\ A = 0.308\ A$$

Current produced by $V_{S2} = I_{R1} = 0.461\ A; I_{R2} = 0.308\ A;$ and $I_{R3} = 0.769\ A$.
To find I_{VS3}, replace V_{S1} and V_{S2} with short circuits. Calculate R_T and I_T for V_{S3}:

$$R_T = R_2 + \frac{R_1 \times R_3}{R_1 + R_3} = 30\ \Omega + \frac{800\ \Omega}{60\ \Omega} = 43.33\ \Omega$$

$$I_T = \frac{V_{S3}}{R_T} = \frac{10\ V}{43.33\ \Omega} = 0.231\ A$$

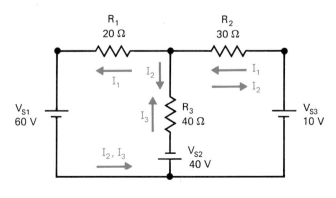

FIGURE 11-18

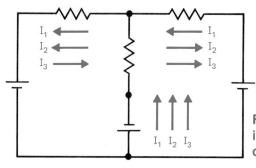

FIGURE 11-19 Each current is shown as a part of the use of the superposition theorem.

Next, calculate I_{R1} and I_{R3}:

$$I_{R1} = \frac{R_3}{R_1 + R_3} \times I_T = \frac{40\ \Omega}{60\ \Omega} \times 0.231\ \text{A} = 0.154\ \text{A}$$

$$I_{R3} = I_T - I_{R1} = 0.231\ \text{A} - 0.154\ \text{A} = 0.077\ \text{A}$$

Current produced by $V_{S3} = I_{R1} = 0.154$ A; $I_{R2} = 0.231$ A; $I_{R3} = 0.077$ A.

The final step in this process is to add the three currents algebraically for each resistance. The direction of current flow for each of the currents is shown in Figure 11-19. These are added as follows:

$$I_{R1} = +I_{VS1} + I_{VS2} - I_{VS3} = +1.615\ \text{A} + 0.462\ \text{A} - 0.154\ \text{A} = 1.923\ \text{A}$$

$$I_{R2} = +I_{VS1} - I_{VS2} - I_{VS3} = +0.923\ \text{A} - 0.308\ \text{A} - 0.231\ \text{A} = 0.384\ \text{A}$$

$$I_{R3} = +I_{VS1} + I_{VS2} + I_{VS3} = +0.692\ \text{A} + 0.769\ \text{A} + 0.077\ \text{A} = 1.538\ \text{A}$$

Voltage source *Power supply providing operating voltage values.*

Current source *Power supply providing operating current values.*

Converting Voltage Sources and Current Sources The process of circuit analysis can often be made easier when a **voltage source** is converted into an equivalent **current source**. This is also true when there is a need to convert a current source into an equivalent voltage source. The process for doing this conversion involves the use of Ohm's law. The conversion of the voltage source into an equivalent current source is done in this manner:

$$I_S = \frac{V_S}{R_S}$$

This process is illustrated in Figure 11-20. The proof of this conversion is

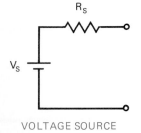

VOLTAGE SOURCE

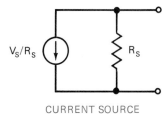

CURRENT SOURCE

FIGURE 11-20 A voltage source (Thévenin) may be converted to a current source (Norton).

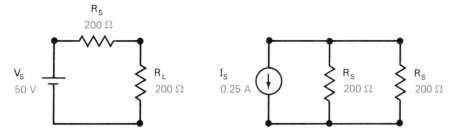

FIGURE 11-21 The ability to convert between a voltage source and a current source is proved using these two circuits.

established by connecting a load resistance across the open terminals in both circuits (Figure 11-21). The value of load resistance is 200 Ω for this conversion.

$$I_L = \frac{V_S}{R_S + R_L}$$

$$= \frac{50 \text{ V}}{200 \text{ Ω} + 200 \text{ Ω}} = \frac{50 \text{ V}}{400 \text{ Ω}} = 0.125 \text{ A}$$

This is the current value for the voltage source circuit. The current source circuit is analyzed as follows:

$$I_L = \frac{R_S}{R_S + R_L} \times \frac{V_S}{R_S} = \frac{V_S}{R_S + R_L}$$

$$= \frac{200 \text{ Ω}}{200 \text{ Ω} + 200 \text{ Ω}} \times \frac{50 \text{ V}}{200 \text{ Ω}} = 0.5 \text{ A} \times 0.25 \text{ A} = 0.125 \text{ A}$$

The results of both sets of calculations are the same. This proves that the conversion formulas are equal.

EXAMPLE 11-12

Convert the voltage source in Figure 11-22 into an equivalent circuit source.

Solution

$$I_S = \frac{V_S}{R_S} = \frac{200 \text{ V}}{50 \text{ Ω}} = 4 \text{ A}$$

The equivalent circuit for this conversion is shown in Figure 11-23.

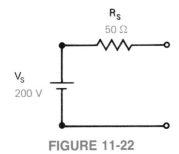

FIGURE 11-22

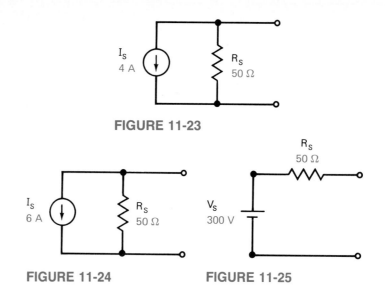

FIGURE 11-23

FIGURE 11-24   FIGURE 11-25

EXAMPLE 11-13

Convert the circuit shown in Figure 11-24 into a voltage source.

Solution

$$V_S = I_S R_S = 6 \text{ A} \times 50 \ \Omega = 300 \text{ V}$$

The circuit drawing for this conversion is shown in Figure 11-25.

REVIEW PROBLEMS

11-11. Explain in your own words how to convert a voltage source into a current source.

11-12. Write the formula for converting a current source into an equivalent voltage source.

11-13. Convert the voltage source in Figure 11-26 into an equivalent current source.

11-14. Convert the current source in Figure 11-27 into an equivalent voltage source.

FIGURE 11-26

FIGURE 11-27

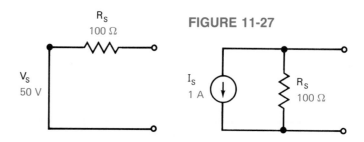

11-4 Millman's Theorem

The advantage of using the theorems presented in this chapter is the ability to analyze a circuit for its mathematical values under a variety of conditions. The theorem presented by Millman is still another process to be used to analyze a complex circuit. This theorem is used for circuits having multiple voltage sources connected in parallel with each other. **Millman's theorem** permits the reduction of any quantity of parallel voltage sources to one equivalent voltage source. It is represented as a single voltage source in series with a load. This theorem is used to find the current through a load or the voltage drop across a load.

Millman's theorem *Process used to reduce multiple parallel voltage sources into one equivalent voltage source.*

The process required to analyze a circuit for the equivalent voltage source is:

Find the total resistance for the resistance values in the circuit. Use either the reciprocal or the conductance rule: $R = 1/G$ or $G = 1/R$.

1. Convert all of the voltage sources in the system into current sources (Figure 11-28).
2. Calculate the total current: $I_T = I_1 + I_2 + I_3$, and so on.
3. Find the total conductance: $G_T = G_1 + G_2 + G_3$, and so on.
4. Convert the total conductance into total resistance. This value is also the equivalent resistance in the circuit. $R_{EQ} = 1/G_T$.
5. Convert each value into an equivalent voltage value:

$$I_T = I_1 = \frac{1}{R_1}, \quad \text{etc.}$$

$$I_T = \frac{V_1}{R_1} + \frac{V_2}{R_2} + \frac{V_3}{R_3}, \quad \text{etc.}$$

6. Use the following formula to find the equivalent circuit voltage:

$$V_{EQ} = \frac{\dfrac{V_1}{R_1} + \dfrac{V_2}{R_2} + \dfrac{V_3}{R_3}}{\dfrac{1}{R_1} + \dfrac{1}{R_2} + \dfrac{1}{R_3}}$$

The resolution of this circuit identifies a voltage value that has the same polarity as the original circuit. It produces a current flow in the same direction as the original currents.

FIGURE 11-28 All voltage sources are converted into equivalent current sources when using Millman's theorem.

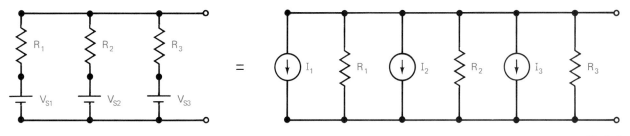

EXAMPLE 11-14

Use Millman's theorem in the circuit shown in Figure 11-29 to find V_L and I_L.

Solution

Find equivalent resistance. Convert the circuit into a current source circuit.

$$R_{EQ} = \cfrac{1}{\cfrac{1}{R_1} + \cfrac{1}{R_2} + \cfrac{1}{R_3}}$$

$$= \cfrac{1}{\cfrac{1}{12\ \Omega} + \cfrac{1}{12\ \Omega} + \cfrac{1}{24\ \Omega}}$$

$$= 4.8\ \Omega$$

Find the equivalent voltage:

$$V_{EQ} = \frac{(V_1/R_1) + (V_2/R_2) + (V_3/R_3)}{(1/R_1) + (1/R_2) + (1/R_3)}$$

$$= \frac{8\text{ A} + 2\text{ A} + 2\text{ A}}{0.208\text{ S}} = \frac{12}{0.208\text{ S}}$$

$$= 57.6\text{ V}$$

The equivalent circuit is shown in Figure 11-30. The load current and load voltage values can now be calculated:

$$I = \frac{V_{EQ}}{R_{EQ} + R_L} = \frac{57.6\text{ V}}{54.8\ \Omega} = 1.05\text{ A}$$

$$V_L = I_L \times R_L = 1.08\text{ A} \times 50\ \Omega = 52.55\text{ V}$$

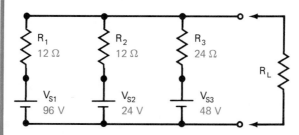

FIGURE 11-29

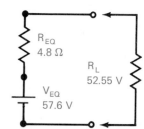

FIGURE 11-30 This circuit is equivalent to the original one shown in Figure 11-28.

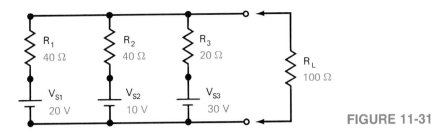

FIGURE 11-31

REVIEW PROBLEMS

11-15. Millman's theorem is applied to what circuit configuration?
11-16. Write the Millman formula for V_{EQ}.
11-17. Write the Millman equivalent for R_{EQ}.
11-18. Determine the load current and voltage for the circuit shown in Figure 11-31.

11-5 Delta (Δ) and Wye (Y) Networks

The resistive networks shown in Figure 11-32 are the same. The only difference between them is in the way they are drawn. This circuit is called a *wye* or *Y* circuit configuration. This circuit is also known as a *tee* or *T* network because of its shape. Figure 11-33 shows the *delta* [part (a)] and *pi* [part (b)] configurations. The reason for these names is, again, based on shape.

The delta and wye circuit configurations are used primarily in the electric power industry. Their use is found in three-phase electric generation and distribution systems. In the delta configuration, instantaneous voltage values around the three sides of the triangle equal zero. This system does not have a neutral wire and it uses three wires for the transmission of electric power. The wye system is used in a similar manner, except the three wires are connected in the center of the wye and this point is the common or neutral point of the circuit. The wye circuit requires four wires.

Delta network *Circuit configuration using three legs connected in series in triangular form.*

Wye network *Circuit configuration using three legs connected at a common point in the shape of the letter Y.*

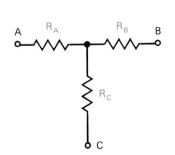

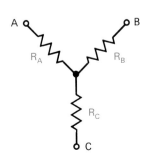

FIGURE 11-32 The wye (Y) network is also known as a tee (T) network, due to its configuration.

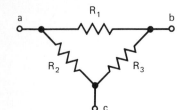 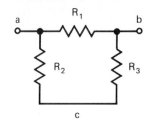

FIGURE 11-33 The delta (Δ) network is also known as a pi (π) configuration, due to its configuration.

When working with electrical or electronic circuit design or analysis it may be convenient to convert from one configuration to the other type of circuit, because conversion often makes it simpler to find the solution. It may even be impossible to analyze the circuit for its mathematical value without first making the conversion.

Formula for Conversion All the resistances shown in Figures 11-32 and 11-33 are identified. The three resistances in the wye circuit are called R_A, R_B, and R_C. Resistances in the delta configuration are identified as R_1, R_2, and R_3. The three output points are identified as A, B, and C in both of the circuits.

The basic formula for the conversion from a wye to a delta circuit configuration is

$$R_\Delta = \frac{\text{sum of all cross products in Y}}{\text{opposite } R_r}$$

The Greek letter sigma (Σ) is used to indicate the term "the sum of" in a formula. When the formula is applied to the circuit shown in the two figures, it looks like this:

$$R_1 = \frac{R_A R_C + R_C R_B + R_B R_A}{R_A}$$

$$R_2 = \frac{R_A R_C + R_C R_B + R_B R_A}{R_B}$$

$$R_3 = \frac{R_A R_C + R_C R_B + R_B R_A}{R_C}$$

The formula for converting a delta to a wye circuit configuration is

$$R_Y = \frac{\text{product of two adjacent } R \text{ in delta}}{\text{sum of all } R \text{ in delta}}$$

This is applied for each of the three components making up the circuit:

$$R_C = \frac{R_2 R_3}{R_{A1} + R_{B2} + R_{C3}}$$

$$R_B = \frac{R_1 R_3}{R_1} + R_1 + R_2 + R_3$$

$$R_A = \frac{R_2 R_1}{R_1 + R_2 + R_3}$$

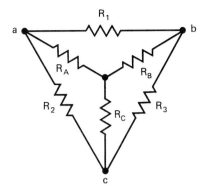

FIGURE 11-34 A combined delta–wye network showing all equivalent resistance values.

Often, a visual image makes the comprehension of the circuit or its analysis a lot simpler. The combined circuits are shown in Figure 11-34. The wye circuit consists of three open-ended arms extending from a central point. The delta circuit is triangular in shape. Notice how points a, b, and c are identical for both of the configurations. The relationship of the two circuits is much easier when illustrated in this way. The terminology used to describe the conversion formulas is simpler. Notice that resistance R_2 is opposite side R_B; R_1 is opposite R_C and R_3 is opposite side R_A. Each of the unterminated resistances in the wye configuration has two adjacent resistances in the delta form. R_A has R_1 and R_2 adjacent; R_B has R_1 and R_3 and R_C have R_2 and R_3 adjacent.

In summary, the conversion process from wye to delta form is as follows:

1. Find the three products of the arms of the Y.
2. Find the sum of the three products.
3. Divide this sum by the value of the opposite arm of the wye.
4. In this conversion the value of the numerator remains the same. The value of the denominator is equal to the value of the opposite arm of the circuit.

In summary, the conversion formula from delta to wye is as follows:

1. Find the product of two adjacent sides.
2. Divide this value by the sum of the three sides of the delta.
3. Repeat for each of the values of the three arms.
4. In this conversion the denominator is a constant value. The value for the numerator is the product of the two adjacent side resistances.

EXAMPLE 11-15

Convert the circuit in Figure 11-35 from delta form to wye form.

Solution

$$R_A = \frac{R_1 \times R_2}{R_1 + R_2 + R_3} = \frac{8\ \Omega \times 20\ \Omega}{40\ \Omega} = \frac{160\ \Omega}{40\ \Omega} = 4\ \Omega$$

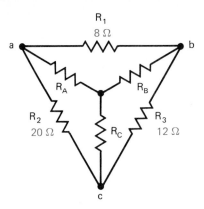

FIGURE 11-35

$$R_B = \frac{R_1 \times R_3}{R_1 + R_2 + R_3} = \frac{8\,\Omega \times 12\,\Omega}{40\,\Omega} = \frac{96\,\Omega}{40\,\Omega} = 2.4\,\Omega$$

$$R_C = \frac{R_2 \times R_3}{R_1 + R_2 + R_3} = \frac{20\,\Omega \times 12\,\Omega}{40\,\Omega} = \frac{240\,\Omega}{40\,\Omega} = 6\,\Omega$$

These values are checked by reversing the process:

$$R_3 = \frac{R_{ARC} + R_{CRB} + R_{ARB}}{R_A} = \frac{24\,\Omega + 9.6\,\Omega + 14.4\,\Omega}{4\,\Omega} = \frac{48\,\Omega}{4\,\Omega} = 12\,\Omega$$

$$R_2 = \frac{48\,\Omega}{R_B} = \frac{48\,\Omega}{2.4\,\Omega} = 20\,\Omega$$

$$R_1 = \frac{48\,\Omega}{R_A} = \frac{48\,\Omega}{6\,\Omega} = 8\,\Omega$$

The results indicate that the values are correct in the conversion of this circuit from delta form to wye form.

Simplifying a Complex Circuit The bridge circuit illustrated in Figure 11-36(a) is very complex. When one wishes to calculate the total source current, the total circuit resistance value needs to be identified. The method of doing this requires a partial circuit conversion. The bridge circuit can be considered as a partial delta configuration with the addition of two series resistances. These two series components are resistances R_4 and R_5. Resistances R_1, R_2, and R_3 form a delta circuit. When this portion of the bridge is converted into a wye circuit, the total resistance value is much easier to calculate.

The steps for this conversion are illustrated in the remaining three sections of this illustration. First, draw the Y configuration on the circuit and identify its components [part (b)]. Then calculate the wye values for the three components. Their values are then added to the redrawn circuit as it now appears in part (c). The next step is to simplify the series–parallel circuit into an equivalent parallel bank and its series component R_C [part (d)]. The final step is to add R_{EQ} to R_C and use this value with Ohm's law to calculate circuit current, I_T.

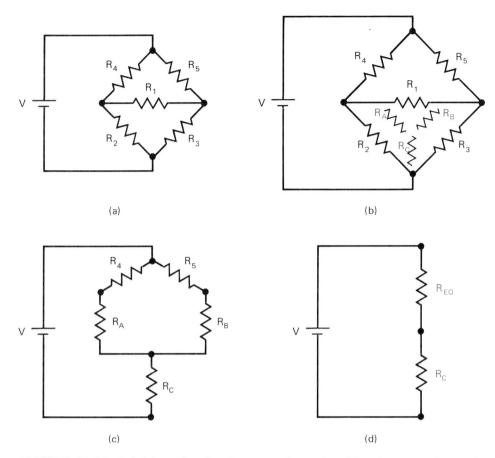

(a)

(b)

(c)

(d)

FIGURE 11-36 A bridge circuit often requires simplification, as shown in these four steps.

Resistances R_1, R_2, and R_3 are first replaced with the wye configuration. This configuration [Figure 11-36(b)] consists of resistances R_A, R_B, and R_C. The delta–wye conversion format is used to do this:

$$R_A = \frac{R_2 R_3}{R_1 + R_2 + R_3} = \frac{12\ \Omega \times 8\ \Omega}{4\ \Omega + 8\ \Omega + 12\ \Omega} = \frac{96\ \Omega}{24\ \Omega} = 4\ \Omega$$

$$R_B = \frac{4\ \Omega \times 8\ \Omega}{24\ \Omega} = 1.33\ \Omega$$

$$R_C = \frac{12\ \Omega \times 4\ \Omega}{24\ \Omega} = \frac{48}{24} = 2\ \Omega$$

The resulting circuit [part (c)] consists of two series resistances in a parallel bank and the two banks connected in series with RC. The total resistance in one branch is

$$R_A + R_4 - 4\ \Omega + 4\ \Omega = 8\ \Omega$$

The total resistance in the other branch is

$$R_B + R_5 = 1.33\ \Omega + 12\ \Omega = 13.33\ \Omega$$

The equivalent resistance for this parallel branch is

$$R_{EQ} = \frac{8\ \Omega \times 13.33\ \Omega}{8\ \Omega + 13.33\ \Omega} = \frac{106.64\ \Omega}{21.33\ \Omega}$$

$$= 4.9995\ \Omega \text{ or } 5\ \Omega \text{ (rounded)}$$

$$R_T = R_{EQ} + R_C = 5\ \Omega + 2\ \Omega + 7\ \Omega$$

The total circuit current, therefore, is

$$I_T = \frac{V_T}{R_T} = \frac{120\ \text{V}}{7\ \Omega} = 17.143\ \text{A}$$

There is another way of recognizing this circuit as a wye–delta combination. The delta could be formed by R_1, R_4, and R_5 just as easily. The solution for circuit values will arrive at the same conclusion using either of the formats.

REVIEW PROBLEMS

11-19. In the wye configuration, which resistance in this circuit is opposite R_2 in the delta?

11-20. In the delta configuration, which two resistances are adjacent to R_C in the wye?

11-21. Explain why the delta-to-wye conversion system is used.

CHAPTER SUMMARY

1. *Thévenin's theorem.* A linear dc circuit, no matter how complicated, can be replaced by an equivalent circuit containing one dc voltage source (V_{TH}) in series with one resistance (R_{TH}).

2. *Norton's theorem.* Any dc circuit, no matter how complicated, can be reduced to a single current source (I_N) and a parallel resistance value (R_N).

3. *Superposition theorem.* Currents in any branch of a multisource circuit are identified by determining the current in each source individually when all other sources are replaced by their internal resistance values. The current flow in a specific branch is the algebraic sum of the current from each source in the specific branch.

4. *Millman's theorem.* A circuit containing multiple source voltages connected in parallel with each other can be reduced to one equivalent voltage in series with a load resistance.

5. Using Ohm's law will convert a voltage source into a current source, and vice versa.

6. A delta configuration may be connected into a wye configuration when a circuit solution is required. It is also possible to reverse the process and convert wye circuits into delta circuits.

Answer true or false; change all false statements into true statements.   **SELF TEST**

11–1. A voltage source has a parallel resistance.

11–2. A current source has a parallel resistance.

11–3. Delta-to-wye conversions are done using Ohm's law.

11–4. A delta network has a triangular form.

11–5. V_{TH} is a short-circuit voltage.

11–6. I_N is a short-circuit current.

11–7. The superposition theorem is used with multiple-path parallel circuits with two or more voltage sources.

11–8. A current source is effectively removed when it is short circuited in the circuit.

11–9. A delta network and a pi network are the same.

11–10. A wye and a pi network are the same.

11–11. A Thévenin equivalent circuit is equal to   **MULTIPLE CHOICE**
  (a) a series load and source
  (b) a parallel load and source
  (c) open source and series load
  (d) shorted source and parallel load

11–12. A Norton equivalent circuit is equal to
  (a) a series load and source
  (b) a parallel load and source
  (c) open source and series load
  (d) shorted source and parallel load

11–13. A Millman equivalent circuit is equal to
  (a) a series load and source
  (b) a parallel load and source
  (c) open source and series load
  (d) shorted source and parallel load

11–14. A voltage source may be converted into a current source by use of
  (a) Millman's theorem
  (b) Norton's theorem

(c) Ohm's law

(d) Thévenin's theorem

11–15. Circuit solution using the superposition theorem is accomplished by

(a) first solving for individual current paths

(b) removing all sources and solving for total resistance

(c) adding all voltage sources

(d) none of the above

ESSAY QUESTIONS

11–16. Explain the use of the term *mesh*.

11–17. Explain why it is possible to solve for circuit values using more than one of the theorems explained in this chapter.

11–18. What is the difference between a delta and a pi network?

11–19. What is the difference between a wye and a tee network?

11–20. Why is network analysis used instead of solving for individual values in each circuit?

PROBLEMS

Use Figure 11-37 to answer problems 11-21 through 11-23.

11–21. Find I_{R3} using Thévenin's theorem. (*Sec. 11–1*)

11–22. Find V_{R3} using Thévenin's theorem. (*Sec. 11–1*)

11–23. Find I_{R1} when $V_s = 150$ V. (*Sec. 11–1*)

Use Figure 11-38 to answer problems 11-24 through 11-26.

11–24. Find current through R_1, using Thévenin's theorem. (*Sec. 11–1*)

11–25. Show the Thévenin equivalent of this circuit. (*Sec. 11–1*)

FIGURE 11-37

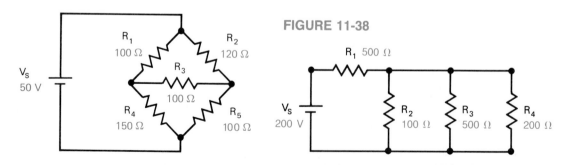

FIGURE 11-38

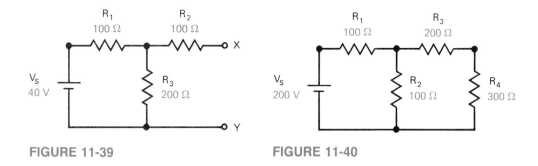

FIGURE 11-39   FIGURE 11-40

11–26. Show the Norton equivalent of this circuit. (*Sec. 11–2*)

11–27. Calculate the current through R_3 in Figure 11-39 using Norton's theorem. (*Sec. 11–2*)

Use Figure 11-39 to answer problems 11-28 and 11-29.

11–28. Compute R_{TH} and V_{TH} for terminals X and Y. (*Sec. 11–1*)

11–29. Compute R_{TH} and V_{TH} for terminals X and Y when R_3 is changed to 400 Ω. (*Sec. 11–1*)

Use Figure 11-40 to answer problems 11-30 and 11-31.

11–30. Draw the Norton equivalent circuit. (*Sec. 11–2*)

11–31. Convert V_s and R_1 into a current source for Figure 11-39. (*Sec. 11–2*)

Use Figure 11-41 to answer problems 11-32 through 11-34.

11–32. Redraw the circuit to show the Millman circuit. (*Sec. 11–4*)

11–33. Calculate V_{AB} using Millman's theorem. (*Sec. 11–4*)

11–34. Calculate V_{AB} using superposition. (*Sec. 11–3*)

Use Figure 11-42 to answer problems 11-35 through 11-37.

11–35. Calculate V_{R2} using Millman's theorem. (*Sec. 11–4*)

11–36. Redraw the circuit to show the Thévenin's equivalent. (*Sec. 11–1*)

11–37. Calculate V_{R2} when R_2 is considered the load resistance. (*Sec. 11–1*)

FIGURE 11-41

FIGURE 11-42

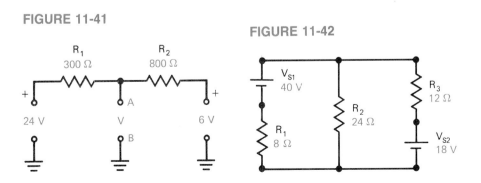

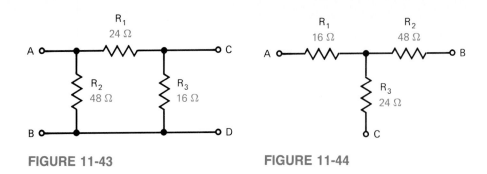

FIGURE 11-43 **FIGURE 11-44**

Use Figure 11-43 to answer problems 11-38 through 11-40.

11–38. Convert the circuit into a wye form. (*Sec. 11–5*)
11–39. Calculate R_T for terminals C_D. (*Sec. 11–5*)
11–40. Calculate R_T for terminals A_C. (*Sec. 11–5*)

Use Figure 11-44 to answer problems 11-41 through 11-45.

11–41. Convert the circuit into delta form. (*Sec. 11–5*)
11–42. Calculate R_T at terminals AC. (*Sec. 11–5*)
11–43. Calculate R_T when measured across terminals AB. (*Sec. 11–5*)
11–44. What current flows through R_3 when a V_s of 48 V is connected to terminals AB? (*Sec. 11–5*)
11–45. What is the current through R_3 when a 48-V source is connected to terminals A and C? (*Sec. 11–5*)

ANSWERS TO REVIEW PROBLEMS

11–1. A linear dc circuit can be replaced with V_{TH} and R_{TH}.
11–2. V_{TH} and R_{TH}
11–3. Short V_{SOURCE}, open load, calculate R in system.
11–4. Open load, calculate V_{TH}.
11–5. $V_{TH} = 50$ V, $R_{TH} = 25$ Ω
11–6. Current source in parallel
11–7. I_N and R_N
11–8. Use R_N and V to calculate I_N (or short load and calculate I_N).
11–9. Short supply, calculate R_N.
11–10. $I_N = 0.01$ A, $R_N = 50$ Ω
11–11. Use Ohm's law, $I_s = V_s/R_s$.
11–12. $V_s = I_s \times R_s$
11–13. $I_s = 50/100 = 0.5$ A
11–14. $V_s = 1 \times 100 = 100$ V
11–15. Multiple voltage sources in parallel
11–16. $V_{EQ} = [(V_1/R_1) + (V_2/R_2) + (V_3/R_3)]/[1/R_1 + 1/R_2 + 1/R_3]$

11–17. $R_{EQ} = 1/(1/R_1 + 1/R_2 + 1/R_3)$

11–18. $V = 20.45$, $I_L = 0.205$ A

11–19. R_B

11–20. R_2 and R_3

11–21. Circuit simplification

CHAPTER 12

DC Meters

One of the major testing instruments for electrical circuitry is the meter. Meters are able to measure almost any electrical quantity. Many of these meters use the principles of direct-current flow to accomplish this. Often one utilizes a meter to measure a variety of electrical values. One meter movement, or unit, may be used with an additional switching arrangement and configured to measure almost all electrical quantities.

KEY TERMS

Analog	Meter Shunt
Autopolarity	Multiplier Resistor
Autoranging	Nonlinear Scale
Continuity Tests	Ohmmeter
D'Arsonval Meter	Ohms Per Volt
Digital	Sensitivity Rating
Full Scale Deflection	Voltmeter
Linear Scale	Voltmeter Loading
Loading	Zero ohms

OBJECTIVES

Upon completion of the material in this chapter, you will

1. Understand the different types of basic meters.
2. Be able to use a meter to measure current, voltage, and resistance values.
3. Be able to explain the shunt and multiplier.
4. Be able to calculate values for shunts and multipliers.
5. Understand the safety rules for meter use.
6. Recognize digital and analog meter movements.

The ability to design an electrical or an electronic circuit is important. Equally important is the ability to use the proper type of testing and measuring equipment to evaluate a working model of the design. The ability to select and use the correct piece of test equipment is necessary if we are to be able to obtain quantitative and meaningful results. Measurements of voltage, current, and resistance are required before the final evaluation of a circuit can be accomplished. We are unable to see electricity. We do have the ability to use instruments to be able to measure different aspects of electrical energy. The industry has developed some basic types of electrical measuring devices. These are known as the voltmeter, the ammeter, and the ohmmeter. In many instances these three types of measuring devices are combined into one composite unit. This unit is commonly called a V-O-M, the letters representing volts, ohms, and milliamperes. Components inside the VOM unit are added to a single measuring meter by use of a selector switch. This reduces the need for individual measuring devices and provides a versatile tool for our use.

12-1 The Meter Movement

D'Arsonval meter *Meter using a rotating armature to indicate current strength.*

There are several types of electrical meters that have been used for measuring purposes. The most popular type of meter is called the moving-coil, or **D'Arsonval, meter** (Figure 12-1). This meter, which uses the principle of electromagnetism, has a semicircular permanent-magnet frame. It also has a small armature suspended between the ends, or poles, of the magnet frame. A pointer is attached to the armature. The armature is held in place by an axle. Coil springs on the axle provide a force that is used to keep the pointer on the left side of the assembly, in an "at rest" position. The coil springs are also used to connect wires from the armature to a set of terminals on the case of the meter. This permits armature rotation while establishing an excellent electrical connection between the moving-coil armature and the nonmoving terminals of the meter.

One other component that is standard with all meter movement assemblies is a calibrated dial, or faceplate. The dial scale is marked with a zero reference place on one side and a numeric value on the right-hand side. Almost all meter dial scales are linear in their calibration. The term **linear scale** indicates that the spacing between each mark is equal. This type of marking is shown in Figure 12-2(a). A **nonlinear scale** is illustrated in Figure 12-2(b). This type of scale has different values for equal distances. Its use is limited primarily to the measurement of resistance values.

Linear scale *Meter dial face having equal spacing between each marking or division.*

Nonlinear scale *Unequal divisions between markings on dial face of meter.*

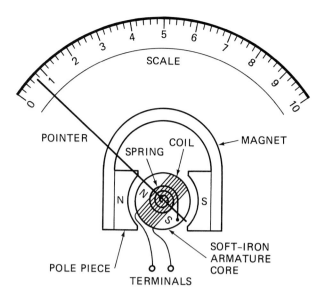

FIGURE 12-1 One of the most popular analog meter movements is known as the D'Arsonval, in honor of its inventor. (From Joel Goldberg, *Fundamentals of Electricity*, 1981. p. 252. Reprinted by permission of Prentice-Hall, Englewood Cliffs, New Jersey.)

The D'Arsonval meter movement uses the principles related to the interaction of magnetic fields. The fixed poles of the magnet establish a magnetic field. The field follows the metal in the magnet and also exists in the space between the pole pieces. When an electrical current flows through the wires on the armature of the meter, a magnetic field is established around the armature. This field reacts with the field from the magnet poles. The result of this is a rotating action. The armature moves in a clockwise manner in the magnetic field. The amount of this movement is dependent on the relative strengths of the two sets of magnetic fields. The pole magnetic field has a constant value, whereas the field in the armature varies in strength. This is due to the quantity of current that flows in the armature winding. The level of current flow rotates the armature from its left-side "at rest" position to some other point until it reaches the extreme right-side part of the meter

All of the action just described occurs when a current flows through the armature of the meter movement. Each meter movement will have specific electrical characteristics. They include a current rating and a resistance. We know that a voltage is required in order to create an electron flow. A voltage

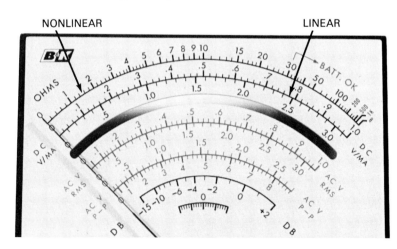

FIGURE 12-2 A linear scale has equal values for each space between each division. The nonlinear scale places different values for each space. (Courtesy B & K-Precision/Dynascan Corporation.)

rating may also be found as one of the ratings for a specific electric meter movement. The D'Arsonval meter movement is often rated with a **full-scale deflection** value. This value states the amount of current that will provide a full-scale deflection of the needle. It is also called the meter's **sensitivity rating**. It is normal to use the terms 1 mA, 5 A, 100 μA meter. These terms report the amount of current required for full-scale deflection. The inference, and unstated, part of these descriptions is a 0 to 1 mA, 0 to 5 A, or 0 to 100 μA meter movement.

The quantity of current that can flow through any armature winding is limited by the physical size of the armature and the size and quantity of wire that can be used for its coil. This limitation resulted in the manufacture of meter movements with rather low current ratings. Most of the basic meters in use have deflection sensitivities of 1 mA or less. All wires exhibit the property of resistance. The wire used to form the armature of the meter is no exception. The resistance of the armature winding of a meter is often less than 1000 Ω.

When the armature of the meter is rated with a deflection sensitivity of 1 mA and its coil resistance is 1000 Ω, only 1 V is required to produce a full-scale deflection. Keep in mind that when less than 1 V is applied to the terminals of the meter, there will be less current flow. The exact amount of current is determined by the use of Ohm's law. When 1 V is applied and there is some resistance in addition to the meter in the circuit, the current flow will also be less than 1 mA. Any current value of less than 1 mA will produce a reading of less than 1 mA on the meter scale. The percent of deflection is directly proportional to the percent of current flow.

The armature of the meter has a definite magnetic relation to the poles of the meter. The polarity of the two must be proper in order to create a *positive* meter deflection. This term indicates a meter needle movement from the left-hand side toward its right-hand side. When, or if, the leads from the meter's terminals have a reverse polarity, the armature and needle attempt to give a *negative* reading. The needle tries to move farther left on the scale. Usually, this movement is limited due to the use of a small mechanical stop that is a part of the meter assembly. A reversal of the two leads to the meter corrects this problem. Almost all basic meters have a + polarity sign on one of the two terminals that connect the movement to the rest of the circuit. This terminal should always be connected to the positive point of the circuit, and the plain, or (−), terminal should be connected to the negative side of the point of measurement. Details on how to connect the meter into a circuit are included in this chapter.

REVIEW PROBLEMS

12-1. Name the parts of a D'Arsonval meter movement.

12-2. Describe the effect of reverse polarity connections.

12-3. Define the following terms: (a) deflection sensitivity; (b) full-scale deflection; (c) negative reading.

12-2 Current Measurement

Almost all moving-coil meters, like the D'Arsonval movement, measure current as a basic function. This type of meter is shown in Figure 12-3, along with the basic symbol used to indicate a meter. The flow of current establishes the size of the magnetic field in the meter's armature. Therefore, whether we realize it or not, whatever quantity being measured, including voltage and resistance, is based on the amount of current flowing through the meter armature's winding.

The ammeter must be connected in series in order to measure circuit current (Figure 12-4). In a series circuit such as this one the current is equal in all parts of the circuit. All of the circuit current must flow through the meter if one expects to obtain an accurate measurement. In theory the ammeter may be connected at any point in the circuit and will display the quantity of circuit current. These points are indicated by an "X" on the drawing. In practice a good habit to form is to place the meter as close to the negative or ground of the system as possible. The circuit in Figure 12-4 has a low voltage value of 24 V from its source. If you would accidentally touch one of the meter leads and touch circuit ground at the same time, you might receive a mild electric shock. If the circuit being tested had a source voltage of 500 or 600 V, the shock you would receive would be greater. When the ammeter is connected near the positive terminal of the source, the possibility of an electrical shock is greater. This is because the meter and its leads are carrying the high-voltage potential. The voltage drop from either of the meter leads to ground is close to 500 or 600 V. When the meter is connected near the ground of the system, the voltage drop between the meter and its leads and ground is a very low value, usually less than 10 V. Any accidental electrical shock would not do much damage when the voltage is this low.

The polarity of the ammeter is important. When one wants a positive reading, the positive (+) terminal of the ammeter is connected to the positive-going side of the point of insertion and the negative (−) lead is connected to

(b)

FIGURE 12-3 Basic meter movement (a) and schematic symbol for a meter (b). A letter inside the circle indicates either a basic meter or the type of measurement the meter is used to indicate. [(a) Courtesy of Simpson Electric Company.]

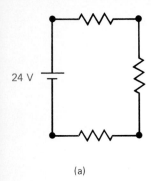

(a)

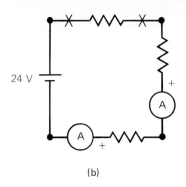

(b)

FIGURE 12-4 Placement of the current measuring meter, or ammeter, in the circuit.

the negative-going side of the insertion point. This connection is necessary no matter where the meter is added to the system.

Extending the Measurement Range Often it is necessary to use a low current meter in a circuit that has a current flow greater than the meter is able to measure. One way of solving this problem is to obtain a meter with the proper sensitivity range. This is not always possible because of cost and size factors. A better solution for a problem of this type is to extend the range of the basic meter movement. The ammeter's range is extended by the addition of a current bypass device. This device is called a **meter shunt** (Figure 12-5). Its purpose is to permit a measured percentage of the total circuit current to flow around the meter movement. The meter shunt is connected in parallel with the meter movement. A parallel resistance bank is created by the two devices. Current flow in a parallel bank is equal to the total circuit current. Current in each of the branch circuits is dependent on the resistance in that branch.

Meter shunt *Resistor placed in parallel with meter movement in order to bypass part of current flow and increase value of indication.*

These rules also apply to the meter and its shunt. In general the value of the shunt resistance affects the total current flow. As the shunt resistance is reduced, the total circuit current increases. Some meters have a built-in shunt. It is part of the meter package. Other meters require the addition of an external shunt resistance in order to extend the range of the basic meter movement.

Calculating Shunt Resistance Calculations can be performed using the rules for parallel circuits. The same rules apply to the meter and its shunt resistance. The application of Ohm's law is required to do this. Ohm's law

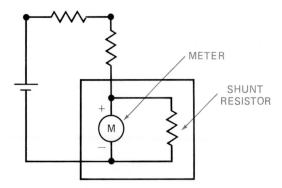

METER

SHUNT RESISTOR

FIGURE 12-5 An ammeter's basic range may be extended by the addition of a parallel "shunt" resistance across the meter movement.

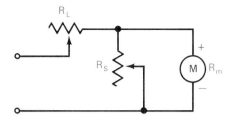

FIGURE 12-6 Determination of the internal resistance of any meter movement is accomplished using this circuit.

requires knowledge of the voltage and current to determine resistance. The resistance of a current meter cannot normally be measured with an ohmmeter. This is because the ohmmeter has an internal power source that will attempt to force more current through the meter armature than it can properly handle. As a result, the needle of the meter deflects to the far right, past the edge of the scale. When this happens the meter is said to be *pinned* and can be damaged internally.

A better method of measuring meter armature resistance, R_M, is shown in Figure 12-6. It requires a series variable resistance, R_L, and a parallel-wired variable resistor, R_S. The first step is to connect the circuit without using R_S. The variable resistance R_L is adjusted to a point where the meter indicates a full-scale deflection. The next step requires the addition of the second variable resistor, R_S, to the circuit. This resistance is adjusted until the needle on the meter indicates a half-scale reading. The circuit is then disconnected and the ohmic value of R_S is measured. The measured value of R_S is equal to the resistance of the meter.

The reason for this is that a current will divide when it has two branches to follow. When the circuit was first built the series resistance was adjusted to indicate a full-scale deflection. Adding a parallel resistance with an ohmic value equal to the resistance of the meter divides the current in the circuit equally. Half of the current flows in each branch. The variable resistance value can be measured without damage to it. Since its resistance equals the meter resistance, the resistance of the meter can safely be identified. When the resistance of the meter is known, it is possible to apply Ohm's law to find the value of voltage required for a full-scale deflection. This is the value of voltage that is applied to both the meter and the shunt.

Circuit Loading The introduction of a current reading meter, or ammeter, should have very little effect on the original circuit if the readings are to be accurate. Generally speaking, the resistance of the ammeter should be under 0.01% of the total circuit resistance. An ammeter used in a circuit whose total resistance is 2500 Ω should have a resistance value of 25 Ω or less. Usually, the shunt in the meter reduces the ammeter's resistance to under 2 Ω. If the resistance of the ammeter is too high, this value of resistance added to the circuit offers a false reading. The meter is **loading** the circuit by adding an additional and unwanted resistance (Figure 12-7). The original circuit contains two 2.5-kΩ resistances connected in series to a 20-V source. This produces a current flow of 4 mA and a voltage drop of 10 V across each resistance. When an ammeter with a resistance of 2.5 kΩ is added to this circuit, all the voltage, current, and resistance values change. The addition of another 2.5 kΩ reduces current flow and reduces the voltage drops across each of the components.

Loading Term used to describe effect of adding meter's resistance to circuit when making measurements.

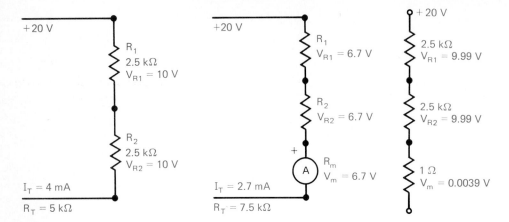

FIGURE 12-7 The addition of a meter to a circuit may "load" the circuit, or change its actual values, since the meter has resistive qualities.

FIGURE 12-8 A low value of meter and shunt resistance has little effect on the circuit.

When the meter and shunt combined resistance is around 1 Ω, there is very little effect on the circuit or its values (Figure 12-8). A total current change of 0.0001 A (100 μA) occurs and the voltage drops across the two resistances remains very close to the values before the meter was added to the circuit.

Shunt Design Design is accomplished by using the rules for parallel circuit. The first step in the design is to determine the value of shunt current, I_S. This value is found by subtracting the full-scale current value from the total current range. The example shown here will use the values shown in Figure 12-9. Total circuit current is 100 mA. The meter maximum current is 1 mA. The shunt current is

$$I_S = I_T - I_M \qquad (12\text{-}1)$$
$$= 100 - 1 = 99 \text{ mA}$$

Find the voltage required to give a full-scale deflection of 1 mA throught the resistance of the meter. Use a resistance value of 1 kΩ for the meter. It represents a typical value for a meter of this type.

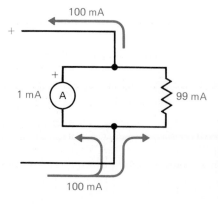

FIGURE 12-9 A meter shunt is designed to permit the by-passing of a predetermined quantity of current. The meter then reads a percentage of the total current in the branch circuit.

$$V_M = I_M \times R_M$$

$$= 0.001 \text{ A} \times 1000 \text{ }\Omega \qquad (12\text{-}2)$$

$$= 1.0 \text{ V}$$

When the voltage, V_M, and the shunt current, I_S, are known, these values are used to determine the value of the shunt resistance, R_S.

$$R_S = \frac{V_S}{I_M}$$

$$= \frac{1 \text{ V}}{0.99 \text{ A}} \qquad (12\text{-}3)$$

$$= 1.01 \text{ }\Omega$$

The value of the shunt resistance is about 1 Ω.

When this value is connected in parallel with the movement of the meter, a current will flow through both the shunt and the armature. The armature has a resistance of 1000 Ω, while the shunt's resistance is just over 1 Ω. When a current of 100 mA is measured, 99/100 of the current flows through the shunt resistance. The needle of the meter will indicate a full-scale deflection. The dial should be marked to indicate a value of 100 mA for those conditions. When a current of 60 mA is measured, the current will divide proportionally and the needle will indicate a value of 60% of the total current.

EXAMPLE 12-1

Design a meter shunt that will extend the range of a 100-Ω meter from 10 mA to 500 mA (Figure 12-10).

Solution

Two values are necessary to solve this problem. First, the voltage value is needed. The second item is the shunt current.

$$V_M = I_M \times R_M$$

$$= 0.01 \text{ A} \times 100 \text{ }\Omega = 0.1 \text{ V}$$

$$I_S = I_T - I_M$$

$$= 500 \text{ mA} - 10 \text{ mA} = 490 \text{ mA} \quad \text{or} \quad 0.49 \text{ A}$$

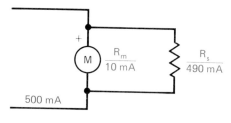

FIGURE 12-10 Circuit used to show the design of a shunt, permitting a 10-mA meter movement to measure up to 500 mA.

The shunt value is determined by use of Ohm's law.

$$R_S = \frac{V_S}{I_S}$$

$$= \frac{0.1 \text{ V}}{0.490 \text{ A}} = 0.2 \text{ }\Omega$$

Shunt resistance is 0.2 Ω for this requirement.

A second method of determining the value of a shunt resistance uses a variation of Ohm's law. This variation is based on the fact that one equality may be subtracted for another and still maintain an equation. It is shown as follows:

$$V_{SHUNT} = V_{METER}$$

$$I_{SH}R_{SH} = I_M R_M$$

Therefore

$$R_{SH} = \frac{I_M R_M}{I_{SH}}$$

This procedure does not require finding the voltage across the meter and the shunt. It may be used instead of the process requiring determination of V_M and then using Ohm's law.

REVIEW PROBLEMS

12-4. Find the shunt current for a meter having a basic movement of 1 mA when it is to be used to measure the following values.
(a) 200 mA full scale
(b) 600 mA full scale
(c) 5 A full scale

12-5. Find the shunt resistance value for the shunt when the meter movement current is 10 mA and the armature resistance is 1 kΩ.
(a) 200 mA full scale
(b) 600 mA full scale
(c) 5 A full scale

12-6. Use the alternative method to find the shunt resistance value required to extend a 0- to 1-mA meter with a 100-Ω coil to read 1 A of current.

When more than one range of current is desired, the system shown in Figure 12-11 is used. A mechanical switch is used to select different values of shunt resistance. The number of shunts is limited only by the quantity of ranges desired. This circuit is used to make one meter movement more versatile. A switching circuit is used in commercial multimeters to extend the range of current it is able to measure.

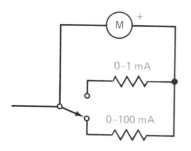

FIGURE 12-11 Use of a switch to select different values of shunt resistance makes the meter more adaptable to a variety of measurement requirements.

12-3 The Voltmeter

The basic meter movement is a current-sensitive device. It reacts to the flow or current through its armature winding. We have seen how the current is shunted, or bypassed, around the armature when a quantity greater than the capabilities is measured. In this type of mesurement the voltage applied to the parallel bank of the meter and its shunt is a constant value. It is also a very low value. The voltage required for full-scale deflection of an ammeter is usually less than 1 V.

   The basic meter movement is also used to measure voltage. If placing a resistance in parallel extends the current range of the meter, placing a resistance in *series* with the armature of the meter will extend the voltage range. A meter used to measure voltage is shown in Figure 12-12. This meter is placed in parallel with the points in the circuit where the measurement is to be taken. The voltmeter is used to measure the difference in electrical potential, or voltage, between two points in a circuit. The series resistance establishes a voltage drop across the series string consisting of the meter (R_M) and the series resistance (R_{mult}). The resistance placed in series is called a **multiplier resistor**. It multiplies, or extends, the voltage range of the meter movement. In this illustration a basic meter movement is used to measure a value of 20 V. The meter movement has a voltage rating of 10 mV, or 0.01 V. The multiplier resistance required to measure 20 V must establish a voltage drop of 19.99 V. These two voltage drops equal 20 V when added together. When a current of 0.001 A flows in this circuit the meter will indicate a 20-V value. Less voltage will cause a smaller amount of current to flow and offer a less-than-full-scale reading on the meter.

   The determination of the value of the multiplier resistance is done by use of Ohm's and Kirchhoff's laws. It may be done as a two-step operation or as a combined one-step calculation. The two-step operation is accomplished by first finding the voltage drop of the meter and subtracting it from the value to

Voltmeter *Used to measure circuit voltage values.*

Multiplier resistor *Resistor placed in series with meter movement in order to use it as a voltmeter.*

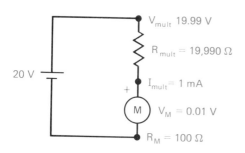

V_{mult} 19.99 V

$R_{mult} = 19,990\ \Omega$

20 V

$I_{mult} = 1$ mA

$V_M = 0.01$ V

$R_M = 100\ \Omega$

FIGURE 12-12 A basic meter movement is used as a voltmeter when a series "multiplier" resistor is included.

be measured. Next, the use of Ohm's law will identify a value of multiplier resistance. The circuit shown in Figure 12-12 is used to show this.

$$V_M = I_M R_M$$

$$= 0.001 \text{ A} \times 100 \text{ }\Omega = 0.1 \text{ V}$$

$$V_{MULT} = V_T - V_M \qquad (12\text{-}4)$$

$$= 20 \text{ V} - 0.1 \text{ V} = 19.9 \text{ V}$$

$$R_{MULT} = \frac{V_{MULT}}{I_{MULT}}$$

$$= \frac{19.90 \text{ V}}{0.001 \text{ A}} = 19,900 \text{ }\Omega$$

The alternative method of finding this value uses a combination formula:

$$R_{MULT} = \frac{V_{FULL\ SCALE}}{I_{FULL\ SCALE}} - R_{METER}$$

$$= \frac{20 \text{ V}}{0.001 \text{ A}} - 100 \text{ }\Omega \qquad (12\text{-}5)$$

$$= 20,000 \text{ }\Omega - 100 \text{ }\Omega = 19,900$$

Either method will produce the same result. This is the value of the resistance to be placed in series with a meter's movement. Its purpose is to extend the voltage reading range of the meter.

EXAMPLE 12-2

Determine the value of multiplier resistance required to measure 500 V with a meter with values of 1000 Ω and 100 μA.

Solution

$$R_{MULT} = \frac{V_{FULL\ SCALE}}{I_{FULL\ SCALE}} - R_M$$

$$= \frac{500 \text{ V}}{0.0001 \text{ A}} - 1000 \text{ }\Omega$$

$$= 5,000,000 \text{ }\Omega - 1000 \text{ }\Omega$$

$$= 4,999,000 \text{ }\Omega$$

The multiplier resistor value is 4,999,000 Ω.

Range Extension The voltmeter's range is extended by changing the value of the multiplier resistance. In Figure 12-13 three multiplier resistances are used. One is selected for each specific range. Since each multiplier resistance

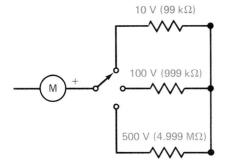

FIGURE 12-13 The basic meter range is extended for voltage readings when a switch and series resistances are included in its circuit.

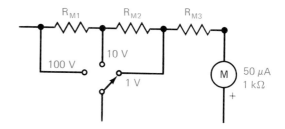

FIGURE 12-14 Use of a series voltage-dividing string instead of individual multiplier resistances is another method of range extension for the voltmeter circuit.

has a different ohmic value, the meter will also measure different values for each range.

A second method of connecting the multiplier resistances is shown in Figure 12-14. This method uses a series string of resistances. The switching arrangement literally *taps* into the series string to select the proper value of resistance. Resistances R_{M1}, R_{M2}, and R_{M3} are selected so that their values, when added, make up the proper multiplier value for a specific voltage range. The system has unlimited range possibilities. Its only limitations are general cost and size of the housing containing the meter movement, switches, and the associated resistance values.

Reading the Dial Most multirange meters use either one or two scales for the voltmeter range. One of these is shown in Figure 12-15. It is calibrated, or marked, with values between 0 and 1. Each major division is marked from 0 on the left and increasing in units of 0.1 V. There are five minor divisions, or markings, between each number. Each of these has a value of one-fifth of the difference between the major values. For example, the difference between 0.3 and 0.2 is 0.1. When 0.1 is divided by 5, the minor divisions have a value of 0.02 each. This relates to 20% of the difference.

The scale is *read* using this format. When the needle rests above one major division, such as 0.2, the measured value is 0.2 of the total scale reading. If the needle rests above 0.2 and two minor divisions, the value is 0.24 (0.2 + 2 × 0.02). This type of interpretation of the dial reading is typical for an analog type of meter. The readings are open to the interpretation of where the needle stops.

The need to expand the range of voltmeter readings requires some method of reading the scale of the meter. The usual method used for this is the mark the switch used to add multiplier resistors with the values for ech position of the switch. When the switch knob is pointed at the 1-V range, the meter dial is read directly as it is marked. The range is 0 to 1 V and each of the markings relates to values between 0 and 1.

When the switch is moved to a higher range, 100 V, the values of the dial scale have to be increased. There is no convenient method of actually changing

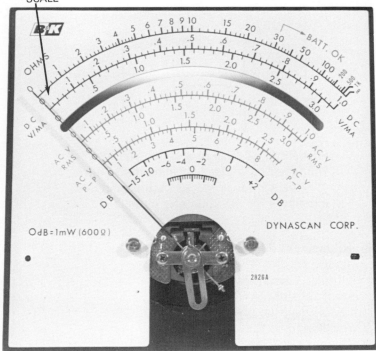

FIGURE 12-15 Typical volt-meter dial scale. Range extension requires the mental multiplication of these values to determine the actual values being measured. (Courtesy of B & K-Precision/Dynascan Corporation.)

the numbers and markings on the dial scale. The method used by most people using this type of meter requires a little mental multiplication. The values of 0 to 1 are now actually indicating a reading between 0 and 100. Each of the marked values has to be mentally multiplied by the new value. In this example each value is multiplied by 100. This will extend the range to the proper reading for the higher scale. When the needle is on the 0.3 mark and the range switch of the meter is on the 100-V setting, the meter is indicating a reading of 30 V.

The method of determining the proper value for the reading requires the division of the range by the full-scale marking of the meter dial. This value is the true reading of the meter.

EXAMPLE 12-3

A meter dial reads from 0 to 1. The meter range switch is on the 50-V position. What is the value for each major division?

Solution

Divide the extended range value by the basic meter reading maximum value:

$$\text{meter multiplier} = \frac{V_{\text{RANGE}}}{V_{\text{FULL SCALE}}}$$

$$= \frac{50 \text{ V}}{1 \text{ V}} = 50 \text{ V}$$

Each value of the scale on the meter is multiplied by 50 to interpret the reading properly.

REVIEW PROBLEMS

12-7. Indicate the multiplication factor for each range setting using a 0-to-10 meter scale.
(a) 100 V
(b) 250 V
(c) 500 V
(d) 25 V

12-8. Fill in the value of voltage indicated when the conditions on the chart are present. The meter is a 0-to-1 dial reading.

Range	Dial Needle Indication	Actual Reading
(a) 10 V	0.4	_____
(b) 25 V	0.6	_____
(c) 100 V	0.21	_____
(d) 500 V	0.35	_____

When the range is selected, the value selected is a *maximum* reading for that range. If the 500-V range is used, the meter will measure any value between 0 and 500 V. This is true for each of the ranges for the voltmeter.

Measuring Voltage The two leads of the voltmeter are placed in parallel with the circuit or components (Figure 12-16). Often the voltmeter is used to measure a value using circuit ground as a reference [part (a)]. When this is done, the negative lead of the meter is connected to circuit ground and the positive lead is connected to the point in the circuit to be measured. This procedure shows the value of voltage between ground and the test point. In many situations the fact that voltage is measured from the ground reference is assumed. The phrase "We have 400 V at this point" is stated when the ground reference is assumed.

A voltage may also be measured as a difference in potential between two points in the circuit. Under the conditions shown in part (b), neither of the meter leads are connected to circuit common. When a measurement of this type is made, the point of reference is included in the statement of the voltage value. A measurement of this type is stated: "There is a voltage drop of 25 V across R_2 in the circuit." This means that the difference in voltage or potential from one end to the other end of that resistor is 25 V.

Both methods of measuring are correct. Both methods are also commonly used. The difference is the reference point. Unless otherwise stated, the reference for voltage measurement is assumed to be circuit common, or ground.

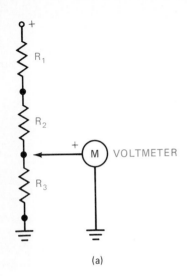

(a)

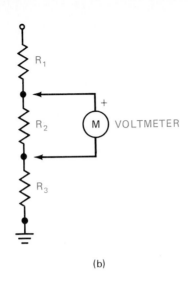

(b)

FIGURE 12-16 Voltage measurements are made when the meter is connected in *parallel* with the circuit components.

Voltmeter Resistance Voltmeter resistance is a term used to describe the total resistance for a voltmeter circuit. This value is the ohmic value for the meter and its multiplier resistance. This value will be different for each range of the multiple-range voltmeter. This is illustrated in Figure 12-17. Assume a 0- to 1-mA meter with an internal armature resistance of 100 Ω. The multiplier resistance for each range is shown. An increase in the voltmeter's range is an increase in the *total* resistance of the voltmeter unit. This total value of resistance may affect the measurement.

Voltmeter loading *Effect of internal resistance of voltmeter on circuit values when making measurements.*

Voltmeter Loading Keep in mind that the voltmeter is a resistance as far as the circuit is concerned. A voltmeter with a total resistance of 10 kΩ placed in parallel with a 10-kΩ resistance develops a parallel circuit with an equivalent resistance of 5 kΩ. A voltage drop across a 10-kΩ resistance that is part of a series string will be reduced when the additional 10 kΩ is placed in parallel with it. The resulting decrease in total resistance will offer a false voltage reading under these conditions. A better method requires the use of a voltmeter having a much higher internal resistance value.

Let us explore the electrical theory behind the loading effect. The two illustrations in Figure 12-18 will help explain this effect. Both circuits have two 5-kΩ resistances connected in series to a 24-V source. Current flow in the circuit before the meter is connected is 0.0024 A. The *IR* voltage drop across each resistor is 12 V since they are equal. This is also found by use of Ohm's law.

$$V_{R1} = I_{R1} \times R_1 = 0.0024 \text{ A} \times 5000 \ \Omega = 12 \text{ V}$$

Range	R_{meter}	R_{mult}	R_{total}
10 V	100 Ω	9,900	10 kΩ
100 V	100 Ω	99,000	100 kΩ
500 V	100 Ω	499,000	500 kΩ
1 kV	100 Ω	990,000	1 MΩ

FIGURE 12-17 Chart showing the values of resistance for multiplier resistors used in the voltmeter.

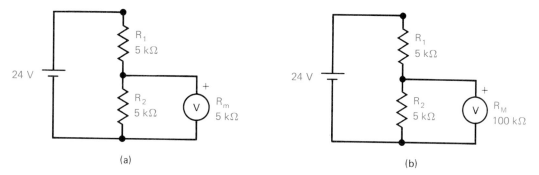

(a)   (b)

FIGURE 12-18 A voltmeter having a low internal resistance placed across a resistance may change the equivalent resistance due to the internal resistance of the meter movement. This also changes the value of voltage being measured and provides a false reading.

A voltmeter having an internal resistance of 5 kΩ is usd to measure the IR voltage drop across R_2, as seen Figure 12-18(a). The equivalent resistance across R_2 is changed by the addition of the meter. The equivalent resistance is now

$$R_{EQ} = \frac{R_2}{2} = \frac{5000 \ \Omega}{2} = 2500 \ \Omega$$

Current flow in the circuit increases:

$$I_T = \frac{V_T}{R_T} = \frac{24 \ V}{7500 \ \Omega} = 0.0032 \ A$$

The IR voltage drop across R_{EQ} decreases:

$$V_{R2} = I_{R2} \times R_{EQ} = 0.0032 \ A \times 2500 \ \Omega = 8 \ V$$

This value of 8.0 V is a false value. It is caused by the addition of the low-resistance meter.

Figure 12-18(b) shows the use of a meter with an internal resistance of 100 kΩ. When this meter is placed in parallel with R_2, the equivalent resistance is

$$\frac{1}{R_{EQ}} = \frac{1}{5000 \ \Omega} + \frac{1}{100,000 \ \Omega} = 0.0002 \ S + 0.00001 \ S = 0.000021 \ S$$

$$R_{EQ} = \frac{1}{0.000021 \ S} = 4762 \ \Omega$$

Current flow in this circuit is

$$I_T = \frac{V_S}{R_T} = \frac{24 \ V}{9762 \ \Omega} = 0.0025 \ A$$

The IR voltage drop across R_2 is

$$V_{REQ} = I_T \times R_{EQ} = 0.0025 \ A \times 4792 \ \Omega = 11.98 \ V$$

The measurement of 11.98 V is still not a true reading. It is so close to the actual 12.0 V that it is acceptable. In a real situation if one used an analog dial meter it would be almost impossible to derive the exact reading of 11.98 V. In practice any reading within 5 or 10% of the expected measurement is usually acceptable.

Voltmeter Ratings Ratings are an essential part of the knowledge one requires if proper test and measurements are to be made. Almost all nonelectronic voltmeters have a resistance rating. The rating is related to the amount of internal resistance for each volt. The term used to describe this value is **ohms per volt**. A voltmeter having a rating of 5000 Ω/V adds 5000 Ω of meter resistance for each volt for a specific range. The use of multiplier resistances for extending the measuring range of the voltmeter also increases the internal resistance of the voltmeter. When a 5000-Ω/V-rated meter is used, an additional 5000 Ω of resistance is added for each volt of that range. A 10-V range has 10 × 5000 Ω or 50 kΩ of internal resistance and a 100-V range has 500 kΩ of internal resistance.

Ohms per volt Term describing internal resistance value of voltmeter.

The ohms per volt rating is determined by taking the reciprocal of the current range of the meter movement. A 100-μA meter movement's internal resistance is calculated by use of the formula $1/I$ or $1/0.0001 = 10$ kΩ. The value of 1000 Ω/V was used for many years as a standard for meter movements. Today, a value of 20 or 30 kΩ/V is standard. The advantage of a higher ohms-per-volt rating is less circuit loading and the ability to measure lower voltage values. Most voltmeters or multiple-function meters have the ohms-per-volt value exhibited on the faceplate of the meter dial. The ohms-per-volt rating of a meter is called its *sensitivity* rating.

12-4 The Ohmmeter

Ohmmeter Meter used to measure resistance values.

A third practical use for the basic meter movement is as an **ohmmeter**. This device is used to measure resistance values. The technician needs to test circuits to measure the value of resistance present. The ohmmeter is used to locate open and short circuits in addition to the measurement of a component's ohmic value.

A schematic diagram for a basic ohmmeter is shown in Figure 12-19. The components required for an ohmmeter are a meter movement, a power source, and a calibrating potentiometer. The theory related to the ohmmeter is that 1 V produces a current of 1 A through a resistance of 1 Ω. Using this formula, 1 V will produce a current of 0.001 A through a resistance of 1000 Ω. Any increase in resistance produces a reduction of current. The ohmmeter uses this principle. When it is used the first step requires short circuiting the ohmmeter's

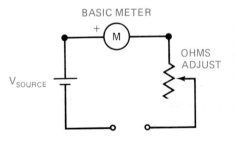

FIGURE 12-19 The basic meter movement is used to measure resistance (ohms) when an internal power source and current-limiting control are included in the circuit.

terminals. This completes the series circuit and current flows. The ohms-adjust potentiometer is adjusted until a full-scale current is obtained. This reading is usually on the right-hand side. It indicates *no resistance* or a *zero ohms* reading. The term **zero ohms** is used to represent an extremely low value of resistance. The short circuit is removed, opening the circuit and stopping all current flow. The meter needle returns to its at-rest position. This indicates an infinitely high resistance. This side of the scale is marked with an infinity symbol (∞). This indicates a value that is beyond the capability of the meter to measure.

Zero ohms *Term describing a very low resistance value, usually too low to measure accurately.*

When the extremes of zero and infinity are established, the next step is to connect a resistance across the meter terminals. This is usually accomplished by connecting the meter's test leads to the component to be measured. The addition of the component completes the circuit. This time, however, the amount of resistance in the series circuit has increased. The source voltage still produces a current flow. The current flow will be less because of the insertion of the additional resistance in the circuit. The meter needle will now indicate a current value of less than full scale. The meter scale (Figure 12-20) is marked in units of the ohm. This scale is nonlinear due to the action of the basic meter. The calibrating, or markings, of the dial represent values obtained by the application of Ohm's law for the current and source voltage.

Range Extension Range extension for the ohmmeter is accomplished as illustrated in Figure 12-21. A resistance for each range is placed in parallel with the meter movement. This permits a portion of the total current flow through the meter. The values of these range extending shunt resistances are selected

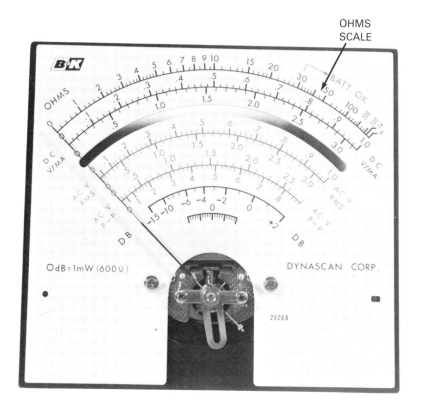

FIGURE 12-20 The dial scale for an ohmmeter is nonlinear. (Courtesy of B & K-Precision/ Dynascan Corporation.)

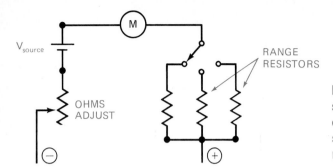

FIGURE 12-21 Range exten-
sion for the ohmmeter is accomplished by use of a
switch and current-limiting
resistances.

to develop a scheme for simplifying meter readings. Normally, the range values for an ohmmeter are in multiples of 10. This is the simplest method of determining the extended range. It requires the technician to multiply the dial reading by units of 10,000,000, and so on. In reality, all that is required is the addition of zeros on the end of the reading value.

The usual set of multipliers used for the ohmmeter are 1, 10, 100, 1000, 10K, 100K, and 1M. The range switch is marked R × 1, R × 10, and so on. This is read by multiplying the R value on the dial by the number after the times in the range. For example, a reading of 30 on the R × 10 range indicates an ohmic value of 300. This format is different from those used to measure voltage and current. The process for those readings requires identifying a maximum value and multiplying the specific reading by that rate. The ohmmeter readings require using the multiplier and the specific reading.

Using the Ohmmeter When using an ohmmeter, one must be certain that the only power source in use is the one contained on the meter. When this source is used it is possible to control the current flow in the circuit. If an external power source is left on, it will produce an undesirable current flow through the meter circuit. If this does not destroy the meter, it will produce a false reading. In addition to this factor, one must be certain that there is only one path for current flow when the ohmmeter is inserted into the circuit. The normal procedure requires the isolation of one end or terminal of the device whose resistance is being measured. A second or third current path in the measured circuit produces an added current flow and, again, a false reading of the circuit resistance.

REVIEW PROBLEMS

12-9. Give the ohmic values for each of the following conditions.

	Needle	Range	Ohmic Reading
(a)	40	R × 10	_____
(b)	180	R × 10 K	_____
(c)	0.8	R × 100	_____
(d)	1.3	R × 1 M	_____
(e)	6.9	R × 1	_____

12-10. What is the purpose of the internal battery in the ohmmeter?
12-11. What is the resistance value for "zero ohms"?
12-12. What is the resistance value for infinity?

12-5 The Multimeter

The three basic functions of measuring voltage, current, and resistance all use the same principle. This principle is the measurement of current flow through the meter movement. Only one meter movement is required for this purpose. A system in use presently uses a basic meter movement and two accessory switches to perform the measurement of voltage current or resistance with one meter. One of the two switches is used to select the *function* of operation. The functions shown in Figure 12-22 are commonly used for a multifunction volt-ohm-milliammeter. This switch is used to select the mode of operation for the multimeter, as it is known. (Often the phrase "VOM" is used to identify the device.) The correct set of wiring connections establishes the meter movement connections for the measurement of the desired circuit component. Usually, the selection includes dc voltage, dc current, ac voltage, and resistance.

The second switch is used to select a suitable *range* for the meter reading. This switch connects the proper shunt or multiplier resistance for range extension of the basic meter movement.

Safety and Precaution The use of any measuring device should be done *after* one considers the aspects of safety. Safety is used here to describe both personal safety and equipment safety. Always remember that electricity is an unseen quantity. Electricity has the potential to do damage to the human body

FIGURE 12-22 The function switch on a multimeter permits the selection of a variety of measurements. (Courtesy of B & K-Precision/Dynascan Corporation.)

when certain basic precautions are not observed. Briefly, the following factors must be kept in mind when working with electricity:

1. Turn off the circuit power when making any connections to circuit components. To obtain the measurement, turn off the power, make the test connection, and then turn the power back on.
2. Turn off the circuit power when changing or replacing any components. The possibility of electrical shock or component damage occurs when parts are changed with the power left on in the circuit.
3. Work with one hand whenever possible. Keep the other hand in your pocket or behind your back. If you should become a part of the circuit (accidentally, of course), current cannot flow through the chest cavity and do major harm to your body when only one hand is used to make circuit connections or modifications.

The second safety-related precaution is conceived with the proper use of equipment. Properly used, test equipment will last for many years. Improperly used, the test equipment will often fail. The failure could be a total loss or it could be a partial loss. This, of course, depends on where the failure occurs in the tester. A loss requires that the instrument be removed from use and repaired. The loss of use is costly in terms of the price of repairs or the productive time lost when the repair is made by you. A partial loss may not be realized immediately, leading to incorrect values of measurement. The time involved in looking for a problem in a circuit that is functioning properly is wasted. Certain basic rules will minimize equipment failure.

1. The range selector switch should be on a range that is higher than the value of voltage or current to be measured. The correct values may be obtained from the schematic diagram or the service literature for the device.
2. When an unknown voltage or current value is to be measured, the range switch is placed in its highest-value position. The switch is moved to lower values until a reasonable reading on the meter is obtained.
3. The positive lead of the voltmeter or ammeter is connected to the most positive of the points being measured.
4. Voltmeters are connected in *parallel* and ammeters connected in *series* in the circuit.
5. Do not switch functions with the meter connected to the circuit and circuit power on.
6. All resistance measurements are done with circuit power off. If possible, one lead of the component should be isolated from the circuit when making resistance measurements.

Types of Multimeters There are several different types of multiple-function meters. Some of these use a meter movement as an output indicator. Such devices are classified as an **analog** unit. They are also classified as VOMs when they are portable and self-contained. One of these VOMs is shown in Figure 12-23. Another type of analog meter uses a more complex circuit for comparison

Analog *Term describing meter dial scale using individual markings for values.*

FIGURE 12-23 Typical VOM (volt-ohm-milliammeter) units available from test equipment manufacturers. (Courtesy of Simpson Electric Company.)

and meter operation. Originally, these meters used vacuum tubes in a bridge circuit. These units are called *vacuum-tube voltmeters* (VTVMs). They are able to measure voltage and resistance, but not current. The VTVM has to be connected to an ac power source to be operated. This does limit its use in areas other than the laboratory or out-of-doors. Technological advances replaced the vacuum tubes with solid-state components such as transistors and integrated circuits. The reduced power requirements of these solid-state devices makes them portable. They operate successfully using internal batteries for their power. These meters are known as transistorized voltmeters (TRVMs) and electronic multimeters (EVMs). Many of them have circuits permitting the measurement of current in addition to voltage and resistance.

Further technological advances have provided us with a solid-state electronic instrument having a **digital** readout. Some of these are shown in Figure 12-24. The analog readout device, or meter, is replaced with a digital readout display. The values given in this readout are usually very accurate. In some instances they are too accurate. We have a tendency to feel that a value of,

Digital *Term describing meter output indicator displaying numbers instead of a calibrated dial scale.*

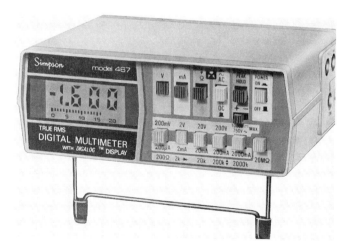

FIGURE 12-24 The digital readout meter similar to this requires an internal power source for the display. (Courtesy of Simpson Electric Company.)

say, 19.8 V is not acceptable when the circuit diagram calls for a value of 20.0 V. The inaccuracy may be due to tolerance values of the meter or the circuit and all may be working correctly. A judgment factor of 2 to 5% error should be included when making comparisons between the circuit and the circuit diagram.

The term *electronic voltmeter* (EVM) is often used to describe this type of meter. Sometimes the reference this to a digital voltmeter (DVM). The DVM or EVM actually measures all of the circuit values, even though the name implies otherwise.

One problem occurs when using a DVM. This problem is related to the lack of an instant response, or readout, when attempting to adjust a circuit for a maximum or minimum value. It is very possible to adjust the system to a point either before or after the peak, due to the time lag between the measurement and the display on the meter. Some instruments combine an analog readout device with the digital readout to compensate for this problem. One such instrument is illustrated in Figure 12-25.

Automatic Meters Some meters have the ability to shift their range automatically. These instruments are classified as **autoranging**. Switching circuits inside the meter connect the proper range resistance. This eliminates the need to make preliminary range switch adjustments.

Another feature in use is **autopolarity** (Figure 12-26). When one connects the leads of the meter "backwards," the needle attempts to move to the left, off the face of the dial. An autopolarity meter corrects for this effect. It always provides a reading. The leads between the meter and the test circuit are switched inside the meter automatically to provide a correct deflection. The digital meter displays either a (+) or a (−) sign as a part of its readout. This indication often eliminates the need for predetermination of circuit polarity when making a measurement.

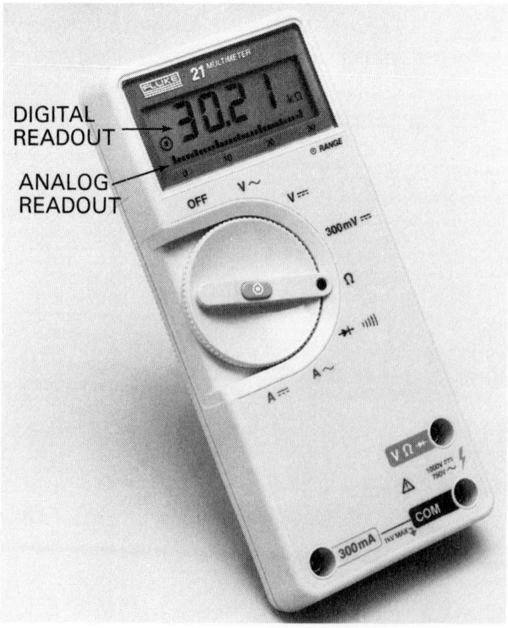

Autoranging Describes type of meter having automatic range switching capabilities.

Autopolarity Describes type of meter having automatic polarity reversal capabilities.

FIGURE 12-25 Some meters have both an analog and a digital display. The analog portion is useful when one has to "peak" or "null" a circuit. (Reproduced with permission from the John Fluke Mfg. Co., Inc.)

FIGURE 12-26 This digital multimeter has both auto-ranging and autopolarity capabilities. (Reproduced with permission from Radio Shack Div., Tandy Corp.)

The selection of a specific piece of measuring equipment often depends on the user. Some people are very comfortable using a VOM, while others prefer the DVM. It often makes little difference. The major factor to be considered is whether the instrument will provide an accurate recording without loading the circuit being measured. Another consideration is, of course, the ability to purchase one of the more expensive meters available today.

Instrument Resistance Resistance is an important factor in the selection of any measuring device. The VOM has an ohms-per-volt rating for its input sensitivity. When this value is 20 kΩ/V or higher, the meter has a minimum effect related to circuit loading. Some EVM units have a transistor circuit as part of the input to the unit. This circuit raises the sensitivity to a value of 1 MΩ or greater. DVMs and VTVMs have an input sensitivity between 5 and 10 MΩ. This is a fixed value for all ranges. It minimizes any circuit loading that may occcur.

REVIEW PROBLEMS

12-13. What does a multimeter measure?
12-14. What is the purpose of the function switch?
12-15. What is the purpose of the range switch?
12-16. Why should circuit power be turned off when connecting a meter to the circuit?
12-17. What is a VOM?
12-18. What is a DVM?

Use of the Meter Meters have a practical use in addition to the ability to measure values of voltage current and resistance. The proper methods of performing these measurements were described earlier in this chapter. A review of them provides these points:

1. Voltmeters are connected in parallel to the points in the circuit where a voltage is to be measured.
2. Current is measured by inserting the ammeter in series with the circuit.
3. Resistance is measured by connecting the ohmmeter in parallel with the component. Power to the circuit must be removed for a valid measurement.

Alternative Current Measurement Current may be measured by an application of Ohm's law. One of the *last* tasks that should be attempted when measuring any circuit value is removal of a component. The normal method of measuring current requires the insertion of the meter leads into the circuit. This procedure is not only inconvenient but requires a great deal of time. A better method to use, one that is equally good, is to measure the voltage drop across a known resistance and then use Ohm's law to calculate the current. This method is illustrated in Figure 12-27. If one has the good fortune to locate a resistance whose value is in units of 1, 10, 100, or 1000, the calculation is very simple. All that is required is to make the measurement and divide the voltage by the value of resistance.

Another method using this principle is even simpler. Consider the relationship of voltage, current, and a fixed value of resistance. Ohm's law shows us that when 10 V is applied to a 100-Ω resistance a current of 0.1 A flows. Using the same concept, when 6 V is applied to the same 100-Ω resistance, a current of 0.06 A flows. The current flow is directly proportional to the value of applied voltages. Since this factor is workable, it is possible to set the range switch of the meter on the 10-V range. A voltage ranging between 0 and 10 V can be "read" as a current of 0 to 0.1 A, or 0 to 100 mA. When this principle is utilized the voltmeter is automatically converted into an ammeter. Current is measured without disturbing the physical circuit.

This concept may be used with a variety of voltages and resistance values. Figure 12-28 illustrates this fact. The horizontal values represent the voltmeter

FIGURE 12-27 The voltmeter may be used in conjunction with an Ohm's law calculation to measure current without disturbing the wiring of the circuit.

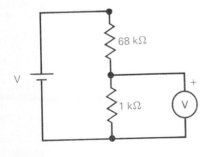

FIGURE 12-28 Chart showing how to use the voltage ranges on a multimeter to determine the value of current flow in a circuit.

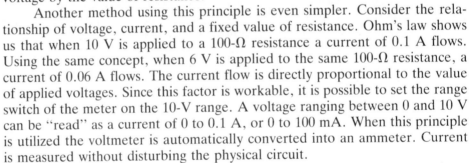

RESISTOR VALUES	VOLTMETER RANGE SWITCH POSITION				
	0–1000 V	0–100 V	0–10 V	0–1.0 V	0–0.1 V
1000 Ω	0–1 A	0–0.1 A	0–0.01 A	0–0.001 A	0–0.0001 A
100 Ω	0–10 A	0–1 A	0–0.1 A	0–0.01 A	0–0.001 A
10 Ω	0–100 A	0–10 A	0–1 A	0–0.1 A	0–0.01 A
1 Ω	0–1000 A	0–100 A	0–10 A	0–1 A	0–0.1 A

range switch settings. The vertical values are the ohmic ratings of the resistors where the voltage reading is measured. The boxes where the two values intersect contain the range of current readings available under the given conditions. This method, once learned and understood, is an easy and timesaving process. It is used by all who really understand electrical theory and its applications.

Continuity Testing Continuity testing is still another practical use for the multimeter. Often, a technician needs to determine if a current path is good. A variation of this requirement is the identification of a specific device or conductor in a multiple-conductor cable. The use of an ohmmeter for this purpose is a standard procedure. **Continuity tests** using an ohmmeter are simple. When both ends of the cable are reasonably close to each other, the ohmmeter's leads are used. One lead is connected to one end of the desired wire. The other lead of the ohmmeter is connected to each of the wires in the cable. Continuity is identified when a 0-Ω reading is obtained. This low reading (0 or close to 0 Ω) indicates a current flow through the meter, its leads, and the wire. The process is repeated using other wires until all are identified and labeled.

TROUBLESHOOTING APPLICATIONS

Continuity tests *Type of measurement made to insure a complete low resistance path for current.*

When one wishes to identify pairs of wires in a cable or any other grouping the continuity test may also be used. The use of this procedure is not limited to having both ends of the cable or wires close together. One end of the pair of wires is short circuited. This is done by connecting a jumper wire between the two wires or simply by twisting them together. Pairs of wires at the other end of the cable are measured for continuity. A very low resistance measurement identifies the desired pair of wires. As before, the process is repeated until all pairs of wires are identified.

Continuity testing procedures are also used to locate opens in the system. An open, by definition, is an infinitely high resistance. An ohmmeter reading of infinity indicates a zero-current-flow condition. When a circuit is being tested for continuity and one cannot obtain a resistance reading, the circuit is open.

Another method of testing for continuity in a circuit may be used if the circuit is a series type. The measurement of the *IR* voltage drop across any component in the circuit will readily identify an open. An open circuit is also a no-current-flow type of circuit. Voltage drops in a series circuit occur only when current flows in the circuit. A reading of 0 V or source voltage is an immediate indication of an open or lack of continuity in the circuit.

The methods described in this section are standard procedures. Their use is limited only by the creativeness of the person wishing to employ them. These and others like them are valid applications of the basic electrical theories being studied. The use of these techniques indicates an excellent understanding of electricity and electronics from the standpoint of being able to apply the learned concepts to practical applications.

1. Meters commonly employed in electrical measurements use the principles of magnetism.
2. The moving-coil meter is called the D'Arsonval meter.
3. Deflection in the moving-coil meter is due to current flow through the coil or armature of the meter.

CHAPTER SUMMARY

4. Current-reading meters are series connected to the circuit.

5. Range extension of the current meter is done by placing a shunt resistance in parallel with the coil of the meter.

6. Voltage-measuring meters are parallel connected to the circuit.

7. Range extension of the voltmeter is done by connecting multiplier resistors in series with the coil of the meter.

8. Meter sensitivity is the ohms-per-volt rating of the coil of the meter. A high ohms-per-volt reading is desirable.

9. Resistance readings are done with an ohmmeter. It consists of the meter coil assembly, an internal battery, and a variable potentiometer.

10. Resistance readings are performed with circuit power off.

11. A meter capable of measuring voltage, current, and resistance is called a VOM.

12. Electronic meters called the VTVM, TRVM, EVM, and DVM are used to measure circuit values.

13. Electronic meters require a power source.

14. Continuity tests for a working current path have a low resistance value, close to 0 Ω.

15. Open circuits have infinite resistance.

Answer true or false; change all false statements into true statements.

12-1. Ammeter multipliers are parallel wired.

12-2. The armature coil winding of the meter determines its resistance.

12-3. Meter sensitivity is described in units of volts per ohm.

12-4. The measurement of voltage is done by placing the meter leads in parallel with the points to be measured.

12-5. A meter display that shows numbers and does not have a calibrated dial is called a digital meter.

12-6. The sum of the voltage drops across the meter and its multiplier resistor determine the value of the voltage.

12-7. The difference between total current and meter movement current is the value of shunt current.

12-8. Resistance measurements require that the system power be turned on.

12-9. An autorange meter automatically displays the polarity of the voltage being measured.

12-10. All voltages are measured from some circuit point of reference.

Identify the values of multiplier resistances required for the meters and voltage values given in problems 12-11 through 12-14.

12-11. 15 V, using a 1-mA, 100-Ω movement.
(a) 1.4 kΩ
(b) 14.9 kΩ

(c) 149.9 kΩ

(d) 1.499 MΩ

12–12. 50 V, using a 500-Ω, 100-μA movement.

    (a) 4.99 MΩ

    (b) 499.5 kΩ

    (c) 49.95 kΩ

    (d) 4.995 kΩ

12–13. 750 V, using a 100-Ω, 1-mA movement.

    (a) 7.49 MΩ

    (b) 0.749 MΩ

    (c) 74.9 kΩ

    (d) 0.00749 kΩ

12–14. 5 V, using a 1000-Ω, 10-μA movement.

    (a) 4.99 MΩ

    (b) 499.5 kΩ

    (c) 49.95 kΩ

    (d) 4.995 kΩ

Identify the values of shunt resistances required for the meters and current values given in problems 12-15 through 12-18.

12–15. 50 mA, using a 1-mA, 100-Ω movement.

    (a) 20.4 Ω

    (b) 204 Ω

    (c) 2.04 Ω

    (d) 0.204 Ω

12–16. 1 A, using a 10-mA, 100-Ω movement.

    (a) 1.01 Ω

    (b) 0.101 Ω

    (c) 10.1 Ω

    (d) 0.009 Ω

12–17. 20 A, using a 100-μA, 1000-Ω movement.

    (a) 0.05 Ω

    (b) 0.50 Ω

    (c) 0.005 Ω

    (d) 5.0 Ω

12–18. 500 mA, using a 1-mA, 200-Ω movement.

    (a) 4.08 Ω

    (b) 0.408 Ω

    (c) 40.8 Ω

    (d) 0.0408 Ω

12–19. Why is it necessary to turn off circuit power when measuring resistance?   **ESSAY QUESTIONS**

12–20. Why is a voltmeter connected in parallel?

12–21. How much current would flow through a 100-mA, 1000 Ω meter when it is accidentally connected across a 100-V source?

12–22. Explain why personal safety is necessary when working with electricity.

12–23. What is the purpose of the zero-ohms control on an ohmmeter?

12–24. What is meant by the term *linear scale*?

12–25. Explain the term *loading* when using a meter.

12–26. What devices are used to convert a basic meter into a multirange ammeter?

12–27. What are the disadvantages of using a digital readout meter?

12–28. What precautions are necessary when using a voltmeter to measure an unknown voltage value?

PROBLEMS

12–29. A meter has an accuracy of 2%. What is the actual voltage range possible when the reading on the meter is 27 V? (*Sec. 12–1*)

12–30. What ohmic value is shown when the needle is at the 40 mark and the range switch is set at R×10? (*Sec. 12–4*)

12–31. A 100-μA, 1000-Ω meter movement is used as an ammeter. What is the maximum shunt current for each of the following ranges? (*Sec. 12–2*)

    (a) 100 μA
    (b) 1 mA
    (c) 100 mA
    (d) 1 A
    (e) 10 mA

Use Figure 12-29 to answer problems 12–32 through 12–35.

12–32. What voltage is measured when the meter is used? (*Sec. 12–3*)

12–33. What current flows through the meter when it is connected to the circuit? (*Sec. 12–2*)

12–34. What multiplier resistance is required for the meter? (*Sec. 12–3*)

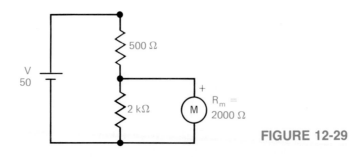

FIGURE 12-29

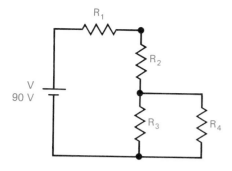

FIGURE 12-30

12–35. What is the needle position for a 0- to 10-mA meter when 00.005 A flows through the meter? (*Sec. 12–2*)

Use Figure 12-30 to answer problems 12-36 through 12-39.

12–36. Redraw the circuit, showing where you would install a current meter in order to measure I_T. (*Sec. 12–2*)

12–37. Redraw the circuit, showing where you would install a voltmeter to measure V_{R1}. (*Sec. 12–3*)

12–38. Redraw the circuit, showing where you would install an ammeter to measure I_{R3}. (*Sec. 12–2*)

12–39. R_1, R_2, and R_3 have values of 5 kΩ each. R_4 represents a voltmeter with a sensitivity of 1000 Ω/V. What is the value of current flow through the meter? (*Sec. 12–1*)

Use Figure 12-31 to answer problems 12-40 and 12-41.

12–40. What voltage is measured across R_4 with a meter having a 100-Ω, 0.1-A movement? (*Sec. 12–3*)

12–41. What voltage is measured across R_3 with a meter having an input resistance of 10 MΩ? (*Sec. 12–3*)

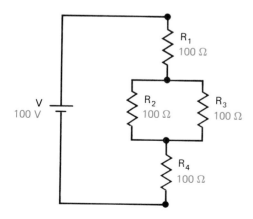

FIGURE 12-31

12-42. Rotation, or adjustment, of an ohmmeter's 'Ohms Adjust' control does not bring the meter's needle to the 0 Ω marking on the meter scale. If all resistances in the circuit are within their tolerances, what could be the problem?

12-43. Connecting a meter in order to measure the voltage in the circuit as shown in Figure 12-7 causes the meter needle to quickly move from its rest position to the extreme right hand side of the meter dial. What is wrong?

12-44. A 1000-Ω resistance is connected in parallel with a meter movement having an internal resistance of 10 Ω. The purpose of the additional resistance is to extend the range of the meter's current reading. The result of the addition of the resistance is a less than normal reading of current. What is wrong?

12-45. A 1000 Ω per volt voltmeter is connected across a 500-Ω resistance in a series circuit. The voltage reading does not agree with the values on the schematic diagram of the tested device. What is wrong?

12-46. A voltmeter designed to measure from 0 to 100 V is placed in series with two 500 Ω resistors. The reading on the meter is very low and you expect to measure about 50 V. What is wrong with the setup?

**ANSWERS TO
REVIEW
PROBLEMS**

12-1. Coil, springs, field magnet, needle, and dial scale

12-2. The needle deflects in the wrong direction.

12-3. (a) Current required for full-scale deflection; (b) total current; (c) a reverse direction reading (right to left).

12-4. (a) 199 mA; (b) 599 mA; (c) 4.999 mA

12-5. (a) 52.6 Ω; (b) 16.95 Ω; (c) 2 Ω

12-6. 0.1 Ω

12-7. (a) 10; (b) 25; (c) 50; (d) 2.5

12-8. (a) 4.0 V; (b) 15 V; (c) 21 V; (d) 175 V

12-9. (a) 400; (b) 1.8 M; (c) 80; (d) 1.3 M; (e) 6.9

12-10. To provide a current flow

12-11. No resistance, maximum conductance

12-12. Extremely high, beyond measurement

12-13. Resistance, voltage, and current

12–14. To select the desired quantity to be measured

12–15. To extend the measurement range

12–16. To protect the operator and the equipment

12–17. A volt-ohm-milliammeter

12–18. A digital multimeter

Chapters 10-12

Often there is a need to analyze a circuit rapidly. This can be accomplished by use of some of the circuit laws presented by Thévenin, Millman, and Kirchhoff. The ability to recognize the need to use these laws and the ability to use them enhances our ability to design and resolve complex circuit problems and values. In addition, the ability to select and use electrical test equipment correctly is mandatory for anyone making circuit tests. How and when to extend the range of a basic meter also makes the work easier as well as more successful. This material is covered in this section and reviewed here.

1. *Kirchhoff's voltage law*. The sum of the voltage drops in a closed loop are equal to zero. When a source voltage is present, the sum of the voltage drops equal the source voltage.

2. *Kirchhoff's current law*. The sum of the currents entering a junction are equal to the currents leaving that junction.

3. Polarity values can be assigned to each component in a circuit depending on the polarity of the source voltage.

4. "Closed loop" is another term for a series circuit.

5. Complex circuit analysis may be accomplished by using loop, node, or mesh analysis.

6. *Thévenin's theorem*. A linear dc circuit, no matter how complicated , can be replaced by an equivalent circuit containing one dc voltage source (V_{TH}) in series with one resistance (R_{TH}).

7. *Norton's theorem*. Any dc circuit, no matter how complicated , can be reduced to a single current source (I_N) and a parallel resistance value (R_N).

8. *Superposition theorem*. Currents in any branch of a multisource circuit are identified by determining the current in each source individually when all other sources are replaced by their internal resistance values. The current flow in a specific branch is the algebraic sum of the current from each source in the specific branch.

9. *Millman's theorem*. A circuit containing multiple source voltages connected in parallel with each other can be reduced to one equivalent voltage in series with a load resistance.

10. Using Ohm's law will convert a voltage source into a current source, and vice versa.

11. A delta configuration may be converted into a wye configuration when a circuit solution is required. It is also possible to reverse the process and convert wye circuits into delta circuits.

12. Meters commonly employed in electrical measurements use the principles of magnetism.

13. The moving-coil meter is called the D'Arsonval meter.

14. Deflection in the moving-coil meter is due to current flow through the coil or armature of the meter.

15. Current-reading meters are series connected to the circuit.

16. Range extension of the current meter is done by placing a shunt resistance in parallel with the coil of the meter.

17. Voltage-measuring meters are parallel connected to the circuit.

18. Range extension of the voltmeter is done by connecting multiplier resistors in series with the coil of the meter.

19. Meter sensitivity is the ohms-per-volt rating of the coil of the meter. A high ohms-per-volt reading is desirable.

20. Resistance readings are done with an ohmmeter. It consists of the meter coil assembly, an internal battery, and a variable potentiometer.

21. Resistance readings are performed with circuit power off.

22. A meter capable of measuring voltage, current, and resistnce is called a VOM.

23. Electronic meters called the VTVM, TRVM, EVM, and DVM are used to measure circuit values.

24. Electronic meters require a power source.

25. Continuity tests for a working current path have a low resistance value, close to $0\ \Omega$.

26. Open circuits have infinite resistance.

CHAPTER 13

Chemical Energy Sources

One major source of electrical energy used in portable equipment is the chemical cell, or battery. There are many types of chemical cells, but they each accomplish the same purpose: conversion of chemical energy into electrical energy. Each type of chemical cell has a specific set of ratings. The designer or operator of portable electrical equipment needs to recognize these differences and be able to select the correct cell for use in equipment.

There are some types of cells that are rechargeable. The use of rechargeable cells is becoming more common. We need to be able to determine the cost-effectiveness of the rechargeable cell as compared to the standard chemical cell. A refinement of the chemical cell has produced cells having long shelf, or storage, life. These are being used more as electronic units are miniaturized.

KEY TERMS

Alkaline-Manganese-Zinc Cell	Maximum Power Transfer
Ampere-hour	Nickel-cadmium Cell
Battery	Plates
Carbon-Zinc Cell	Primary Cell
Chemical Cells	Secondary Cell
Constant-Current Charging	Trickle Charge
Dry Cell	Voltaic Cell
Internal Resistance	Wet Cell
Lithium Cell	Zinc Chloride Cell
Load Current	

OBJECTIVES

Upon completion of the material in this chapter, you will

1. Understand the concept of voltage creation by chemical means.
2. Be able to explain the concept of internal resistance.
3. Understand cell recharging.
4. Be able to combine cells in series and parallel.
5. Be able to explain power transfer.

All electrically operated equipment requires a source of energy or power. This power may be created in several ways. Two of the more common methods in use today are chemical energy and electromagnetic energy. A discussion about power sources must start with a statement about energy and matter. In our physical world, matter cannot be created, nor can it be destroyed. All matter has energy of some form. The form of energy may be changed. This will not create more energy than was in the original form. Let us say that energy can be found in many different forms. The form may be changed into another type of energy, but we are not able to increase the level of energy beyond that in its original form.

Matter often has levels of energy that are unmeasurable. When the form of the matter is changed, it is often possible to measure the energy in its newer form. We say that the original form of the matter has the *potential* to produce a useful level of energy. The useful level may require further processing before we are able to use it to perform some quantity of work. An example of this concept is related to the use of coal or oil. Both of these have the potential to create useful quantities of energy. In the form in when they originate, neither of these will perform much work. When coal or oil is burned, however, the basic matter is converted into heat. The heat is then used in one form to operate a device directly, or it may be used to change water into steam. The steam is then used to operate some form of motor. The motor, in turn, operates a machine.

These are two examples of how matter is transformed into different forms. Under ordinary circumstances a loss of energy occurs at each step of the process. This loss is related to the efficiency of the process. The loss, as you will find out later in this chapter, is a portion of the energy. It must be minimized to maximize the transfer of the energy from one form to another form.

This chapter's topic is chemical sources for electrical energy. The chemical sources we use are called *batteries*. The **battery** is a combination of chemical cells. The **chemical cells** create a voltage. The voltage has the potential of creating a current flow when the cells are connected to a load. There are many different types of chemical cells. The material presented in this chapter explains how several of the more commonly used cells are constructed, their applications, and some of their limitations.

13-1 Features of Batteries

One or more chemical cells are used to produce a finished product known as a battery. The battery is housed in a container containing some chemicals. The

Battery *Group of chemical cells connected to provide desired operating power.*

Chemical cell *Unit using chemical energy in order to create electrical energy.*

chemicals interact to produce a voltage. This voltage level can be as low as 1.0 V for a single-cell battery to well over 100 V for a multiple-cell battery.

Batteries are used to produce a direct-current flow through the load. The output connections of the battery have commonly accepted terminals to connect them to the load. These include clips, buttons, and snap fittings. Some of these are shown in Figure 13-1. You will also note the variety of sizes for batteries. All batteries have a voltage rating.

Batteries are generally classified into two groups: *primary cells* and *secondary cells*. The difference between the two groups is simple. **Primary cells** are not normally rechargeable. Once their chemistry is exhausted, the cell must be replaced. **Secondary cells** are rechargeable. They may be used over a long period of time by keeping the charge "up" in the cells. Typical primary cells include a large group of cells made primarily of a combination of lead and zinc. Secondary cells include lead-acid automobile batteries, nickel-cadmium, and some types made of a manganese dioxide mixture.

In the early days of radio, batteries were used to supply the operating power. There were three categories for these batteries. They were identified as A, B, and C batteries. The A battery was used to heat the filaments of the vacuum tubes in the radio. B batteries were used for the dc plate voltage of the tubes, and the C battery established a point of operation for the control elements of the tube. Almost no tube-type radios requiring batteries are in general use. The terminology relating to the battery class function is still partially in use. The A supply is a low-voltage source still identified for its original purpose of heating vacuum tube filaments. The term "B +" refers to the positive voltage obtained from an electronic power source, while "B −" is used to indicate circuit ground. These terms are not to be confused with an A cell or type A cell, which is a specific type or size of battery, not a power source.

Another method of classifying batteries relates to the type of the chemical mix used to create the terminal voltage. In a **dry cell** the mix is moist but not in liquid form. A **wet cell**, on the other hand, contains a form of liquid as its electrolyte.

Primary cells *Nonrechargeable chemical cells.*

Secondary cells *Rechargeable chemical cells.*

Dry cell *Cell containing semi-dry electrolyte.*

Wet cell *Cell containing liquid electrolyte.*

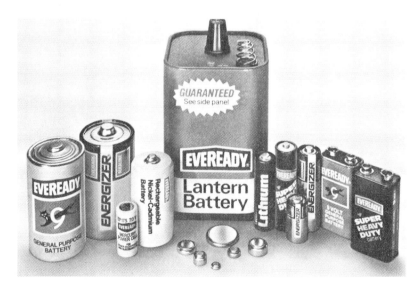

FIGURE 13-1 Typical commercial batteries used in electronic equipment are available in a variety of sizes and shapes. (Courtesy Eveready Battery Company, Inc.)

13-2 The Voltaic Cell

A typical cell is shown in Figure 13-2(a). This cell has three major components: the positive electrode, the negative electrode, and the electrolyte. The cell is called a **voltaic cell**. It was invented by Galvani and is also called a *galvanic* cell, in his honor. The process of creating a potential difference between the electrodes of the cell is due to the chemical action between the electrolyte and each of the electrodes. The electrodes are also known as **plates** of the cell. The plates serve two purposes in the cell. One purpose is to provide an electrical connection from the cell to the load. The other purpose is their use as chemical reactants to create a positive and a negative charged plate.

Current flow *inside* the cell is from the positive plate to the *negative* plate. Current flow *outside* the cell starts at the negatively charged plate, continues through the load, and then goes to the positively charged plate. The chemical action between the electrolyte and the plates is a continuous effort to keep the plates properly charged. The difference in charges on the plates is the potential difference or voltage required to move electrons through the load. The action inside the cell is an ionization process, and the activity external to the cell is an electron flow activity.

13-3 Cell Internal Resistance

All sources of electrical energy have some amount of **internal resistance.** The source internal resistance limits the amount of current flow in the system. The internal resistance of a fully charged cell is around 1 Ω. When the cell starts to "run down," its internal resistance will increase. The result is a drop in the voltage available at the terminals of the cell. In Figure 13-3 the cell has an output voltage of 2.0 V. The internal resistance is 1.0 Ω and the load resistance is 15 Ω. A series circuit formed when the components are connected to each other. The total system resistance is the sum of all resistances in the circuit, or 16 Ω. The total current flow is

$$I_T = \frac{V_T}{R_T} = \frac{2\ \text{V}}{16\ \Omega} = 0.125\ \text{A}$$

FIGURE 13-2 (a) The basic chemical cell consists of two electrodes and an electrolyte connected to a load. (b) The circuit is redrawn using standard schematic symbols.

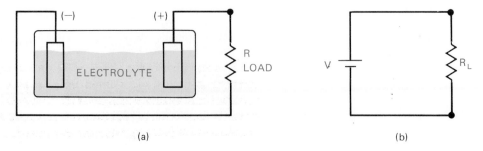

(a)   (b)

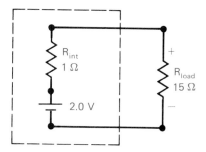

FIGURE 13-3 All chemical cells have a value of internal resistance.

As the voltage drop across the cell and its internal resistance is determined when the source and its internal resistance are added:

$$V_T = V_{cell} + V_{R_{int}}$$

$$= 2.0 \text{ V} + (-0.125) \text{ V} = 1.875 \text{ V}$$

The terminal voltage, as it is called, is not 2.0 V, but it is 1.875 V.

The internal resistance of the cell increases as the cell deteriorates. The internal resistance for this cell could rise to 100 Ω. Now the total circuit resistance is

$$R_T = R_{int} + R_{load} \tag{13-1}$$

$$= 100 \text{ Ω} + 15 \text{ Ω} = 115 \text{ Ω}$$

The total current flow in this circuit now drops:

$$I_T = \frac{V_T}{R_T}$$

$$= \frac{2 \text{ V}}{115 \text{ Ω}} = 0.0174 \text{ A}$$

The voltage drop across R_{int} under these conditions is

$$V_{R_{int}} = R_{in} \times I_T$$

$$= 100 \text{ Ω} \times 0.0174 \text{ A} = 1.74 \text{ V}$$

When this value is subtracted from the 2.0 V created by the cell, the terminal voltage is

$$V_{term} = V_{cell} + (-V_{R_{int}}) \tag{13-2}$$

$$= 2.0 \text{ V} - 1.74 \text{ V} = 0.26 \text{ V}$$

The cell is exhausted and will not perform adequately under these conditions.

13-4 Measuring Cell Voltage

The operating characteristics of the cell must be tested under loaded conditions. The terminal voltage of a cell or battery that is not connected to its normal load will provide a false reading. When the open terminal voltage is measured, the terminal voltage will reflect the cell voltage without any current flowing in the circuit. When there is no current flow, there are no voltage drops across

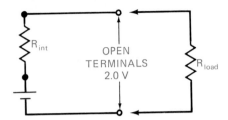

FIGURE 13-4 Voltage drop across the internal resistance of the cell can be determined only when the cell is connected to a load and its terminal voltage measured.

any resistances in the circuit. Consequently, the value of R_{int} is not reflected in the terminal voltage reading. This open circuit test is shown in Figure 13-4. Connecting the load to the terminals of the cell completes the circuit. Current will now flow and we are able to measure accurately the terminal voltage of the cell. When the terminal voltage is close to the design value (2.0 V in this example), the cell is usable. When the terminal voltage drops to less than about 60 to 65%, it should be replaced. The exact value for replacement depends on the manufacturer's specific action.

REVIEW PROBLEMS

Answer true or false; change all false statements into true statements.

13-1. The form of matter cannot be changed.
13-2. Batteries are formed from individual cells.
13-3. A primary cell is rechargeable.
13-4. Internal resistance of a cell should be very low value.

13-5 Cells in Combination

All cells are rated in units of optimum terminal voltage and current. When the voltage requirement for a power source is greater than the capability of a single cell, several cells may be connected in series. This series combination will create a terminal voltage *for the group of cells* that is equal to the sum of the individual cell voltages. For example, when one requires a 9.0-V source, the battery manufacturer builds a battery containing six 1.5-V cells, since $6 \times 1.5 = 9.0$. Cells are also wired in parallel when current values larger than those available from one cell are required. The rules established by Kirchhoff for series voltage drops and parallel current paths are applied here. Since series voltage drops add, we can also state that series voltage sources also add in a closed-loop circuit. The same is valid for parallel circuits: Parallel current sources also add to equal total current flow.

The circuit of Figure 13-5 illustrates a series cell connection. Each cell has a value of 1.5 V and 0.5 A. When they are wired in series the voltage produced by each cell adds to equal 6.0 V. The current flow in this configuration has only one path. It is limited to the maximum current flow of any one cell. In this example if one cell had a current rating of 0.1 A, the maximum current flow in the series circuit would also be limited to 0.1 A.

When cells are wired in parallel (Figure 13-6), their terminal voltages are

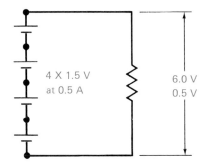

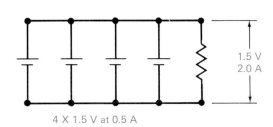

FIGURE 13-5 Series cells produce the sum of the individual cell voltages and a current value limited to the lowest single-cell current value.

FIGURE 13-6 Parallel cells produce the voltage of a single cell and the sum of the individual cell current values.

the same value. In this circuit the voltage is 1.5 V for the parallel bank. The current capabilities for this parallel cell bank are increased by the number of parallel current sources in the system. When four cells, each having the capacity of producing a 0.5-A current, are placed in parallel, their total current capacity is equal to the sum of the value of each. If the cells have the same terminal voltage but have different current ratings, the total current capacity is still the sum of the current capacity of each cell.

Cells are rated in terms of **ampere-hours**. This term is used to identify the amount of current available for one hour of operation. The ampere-hour capacity of the battery is its current rating for a specific number of hours of operation.

Cells may also be connected in series–parallel combinations. One such arrangement is shown in Figure 13-7. This figure shows three series strings. Each string has two 1.5-V cells. The total voltage and current for each string is 3.0 V and 0.5 A. When three of the strings are connected as a parallel bank, the total voltage and current of the package is 3.0 V and 1.5 A. Any number of different combinations of cells is possible. Cell combinations are formed into battery packages by the cell manufacturer. The package rating has a total voltage and current capacity listed. Cells and individual batteries are also placed in series, parallel, or series–parallel combinations by the user. When a specific voltage is not readily available from the battery manufacturer, the voltage can be created by the user. If one required an 18-V source, it is possible to series

Ampere-hour *Amount of current available for one hour of operation of a battery.*

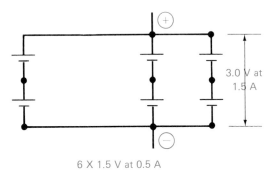

6 X 1.5 V at 0.5 A

FIGURE 13-7 Cells may be combined in both series and parallel to achieve a specific value of voltage and current.

connect two 9.0-V batteries. The popular 9.0-V batteries used in portable radios and calculators are readily available and low in cost. An 18-V battery could be much more expensive and more difficult to obtain than two 9.0-V batteries—thus their use in this application. Any type of cell may be used in combination to create a battery. Section 13-6 explains the construction of the various cells in common use.

REVIEW PROBLEMS

13-5. What is the terminal voltage when six 1.5-V cells are series connected?

13-6. What is the current capacity of four 2.0-V 0.1-A cells in series?

13-7. What is the current capacity when cells rated at 0.5, 0.1, 1.2, and 1.0 A are parallel connected?

13-8. What is terminal voltage and current for this series cell combination: 1.2 V at 1 A, 1.55 V at 0.75 A, and 1.25 V at 1.25 A?

13-6 Cell Construction

A variety of chemical cells are available today. The development of the various cells is due to the special requirements of their user. Primary, nonrechargeable cells are commonly used in flashlight applications. The promotion of portable tools and light sources in recent years has led to the need for a rechargeable secondary cell. The most common of these is the nickel-cadmium, or ni-cad, cell. A review of the construction and typical application of these and other cells will aid in their selection and proper care when they are used.

The **carbon-zinc cell** has been one of the most often used types. It has a formal name, the Leclanche cell, named after its inventor. A cutaway view of the carbon-zinc cell is shown in Figure 13-8. This cell has an anode made of zinc. Its cathode is made of a manganese dioxide mixture, and the electrolyte is a mix of ammonium chloride and zinc chloride dissolved in water. A carbon rod is used as a means of improving conductivity and to retain moisture in the cell. The case of this cell is a zinc can. The terminals for its electrodes are located at the top and at the bottom of the case. The positive (+) terminal is the upper one and has a formed button-like terminal. The negative (−) terminal is also a formed metal piece. It normally has an identation in its center.

The terminal voltage of the carbon-zinc cell is 1.5 V. It may be placed in combination to obtain voltages as high as 510 V. The Leclanche cell is available in two basic forms. One of these is the previously shown round cell. These are available in a variety of sizes. They may be assembled and wired in combination to create a specific battery. The second form for this cell is called a flat cell. The flat-cell construction is shown in Figure 13-9. Flat cells are available only as multiple-cell batteries. The major difference between round and flat cells is physical. They both have the same chemical composition.

The carbon-zinc cell has the following major characteristics:

Cell voltage: 1.5 V

Type: primary

Carbon-zinc cell Chemical cell using carbon and zinc electrodes.

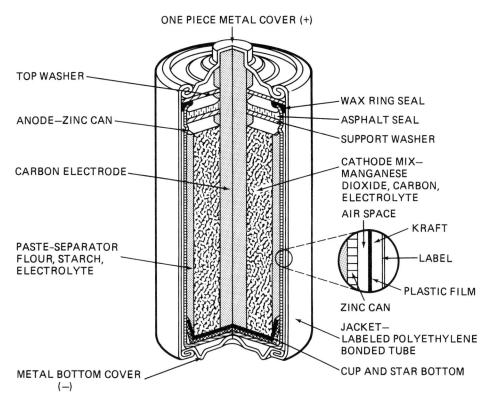

ONE PIECE METAL COVER (+)

TOP WASHER

ANODE—ZINC CAN

CARBON ELECTRODE

PASTE-SEPARATOR
FLOUR, STARCH,
ELECTROLYTE

METAL BOTTOM COVER
(−)

WAX RING SEAL
ASPHALT SEAL
SUPPORT WASHER

CATHODE MIX—
MANGANESE
DIOXIDE, CARBON,
ELECTROLYTE

AIR SPACE

KRAFT

LABEL

PLASTIC FILM

ZINC CAN

JACKET—
LABELED POLYETHYLENE
BONDED TUBE

CUP AND STAR BOTTOM

FIGURE 13-8 Cutaway view of a carbon-zinc cell, showing its internal construction. (Courtesy Eveready Battery Company, Inc.)

Rechargeability: poor
Number of cycles: 10 to 20
Capacity: 60 mA to 30 Ah (ampere-hour)
Temperature: poor low-temperature efficiency
Operating cost: low

Recharging of carbon-zinc cells is not a normally accepted procedure. One may purchase a recharging unit for these cells but the cost-effectiveness of

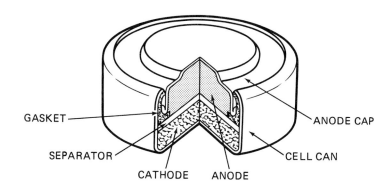

GASKET

SEPARATOR

CATHODE   ANODE

ANODE CAP

CELL CAN

FIGURE 13-9 Cutaway view of a flat cell, showing its internal construction. (Courtesy Eveready Battery Company, Inc.)

recharging carbon-zinc cells is limited. A term that better describes this process is "rejuvenating" rather than "recharging." If one wishes to utilize one of these rejuvenating units, the U.S. National Bureau of Standards has made the following statement (*National Bureau of Standard LC 965*):

The dry cell is considered a primary battery. It may be recharged for a limited number of cycles under certain conditions:

1. Open-circuit voltage of the cell is not under 1.0 V.
2. The battery must be placed on charge very soon after its removal from service.
3. Recharging times should be between 16 and 20 hours.
4. Cells must be returned to service (use) soon after they are recharged, as they have a very short storage, or shelf life.

As you can see, the practicality of recharging the carbon-zinc cell is very limited. One also has to be aware that gases are formed during the recharge period. When the cell's container is sealed, a large quantity of gas will produce a rupture of the container. This could, in extreme cases, produce a minor explosion and result in parts of the case flying through the air, damaging people or property near them.

REVIEW PROBLEMS

13-9. Name the major components used in the construction of a carbon-zinc cell.

13-10. Is this cell rechargeable?

13-11. What is the open-circuit terminal voltage of this cell?

Zinc chloride cells Type of chemical cells.

Zinc chloride cells are very similar in construction to the carbon-zinc cell. The construction of this cell is shown in Figure 13-10. A different internal construction is required because of the change in the electrolyte for this cell. Removal of the ammonium chloride from the electrolyte permits a higher current output for a longer time period than is available from the carbon-zinc cell. Another feature is the cell's ability to maintain its operating voltage under load for a greater time. The characteristics of this cell are similar to those of the carbon-zinc cell, except for the following:

Capacity: several hundred mA to 9 Ah

Temperature: good low-temperature operation

Cost: low to medium

REVIEW PROBLEMS

13-12. Name the major component used in the construction of a zinc chloride cell.

13-13. Is this cell rechargeable?

13-14. What is the open-circuit terminal voltage of this cell?

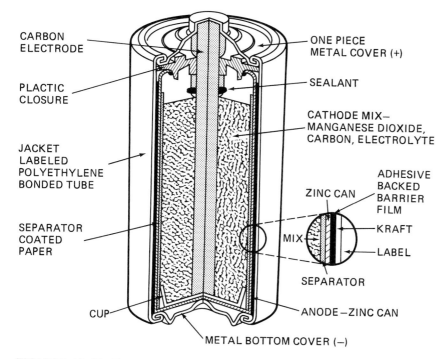

FIGURE 13-10 Cutaway view of a zinc-chloride cell, showing its internal construction. (Courtesy Eveready Battery Company, Inc.)

FIGURE 13-11 Cutaway view of an alkaline-manganese cell, showing its internal construction. (Courtesy Eveready Battery Company, Inc.)

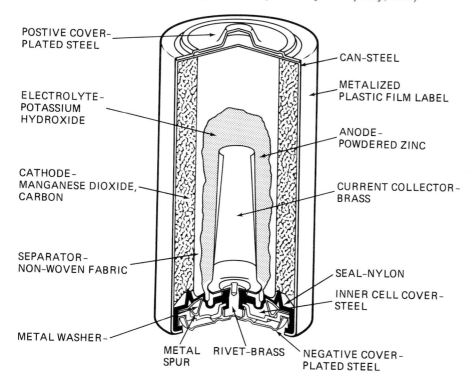

Alkaline-manganese-zinc cells are produced as both primary and secondary types. The composition of the cell (Figure 13-11) is a zinc anode, a potassium-hydroxide electrolyte, and a manganese-dioxide cathode. Its development is a major step in the improvement of operating conditions for the basic Leclanche type of cell. This cell has a very low internal resistance factor and a high service capacity. The primary cell voltage is 1.5 V. The application for this cell is in devices requiring more power or longer life than can be obtained from the carbon-zinc cell. The total energy contained in this alkaline cell is between 50 and 100% greater than from the carbon-zinc cell of the same size. Applications for alkaline cells include radios, portable tape players, toys, model boats, and electronic flash units for cameras. Almost any high-drain, heavy-discharge application requires the use of this alkaline cell.

Many of the batteries used in electronic devices are composed of several individual cells. These cells are housed in a standard size container. The internal wiring of the individual cells creates the desired terminal voltage for the battery. The 9 volt alkaline cell, shown in Figure 13-12, is composed of six 1.5 V units.

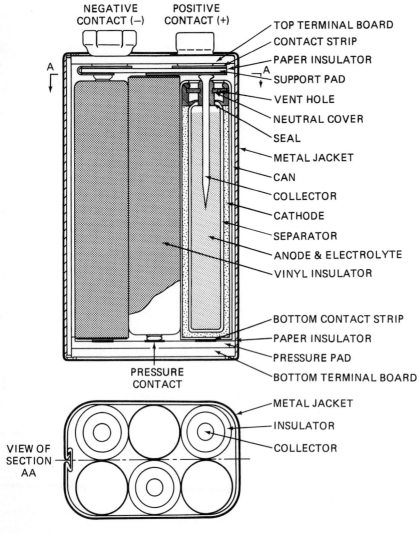

FIGURE 13-12 Cutaway view of an alkaline 9 volt cell, showing its internal construction. (Courtesy Eveready Battery Company, Inc.)

The six cells are series connected in order to produce the desired 9.0 V at the terminals of the case.

Nickel-cadmium cells have the reputation of being one of the best of the rechargeable cells. The cells are assembled into a variety of packages. The output voltage for the packages ranges from 1.2 V to 14.0 V. The construction of the ni-cad cell, as it is called, is shown in Figure 13-13. It uses a rolled sandwich of electrodes with the electrolyte sandwiched between the two elec-

Nickel-cadmium cells *Rechargeable chemical cells.*

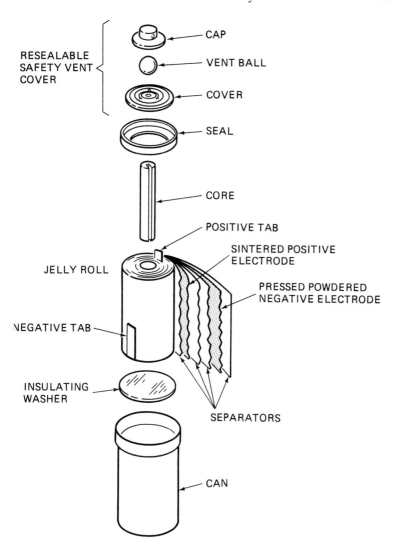

FIGURE 13-13 Cutaway view of a nickel-cadmium cell, showing its internal construction. (Courtesy Eveready Battery Company, Inc.)

trodes. Terminal connections may be solder tabs or the package may use the standard flashlight cell configuration.

The chemistry of the ni-cad cell uses nickelous hydroxide for the positive electrode. The negative electrode is cadmium hydroxide and the electrolyte is potassium hydroxide. The cell creates a nominal terminal voltage of 1.2 V. These cells represent the best features of a rechargeable chemical cell.

All good things seem to have at least one negative characteristic. This is also true for the ni-cad cell. When the cell is connected in a series voltage source string the rate of discharge for each of the cells in the string is slightly different. When the cells are completely discharged during operation this factor may create a reverse polarity condition in the first cell to be fully discharged. A negative secondary effect resulting from polarity reversal is the formation of gas in the cell as current is forced through it. This could result in a cell rupture. The answer for this problem is not to let the cells become fully discharged during their operation.

The characteristics for the nickel-cadmium cell are:

Voltage: 1.2 V
Type: secondary
Rechargeable: yes
Number of cycles: 300 to 2000
Capacity: 20 mA to 4 Ah
Temperature: very good efficiency at low temperatures, poor at high temperatures
Operating cost: low

REVIEW PROBLEMS

13-18. Name the major components used in the construction of a nickel-cadmium cell.

13-19. Is this cell rechargeable?

13-20. What is the open-circuit terminal voltage of this cell?

Other Cell Types The cells described in the preceding section are only a few of the types of cells in use today. They are by far the most popular. Several other cells are also used. Their use is limited, but they do require a brief description. The cells in this category include the mercuric oxide, silver oxide, and lithium types. They are described in this section.

Mercuric oxide cells are made using a mercuric oxide cathode, a zinc anode, and an electrolyte of either potassium hydroxide or sodium hydroxide. The terminal voltage for this cell is 1.35 V. The construction of this cell is shown in Figure 13-14. These cells are commercially packaged in batteries with voltages ranging up to 97.4 V. A characteristic of this cell is that its output voltage remains almost constant until it is fully discharged. The mercuric oxide

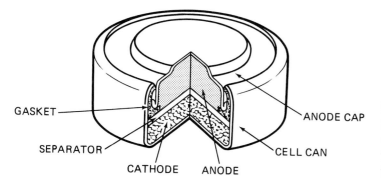

GASKET

SEPARATOR

CATHODE   ANODE

ANODE CAP

CELL CAN

FIGURE 13-14 Cutaway view of the mercuric oxide cell, showing its internal construction. (Courtesy Eveready Battery Company, Inc.)

cell is classified as a primary, nonrechargeable type. Its major characteristics are:

Cell voltage: 1.35
Type: primary
Rechargeable: no
Capacity: 16 mA to 28 Ah
Temperature: poor efficiency at low temperature good at high temperature
Operating cost: high

Batteries made of mercuric-oxide cells are used in television receivers, radios, test equipment, hearing aids, and electronic watches. These are just a few of the wide range of applications for this cell.

Silver oxide cells are also classified as primary cells. Their construction is the same as the mercuric oxide cell except that they have a silver oxide anode. This cell is capable of a terminal voltage of 1.5 V. Almost all of these cells are packaged in either 1.5- or 6.0-V output batteries. This cell is very similar in many ways to the mercuric oxide cell. It has a few characteristics that differ. These characteristics are:

Cell voltage: 1.5 V
Capacity: 35 mA to 210 mAh
Temperature: good low-temperature efficiency
Operating cost: high

Silver oxide cells are used in a number of small electronic devices. These include photoelectric exposure units, hearing aids, electronic watches, and test instruments.

The lithium cell is a rather recent development in the field of primary cells. Its characteristics include a high output capacity, about five times that for a carbon-zinc type of cell. The output voltage for the lithium cell is between 2.9 and 3.7 V. The specific value depends on the electrolyte used in its construction. This cell has an extremely long shelf life. It can be stored for close to 10 years without losing any of its charge. No other cell has a shelf life even close to this. The lithium cell is used in electronic calculators, watches, and other devices requiring small size, long life, and high-capacity output.

Lithium cell *Small chemical cell often used in calculators and watches.*

13-7 The Wet Cell

One of the most commonly used cells where large capacities of current are required is the wet cell. This cell is commonly used in the automobile. Its construction is illustrated in Figure 13-15. The plates of this cell are made of an alloy of lead and antimony. The electrolyte is concentrated sulfuric acid in water. The positive plate is lead peroxide and the negative plate is a spongy lead material. The chemical interaction of these materials forms a positive electrode and a negative electrode. The discharge of the cell dilutes the water-acid electrolyte. The cell has to be recharged when this occurs. The recharging process reverses the chemical action and the plates are once again charged properly. The terminal voltage for a wet cell is 2.0 V.

Typical cell combinations for the wet cell are in groups of three or six units. These are series connected and the resulting battery terminal voltages are 6 and 12 V, respectively. The ratings for these cell combinations are in terms of the ampere-hour (Ah). This terminology describes the quantity of current available for 1 h of continuous use. A battery capable of producing its rated voltage for 1 h might be rated as having a 100-Ah capacity. This battery could also produce 200 A for $\frac{1}{2}$ h or 25 A for 4 h before being discharged.

Charging the Wet Cell When discharged the battery is connected to a charging system (Figure 13-16). The voltage source of the charger will create a reverse current flow through the cells of the battery. This action reforms the original chemical composition of the cells. The value of the charging voltage has to be slightly higher than the terminal voltage of the battery when it is fully charged or this system will not work properly. Usually, a value of around 15 V will recharge a 12-V battery. The system for recharging will operate successfully using either a dc source or an ac-operated source. The dc source most often used is the generating system of the automobile since most wet cell batteries are used in applications with an internal combustion engine. It is possible, and often done, to use a charging system operated from the home or business power lines. The charger is actually an electrical system used to convert the alternating current from the electric company into a direct current and for charging the battery.

The ability to recharge a wet cell or any other rechargeable cell is dependent on its construction. The automobile battery in recent years has undergone a design change. Batteries in new production cars require a high current output for initial motor cranking. The battery's requirements after this startup

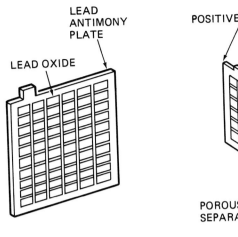

LEAD
ANTIMONY
PLATE

LEAD OXIDE

(a) SINGLE LEAD PLATE

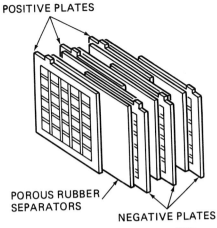

POSITIVE PLATES

POROUS RUBBER
SEPARATORS

NEGATIVE PLATES

(b) ARRANGEMENT OF PLATES
AND SEPARATORS

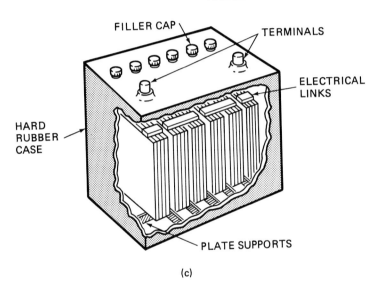

FILLER CAP

TERMINALS

ELECTRICAL
LINKS

HARD
RUBBER
CASE

PLATE SUPPORTS

(c)

FIGURE 13-15 Cutaway view of a wet cell, showing its internal construction. (From Joel Goldberg, *Fundamentals of Electricity*, 1981, p.17. Reprinted by permission of Prentice-Hall, Englewood Cliffs, New Jersey.)

are minimal. The construction of this battery does not permit its recharge when it is fully discharged. When this condition occurs, the battery must be replaced.

Batteries designed for truck, boat, or recreational vehicles do not have the same limitations as automobile batteries. These batteries may be run down, and then when recharged, they will accept a full charge. The best thing to do

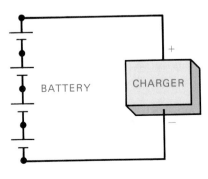

BATTERY

CHARGER

FIGURE 13-16 Use of a charging system to maintain terminal voltage in the wet-cell battery.

is to check the manufacturer's literature to see what conditions are required for recharging the battery.

REVIEW PROBLEMS

13-24. Name the major components used in the construction of a wet cell.

13-25. Is this cell rechargeable?

13-26. What is the open-circuit terminal voltage of this cell?

13-8 Charging the Dry Cell

The purpose of any cell or battery is to develop a potential voltaic source. The cells use chemical energy principles to accomplish this purpose. The electrical energy is then used to perform some form of useful work. The work is performed when a load is connected to the terminals of the cell. Some of the dry cells described in this chapter are constructed from chemical systems that may be reversed. The reversal process will restore the original voltaic charge to the cell. We know this process as battery or cell recharging.

The primary requirements for any battery recharging system is a dc voltage source. The source must be capable of producing a voltage that is higher than the voltage obtained from a fully charged battery. The purpose of this dc voltage is to force a current through the battery in a reversal from normal direction. The electrical circuit for this was shown in Figure 13-16.

The charging time for a dry cell will depend on the specific chemical makeup of the cell or battery. Typically, the charging rate for a ni-cad and alkaline cell is around 20 h. The methods of charging include a constant voltage, constant current, and taper current processes. Specific procedures are discussed in the technical material available from battery manufacturers.

In general, the constant-voltage process indicates the presence of a fixed-voltage value from the charger. This voltage is maintained during the entire charging period, usually within 1%. Current flow will be heavy during the entire charge period. This type of charger is not normally used because of its high cost and circuit complexity.

The **constant-current charging** source is the simplest method to use. Current between the charging source and the battery is limited by placing a large resistance in series with the cell being charged (Figure 13-17). The addition of the series resistance will aid in the maintenance of a constant current flow from the charger to the cell.

Constant-current charging *Method of charging cells using a constant-current charging unit.*

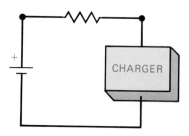

FIGURE 13-17 Addition of a current-limiting series resistance reduces current flow during the charging period.

The final charging method is the taper current process. This process has a high current at the beginning of the charge. Current tapers, or declines, during the charging period. The characteristic of the taper, its amount and charge rate, depend on the requirements of the cell or battery. Many of the less expensive types of taper chargers require adjustments by the person doing the charging.

Another method of maintaining battery voltage is called the **trickle charge** process. Most batteries have a tendency to discharge when they are not in use. This effect is true for batteries installed in equipment as well as new batteries sitting in a box or on a shelf waiting to be used. In some situations the batteries are on a "standby" basis. Equipment such as emergency lighting systems or two-way radios need to have these batteries kept in a fully charged condition. A trickle charger is connected to the terminals of the battery. A low value of voltage or current charger is placed across the battery terminals. This slow charge will maintain the battery's full capabilities and keep it ready for immediate use at all times.

Trickle charge A method of slow charge used to maintain voltage in cell.

REVIEW PROBLEMS

13-27. Should the charging voltage for a dry cell be higher or lower than that of a fully charged cell?

13-28. What is the recharge time for a dry cell?

13-29. What is a trickle charger?

13-9 Voltage and Current in a Power Source

The concept of an ideal power source is easy to understand. The ideal power source has a constant voltage output. This source also has a constant-current output. The ideal power source also has unlimited quantities of both voltage and current available. This condition exists only in textbooks. Almost all power sources have voltage and current limitations. The term *ampere-hour* was introduced earlier in this chapter. What occurs when the limits of the source are reached and what occurs when load demands go beyond this rating are topics of this section.

Load current values will vary depending on the value of load resistance (Figure 13-18). This factor is based on the relationships identified by Ohm. Current is 25 mA with a load of 240 Ω connected to a 6.0-V source [Figure 13-18(a)]. The current rises to a value of 2.5 A when the load resistance is

Load current Current requirements of operating load in system.

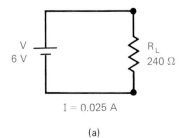

$I = 0.025$ A

(a)

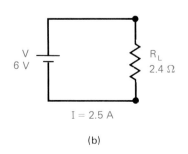

$I = 2.5$ A

(b)

FIGURE 13-18 Load current in a battery circuit is determined by load resistance.

reduced to 2.4 Ω [Figure 13-18(b)]. If the source is rated at 10.0 Ah, the 240-Ω load can be used for 400 h. We calculate this by dividing the load current into the Ah rating. A load of 2.4 Ω will operate for slightly more than 4 h. The useful life of this battery is determined by its rating and the demands of the loads it services.

The limitations of batteries or any other type of power source include a maximum power rating factor. Power in electrical term is the relationship of voltage and current. The power formula is

$$P = V \times I$$

Exceeding the capabilities of the power source will create a decrease in the output level of voltage from the source. This factor is explained when one considers the internal resistance. The circuit illustrated in Figure 13-19 helps show this concept. A load is connected to a 50-V source. The internal resistance (R_{int}) of the source is a value of 100 Ω. The effect of varying loads on the output voltage of the source is described as follows: When the load resistance is 10 kΩ, the current flow in the circuit is 495 mA. Voltage drop across the internal resistance of the source is 0.495 V and the terminal voltage is 49.01 V, almost the same value as the unloaded source.

When a heavy load (low resistance–high current demand) of 10 Ω is used, the circuit current rises to 0.45 A. Under this condition the *IR* voltage drop across the internal source resistance is 45 V. Since this is a series circuit the voltage drop across the load is the difference between the source voltage and all other voltage drops. The 50-V source and its −45-V internal *IR* drop leave a value of 5.0 V for the load. Should the load resistance decrease to 1 Ω, the output voltage of the source is reduced further. The new value of output voltage is now less than $\frac{1}{2}$ V.

The concept of a voltage drop due to excess circuit current flow is one requiring acceptance by those learning all about electricity. All power sources presently known to humankind are affected by this concept. The design of a power source for any electrical device includes its normal operating conditions. When the internal resistance of the load exceeds the design factors for the source, the output voltage of the source decreases. The effect will occur when the load circuit has a total or partial short circuit. The result is a drop in the terminal voltage from the source.

REVIEW PROBLEMS

13-30. What is an ideal power source?
13-31. How does load resistance affect current from the source?
13-32. Write a formula showing V_S, $V_{R_{int}}$, and V_L.

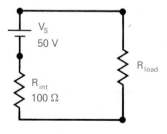

FIGURE 13-19 The internal resistance of the battery source is a factor in the availability of current for load operation.

When attempting to diagnose and repair a system having a low operating voltage, one has first to determine the source of the trouble. The load and source need to be isolated from each other. When the source is a power supply converting ac into dc, it is possible to measure the resistance of the load. An extremely low reading usually indicates a short-circuit path in the section of the device having the below-normal resistance value.

TROUBLESHOOTING APPLICATIONS

Power Transfer A voltage drop due to the value of internal resistance of the source can have an adverse effect. The purpose of the power source is to provide a maximum quantity of power to the load. When the source resistance is equal to the load resistance, a maximum quantity of power is transferred from the source to the load.

Maximum power transfer occurs when the load resistance and the source resistance are equal. This, of course, has to be some realistic value. A battery having an internal resistance on the order of 1 Ω would soon be depleted when a 1-Ω load was placed across its terminals. Any difference in the values of the two resistances also demands the use of power. The power consumed by the "extra" resistance is subtracted from the total power value of the circuit. If the mismatch, or unbalance, occurs in the source, this difference is subtracted from the total power. The difference is the quantity remaining to perform the load's work. When the imbalance is in the load, it, too, dissipates power and is not available to perform the required work. The undesirable power loss is often in the form of heat. The mismatched units create heat as a form of energy. This heat is dissipated into the air and is not available for useful work in the load.

Maximum power transfer Ability of system to transfer output power of one stage or unit to another with minimum loss.

One example of this is the matching of the output of a radio transmitter to its antenna system. A properly matched system delivers all of the power created in the transmitter to the antenna. An improperly matched system "reflects" the power created by the mismatch back into the transmitter. It is a negative force and its value opposes the delivery of all of the power to the antenna system. The results are a loss in the amount of power radiated from the antenna and the possibility of overheating the output stage of the transmitter. Overheating the output stage often destroys the transistors comprising this section of the transmitter.

The process of maximum power transfer is necessary for all types of machinery and electrical devices. A power loss in a machine is normally created by the friction between moving parts. Machine designers attempt to minimize friction, as it produces heat. Energy is required to overcome the effects of friction. This energy cannot be considered as doing useful work under these circumstances.

This effect is seen in electrical circuitry in a very similar manner. Energy is required to force electrons through a conductor. When the conductor is small the electrons bump into each other and create heat. The heat is a result of wasted energy. Often a larger conducting path will minimize this effect. The use of a bigger conductor minimizes conductor resistance and reduces heat buildup due to friction.

A final example of this concept deals with the addition of speakers to stereo or audio amplifying systems. Both the audio amplifier and the speakers exhibit the qualities of resistance. When the values of the two devices are close to matching, the sound reproduced by the speakers is at its loudest level. Any mismatching of values between speakers and amplifier results in a loss of level of audio output. An excess in mismatch can result in destroying the output stage transistors in the amplifier unit. What is being said can be stated in the form of a formula:

$$P_{out} = P_{in} \qquad (13\text{-}3)$$

Either of the power (P) units must include any losses due to friction. This formula could be modified as follows and still be correct:

$$P_{out} + P_{loss} = P_{in} + P_{load} \qquad (13\text{-}4)$$

the loss in this example being the power required to overcome electrical friction.

REVIEW PROBLEMS

13-33. What is meant by the term maximum power transfer?
13-34. What is the undesired by product of a poor power transfer?
13-35. What is the result of a mis-match between stages when power is to be transferred?

CHAPTER SUMMARY

1. Matter cannot be created or destroyed. Its state can be changed from one form to another form.
2. Chemical energy may be converted into electrical energy.
3. A battery is constructed from one or more chemical cells.
4. Primary cells are not rechargeable.
5. Secondary cells are rechargeable.
6. A cell consists of two electrodes, or plates, and an electrolyte mixture.
7. Current flow inside the cell is from positive to negative. Outside the cell current flow is from negative to positive.
8. All cells have an internal resistance. Ideally, this value should be very low.
9. An increase in the internal resistance of the cell decreases its terminal voltage when it is operating with a load.
10. Cell terminal voltage should be measured when the cell is connected to its normal load.
11. No primary cell is rechargeable. Some may be rejuvenated, but they are really not recharged.
12. Cells may be wired in series, in parallel, and in series–parallel combinations.

13. The rules for resistive circuits also apply to cells in combination.

14. A variety of dry cells are available. Each has its own set of characteristics. The cell types include carbon-zinc, alkaline, zinc chloride, nickel-cadmium, mercury, lithium, and silver oxide.

15. The wet cell is a larger-capacity type. It can be recharged hundreds of times.

16. Terminal voltage for the wet cell is 2.0 V.

17. Cells are rated in output voltage and ampere-hour capacities.

18. Cells may be recharged when the charger voltage is greater than the full charge voltage of the cell.

19. Output current from a power source is inversely proportional to the output voltage.

20. All power sources have some design limitations for output voltage and current.

21. Maximum power transfer occurs when the electrical characteristics of the power sources are equal to the input characteristics of the load.

Answer true or false; change all false statements into true statements.　　**SELF TEST**

13–1. Batteries do not have any internal resistance.

13–2. Terminal voltage for a wet cell is 2.0 V.

13–3. Four dry cells connected in series have a total voltage of 6.0 V.

13–4. Nickel-cadmium cells are rechargeable.

13–5. Primary cells are not rechargeable.

13–6. Recharge voltages must be higher than the terminal voltage of a new cell.

13–7. Electrical friction generates heat.

13–8. Output power is the sum of the output system and friction power.

13–9. A 6-Ah cell will deliver 6 A for 6 h.

13–10. Current flow inside a cell occurs from negative terminal to positive terminal.

13–11. Three 1.5-V carbon-zinc cells are series connected to a 1.5-kΩ load.　　**MULTIPLE CHOICE**
The load voltage and current are
(a) 1.5 V and 0.001 A
(b) 4.5 V and 0.001 A
(c) 4.5 V and 0.003 A
(d) 1.5 V and 0.003 A

13–12. Three 1.5-V 0.5-A cells are parallel connected. The available load voltage and current are

(a) 1.5 V and 0.5 A
(b) 4.5 V and 0.5 A
(c) 1.5 V and 1.5 A
(d) 4.5 V and 1.5 A

13–13. A battery has an open terminal voltage of 12.0 V. The voltage drops to 10.0 V when a 100-Ω load is connected to the battery. The R_{in} of the battery is

(a) 1.67 Ω
(b) 16.7 Ω
(c) 166.7 Ω
(d) 0.167 Ω

13–14. Four cells are series connected. Their ratings are 1.5 V at 0.5 A, 1.2 V at 0.3 A, 1.5 V at 0.1 A, and 1.5 V at 0.5 A. The terminal voltage of the combination is

(a) 1.5 V
(b) 3.0 V
(c) 4.5 V
(d) 5.7 V

13–15. The maximum current flow for the battery combination of problem 13–33 is

(a) 1.4 A
(b) 0.5 A
(c) 0.3 A
(d) 0.1 A

13–16. A 18-V source is connected to a 40-Ω load. Voltage measured at the load is 14 V. Its load current is

(a) 3.5 A
(b) 0.35 A
(c) 4.5 A
(d) 0.45 A

13–17. The value of R_{int} for problem 13–16 is
(a) 88 Ω
(b) 8.8 Ω
(c) 7.7 Ω
(d) 11.42 Ω

ESSAY QUESTIONS

13–18. Draw a schematic diagram showing a series–parallel connection using six 1.5-V cells and creating a 4.5-V output.

13–19. Draw a schematic diagram showing a 40-V source, an internal resistance, and a 80-Ω load.

13–20. What are the limitations of cells connected in series?

13–21. Why should the R_{int} of a source have a low resistance?

13–22. What is meant by maximum power transfer?

13–23. Why is it necessary to match the load resistance to the source resistance?

13–24. What advantage is there where cells are connected in parallel?

13–25. Why is it necessary to test a battery under load conditions?

13–26. A power source has a 24-V output. It is rated at 60 W. How much power is available for use at the load?   **PROBLEMS**

13–27. A source has an open terminal voltage of 100 V. When a 500-Ω load is connected to the source, the output voltage drops to 60 V. What is the internal resistance of the source?

13–28. A source creates 60 W of power. Its internal resistance consumes 10 W. What power is available for the load?

13–29. A load is rated at 85 W. It creates 15 W of heat during operation. What amount of power is required for its operation?

13–30. A battery is rated at 40 Ah. How long can it produce a 5-A current at its rated voltage?

13–31. A 12-V battery has an internal resistance of 0.5 Ω. What is the short-circuit current? What is the open-circuit current?

13–32. Six 1.5 V cells rated at 1.0 A are series connected to a 100-Ω load. The voltage measured across the load is 4.5 V. What is wrong?   **TROUBLESHOOTING PROBLEMS**

13–33. Six 1.5 V cells rated at 1.0 A each are parallel connected to a 100 Ω load. The voltage measured across the load is 1.5 V. What is wrong?

13–34. Four 1.5 V cells are rated as follows: 1.5 V at 1.0 A, 1.5 V at 0.75 A, 4.5 V at 0.50 A, and 1.5 V at 0.50 A. Can these four cells be series connected to provide a current of 1.0 A?

13–35. The same four cells are parallel connected in an attempt to provide 4.5 V at 2.75 A. Will this arrangement work?

13–36. The voltage measured across the terminals of a battery without its load is 45.0 V. When connected to a load the voltage measures 32 V. Where is the remaining 13 V?

13-1. False

13-2. True

13-3. False

13-4. True

13-5. 9.0 V

13-6. 0.1 A

13-7. 2.8 A

13-8. 4.0 V at 0.75 A

13-9. Zinc anode, manganese dioxide cathode, ammonium chloride, and zinc chloride electrolyte

13-10. No

13-11. 1.5 V

13-12. Zinc anode, manganese dioxide cathode, ammonium chloride, and zinc chloride electrolyte

13-13. No

13-14. 1.5 V

13-15. Zinc anode, potassium hydroxide electrolyte, and manganese dioxide cathode

13-16. Yes

13-17. 1.5 V

13-18. Nickelous hydroxide anode, cadmium hydroxide cathode, and potassium hydroxide electrolyte

13-19. Yes

13-20. 1.2 to 1.4 V

13-21. *Mercury*: mercuric oxide cathode, zinc anode, and potassium hydroxide or sodium hydroxide electrolyte; *silver*: silver oxide anode, zinc cathode, and potassium hydroxide or sodium hydroxide electrolyte; *lithium*: not given

13-22. Mercury, no; silver, no; lithium, no

13-23. 1.35; 1.35; 2.9 to 3.7 V

13-24. Lead-peroxide anode, lead cathode, and sulfuric acid electrolyte

13-25. Yes

13-26. 2.0 V

13-27. Higher

13-28. About 20 h

13-29. A trickle charger is a slow, low-current charger used to maintain a battery's charge during standby periods.

13-30. One with zero internal resistance

13–31. Low values of R_L draw more current.

13–32. $V_s + V_{RI} = V_L$

13–33. Resistance of source and load should be matched.

13–34. Heat

13–35. Reduces power available for performance of work

CHAPTER 14

Magnetism

It has been stated that "The one who controls magnetism is able to control the world." The reason this statement is true is that almost all of our commercial electrical energy is obtained by the use of magnetic principles. The earth is polarized around magnetic poles. The compass used to indicate direction is a magnetic device. Equipment presently utilized to set type, process, and store data and communicate utilizes magnetic principles. Each chapter of this book was stored after it was written on a magnetized medium known as a "floppy disk." The properties of magnetism and their effect on our life-style is important knowledge.

KEY TERMS

Electromagnet	Magnetic Strength
Ferrite Material	Magnetically Hard
Gauss	Magnetically Soft
Magnetic Field	North Pole
Magnetic Density	Relative Permeability
Magnetic Flux	Permanent Magnet
Magnetic Flux Density	Saturation
Magnetic Induction	South Pole
Magnetic Shield	Toroid Coil
	Transformer

OBJECTIVES

Upon completion of the material in this chapter, you will

1. Be able to explain the concept of a magnetic field.
2. Understand magnetic flux and density.
3. Be able to explain magnetic terms.
4. Be able to explain shielding and the Hall effect.
5. Understand the relationship of core materials to an operating frequency.

The world as we know it today could not exist without the application of magnetic principles. All of the major production of electrical energy is dependent on these principles. The entire business of communication is based on the application of magnetic principles. Almost all of our commercial transportation systems requires the use of magnetic principles. The author of a famous comic strip once stated that the person controlling magnetism will control the world. When one considers all the devices that use magnetic principles, this statement makes a lot of sense.

Most of us are aware of the effects of magnetism in one form or another. Perhaps the most common application we can visualize that uses magnetic principles is the compass. The north-seeking needle of the compass will attempt to point toward the magnetic north pole. Another device that is usually familiar to us is the bar magnet. We know that the magnet will pick up materials containing iron. Many of us are aware that two magnetic poles having the same charge will repel each other. We also know that two magnetic poles having opposite charges will attract each other.

These concepts are used in the generation of electrical energy. Principles of magnetism are applied to electrical theories. This results in a group of theories related to electromagnetism. In this chapter we describe the permanent magnet, how it functions, and the terminology describing its operation. In Chapter 15 we describe electromagnetism, its units, and its application.

14-1 The Magnetic Field

Magnetic field *Area of magnetic lines of force surrounding magnet.*

The concept of a magnetic field is theoretical. The **magnetic field** created by a magnet cannot be seen or felt. It is a concept used to help us comprehend the action and interaction of the devices we identify as magnets. Whether these fields actually exist or whether they have become almost real, but imaginary factors is a topic that goes beyond the scope of this book. Here we will use the assumption that these fields actually exist for our purposes.

There are some specific terms we use to describe these magnetic lines and the magnetic field created by them. These terms include *direction*, *density*, and *strength*. The *direction* of the magnetic field is used to describe the way it is moving. This movement is in space and we are unable to see it or to feel it. Lines with arrowheads are used to indicate a magnetic field and its direction. The arrowheads, of course, indicate the direction of the movement of the magnetic field. The second term used to describe the magnetic field is its **magnetic density**. The density of the magnetic field describes the number of

Magnetic density *Quantity of lines of force in given area.*

individual magnetic lines contained in a given area of a square centimeter or a square inch. The third term, **magnetic strength**, refers to how strong the magnetic field is in the space containing it.

Magnetic strength *Power of magnetic field.*

The *magnet* as we know it consists of a piece of metal. This metal contains iron particles. It will accept the influence of another magnet in order to become a magnet in its own right. Each magnet has two distinct poles. One of these is the north-seeking pole. It is identified as the **north pole** of the magnet and by the letter N. The other pole is the south-seeking pole. It is identified as the **south pole** and by the letter S. When the magnet is formed as a straight bar of metal (Figure 14-1), the poles are at opposite ends of the metal bar. The magnetic pole is usually defined as that portion of the end of the magnet where there is a high concentration of lines of force. These lines of force flow through the air between the two poles of the magnet.

North pole *Magnetic pole containing north seeking lines of force.*

South pole *Magnetic pole containing south seeking lines of force.*

A magnetic field surrounds the magnet. Invisible lines of force emanate from one magnetic pole and are attracted to the other magnetic pole. This occurs through the metal of the magnet. The lines of force form a magnetic circuit. They also flow through the air. Their direction is then from the north pole to the south pole (Figure 14-2). The lines of force surround the magnet. This is difficult to show in a drawing, but the right-hand part of the illustration shows how the lines of force form a field surrounding the magnet.

Magnetic lines of force have similar strength and polarity as they travel between the magnetic poles. Because of this they do not touch each other. These lines of force are concentrated at the poles of the magnet. This concentration provides the magnet with polar strength.

The lines of force used to show a magnetic field are not necessarily straight. These lines often assume a curved shape. The four illustrations shown in Figure 14-3 are typical of magnetic line formation. Parts (a) and (b) show magnetic lines of force that exist between the poles of a permanent magnet. Notice how the lines of force concentrate at the polar ends of the magnets in these two illustrations. The horseshoe-shaped magnet exhibits a stronger magnetic field between its poles than does the straight bar magnet. The reason for this is that the poles of the horseshoe-shaped magnet are closer to each other. This pro-

FIGURE 14-2 Magnetic lines of force flow through the air and between the poles of the magnet.

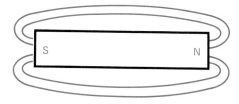

FIGURE 14-1 The bar magnet with its poles identified.

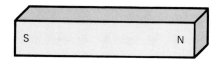

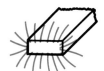

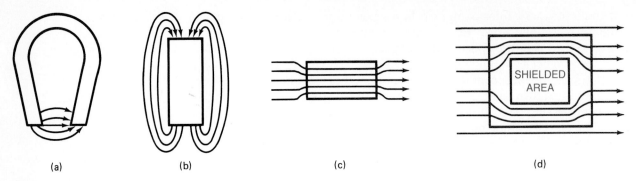

(a)   (b)   (c)   (d)

FIGURE 14-3 Magnetic lines of force may be concentrated when the magnet poles are close together (a), directed through a metal (c), or directed around a specific area (d).

duces shorter lines of force between the poles and also increases the density of the field between these poles.

Parts (c) and (d) illustrate how part of a magnetic field is directed due to the influence of an object containing iron particles. The lines of force can be concentrated when an iron-containing metal is placed in their field. Iron offers less opposition to lines of force than does air; thus the lines of force take the easier path for their travels. This effect is similar to electric current flow through parallel resistances. The smaller resistance offers less resistance, hence the great current flow through it.

Part (d) shows how a magnetic field can be directed around a specific area by use of the same type of iron-containing metal. It is used to concentrate, or direct, the lines of force. An iron type of box is used as a shield, or protection, in this application. Some types of electrical and electronic equipment are adversely affected by magnetic fields. The **magnetic shield** protects the equipment from the influence of the magnetic field.

Magnetic shield Metal used to protect an area or device from the effects of magnetic lines of force.

In summary, magnetic lines of force form a magnetic circuit. This circuit is similar to a series electrical circuit. The lines of force emanate from one magnetic pole of the magnet and flow through the air to the other pole. The strongest concentration of these lines of force is at the polar ends of the magnet. They have the greatest density at these points. Any substance containing iron may be used to direct the lines. The direction can either be toward a specific area when the lines will be concentrated or it can be away from an area when the area is shielded from the influence of the lines. These lines of force never touch each other. Each line exhibits the same magnetic properties as other lines as they move between the poles of the magnet.

Magnetic interaction occurs between two magnets (Figure 14-4). Magnetic lines of force exist around a permanent magnet. The lines of force will complement each other when two magnets are placed end to end and the middle of the two magnets has opposite poles. When like magnetic poles are facing each other, the lines of force of each magnet are repelled. The two magnets do not complement each other under these conditions. In fact, it is almost impossible to hold the two ends together. From these examples we can form these basic magnetic statements:

1. Like magnetic fields oppose each other.
2. Opposite magnetic fields attract each other.

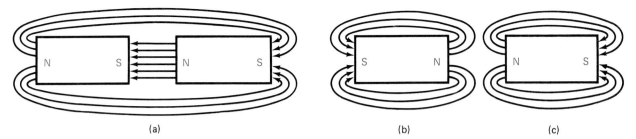

FIGURE 14-4 Magnetic fields will aid each other (a) or oppose each other (b), depending on how their poles are placed.

REVIEW PROBLEMS

14-1. The north poles of two magnets are placed facing each other. Will they be attracted or repelled?

14-2. The north pole of one magnet is placed facing the south pole of another magnet. Are they attracted or repelled?

14-3. Will magnetic lines of force flow through a magnetic shield?

14-2 Magnetic Properties

Magnetic flux is the quantity of lines of force that occupy a specific space. It is symbolized by the Greek lowercase letter phi (ϕ). The strength of the magnetic field depends on the number of magnetic lines. The greater the quantity of lines, the stronger the field and the greater is the magnetic flux of that field.

Magnetic flux *Number of lines of magnetic force occupying a specific space.*

A term used to describe one line of magnetic force is the *maxwell* (Mx). Since most magnets have many lines of force and it is nearly impossible to count them, this term is not often used. A unit capable of describing a greater quantity of magnetic lines of force is used in its place. The maxwell is a unit developed prior to the adoption of the international SI units of measure.

Magnetic flux density is the amount of flux in a given area of the magnetic field. The symbol for flux density is B. The unit of flux density is the *tesla* (T). The tesla is equal to 1 weber per square meter (Wb/m²). Flux density is expressed by the formula

Magnetic flux density *Quantity of magnetic flux in a given area.*

$$B = \frac{\phi}{A}$$

where A is equal to the area.

EXAMPLE 14-1

Find the flux density of a field when the flux in 0.5 m is 400 μWb.

Solution

$$B = \frac{\phi}{A}$$

$$= 400 \ \mu\text{Wb}/0.5 \ \text{m}^2$$

$$= 16 \times 10^{-4} \ \mu\text{Wb}$$

Gauss *Term used to describe a magnetic strength.*

Gauss is a term used to describe a magnetic field strength. This term is also not considered one of the SI units now in use in the scientific field. Magnetic fields are measured by a *gaussmeter*. The video display, or picture tube, of a color television receiver requires the use of a *degaussing coil* to neutralize any undesirable magnetic fields around it. It is the same type of demagnetizing unit as that used on video and audio tape recorders to neutralize any residual magnetism in their recording or playback heads. It does seem to be more comfortable to use the term "degaussing" rather than a term "de-teslaing" when describing this activity.

Magnetic Induction *Ability of one magnetic field to influence another area.*

Magnetic induction is the principle used to describe the influence of the field from a magnet to another material. Induction occurs without the two bodies touching each other. This influence was partially described earlier in the example of an iron-containing material being used to direct lines of force [Figure 14-3(c)].

When a magnetic field is induced onto another material, the particles of the second material will align themselves along the magnetic lines of force from the magnet. Each particle's poles act as a small magnet under these conditions. This results in the formation of a temporary magnet. It has a north pole and a south pole. The magnetic poles in the material are arranged to complement the permanent magnet's poles (Figure 14-5). This phenomenon occurs only in material containing iron particles or any other material capable of being magnetized. The effect of acceptance of magnetic induction is rated in units of *relative permeability* of the material.

Relative permeability *Rating of the ability of a material to accept magnetic properties.*

Relative permeability is a numeric value. The rating numbers used in permeability charts indicate the ability of the material to accept a magnetic field. The symbol for magnetic permeability is the Greek lowercase letter mu (μ). Typical values for permeability range up to around 2000, the value for

FIGURE 14-5 A magnet's influence may be included in a nonmagnet having iron properties.

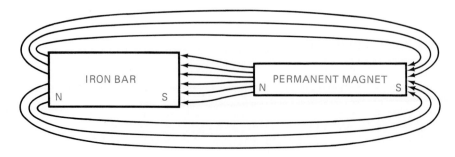

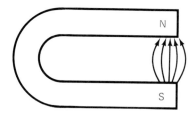
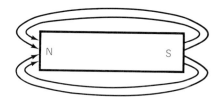

FIGURE 14-6 The relationship between the two poles of a magnet determines the direction and strength of its magnetic field.

iron. Relative permeability is a ratio, the ratio between air and the material being rated. A relative permeability rating of 2000 for iron indicates that a magnetic field will induce a field 2000 times stronger than a similar field in air.

REVIEW PROBLEMS

14-4. What is the relative permeability of air?
14-5. What is the polar alignment when magnetic fields oppose each other?
14-6. What is the name of the process of aligning iron particles in a nonmagnet?

The magnetic field of any magnet flows from one pole to the other pole. When the field is directed by a magnetic material such as iron, the lines of force are concentrated. When the lines of force move through air they tend to spread out. This is because they have the same magnetic polarity and like charges will repel each other. A magnet that is bent into a horseshoe form has its two poles much closer together than when the magnet is in bar form (Figure 14-6). The magnetic field is more concentrated between the poles of the horseshoe magnet due to the closeness of its poles. We can state that the smaller air gap will produce a stronger magnetic field between the magnet's poles. This effect is used for the magnetic field of a tape recorder's recording and playback heads.

The strongest magnetic field has no air gap or break in it. Magnets having this shape are used in places requiring a highly concentrated magnetic field. The doughnut-shaped form shown in Figure 14-7 is called a *toroid*. The toroid is used with wires wound around its form. The combination of the wire and the toroid form core is called a **toroid coil**. Toroid coils are used in electronic

Toroid coil   Round magnetic form used as base for winding wire. Makes a strong magnetic field.

WIRE AROUND CORE

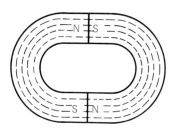

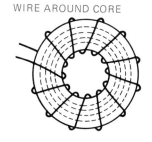

FIGURE 14-7 A toroid, or doughnut-shaped, coil has a very strong magnetic field because all lines of force are contained by its shape.

equipment where a high-permeability core for a coil of wire is required. The toroid coil is described in further detail in the section describing electromagnetic induction.

14-3 Types of Magnets

Permanent magnet *Metal having constant magnetic properties.*

A general statement describes magnets as being either *permanent* or *electromagnetic* devices. A **permanent magnet** is one whose magnetic field can be held at a constant value. It requires no outside force to change its state into a magnet on a temporary basis. The only major external force affecting its magnetism is extreme heat. This force will disturb its magnetic properties, making it a non-magnet.

Electromagnet *Device requiring current flow through coil of wire in order to act as a magnet.*

The **electromagnet** is a device requiring a flow of electron current to make it magnetic. It consists of an iron core and a coil of wire. The coil of wire is wound around the iron core. When an electric current flows through the coil of wire, an electromagnet is formed. The magnetic field will remain as long as the external current flow is maintained. The amount of electron current and the type of core (if any) on which the coil of wire is wound determine the magnetic strength of the electromagnet.

The electromagnet is shown in Figure 14-8. Iron, having a high permeability rating, will increase the strength of the magnetic field created in the unit. The addition of a switch allows the electromagnet to be turned on or off, as required. The iron-core material is made of a soft iron. Soft iron is used because it does not hold its magnetic properties when the current is turned off.

Magnetically hard *Classification of material able to maintain its magnetic properties.*

Magnetically soft *Classification of material not able to maintain its magnetization when influence is removed.*

Permanent magnets are constructed from a magnetic material designed to maintain its magnetic properties. Materials such as cobalt steel are classified as being **magnetically hard**. Other materials, such as soft iron, are considered to be **magnetically soft**. One of the more common magnetically hard materials is called *alnico*. Alnico is an alloy containing primarily aluminum, nickel, and iron. Other materials, such as cobalt, titanium, and copper, are added to this mixture. This results in several levels, or grades, of alnico steel. One of these, called alnico V, is used for the permanent-magnet component of a loudspeaker.

REVIEW PROBLEMS

14-7. Name the parts of an electromagnet.
14-8. Is alnico steel magnetically soft or hard?
14-9. What causes a permanent magnet to lose its magnetic properties?

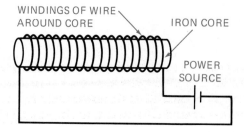

WINDINGS OF WIRE AROUND CORE   IRON CORE

POWER SOURCE

FIGURE 14-8 An electromagnet is formed when current flows through a wire. The strength of this magnet is increased by the addition of the iron core.

14-4 Ferrite Materials

A group of nonmagnetic materials have been developed in recent years. These materials do not contain any iron particles, yet they have the properties of magnetic materials. These materials are called *ferrites*. They are made of a ceramic material. The permeability of **ferrite material** is very high, usually on the order of up to 3000. One advantage of ferrite over iron is that ferrite is an insulator, whereas iron is a conductor. Ferrite is used as the core material for many electromagnetic devices. Usually, the ferrite core is used in radio, television, and similar communication equipment as a core for coils. It is also used as a core material in certain types of transformers. The term **transformer** refers to an electrical device consisting of a magnetic core and one or more windings of wire.

Ferrite materials *Group of nonferrous materials able to act as magnets*

Transformer *Electrical device with magnetic core and windings of wire using induction to transfer energy from one winding to another winding.*

There are many different *mixes* of ferrite materials. Each mix is designed to be used in a specific range of frequencies. The term mix refers to the specific composition of the ferrite material. One major advantage of ferrite types of materials is their ability to rapidly change their magnetic field strength. This is useful in electronic circuits requiring rapid value changes during operation. The subject of ferrite materials and related characteristics is a topic unto itself. It is beyond the scope of this book to do more than briefly state basic facts.

14-5 Shielding

Earlier in this chapter the concept of magnetic shielding was introduced. Magnetic shielding is used to protect some component or components from the effects of magnetic field radiation. Both permanent magnets and electromagnets produce external magnetic fields. These fields may interact and interfere with the normal operation of the unit or one of its components. Magnetic shields consist of boxes, metallic tubes, and a braid woven around the insulation of a wire conductor. Some types of magnetic shields are shown in Figure 14-9. These are used with electrical and electronic devices to insulate whatever is inside them from the influence of the magnetic field. Each of these directs the magnetic field around the area it is protecting.

REVIEW PROBLEMS

14-10. What is the basic composition of ferrite?
14-11. Why is ferrite used in electrical devices?
14-12. What is the purpose of a magnetic shield?

14-6 Saturation

One other term used to describe magnetic properties is **saturation**. Magnetic saturation occurs when the magnetic material is unable to accept any more lines of force. The external force creating the magnet may increase its magnetic field, but the material being magnetized or directing the lines of force is unable

Saturation *Term describing maximum condition where an increase in the influence does not increase the condition of operation.*

FIGURE 14-9 Magnetic shields used in electronic equipment to protect a specific circuit or component from the effects of a magnetic field. [Courtesy of Magnetic Shield Corp. (Perfection Mica Company.)]

to accept them. This effect is called saturation. It is similar in action to the process of dissolving salt or sugar in water. The liquid will absorb only a certain amount. All additional salt or sugar falls to the bottom of the container. The point at which the absorption stops is identified as the point of saturation. The liquid is said to be saturated and will not accept more even though it is still being added to the mixture. This is the same effect as magnetic saturation.

CHAPTER SUMMARY

1. The production of major qualities of electricity is dependent on magnetic principles.
2. Like magnetic charges repel each other.
3. Opposite magnetic charges attract each other.
4. A magnetic field surrounds all magnets.
5. Magnetic fields flow from one magnetic pole of a magnet, through the air, to the other magnetic pole of the magnet.
6. The strongest part of the magnetic field exists close to the poles of the magnet.
7. Magnetic lines of force can be directed either through or around an object.
8. Magnetic flux describes the number of lines of force occupying a specific space.

9. Gauss is a term used to describe a magnetic flux field's strength.
10. Magnet induction occurs when the lines of force from a magnet create a magnetic field in an adjacent piece of metal. This occurs even though the magnet is not touching the adjacent metallic material.
11. Permeability is a term used to describe the ability of a substance to become magnetized.
12. A horseshoe-shaped magnet's strength develops from the closeness of its poles to each other.
13. Permanent magnets are metals whose magnetic field is constant without any outside influence.
14. An electromagnet requires an external power source to create its magnetic field.
15. Materials can be either soft or hard magnetically. Soft materials do not retain their magnetism, while hard materials do retain their magnetism.
16. A toroid is a circular magnet and has no air gap.
17. Magnetic shielding is used to direct magnetic lines of force around a specific object or conductor.

Answer true or false; change all false statements into true statements.   SELF TEST

14–1. All magnets have a north pole and a south pole.
14–2. Permeability is the term used to describe magnetic strength.
14–3. The electromagnet does not require an external power source.
14–4. Cobalt iron is a magnetically soft material.
14–5. Alnico is an alloy used to make magnets.
14–6. Heat will destroy a magnet.
14–7. Ferrite is a noniron material having magnetic characteristics.
14–8. Permeability ratings for ferrite can be higher than those for iron.
14–9. The toroid has two air gaps.
14–10. Like magnetic poles attract each other.

14–11. Magnetic lines of force flow   MULTIPLE CHOICE
  (a) from south pole to north pole of the magnet
  (b) from north pole to south pole of the magnet
  (c) from both north to south and south to north
  (d) only in the magnet's metal form
14–12. Opposite magnetic poles
  (a) repel each other
  (b) attract each other
  (c) have no effect on each other
  (d) force crossing of magnetic lines of force

14–13. The electrical property describing the influence of a magnetic field on a second material is magnetic
    (a) force reaction
    (b) attraction
    (c) induction
    (d) reactance

14–14. The ability of a material to accept a magnetic field is known as its
    (a) permeability
    (b) acceptability
    (c) inductance factor
    (d) permanance

14–15. When a magnetic field is at its maximum, the magnet is said to be
    (a) inflated
    (b) saturated
    (c) supersaturated
    (d) semisaturated

ESSAY QUESTIONS

14–16. Describe the action of an electromagnet when external power is applied and when external power is removed from the coil of the electromagnet.

14–17. Describe the movement of magnetic particles in a substance when it is under the influence of an external magnetic field.

14–18. Draw the lines of force from a bar magnet as they affect a piece of bar iron stock placed in line with the north pole of the bar magnet.

14–19. What is magnetic saturation?

14–20. Draw a picture of a bar magnet, showing poles and direction of magnetic field flow.

PROBLEMS

14–21. What is the flux density of a magnetic field having 1000 μWb and an area of 0.038 m^2?

14–22. A magnet having an air core has a flux density of 0.0025 T. When an iron core is added, the flux density increases to 0.35 T. What is the permeability of the iron core?

14–23. What is the flux density of a magnetic field for a 180-μWb flux in an area that is 3 cm $\times$ 15 cm?

ANSWERS TO REVIEW PROBLEMS

14–1. Repelled

14–2. Attracted

14–3. No; around it

14–4. One

14–5. N to N or S to S

14–6. Magnetic induction

14–7. Iron core, coil of wire, and external power source

14–8. Magnetically hard

14–9. High temperature

14–10. Ceramic

14–11. Higher permeability rating, nonconductor

14–12. Protects a given area from magnetic influence

CHAPTER 15

Magnetic Units

Practical applications of magnetic principles are increased by the use of an electrical magnet, or electromagnet. This device uses an electric current to create a magnetic field. The magnetic field is used to transfer, or induce, a voltage from one circuit to a second circuit. All of the principles of radio and television broadcasting are based on the concept of electromagnetism. Computer use and machine control use the same principles. One of the more basic uses of this concept is the generation of electrical energy. Almost all of our commercial electric industries used the principle of the motion of a wire and a magnetic field to produce the electrical energy required to survive in our present world.

KEY TERMS

Ampere-turns

B-H Magnetism Curve

Electromagnetic Induction

Hysteresis

Hysteresis Loss

Induced Voltage

Inductor

Lenz' Law

Magnetic Retention

Magnetic Saturation

Magnetomotive Force

Permeability

OBJECTIVES

Upon completion of the material in this chapter, you will

1. Be able to explain common magnetic terms.
2. Be able to apply Ohm's law for magnetic fields.
3. Be able to explain electromagnetic induction.
4. Be able to explain the polarity of induced voltage.
5. Be able to explain the effect of induced current.

The strength of any magnet is limited by its physical size and its ability to be magnetized. The development of the electromagnet enables us to overcome these limitations. It is now possible to reduce the physical size of the magnetic device without reducing its magnetic properties. The use of an electromagnet also permits us to turn it on and off as required.

15-1 Magnetic Units

Electric current flow through a conductor creates a magnetic field around the conductor. This permits us to consider any similarities occurring between a magnetic circuit and a dc circuit. A magnetic circuit consists of two poles and a conductive path between the poles. The conductive path occurs in the magnet as well as outside of the magnet (Figure 15-1). Magnetic charges move from the south pole to the north pole in the magnetic material. These charges also move through the field around the magnet. Their direction outside the magnet is from the north pole to the south pole. This forms a magnetic circuit.

An electromagnet consists of a coil of wire wound around a steel core. The electromagnet also requires a power source in order to create a current flow through the coil of wire. The strength of the electromagnetic field depends on three basic factors. These are the type of core material, the quantity of current flowing through the coil, and the number of turns of wire used to create the coil. If we are willing to use a standard type of core, the other factors that influence the strength of the magnetic field are the number of turns of wire in the coil and the quantity of current flow. These factors are described as the **ampere-turns** of the coil. The letters *NI* are used to represent ampere-turn. The letter *N* describes the number of complete turns of wire and the letter *I* describes the quantity of current in units of the ampere. The letters At are used to represent the term *ampere-turns*.

The quantity *NI* produces a magnetic effect. This effect is commonly known as the **magnetomotive force (mmf)**, magnetic potential, or magnetizing

Ampere-turns *Term describing number of turns of wire and quantity of current flow in coil.*

Magnetomotive force (mmf) *Force producing magnetic field.*

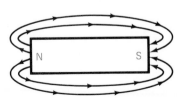

FIGURE 15-1 **Magnetic lines of force flow from one pole of the permanent magnet to the opposite pole of the magnet.**

force. It is analogous to voltage, or electromotive force, in a wired current circuit. The formula for ampere-turns is stated as

$$At = NI$$

(15-1)

EXAMPLE 15-1

Calculate the ampere-turns in a coil of wire having 1500 turns and a current of 6 A.

Solution

$$NI = 1500 \times 6$$
$$= 9000 \text{ At}$$

The formula may also be used to determine either the number of turns of wire or the current.

EXAMPLE 15-2

Find the current through a coil of wire with 600 turns and a magnetizing force of 400 At.

Solution

$$I = \frac{NI}{N} = \frac{400}{600} = 0.67 \text{ A}$$

EXAMPLE 15-3

A current of 8 A is required to produce a magnetizing force of 1200 At. How many of these turns of wire are required?

Solution

$$N = \frac{NI}{I}$$
$$= \frac{1200 \text{ At}}{8 \text{ A}}$$
$$= 150 \text{ turns}$$

This formulation may also be used in conjunction with Ohm's law when required.

EXAMPLE 15-4

A coil of wire has 400 turns and a resistance of 8 Ω. Determine current when the coil is connected to a 12-V source. Also calculate the ampere-turns of the coil.

Solution

$$I = \frac{V}{R} = \frac{12 \text{ V}}{8 \text{ Ω}} = 1.5 \text{ A}$$

$$At = NI$$

$$= 400 \text{ t} \times 1.5 \text{ A}$$

$$= 600 \text{ At}$$

REVIEW PROBLEMS

15-1. How much current is required to increase a magnetizing force from 100 to 800 in a coil having 600 turns of wire?

15-2. A magnetizing force is created with 5000 turns of wire and 2 A of current. What force is created when the current is tripled?

15-2 Field Intensity

Another major factor for an electromagnet is the intensity of its magnetic field. The intensity of the magnetic field depends on the length of the coil of wire used to create it. The capital letter *H* is used in the SI system to indicate the intensity of a magnetic field. The formula for this is:

$$H = \frac{NI}{l} \tag{15-2}$$

EXAMPLE 15-5

Find the field intensity for the electromagnet shown in Figure 15-2.

Solution

$$H = \frac{NI}{l}$$

$$= \frac{500 \text{ At}}{2} = 250$$

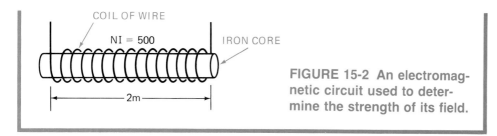

FIGURE 15-2 An electromag-netic circuit used to deter-mine the strength of its field.

where *l* is the length between the magnetic poles in units of the meter. The method of determining this is shown in the following example.

The field intensity of any coil is considered to be equal to the magneto-motive force divided by the length of the coil. This is also expressed as ampere-turns per meter.

15-3 Permeability

The **permeability** of a magnet describes its ability to concentrate a magnetic flux field. Units of permeability are indicated by the Greek lowercase letter mu (μ). The ratio of *B* (flux density/m²) to *H* (At/m²) will provide a permeability value.

The **B–H magnetism curve** is often used to indicate the quantity of flux density resulting from increasing the intensity of the magnetic field. A graphic representation of this curve is shown in Figure 15-3. The vertical component of the graph represents the quantity of flux density in a given material. The horizontal component represents the magnetizing force. The ability to impress magnetic properties on a metal is not a linear factor, as seen in this graph. As the extremes of the graph are reached there is a tendency to have a lesser effect. An increase in the magnetizing force does not create a linear increase in the magnetic field. This effect is called **magnetic saturation**. The saturation effect produces a nonlinearity to the curve.

Magnetic retention occurs when any magnetic field is imposed on a mag-netic core. The magnetic field in the core retains some of its magnetic properties, and this results in a lag in time between the application of the magnetic field producing force and the actual magnetic field. This effect occurs when the force

Permeability *Ability of magnet to concentrate a magnetic flux field.*

B-H magnetism curve *Graph indicating magnetic flux density as intensity of magnetic field is increased.*

Magnetic saturation *Increase of magnetizing influence does not increase magnetic properties.*

Magnetic retention *Ability of material to maintain magnetism when magnetizing influence is removed.*

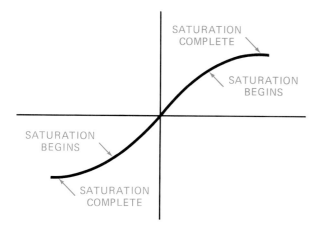

FIGURE 15-3 A magnetiza-tion curve known as a *B–H* curve shows magnetic strength in relation to time to produce it.

is applied as well as when the force is removed. The result of this is a lagging effect between the force and the field.

The results of the lagging effect are known as **hysteresis**. Hysteresis is created by the retention of magnetic properties in the core material. A typical hysteresis curve is shown in Figure 15-4. The effects of hysteresis will be better understood as you follow its development. The center point of the graph, O, is the state of the core when all magnetic fields are neutralized. The initial force creates the line between O and point P, the point of saturation.

When the magnetizing force is removed, the magnetism in the core drops from point P to point Q. The lag in time between removal of the force and the reduction of the field strength is due to hysteresis. When a reverse polarity force, such as an alternating current, is applied, the curve becomes negative. This is shown in the lower left quarter of the graph. The reverse saturation point is marked S on the graph. As the force is again reversed, the curve rises. The rise does not follow the original path from O to P, due to the retention of the magnetic field in the core material. This provides us with a hysteresis curve. The values of the curve will vary due to the strength of the magnetizing force and the type of material used in the core.

The effects of hysteresis create heat in the core material. This is due to the movement of the alignment of the magnetic particles in the material. The speed of the motion is a factor in the quantity of heat being created. When a slow rate of change, or low-frequency rate, is used, the quantity of heat is minimal. The amount of heat increases with the frequency rate of the charge. The energy required to create the heat is a nonproductive force. This energy is known as **hysteresis loss**. The hysteresis loss increases with frequency. The type of core material affects the amount of hysteresis loss. Proper core material solution will minimize hysteresis losses. Manufacturers of core materials will provide information relative to minimizing these losses at a variety of operating frequencies.

REVIEW PROBLEMS

15-3. What is the relationship between hysteresis losses and operating frequency?

15-4. What does the term *hysteresis* mean?

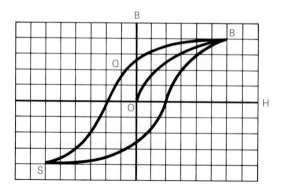

FIGURE 15-4 When plotted as shown, the magnetization curve represents a value known as hysteresis.

15-4 Electromagnetic Induction

The interaction between a conductor and a magnetic field is called **electro-magnetic induction**. The result of electromagnetic induction is the production of a voltage on the conductor. This voltage is created on the conductor when there is a movement of either the magnetic field or the conductor.

The method of producing a voltage by electromagnetic induction is illustrated in Figure 15-5. A strong magnetic field exists between two poles of a permanent magnet. When a conductor is moved through the magnetic field at right angles to the lines of force, a voltage is induced on the conductor. Conversely, when the conductor is stationary and the magnetic field is moved, the same effect results. Also, when both the magnet and the conductor are stationary, electromagnetic induction produces a voltage in the conductor if the strength of the magnetic field is altered and the lines of force move. The effect of electromagnetic induction occurs as long as either the conductor or the magnetic line of force are in motion. When both are stationary, the process of electromagnetic induction cannot occur.

The quantity of induced voltage on the conductor is also influenced by the speed of the motion of either the field or the wire. The amount of induced voltage is dependent on the rate, or frequency, of the interaction during a standard time period of 1 s. The amplitude of the induced voltage is also influenced by the number of positive peaks occurring during 1 s. The greater quantity of occurrences, the higher the average of the voltage during the time period (Figure 15-6). The letters assigned to each quarter of the graph in Figure 15-6 represent specific points of action during the electromagnetic induction cycle. *A* illustrates the "at rest," or no-movement, condition. The line between *A* and *B* represents an increase in voltage as motion starts. *B* is the maximum voltage value and is also representative of the fastest rate of movement. As the interaction decreases, the voltage drops, as seen between *B* and *C*. The induction process occurs once during a second. The average voltage induced during this period is zero.

FIGURE 15-5 A voltage is produced when a wire moves through a magnetic field (a) or when the field moves past the wire (b).

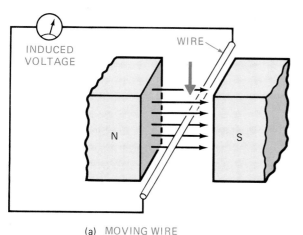

(a) MOVING WIRE

(b) MOVING FIELD

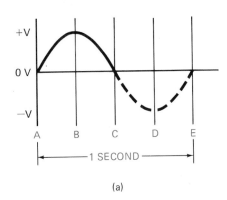
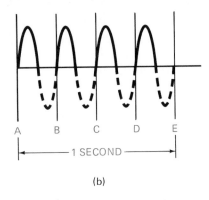

(a)

(b)

FIGURE 15-6 The value and polarity of the induced voltage is dependent on the rotational frequency of the armature.

Figure 15-6(b) shows the induced voltage when the frequency of operation is increased. The higher frequency rate produces more waveforms in the same period of time. The result is a higher output voltage. This action confirms the statement that the higher frequency rate induces a higher voltage. This factor is the basis of the design of modern power supplies.

15-5 Polarity of the Induced Voltage

The polarity of any induced voltage depends on the direction of the movement between the conductor and the magnetic field (Figure 15-7). When the conductor is moved downward through the magnetic field, the induced voltage has a polarity [part (a)]. A reversal of the direction of the conductor creates a voltage having the opposite polarity [part (b)]. In general, the polarity of the induced motion depends on the direction of the conductor's motion as it relates to the magnetic field.

This effect is also true when the direction of the movement of the lines of force change. The polarity of the induced voltage will create one polarity as the force fields strength increase. When the force field decreases and its lines of force collapse, the induced voltage has the opposite polarity.

FIGURE 15-7 The direction of motion between the wire and the field determines the polarity of the voltage induced on the wire.

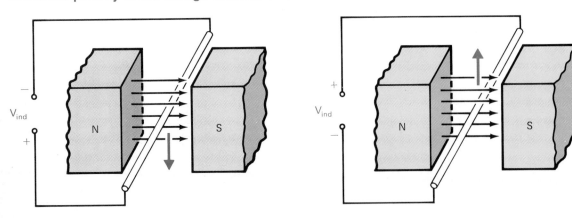

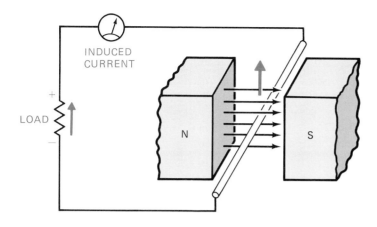

FIGURE 15-8 The addition of a load on the wires of the armature moving in a magnetic field will **cause** current flow through the wire.

15-6 Induced Current

When a load is connected to a conductor, a complete electrical circuit is created. Any induced voltage on the conductor will then force a movement of electrons through the conductor and the load (Figure 15-8). The current flow in this circuit is due to the effect of the voltage induced on the conductor. This action is the fundamental basis for electrical generators.

15-7 Forces on Conductors

A force field is created around any conductor as a current flows through the conductor. This force field will interact with the magnetic lines of force between the poles of any magnet. The result of this interaction moves the conductor. The direction of the current flow in the conductor determines the direction of its motion. This factor is the basis for the electrical motor and the moving-coil electric meter movement (Figure 15-9). The direction of current flow in the conductor and the direction of the magnetic field both determine the direction of rotation in the magnetic field.

FIGURE 15-9 Motor action occurs when a current-carrying wire reacts to a fixed magnetic field: (a) upward force; (b) downward force.

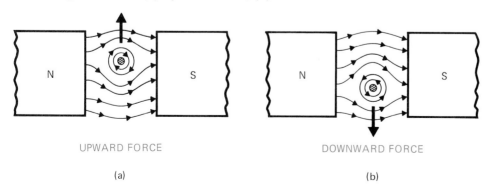

15-8 Conductors and Coils

Induction is a process occurring any time a magnetic field and a conductor interact. The maximum amount of induction occurs when the conductor and the field are at right angles to each other. When the two are parallel to each other, there is minimum interaction. The reason for this becomes evident when we look at Figure 15-10. The number of lines of force cutting across the conductor in this illustration depends on the intensity of the magnetic field. Each line of force cuts across the conductor and each induces a voltage on the conductor. When the conductor is oriented so that it is parallel with the lines of force, only a few of the lines cut across the conductor and the resulting induced voltage is minimal.

The strength of the induced voltage, therefore, is dependent on the relationship of the lines of force and the conductor. Higher levels of induced voltage will occur when more of the conductor is influenced by the lines of force. This is accomplished by use of a coil of wire instead of a single straight conductor wire.

Figure 15-11 shows this effect. A coil of conductive material, such as copper wire, is wound. The coil has many turns. Its total length depends on the diameter of the wire, any spacing between its turns, and the diameter of the coil. A 1-in. length of a straight piece of wire placed in a flux field having

FIGURE 15-10 The amount of voltage induced on wire depends on the strength of the magnetic field through which the wire passes.

FIGURE 15-11 A coil of wire has a greater length than a single straight piece of wire; hence it will have a higher voltage induced on it.

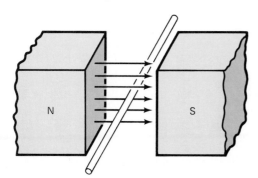

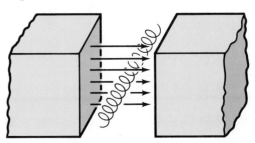

a density of 800 lines of force will accept only 800 lines of force. A coil of wire with a diameter of 1 in. has a length of 3.14 in. If the coil of wire has 200 turns per inch, the total length of wire being influenced by the 800 lines of force is over 600 in. The voltage induced for a 600-in. length of wire is much greater than for the 1-in. length. The reason for using coils of wire instead of straight conductors is that they increase the amount of induction.

15-9 Basic Inductor

A coil of wire is considered to be a basic **inductor**. Scientific work done by Michael Faraday in 1831 identified the principle we know as electromagnetic induction. Faraday discovered that moving a magnet through a coil of wire produced a current flow in the coil. This, of course, occurred only when a load was connected to the coil in order to complete the electrical circuit.

Simply stated, Faraday's law tells us that when there is motion between a magnetic field and a coil of wire a voltage is impressed, or induced, onto the coil. This voltage is called an **induced voltage**. The quantity of induced voltage is directly proportional to the rate of change between the magnetic field and the coil of wire. This is the principle for the electric generator and for a device called an electric transformer.

Inductor *Conductor able to induce inductance into circuit. Also called a coil.*

Induced voltage *Electromotive force impressed on conductor by magnetic induction.*

REVIEW QUESTIONS

15-8. How does electromagnetic induction occur?
15-9. Why is a coil of wire used instead of a straight piece of wire for a inductor?
15-10. What effect does frequency have on induction?

15-10 Lenz's Law

One of the basic rules related to the process of producing an induced voltage and current is called **Lenz's law**. This law describes the polarity of the induced current and voltages. Lenz's law states that the direction of an induced current in a coil is opposite the polarity of the field that created it. Another way of stating this is to say that the polarity of an induced current and voltage is opposite to the polarity of the magnetic field that produced it.

Lenz's law *Relates polarity of induced voltage to direction of motion between lines of force and conductor.*

The process illustrating Lenz's law is illustrated in Figure 15-12. When switch S is closed, current flows through the left-hand circuit. Current flow on this circuit develops a magnetic field around the conductor. As the current rises to its Ohm's law value, the magnetic field around the conductor increases. This magnetic field moves across the conductor in the right-hand circuit and induces a voltage on the wire in the right-hand circuit. The polarity of the induced voltage is opposite that of the circuit producing the inductive effect.

This should help illustrate the action described by Lenz. The polarity of the induced voltage in the right-hand circuit is opposite to the polarity of the current that created it. When the polarity of the induced voltage is opposite, the current in the right-hand circuit also has the opposite polarity.

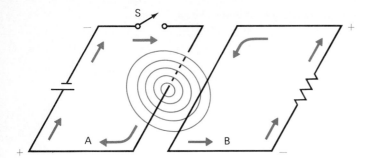

FIGURE 15-12 The current flow in circuit *A* creates a magnetic field around the wire of this circuit. The field crosses wires in circuit *B* and induces a voltage on this second set of wires.

A dc source is used in Figure 15-12. The effect of an induced voltage and current will occur only during the period when current is *changing* in the source, or left-hand circuit. This limits the process of voltage induction in a circuit having a dc source. Most voltage inductive circuits use a source having the capability of changing its magnetic field with some regularity. Details about the action involved as the source current is varied are given in Chapter 20.

CHAPTER SUMMARY

1. Magnetic strength of a permanent magnet is limited by its size.
2. Current flow through a conductor will create a magnetic field around the conductors.
3. An electromagnet consists of a coil of wire, a core, and a power source.
4. The strength of the electromagnet depends on its core, the power source, the number of turns of wire, and the amount of current flow.
5. Ampere-turns (*NI*) describe the number of turns of wire in a coil and the current in the coil.
6. The intensity of the magnetic field is also a major factor for an electromagnet.
7. The ability to concentrate a magnetic field is called permeability.
8. A magnetization curve is known as the *B–H* curve.
9. Magnetic saturation occurs when the electromagnet cannot accept further magnetic properties.
10. Hysteresis is the lag between the application of the magnetizing force and the magnet's reaction to the application.
11. Hysteresis loss creates heat in a coil or magnet.
12. The ability to impress a voltage and current on a secondary wire is called electromagnetic induction.
13. The polarity of an induced voltage is opposite to the polarity of the force creating it.
14. The quantity of the induced voltage is directly proportional to the speed of the changing field creating it.
15. An induced voltage produces an induced current in a secondary circuit.

16. A magnetic field produces a force, or motion, on a current-carrying conductor in the magnetic field.

17. Maximum interaction between a magnetic field and a conductor occurs when the two are at right angles to each other.

18. A coil of wire will accept a larger induced voltage than will a straight piece of wire.

19. Faraday's law states that when there is motion between the magnetic and the conductor, a voltage is induced onto the conductor.

20. The polarity of the induced voltage is opposite to the force that created it.

Answer true or false; change all false statements into true statements.   SELF TEST

15–1. The frequency of an electromagnet current change is directly proportional to the quantity of induced voltage.

15–2. Current flow in an induced circuit is independent of the size of the load.

15–3. An induced voltage has the same polarity as that of the force creating it.

15–4. Electromagnetic induction is a constant factor in a dc circuit.

15–5. Maximum interaction between magnetic fields occurs when they are parallel to each other.

15–6. Inductors at right angles to each other have maximum interaction of their magnetic fields.

15–7. An electromagnet's strength is increased as its turns of wire are decreased.

15–8. Heat loss is a cause of hysteresis in an inductor.

15–9. The ability of a material to retain its magnetism is called magnetic retentivity.

15–10. Core material has no effect on the magnetic field around a magnet.

15–11. The term *ampere turns* refers to the   MULTIPLE CHOICE
    (a) size of the coil wires and the number of windings
    (b) number to windings and the voltage applied
    (c) voltage applied and the number of winding turns
    (d) number of winding turns and the current in the coil

15–12. The ability of a magnet to concentrate its magnetic field is called its
    (a) retentivity

(b) permanence
(c) reluctivity
(d) permeability

15–13. Hysteresis loss results in _____ in the magnet.
(a) current loss
(b) heat
(c) voltage drop
(d) loss of magnetism

15–14. Magnetic induction occurs when
(a) a wire moves across a magnetic field
(b) the magnetic field moves across a wire
(c) there is no motion between the wire and the magnetic field
(d) both (a) and (b)

15–15. Lenz's law states that the
(a) induced current is in phase with the originating current
(b) induced voltage is out of phase with the originating current
(c) polarity of the induced voltage and current are in phase with the magnetic field producing it
(d) polarity of the induced voltage and current are out of phase with the magnetic field producing it

ESSAY QUESTIONS

15–16. Describe the effect of an increasing electrical current flowing in a coil of wire.

15–17. Describe the effect of a decreasing electrical current flowing in a coil of wire.

15–18. Describe the effect of a stationary (constant level) current flow and its ability to induce a voltage in another conductor.

15–19. Describe the effect on the value of inductance of moving the core of an inductor in and out of the center of the inductor.

15–20. How does the size or spacing of the air gap of an electromagnet affect its magnetic effect?

PROBLEMS

15–21. What is the MMF of a coil having 500 turns of wire and 0.75 A of current?

15–22. What is the field intensity for a coil having 600 mA and a length of 1.6 m?

15–23. A coil having 1500 turns of wire has a current of 80 mA. What is its magnetomotive force?

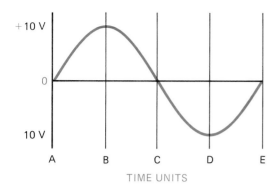

+ 10 V

0

10 V

A   B   C   D   E

TIME UNITS

FIGURE 15-13

15–24. What is the current for a 500-turn coil required to produce an MMF of 16 At?

15–25. How many turns of wire are required to produce an MMF of 6500 when a current of 6 A is used?

15–26. What is the magnetic field intensity for an electromagnet having 4500 At and a length of 1.5 m?

15–27. What is the permeability of a magnet having a B of 40 and an H of 60?

15–28. A coil has a current flow that alternates at a rate of 80 times a second. Increasing the rate of its alternator to 160 has what effect on the induced voltage?

15–29. What relation is there between the magnetic field and the induced voltage?

15–30. Using Figure 15-13, identify the quantity of induced voltage produced by the current for each time unit. Use the terms "maximum" or "minimum" for your answer.
  (a) A
  (b) B
  (c) C
  (d) D
  (e) E

15–1. 1.16 A

15–2. 30,000 At

15–3. Directly proportional

15–4. Lagging

15–5. None

15–6. Directly proportional

ANSWERS TO REVIEW PROBLEMS

15-7. It creates a complete circuit and current flow.

15-8. Magnetic field moves across a conductor.

15-9. More lines of force cut the conductor for a given overall length of the coil.

15-10. Directly proportional

Chapters 13-15

The material in this section is related to the production of electrical energy. Both chemical and magnetic concepts are presented. The principles of each of these topics are also related to basic, required knowledge for those using electricity and electronics.

1. Matter cannot be created or destroyed. Its state can be changed from one form to another form.

2. Chemical energy may be converted into electrical energy.

3. A battery is constructed from one or more chemical cells.

4. Primary cells are not rechargeable.

5. Secondary cells are rechargeable.

6. A cell consists of two electrodes, or plates, and an electrolyte mixture.

7. Current flow inside the cell is from positive to negative. Outside the cell current flow is from negative to positive.

8. All cells have an internal resistance. Ideally, this value should be very low.

9. An increase in the internal resistance of the cell decreases its terminal voltage when it is operating with a load.

10. Cell terminal voltage should be measured when the cell is connected to its normal load.

11. No primary cell is rechargeable. Some may be rejuvenated, but they are really not recharged.

12. Cells may be wired in series, in parallel, and in series–parallel combinations.

13. The rules for resistive circuits also apply to cells in combination.

14. A variety of dry cells are available. Each has its own set of characteristics. The cell types include carbon-zinc, alkaline, zinc chloride, nickel-cadmium, mercury, lithium, and silver oxide.

15. The wet cell is a larger-capacity type. It can be recharged hundreds of times.

16. Terminal voltage for the wet cell is 2.0 V.

17. Cells are rated in output voltage and ampere-hour capacities.

18. Cells may be recharged when the charger voltage is greater than the full charge voltage of the cell.

19. A change in output current from a power source is inversely proportional to changes in output voltage.

20. All power sources have some design limitations for output voltage and current.

21. Maximum power transfer occurs when the electrical characteristics of the power source are equal to the input characteristics of the load.

22. The production of major quantities of electricity is dependent on magnetic principles.

23. Like magnetic charges repeal each other.

24. Opposite magnetic charges attract each other.

25. A magnetic field surrounds all magnets.

26. Magnetic fields flow from one magnetic pole of a magnet, through the air, to the other magnetic pole of the magnet.

27. The strongest part of the magnetic field exists close to the poles of the magnet.

28. Magnetic lines of force can be directed either through or around an object.

29. Magnetic flux describes the number of lines of force occupying a specific space.

30. Gauss is a term used to describe a magnetic flux field's strength.

31. Magnetic induction occurs when the lines of force from a magnet create a magnetic field in an adjacent piece of metal. This occurs even though the magnet is not touching the adjacent metallic material.

32. Permeability is a term used to describe the ability of a substance to become magnetized.

33. A horseshoe-shaped magnet's strength develops from the closeness of its poles to each other.

34. Permanent magnets are metals whose magnetic field is constant without any outside influence.

35. A electromagnet requires an external power source to create its magnetic field.

36. Materials can be either soft or hard magnetically. Soft materials do not retain their magnetism, while hard materials do retain their magnetism.

37. A toroid is a circular magnet and has no air gap.

38. Magnetic shielding is used to direct magnetic lines of force around a specific object or conductor.

39. Magnetic strength of a permanent magnet is limited by its size.

40. Current flow through a conductor will create a magnetic field around the conductor.

41. An electromagnet consists of a coil of wire, a core, and a power source.

42. The strength of the electromagnet depends on its core, the power source, the number of turns of wire, and the amount of current flow.

43. Ampere-turns (NI) describe the number of turns of wire in a coil and the current in the coil.

44. The intensity of the magnetic field is also a major factor for an electromagnet.

45. The ability to concentrate a magnetic field is called permeability.

46. A magnetization curve is known as the B–H curve.

47. Magnetic saturation occurs when the electromagnet cannot accept further magnetic properties.

48. Hysteresis is the lag between the application of the magnetizing force and the magnet's reaction to the application.

49. Hysteresis loss creates heat in a coil or magnet.

50. The ability to impress a voltage and current on a secondary wire is called electromagnetic induction.

51. The polarity of an induced voltage is opposite to the polarity of the force creating it.

52. The quantity of the induced voltage is directly proportional to the speed of the changing field creating it.

53. An induced voltage produces an induced current in a secondary circuit.

54. A magnetic field produces a force, or motion, on a current-carrying conductor in the magnetic field.

55. Maximum interaction between a magnetic field and a conductor occurs when the two are at right angles to each other.

56. A coil of wire will accept a larger induced voltage than will a straight piece of wire.

57. Faraday's law states that when there is motion between the magnetic and the conductor, a voltage is induced onto the conductor.

CHAPTER 16

Inductance

A coil of wire is known as an inductor because its magnetic field is able to transfer, or induce, a voltage from the source circuit onto a secondary circuit. The fundamental inductor consists of a coil of wire. A core material may be used to increase the quantity of magnetic field in the inductor. Inductors are used as one of the components in a resonant, or tuned, circuit. Inductors are also used as transformers. The transformer circuit is able to step up, step down, or isolate one voltage value from a second voltage value. One practical application for the transformer is found in much of our electronic equipment designed to operate from the commercial power companies. The value of this voltage is too high for most solid-state equipment. It is dropped, or stepped down, by the action of a transformer to a level adequate for circuit operation.

KEY TERMS

Choke	**Motor**
Coil Resistance	**Radio Frequencies**
Core	**Resonant Circuit**
Filter Choke	**RF Choke**
Generator	**Solenoid**
Henry	

OBJECTIVES

Upon completion of the material in this chapter, you will

1. Understand the basic inductor.
2. Be able to explain mutual and self-inductance.
3. Be able to analyze series and parallel inductive circuits.
4. Understand transformer action.
5. Understand motor action due to inductance.
6. Be able to apply Lenz's law.
7. Be able to apply Faraday's law.

The term *inductance* has little meaning until one has some understanding of what an inductor is and what purpose it serves in an electrical or electronic circuit. The description of a basic inductor in Chapter 15 is very general. It requires both pictures and a more thorough description to provide a valid meaning to the term. The material in this chapter will accomplish both of these ends.

16-1 Inductor Characteristics

There are some physical values of inductors used to provide a better description of their characteristics. These include the units of inductance, types of core materials, shape, and form. Figure 16-1(a) shows some of the large variety of inductors commercially available. This is only a sample of the types found in electronic equipment. Figure 16-1(b) shows the schematic symbol for the basic inductor, together with some variations for this symbol. A more thorough set of graphic symbols for inductors is provided in Appendix B.

Coils, or inductors, are constructed by forming a length of wire into a coil. The coil may be wound on a form of some type for purposes of support for the windings. Normally, this form for the coil is tubular and has a hollow center, or **core**. The coil form does not normally have any effect on the inductance of the coil. Its sole purpose is support. The area inside of the coil form may be used to modify the quantity of inductance in the coil. The modification is created by the use of one of a variety of materials that may be inserted into the core of the coil form.

Core *Center of inductor, may be air or other material.*

The type of *core material* used with the coil is one determinant for the amount of inductance. A basic inductor will have an air core. This is the same as stating that this inductor has no core at all (Figure 16-2). The inductor may be formed, or wound, on a cardboard form in order to support the wires of the coils. Some large-diameter air-core inductors use plastic strips to separate adjacent turns and to maintain the shape of the coil.

Other core materials affect the magnetic quality of the coil. When the core material is a ferrous material, the lines of force are given more direction. The inductance value of the coil will be greater when a ferrous core material is used. This value will often be many times higher than that of the air reference coil, or inductor.

The type of core material and its relation to the quantity of inductance is directly related to the permeability of the core material. The permeability factors of different materials vary. Those having ferrous, or iron, bases will

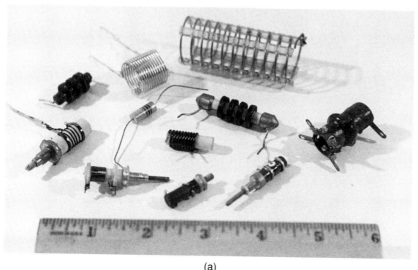

(a)

(b)

increase the magnetic properties of the coil. Others, made of nonferrous materials, such as copper or brass, will decrease the magnetic properties of the coil. The permeability of an air core is 1. The permeability of the inductor changes by the core's permeability factor.

Another factor related to ferrous types of core materials is the frequency of operation of the inductor. Inductors functioning at a low range of frequencies, between 20 Hz and about 600 Hz, will utilize a laminated steel core material or a solid steel rod. When the operational frequency increases beyond 600 Hz or so, the core material must be changed in order to have effective operation. The higher-frequency coils utilize a core material of a powdered ferrite. This material is formed into a solid shape. Some of these are shown in Figure 16-3.

A second characterisitic for an inductor is the length of the coil. In general, the length of the coil is inversely proportional to the quantity of inductance. The reason for this is that the concentration of the magnetic field is less when the coil is stretched out.

A third factor affecting the amount of inductance is the diameter of the coil of wire. This is related to the cross-sectional area of the windings. The inductance is directly proportional to the area of the coil.

The final factor involved in the amount of inductance is the number of turns of wire in the coil. The square of the number of turns of wire is directly proportional to the quantity of inductance. If all other factors are equal, a coil having 100 turns will have more inductance than a coil having 25 turns of wire.

FIGURE 16-2 The basic inductor does not have a core material.

FIGURE 16-3 Ferrite cores
are used for increasing the
inductance in high-frequency
circuit components.

The general factors related to the value of inductance for any coil of wire can be summarized in this manner:

1. Inductance L increases as the square of the number of turns of wire in the coil increase.
2. Inductance L decreases as the length of the coil is increased.
3. Inductance L increases with the diameter of the coil of wire.
4. Inductance L increases when the permeability value of the core material increases.

The characteristics of the coil and its inductance are expressed using this formula:

$$L = \frac{N^2 \mu A}{l} \times (1.26 \times 10^{-6}) \qquad (16\text{-}1)$$

where L = quantity of inductance (H)
N = number of turns
μ = permeability of the core material
A = cross-sectional area (m^2)
l = length of the core (m)

EXAMPLE 16-1

Find the inductance of a coil having 25 turns of wire with a diameter of 0.2 m wound on an air core having a length of 0.5 m (see Figure 16-4).

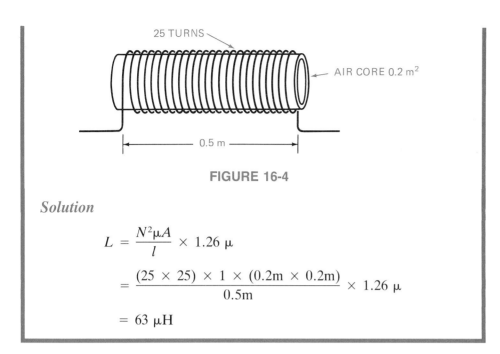

25 TURNS

AIR CORE 0.2 m²

0.5 m

FIGURE 16-4

Solution

$$L = \frac{N^2 \mu A}{l} \times 1.26\,\mu$$

$$= \frac{(25 \times 25) \times 1 \times (0.2m \times 0.2m)}{0.5m} \times 1.26\,\mu$$

$$= 63\,\mu H$$

REVIEW PROBLEMS

16-1. What is the inductance value for a coil having six turns of wire 0.3 m² in diameter, 0.8 m long, on a core with a permeability factor of 3?

16-2. What is the inductance value when the coil in problem 16-1 has 100 turns?

16-3. Name three factors influencing inductance.

16-4. What factors reduce the value of inductance for a 0.01-m coil with 10 turns, 10 cm in diameter?

16-2 Units of Inductance

The unit of inductance, as stated earlier, is the **henry**. By definition, 1 henry is the quantity of inductance required to induce 1 V when the current changes at a rate of 1 A per second. In the field of electronics this unit is often too large to use. Often, the subunits millihenry (mH) and microhenry (μH) are used to describe small values of inductance.

Henry *Unit of inductance.*

REVIEW PROBLEMS

16-5. Convert 0.048 H into millihenries.

16-6. Convert 0.03794 H into microhenries.

16-7. Convert 283 μH into henries.

16-3 Types of Inductors

Inductors are classified in several different ways. The most common describe the type of core material, its use, and whether the inductor has a constant or variable value of inductance. Let us examine these characteristics.

The inductor will have either an air core, an iron core, or a ferrite *core material*. The schematic symbols for these three core materials are shown in Figure 16-5. An iron core is illustrated with two parallel solid lines drawn along the length of the basic coil symbol. When the core is ferrite the solid lines are replaced by two parallel dashed lines. A lack of any core material is illustrated by using no parallel lines.

Some inductors are produced in a manner that permits them to have a range of inductance values. These are known as *variable* inductors. The schematic symbols for these are shown in Figure 16-6. A general symbol shows an arrow crossing through the winding of the coil wires. When an adjustable core is utilized the lines drawn showing the type of core material have arrowheads. Some of the more common variable inductors are shown in Figure 16-7.

Inductor use is another way of describing inductors. The term **choke** is often used to describe an inductor. The reason for the use of this term is due to the action of an inductor as it relates to current flow in a circuit. The inductor attempts to maintain circuit current; hence it *chokes* off any changes in current. When used with a dc power supply the inductor and other components remove, or filter out, changes in current. This application earns the name of **filter choke** for the inductor.

When the application is in circuits operating in a range known as **radio frequencies**, the inductor is often called an **RF choke**. The application of the inductor's function to attempt to remove current changes may be accomplished at any frequency. The design of the core and the methods of winding the coils are important factors in determining the use of the inductor.

Finally, inductors are often used in conjunction with capacitors and resistors. When combined in either a series or a parallel circuit, the result is called a **resonant circuit**. Details about resonant circuits are presented in Chapter 29.

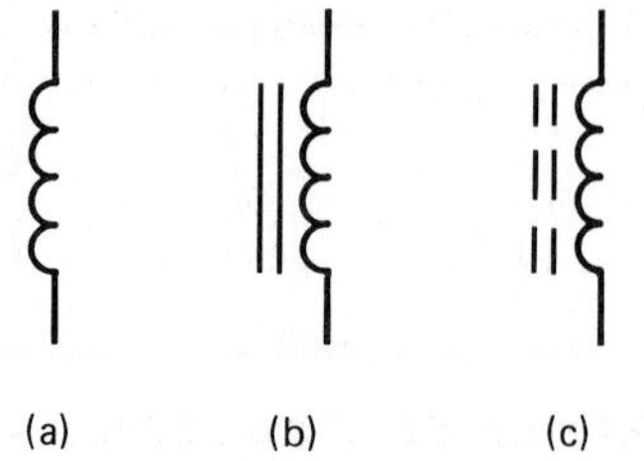

Choke *Term used to identify an inductor.*

Filter choke *Inductor used in power supply section to attenuate variations in current.*

Radio frequencies *Group of frequencies ranging from 20,000 cycles to millions of cycles.*

RF choke *Inductor used to attenuate currents in the radio frequency range.*

Resonant circuit *Circuit containing inductance, capacitance, and resistance having values able to attenuate or accept a specific frequency.*

FIGURE 16-5 Schematic symbols for inductors having different core materials (a) air core; (b) iron core; (c) ferrite core. The type of core material is indicated by the lines included with the inductor symbol.

FIGURE 16-6 The addition of arrowheads on the lines in the schematic drawing for an inductor indicates that it is adjustable or variable: (a) general symbol; (b) iron core; (c) ferrite core.

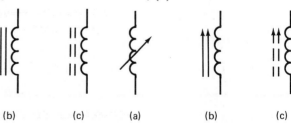

(a)   (b)   (c)   (a)   (b)   (c)

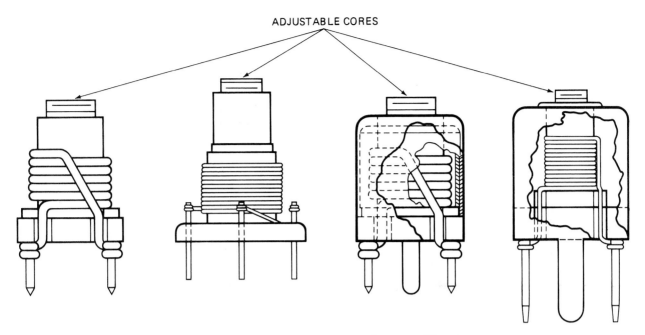

ADJUSTABLE CORES

FIGURE 16-7 Variable inductors as used in electronic equipment. (Courtesy Communication Associates, Inc., Anniston, Alabama.)

REVIEW PROBLEMS

16-8. Name three materials used as cores for inductors.

16-9. What is a choke?

16-10. What is the difference in schematic symbols between a fixed iron-core inductor and a variable unit?

16-4 Series Inductance

Inductors may be connected to each other to develop a total inductance value. One method of doing this is to connect them as a series string (Figure 16-8). When inductors are series connected the total value of inductance L_T is equal to the sum of the individual inductances. The formula for this is stated

$$L_T = L_1 + L_2 + L_3 + \cdots + L_n$$

where L_n represents the final inductor in the circuit. The rule for series inductances is the same rule used for series resistance.

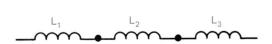

L_1 L_2 L_3

FIGURE 16-8 Series inductors are connected in an end-to-end manner.

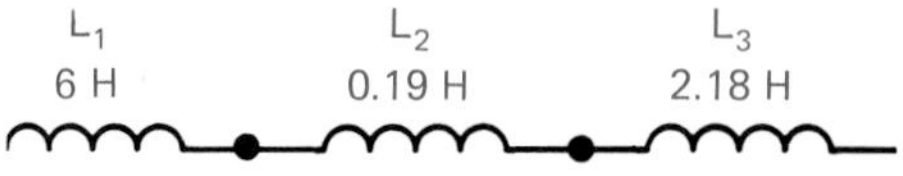

FIGURE 16-9 The total value of inductance is equal to the sum of the individual values.

L_1 6 H L_2 0.19 H L_3 2.18 H

EXAMPLE 16-2

Find total inductance for the circuit shown in Figure 16-9.

Solution

$$L_T = L_1 + L_2 + L_3$$
$$= 6 \text{ H} + 0.19 \text{ H} + 2.18 \text{ H}$$
$$= 8.37 \text{ H}$$

EXAMPLE 16-3

Find total inductance when three 4.1-H inductors are series connected.

Solution

$$L_T = L_1 + L_2 + L_3$$
$$L_1 = L_2 = L_3$$
$$L_T = 3 \times 4.1 = 12.3 \text{ H}$$

REVIEW PROBLEMS

16-11. What is the rule for series inductances?
16-12. What is L_T when a 100-mH, a 800-μH, and a 5-mH inductor are series connected?

16-5 Parallel Inductance

Inductors connected in parallel present a different set of circumstances. Total inductance for a parallel inductive circuit is less than the lowest single value of inductance used in the circuit. This factor is similar to that used to describe total resistance in a parallel resistor bank.

The formula for this circuit states that the reciprocal of the total inductance is equal to the sum of the individual reciprocal values. It is written

$$\frac{1}{L_T} = \frac{1}{L_1} + \frac{1}{L_2} + \frac{1}{L_3} + \cdots + \frac{1}{L_n} \tag{16-3}$$

Another method of writing this formula is

$$L_T = \frac{1}{1/L_1} + \frac{1}{1/L_2} + \frac{1}{1/L_3}$$

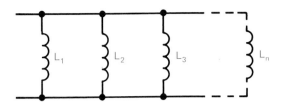

FIGURE 16-10 Parallel inductors are connected as shown.

The basic parallel inductive circuit is shown in Figure 16-10. Any number of inductors may be parallel connected to achieve a single total value of inductance.

When two unequal value inductors are used, we may desire to utilize a modification of the basic parallel inductive formula. This modification is referred to as the *product over the sum* formula. It is stated

$$L_T = \frac{L_1 \times L_2}{L_1 + L_2} \tag{16-4}$$

This formula is limited to two inductor values.

Another unique situation is when two or more equal-value inductors are used. Under this condition the value of one inductor is divided by the number of inductors in the circuit. This formula is

$$L_T = \frac{L}{n} \tag{16-5}$$

This formula may only be used with equal-value inductors.

EXAMPLE 16-4

Find the total inductance for the circuit shown in Figure 16-11.

Solution

$$L_T = \frac{1}{1/L_1} + \frac{1}{1/L_2} + \frac{1}{1/L_3}$$

$$= \frac{1}{1/0.5 \text{ H}} + \frac{1}{1/1.3 \text{ H}} + \frac{1}{1/0.9 \text{ H}}$$

$$= \frac{1}{2 + 0.77 + 1.11}$$

$$= \frac{1}{3.88} = 0.26 \text{ H}$$

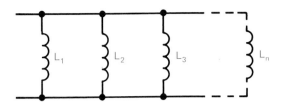

FIGURE 16-11

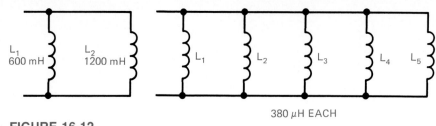

FIGURE 16-12

380 µH EACH

FIGURE 16-13

EXAMPLE 16-5

Find the total inductance for the components in the circuit shown in Figure 16-12.

Solution

$$L_T = \frac{L_1 \times L_2}{L_1 + L_2}$$

$$= \frac{600 \times 1200}{600 + 1200}$$

$$= \frac{720,000}{1800}$$

$$= 400 \text{ mH}$$

EXAMPLE 16-6

Find total inductance for the parallel-connected inductors in Figure 16-13.

Solution

$$L_T = \frac{L_1}{n}$$

$$= \frac{380 \times 10^{-6}}{5}$$

$$= 76 \text{ µH}$$

REVIEW PROBLEMS

16-13. How does total inductance in a parallel circuit compare to the smallest inductor in the circuit?

16-14. What other components utilize a formula similar to the one used for parallel inductors?

16-6 Inductors in DC Circuits

The electrical property known as inductance functions only when either the magnetic field or the conductors in the magnetic field are moving. When both are stationary the ability to induce a voltage does not occur. This limits the process to those moments in time when there is movement. When a dc voltage is applied to an inductive circuit, the magnetic field movement occurs only when the dc source is first connected to the inductor. When connection is complete the current in the circuit will rise to the Ohm's law value. The magnetic field increases from zero to the level established by the rising current flow. When this level is reached the size of the flux field becomes stationary. At this point the process of inducing a voltage stops.

Under ordinary circumstances the action known as induction does not work well in a dc circuit. The exception to this occurs when some method is used to break up the constant current flow in the dc circuit. This is accomplished by use of an electronic or mechanical switching circuit. The electronic switch turns the current on and off at a constant repetitive rate. The magnetic field is then able to rise and fall. It acts in a manner similar to the effect of using an ac source for the circuit. This process is used in systems operating from a dc source and requiring inductive action.

16-7 Applications of Inductors

The process of electromagnetic induction has many practical uses. Some of the more common uses include the solenoid coil, the transformer, choke coils, electric motor and generator, and a filtering device and resonant circuit.

Solenoids are coils of wire wound on a tubular form (Figure 16-14). The tubular form is hollow. A steel core is used with the solenoid. The core is

Solenoid *Electromagnet having moveable core used to perform mechanical motion when energized.*

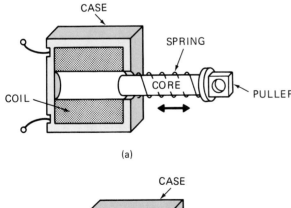

(a)

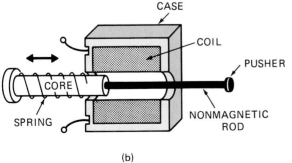

(b)

FIGURE 16-14 A solenoid coil has a fixed coil and a moving core. (From Joel Goldberg, 1981, *Fundamentals of Electricity*, p. 241. Reprinted by permission of Prentice-Hall, Englewood Cliffs, New Jersey.)

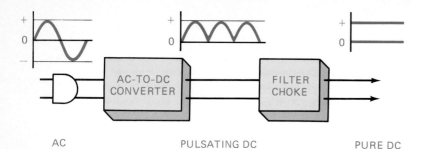

FIGURE 16-15 Use of an inductor known as a "filter choke" to smooth out any variations of the dc voltage produced by the ac-to-dc converter.

designed to move in the hollow tube. Usually, the core material is not centered. The core is attached to a mechanism. When the coil of the solenoid is energized, the core material will try to center itself inside the coil. This results in the activation of the mechanism. The device connected to the solenoid then moves and some work is performed. The mechanism may be designed to either pull or push, as required. The spring used in the assembly will return the core to its "at rest" position when the power to the coil is removed.

Choke coils are used to smooth out any variations in the current in a circuit. Often, the choke coil is used in a system converting ac into dc. The output of the system requires a constant value of current. Devices used to change ac into dc often produce a pulsating direct current (Figure 16-15). When a choke coil is connected in series with the output the changes in current are minimized. The waveforms shown at each section in Figure 16-15 are similar to those observed with electrical measuring equipment. The straight lines on the right-hand side represent a pure, or constant value of dc.

Transformers are devices utilizing the principle of electromagnetic induction. These are used to provide isolation between circuits and to couple voltages between electrical circuits. This topic is described fully in Chapter 19.

REVIEW PROBLEMS

16-15. What is required when inductive action uses a dc source?

16-16. What is meant by *solenoid action*?

16-17. What two factors result from transformer action?

Motor *Device used to convert electrical energy into rotary motion.*

Generator *Device used to convert rotary motion into electrical energy.*

Motors and **generators** use the principle of electromagnetic induction. The effect of rotary motion is common when a current-carrying wire or conductor is placed in a magnetic field. This principle is used to convert electrical energy into mechanical energy in the rotating electrical motor. This same principle is currently being developed for use in linear electric motors. These motors are capable of powering transportation vehicles at high speeds. The ideal part of this is that there are no moving parts to the linear motor; thus maintenance is minimal.

When a mechanical device is used to rotate an armature or coil of wire in a magnetic field, a voltage is induced on to the coil's wires. This principle is used in the process of electrical voltage generation.

Resonant circuits consist of inductors, resistors, and capacitors. The values of the L and C components determine the specific frequency or point of resonance. These circuits may be used to enhance a frequency or to attenuate it.

The specific action is dependent on how the components are connected to each other. This subject is covered in detail in Chapters 29 and 30.

16-8 Coil Resistance (DC)

Since any coil is constructed using conductive materials, all coils will have a value of dc resistance. When the coil is wound using copper wire, its resistance may be found from the information provided in the copper wire table in Appendix D. A second method of determining the dc resistance of a coil is to measure it using a dc ohmmeter.

Coil resistance *DC resistance value of an inductor or coil.*

The dc resistance of a coil is proportional to the gage of the wire. A high-gage-number wire has a very small diameter. Its resistance is relatively high. When a coil is constructed using a low-gage-number wire, its resistance is low. The dc resistance of any coil is a factor related to its total value. This value may be used to test the integrity of the coil.

REVIEW PROBLEMS

16-18. What is the relationship, or action, between a current-carrying wire and a magnetic field?

16-19. What three components make up a resonant circuit?

16-20. What is meant by the term *dc resistance of a coil*?

16-9 Testing Inductors

Since the inductor has a dc resistance value, it is easily tested with an ohmmeter. There are two extreme conditions for failure of the inductor. These are the open circuit and the short circuit. When testing for resistance it is convenient to know the normal dc resistance value of the inductor. This information is often provided by the manufacturer or included as part of service literature.

The ideal method to use to test the resistance of an inductor, or coil, is to isolate at least one end from the circuit to which it is connected. This procedure ensures that the resistance measurement does not include any other components connected in parallel to the coil. In practice this should not be done until *after* a preliminary measurement is made. If the preliminary measurement is not conclusive, the coil should be isolated from the circuit. Before any resistance measurement is done, the schematic diagram should be checked. This will show if any components are connected in parallel with the coil. The resistive value of the component will be an indication of the total resistance across the terminals of the coil. This will then aid in determining if the coil is either open or short circuited. Some very small inductors are wound on a high-value resistor body. This is done for support of the coil. Total resistance in these instances should be very low, usually just under the resistance value of the coil.

EXAMPLE 16-7

Find the total resistance of a 3-Ω coil wound on a 100-kΩ resistor.

Solution

$$R_T = \frac{R_1 \times R_2}{R_1 + R_2}$$

$$= \frac{3 \ \Omega \times 100 \ \text{k}\Omega}{3 \ \Omega + 100 \ \text{k}\Omega}$$

$$= \frac{300 \ \text{k}\Omega}{100,003 \ \Omega}$$

$$= 2.9999 \ \Omega$$

In this example the ohmmeter would indicate a value of 3 Ω.

The *open coil* is easily identified by a resistance reading of infinity. This is the same as stating that the resistance is too high to measure. Often, the open in the coil is located where its wires are connected to the form used to support it. A close inspection of the coil will show the point of the break. It may be possible to clean the wire and resolder or reconnect it to the form. This should restore the integrity of the coil and the circuit should be operational. Another solution, although temporary, is to unwind one turn of wire from the coil and solder it to the base connection. This will often work until the proper replacement can be installed.

A second common problem for an open coil is located at the place where the coil is connected to the circuit. This is very often true when the coil is mounted on a printed circuit board. There are times in the automated fabrication process when a relatively poor solder connection is made. This poor connection may not create troubles until after the board is in service for a period of time. This type of failure is corrected by resoldering the connection at the board.

TROUBLESHOOTING APPLICATIONS

The open-circuited coil will interrupt current flow in the circuit. The result of this is usually a failure of the function of the circuit. Open-circuit coils often do not exhibit any visual indications. The major indication of the open coil is a circuit malfunction, probably a total failure of operation.

A *short circuit* in a coil may be a total short or it may be a partial short. The total short will usually result in an excess current flow. Current will bypass the shorted windings in the coil in this situation. The excess current flow will probably cause some other component to overheat and ultimately fail. The short circuit in the coil is located by measuring its ohmic value. A total short circuit will have a very low resistance value. This measurement will be very close to 0 Ω. The fact that the coil has a total short circuit verified when the coil is removed from the circuit and its resistance measured. A total short circuit in a coil is one of the less common faults.

A partial short circuit in a coil is more common. When an excess amount of current flows through the coil its wires will heat. The wires are coated with an insulating material before being formed into the coil. When the insulation is too hot it will melt. The result is the short circuiting of adjacent turns of wire on the coil. This effectively reduces total coil resistance to a value less than a new coil would have. Often, a partially shorted coil is easily identified by having an overheated, cracked, or charred exterior.

Another method of testing the validity of an inductor is to measure the voltage that develops across its terminals during circuit operation. The voltage drop across a good inductor will be provided in a service manual. When the inductor is open its voltage drop is equal to the voltage applied to the circuit. This factor is the same for series resistances and series inductors. When one has an open there is no current flow. A no-current condition will not produce a voltage drop across the series components in the circuit.

If the inductor is shorted, the voltage drop measured across its terminals will be close to, or equal to, zero. This is also related to the voltage drop across series-connected resistances. The very low resistance offers a very low voltage drop across the short-circuited component. This method of testing for a defective inductor may be used as a preliminary step prior to testing for resistance.

When an open or a shorted coil is discovered, it is wise to look further in the circuit for a possible cause. In normal use coils do not exhibit either full or partial shorts. The cause for these is often the failure of another component in the same circuit. This may also be true when an open-circuited coil is discovered. Excess current could easily produce any of these failures, including the open. If the coil is one component in a series-connected string, the failure anywhere in the string will produce a break at its weakest point. This could be either the coil or its connection to the rest of components. The experienced repair technician learns to look for this kind of problem during the analysis and repair process.

REVIEW PROBLEMS

16-21. What is the measured resistance of a short-circuited coil?
16-22. What is the measured resistance of an open-circuited coil?
16-23. What physical signs can one look for to indicate a partially shorted coil?
16-24. What physical signs can one look for to indicate an open-circuit coil?

1. The characteristics of an inductor are its size, shape, and type of core.
2. Core material may be air, laminated steel, and ferrite forms.
3. The core material will vary, depending on the frequency of operation for the coil.
4. Inductance of the coil is directly proportional to the square of the number of turns of wire.

CHAPTER SUMMARY

5. Inductance of the coil is inversely proportional to the length of the coil.
6. The unit of inductance is the henry (H). The letter L is used to represent an inductor.
7. Subunits of the henry are the millihenry and the microhenry.
8. Inductors may be fixed, variable, or adjustable.
9. Inductors are often called *chokes* because they choke any current changes in the circuit.
10. Inductors, capacitors, and resistors can form resonant circuits.
11. Inductors may be series connected. Total L is the sum of each inductance in the circuit.
12. Inductors may be parallel connected. Total L is less than the value of the lowest single value.
13. Inductors do not work well in a pure dc circuit.
14. Practical uses for inductors include solenoids, transformers, resonant circuits, motors and generators, and filter circuits in power supplies.
15. Inductors have a dc resistance value.
16. Inductors, when defective may be open, shorted, or partially shorted.

SELF TEST

Answer true or false; change all false statements into true statements.

16–1. An inductor is a coil of wire.
16–2. Core material has no effect on amount of inductance.
16–3. Reducing the number of turns of wire will increase inductance in a coil.
16–4. The open coil has a resistance close to 1 Ω.
16–5. A ferrite core is used for high-frequency inductors.
16–6. Changing the position of the core has no effect on the value of inductance.
16–7. The unit of inductance is the henry.
16–8. Inductance is represented by the letter L in formulas.
16–9. A solenoid is a coil having a movable core.
16–10. An open-circuited inductor may be caused by the failure of a component connected in series with the coil.

MULTIPLE CHOICE

16–11. Reducing the number of turns of wire in a coil will
  (a) increase inductance
  (b) reduce inductance
  (c) increase applied voltage to the coil
  (d) decrease the applied current to the coil
16–12. The electromagnet inductor having a moving coil is known as a(n)
  (a) solenoid
  (b) relay
  (c) electromagnetic switch
  (d) switchable coil unit

16–13. Inductors operating in the high-frequency range use cores of
    (a) hard steel
    (b) soft steel
    (c) brass
    (d) ferrite

16–14. A core having a resistance of less than 1 Ω is probably
    (a) open circuited
    (b) normal
    (c) short circuited
    (d) usable

16–15. When a steel rod is moved into the core of an inductor
    (a) its inductance decreases
    (b) its inductance increases
    (c) the coil resistance increases
    (d) the inductance remains the same

ESSAY QUESTIONS

16–16. Explain why two inductors, when series connected, will provide an increase in total inductance.

16–17. Explain why two inductors, when parallel connected, will provide a decrease in total inductance.

16–18. Explain why the type of core material affects the amount of inductance in a coil.

16–19. List the four major factors affecting the amount of inductance in a coil. Explain in your own words why these affect the inductance.

16–20. Explain the term *choke*.

PROBLEMS

16–21. Find L_T for two 8-H inductors connected in series. (*Sec. 16–4*)

16–22. Find L_T for series connected inductors of 0.8 mH, 0.93 H, and 0.0445 H. (*Sec. 16–4*)

16–23. Find L_T for the inductors in problem 16–21 when they are connected in parallel. (*Sec. 16–5*)

16–24. Find L_T for the inductors in problem 16–22 when they are wired in parallel. (*Sec. 16–5*)

16–25. Find L_T for the circuit in Figure 16-16. (*Sec. 16–5*)

16–26. Find L_T for the circuit in Figure 16-17. (*Sec. 16–4*)

FIGURE 16-17

FIGURE 16-16

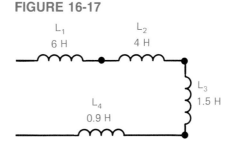

FIGURE 16-16: 1 H, 0.96 H, 2.1 H

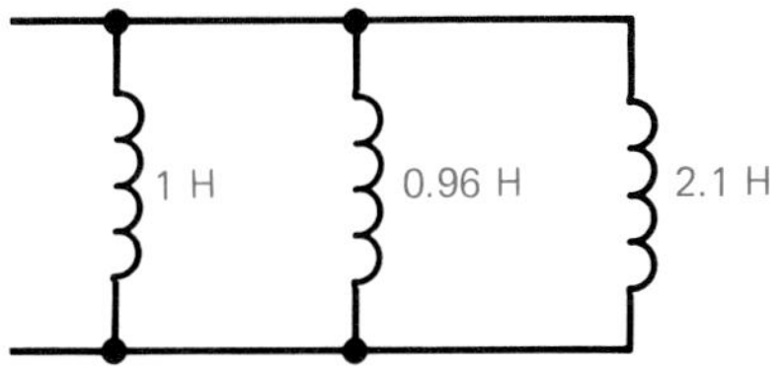

FIGURE 16-17: L_1 6 H, L_2 4 H, L_3 1.5 H, L_4 0.9 H

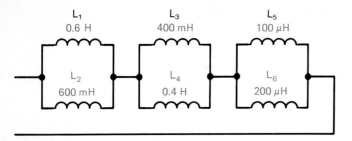

FIGURE 16-18

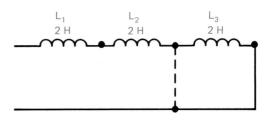

FIGURE 16-19

16–27. What is L_T for problem 16–26 when L_4 is open? (*Sec. 16–9*)

16–28. What is L_T for problem 16–26 when L_1 is shorted? (*Sec. 16–9*)

Use Figure 16-18 to answer problems 16–29 and 16–30.

16–29. Find L_T. (*Sec. 16–4,5*)

16–30. Find L_T when L_1 and L_3 are removed. (*Sec. 16–4,5*)

Use Figure 16-19 to answer problems 16–31 through 16–33.

16–31. The inductors all have the same resistance. What is the voltage drop across L_2 when 60 V is applied to the circuit (*Sec. 16–4*)

16–32. What is the voltage drop across L_1 if L_3 is removed, as indicated by the dashed line? (*Sec. 16–4*)

16–33. What is total inductance when L_3 is connected in parallel across both L_1 and L_2? (*Sec. 16–5*)

16–34. What is the inductance of coil having 100 turns of wire wound on a 0.75-m^2 core 0.5 m in length having an air core? (*Sec. 16–2*)

16–35. What is the inductance for the coil in problem 16–34 when a 300 permeability core is used? (*Sec. 16–2*)

Calculate total inductance for the coils described in problems 16–36 through 16–38.

16–36. Air core, 200 turns, 6 cm diameter, 80 cm long (*Sec. 16–2*)

16–37. 4000 core, 100 turns 15 cm diameter, 0.2 m long (*Sec. 16–2*)

16–38. 10000 core, 600 turns, 40 cm diameter, 0.2 m long (*Sec. 16–2*)

16–39. What is L_T for the circuit in Figure 16-20? (*Sec. 16–4,5*)

16–40. What is L_T for Figure 16-21? (*Sec. 16–4,5*)

FIGURE 16-20

FIGURE 16-21

EACH INDUCTOR = 10 mH AND
200 Ω dc RESISTANCE

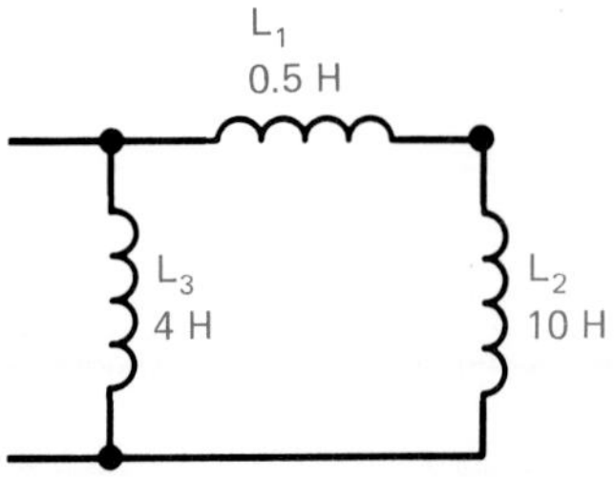

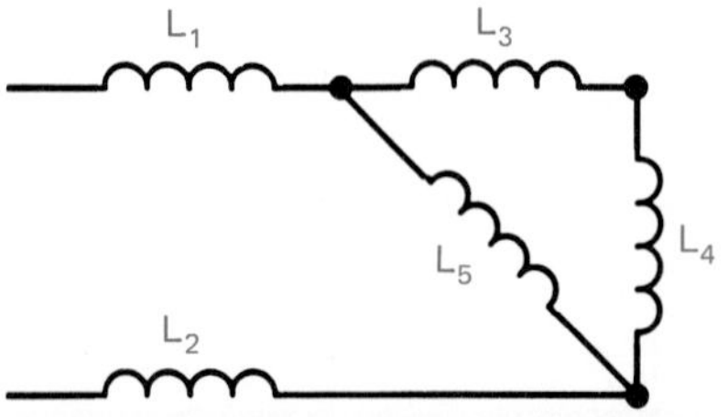

Use Figure 16-21 to answer problems 16–41 through 16–46.

16–41. What is L_T when L_5 is open?

16–42. What is L_T when L_5 is shorted?

16–43. What is R_T when all inductors are good?

16–44. What is the voltage across L_1 when 60 V dc is applied to the circuit?

16–45. What voltage would be measured across L_2 when it is open?

16–46. What voltage would be measured across L_2 when it is shorted?

16–47. Is it possible to replace a 100-mH inductor with a 40-mH and a 60-mH unit? Explain your answer.

16–48. An inductance value of 600 mH is required. The only inductors available have a value of 1.2 H. Can these be used? Explain your answer.

16–49. What effect, if any, does moving the core out of an inductor have on its value of inductance?

16–50. What effect on total inductance is there when a laminated steel core is removed from the unit?

16–1. 50.89 µH

16–2. 14.14 mH

16–3. Core material, length, and cross-sectional area of the coil.

16–4. A longer coil, smaller diameter, or a lower-permeability core material

16–5. 48 mH

16–6. 37,940 µH

16–7. 0.000283 H

16–8. Air, steel, and powdered iron

16–9. An inductor

16–10. The variable unit has arrowheads on its lines.

16–11. Series inductances add.

16–12. 105.8 mH

16–13. It is less than the lowest single value.

16–14. Parallel resistance.

16–15. It must be turned on and off at a regular rate.

16–16. A core is pulled into the center of a coil when the coil is energized.

16–17. Induced voltage and circuit isolation.

16–18. The wire will try to move at right angles to the magnetic field.

16–19. Resistors, inductors, and capacitors

16–20. The value of dc wire resistance in the coil

16–21. Almost 0 Ω

16–22. Infinite resistance; highest possible ohmic reading

16–23. Burned wrapper or insulation, broken wire

16–24. Broken wire

CHAPTER 17

Alternating Voltage and Current

The utilization of electrical energy during the early years was limited by the ability to transport it over any distance. Generating plants had to be located close to the user of the energy. One significant discovery that changed all of this was alternating current. Alternating current (ac) could be carried over wires for long distances. The principle of electromagnetic induction was used to accomplish this. Alternating-current electricity is our principal source of power for operation of almost all equipment except that using a chemical cell or solar cell voltage source.

The rules for ac are similar to those developed for direct current. There are certain characteristics of ac that make it unique. Included in these is an additional set of terms and phrases that are used by those working in the industry.

KEY TERMS

Amplitude	Period
Average Value	Phase
Effective Value	Polarity
Frequency	Root Mean Square Value
Harmonic Frequencies	Sawtooth Wave
Hertz	Sine Wave
Instantaneous Value	Square Waves
Oscilloscope	Time-Base Generator
Peak-to-Peak Value	Waveform
Peak Values	Zero-Volt Reference Line
	Zero Volts

OBJECTIVES

Upon completion of the material in this chapter, you will

1. Recognize the sine wave.
2. Understand terminology related to ac sine waves.
3. Be able to use Ohm's law for ac circuits.
4. Understand voltage and current values for ac circuits.
5. Understand harmonics.
6. Be able to explain and use amplitude-related values for ac.

So far in this book we have dealt with direct-current (dc) circuits. In a dc circuit the current has a constant value. This value is determined by the resistance in the circuit and the magnitude of the applied voltage. The flow of current is always in the same direction through the circuit. The use of dc is normally limited to devices where the power source and the load are reasonably close to each other. It is also limited to use in components requiring a constant-polarity voltage source. Perhaps the greatest limitation to the use of dc is that it cannot be carried over long lengths of wire without a large voltage drop. The development and use of an alternating-current source has overcome these limitations.

Our homes are powered by an ac source, as are our businesses and industries. A modified form of ac is also used in another manner. This form is called a *signal*. A signal is a voltage carrying some form of intelligence. When discussing the ac voltage or current the reference is to a *sine-wave* ac current or voltage. The signature, or wave shape, for this is illustrated in Figure 17-1. The sine wave is used to represent the ac voltage. It is also used as a general shape for an intelligence-carrying signal. The major difference is in the quantity of power and in the application. The sine wave has two characteristics. One of these is its amplitude and the other is its repetition rate. Note in Figure 17-1 how both halves of the wave have the same shape. Also note how both halves of the wave take the same amount of time. This is typical of all sine-wave forms. There are several characteristics of the **sine wave** that one must know. These may be divided into two major groups. One group describes the time-related values. The second group describes the amplitude-related values of the sine wave.

Sine wave *Electrical wave having values expressed as a sine of a linear function of time.*

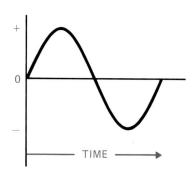

FIGURE 17-1 The signature of the ac wave, known as a *sine wave.*

17-1 Time-Related Values

The time-related values of the sine wave are identified as polarity, period, frequency, and rate of change. Let us look at each of these terms as well as any interrelationship they may have.

Polarity **Polarity** refers to the wave's value in reference to a common point in the system. Figure 17-2 illustrates polarity. The common reference for this description is the **zero-volt reference line**. This line represents a value of **zero volts.** In this illustration the voltage is positive during its first half-cycle. The voltage value starts at time zero with a value of 0 V. The voltage then rises to a maximum value during the first quarter of its cycle. This is considered to be positive because of its rise *above* zero-voltage reference line. The voltage remains positive during the second quarter of the cycle even though it falls back to the zero-voltage reference line.

The third and fourth quarters of the cycle are considered to be negative in this example. This is because the voltage value is *below* the zero-voltage reference line. In general, all voltage values (as well as parts of the waveform) that rise above the zero-voltage reference line are considered to be positive in their value. All voltages falling below the zero-voltage line are considered to be negative.

The reason for the creation of the positive and negative voltage values may be clearer when Figure 17-3 is explained. This drawing represents the ac generator. The generator consists of a rotating armature and a stationary magnetic field. The armature is rotated through this field. One side of the armature is shaded and this side is used as a reference for this explanation.

The electrical concept used to explain generator action is: When a wire moves through a magnetic field a voltage is induced on the wire. The concentration and polarity of the magnetic field determine the strength and polarity of the induced voltage. In Figure 17-3(a) the shaded part of the armature is equidistant between the two poles of the field. In this position the effects of the opposing magnetic poles cancel each other. Under these conditions no measurable amount of voltage is induced onto the armature wire.

When the armature is rotated 90°, the shaded part of the winding [Figure 17-3(b)] is close to the north (+) pole of the field. As it moves through this position, this winding part crosses the maximum number of magnetic lines of

Polarity *Relationship of the waveform value being either positive (+) or negative (−).*

Zero-volt reference line *Line on oscilloscope used as a zero, or no-voltage reference.*

Zero volts *Value of voltage at a reference line.*

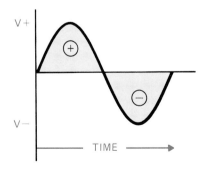

FIGURE 17-2 Polarity of the ac wave is considered positive when the wave is above the horizontal line and negative when it falls below this line.

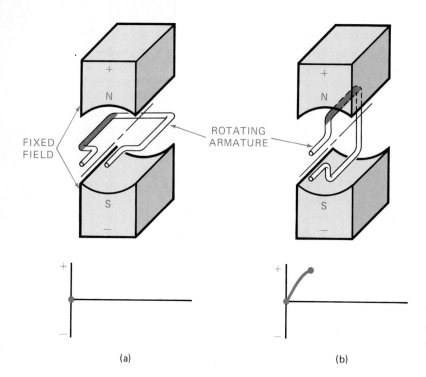

FIXED
FIELD

ROTATING
ARMATURE

(a)

(b)

FIGURE 17-3 Creation of the varying-polarity ac wave is due to the relationship of the armature and the magnetic fields of the generator.

force. The lines of force are strongest here and the induced voltage is the highest possible value, or at its peak. Since the shaded winding is moving through the field created by the north (+) pole, the induced voltage on this part of the armature is positive in polarity.

As the armature continues to rotate past the first 90°, the shaded part of the armature moves away from the intense positive field. It rotates until it has completed one-half of its full circle rotation. At the halfway point the shaded part of the armature is again back at the midpoint between the positive and negative fields. The induced voltage drops to zero again.

As the armature continues to rotate the shaded part of its winding is now close to the south (−) pole of the field. The voltage that is induced is negative in nature during this second half of the cycle. The induced voltage rises to its peak negative value and then falls back to the zero level. Thus, as the armature rotates through one full 360° circle, the voltage rises to a maximum positive value, returns to zero, then rises to its maximum negative value and again falls to zero. This completes one cycle of the generation of the voltage. Its form is that of a sine wave. The polarity of the sine-wave voltage reverses once each full cycle.

Current Flow Current flow in any electrical circuit is produced by connecting a voltage source to the circuit. The direction of current flow, the movement of electrons, is from source negative, through the load, to source positive. The direction of current flow in a circuit with a dc source is always in the same direction in the circuit. When an ac source is used the direction of the electron flow is reversed during each cycle (Figure 17-4). Current flow does not occur at the start of the cycle. This is because the source voltage is zero. The current

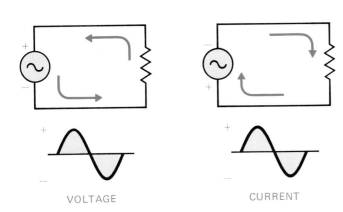

VOLTAGE   CURRENT

FIGURE 17-4 Current flow in an ac circuit is directly related to the polarity of the voltage created in the generator.

rises as the source voltage is greater [part (a)]. When the voltage reaches its peak the current is at its maximum. After the voltage falls to zero the polarity of the source is reversed. This produces a reversal of the current flow in the circuit [part (b)]. Current flow is equal to that produced during the first half-cycle, only in the opposite direction.

REVIEW PROBLEMS

17-1. How many positive peaks are there in one sine wave?
17-2. How often does the polarity change during one cycle of a sine wave?
17-3. What happens to the circuit current when the polarity of the sine wave reverses?

Period The **period** of any waveform is the time required for the wave to complete one cycle. The period for one cycle is illustrated in Figure 17-5. This is the total time required for the wave to start, rise to a maximum positive value, drop back to zero, fall to a maximum negative value, and return again to the zero level. The definition for a period applies to all wave shapes and forms. It also applies when the polarity of the wave is reversed. The wave can start on the negative half of the cycle first instead of the positive half. One cycle is identified when the wave has completed its shape and is ready to start

Period *Time required for the waveform to complete one cycle.*

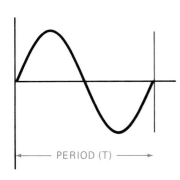

PERIOD (T)

FIGURE 17-5 The period of the ac wave is related to a specific unit of time, usually 1 second.

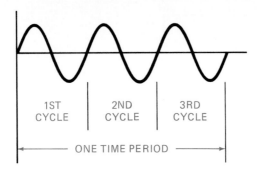

1ST CYCLE   2ND CYCLE   3RD CYCLE

ONE TIME PERIOD

FIGURE 17-6 Three cycles or complete waves are created in one time period.

a second reproduction of the shape. A multiple-cycle waveform is shown in Figure 17-6. Each complete waveform is identified as one cycle. The illustration shows three cycles of a repetitive sine wave.

The time interval required to complete one cycle is the period of the wave. In electrical and electronic work the standard unit for time measurement is the second (s). A test instrument called an **oscilloscope** is used to display the **waveform**. A high-quality oscilloscope has circuits enabling it to display one, two, or three or more cycles of the waveform. This circuit is called its *time-base* generator. The **time-base generator** is adjusted so that its time rate is close to that of the electrical signal. This permits the display of one or a few of the cycles of the wave. The time-base generator control (Figure 17-7) is calibrated in units of the second. Many of the controls are calibrated in smaller units. These units are the millisecond (0.001 s) and the microsecond (0.000001 s). The scale on the face of the oscilloscope is usually divided into units of the centimeter. The time-base control is calibrated in values related to seconds per centimeter. When one needs to know the time for one wave the procedure is to count the number of centimeters (and fractions) and multiply this figure by the time per centimeter (s/cm) marked on the position of the knob controlling

Oscilloscope Test instrument used to display the waveform or shape on a cathode ray tube.

Waveform Term used to describe the visual display of the electrical wave.

Time-base generator System used to create synchronizing signals in the oscilloscope.

FIGURE 17-7 The time-base switch on an oscilloscope is used to establish the rate of movement of the displayed waveform. (Courtesy Leader Instruments Corp.)

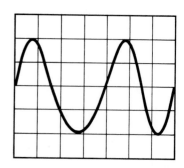

(b)

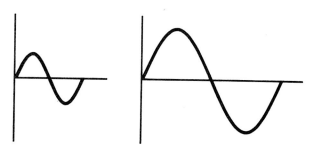

FIGURE 17-8 Amplitude of any wave describes its height, or voltage value.

the time-base circuit. The result is the *total* time required to complete one waveform.

The period of the wave is related to the time required for one complete wave. Time is determined by the horizontal movement of the wave. Time has nothing to do with the height, or amplitude, of the wave. This is another factor and is discussed later in this chapter. The two waveforms shown in Figure 17-8 the same time period, even though their amplitudes are not equal. Do not confuse these two factors. Both help to describe the electrical wave. They are independent values used for this purpose.

EXAMPLE 17-1

A waveform occupies 3 cm on the face of an oscilloscope. The time-base control is set on 0.5 s/cm. What is the time period?

Solution

$$\text{period} = \text{total time}$$
$$= \text{time(s)} \times \text{number of centimeters}$$
$$= 0.5 \text{ s} \times 3 \text{ cm}$$
$$= 1.5 \text{ s}$$

EXAMPLE 17-2

Find the period for each of the following set of conditions.

Time base	Space
(a) 5 ms/cm	4 cm
(b) 50 μs/cm	6 cm
(c) 0.1 s/cm	3.8 cm

Solution

(a) 5 × 4 = 20 ms or 0.02 s
(b) 50 × 6 = 300 μs or 0.0003 s
(c) 0.1 × 3.8 = 0.38 s

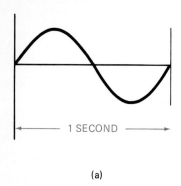

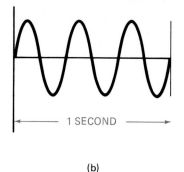

(a)   (b)

FIGURE 17-9 **Frequency of any wave describes the number of repetitions during a standard time period of 1 s.**

Frequency *Repetition rate of the waveform in time units of the second.*

Frequency The **frequency** of any occurrence is the number of times the event occurs during one time period. We know that the sun is supposed to shine once during a 24-h period. Therefore, the frequency of sunshine is expected to be once per day. In electrical and electronic terms the time period of reference is the second. All measurements are related to seconds or fractions of seconds.

The frequency of an electrical wave is the number of times the wave repeats itself during a 1-s period. We discuss the frequency in units of complete cycles per second (Figure 17-9). Waveform (a) is shown as occurring once during 1 s. Waveform (b) occurs three times during the same period. Waveform (b) is identified as having a frequency of three cycles per second. The unit used to describe frequency is the **hertz** (Hz). One hertz is equal to one cycle per second. The SI symbol for frequency is f. Waveform (b) has a frequency of 3 Hz.

Hertz *Term describing the number of cycles of a wave in a time period of one second.*

REVIEW PROBLEMS

17-4. What is the frequency of a waveform displaying 6 cycles in 1 s?
17-5. What is the frequency of a waveform displaying 10 cycles in 5 s?

Frequency and Period Relationship Frequency and period have a definite relationship. The equations for this relationship are

$$f = \frac{1}{T} \tag{17-1}$$

$$T = \frac{1}{f} \tag{17-2}$$

The relationships are reciprocal; it is possible to convert from time to frequency, and vice versa, as long as either time or frequency is known. Since the two factors are reciprocals, when the frequency increases the length of time for one wave will decrease. Also, when the time increases the frequency will decrease.

EXAMPLE 17-3

Using Figure 17-10, what is the frequency when the total time of the wave displayed is 3 s?

Solution

$$f = \frac{1}{T}$$

$$= \frac{1}{3 \text{ s}} = 0.33 \text{ Hz}$$

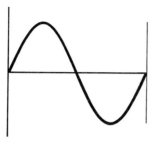

FIGURE 17-10

When one has a multiple waveform display, the method of determining the frequency is to evaluate one cycle of the wave.

EXAMPLE 17-4

In Figure 17-10, what is the frequency of the waveform when it occupies 2 cm and the time for each cm is 10 ms?

Solution

First determine the time for one wave. In this example, each complete cycle occupies 2 cm. Each cm has a value of 10 ms, or 0.01 s. The total time for one complete cycle is

$$2 \times 0.010 \text{ or } 0.02 \text{ s}$$

The final step is to apply the frequency formula

$$f = \frac{1}{T}$$

$$= \frac{1}{0.02 \text{ s}}$$

$$= 50 \text{ Hz}$$

17-2 Amplitude-Related Values

Amplitude *Vertical or voltage value of a wave.*

The second set of values for the sine wave is related to the description of its amplitude. The **amplitude**, or height, of the sine wave is a voltage or current value. The descriptions for this value include *instantaneous*, *peak*, *peak-to-peak*, *rms*, and *average* units. Each of these values are commonly used to describe the ac sine wave.

Instantaneous value *Voltage value of electrical wave at one given instant of time.*

Instantaneous Value This term is used to describe the amplitude of the ac wave at a given moment, or instant, in time. The voltage or current values of the sine wave are constantly changing. This requires the description of the amplitude to be referenced with a time value. An example of this is shown in Figure 17-11. Each time interval is identified by the lowercase letter t. The voltage values v are also indicated. The instantaneous value of voltage at t_2 has a value of v_2 in this example. Actual voltage values are used in practice instead of reference to v_1, v_2, and so on.

Peak value *Value of wave from the zero volt value to either its positive or its negative peak.*

Peak Value A second way of describing the ac wave is related to its maximum amplitude during one-half of its full cycle. This relation may be either to the positive half-cycle or to the negative half-cycle. An example of this is shown in Figure 17-12. When describing the peak value of a voltage or current the zero line for the wave is used as the reference. Peak values may be either positive values or negative values. This, of course, depends on the direction above or below the zero line taken by the wave. Peak values are a *maximum*

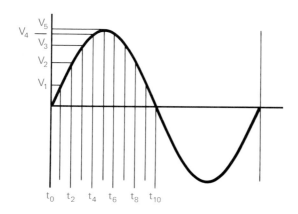

FIGURE 17-11 Instantaneous values are those values of voltage measured at specific instances of time during the period of the wave.

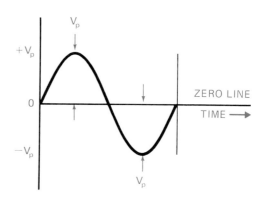

FIGURE 17-12 The peak value of a wave is its amplitude from the zero reference line to one of its peaks.

value, using the zero line as the point of reference. Peak values for voltage are described using the expression V_p or V_{pk}.

Peak-to-peak Value

Another method of describing the sine-wave voltage or current is to use the total excursion from maximum positive to maximum negative. This is its peak-to-peak value (Figure 17-13). The peak-to-peak value is the difference in voltage or current between the maximum positive value and the maximum negative value. The description for this voltage is V_{pp} or V_{p-p}. The peak-to-peak value is twice that of an equivalent peak value. This value is usually observed and measured with an oscilloscope. Most oscilloscopes can display only a voltage value. Thus the peak-to-peak values described are usually voltage values.

Peak-to-peak value *Value of wave from its most positive value to its most negative value.*

Root-Mean-Square (RMS) Value

Still another way of describing the ac wave is to use its rms value. The term rms means **root mean square**. This term identifies the mathematical process for this value. Rms values are developed from a complex formula.

A second way of considering the rms value of the wave is to identify it as the **effective value**. In its truest form the rms value is basically a measure of the heating effect of the sine-wave value. This value is derived by comparing

Root-mean-square value *Value of sine-wave voltage developed by mathematical means.*

Effective value *Reference to the quantity of heat produced by an ac current compared to an equal amount of dc current.*

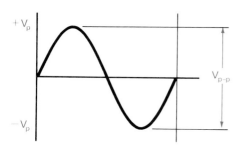

FIGURE 17-13 The peak-to-peak values of a wave is its amplitude from the most positive value to its most negative value.

the quantity of dc voltage required to heat a resistance to the quantity of *ac* voltage required to generate the same amount of heat. This ac sine-wave value producing the same amount of heat as the dc value is the rms value of the voltage. This value can be determined mathematically to be 70.7% of the peak value of the ac voltage or current.

The formulas for describing the rms value are

$$V_{rms} = 0.707V_p \qquad (17\text{-}3)$$

$$I_{rms} = 0.707I_p \qquad (17\text{-}4)$$

The same values may be used to reverse the formula process:

$$V_p = \frac{V_{rms}}{0.707} = \frac{1}{0.707}V_{rms} \qquad (17\text{-}5)$$

$$= 1.414V_{rms}$$

$$I_p = 1.414I_{rms} \qquad (17\text{-}6)$$

If describing peak-to-peak values instead of peak values, the relationship is doubled:

$$V_{p-p} = 2.828V_{rms} \qquad (17\text{-}7)$$

$$I_{p-p} = 2.828I_{rms} \qquad (17\text{-}8)$$

Ac voltage and current values are always expressed in rms units *unless otherwise noted* in the technical literature. Any voltage or current values displayed on schematic diagrams or in service literature are shown as rms values. Meters used to measure ac voltage or current display these values in rms units. Oscilloscopes are designed to display either peak or peak-to-peak values. The reason for this difference is that meters cannot relate to time units and the oscilliscope is designed to do this.

Average value *The mathematical average of one-half of the complete cycle of the ac wave.*

Average Value The final method of describing the ac sine wave uses its **average value**. If we average both the positive and negative halves of the wave, the result is zero. This is because the values will cancel each other. The average value of the rms wave is determined using only one-half of the wave, or one half-cycle. It is determined by adding each instantaneous value and then finding the average of all of these values. This also requires a complex mathematical derivation. Simply stated, the average value (Figure 17-14) is equal to 0.637

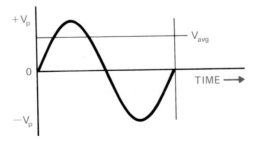

FIGURE 17-14 The average value of one half-cycle of the wave is equal to an equivalent quantity of dc voltage.

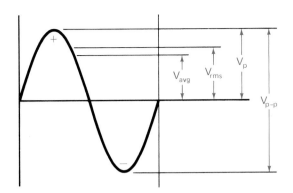

GIVEN:	TO FIND:	USE:
V_{peak}	V_{rms}	$V_{pk} \times 0.707$
V_{peak}	V_{avg}	$V_{pk} \times 0.636$
V_{rms}	V_{peak}	$V_{rms} \times 1.414$
V_{rms}	V_{avg}	$V_{rms} \times 0.899$
V_{avg}	V_{peak}	$V_{avg} \times 1.57$
V_{avg}	V_{rms}	$V_{avg} \times 1.11$
V_{peak}	V_{p-p}	$V_{peak} \times 2.0$
V_{rms}	V_{p-p}	$V_{peak} \times 2.828$

FIGURE 17-15 Various levels of voltage and their relationship to one cycle of the ac wave.

FIGURE 17-16 Comparisons of the various methods of describing the ac wave. Included are the formulas required to convert from one value to another value.

of the peak value of the wave. This is expressed as

$$V_{avg} = 0.637V_p \qquad (17\text{-}9)$$

$$I_{avg} = 0.637I_p \qquad (17\text{-}10)$$

A comparison of the methods of describing the ac sine wave is shown in Figure 17-15. The most commonly used terms of this group are rms, peak, and peak to peak. The ability to relate the different values and to be able to interchange them is important. This ability aids in the understanding of ac voltage and current values. It is also important when learning how the ac voltage and current is changed into dc values for use in the operation of electrical and electronic devices. All of the major conversion formulas are shown in Figure 17-16.

REVIEW PROBLEMS

17-9. Convert the following into rms values.
  (a) $356V_{p-p}$
  (b) $400V_p$
  (c) $80V_{avg}$
17-10. Convert the following into p-p values.
  (a) $120V_{rms}$
  (b) $24V_{rms}$
  (c) $60V_{avg}$
  (d) $40V_p$
17-11. Convert the following into average values.
  (a) $65V_{rms}$
  (b) $80V_p$

17-12. Convert the following into peak values.
(a) $480V_{p-p}$
(b) $48V_{rms}$

17-3 Ohm's Law for ac Circuits

Ohm's law describes the relationship of voltage, current, and resistance in an electrical circuit. The rules set forth by Ohm apply equally to circuits having dc sources and circuits having an ac source. All that is required is to use the same type of ac values when applying Ohm's law. When attempting to find the peak-to-peak current, be sure to use the peak-to-peak voltage values in the formula. Ohm's law for ac is presented as

$$V = IR$$

$$I_p = \frac{V_p}{R} \tag{17-11}$$

$$I_{p-p} = \frac{V_{p-p}}{R} \tag{17-12}$$

$$I_{rms} = \frac{V_{rms}}{R} \tag{17-13}$$

$$V_{avg} = I_{avg} \times R \tag{17-14}$$

All of the above are expressed in terms of sine-wave values.

REVIEW PROBLEMS

17-13. An ac voltage of $6V_{rms}$ is applied to a 10-Ω load. What is the current flow?

17-14. $20V_{p-p}$ is applied to a 100-Ω load. What is the current?

17-15. A current of $6A_{p-p}$ flows through a load of 180 Ω. What ac source voltage is required to do this?

17-4 Phase Relationship

The production of an ac wave from a rotating electrical machine, or generator, requires the armature to revolve through a 360° circle. This rotation produces one cycle of the ac sine-wave voltage. Using this as a reference, it is possible to relate points on the sine wave to degrees of rotation of the armature. This relationship is called the angular of **phase** measurement of the sine wave. It is related to one cycle and is independent of the frequency of the wave.

Phase Angular relationship between two waves in the same electrical circuit.

The relationship of the sine wave to rotation is illustrated in Figure 17-17. The wave starts on the zero line. The point of reference is considered as 0° of

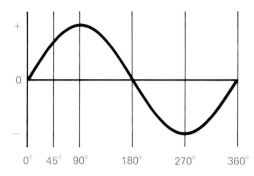

FIGURE 17-17 The phase relationship of the ac wave refers to when it starts its cycle.

0° 45° 90° 180° 270° 360°

rotation. One-eighth ($\frac{1}{8}$) of the rotation is identified as 45°, one-fourth ($\frac{1}{4}$) rotation is 90°. The midpoint, or half of the cycle, is 180° and the full cycle point is identified as 360°.

The relationship of two sine waves is often described by discussing their respective phase relationships. This relationship can be described only when both waves have the same frequency. The waves may both be voltage or they both may be current. They can also represent both voltage and current (Figure 17-18).

When both waves start at exactly the same time reference, they are considered to be *in phase* with each other. If either of the waves should start at a different moment in time, the waves are no longer in phase, or step, with each other. They are now described as being *out of phase* with each other. The specific amount of phase difference is described using the number of degrees of difference. The out-of-phase waveforms shown in Figure 17-19 have a difference in phase of 60°. They are out of phase by 60°.

An additional method of describing the phase difference of two identical waves is to identify one wave as *lagging* behind the other. In Figure 17-19 the current wave, *I*, lags behind the voltage wave, *V*, by 60°. It is necessary to use the same time reference when describing phase differences. The current waveform, *I*, is drawn in the same time reference by the addition of the dashed-line

FIGURE 17-18 Waveforms may be used to represent both voltage and current values.

FIGURE 17-19 When two waves start at different moments in time, they are said to be "out of phase" with each other.

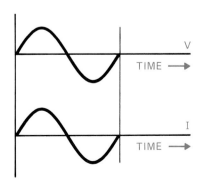

V

TIME →

I

TIME →

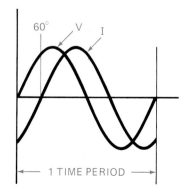

60° V I

1 TIME PERIOD

portion under the first 60° of the voltage wave in this illustration. A general rule to use when describing a phase relationship is to use as a reference the wave starting at the beginning of the time period.

REVIEW PROBLEMS

17-16. What is the phase relationship of two waves that both start at the same time?

17-17. What is the phase relationship for two waves when the second wave starts at the halfway point of the first wave?

17-5 Non-Sine-Wave Forms

Square waves Rectangular electrical wave having identical positive and negative values.

Much of the circuitry used in computers, machine controls, and the latest technology communications equipment does not use a sine-wave format. Most of the circuits require some form of square-shaped wave (Figure 17-20). These waves are often called **square waves,** due to their general shape. The reason for the use of this shape of wave is that the circuits require an instantaneous change. The change occurs from one peak of the wave to the opposite peak. This sharp and rapid change is required to move from one voltage level to another level. A sine-wave change is gradual and the point marking the shift from a positive value to a negative value can vary. When a computer or machine requires an instantaneous change or command, the gradual change that occurs with a sine-wave-shaped voltage is not accurate for this purpose. The moment in time for the change may be accurate on a display or in a service manual, but variances in components and actual circuits permit some latitude for this point. The use of a square-shaped wave having an instantaneous change from one peak to the other permits better control for the timing of circuit action.

The square-shaped waves illustrated in Figure 17-20(a) have equal amplitude for both the positive and the negative halves of its cycle. Note that one complete cycle includes both halves of the wave. The wave repeats at the end of its negative incursion. The wave may have any one of several different shapes and still be generally classified as a square wave. Figure 17-20(b) and (c) show square waves having different positive and negative values. These types of square waves often are referred to as pulse wave forms. Part (b) illustrates a

FIGURE 17-20 The square wave has some characteristics that are similar to those of the sine wave.

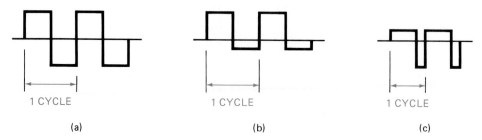

| (a) | (b) | (c) |

  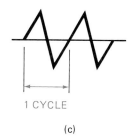

(a)　　　　　　　　　　(b)　　　　　　　　　　(c)

FIGURE 17-21 The sawtooth, or triangular, wave also has some characteristics similar to those of the sine wave.

wave having equal time for both the positive and the negative halves of its cycle. In this illustration the amplitudes of the two halves are different.

Part (c) shows a square wave having different directions for the positive and the negative halves of its cycles. The specific values for both amplitude and time are normally provided by the manufacturer of the equipment using these waveforms. Usually, the information will also give an amplitude value. This is given in units of peak-to-peak values since the waveform is observed on an oscilloscope and the "scope's" calibration uses peak-to-peak values.

Not all nonsinusoidal ac waves use a square wave, or pulse wave, form. Some use a triangular, or sawtooth, form (Figure 17-21). The **sawtooth wave** shown in part (a) has a linear rise time as it increases from its negative value to its most positive point. The rapid drop at the end of its cycle does require some time. This time is so short that the waveform appears to have an instantaneous drop time at the end of its cycle. The waveform can also be reversed, as seen in part (b). The specific waveform, of course, depends on the needs of the circuit. In other applications the rise time and the fall time of the wave, as seen in part (c), are equal. This type of waveform is called a triangular wave.

Sawtooth wave Electrical wave having shape of the tooth of a saw blade, or triangular in form.

These waves are also described in terms of peak-to-peak amplitude. One cycle of the wave occurs from its start and continues until the wave starts to repeat itself. The sawtooth wave is used in oscilloscopes as a means of producing the horizontal movement of the display. In the video display terminal this waveform is used for both horizontal and vertical movement of the display.

Harmonic Frequencies The term *harmonic frequency* refers to exact multiples of the frequency of the waveform. **Harmonic frequencies** are described as the multiple of the fundamental, or basic, frequency of the wave. The second harmonic of a 1-kHz wave is 2 kHz. The fifth harmonic of the wave would be a frequency of 5 kHz.

Harmonic frequencies Multiples of the basic, or fundamental, frequency of an electrical wave.

It is possible to describe the harmonic frequency further in terms of either *even* harmonics or *odd* harmonics. These terms refer to even or odd numbers used as multiples of the fundamental frequency.

The square-shaped wave is developed from the addition of the fundamental frequency and harmonics of this frequency (Figure 17-22). The addition of the third, fifth, and seventh harmonics of the fundamental sine wave will create a square-shaped waveform. This process creates an almost ideal square wave. The use of harmonics is only one way to create a square wave. These wave shapes may also be created by use of an electronic switching circuit.

The creation of a square wave is one use for harmonics. Another use for harmonics is the development of very high and ultrahigh frequencies for broad-

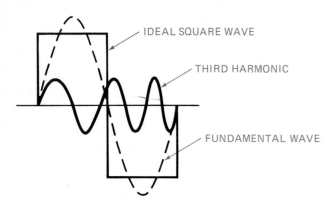

IDEAL SQUARE WAVE

THIRD HARMONIC

FUNDAMENTAL WAVE

FIGURE 17-22 A square wave is formed by the addition of several harmonics, or multiples, of the fundamental wave.

cast transmitters. When used for these purposes the integrity of the sine wave is maintained and only the harmonic frequency is utilized.

REVIEW PROBLEMS

17-18. What term is used to describe the amplitude of a square wave?

17-19. Name three types of non-sine-wave forms.

17-20. How does a pulse wave differ from a square wave?

17-21. What harmonics are used to create a square-wave shape from a sine wave?

17-22. What is the sixth harmonic of a 1.5-MHz wave?

17-6 60-Hz AC Power

Almost all the homes and industries in the United States utilize 60-Hz ac power. The frequency is a standard for this country. Many electric clocks use this frequency to maintain accurate time. Timing circuits in television receivers have used a 60-Hz frequency as a time reference for the operation of part of their circuits. The voltage created by the ac power generator may vary between 115 and 125 V, but the frequency of the generators used by the power companies is locked to 60 Hz by use of a computer. This rigid control permits power companies either to buy or to sell power to other companies, since the frequency is constant.

The voltage created by local power companies for use in homes and offices has a value of 120 V and 240 V. The power is delivered to local users in 240-V form. A three-wire system is used. One of the wires is considered to be 0 V, or ground. Each of the other two wires is rated at 120 V above ground. The difference between the two 120-V lines is 240 V. The difference between each 120-V line and ground is 120 V. The 240-V system is divided in the electrical distribution panel in the home or office into 120-V branch circuits.

The practical advantage of this system is the ability to transport ac without much loss of power. The process of generating ac power creates a high voltage

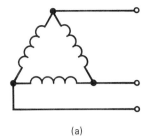
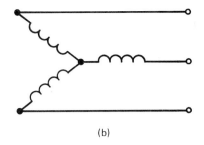

(a)   (b)

FIGURE 17-23 Delta (a) and wye (b) electrical connections used for three-phase electric power generator.

and a low current. Many power companies create voltages on the order of 18 kV or higher. This voltage is "stepped up" to around 345 kV by use of a transformer. It is sent over the power lines in this form until it reaches a local distribution center. In this center the 345 kV is stepped down to the levels required for local use.

The reason for transporting the high levels of voltage is apparent when one considers that the I^2R losses due to wire resistance are very high. When the voltage is stepped up, the line current drops and the same power level is maintained. The low current–high voltage condition has a much smaller I^2R loss factor. The result is a more efficient delivery system.

Much of the electrical power generated by local power companies is three-phase power. The reason for this is an increased efficiency for the generator. Large industrial motors operate more efficiently on three-phase ac power. The three-phase power may be connected in the form of a delta or a wye (Figure 17-23). These are connections for the individual windings of a generator. The same connections are used in motors, for maximum efficiency.

REVIEW PROBLEMS

17-23. What is the standard frequency for ac power in the United States?
17-24. What is the voltage delivered to homes and businesses?
17-25. Why does the power company create a higher voltage?

1. The ac voltage and current used for operating power in homes, businesses, and industry is sinusoidal.

2. The sine wave has both time and amplitude characteristics.

3. Time characteristics are independent of amplitude wave characteristics, and vice versa.

4. The sine wave is positive during its upward half-cycle and negative during its downward half-cycle.

5. One full 360° rotation of the ac generator creates the sine-wave voltage.

CHAPTER SUMMARY

6. Ac current flow is produced when an ac voltage is connected to a load.
7. The direction of ac current flow reverses during each full cycle of the wave.
8. The period of the ac wave is the time required to complete one full cycle.
9. The standard unit for measurement of the period of the ac wave is 1 s.
10. The frequency of the sine wave is the number of cycles occurring during 1 s.
11. Frequency is measured in units of the hertz (Hz). This term means cycles per second.
12. Wave frequency is inversely proportional to the period of the wave.
13. The amplitude of the sine wave may be described as peak-to-peak, peak, rms, average, and instantaneous values.
14. The amplitude of the ac wave at a given instant is called its instantaneous value.
15. The peak value is the amplitude of one-half of the ac wave. It is measured from the zero line to the maximum, or peak, voltage value.
16. The peak-to-peak value describes the voltage amplitude from the maximum positive point to the maximum negative point of the wave.
17. Rms, or root-mean-square, value is the same as the effective value. This is based on the value of ac voltage equal to an amount of dc voltage used to create a specific quantity of heat.
18. Ac voltage values are expressed in rms units unless otherwise specified.
19. Average values of the ac wave are determined by averaging all the instantaneous values during one-half of the cycle.
20. Ohm's law is the same for both ac and dc circuits.
21. The phase relationship of two ac waves describes their starting times. It is figured on a 360° basis.
22. Nonsinusoidal waves include square, triangular, and sawtooth forms.
23. Amplitude and frequency for nonsinusoidal waves is computed in the same manner as sine waves.
24. Harmonics are multiples of the fundamental frequency of the wave.
25. 60-Hz power is used almost exclusively in the United States for operating power.
26. Ac is used for operating power because it is easy to transport between generator and consumer and has a minimum I^2R loss.

SELF TEST   *Answer true or false; change all false statements into true statements.*

17–1. The period of an ac voltage is expressed in units of peak voltage.

17–2. One cycle of an ac wave represents a 360° generator rotation.

17–3. The rms value of a sine wave is 0.707 of its peak value.

17–4. The average value of the RMS wave is 1.414 times its peak-to-peak value.

17–5. The sine wave can only be used to represent an ac power voltage.

17–6. The third harmonic of a 600-Hz wave is 1.8 kHz.

17–7. Convert frequency into time by multiplying the two values together.

17–8. Two ac waves of different frequencies that start together are said to be *in phase*.

17–9. The ac voltage always has the same polarity.

17–10. The current produced by an ac voltage reverses during one cycle.

17–11. The voltages used in the electric power industry are

    (a) dc
    (b) RF
    (c) ac
    (d) signals

17–12. One full rotation of the generator armature is called its

    (a) hertz
    (b) frequency
    (c) alteration
    (d) cycle

17–13. The difference in starting times of two similar ac waves is called the

    (a) rotation sequence
    (b) phase angle
    (c) starting difference
    (d) lead time

17–14. Peak values are converted into rms values when multiplied by

    (a) 1.414
    (b) 2.828
    (c) 0.707
    (d) 0.634

17–15. The frequency of a 3-μs period is

    (a) 3.3 kHz
    (b) 3.3 Hz
    (c) 33 kHz
    (d) 0.33 MHz

17–16. The time period for a frequency of 28 kHz is

    (a) 35 ms
    (b) 35 μs
    (c) 3.5 ms
    (d) 3.5 μs

17–17. The peak-to-peak value for a wave that has a $+3$ positive peak and a -1 negative peak is

    (a) 3 V p-p
    (b) 4 V p-p

(c) 1 V p-p
(d) 2 V p-p

17–18. A rms voltage value of 75 V is equal to

(a) 47.7 V p-p
(b) 212.1 V p-p
(c) 53.025 V p-p
(d) 106.05 V p-p

17–19. The fifth harmonic of a 160-kHz wave is

(a) 0.8 MHz
(b) 80 kHz
(c) 40 kHz
(d) 32 kHz

17–20. The rms voltage for a circuit having 1 k Ω resistance and an ac current of 100 mA

(a) is 10 V
(b) is 100 V
(c) is 1000 V
(d) cannot be determined

ESSAY QUESTIONS

17–21. How does the generated ac voltage relate to the rotation of the armature of the generator?

17–22. How does generated current relate to the rotation of the armature of the ac generator?

17–23. What is meant by the term *period* of the ac wave?

17–24. How is an oscilloscope used to measure the frequency of the ac wave?

17–25. What is the relationship of frequency to time?

17–26. What is the relationship of the effective value of the ac wave to a dc voltage?

17–27. How does the peak-to-peak value of the ac wave relate to its effective value?

17–28. Is there any difference in using Ohm's law in either a dc circuit or an ac circuit?

17–29. Explain the term *phase relationship*.

17–30. Why is ac presently used for generation of electrical power by the commercial power companies?

17–31. A square wave with a period of 20 μs has what frequency? (*Sec. 17–1*)

17–32. A 208-V rms value has what peak-to-peak value? (*Sec. 17–2*).

17–33. A 12-V rms voltage has what peak value? (*Sec. 17–2*)

17–34. A 356-V p-p voltage has what rms value? (*Sec. 17–2*)

17–35. A 4-ms triangular wave has what frequency? (*Sec. 17–1*)

17–36. What is the time period of a 10-kHz voltage? (*Sec. 17–1*)

17–37. What is the average value of a 100-V rms value? (*Sec. 17–2*)

17–38. A signal having a time of 0.02 s has what frequency? (*Sec. 17–1*)

17–39. What is the ac power-line frequency? (*Sec. 17–1*)

17–40. What is the ac power-line period? (*Sec. 17–1*)

17–41. What is the current flow when a 100-V rms value is connected to a 500-Ω load? (*Sec. 17–3*)

17–42. What is the rms current when a 60-W bulb is connected to a 120-V source? (*Sec. 17–3*)

17–43. What is the load resistance when a 480-V rms voltage source produces a current of 1.6 A? (*Sec. 17–3*)

17–44. A sine wave completes 10 cycles during a 25-s period. What is its frequency? (*Sec. 17–1*)

17–45. Find the peak, average, and peak-to-peak values for a 24-V rms sine wave. (*Sec. 17–2*)

17–46. A sine wave completes its positive half-cycle in 25 ms. What is its frequency? (*Sec. 17–1*)

17–47. An ac current starts at the quarter-wave point of an ac voltage. What is the degree of phase difference? (*Sec. 17–4*)

17–48. The peak value of an ac voltage is 48 V. What is its effective voltage value? (*Sec. 17–2*)

17–49. The peak amplitude of an ac wave is 4 V. What is the rms current through a 10-kΩ resistance? (*Sec. 17–3*)

17–50. An ac wave has a period of 25 s. What is its third harmonic? (*Sec. 17–5*)

17–51. What is the fundamental frequency of a 360-Hz third harmonic? (*Sec. 17–5*)

17–52. What are the second, third, and fifth harmonics of a wave having a fundamental frequency of 1.8 kHz? (*Sec. 17–5*)

17–53. What current flows through two series connected 2.2-kΩ resistors when 24 V rms is connected to them? (*Sec. 17–3*)

17–54. A 5-kΩ and a 10-kΩ resistor are connected in parallel to a 120-V source. What current flows through each resistor? (*Sec. 17–3*)

17–55. Three 5-kΩ resistors are wired as a parallel bank and connected to a 48-V source. What is the power consumption for the total circuit? (*Sec. 17–3*)

17–56. A series string consisting of a 1-kΩ, 5-kΩ, 2.5-kΩ, and a 500-Ω resistor is connected to a 100-V p-p source. What is the p-p current in the circuit? (*Sec. 17–3*)

17–57. In problem 17–56, what rms current flows through the 5-kΩ resistor? (*Sec. 17–3*)

17–58. In problem 17–56, the source current frequency is 120 Hz. What is the frequency of the current through the 500-Ω resistor? (*Sec. 17–1*)

17–59. An ac wave has an amplitude of 18 V p-p and a time of 6 s. What is its rms voltage value and its frequency? (*Sec. 17–1* and *Sec. 17–2*)

17–60. A sine wave has an amplitude of 6 V p-p. A second sine wave has an amplitude of 4 V p-p and the same frequency. The second wave starts at the 60° point of the first wave. What is the phase angle of the second wave? Is it leading or lagging the first wave? (*Sec. 17–4*)

ANSWERS TO REVIEW PROBLEMS

17–1. One

17–2. Once; there are two polarities, and two changes per cycle.

17–3. It also reverses.

17–4. 6 Hz

17–5. 2 Hz

17–6. (a) 6.25 Hz; (b) 2.5 Kz; (c) 26.32 Hz; (d) 0.5 Hz

17–7. (a) 0.1 s; (b) 0.0025 s; (3) 0.000067 s or 67 μs; (d) 0.167 μs

17–8. (a) 25 Hz; (b) 12.5 kHz; (c) 1.67 MHz; (d) 0.1 Hz

17–9. (a) 125.8 V; (b) 282.8 V; (c) 88.8 V

17–10. (a) 339.36 V p-p; (b) 67.874 V p-p; (c) 188.4 V p-p; (d) 80 V p-p

17–11. (a) 58.43 V; (b) 50.88 V

17–12. (a) 240 V; (b) 67.87 V

17–13. 0.6A

17–14. 0.2A p-p

17–15. 1080 V p-p

17–16. In phase

17–17. 180° out of phase

17–18. Peak-to-peak value

17–19. Square, sawtooth, and triangular

17–20. The shape is basically square, but the time and or amplitude are not symmetrical for both halves of the wave.

17–21. Third, fifth, and seventh minimum

17–22. 9.0 MHz

17–23. 60 Hz

17–24. 240 V

17–25. To use less current, or to send an equal amount of power more efficiently

CHAPTER 18

60-Hz AC Power

The standard power generation values in the United States are a frequency of 60 Hz and a voltage of 120 (rms). Ac power is used because the equipment required to create it is more efficient than that of the direct-current counterparts. The frequency of the ac voltage is computer regulated to exactly 60 Hz. This frequency is used as a time reference in a good deal of electronic equipment. One major example is the clock. The motors of electromechanical clocks are synchronized to rotate at a rate related to 60 Hz. Electronic clocks use an internal circuit having a reference to the power-line frequency. Both types of clocks stay fairly accurate using this time-base reference. The television receiver's signal processing includes a signal having a reference to the 60-Hz power-line frequency as well.

KEY TERMS

Alternator	Power Factor
Brush	Rotating Field
Bus Bar	Sine Function
Excitation	Single Phase
Line Current	Slip Rings
Line Voltage	Three Phase

OBJECTIVES

Upon completion of the material in this chapter, you will

1. Understand why the 60-Hz power system is used.
2. Understand basic motor and generator theory.
3. Understand the basics of residential wiring systems.
4. Understand basic three-phase power systems.

The production and distribution of 60-Hz alternating-current power is a major industry in the United States. Almost all of our homes use 60-Hz ac power. Office buildings and stores also use this type of power. The power used by industry to operate a wide variety of production machinery is also 60-Hz alternating current.

This almost universal use is based on both practical and economical reasons. The efficiency of an ac generator is higher than a comparable dc generator. In addition, the ability to transport the ac power from its source to the consumer is better. Finally, the motors used by the consumer of ac power are smaller and less expensive than their dc-operated counterparts.

These are some of the basic reasons for our use of ac power. Persons using ac power have a need to understand how it is produced, how it is transported, and how it is distributed for use by the ultimate consumer. Material in this chapter covers these topics.

18-1 The Ac Generator

The device used to create ac power is a rotating machine. This machine is called a *generator*. The generator relies on the principles of inductance. The basic concept is that a wire moving through a magnetic field will have a voltage induced on it. The same concept is true when the magnetic field is moved past a stationary wire. The device used to create an ac voltage using these principles is called an *alternator*. The **alternator** is a rotating machine having the capability of creating an alternating-current (ac) voltage.

Alternator *A generator used to create alternating current (ac) voltage.*

Ac power in the western hemisphere uses a standard frequency of 60 Hz. This standard frequency is computer controlled. One reason for this control is that when all of the power companies generators use a common frequency it is easy to send excess power from one system to another system. An example of this is that when a major city needs more power than it can economically produce, it may purchase extra power from a system in another city. Since the two systems have exactly the same standards, the two sources are compatible. The result is the power users in the first city have adequate power for their requirements.

The alternator, or ac generator, is illustrated in Figure 18-1. It consists of a rotating armature, a stationary field coil, and a pair of slip rings. The armature winding is made of many turns of wire wound on a laminated steel core. Laminated steel is used for the armature core because it will develop a more concentrated magnetic field than will a solid steel core.

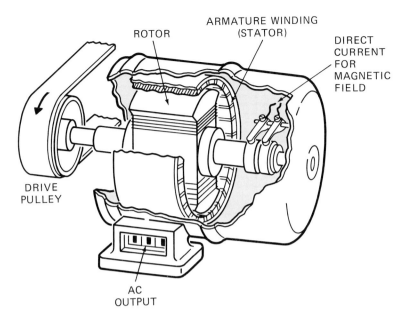

ARMATURE WINDING
(STATOR)

DIRECT
CURRENT
FOR
MAGNETIC
FIELD

DRIVE
PULLEY

AC
OUTPUT

FIGURE 18-1 Cutaway view of the ac generator, or alternator.

The field coil actually is a pair of windings. These are also constructed using a core material capable of producing a very concentrated path for the magnetic lines of force. The slip rings are connected to the winding of the armature. One slip ring is connected to the starting point of the winding. The other slip ring is connected to the ending of the same winding. Brushes, made of a carbon material, ride on the slip rings. The brushes are used as a connection to carry the ac voltage from the alternator to wherever it is to be used.

The concept of creating an ac voltage by rotating a wire through a magnetic field is shown in Figure 18-2. In this illustration the fixed field is developed by use of an electromagnet. A dc source is used to create a fixed magnetic field in the field coil windings. The armature is illustrated by a single-turn coil of

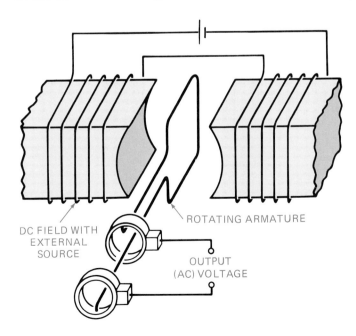

ROTATING ARMATURE

DC FIELD WITH
EXTERNAL
SOURCE

OUTPUT
(AC) VOLTAGE

FIGURE 18-2 Component parts of the ac alternator.

wire. In reality both the field coil and armature windings consists of many turns of wire for their respective windings.

The armature is rotated through the stationary magnetic field. Any power source may be used for this purpose. Alternators used by electric power companies use a turbine to rotate the armature. The turbine may be powered by water from a reservoir. When a water source is not readily available, a steam turbine is used to rotate the machine's armature. Steam is created by use of oil, natural gas, coal, and atomic energy to heat water. The water turns to steam and the steam is piped to the turbine.

An alternative method of creating an ac voltage is shown in Figure 18-3. The process is the same as the action described previously. The difference lies in the reversal of the roles of the armature and the field. In this example the rotating part is the field and the stationary windings are considered to be the armature. This process is called a **rotating field** method. The dc source is connected to the rotating part of the alternator. This is identified as the field. It rotates through the outside windings and induces a voltage onto these windings. Since the field is the part of the alternator that rotates, this device is called a *rotating field alternator*. This type of alternator construction is typical for the alternators used in automobiles.

All rotating machines of this type require some method of getting the generated voltage out of the alternator. The method used for ac generators requires two metal disks and a pair of contact brushes. The metal disks are mounted on the same shaft as the armature. They are called **slip rings**. A contact called a **brush** rides against each of the slip rings. The brushes are stationary. They are connected to terminals on the alternator. The terminals are used to connect the load wires to the rotating armature.

Rotating field *Type of generator where the field rotates instead of the more common method or armature rotation.*

Slip rings *Circular rings on shaft or generator connected to windings. Contacts on these connect to output terminals of unit.*

Brush *Device used to conduct electrical energy from rotating armature of generator to output terminals.*

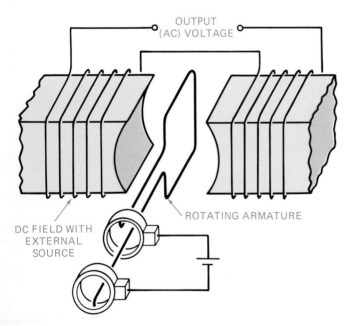

OUTPUT (AC) VOLTAGE

ROTATING ARMATURE

DC FIELD WITH EXTERNAL SOURCE

FIGURE 18-3 The alternator may have a fixed field and a rotating armature, or the two major parts may be reversed.

18-1. Name the basic parts of an alternator.
18-2. What is the name of the device used to connect the armature wind-ings to the brushes?
18-3. What electrical principle is used to generate a voltage?

18-2 Ac Voltage Production

The voltage created in the alternator is directly related to the position of the armature and the fixed magnetic field. This voltage is identified as having a specific value during each instant of time as the armature moves through one 360° rotation. The value of voltage created is called the *instantaneous* value. This value is determined by use of the trigonometric value called the **sine function.** Sine voltage values are based on the sine of the angle created by the rotation of the armature. A trigonometric table may be used to find the value of the sine of each angle for a 360° circle. The sine of the angle is the ratio of the side opposite to the hypotenuse of a triangle [Figure 18-4(a)]. The hypo-tenuse is marked by the letter h. The formula for the sine of the angle is

Sine function *Trigonometric value based on relationship of the side opposite to the hypotenuse of a right triangle.*

$$\sin \phi = \frac{\text{opposite}}{\text{hypotenuse}}$$

This provides a numeric value. Values for the sine of every angle for a 360° circle may be found in almost any mathematics book. A sample of some of these values follows.

FIGURE 18-4 Values of generated ac voltage are determined by relating the values of voltage to one 360° rotation of the generator's armature.

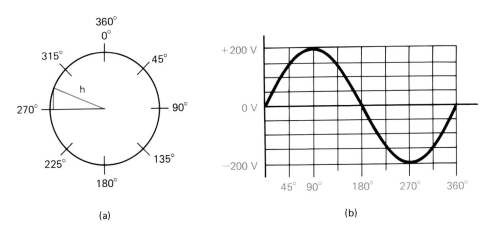

(a)   (b)

Degree	Sine	Degree	Sine
0/360	0.0000	200	−0.3420
20	0.3420	220	−0.6428
40	0.6428	260	−0.8660
60	0.8660	240	−0.9848
80	0.9848	270	−1.000
90	1.000		
100	0.9848	280	−0.9848
120	0.8660	300	−0.8660
140	0.6428	320	−0.6428
160	0.3420	340	−0.3420
180	0.0000	360	0.0000

The sine value is used to represent the exact value of voltage generated for the specific amount of rotation. The sine values may be either positive or negative. The specific value depends on the position of the armature in the circle. Usually, the upper half of the circle will have positive values and the lower half will have negative values.

Instantaneous Voltage values are found by use of the sine value and the maximum voltage value for the generator. Figure 18-4(b) is used with this formula:

$$V_{instantaneous} = V_{maximum} \times \sin \phi \qquad (18\text{-}1)$$

In this example, the maximum peak voltage is 200 V. Using the formula $V_i = V_{max} \times \sin \phi$, the voltage for a 40° rotation is

$$V_i = 200 \times 0.6428$$

$$= 128.56$$

This formula is used to establish voltage at every instant of time for the rotation of the armature. Usually, a sample value set is utilized to show the creation of the sine wave during one alternation.

The development of sine-wave voltage is illustrated in Figure 18-5. In this example a maximum peak voltage of 200 V is used. The instantaneous values for rotational degrees are plotted on a graph. The points are then connected to each other. The result is the sine-wave shape.

SIN ϕ	V_{INST}
0° = 0	0
45° = 0.707	141.4
90° = 1.0	200 V
135° = 0.707	141.4
180° = 0	0°
225° = −0.707	−141.4
270° = −1.0	−200
315° = −0.707	−141.4
360° = 0	0

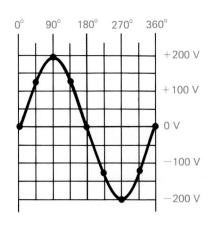

FIGURE 18-5 The specific value of voltage created at any instant of time is determined by use of the trigonometric functions.

REVIEW PROBLEMS

18-4. What is meant by the term *instantaneous voltage*?
18-5. What is the instantaneous voltage for a 40° rotation and a maximum voltage of 150 V?
18-6. What is the instantaneous voltage for a 220° rotation and a maximum voltage of 360 V?

18-3 Single-Phase Voltage

The voltage produced by the alternator described earlier in this section would be classified as a *single-phase voltage*. The label **single phase** refers to the number of waves created by the alternator during one cycle. A practical single-phase alternator will have one pair of armature windings and one pair of field coil windings. This type of alternator is not practical for large quantities of electrical energy. Under these conditions the armature of the alternator is required to rotate at a speed of 60 times a second. A practical alternator design will utilize more pairs of magnetic poles and rotate at a slower rate in order to create the desired voltage at a frequency of 60 Hz. A formula for determining alternator frequency is

$$f(\text{Hz}) = \frac{\text{number of magnetic poles} \times \text{rotational speed (rpm)}}{120} \qquad (18\text{-}2)$$

When a frequency of 60 Hz is desired, the alternator operates at a speed of 1800 or 3600 rpm. The slower speed will require four magnetic poles (two pairs). The higher speed requires two magnetic poles (one set). Either of these two speeds is often used for portable ac alternators.

The ability of an alternator or generator to convert mechanical rotary energy into electrical energy depends on three conditions: (1) a magnetic field must exist; (2) conductors must cut through the magnetic field; and (3) there must be motion between the conductors and the magnetic field. The two methods described in this section capable of meeting these three criteria are called the *rotating field* and the *rotating armature* methods.

The rotating armature uses a fixed magnetic field. The field consists of pairs of coils of wires. The coils are wound onto the fixed frame of the alternator. A dc voltage is used to establish the magnetic field. The term used to describe this field and its dc source is **excitation**. The armature winding is wound on a laminated core. The core is supported by a steel shaft suspended on bearings. The shaft of the armature is rotated by an external power source. The ac voltage is removed from the armature by the use of slip-ring assemblies and brushes.

The rotating field alternator has its field and armature reversed. The dc excitation voltage is connected to the armature by slip rings and brushes. The windings of the rotating coil are used as the field. The fixed windings mounted on the frame of the alternator are used as the armature. The system works in the same manner since all three criteria are met. The advantage of the rotating field system is that the large quantities of current produced do not pass across the brushes. A heavy current on the brushes could produce arcing and an earlier-than-normal failure of the brushes.

Single phase *Term used to describe voltage created during one cycle of the rotation of the generator.*

Excitation *Electromagnetic field required by generator produced by an output power source.*

18-4 Three-Phase Ac Power

Three phase *Describes a system where three voltages are created during one complete alternation of the generator.*

The efficiency of the ac generator may be increased by use of a three-phase system. The **three-phase** system uses three sets of coils on the alternator. The alternator is constructed so that each set of coils is 120° away from the adjacent sets (Figure 18-6). The three pairs of stationary windings are spaced equally around the perimeter of the alternator. Since the perimeter represents a 360° circle, the spacing is 360/3, or 120°. The three pairs of coils are connected to each other internally. They are wired in series, and one pair of wires from each set is identified as either phase 1, 2, or 3. The rotating armature is connected to a dc excitation source. The rotation of the armature produces three voltages (Figure 18-7). The phase shift of 120° is also apparent in this illustration.

Three-phase ac is produced by all electric power companies in the United States. The reason for this is related to the ability to transport the ac from the source to the consumer. Use of one of two transformer connection systems permits using three wires instead of three pairs of wires between source and

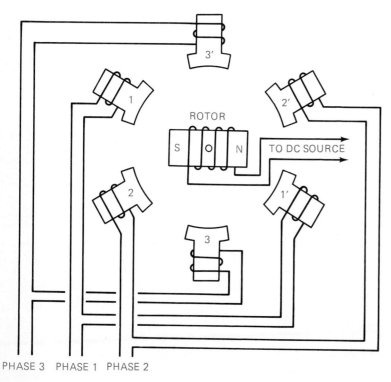

PHASE 3   PHASE 1   PHASE 2

FIGURE 18-6 Schematic diagram of a three-phase ac generator.

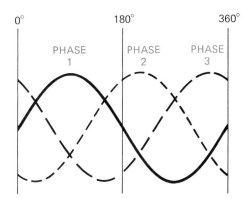

FIGURE 18-7 Waveforms for a three-phase ac voltage-generating system.

consumer. This, of course, reduces the number of miles of wire connecting the generating station to the consumer by one-half.

The two methods of wiring three-phase alternators are called *delta* and *wye*. The wiring for the delta connection is shown in Figure 18-8. The windings of each phase are series connected to each other. These form a triangle. Since the shape of the Greek capital letter delta is triangular this wiring type uses that name. In the delta configuration the terms *line voltage* and *line current* are used. **Line voltage** is determined by using the peak value of the voltage of one phase. The formula for this is

$$V_L = V_p \qquad (18\text{-}3)$$

when L stands for line and p for peak. **Line current** is determined in a different manner. The line current is found by using $\sqrt{3}$, or 1.73, and the peak current value. The formula for this is

$$I_L = I_p \times 1.73 \qquad (18\text{-}4)$$

The wye configuration for alternator windings is illustrated in Figure 18-9. In this configuration one end of each of three windings is connected to the other windings. This forms a *wye* or star shape. The other ends of the three windings are connected to each of the three lines connected to the load. The relationship of voltage and current for the wye is opposite to that of the delta arrangement. Line voltage is

$$V_L = V_p \times 1.73 \qquad (18\text{-}5)$$

Line voltage *Voltage developed from generator and sent, or delivered, to consumer on power lines.*

Line current *Current developed from generator and delivered to consumer.*

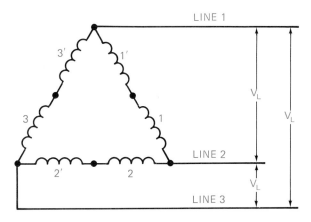

FIGURE 18-8 Delta wiring for a three-phase ac generating system.

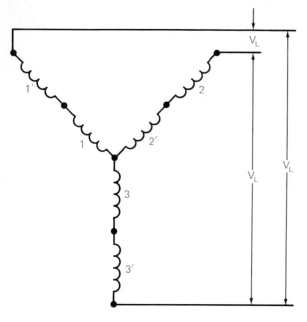

FIGURE 18-9 Wye wiring for a three-phase ac generating system.

and line current is

$$I_L = I_p \tag{18-6}$$

When three-phase electrical power is needed by the consumer, all three lines are connected to the load. This is accomplished by use of a three-phase motor or transformer. When single-phase ac power is required, only one of the three phases available is utilized.

REVIEW PROBLEMS

18-10. How many wires are used to carry three-phase power?
18-11. What are the formulas for line and current voltage in a wye system?
18-12. What are the formulas for line voltage and current in a delta system?

18-5 Electric Power Distribution

The production of great quantities of electrical power is usually done at a site that is a distance away from the ultimate consumer. There are several reasons for this. Among these are the costs and availability of fuel to operate the turbines, the cost of land for the facility, and the situation that few of us want an operation of this magnitude in our back yards. As a result of these factors there is a requirement to carry the electrical power over reasonably long distances.

All electrical conductors have a certain amount of resistance. The specific amount of resistance is described as the resistance for each 1000 ft. A copper wire table provides this information (see Appendix D). Each size of wire is given a numerical assignment, with the smaller-diameter wires having the largest number assignment. In other words, a number 18 wire is larger in diameter than a number 30.

Let us examine the conditions that prevail when a distance of 50 miles exists between the generating plant and the consumer of electricity. In this example we use a number 0 wire. The copper wire table indicates a resistance of 0.1 Ω for each 1000 ft of wire. In the example of 50 miles there are 264,000 ft (5280 × 50). When this is divided by 1000, there are 264 one-thousand-foot sections for this distance. Each 1000-ft section has a resistance of 0.1 Ω. The total resistance for the 50-mile length of wire is 28.4 Ω. This does not seem like a very large resistance factor in itself.

Let us now examine the voltage drop that occurs when a maximum current of 325 A flows through this wire. Using Ohm's law, the voltage drop is calculated:

$$V = I \times R$$
$$= 325 \times 26.4$$
$$= 8580 \text{ V}$$

The generator would have to produce a voltage of 8580 V just to overcome the resistance of the wire. When 240 V is required, the generator has to create a value of 8820 V. The efficiency of this system is counterproductive and too expensive to use.

One method of reducing the ohmic value of the wire is to use several parallel-connected wires. This method requires the use of a stronger tower to hold the wires. In addition, the towers may have to be closer together. Both methods greatly increase the cost of the installation.

There is a better method of solving a problem of this type. This is the method in common use today. The method used requires a series of step-up and step-down transformers. (Transformer details are discussed in Chapter 19.) This is understood when one considers the power losses occurring during the transmission of electrical power. The goal of course, is to reduce any losses to a very small percentage of what is being produced.

The power formula used by electrical power companies is a combination of Ohm's and Watt's laws. We know that

$$V = IR \quad \text{and} \quad P = IV$$

It is possible to substitute one equality for another without changing either equation. Since $V = IR$, the IR portion may replace the V in the formula $P = IV$. The result is

$$P = IIR$$

When like terms are combined, the formula is

$$P = I^2R$$

This formula is used to illustrate how power losses may be minimized.

The example used earlier in this section used a current of 325 A and a voltage of 240 V. These values create a power level of 78 kW (240 V × 375 A). It is possible to increase the voltage level and reduce the current level and still maintain the same power level. If the generated voltage was increased to

3.45 kV, the current would be reduced to

$$I = \frac{P}{V} = \frac{78 \text{ kW}}{3.45 \text{ V}} = 22.61 \text{ A}$$

Referring to the copper wire table, it can be seen that this current can be carried safely by a number 12 wire.

The diameter of the conductor could be reduced from 0.325 in. to 0.081 in. Carrying this idea one further step, increasing the generated voltage to a value of 13.8 kV permits a current flow of 5.65 A. Wire size can be reduced even further with this level of current. It could be as small as 0.061 in. Stepping up the voltage permits a smaller amount of current flow when the power level is held at a constant value.

The efficiency of increasing the voltage and reducing the current may be carried even further. Many power companies now operate at a 1-MW level. Using a value of 13.8 kV, the current created for this power is about 72.5 A. A more efficient method of transporting the power is to step up the voltage to an even-higher value. One common value for this voltage is 345 kV. Using a voltage of 345 kV, the current required for a 1-MW system is

$$I = \frac{P}{V}$$

$$= \frac{1,000,000 \text{ W}}{345,000 \text{ V}}$$

$$= 2.9 \text{ A}$$

The current level is reduced to a low value of less than 3.0 A. This high voltage and low current power is used to carry the required power from the generating station to the local distribution point (Figure 18-10). Electric power is generated at a level of 13,800 V. This value is then stepped up through a transformer to a voltage value of 345 kV. The power level for this system is rated at 1 MW. Current levels at 13.8 kV are 72.5 A. The current value for the higher voltage value is reduced to 2.9 A. This permits the use of smaller, lighter-weight conductors.

The high voltage and low current power is carried over high-tension wires from the generating plant to a series of *substations*. The purpose of the substation is to provide a terminus for further processing. An industrial customer requiring large quantities of electrical power would probably set up a local distribution and stepdown voltage system. The voltage is reduced in amplitude and the current is increased proportionally.

In addition to this departure the power may also be carried to a series of other substations. One of these will be used to reduce the voltage level to 69 kV. This value is also utilized by some industrial customers. Another substation may reduce the voltage level even further. Typical values for this reduction are 13.8 kV, 4.8 kV, 240 V, and 120 V. Usually the values higher than 240 V are three-phase systems. 240-V and 120-V systems use both single-phase and three-phase types of power. The specific voltage values and phase requirements are established by the consumer.

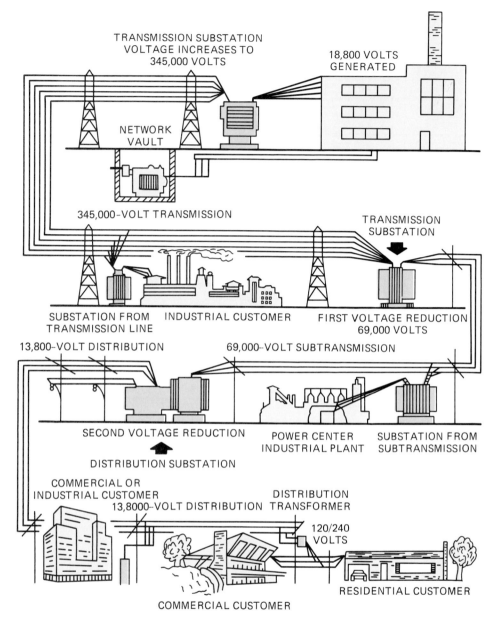

FIGURE 18-10 Typical electric power distribution system. (From Joel Gold-berg, *Fundamentals of Electricity*, 1981, p. 300. Reprinted by permission of Prentice-Hall, Englewood Cliffs, New Jersey.)

Figure 18-11 shows the various levels of voltage and current required to maintain a 1-MW power level. Notice how the levels of voltage and current are inversely proportional to each other. This chart illustrates how electrical power is transported, or delivered, from the generating plant to the user.

POWER	VOLTAGE	CURRENT
1 MEGAWATT	13.8 kV	72.5 A
1 MEGAWATT	34.5 kV	28.9 A
1 MEGAWATT	69 kV	14.5 A
1 MEGAWATT	13.8 kV	72.5 A
1 MEGAWATT	4.8 kV	208 A
1 MEGAWATT	240 V	4167 A

FIGURE 18-11 Relationship of possible combinations of voltage and current used in a commercial electrical production system.

REVIEW PROBLEMS

18-13. Why is the voltage increased for transportation of electrical power from the generating plant to the consumer?

18-14. What is a transformer?

18-15. What is the relationship between voltage and current in a transformer?

FIGURE 18-12 Use of large transformers permits a high-voltage, low-current value for transportation from the generator to the consumer, where another set of transformers reverse the voltage-to-current ratio. (From Joel Goldberg, *Fundamentals of Electricity*, 1981, p. 303. Reprinted by permission of Prentice-Hall, Englewood Cliffs, New Jersey.)

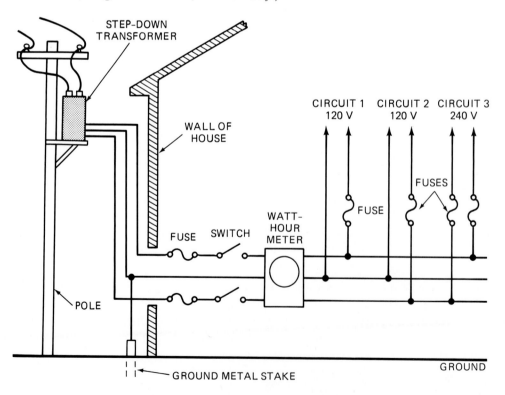

240-Volt Wiring Almost all homes, offices, and small industries have 240-V electrical systems. The system normally uses a 240-V and three-wire format. The method of wiring this is illustrated in Figure 18-12. Two of the three wires are "hot" electrically and the third wire is at ground, or zero volts, potential. It is also called the neutral wire. The third wire is actually connected to the earth. This is often accomplished by connecting a wire from the ground in the electrical distribution panel to a water pipe. Water pipes are normally buried under the ground and thus make a good connection between the power line and the earth. The voltage level for the remaining two wires is 240 V. This is measured from one of the wires to the other. When measuring from either of these wires to the ground connection the voltage is one-half of this value, or 120 V. The wiring is split in the building as shown in Figure 18-12. This split provides the required 120-V circuits. When 240 V is required, both of the wires and the ground are used. The dual-voltage wiring system is shown in Figure 18-13. This is the layout of a typical electrical distribution panel in the home or business. Each of the main 240-V lines is connected to a heavy conductor. This heavy conductor is called a **bus bar.** Each secondary circuit is connected to a line through either a fuse or a circuit breaker. The lines are the wires in the walls having outlets for terminators. Each fuse or circuit breaker is rated as a current level designed to protect the system from overcurrents. The specific value for circuit protection depends on the current requirements of the load.

Bus bar *Large conductor used to carry great amounts of current.*

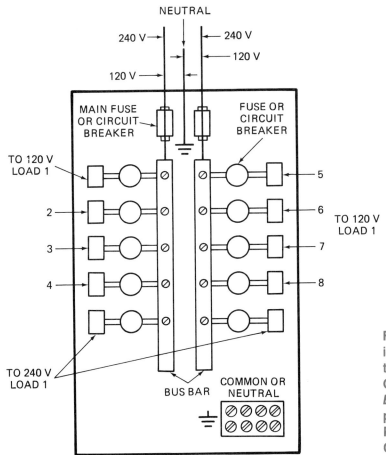

FIGURE 18-13 Electric power is distributed in the home in this manner, (From Joel Goldberg, *Fundamentals of Electricity*, 1981, p. 305. Reprinted by permission of Prentice-Hall, Englewood Cliffs, New Jersey.)

Power Factor All electrical generating systems are designed with the concept of being able to transfer the maximum amount of power from the source to the load or consumer. When an electrical load is purely resistive in nature, the voltage and the current are in phase. The power value is the product of voltage and current. When the values of voltage and current are multiplied (Figure 18-14), the result is a positive value of power. This is true during the total cycle of one wave. This condition is the ideal for all power companies.

In a industrial plant there are often a large quantity of motors and transformers. Both of these are inductive devices. An inductive device introduces a phase shift between voltage and current. A circuit having inductive qualities creates a current lag behind the voltage. This phase shift results in some negative power values (Figure 18-15). These conflict with the delivery of power to the consumer. The negative power factor reduces the efficiency of the generating system and raises the cost of power production. The ideal **power factor** is a 1 : 1 ratio. In an inductive circuit the power company will charge a higher rate than normal to overcome this condition. Often the consumer will be required to add some values of capacitance to compensate for the excess inductance. When capacitance and inductance values of reactance are equal, they cancel each other. The result is a circuit having only resistive qualities.

Power factor *Ratio of actual power created by generator to apparent power used by consumer.*

FIGURE 18-14 Electric power is the product of voltage and current. When they are in phase, a high power rate can be obtained.

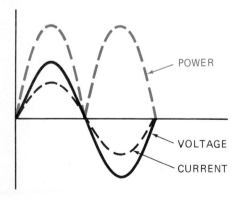

FIGURE 18-15 When voltage and current are out of phase, the total power is less than in an in-phase condition.

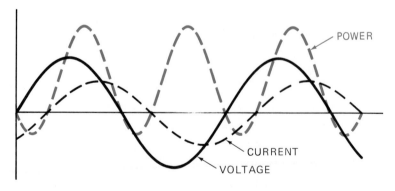

1. Electrical power produced in the United States is 60-Hz alternating-current power.
2. Ac is used because of its ease in transporting between source and consumer.
3. Ac power is created by use of a rotating electrical machine.
4. The ac generator is called an alternator.
5. The alternator consists of an armature, a field, and slip rings.
6. Induced voltage is accomplished when a wire is passed through a magnetic field or the field is passed across a conductor.
7. Alternators use either the rotating field or the rotating armature process for creating voltage.
8. Slip rings are used to connect the stationary wires from the outside to the rotating part of the alternator.
9. The value of an ac voltage created at a specific moment in time is called the instantaneous value.
10. Sine-wave voltage values are determined by using the sine of the angle of rotation and the peak voltage value.
11. Most ac power is produced as three-phase power.
12. Three-phase power is more efficient than its single-phase counterpart.
13. Two types of three-phase electrical connections are called delta and wye.
14. Electrical power is distributed by first stepping the voltage up to 345 kV.
15. Voltage and current are universally proportional when the level of power is held constant.
16. High voltage levels are utilized because the current level is reduced to a low value, permitting the use of smaller wiring.
17. Electrical power losses are described using $P = I^2R$ calculations.
18. Almost all homes and small industries use 240-V three-wire single-phase power.
19. Power factor is an important consideration for production and consumption of electrical energy.

Answer true or false; change all false statements into true statements.

18-1. Electrical power frequency in the United States is 60 Hz.
18-2. The process of creating a voltage on a wire is called capacitance.
18-3. A voltage created at a specific moment in time is called an instant value.
18-4. Electric power is the product of voltage and resistance.
18-5. Electric power in the United States is direct-current power.
18-6. Four-phase power is created at the generating station.
18-7. Home wiring uses 120-V and 240-V.
18-8. An electrical distribution center is called a substation.

18–9. Voltage and current in a purely resistive circuit are out of phase with each other.

18–10. Purely resistive loads produce a low power factor value.

MULTIPLE-CHOICE

18–11. The values of voltage commonly available for use in the home are:
(a) 240 and 177 V
(b) 240 and 100V
(c) 120 and 240 V
(d) 100 and 120 V

18–12. Another name for an ac generator is
(a) dc converting generator
(b) alternator
(c) rotary ac generator
(d) none of the above

18–13. The ac generator
(a) converts rotary energy into electrical energy
(b) converts electrical energy into rotary energy
(c) uses a motor to create electrical current
(d) uses mechanical energy to create rotary energy

18–14. An ac generator uses _____ to remove the created voltage.
(a) a segmented communtator
(b) a rotating commutator
(c) a set of slip rings
(d) fixed field activity

18–15. The quantity of voltage created at a given moment in time is called: _____ , voltage.
(a) degree
(b) sine
(c) maximum time
(d) instantaneous

ESSAY QUESTIONS

18–16. What is meant by the term *out of phase*?

18–17. Explain the term *voltage leads the current*.

18–18. What is the part of one cycle when the voltage crosses the zero line?

18–19. How does the voltage produced during the positive and negative half-cycles compare.

18–20. Why is a rotating field used instead of a rotating armature type of generator?

18–21. Why is a transformer used to step up or step down voltage?

18–22. How is the sine of an angle determined?

18–23. What is the difference between a single-phase generator and a three-phase generator?

18–24. Draw a three-phase waveform.

18–25. Draw a delta and a wye connection.

18–26. What is the operating frequency of a four-pole generator operating at a speed of 1800 rpm? (*Sec. 18–1*)

18–27. Find the operational frequency of a generator rotating at 3600 rpm and having two poles. (*Sec. 18–1*)

Find the instantaneous values for the peaks and rotations given in problems 18-28 through 18-31.

18–28. Peak = 480 V, rotation = 30°. (*Sec. 18–2*)

18–29. Peak = 240 V, rotation = 15°. (*Sec. 18–2*)

18–30. Peak = 120 V, rotation = 260°. (*Sec. 18–2*)

18–31. Peak = 4800 V, rotation = 105°. (*Sec. 18–2*)

18–32. Find the line current for a three-phase delta-connected system having a peak voltage value of 200 A. (*Sec. 18–4*)

18–33. Find the line voltage for a three-phase delta-connected system having a peak voltage value of 200 V. (*Sec. 18–4*)

18–34. What is the line current for a wye-connected three-phase system having a peak voltage of 240 A? (*Sec. 18–4*)

18–35. What is the line voltage for a wye-connected three-phase system having a peak voltage of 240 V? (*Sec. 18–4*)

18–36. What is the current for a 5000-kW system generating 345 kV? (*Sec. 18–2*)

18–37. What is the current for a 1-MW system generating 345 kV? (*Sec. 18–2*)

18–38. What is the power for a system generating 172,500 V and 6 A? (*Sec. 18–2*)

Find the currents for a 250-kW system using the voltages given in problems 18-39 through 18-42.

18–39. 100 kV (*Sec. 18–2*)

18–40. 17.5 kV (*Sec. 18–2*)

18–41. 4.8 kV (*Sec. 18–2*)

18–42. 240 V (*Sec. 18–2*)

18–43. State the three methods of creating a voltage using an alternator. (*Sec. 18–2*)

18–44. What device is used to transfer voltage to and from the rotating part of the alternator? (*Sec. 18–2*)

18–45. How is a magnetic field established in the alternator? (*Sec. 18–2*)

18–1. Field, armature, and slip rings

18–2. Slip rings

18–3. Electromagnetic induction

18–4. Instantaneous voltage is the voltage created at a given instant in time during one rotation of the generator.

18–5. 96.24 V

18–6. -231.4 V

18–7. To establish a fixed magnetic field

18–8. Rotating field and rotating armature

18–9. A magnetic field, conductors cutting the field, and motion between the two.

18–10. Three and a ground, total four

18–11. $E_L = E_p \times 1.73$ and $I_L = I_p$

18–12. $E_L = E_p$ and $I_L = I_p \times 1.73$

18–13. Use of smaller wire

18–14. A device to increase voltage

18–15. A rise in voltage will produce a drop in current, and vice versa.

18–16. 240-V single phase and 120-V single phase

18–17. Use one 240-V line and common.

18–18. To earth ground

18–19. A large metal conductor

Chapters 16-18

The characteristics of inductance are applied to produce an alternating current and voltage. This type of current and voltage is standard throughout the world. The methods of creating this power and its distribution are discussed in this section.

1. The characteristics of an inductor are its size, shape, and type of core.
2. Core material may be air, laminated steel, and ferrite forms.
3. The core material will vary, depending on the frequency of operation for the coil.
4. Inductance of the coil is directly proportional to the square of the number of turns of wire.
5. Inductance of the coil is inversely proportional to the length of the coil.
6. The unit of inductance is the henry (H). The letter L is used to represent an inductor.
7. Inductors are often called *chokes* because they choke any current changes in the circuit.
8. Inductors, capacitors, and resistors form resonant circuits.
9. Inductors may be series connected. Total L is the sum of each inductance in the circuit.
10. Inductors may be parallel connected. Total L is less than the value of the lowest single value.
11. Inductors do not work well in a pure dc circuit.
12. Inductors have a dc resistance value.
13. The ac voltage and current used for operating power in homes, businesses, and industry is sinusoidal.
14. The sine wave has both time and amplitude characteristics.

15. Time characteristics are independent of amplitude wave characteristics, and vice versa.

16. The sine wave is positive during its upward half-cycle and negative during its downward half-cycle.

17. One full 360° rotation of the ac generator creates the sine-wave voltage.

18. Ac current flow is produced when an ac voltage is connected to a load.

19. The direction of ac current flow reverses during each full cycle of the wave.

20. The period of the ac wave is the time required to complete one full cycle.

21. The standard unit for measurement of the period of the ac wave is 1 s.

22. The frequency of the sine wave is the number of cycles occurring during 1 s.

23. Frequency is measured in units of the hertz (Hz). This term means cycles per second.

24. Wave frequency is inversely proportional to the period of the wave.

25. The amplitude of the sine wave may be described as peak-to-peak, peak, rms, average, and instantaneous values.

26. The amplitude of the ac wave at a given instant is called its instantaneous value.

27. The peak value is the amplitude of one-half of the ac wave. It is measured from the zero line to the maximum, or peak, voltage value.

28. The peak-to-peak value describes the voltage amplitude from the maximum positive point to the maximum negative point of the wave.

29. Rms, or root-mean-square, value is the same as the effective value. This is based on the value of ac voltage equal to an amount of dc voltage used to create a specific quantity of heat.

30. Ac voltage values are expressed in rms units unless otherwise specified.

31. Average values of the ac wave are determined by averaging all the instantaneous values during one-half of the cycle.

32. Ohm's law is the same for both ac and dc circuits.

33. The phase relationship of two ac waves describes their starting times. It is figured on a 360° basis.

34. Nonsinusoidal waves include square, triangular, and sawtooth forms.

35. Amplitude and frequency for nonsinusoidal waves is computed in the same manner as sine waves.

36. Harmonics are multiples of the fundamental frequency of the wave.

37. 60-Hz power is used almost exclusively in the United States for operating power.

38. Ac is used for operating power because it is easy to transport between generator and consumer and has a minimum I^2R loss.

39. Power factor is an important consideration for production and consumption of electrical energy.

40. Ac is used because of its ease in transporting between source and consumer.

41. Ac power is created by use of a rotating electrical machine.

42. The ac generator is called an alternator.

43. The alternator consists of an armature, a field, and slip rings.

44. Induced voltage is accomplished when a wire is passed through a magnetic field or the field is passed across a conductor.

45. Alternators use either the rotating field or the rotating armature process for creating voltage.

46. The value of an ac voltage created at a specific moment in time is called the instantaneous value.

47. Sine-wave voltage values are determined by using the sine of the angle of rotation and the peak voltage value.

48. Most ac power is produced as three-phase power because of its greater efficiency.

49. Two types of three-phase electrical connections are called delta and wye.

50. Voltage and current are inversely proportional when the level of power is held constant.

CHAPTER 19

Transformers

Transformer action is the ability to transfer energy from one coil of wire to a second coil of wire. This is done using the concept of electromagnetic induction. All the electrical energy produced by commercial power companies passes through one or more transformers as it is transported from the generating plant to the consumer. Electrical signals are processed from one stage to the next stage of radios and television receivers by use of transformer action. A great many sources of dc power required to operate solid-state equipment require the use of a transformer to provide the correct values of operating voltage and current.

KEY TERMS

Coefficient of
Coupling
Eddy Current
Isolation Transformer
Leakage Flux
Mutual Inductance
Power Transfer
Power Transformers

Primary Winding
Secondary Loading
Secondary Winding
Step-Down Transformer
Step-Up Transformer
Tapped Windings
Turns Ratio

OBJECTIVES

Upon completion of the material in this chapter, you will

1. Understand basic transformer action.
2. Understand secondary loading effect.
3. Be able to troubleshoot basic transformer circuits.
4. Understand the concepts of tapping, impedance matching, and step-up or step-down windings.

The principles of electromagnetic induction are utilized when the magnetic field surrounding the current carrying conductor is in motion. One of the methods of accomplishing this used most often is to utilize an ac source. The action related to electromagnetic induction may be used to induce a voltage on a second wire or inductor. This second inductor does not have to be electrically connected to the circuit producing the magnetic field. This concept is called *mutual inductance*. The symbol L_m is used to represent mutual inductance.

The concept of mutual inductance is illustrated in Figure 19-1. Coil L_1 is connected to a source of power. Current flow through coil L creates a magnetic field around its conductors. A second coil, L_2, is placed close to coil L_1. There is no direct electrical connection between these two coils. The magnetic field developed in L_1 cuts across the conductors in L_2. The moving magnetic field of L_1 will induce a voltage in the conductors of L_2. This same varying field will also induce a voltage on the conductors of L_1. When all of the magnetic flux created in L_1 is able to link all of the conductors in L_2, each turn of wire in L_2 will have the same voltage as its counterpart in L_1. Since a voltage is able to create a current flow, the voltage induced in L_2 will produce a current flow in L_2 when a load is connected to its terminals.

The induced voltage in L_2 and its resulting current flow also produces a voltage back to the conductors in L_1. This concept is called *mutual inductance* because current in either coil can produce a current in the other coil.

The unit of **mutual inductance** is the henry (H). Two coils are said to have a mutual inductance (L_m) of 1 henry when a change in current of 1 A per second in one of the coils will induce 1 V in the other coil.

The schematic symbols used to represent coils of wire having mutual inductance are illustrated in Figure 19-2. This figure shows symbols for units having three different core materials. Part (a) is the symbol for an iron or steel core unit, part (b) for an air core, and part (c) for a ferrite core. Both the iron

Mutual inductance *Property existing between two conductors where the magnetic lines of force from one link with the other.*

FIGURE 19-1 Effects of mutual inductance are used for transformer action.

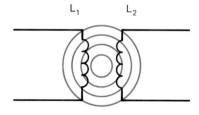

FIGURE 19-2 Schematic symbols for transformers using different core materials: (a) iron or steel; (b) air; (c) ferrite.

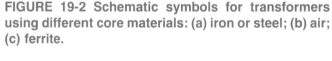
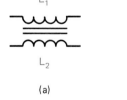

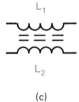

(a)   (b)   (c)

core and ferrite core materials increase the quantity of mutual inductance. This is because the core materials help concentrate the lines of force. In theory almost all the lines of force from one coil will cut across the conductors in the second coil. There are some lines of forces that do not do this. These are classified as *leakage* and since they are a part of the magnetic flux field, they are known as **leakage flux**.

Leakage flux *Quantity of lines of force that do not cut across another conductor.*

19-1 Coupling Coefficient

The quantity of the total flux developed in one coil of conductors that links a second coil is known as the **coefficient of coupling**. The letter k is used to express this value. When perfect coupling exists, this value is 1.0.

Coefficient of coupling *Amount of flux from one coil linked to a second coil.*

   The coefficient of coupling is actually a fraction of the total flux linking two coils of wire. When one-half of the flux of one coil cuts across a second coil, the coefficient of coupling, k, is 0.5. If all the flux links the two coils, the coefficient of coupling is 1. The formula for coefficient of coupling is

$$k = \frac{\text{flux lines between } L_1 \text{ and } L_2}{L_1 \text{ flux}} \qquad (19\text{-}1)$$

   The coefficient of coupling is stated using a numeric value. There is no unit used to express this fraction.

19-2 Coupling Methods

The specific method used to transfer energy from one coil to another is determined by the needs of the circuit. In general, the coupling is described as either *loose* coupling or *tight* coupling. Tight coupling has a high value of k, whereas loose coupling has a low value. Coupling may be accomplished by several methods. Figure 19-3 (a) uses an iron core to concentrate the lines of force. Each coil, L_1 and L_2, is wound onto the core material. Cores of this type may be either rectangular or doughnut shaped.

   Other methods of coupling are shown in Figure 19-3 (b) and (c). Part (b) shows a loose type of coupling. The two coils are wound on a common form.

FIGURE 19-3 Primary and secondary windings on any transformer depend on the shape of the core material. The amount of coupling depends on the physical relationship of the windings.

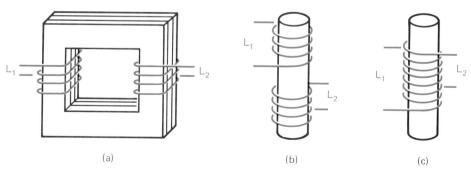

(a)   (b)   (c)

The form material may be any of the various types of cores. This, of course, depends on the requirements of the circuit using the coil. Part (c) illustrates a tight coupling method on the same type of core. The two windings are wound with turns of wire of one coil next to wires from the other coil. This is similar to holding the two wires next to each other and winding them onto the core.

When the coefficient of coupling of zero is required, the two coils should be placed at right angles to each other. When placed in this position the lines of force from one coil exert minimum interaction on the second coil. This is because the lines of force cut across fewer wires in the second coil.

EXAMPLE 19-1

Coil L_1 has a magnetic flux of 120 W, 95 W of which cuts cross coil L_2. Find the flux between L_1 and L_2.

Solution

$$k = \frac{L_2 \text{ flux}}{L_1 \text{ flux}}$$

$$= \frac{95}{120} = 0.79$$

Mutual inductance may be calculated when one considers the three factors involved in this process. These factors are the flux of L_1 and L_2 and the coefficient of coupling. The formula for mutual inductance is

$$L_m = k\sqrt{L_1 L_2} \tag{19-2}$$

EXAMPLE 19-2

Two coils are wound on a common core. Their coefficient of coupling is 0.6. Coil L_1 has an inductance of 20 mH and coil L_2 has an inductance of 30 mH. Find L_m.

Solution

$$L_m = k\sqrt{L_1 L_2}$$

$$= 0.6 \sqrt{0.02 \times 0.03}$$

$$= 0.6 \sqrt{0.0006}$$

$$= 0.6 \times 0.0245$$

$$= 0.0147$$

19-3 The Basic Transformer

A transformer is an electrical device used to transfer electrical energy between two circuits. The transformer utilizes the principles of electromagnetic induction to accomplish this action. A basic transformer has two coils of wire. One of these is identified as the **primary winding**. This winding is considered the *input* winding. Electrical power is connected to the primary from a source. The voltage from the source produces a current flow in this winding. An electromagnetic flux field is formed by this action.

Primary winding Input winding of transformer.

The schematic diagram and some drawings of transformers are illustrated in Figure 19-4. The magnetic flux lines must cut across some other conductors to produce transformer action. A second winding is required. This winding is called the **secondary winding**. It is connected to a load. When wired in this manner the voltage created on the secondary wires produces a current flow through the series circuit consisting of the windings and the load. The result is the production or accomplishment of an amount of work in the secondary circuit. An analysis of this action tells us that the electrical power in the primary is transformed with a value of electrical power in the secondary of the transformer.

Secondary winding Output winding of transformer.

Transformers are usually wound on a common core material. The specific core material is dependent on the frequency of operation and the types of power being transferred. The schematic symbols shown in Figure 19-4 are typical for all transformers. Some more detailed drawings are provided in Appendix B.

19-4 Transformer Terminology

There are several specific terms used in the description of transformers. These include applications, turns ratios between windings, voltage relationships be-

  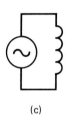 

(a)   (b)   (c)   (d)

FIGURE 19-4 The two basic windings of any transformer are called (a) the primary, or input, winding and (b) the secondary, or output, winding. (c) source; (d) load.

tween windings, winding polarity, impedance relations, and other features. Many of these are described in the following section.

Turns ratio *Relationship of number of windings, or turns, of wire on the primary to those on the secondary of the transformer.*

Turns Ratio The **turns ratio** is directly related to the number of coil loops, or turns around the coil, of the individual windings of the transformer. This ratio is important when we consider that this relationship is directly related to the quantity of induced voltage. The formula for turns ratio is

$$\text{turns ratio} = \frac{N_P}{N_S} \qquad (19\text{-}3)$$

where N_S is the number of turns of wire in the secondary winding and N_P is the number of turns of wire in the primary winding.

EXAMPLE 19-3

What is the turns ratio of a transformer having 600 turns of wire on its primary and 150 turns of wire on its secondary?

Solution

$$\text{turns ratio} = \frac{N_P}{N_S}$$

$$= \frac{600}{150}$$

$$= 4{:}1$$

REVIEW PROBLEMS

19-4. What are the components of a basic transformer?
19-5. What is a turns ratio?

Winding Direction Winding direction is very important when a transformer is being constructed or when the polarity of the voltages is created in the secondary winding. The polarity of the induced voltage is directly related to the direction of the primary and secondary windings (Figure 19-5).

The induced voltage will have either an in-phase or an out-of-phase relationship with the primary winding. For many ac applications the phase relationship is not very critical. This is true because the polarity of the ac voltage changes at a rapid rate, depending on the specific frequency of operation. When

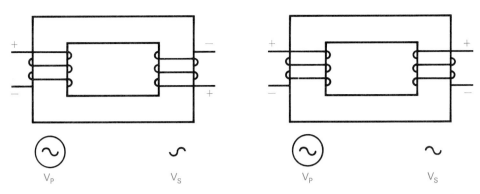

FIGURE 19-5 The polarity relationship between the input and output windings of any transformer depends on how each winding is wound on the core.

the polarity of the windings is critical, a method of identifying it is required. A commonly used method is shown in Figure 19-6. The black dots added to the schematic symbols indicate the common polarity of the winding. These may identify either the positive or the negative connection. The graphic standards of the company publishing the schematic diagram determine which of the two systems are used.

Voltage Relationships   Another way of identifying transformers is to identify the voltage relationship between the primary and the secondary windings. They are *step-up*, *step-down*, and *isolation*. One formula is used to identify the voltage ratio of any transformer. The formula states that the ratio of primary voltage to secondary voltage is equal to the ratio of the number of turns of wire on the primary to the number of turns of wire on the secondary winding. It is expressed

$$\frac{V_P}{V_S} = \frac{N_P}{N_S} \qquad \frac{V_P}{N_P} = \frac{V_S}{N_S}. \tag{19-4}$$

Another way of expressing this formula is

$$V_S = \frac{N_S}{N_P} \times V_P \tag{19-5}$$

This adaptation will provide the secondary voltage value. Since a great many power transformers operate using the 120-V ac power-line value, this adaptation will determine the value of secondary voltage.

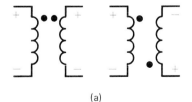
(a)

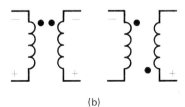
(b)

FIGURE 19-6 Often the polarities of the specific windings of a transformer are indicated by the use of a dot to show the start of each winding.

EXAMPLE 19-4

A transformer has 80 turns on its primary winding and 400 turns on its secondary. What is the voltage at its secondary when it operates from a 120-V source?

Solution

$$\frac{V_P}{V_S} = \frac{N_P}{N_S}$$

$$\frac{120}{V_S} = \frac{80}{400}$$

$$80V_S = 48,000$$

$$V_S = \frac{48,000}{80}$$

$$= 600 \text{ V}$$

or

$$V_S = \frac{N_S}{N_P} V_P$$

$$= \frac{400}{80} \, 120$$

The formulas used in these examples assume a coefficient of coupling having a value of 1. Most low-frequency power transformers have an iron core. Their coupling coefficient is close to this value.

The example shown above provides a secondary voltage higher than the primary value. This type of transformer is known as a *voltage step-up* type. In practice it is called a **step-up transformer**; the word *voltage* is assumed. The secondary voltage in a step-up transformer is always higher than the primary voltage.

Another common use for a transformer is to reduce the value of secondary voltage. This type of transformer is called a *voltage step-down*, or **step-down transformer**. Its secondary voltage is always less than the primary voltage. The same formulas are used to determine the value of secondary voltage. They require the turns ratio and the voltage of the primary.

Step-up transformer *Type of transformer where secondary voltage is greater than the primary voltage.*

Step-down transformer *Type of transformer where secondary voltage is less than the primary voltage.*

EXAMPLE 19-5

What is the secondary voltage available from a transformer having 800 primary turns and 200 secondary turns when it is connected to a 120-V source?

Solution

$$V_S = \frac{N_S}{N_P} V_P$$

$$= \frac{200}{800} \times 120$$

$$= 30 \text{ V}$$

It may be noted that the turns ratio for a step-down transformer is always less than the value of 1.

The third type of relationship between transformer windings is called an *isolation* type of transformer. The isolation transformer primary and secondary have identical winding values. The function of the **isolation transformer** is to isolate the primary circuit from the secondary load circuit while maintaining the same value of the secondary voltage. This type of transformer is usually used as a safety device between circuits that do not have the same common ground or point of reference.

Isolation transformer One having equal voltages at primary and secondary. Primarily used to provide electrical isolation between circuits.

REVIEW PROBLEMS

19-6. What is the purpose of a step-up transformer?
19-7. What is the purpose of an isolation transformer?
19-8. What is the purpose of a step-down transformer?
19-9. A step-down transformer connected to a 120-V source has a turns ratio of 6 : 1. What is the secondary voltage?
19-10. A step-up transformer has a turns ratio of 1 : 4. What is the secondary voltage when it is connected to a 12-V ac source?

Secondary Loading The secondary circuit is completed when a load is connected to its windings (Figure 19-7). The load connection permits the flow of current through the secondary winding. Load current is the same as current in the secondary winding since they are connected as a series, or loop, circuit. The quantity of current is dependent on the resistance of the load as well as the secondary voltage, V_S. In the basic transformer the primary current has a relationship to the secondary current. These two values are directly related to the relationship of primary turns to secondary turns of wire in the transformer.

Secondary loading Drop in secondary voltage due to connection of load to transformer secondary winding.

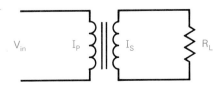

FIGURE 19-7 Current flow in the secondary of a transformer occurs when its winding is connected to a load.

This is expressed by the formula

$$\frac{I_P}{I_S} = \frac{N_S}{N_P} \qquad I_P \, N_P = I_S \, N_S \qquad \text{(19-6)}$$

A modification of the formula above, which may be more practical, is

$$I_S = \frac{N_P}{N_S} I_P$$

These formulas indicate the relationship of primary turns, voltage, and current to their counterparts in the secondary winding. The turns ratio is directly proportional to the voltage values that when a voltage step-down transformer is used, the secondary current will be high. The opposite is also true: When a voltage step-up transformer is employed, there is a drop in current between the primary and the secondary windings.

EXAMPLE 19-6

Find secondary current for a voltage step-up transformer having a times ratio of 1 : 5 and connected to a 120-V source using 200 mA of current.

Solution

$$I_S = \frac{N_P}{N_S} I_P$$

$$= \frac{1}{5} \times 0.2$$

$$= 0.2 \times 0.2$$

$$= 0.04 \text{ A} \quad \text{or} \quad 40 \text{ mA}$$

EXAMPLE 19-7

Using the values in Example 19-6, find the secondary current when the transformer has a turns ratio of 6 : 1.

Solution

$$I_S = \frac{N_P}{N_S} I_P$$

$$= \frac{6}{1} \times 0.2$$

$$= 6 \times 0.2 = 1.2 \text{ A}$$

Examples 19-6 and 19-7 prove the statements relating to the relationship of primary current to secondary current. These values are inversely proportional to each other. Thus a voltage step-up transformer has a reduction in secondary current and a voltage step-down transformer has an increase in secondary current over the primary current.

One also has to consider the effect of a load on secondary current. Users of Ohm's law will identify the current in the secondary of any transformer winding.

$$I_S = \frac{V_S}{R_L}$$

where V_S is the secondary induced voltage and R_L is the ohmic value of the load.

EXAMPLE 19-8

Find secondary current when a 400-Ω load resistance is connected to a 12-V secondary.

Solution

$$I_S = \frac{V_S}{R_L}$$

$$= \frac{12}{400}$$

$$= 0.03 \text{ A}$$

REVIEW PROBLEMS

19-11. Is the secondary voltage in a step-up transformer smaller or larger than primary voltage?

19-12. When a voltage step-down transformer is used, what is the usual relationship between primary current and secondary current?

Power Transfer **Power transfer** is a factor in any transformer. If we can consider an ideal transformer, the primary power and the secondary power values are equal. This is explained using the following equations:

$$P_P = I_P \times V_P \quad \text{and} \quad P_S = I_S \times V_S \qquad (19\text{-}7)$$

$$I_S = \frac{N_P}{N_S} I_P \quad \text{and} \quad V_S = \frac{N_S}{N_P} V_P \qquad (19\text{-}8)$$

Substitution of equal terms:

$$P_S = \frac{\cancel{N}_P \, \cancel{N}_S}{\cancel{N}_S \, \cancel{N}_P} V_P I_P$$

$$\qquad (19\text{-}9)$$

$$= V_P I_P = P_P$$

Power transfer *Ability of system to move power from the output circuit of one stage to the input circuit of the next stage or section of a unit.*

The concept of primary power and secondary power is also true when the transformer has more than one primary or secondary winding. There are many different configurations for transformer windings. One of these is illustrated in Figure 19-8. This arrangement shows a transformer with two 120-V primary windings. Many electrical and electronic devices use this arrangement. It is done in order to accommodate either a 120- or 240-V power source. There are many nations in the world using 240-V power lines instead of the 120-V standard used in North America. A universal type of transformer is used rather than one having only one of the two input voltage capabilities.

When the power source is 240 V, the two primary windings are series connected. The series connection permits the use of the higher source voltage. Each of the windings requires 120 V; thus the source voltage divides equally across both of the windings.

When operated from a 120-V source the two primary windings are parallel connected. Each receives 120 V and this permits full operating voltage and

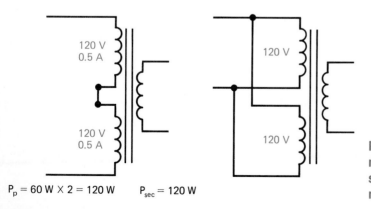

120 V
0.5 A

120 V
0.5 A

$P_p = 60 \text{ W} \times 2 = 120 \text{ W}$ $P_{sec} = 120 \text{ W}$

120 V

120 V

FIGURE 19-8 Two windings may be connected in either series or parallel in order to meet the needs of the circuit.

current from the source. Thus one transformer in the device is capable of operating from a 120- or 240-V source. Usually, the device will have either a voltage selector switch or a means of connecting some internal jumper wires in order to select the required operating voltage value.

When a transformer has two primary windings the sum of the power used by each is the value of total primary power. This is the same amount of power available from the secondary windings.

Transformers also are available having more than one secondary winding. Devices requiring different secondary operating voltage levels often use a multiple secondary winding transformer. The voltages available from a device like this can be either stepped up or stepped down, and in many applications the secondary voltages from one winding will be stepped up and from another winding the voltage will be stepped down. The circuit shown in Figure 19-9 illustrates this application.

The transformer shown has two secondary windings. One of them has an output of 400 V. When connected to a 2-kΩ load the current in this winding is 200 mA. Power for this winding is 80 W. The second winding is a step-down circuit. When connected to a 4-Ω load its output is 12 V with a current of 3 A. The power for this winding is 36 W. The total secondary power for this transformer is the sum of the power from each of its secondary windings.

$$P_T = P_1 + P_2$$

$$= 80 \text{ W} + 36 \text{ W} = 116 \text{ W}$$

This value is the same power as that used by the primary winding since P_S and P_P are equal. The current flow in the primary winding is 0.97 (approximately) when it is connected to a 120-V source.

Transformers used to provide the correct values of operating voltage and current are called **power transformers**. The specific voltage values for the primary and secondary windings is dependent on circuit requirements. A power transformer often has more than two secondary windings. This, again, is dependent on requirements of the system requiring the power.

Power transformer *Type of transformer used to change input voltage and current to required levels for load operation.*

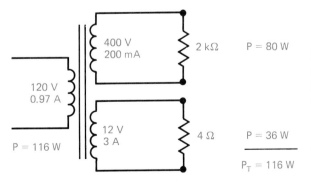

FIGURE 19-9 **Multiple values of voltage and current may be obtained from groups of windings on the transformer. Power input must equal power output, regardless of the levels of voltage and current in each winding.**

REVIEW PROBLEMS

19-13. What is the relationship between primary power and secondary power in a transformer?

19-14. How can a transformer with two 120-V primary windings be operated from a 240-V source?

Winding Connections Transformer windings may be wired either in series or in parallel. When they are connected together one has to be certain of the polarities of the windings. Transformer windings must have the same polarity when parallel connected. If the polarities are not the same, the voltages oppose each other. They tend to cancel and the net result is close to a value of zero. This is the same effect that one obtains when two batteries are series connected and their terminal polarities are opposite.

The general rule for parallel-wired transformer windings is closely related to a parallel resistive circuit. The parallel-wired winding voltages have the same value. Current in the parallel-connected bank is equal to the sum of the current in each of the windings.

When transformer windings are series connected the rules for series circuits apply. First, the polarities of each winding must be arranged so that they add. The total voltage is the sum of the individual voltages of each winding. Current in the series-connected transformer circuit depends on the lowest single current value available from one winding. When two windings of 0.5 A and 1.0 A are series connected, the maximum current available for the series is the lowest value of 0.5 A.

Analysis of a loaded power transformer is accomplished using the following procedure. First, calculate the secondary voltages. Then calculate the secondary currents. Follow this by calculating the total power used by the secondary. Finally, use Watt's law to calculate total power. When more than one secondary winding is utilized, the total power is the sum of the power used by each secondary winding. Determine each value using this procedure and then add the individual power values to find the total secondary powers.

REVIEW PROBLEMS

19-15. What is total secondary power for a transformer having a 6-V 3-A and a 400-V 0.1-A secondary?

19-16. What is the primary power for this transformer?

Unloaded Secondaries Unloaded secondaries of transformers will provide voltages higher than the rated value of the transformer. All transformers are rated by both their operating voltage and current. When an unloaded transformer's secondary voltage is measured the value will be higher than its rating because there is no current flow except a very small value through the meter circuit. For example, a transformer rated at 6 V and 1.0 A will have an unloaded secondary voltage of over 8 V. When connected to a load having a resistance of 6 Ω, the secondary voltage will drop to the rated value of 6 V. This action is similar to the operation of an automobile engine. When started the transmission is not connected to the engine and the engine rpm is high. When one connects the transmission to the engine, it slows down. This is the loading effect of the transmission on the engine.

Tapped Windings Some transformers have connections to their windings at places other than the endings of the windings. These are called *taps*. Figure 19-10(a) shows a tapped secondary winding. Some circuits require a tap at the center of the winding. The secondary voltage is divided into two equal parts. The center tap is a point of reference and each end has a 6-V potential when measured from the center-tap connection. A transformer of this type is described in one of two ways. One of these is called "a 12-V center-tapped transformer" and the other is "6 V on each side of the center tap." Both ways are correct. The specific description is left to the individual's own choice.

Figure 19-10(b) shows a tapped primary winding. In this application the taps are not at the middle of the winding. They are close to one of the ends. A tap situation like this is required when the source of the primary voltage for the transformer requires a one-time adjustment due to different levels of source voltage.

Tapped windings *A winding of a transformer with a connection for an outside circuit located between the start and end of the entire winding.*

Isolation Transformer An isolation transformer is used to separate or isolate two circuits. Usually, the turns ratio of this transformer is 1:1. The secondary voltage is the same value as the primary voltage. Isolation transformers are used to electrically separate, or isolate, two circuits when both circuits require the identical level of voltage. One major application for the isolation transformer is on testing of electrical devices. The device to be tested is connected to the secondary of the transformer. It is operated using the same level of voltage as the source, but there is no direct connection between this circuit and the test circuit. This is a very important safety factor. The isolation transformer should be used to protect both human life and the devices being tested.

REVIEW PROBLEMS

19-17. What is the voltage ratio for an isolation transformer?
19-18. An isolation transformer has a 120-V 3-A primary. What is its secondary voltage, current, and power?

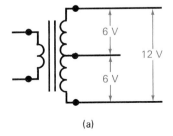

(a)

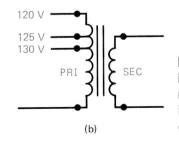

(b)

FIGURE 19-10 Tapped windings: (a) secondary; (b) primary. A tapped primary winding is used to match a variety of input voltage values.

19-5 Efficiency of the Transformer

Transformer efficiency is another factor in the design of these devices. Efficiency in any item is expressed as a percentage. The basic expression of efficiency is

$$\text{efficiency} = \frac{\text{output}}{\text{input}} \times 100 \qquad (19\text{-}10)$$

The efficiency rating for a transformer also uses this formula. Since both voltage and current are required for transformer action, efficiency is normally described by comparing input power and output power. The formula for this is

$$\text{efficiency} = \frac{P_{\text{out}}}{P_{\text{in}}} \times 100$$

Power transformers range in efficiency between 80 and 90%. It is possible to produce transformers having higher efficiency ratings, but this is not economical or always practical. Any primary power that is not used by the secondary of a transformer creates heat in the transformer. The power used to heat the transformer in this manner is not a productive by-product of transformer action. It is a part of the total secondary power and should be kept at a minimum level.

The major reason for this heat in the transformer is related to the reaction of the core and the induced voltage. The transformer core material used for power and audio-frequency application is basically constructed of iron or laminated steel. When a voltage is induced onto this type of core a current is produced in the core. This current is called an **eddy current.** It flows through the cross-sectional area of the core. The result of the eddy current is a heating effect. The eddy current flux and the flux of the coil are opposite polarities. This opposition is known as core resistance. The coil requires a higher-than-normal amount of current to overcome this resistance.

Eddy current *Current flowing through cross-sectional area of conductor creating heat.*

19-6 Transformer Applications

The discussion so far in this chapter is related to power transformer types. The amounts of power being transferred from the primary winding to the secondary are reasonably large. Other applications for transformers also transfer power between circuits. Many of these applications are low powered. Many use frequencies other than those required by power transformers.

Some of the applications of transformers are those used in consumer electronic products. Audio amplifiers may use transformers to transfer, or couple, power between the individual stages of the unit. Radios use several different transformers to couple the signals from one stage to the next. This is also true for television and tape recorders. Almost all consumer products have transformer applications. Many of these are identified by their function. These include audio frequency coupling and intermediate-frequency and radio-frequency applications. The name associated with the transformer indicates the use of the device.

19-7 Toroids

The purpose of the core of any transformer is to direct the lines of magnetic flux from the primary to the secondary of the transformer. When a rectangular or square core is used, some of these lines of force do not have any influence on the secondary coil. The use of a doughnut-shaped core material will produce a maximum coefficient of coupling between primary and secondary windings. Coils having this shape are called *toroids*. The toroid core is usually a ferromagnetic material. A toroid coil and its form are illustrated in Figure 19-11. The magnetic flux is indicated by the dashed lines inside the coil form. The primary and secondary windings are also indicated.

The major advantage of the toroid coil arrangement are more apparent at frequencies above the power and audio spectrum. Toroids are used primarily in radio-frequency circuits. The permeability of the core is so high that few of magnetic flux lines appear outside the core. In practice two toroid coils may be placed next to each other with little, if any, coupling between the two units.

REVIEW PROBLEMS

19-19. What is the characteristic shape of a toroid coil?
19-20. How does the magnetic flux flow in a toroid core?

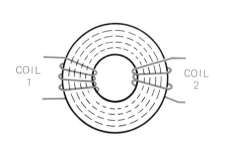

FIGURE 19-11 Typical toroid coil forms and windings.

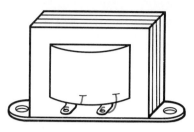

FIGURE 19-12 One problem often encountered is a broken wire at the terminal of the transformer.

19-8 Testing Transformers

TROUBLESHOOTING APPLICATIONS

Problems found in transformers are similar to those found in inductors. The two prime problems are open circuits and short-circuited turns. Opens may be anywhere in the winding. Often, the open occurs at the point where the wire used in the winding is connected to a terminal on the transformer (Figure 19-12). The wires are wrapped around a solder terminal. Often, the break occurs at this point. A close inspection, often with a magnifier, will show this type of break. A piece of wire can be spliced onto the end of the winding wire and reconnected to the terminal. This will make an effective repair.

Shorted turns are caused by an overcurrent through the winding. The result of this is heating of the wires and a breakdown of the insulation on the wire. The only successful repair procedure for this is to replace the transformer.

Transformer windings have a dc resistance value. An ohmmeter may be used to measure this value. The dc resistance should agree with the specifications for the transformer provided by the manufacturer or in the service literature. Shorted turns will be indicated by a lower-than-normal value. An open is indicated by an infinitely high value of resistance in the winding.

When measuring the dc resistance of a transformer winding, be sure to include the dc resistance between the terminal and any connection to a printed circuit board. Too often, these have a way of being imperfect. A bad solder connection between the terminal and the board will act as an unwanted series-connected resistance. This produces an unwanted voltage drop across this connection. The result is a lower-than-normal voltage and current to the winding of the transformer.

CHAPTER SUMMARY

1. The ability to transfer energy from one coil to a second is accomplished with mutual inductance.
2. The unit of mutual inductance is the henry (H).
3. The amount of flux from one coil to a second coil is the coupling coefficient (k).
4. Coupling coefficient is a ratio of the flux from one coil to a second coil.
5. Coupling is classified as being *loose* or *tight*.
6. A zero coupling coefficient occurs when the two coils are at right angles to each other.
7. Mutual inductance includes the coupling coefficient, coil 1 flux, and coil 2 flux.

8. A transformer has an input, or primary winding, an output, or secondary winding, and a core.

9. The value of induced secondary voltage is determined by the turns ratio between primary and secondary windings.

10. Transformers may be classified as voltage step-up, voltage step-down, or voltage-isolation types.

11. When a load device is connected to the secondary winding, it is *loaded* by this device and forms a complete electrical circuit.

12. Secondary current depends on secondary voltage and load resistance values.

13. Secondary power and primary power are equal in the ideal transformer. This is known as power transfer.

14. Transformers may have either multiple primary windings or multiple secondary windings, or in some cases, both.

15. An unloaded secondary will produce a higher-than-rated output voltage.

16. Transformer efficiency is the percentage of output to the input × 100.

17. The efficiency of a transformer is reduced by eddy currents and their heating effect.

18. A toroid is a highly efficient doughnut-shaped core. It is usually utilized at frequencies above the audio range.

19. Major problems in transformers are opens and shorted turns in internal windings.

Answer true or false; change all false statements into true statements. **SELF TEST**

19–1. Primary current is always equal to secondary current in a transformer.

19–2. Transformers can be used only for power applications.

19–3. Transformer windings that are parallel to each other provide the best coupling between windings.

19–4. Iron-core transformers are used for low-frequency applications.

19–5. A turns ratio of 1:6 will produce a drop in secondary voltage.

19–6. Secondary voltage values for a loaded and an unloaded secondary are the same.

19–7. Transformer windings may be connected either in series or in parallel.

19–8. A transformer secondary winding always has more turns or wire than its primary winding.

19–9. Secondary power is always greater than primary power in a step-up transformer.

19–10. A toroid-shaped core is rectangular.

19–11. A transformer may have **MULTIPLE CHOICE**
  (a) only one input
  (b) only two input windings
  (c) unlimited input windings
  (d) only one input and one output winding

19-12. A transformer may have
  (a) only one output
  (b) only two output windings
  (c) unlimited output windings
  (d) only one input and one output winding

19-13. Two transformer secondary windings are series connected. Each is rated at 5.0 V and 1.2 A. The total voltage and current available from this connection is
  (a) 5.0 V and 1.2 A
  (b) 5.0 V and 2.4 A
  (c) 10.0 V and 1.2 A
  (d) 10.0 V and 2.4 A

19-14. The same two transformers as shown in problem 19–13 now have their secondaries connected in parallel. The total voltage and current available from this connection is
  (a) 5.0 V and 1.2A
  (b) 5.0 V and 2.4 A
  (c) 10.0 V and 1.2 A
  (d) 10.0 V and 2.4 A

19-15. If the two transformers described in problem 19–14 have their primary windings connected in parallel, the output voltage and current will be
  (a) 5.0 V and 1.2 A
  (b) 5.0 V and 2.4 A
  (c) 10.0 V and 1.2 A
  (d) 10.0 V and 2.4 A

19-16. If the two transformers described in problem 19–14 have their primary windings connected in series, the output voltage and current will be
  (a) 2.5 V and 1.2 A
  (b) 5.0 V and 2.4 A
  (c) 5.0 V and 1.2 A
  (d) 10.0 V and 2.4 A

19-17. Increasing the number of turns of wire on the secondary of a transformer will
  (a) decrease the output voltage
  (b) increase the output voltage
  (c) have no effect on the output voltage
  (d) increase the primary input voltage value

19-18. When the primary power of a transformer having a 2:1 voltage ratio is 100 W, the secondary power is
  (a) 50 W
  (b) 75 W
  (c) 100 W
  (d) 200 W

19-19. The secondary power of an audio transformer is developed from its operating values of 1.0 V and 10 A. Its primary power is developed from the values
  (a) 2 V and 10 A
  (b) 5 V and 4 A

(c) 6 V and 1.67 A

(d) 4 V and 3 A

19–20. An isolation transformer has two primary windings, each rated at 120 V and 2 A. Both are required for proper transformer operation. The secondary of the transformer has a single winding. It is rated at

(a) 120 V and 2 A

(b) 120 V and 4 A

(c) 240 V and 1 A

(d) 120 V and 1 A

19–21. Why does a transformer having a turns ration of 1:3 provide an increased output voltage?

19–22. What is the relationship between primary voltage and current and secondary voltage and current when a step-up transformer having input values of 120 V and 1.5 A is used?

19–23. Explain the process of electromagnetic induction as it applies to transformer action.

19–24. Describe the effects of electromagnetic induction as they relate to the position of the two sets of conductors in a transformer.

19–25. Describe the effects of the use of the proper type of core material in a transformer.

19–26. What is meant by the term *loose coupling*?

19–27. What effect is there in a transformer when its two windings are in opposite directions?

19–28. Describe the phase relationship between primary and secondary voltage and current of a transformer during both the rise and the fall of primary current.

19–29. Describe the primary function of the isolation transformer.

19–30. What advantages are there to the use of a toroid type of transformer?

19–31. A transformer has 60 turns on its primary and 300 turns on its secondary. What is its turn ratio? (*Sec. 19–4*)

19–32. What voltage is induced on the secondary when a 240-V source is applied to the primary? (*Sec. 19–4*)

19–33. What is the secondary voltage when the primary voltage is reduced to 120 V? (*Sec. 19–4*)

19–34. A transformer has a turns ratio of 4:1 and is connected to a 120-V source. It operates a load having a resistance of 10 Ω. What is its secondary current? (*Sec. 19–4*)

19–35. What is the primary current and power for the transformer in problem 19–34? (*Sec. 19–4*)

Use Figure 19-13 to answer problems 19–36 through 19–40.

19–36. What is primary power? (*Sec. 19–4*)

19–37. What is secondary power? (*Sec. 19–4*)

19–38. What is secondary current? (Sec. 19–4)

19–39. What is the turns ratio? (*Sec. 19–4*)

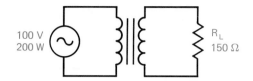

100 V
200 W

R_L
150 Ω

FIGURE 19-13

19–40. What power is delivered to R_L? (*Sec. 19–4*)

19–41. Find the turns ratio for a transformer having a 100-turn primary and a 150-turn secondary. (*Sec. 19–4*)

19–42. A transformer is connected to a 48-V 400-Hz source. Its secondary voltage is 12 V. What is its turns ratio? (*Sec. 19–4*)

19–43. A transformer has an efficiency of 0.83. Find its coefficient of coupling. (*Sec. 19–1*)

19–44. The transformer in problem 19–43 has a primary inductance of 25 H and a secondary inductance of 5 H. Find its L_M. (*Sec. 19–1*)

19–45. A transformer has a turns ratio of 1:5. What is its secondary current when it is connected to a 120-V source and a 500 Ω load? (*Sec. 19–4*)

19–46. A transformer has 60 turns on its primary and 40 turns on its secondary. What number of turns must be on its secondary to triple the secondary voltage? (*Sec. 19–4*)

19–47. A transformer is rated at 24 V center tapped. What is its voltage between the center tap and each end of the winding? (*Sec. 19–4*)

19–48. A transformer has three secondaries. They are rated at 6 V, 1 A, 12 V, 0.9 A, and 18 V, 1A. What is the maximum voltage available when they are series connected? (*Sec. 19–4*)

19–49. What is the maximum current available from the series-connected secondary windings in problem 19–48? (*Sec. 19–4*)

19–50. What tis the dc resistance of an open-circuited transformer winding? (*Sec. 19–8*)

19–51. Calculate the turns ratio for the following secondary voltages. Use a 120-V source. (*Sec. 19–4*)
  (a) 16 V
  (b) 24 V
  (c) 120V

For problems 19–52 through 19–60, consider the following: A power transformer is connected to a 120-V 60-Hz source. The transformer has a design turns ratio on secondary 1 of 1:3, on secondary 2 of 4:1, and on secondary 3 of 8:1.

19–52. What are voltages for each secondary? (*Sec. 19–4*)

19–53. What is the total voltage when secondaries 2 and 3 are series connected? (*Sec. 19–4*)

19–54. What current flows through secondary 3 when a 10-Ω load is connected to it? (*Sec. 19–4*)

19–55. What is the power value of secondary 2 when a 15-Ω load is connected to it? (*Sec. 19–4*)

19–56. What is the current for secondary 1 when a 5-kΩ load is connected to it? (*Sec. 19–4*)

19–57. What is the primary power of this transformer? (*Sec. 19–4*)

19-58. What is the number of turns on secondary 2 when 100 turns are on the primary? (*Sec. 19-4*)

19-59. The output voltage from secondary 2 is measured as 20 V. Is there a problem with the winding? If so, what do you think it is? (*Sec. 19-4*)

19-60. What is the maximum current available when secondaries 2 and 3 are series connected? (*Sec. 19-4*)

19-61. The voltage measured across the full secondary winding of a transformer is 0 V when it is connected to its normal 120-V 60-Hz source. What could be the problem?

19-62. The resistance measurement of the primary winding of a power transformer is infinite. There is no secondary voltage. What is the problem?

19-63. A transformer designed with a 12-V center-tapped secondary winding measures 6.0 V across its full secondary winding. What is wrong?

19-64. A transformer is designed to produce 24.0 V at a current of 5.0 A. A second transformer is designed to produce 12.0 V at a current of 2.0 A. Can these two transformers be connected in series in order to produce 36 V at a current of 5.0 A?

19-65. The resistance measurement of the secondary winding of a transformer is 0 Ω. What could you expect the voltage of this winding to be?

19-1. Mutual inductance is the coupling of two coils of wire due to magnetic flux.

19-2. The coupling coefficient is the quantity of coupling between coils.

19-3. 6.58 H

19-4. Primary and secondary coils wound on a core

19-5. The relationship between number of turns or primary and secondary

19-6. To increase voltage

19-7. To electrically isolate two circuits while maintaining the same voltage and current values

19-8. To reduce voltage

19-9. 20 V

19-10. 48 V

19-11. Larger

19-12. Secondary current is increased.

19-13. They are equal or the ideal transformer.

19-14. Series connect the two primary windings.

19-15. 58 W

19-16. 58 W

19-17. 1:1

19-18. 120 V, 3 A, 360 W

19-19. Round, like a doughnut

19-20. Around the core

CHAPTER 20

Inductive Reactance

A dc circuit uses the quantity known as resistance to describe the opposition to the flow of current. Resistance in the dc circuit is a constant value due to the static value of the direct current. In the ac circuit the current is constantly changing. One must have a different set of terminology to use to describe the activity in an ac circuit. One such term is reactance. Reactance in the ac circuit is the opposition of the movement of current due to the effects of either inductance or capacitance in the circuit. When the circuit has inductive properties this opposition is called inductive reactance.

KEY TERMS

Inductive Reactance   **Out of Phase**
In Phase

OBJECTIVES

Upon completion of the material in this chapter, you will

1. Understand how inductance affects current flow.
2. Be able to apply Ohm's law to inductive reactance circuits.
3. Understand the frequency effect in an inductive reactive circuit.
4. Be able to calculate current flow in the circuit.

Inductance is the property of electricity describing the opposition to a change in the quantity of current flowing in a circuit. The opposition to current flow in a dc circuit is known as resistance. Its value is measured in units of the ohm. Opposition to current flow in a circuit having an alternating current source is called *reactance*. When the circuit has some value of inductance and an ac source the phrase used to describe its opposition to current flow is called *inductive reactance*. It, too, is measured in units of the ohm. The letter X is used to identify reactance. When used in conjunction with the subscript L, the term X_L denotes a value of inductive reactance.

20-1 Inductive Effect on Current

A circuit containing a pure resistance and no inductance will establish a specific level of opposition to current flow. The circuit shown in Figure 20-1 meets this requirement. This circuit contains a fixed-noninductive resistance having a value of 100 Ω. The dc current flow for this circuit is based on Ohm's law:

$$I = \frac{V}{R} = \frac{120\ \text{V}}{100\ \Omega} = 1.2\ \text{A}$$

When an inductance is added to the circuit, or the resistance load has an inductive quality (Figure 20-2), the operating condition may change. The in-

FIGURE 20-2 The addition of an inductance provides reactance to the current flow in the circuit.

FIGURE 20-1 A purely resistive circuit offers no reaction to the flow of current.

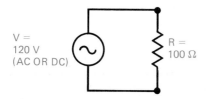

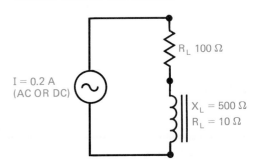

ductive added to this circuit has a dc resistance of 10 Ω. It also has an inductive reactance (X_L) value of 500 Ω. When a dc source is applied to the circuit, only the dc resistance of the inductor is a factor involved in total current flow:

$$R_T = R_L + R_{\text{load}} \qquad (20\text{-}1)$$
$$= 10\ \Omega + 100\ \Omega = 110\ \Omega$$

The total circuit current is determined by use of Ohm's law:

$$I_T = \frac{V}{R} = 120\ \text{V}/110\ \Omega = 1.09\ \text{A}$$

The use of an ac source has a different effect on total circuit current. The ac source requires the recognition of the reactive value of the inductor:

$$Z_T = \sqrt{R^2 + X_L^2} = \sqrt{100\ \Omega^2 + 500\ \Omega^2} = 509.9\ \Omega$$

The total circuit current is calculated using this value:

$$I_T = \frac{V_T}{Z_T} \qquad (20\text{-}2)$$
$$= \frac{120\ \text{V}}{510\ \Omega} = 0.235\ \text{A}$$

In this example the current is reduced when an ac source is used instead of a dc source. This indicates that the changing value of the ac wave has a direct effect on the current values of a circuit containing some value of inductance.

20-2 Derivation of X_L Values

The value of X_L in a circuit is derived from the ratio of applied voltage to the ac current flow. The formula for this is

$$X_L = \frac{V}{I} \qquad (20\text{-}3)$$

The derivation of the value of X_L for the circuit shown in Figure 20-2 is accomplished:

$$X_L = \frac{120\ \text{V}}{0.235\ \text{A}} = 510\ \Omega$$

20-3 Relationship of X_L to Frequency

Earlier you learned that the frequency of the alternations of the sine wave were directly related to the quantity of voltage induced in an inductor. Therefore, a circuit using a 60-Hz source will have less induced voltage than a similar circuit utilizing a 400-Hz ac source. An increase in frequency will induce a higher voltage across the inductance in the same circuit.

An increase in the amount of induced voltage creates more opposition to current (X_L). Therefore, X_L is directly proportional to the frequency of the source. Simply stated: X_L *is directly proportional to f*. In an inductive circuit any increase in the value of induced voltage indicates a greater opposition (X_L). We can state from this that X_L is directly proportional to induced voltage. It is also directly proportional to the amount of inductance in the circuit in addition to the frequency of the source. Therefore, X_L is directly proportional to fL.

A formula stating this relationship is shown below. Since the development of any sine wave is derived from rotary motion, we must include a constant multiplier of 2π, or 2×3.14. This formula is

$$X_L = 2\pi f L \tag{20-4}$$

where f is in units of hertz and L is in units of the henry.

EXAMPLE 20-1

A 120-Hz sine wave is applied to the circuit shown in Figure 20-3. Find X_L.

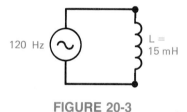

FIGURE 20-3

Solution

$$X_L = 2\pi f L$$
$$= 2 \times 3.14 \times 120 \text{ Hz} \times 0.015 \text{ H}$$
$$= 11.3 \ \Omega$$

Inductive reactance *Opposition to current flow in an ac circuit due to inductance in circuit.*

The **inductive reactance** value of any circuit is directly related to the frequency of the ac source. This statement is supported by the following data, which show the effect of an increasing ac frequency on a fixed value of inductance. The value of inductance in this example has a value of 0.6 H.

Frequency	$X_L = 2\pi f L$
60 Hz	226 Ω
120 Hz	452 Ω
400 Hz	1.507 kΩ
1 kHz	3.768 kΩ
15 kHz	56.5 kΩ
35 kHz	131.9 kΩ

It is also possible to increase the quantity of inductive reactance in a circuit by increasing the value of inductance and maintaining a constant frequency, as illustrated by the following data. The frequency is held at a value of 150 Hz for this example.

Inductance	$X_L = 2\pi fL$
0.1 H	94.2 Ω
0.2 H	188.4 Ω
0.3 H	282.7 Ω
0.4 H	376.9 Ω
0.5 H	471.2 Ω
1.0 H	942.4 Ω
1.5 H	1.413 kΩ
2.0 H	1.884 kΩ

These two examples illustrate the proportional factors of frequency and inductance in an ac circuit having inductive resistance values.

EXAMPLE 20-2

Find the X_L for a circuit having 1.6 H and a frequency of 400 Hz.

Solution

$$X_L = 2\pi fL$$
$$= 2 \times 3.14 \times 400 \text{ Hz} \times 1.6 \text{ H}$$
$$= 4.019 \text{ k}\Omega$$

EXAMPLE 20-3

Find the X_L for Example 20-2 when the frequency is reduced to 60 Hz.

Solution

$$X_L = 2\pi fL$$
$$= 2 \times 3.14 \times 60 \text{ Hz} \times 1.6 \text{ H}$$
$$= 603 \text{ }\Omega$$

Determining L can be accomplished using a variation of the basic X_L formula:

$$L = \frac{X_L}{2\pi f} \tag{20-5}$$

EXAMPLE 20-4

Find L for a circuit having 600-Ω X_L and a frequency of 1 kHz.

Solution

$$L = \frac{X_L}{2\pi f}$$

$$= \frac{600\ \Omega}{2 \times 3.14 \times 1000\ \text{Hz}}$$

$$= 95.5\ \text{mH}$$

EXAMPLE 20-5

Find the value of L when a coil having an X_L of 5 kΩ operates at a frequency of 15,734 Hz.

Solution

$$L = \frac{X_L}{2\pi f}$$

$$= \frac{5000\ \Omega}{2 \times 3.14 \times 15,734\ \text{Hz}}$$

$$= 0.0506\ \text{H or } 50.6\ \text{mH}$$

A third method of using the basic X_L formula may be utilized in order to find the frequency of the circuit. This adaptation is

$$f = \frac{X_L}{2\pi L}$$

EXAMPLE 20-6

Find the frequency for a circuit having an L of 0.45 H and an X_L of 400.

Solution

$$f = \frac{X_L}{2\pi L}$$

$$= \frac{400\ \Omega}{2 \times 3.14 \times 0.45\ \text{H}}$$

$$= \frac{400\ \Omega}{2.826\ \text{H}}$$

$$= 141.54\ \text{Hz}$$

20-4 Circuit Rules for X_L

Inductive reactances may be determined for both series- and parallel-connected inductors. When one considers that reactance is an opposition to current flow and it is measured in units of the ohm, inductive reactances are combined in a manner similar to resistances. The same circuit rules are used with a slight modification. The rule for series inductive reactances is

$$X_{LT} = X_{L1} + X_{L2} + X_{L3} + \cdots + X_{Ln} \qquad (20\text{-}6)$$

The terms representing inductive reactances, X_L, is substituted for the term R, representing resistances in this formula (see Figure 20-4). These inductive reactances are series connected. The total value of X_L is equal to the sum of the individual values

$$X_{LT} = 50\ \Omega + 400\ \Omega + 150\ \Omega$$

$$= 600\ \Omega$$

When reactances are connected in a parallel bank configuration the rule for parallel resistances is utilized (see Figure 20-5). The total reactance due to

FIGURE 20-4 Series induct-ances are equal to the sum of the individual inductance values.

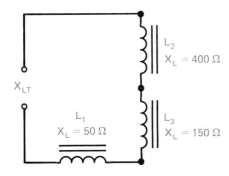

FIGURE 20-5 Total inductance for paral-lel-connected inductances is found by us-ing the parallel resistance formula.

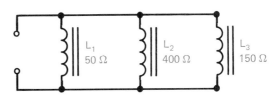

inductance is less than the lowest single value. This is the identical rule used for resistors in a parallel bank.

$$\frac{1}{X_{LT}} = \frac{1}{X_{L1}} + \frac{1}{X_{L2}} + \frac{1}{X_{L3}}$$

$$= \frac{1}{50 \ \Omega} + \frac{1}{400 \ \Omega} + \frac{1}{150 \ \Omega}$$

$$= 0.02 \ \text{S} + 0.0025 \ \text{S} + 0.0067 \ \text{S} \qquad (20\text{-}7)$$

$$= 0.02925 \ \text{S}$$

$$X_{LT} = 34.29 \ \Omega$$

The rules explained earlier related to two resistances in parallel also apply for reactances. So does the rule for total reactance when equal values of reactance are in parallel, being the value of one of the reactances divided by the number of equal units.

REVIEW PROBLEMS

20-6. An X_L of 150 Ω is parallel connected to a second 150-Ω reactance. What is total X_L?

20-7. Four reactances of 180 Ω are series connected. What is total X_L?

20-8. Reactances of 10, 100, and 60 are series wired. Find X_{LT}.

20-9. What is X_{LT} when the three reactances in problem 20-8 are parallel connected?

20-10. Find total reactances for the circuit of Figure 20-6.

20-5 Ohm's Law for Inductive Reactance

The relationship between voltage, current, and reactance is similar to the relationship between these same values in a dc circuit. The formulas for Ohm's law in a circuit having inductive reactance are

$$V = I \times X_L \qquad I = \frac{V}{X_L} \qquad X_L = \frac{V}{I} \qquad (20\text{-}8)$$

The method of solving for circuit values in reactive circuits is similar to that used for solutions to dc circuits.

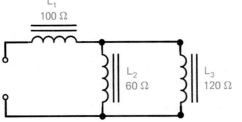

FIGURE 20-6

EXAMPLE 20-7

Find current in the circuit shown in Figure 20-7.

Solution

$$I = \frac{V}{X_L} = \frac{50 \text{ V}}{500 \text{ }\Omega} = 0.1 \text{ A}$$

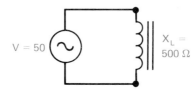

FIGURE 20-7

In a practical application the values of X_L are not provided on inductors. This, of course, is due to the effect of operating frequency on the circuit. We must calculate the value of X_L first when attempting to find the Ohm's law values for a circuit.

EXAMPLE 20-8

Find the current in the circuit shown in Figure 20-8.

Solution

First determine the value of reactance.

$$X_L = 2\pi f L$$

$$2 \times 3.14 \times 60 \text{ Hz} \times 0.5 \text{ H}$$

$$= 188.4 \text{ }\Omega$$

Next, find the Ohm's law value:

$$I = \frac{V}{X_L} = \frac{100 \text{ V}}{188.4 \text{ }\Omega}$$

$$= 0.53 \text{ A or } 531 \text{ mA}$$

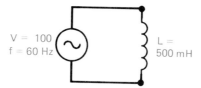

FIGURE 20-8

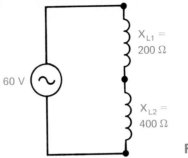

FIGURE 20-9

The applications for Ohm's law solutions as well as Kirchhoff's laws apply to circuits containing inductive reactances. The circuits shown in Figures 20-9 and 20-10 are examples of how these laws are applied.

In Figure 20-9, two reactances are series connected to a 60-V source. The total reactance in this circuit is equal to the sum of the two values, or 600 Ω. Total circuit current is equal to V/X_L, or 0.1 A. The voltage drop across each of the components is determined as follows:

$$V_{L1} = I \times X_{L1} \qquad\qquad V_{L2} = I \times X_{L2}$$

$$= 0.1 \text{ A} \times 200 \ \Omega \qquad\qquad = 0.1 \text{ A} \times 400 \ \Omega$$

$$= 20 \text{ V} \qquad\qquad\qquad = 40 \text{ V}$$

Thus the two voltage drops in this series circuit are equal to the applied voltage. Also note that they are also proportional to their values. The ratio of 200/400 is equal to 1 : 2. The voltage across X_{L2} is twice the value of the voltage across X_{L1}.

When reactances are connected in parallel (Figure 20-10), the rules for parallel circuits apply. In any parallel circuit the voltages are equal across each component. Current values will be different. This depends on the specific values of the reactances. Current for each reactance is determined:

$$I_{L1} = \frac{V}{X_{L1}} \qquad\qquad I_{L2} = \frac{V}{X_{L2}}$$

$$= \frac{50 \text{ V}}{200 \ \Omega} \qquad\qquad = \frac{50 \text{ V}}{400 \ \Omega}$$

$$= 0.25 \text{ A} \qquad\qquad = 0.125 \text{ A}$$

Total circuit is the sum of the currents through each path, or

$$I_T = I_1 + I_2$$

$$= 0.25 \text{ A} + 0.125 \text{ A}$$

$$= 0.375 \text{ A}$$

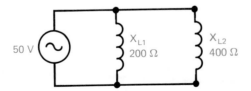

FIGURE 20-10

This may also be determined by first finding total reactance and then using Ohm's law to obtain total circuit current.

$$\frac{1}{X_{LT}} = \frac{1}{X_{L1}} + \frac{1}{X_{L2}}$$

$$= \frac{1}{200 \ \Omega} + \frac{1}{400 \ \Omega}$$

$$= 0.005 \ \text{S} + 0.0025 \ \text{S}$$

$$X_{LT} = \frac{1}{0.0075 \ \text{S}} = 133.3 \ \Omega$$

$$I_T = \frac{V_T}{X_{LT}}$$

$$= \frac{50 \ \text{V}}{133.3 \ \Omega}$$

$$= 0.375 \ \text{A}$$

REVIEW PROBLEMS

20-11. What is the circuit current when two reactances of 100 Ω are parallel connected to a 100-V ac source?

20-12. What is circuit current when the 100-Ω inductors are series connected to the 100-V ac source?

20-6 Phase Relationship in the X_L Circuit

Phase relationships are time relationships when two waveforms start and end at the same time they are **in phase** with each other (Figure 20-11). Should either of the waveforms start at a different moment, the two are said to be **out of phase** with each other. A numeric value is used to describe the difference in time between the waveforms. The system used is related to the rotation of a generator armature as it goes one 360° cycle. First, let us review the in-phase relationships of two circuit values.

In phase *Time relationship of two identical electrical waves, both starting at the same moment in time.*

Out of phase *Relationship of two identical electrical waves having a different starting time.*

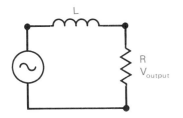

FIGURE 20-11 Circuit containing only resistance and a source has voltage and current in phase with each other.

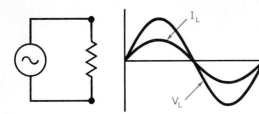

FIGURE 20-12 Ac circuit containing both inductance and resistance.

Figure 20-12 shows an ac circuit containing a purely resistive load. In this circuit the voltage applied to the load resistance provides a direct proportional relationship to circuit current. In this circuit the current rises and falls just as the voltage producing the current flow rises and falls. The voltage and current are in phase in this circuit.

The value of the current in this circuit depends on the method of determining current. In an ac circuit one has three different means of describing the values. These are average, rms, and peak values. Putting these into the proper form, these three formulas are available:

$$I_{avg} = \frac{V_{avg}}{R} \tag{20-9}$$

$$I_{rms} = \frac{V_{rms}}{R} \tag{20-10}$$

$$I_{pk} = \frac{V_{pk}}{R} \tag{20-11}$$

Since there is no phase shift on this circuit, any of the three values may be used to describe circuit current.

When an ac source is applied to an inductor, the phase relationship is not the same. This difference is produced as an effect of the magnetic field in the inductor. This effect is explained using Figure 20-13. In this example the waveforms represent the current flow in the circuit. Current flow is from maximum positive, through the cycle to maximum negative, and then back to the maximum positive value.

When current increases through an inductor, its magnetic field also increases. These two values occur in step, or in phase, with each other. When the circuit current is at its maximum value, its rate of change is considered to be zero. There is a maximum field strength but no movement of the magnetic field at this point. The amount of induced or counter EMF is zero under this condition. Maximum magnetic field and zero-induced EMF occur at the same moment in time.

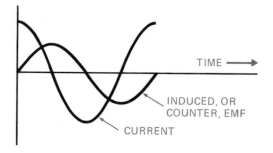

FIGURE 20-13 Effects of the applied voltage, the induced EMF, and the current in a circuit with resistance and inductance.

When the circuit current falls toward the zero line, field movement is toward the center of the coil. The rate of charge of the magnetic field is very rapid and this creates a maximum value of induced voltage. The polarity of the induced voltage is positive because of the direction of movement of the magnetic field.

Current in this circuit now starts to increase in the negative direction. This action produces an expanding magnetic field. The reverse-polarity magnetic field creates a voltage having the same polarity as that of the original voltage used to create the magnetic field. Another way of explaining this is to state that the result of this action is a double reversal. The collapsing magnetic has reversed to an expanding field, and at the same time, the field polarity has also reversed. The result is the same EMF polarity as used in the beginning of this action. When the current reaches its maximum value in the now-negative direction, its rate of change again drops to zero. This, of course, produces a zero-value EMF in the coil because the field is not in motion.

The current now rises from its maximum negative value back to the zero line. It again is in motion, and as it approaches the zero line it is moving at its maximum rate. When the current in the circuit drops to zero, the induced voltage is at its maximum value. This time, however, the voltage has a negative polarity.

One factor remaining is the phase relationship of voltage to current. There is a 90° phase shift between these two values (Figure 20-14). The current and the induced voltage in the coil in Figure 20-14 are constantly out of phase with each other by a factor of 90°. The current lags behind the source voltage by this same value of 90°. Counter EMF is the opposite of the source voltage. It has a 180° phase relationship with the source EMF. Circuit current leads the counter EMF by 90°. This is the same as stating that counter EMF lags circuit current by 90°.

REVIEW PROBLEMS

20-13. At what angle in the circuit is V_L at its maximum?
20-14. What angle is maximum I in the circuit?
20-15. What is the degree of phase shift between I and V in the circuit?

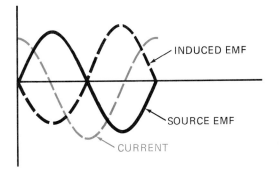

FIGURE 20-14 Voltage and current relationships in an inductive circuit.

1. Opposition to an alternating current is called reactance.
2. Opposition to an alternating current in an inductive circuit is known as inductive reactance.
3. The symbol X_L represents inductive reactance, measured in units of the ohm.
4. An inductive circuit's reactance value affects current flow in the circuit instead of the dc resistance value.
5. Inductive reactance is directly proportional to operational frequency of the circuit.
6. X_L is also directly proportional to the value of inductance; therefore, X_L is proportional to L.
7. The formula for X_L is $X_L = 2\pi fL$.
8. Total X_L in a series circuit is equal to the sum of the individual X_L values.
9. Total X_L in a parallel X_L circuit is expressed by the formula $1/X_{LT} = 1/X_{L1} + 1/X_{L2}$, and so on.
10. The rules for series resistances and series X_L values are identical.
11. The rules for parallel resistances and parallel X_L values are identical.
12. Ohm's law rules may be used for circuits containing inductive reactance values.
13. Kirchhoff's rules for voltage drops and current division apply to inductive reactance circuits.
14. Voltage and current are out of phase in a circuit containing X_L.
15. Voltage leads current by 90° in the inductive reactance circuit.

SELF TEST

Answer true or false; change all false statements into true statements.

20–1. Reactance is the opposition to an alternating current.
20–2. The dc resistance and reactance values for an inductor are equal.
20–3. The symbol for reactance is X.
20–4. Frequency of the ac wave will decrease the value of reactance.
20–5. Reactances in series are similar to resistances in parallel.
20–6. The phase shift between voltage and current in a reactive circuit is 90°.
20–7. In an inductive reactance circuit the current leads the voltage.
20–8. Total reactance for two series-connected inductances is the sum of their values.
20–9. X_L is directly proportional to frequency and inductance.
20–10. The unit of inductive reactance is the ohm.

MULTIPLE CHOICE

20–11. The phase shift between voltage and current in an inductive circuit is
(a) 50°
(b) 70°
(c) 90°
(d) 0°

20–12. In an inductive circuit,
    (a) current leads voltage
    (b) current and voltage are in phase
    (c) voltage leads current
    (d) current is 180° out of phase with the input

20–13. Reactance in the ac circuit is the equivalent of _____ in a dc circuit.
    (a) capacitance
    (b) resistance
    (c) impedance
    (d) conductance

20–14. The inductive reactance in a circuit will _____ as the frequency of operation rises.
    (a) fall
    (b) remain the same
    (c) vary
    (d) rise

20–15. Inductive reactances connected in series
    (a) maintain the same value as a single inductor
    (b) are equal to the sum of the reciprocals of the individual values
    (c) have no effect on circuit current
    (d) are equal to the sum of the individual values

ESSAY QUESTIONS

20–16. Explain how inductive reactance acts to attempt to maintain a constant level of circuit current.

20–17. Explain why Ohm's law may be used in an inductive circuit.

20–18. Explain why there is a phase shift between V and I in an inductive circuit.

20–19. Explain the term *counter emf*.

20–20. Explain why inductive reactance affects the frequency of the applied ac voltage.

PROBLEMS

20–21. What is the total X_L for these reactances having values of 6, 400, and 21 Ω? (*Sec. 20-4*)

Use Figure 20-15 to answer problems 20–22 through 20–25.

20–22. What is the inductive reactance? (*Sec. 20-4*)

20–23. What is the inductive reactance when the frequency is tripled? (*Sec. 20-3*)

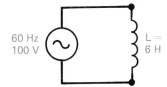

60 Hz
100 V

L = 6 H

FIGURE 20-15

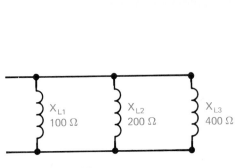

FIGURE 20-16

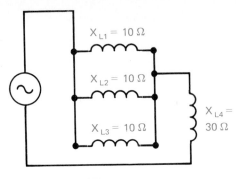

FIGURE 20-17

20–24. What is the inductive reactance when the value of inductance is reduced to one-third of its marked value? (*Sec. 20-4*)

20–25. What is the value of X_L for this circuit when the applied voltage is 100 V dc? (*Sec. 20-4*)

Use Figure 20-16 to answer problems 20–26 through 20–28.

20–26. What is the total X_L? (*Sec. 20-4*)

20–27. What is total X_L when X_{L2} is removed? (*Sec. 20-4*)

20–28. What is total X_L when X_{L1} is shorted? (*Sec. 20-4*)

Use Figure 20-17 to answer problems 20–29 and 20–30.

20–29. What is total reactance? (*Sec. 20-4*)

20–30. What is X_L when an ac voltage of 600 V with a frequency of 400 Hz is applied to the circuit? (*Sec. 20-4*)

Use Figure 20-18 to answer problems 20–31 through 20–39.

20–31. Find X_L when the operational frequency is 60 Hz. (*Sec. 20-4*)

20–32. Find X_L when the frequency of operation is 15 kHz. (*Sec. 20-4*)

20–33. Find the total inductance. (*Sec. 20-4*)

20–34. Find the total reactance. (*Sec. 20-2*)

20–35. Find the total current. (*Sec. 20-5*)

20–36. Find the current through each component. (*Sec. 20-5*)

20–37. Determine the voltage drops across each component. (*Sec. 20-5*)

20–38. Determine the voltage drops across each component when the operational frequency is 1.0 MHz. (*Sec. 20-5*)

20–39. What would happen to current flow if L_1 were short circuited? (*Sec. 20-5*)

20–40. Find the frequency when X_L is 3200 Ω for each of the following inductors. (*Sec. 20-3*)
(a) 3.8 H

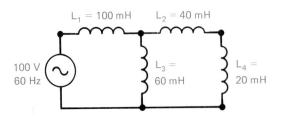

FIGURE 20-18

(b) 420 mH
(c) 600 H
(d) 180 H
(e) 15 H

20–41. A 15-mH inductor is parallel connected to a 20-mH inductor and to a 15-kHz 20-V source. Find the following values. (*Sec. 20-1*)
(a) L_T
(b) X_{LT}
(c) X_{L1T} (15 mH)
(d) X_{L2} (20 mH)

20–42. How much current flows through an X_L of 10 Ω when it is connected to a 120-V 60-Hz source? (*Sec. 20-5*)

20–43. Determine the value of X_L for the circuit in problem 20–42 for the following frequencies. (*Sec. 20-3*)
(a) 120 Hz
(b) 400 Hz
(c) 1 kHz
(d) 10 kHz
(e) 1.5 mHz

20–44. Calculate circuit currents for each frequency in problem 20–43. (*Sec. 20-4*)

20–45. What is circuit current for problem 20–42 when the value of inductance is 10 mH? (*Sec. 20-4*)

20–46. What is the circuit reactance for a 2.5-H inductor at an operational frequency of 30 MHz? (*Sec. 20-4*)

20–47. What is the value of inductance for a reactance of 80 Ω operating at 800 Hz? (*Sec. 20-4*)

20–48. What is the value of inductance for a 10-Ω reactance on a frequency of 15 MHz? (*Sec. 20-4*)

20–49. An inductor is added to a resistive circuit. What is the phase relationship, source voltage, and current across the resistor? (*Sec. 20-6*)

20–1. 490.088 Ω
20–2. 25.12 Ω
20–3. 7957 Hz
20–4. 1018.6 Hz
20–5. 0 Ω
20–6. 75 Ω
20–7. 720 Ω
20–8. 170 Ω
20–9. 7.89 Ω
20–10. 140 Ω
20–11. 2 A
20–12. 0.5 A
20–13. 90°
20–14. 0°
20–15. 90°

Inductive Circuits

Two major applications for inductors are the motor and the transformer. Motors utilize electromagnetic principles to convert the electrical energy into rotational motion. The transformer uses electrical energy at one level to convert the same electrical energy into voltage and current at a different level. Transformers are also used to transfer energy from one stage to the next stage in a signal processing device.

KEY TERMS

Effective Resistance   **Phasor Diagram**
Figure of Merit Q
Impedance   **Quality**
L/R **Rise Time**   **Signal Voltage**
Operating Voltage   **Skin Effect**
Phase Angle

OBJECTIVES

Upon completion of the material in this chapter, you will

1. Understand current and voltage relationships in the inductive circuit.
2. Be able to calculate impedance for the circuit.
3. Be able to calculate L/R time constants.
4. Be able to analyze series and parallel inductive circuits.
5. Understand the terminology related to inductive circuits.

A need to analyze circuits having inductive reactance and resistance exists. People employing electrical concepts need to be able to identify the amount of reactance and its effect on circuit action. In this chapter we review the basic concepts and then present some applications for this concept.

21-1 Phase Analysis

Phasor diagram *Shows relationship of two values in graphic form.*

The use of two comparative sine waves for circuit analysis may at times be confusing. Another method of illustrating the phase relationship is accomplished by use of a **phasor diagram** (Figure 21-1). These diagrams use arrows to represent the forces present in the circuit. The phasor is a quantity having both amplitude and direction. The length of the arrow is used to indicate the magnitude of the ac voltage. This voltage may use any of the accepted forms of expressing an ac value, either rms, peak to peak, average, or any other value. The stipulation is that both values must be expressed in the same terms or format. The angle of the arrow used in reference to the horizontal axis indicates the degrees of the phase angle. It also represents the difference in time between in degrees of rotation the start of the two waves.

In Figure 21-1, one line is used as the reference. In this illustration the arrow on the horizontal line marked V_A is the reference. It represents that moment in time when the generator armature starts to rotate. The second arrow, V_B, represents the difference in time between the two waveforms as presented during the rotation of the generator armature.

The phase angle must be compared to a reference angle in order to make much sense. The specific method of drawing the phasor depends on which of the phasors is used as the reference. Normally, the horizontal phasor is used as the reference in drawings of this type. The direction of the reference phasor

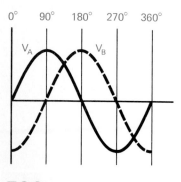

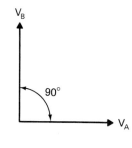

FIGURE 21-1 A phasor diagram illustrates the two forces acting on an ac circuit.

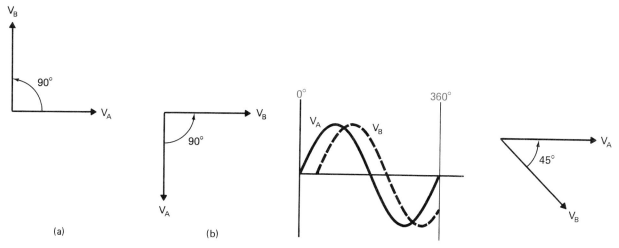

FIGURE 21-2 Phase angles are compared to a reference angle in the phasor diagram.

FIGURE 21-3 Phase angle between two waves.

also shows its polarity and its **phase angle.** In Figure 21-2(a) the horizontal arrow marked V_A is the reference. The second phasor, marked V_B, is separated from V_A by an angle of 90°. It is rotated away from V_A in a counterclockwise direction. This movement is used to indicate a positive direction for that angle. This example illustrates that V_B is leading V_A by 90°.

Both of these examples have the same phase angles. The basic difference between the two is the waveform used as a reference. It really makes little difference whether we state that V_A is leading V_B by 90° or if we say that V_B is lagging V_A by 90°.

In Figure 21-2(b), V_B is the reference. In this portion the angle V_A is 90° in a clockwise direction from V_B. The clockwise direction indicates a negative value. The use of a negative angle, such as this, indicates a waveform that is lagging behind the reference wave. In this example V_A is lagging behind V_B by 90°.

The phase angle between any two waves may be any value from 0 to 360°. In Figure 21-3, waveform V_A is the reference. Its phasor occupies the horizontal plane and is identified as V_A. Waveform V_B lags behind the reference by a value of 45°. This is illustrated as a negative phasor.

When two waves have a phase angle of zero degrees, they both start at the same moment in time. The phasor relationship for this phase angle is shown in Figure 21-4. The arrows are located in the same plane. The lesser force, V_B, is shown first and then the larger phasor, V_A, is added to its end.

Phase angle *Phase relationship in degrees between two electrical forces.*

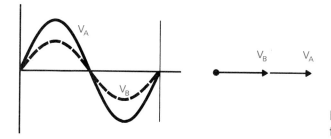

FIGURE 21-4 In-phase relationship.

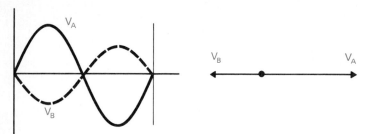

FIGURE 21-5 A 180° phase shift.

A phase difference of 180° is shown in Figure 21-5. The two phasors are opposite to each other. As a result, the phasors are illustrated as moving away from each other in a horizontal plane.

Keep in mind that the phase angle may have any value of degrees. Phase shifts are not limited to 90° displacements. The actual values are determined by the strength of the individual forces acting on the circuit.

REVIEW PROBLEMS

21-1. Identify the phase angle for Figure 21-1.
21-2. Identify the phase angle for Figure 21-3.

21-2 Analysis of Sine-Wave Action

When a sine-wave current produces an induced voltage the current lags the induced voltage by 90°. This is shown in Figure 21-6. The circuit shown in part (a) creates the waveforms shown in part (b). The phasor relationship is shown in part (c). In this circuit V_L leads I_L by 90°. This 90° phase relationship is true for any sine-wave circuit. It is true for a circuit with L in series or when L is in parallel. It is also true for circuits containing components in addition to inductors. The statement that the voltage across an X_L is 90° out of phase with the current through the X_L is always a true one.

FIGURE 21-6 The inductive circuit (a) showing its phase relationship in both waveform (b) and phasor form (c).

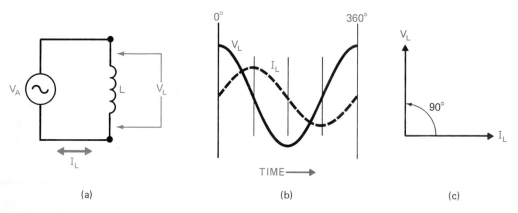

(a) (b) (c)

In the series-resistive circuit current is equal everywhere in the circuit. This is also true for circuits having X_L characteristics. In the circuit shown in Figure 21-6(a) the current flow through the inductor L has exactly the same value as the current in the source generator. The connecting wires between the generator and the load L also have the same value of current flow. The difference in the phase angle between voltage and current does not affect the current flow in any one part of the circuit. Since all of the parts of this system are connected as a series circuit, the current in each part must be the same. The difference in this circuit is the relationship between voltage and current. This is a time reference and must only be considered as such.

In a parallel inductive circuit the voltage developed by the source generator is equal to the voltage applied to the load inductance. This supports the basic rule for voltages in a parallel bank. The voltages, being the same, cannot have any phase shift. As before, the phase difference is developed between voltage and current and represents a difference in time.

Another factor to be considered is the frequency of the source generator voltage and the voltage in the loads. Here, again, the two values must be equal. Any difference in time between voltage and current is a constant value. The frequency of the source is always equal to the frequency at the load.

REVIEW PROBLEMS

21-3. In Figure 21-6, what is the phase relationship of V_L to I_L?
21-4. What is the phase relationship of V_A to V_L?

21-3 Series X_L and R Circuits

A current will flow in a circuit containing both an inductance and a resistance. This is true for both an alternating-current source and a direct-current source. If this circuit contained only a resistance, the voltage and current would be in phase with each other. In addition, all the power in the circuit would then be converted into heat. The formula for Ohm's law, $V = IR$, would be in effect.

If this same circuit contained only an inductance, there would be a 90° phase shift between voltage and current. The current would lag the voltage by this value. In this circuit the power during the first half of the cycle would be used to expand the magnetic field surrounding the inductor. During the second half of the cycle the power would be returned to the source as the magnetic field collapsed. There would be no heat loss for this circuit.

21-4 Impedance

When both the frequency of the ac voltage and the value of inductance are known, one can use the formula $X_L = 2\pi fL$ in order to find the value of inductance reactance. The load in a circuit similar to the one shown in Figure 21-7 is neither purely inductive nor purely resistive. Since it has both features, it has a value of opposition that reflects both of these quantities. This value of

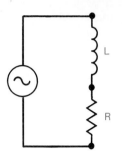

FIGURE 21-7 Circuit containing both resistance and inductance.

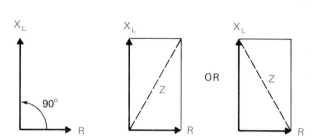

FIGURE 21-8 Impedance diagram for the inductive circuit.

opposition is known as the **impedance** of the circuit (see Figure 21-8). The lengths of the arrows represent the values of reactance and resistance. A diagonal line connecting the arrowheads represents the value of impedance for the circuit. The diagonal line may be drawn in either of the two methods shown in Figure 21-8.

The specific value of impedance is determined using one of two methods. One way to do this is to use a ruler or scale to draw the lengths of the arrows. This method is followed by a measurement of the diagonal line length connecting the two arrow heads. When the values of X_L and R are low (Figure 21-9), this method is easy to utilize. A better method is to perform a mathematical analysis for this circuit. The two arrows 90° apart form a right triangle. The connecting impedance value forms the hypotenuse of this triangle. The Pythagorean theorem is used to solve for this value.

The formula for the Pythagorean theorem is adapted as follows:

$$Z^2 = R^2 + X_L^2$$

When the square root for both sides is taken, the formula is

$$\sqrt{Z^2} = \sqrt{R^2 + X_L^2}$$

The formula is simplified in this manner:

$$Z = \sqrt{R^2 + X_L^2}$$

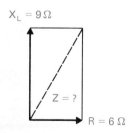

$X_L = 9\,\Omega$

$Z = ?$

$R = 6\,\Omega$

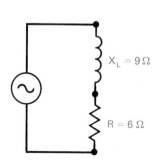

$X_L = 9\,\Omega$

$R = 6\,\Omega$

FIGURE 21-9 Use of the impedance diagram to determine circuit impedance.

This formula may then be used to find the impedance for this circuit:

$$Z = \sqrt{R^2 + X_L^2}$$
$$= \sqrt{6^2\ \Omega + 9^2\ \Omega}$$
$$= \sqrt{36\ \Omega + 81\ \Omega}$$
$$= \sqrt{117\ \Omega}$$
$$= 10.8\ \Omega$$

When the value of impedance is known, it is easy to use Ohm's law to determine circuit values. This is accomplished in the following example.

EXAMPLE 21-1

Find circuit current and voltage drops across each component in Figure 21-10.

Solution

First find circuit impedance.

$$Z = \sqrt{R^2 + X_L^2}\ \sqrt{100\ \Omega + 2500\ \Omega} = \sqrt{2600\ \Omega} = 50.99\ \Omega$$

Next determine circuit current.

$$I = \frac{V}{Z} = \frac{100\ V}{50.99\ \Omega} = 1.96\ A$$

The final step is to determine the voltage drop across R and X_L.

$$V_R = I \times R = 1.96\ A \times 10\ \Omega = 19.6\ V$$
$$V_{XL} = I \times X_L = 1.96\ A \times 50\ \Omega = 98\ V$$

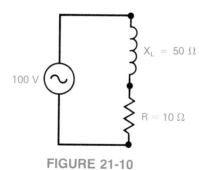

100 V $X_L = 50\ \Omega$ $R = 10\ \Omega$

FIGURE 21-10

Example 21-1 also illustrates another concept. In a purely resistive circuit the sum of the voltage drops that develop as a result of current flow equal the source voltage. This is because they are in phase with each other and the current is also in phase. When one has a circuit with a phase shift the individual voltage

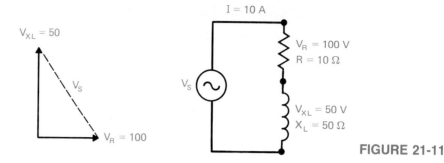

FIGURE 21-11

drops do not directly add to equal the source voltage. In Example 21-1 the sum of two voltage drops produce a value that is greater than the source voltage.

The actual values of the voltage drops across X_L and R can be used to determine the source voltage values. This, of course, is a reverse method when compared to Example 21-1.

In the circuit shown in Figure 21-11 the value of source voltage is to be determined. The use of a right triangle and the Pythagorean theorem will accomplish this.

The first steps require the determination of the voltage drop across each of the components in the circuit. The current flow of 10 A is used with the values of X_L and R.

$$V_{XL} = 10 \text{ A} \times 5 \text{ } \Omega = 50 \text{ V} \quad \text{and} \quad V_R = 10 \text{ A} \times 10 \text{ } \Omega = 100 \text{ V}$$

These two values can then be used with the Pythagorean theorem in order to find the source voltage V_S.

$$\begin{aligned}
V_S &= \sqrt{V_{XL}^2 + V_R^2} \\
&= \sqrt{(50^2 \text{ V} + 100^2 \text{ V})} \\
&= \sqrt{2500 \text{ V} + 10,000 \text{ V}} = \sqrt{12,500 \text{ V}} \\
&= 111.8 \text{ V}
\end{aligned}$$

This method of using the phasor values will easily determine the specific values of impedance and voltage drops in circuits containing X_L and R values.

REVIEW PROBLEMS

21-5. Find Z for an X_L of 10 Ω and an R of 15 Ω.
21-6. Find Z for an R of 4 Ω and an X_L of 3 Ω.
21-7. Find V_S for a circuit having 2 A of current, X_L of 5 Ω, and R of 8 Ω.
21-8. Find V_{XL} and V_R for the circuit of problem 21-7.

21-5 Parallel Reactive Circuits

The solution of circuit values for parallel connected X_L and R components is quite different from the one used to determine series-connected values. All parallel circuits have a common source voltage. They may have also different values for each of the branches. Figure 21-12 will be used to illustrate the methods of determining branch values.

In this circuit the source voltage V_S is applied to both of the branches. Current through each branch is determined by use of Ohm's law.

$$I_R = \frac{V_R}{R} = \frac{20 \text{ V}}{15 \text{ }\Omega} = 1.33 \text{ A}$$

$$I_{XL} = \frac{V_{XL}}{X_L} = \frac{20 \text{ V}}{30 \text{ }\Omega} = 0.67 \text{ A}$$

Current in the resistive branch is in phase with the source voltage since this is a purely resistive circuit. Current in the inductive branch lags the applied voltage by a value of 90°.

In this circuit the source current is equal to the sum of the current through each branch. This value can be determined only by use of a phasor diagram and the application of the Pythagorean theorem (Figure 21-13). The method of solving for this value is similar to the method used to find impedance and source voltage in the series $L + R$ circuit.

$$I_T = \sqrt{I_R^2 + I_{XL}^2}$$
$$= \sqrt{1.33^2 \text{ A} + 0.67^2 \text{ A}} \qquad (21\text{-}2)$$
$$= \sqrt{1.77 \text{ A} + 0.45 \text{ A}}$$
$$= \sqrt{2.22 \text{ A}} = 1.49 \text{ A}$$

Impedance in the parallel circuit is found by use of Ohm's law.

$$Z_T = \frac{V_T}{I_T}$$

$$= \frac{20 \text{ V}}{1.49 \text{ A}} \qquad (21\text{-}3)$$

$$= 13.42 \text{ }\Omega$$

FIGURE 21-12 A parallel re-active circuit also requires analysis.

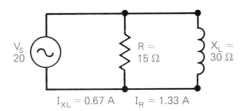

FIGURE 21-13 Phasor dia-gram for the circuit in Figure 21-12.

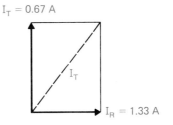

21-6 The Q of a Coil

Quality *Term describing ratio of reactance to effective series resistance at a specific frequency.*

Q *Term used to represent quality of resonant circuit.*

Figure of merit *Term describing Q of a coil.*

A coil's ability to create a self-induced voltage is indicated by its value of inductive reactance, X_L. This is true because it includes the two values of frequency and inductance. Another factor to be considered is the value of the internal resistance of the coil. This value is produced by the quantity of resistance of the wire used to wind the coil. The term **quality** or **figure of merit**, or **Q**, is used to describe the ratio of reactance to its effective series resistance at a specific frequency. This is determined by the formula

$$Q = \frac{X_L}{r_i} = \frac{2\pi f L}{r_i} \qquad (21\text{-}4)$$

where r_1 is the internal resistance of the wire in the coil (Figure 21-14). A demonstration of how the Q value is determined by using the values is shown in this illustration. The Q for this component is

$$Q = \frac{X_L}{r_i} = \frac{300 \ \Omega}{20 \ \Omega} = 15$$

The value of Q is expressed as a number and has no units. In this example the value of X_L is 15 times greater than the internal resistance of the coil.

Values of Q will range from a low of around 5 or 10 to a high of well over 1000. Based on this statement one would expect the Q of a coil to increase in a linear manner as the operational frequency of the circuit increased. This is *not* a true statement. The reason for this is that there are some other factors affecting circuit Q at higher frequencies. These are known as the *skin effect* and *eddy currents*.

Skin effect *Current flow along surface of conductor at radio frequency wavelengths.*

Skin effect is a phenomenon occurring with current flow in a conductor. At high frequencies the electron flow tends to be along the surface of the conductor. There is little, if any, electron movement at the center of the conductor. One result of skin effect is an increase in the internal resistance value of the conductor since there is actually a much smaller cross section in the conductor for current flow.

Eddy currents are a result of inductance in a wire. Inductance creates

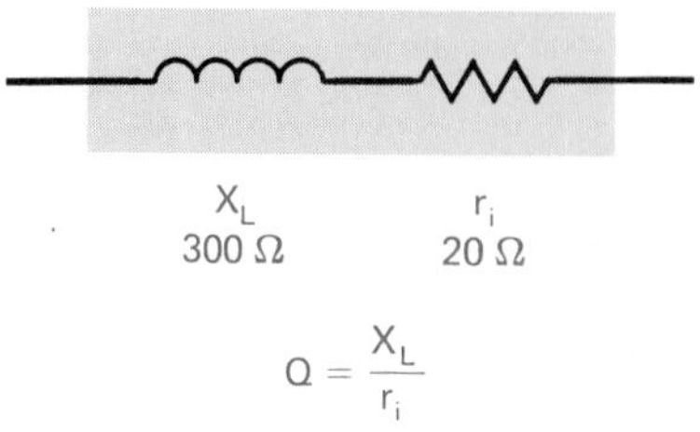

X_L
300 Ω

r_i
20 Ω

$Q = \dfrac{X_L}{r_i}$

FIGURE 21-14 The Q of any coil includes its internal resistance and its inductive reactance values.

current flow inside the conductor. This flow is not in the normal direction along the conductive path. The eddy currents create an opposition to normal current flow.

The losses due to skin effect and eddy currents act to limit the Q of the coil as the operational frequency increases. The Q of a coil can be defined as being a ratio of reactive power of the inductance to the real power dissipated by the resistance. This may be expressed by the following:

$$Q = \frac{P_L}{P_{ri}} = \frac{I^2 X_L}{I^2 r_i} = \frac{X_L}{r_i} = \frac{2\pi f L}{r_i} \qquad (21\text{-}5)$$

Any one of these equations may be utilized to determine the Q of an inductor.

EXAMPLE 21-2

An inductor has an X_L of 600 Ω and an internal resistance of 40 Ω. What is its Q?

Solution

$$Q = \frac{X_L}{r_i} = \frac{600\ \Omega}{40\ \Omega} = 15$$

Effective Resistance A coil has a dc resistance value. This value is measured with an ohmmeter using an internal dc battery. The factors of skin effect and eddy currents create an ac resistance for the same coil. A coil having a dc resistance of 2 Ω may have an ac resistance value of 12 to 15 Ω. This is known as the **effective resistance** of the coil. This value will actually reduce the Q of the coil.

The value of Q may be used to calculate the effective resistance of a coil. This value is known as R_e and is also measured in units of the ohm. It is possible to substitute R_e for r_i of a coil since these values relate to the same factor. The formula

$$Q = \frac{X_L}{R_e} \qquad (21\text{-}6)$$

is used to express this.

Effective resistance *The ac resistance value of a coil of wire, or inductor.*

EXAMPLE 21-3

Find the Q for a coil with an X_L of 100 Ω and an R_e of 10 Ω.

Solution

$$Q = \frac{X_L}{R_e}$$

$$= \frac{100\ \Omega}{10\ \Omega} = 10$$

This formula may be applied in an alternative way to determine the R_e of a coil.

EXAMPLE 21-4

Find the R_e of a coil having a Q of 50 and an X_L of 300 Ω.

Solution

$$R_e = \frac{X_L}{Q}$$

$$= \frac{300 \ \Omega}{50} = 6 \ \Omega$$

A general review of the relationship of these two factors indicates that a low internal resistance can produce a high-Q coil.

REVIEW PROBLEMS

21-11. Find the Q for a coil with an X_L of 400 Ω and an r_1 of 30 Ω.
21-12. Find the R_e for a coil having an X_L of 100 Ω and a Q of 6.

21-7 Operating and Signal Voltages

Operating voltage *Dc voltage used to establish operational conditions of semiconductors or vacuum tubes.*

Signal voltage *Variations in amplitude of operating voltage produced by information being processed in system.*

Electronic devices require an internal level of operational voltage. This voltage is direct current. It is produced from the power source section of the device. Almost all transistors and integrated circuits require a dc **operating voltage**. The voltage establishes the operational conditions for the systems. Once the dc voltage levels are established the device requires some additional voltage in order to function. This additional voltage is called a **signal**.

Electronic signals are composed of a varying voltage. When the signal is introduced to the system, it often is an ac voltage. Once inside the device the signal becomes a variation in the amplitude of the operating voltage (Figure 21-15). The operating voltage established with no signal applied takes on the characteristic shape of the ac signal voltage. This is processed at each stage of the device. In practice the signal variations from one stage control the operation of the following stage in the system. The variations of the operating voltage may be observed on an oscilloscope screen. Technically, the variations are called the *signal*, but in reality they are the operating voltage for a specific stage or stages of the device.

The frequency of a dc power source is 0 Hz. Frequencies of electronic signals will range between 20 Hz and 20 kHz for audio frequencies. Digital data processing signals often range in the middle- to high-megahertz area. The processing of these, or any other, signals requires that they follow a specific path through the electronic device. Signal voltages that stray from this path

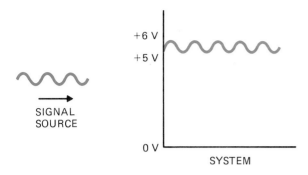

FIGURE 21-15 The electrical signal is actually a variation in the amplitude of the operating voltage in a circuit.

may adversely affect other sections of the system. Some type of device or circuit is required to keep this from occurring.

21-8 The Choke

One type of device used to keep the varying signal voltage from interfering with other circuits is called a *choke*. Inductors will provide many ohms of reactance at high frequencies. Resistance, as we know, has the same value, and there is no effect on the frequency of operation. Inductors having a high value of X_L will act as a *choke* to signal frequencies. An inductor placed in series with any conductive path processing dc and signals will act as a very high impedance to the signal. This prevents the signal from passing through the inductor. A sample circuit illustrating the action of a choke is shown in Figure 21-16. A common power source line is connected to each amplifier block. A choke is connected in series between the power source and the amplifier. Signal voltage variations are processed from the left side of the system to the right side in a horizontal flow path. The signal voltage variations are necessary for correct operation. What is not desired is to have the signal flow from one amplifier block to another through the power source. A pair of chokes are used between the amplifier and the dc source. These chokes present a high reactance to the signal and permit the zero frequency dc to pass through to the amplifiers.

REVIEW PROBLEMS

21-13. What is the basic electrical name for a choke?
21-14. Is the dc resistance of a choke high or low?
21-15. Is the reactance of a choke high or low?

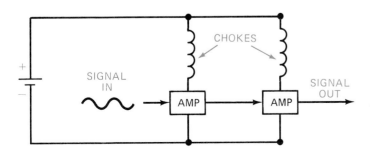

FIGURE 21-16 Use of small inductors, known as chokes, to attenuate signals between the circuit and the common power source feed.

21-9 Induced Voltage

The factor of inductive reactance is applied only to circuits using sine-wave voltages. This factor we call X_L cannot be applied to non-sine-wave circuits. The ability to induce a voltage, however, does apply to circuits having other wave-shape forms. Two of the more common circuits using inductance and a non-sine wave are the sweep circuits in an oscilloscope or a video display terminal and in the automobile electrical systems spark coil circuits. Both of these systems use a sawtooth waveform (Figure 21-17).

The rise in this sawtooth wave is 8 μs. The maximum current through the inductor is 80 mA during this period. A formula comparing the change in current to the change in time is used to determine the voltage across the inductor. This formula is

$$V_L = L \frac{\Delta i}{\Delta t} \tag{21-7}$$

where Δi is the change in current in amperes and Δt is the change in time in seconds.

In this illustration the value of V_L is equal to a change of 10 mA for each microsecond. Since the waveshape is a sawtooth form, the rise in current is equal for each subperiod of time. The factor of 10 is a constant value for the part of the wave when the current rises. The time shown for the drop in the current is much faster. Only 1 μs is required for this current to fall back to zero. This gives a ratio of 80 : 1 for this portion of the wave's cycle.

When a current is rising in a conductor the inducted voltage opposes the applied voltage. Therefore, the induced voltage is negative with respect to a positive applied or source voltage. In this circuit the value of induced voltage during the first 8 μs, or during the rise time, is

$$-V_L = L \frac{\Delta i}{\Delta t}$$

$$= 0.5 \text{ H} \times \frac{0.08}{0.000008}$$

$$= 500 \times 10^{-3} \times \frac{80 \times 10^{-2}}{8 \times 10^{-6}}$$

$$= 5000 \text{ V}$$

The polarity of induced voltage created during the collapse of a magnetic

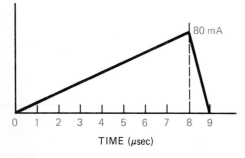

80 mA

TIME (μsec)

$L = 500 \text{ mH}$

+

FIGURE 21-17 Rise and decay of the sawtooth wave. Normally, there is a relatively long rise time and a very short decay period.

field is the same as the polarity of the voltage originally used to enlarge the magnetic field. In this circuit the polarity of this voltage is positive. The determination of the value of the voltage shows it to be much higher than the negative voltage created during the rise time of the wave. This value is

$$+V_L = L\frac{\Delta i}{\Delta t}$$

$$= 500 \times 10^{-3}\frac{80 \times 10^{-2}}{1 \times 10^{-6}}$$

$$= 40,000 \text{ V}$$

This method of utilizing a sawtooth wave form with a relatively long rise time and a very short drop time is useful when some voltages of a very high order are required. It is not unusual to create voltage values of up to 30 kV in this manner. The spark plug system in an automobile requires values up to 20 kV. The cathode ray tube used in a color video display terminal or television receiver could have voltage values of 24 to 30 kV, depending on the size and electrical requirements of the display tube.

21-10 Determination of L/R Rise Time

The time required for current rise or fall is a factor often used in electrical calculations. This factor is the relationship of the inductance to the resistance in a circuit, or *L/R*. This value is known as the *time constant* of the circuit. The time constant for an inductor is the time required for current in a circuit to rise from zero to a value of 63.2% of the total current. The formula for this is

L/R rise time *Time for current to rise to 63.2% of maximum in an inductive circuit.*

$$T = \frac{L}{R}$$

where T = time (s)

L = inductance (H)

R = resistance (Ω)

The wave shown in Figure 21-18 has risen to 63.2% of its possible value. The rise has occurred during a period of 1 s. This value is equal to one time constant.

Figure 21-19 illustrates how this is calculated. A value of 100 Ω of resistance and 100 H are series connected to a dc source of 100 V. Maximum dc

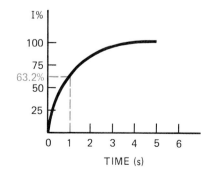

FIGURE 21-18 The time constant for changes in voltage or current in the inductive circuit.

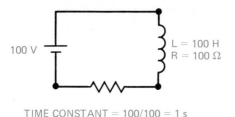

FIGURE 21-19 Determination of the time constant for a circuit.

TIME CONSTANT = 100/100 = 1 s

current flow on this circuit is determined by Ohm's law. The value of inductance is not a factor in this calculation since the inductor has zero reactance at dc. Current flow is 100/100, or 1 A maximum. The time constant for this circuit is calculated as $T = L/R$ or 100/100 or 1 s.

When these values are plotted on the graph in Figure 21-18, the rise time during 1 s is 63.2% of the total possible current in the circuit. This graph shows that five time periods are required for current to rise to close to 100% of its value.

This factor is also true when current falls from a maximum value toward zero in an inductive circuit. The 63.2% drop value is valid under these circumstances as well. Further details about time constants for circuits containing inductors as well as circuits with capacitance are presented in Chapter 26.

REVIEW PROBLEMS

21-16. What is the time constant for a circuit having an R of 60 Ω and an L of 400 mH?

21-17. A circuit has a time constant of 30 μs. What time is required for current to rise to 63.2% of its full value?

CHAPTER SUMMARY

1. A phasor diagram is used to show the phase relationships of two electrical waves.

2. An inductive circuit has a 90° phase shift between voltage and current.

3. Voltage leads current in an inductive circuit.

4. Rules for current are the same for ac and dc circuits.

5. Circuit laws for series circuits are the same for dc and inductive ac circuits.

6. Circuit laws for parallel circuits are the same for dc and ac inductive circuits.

7. Impedance is opposition to an ac wave in a circuit containing reactance and resistance.

8. Impedance is represented by the letter Z and is measured in units of the ohm.

9. Impedance values may be calculated by use of the Phythagorean theorem: $Z^2 = A^2 + B^2$.

10. The Q of a coil is a measure of its quality.

11. Q is determined as a ratio of X_L to internal resistance.
12. Electron flow in conductors occurs at or near the surface at high frequencies.
13. The effective resistance of an inductor (R_e) is the ratio of X_L to Q.
14. A signal voltage is a varying value of the dc operating voltage.
15. An inductor will choke a signal on an operating voltage conductor. This offers a high impedance to signal variations.
16. Induced voltage can create very high values in an inductor when the fall time of current has a very short time period compared to its rise time.
17. A sawtooth waveform is used to induce high voltages.
18. Time required for current or voltage to rise to 63.2% of its full value is called the time constant.
19. Five time-constant periods are required for a rise to 100% of the value.

Answer true or false; change all false statements into true statements. **SELF TEST**

21-1. Voltage induced when current rises has the same polarity as the source voltage.
21-2. Current and voltage are out of phase in an inductive circuit.
21-3. Current leads voltage in an inductive circuit.
21-4. A phasor drawing shows the phase relationship between voltage and resistance.
21-5. The frequency of a dc voltage is 0 Hz.
21-6. Q is the term used to identify the quality of a coil.
21-7. Coil Q is related to the frequency of operation of the coil.
21-8. A choke impedes dc current in a series circuit.
21-9. Impedance is measured in ohms.
21-10. Rules for series and parallel inductive circuits are similar to series and parallel resistive circuits.

21-11. Two waves starting at the same moment in time have a phasor angle **MULTIPLE CHOICE**
of
(a) 0°
(b) 90°
(c) 180°
(d) 270°
21-12. In an inductive circuit, current flow is
(a) equal throughout
(b) higher at the + connection of the source
(c) lower at the + connection of the source
(d) dependent on the wiring connections of the source and the loads

21-13. In a series circuit containing both X and R, the current phase relationship between the two components is

(a) an in-phase condition
(b) 45° shift, I lagging
(c) 90° shift, I lagging
(d) 90° shift, V lagging

21-14. The ratio of reactance to series resistance at a specific frequency is known as the

(a) Q of a coil
(b) meritorious rating of the coil
(c) specific frequency quality of the coil
(d) none of the above

21-15. A variation of the operating voltage of a circuit due to information processing is known as

(a) information voltage
(b) processing information voltage
(c) variation voltage
(d) signal voltage

21-16. The choke in an electrical circuit

(a) opposes changes in current
(b) opposes changes in voltage
(c) creates an impedance to the operating frequency
(d) reacts to output signal information

21-17. The total time for a voltage or current to rise to 63.2% of its full value in an inductive circuit is called its

(a) inductive rise time
(b) resistive rise time
(c) L/R rise time
(d) LR rise time

21-18. The number of rise time periods required to achieve close to 100% is

(a) 5
(b) 4
(c) 6
(d) 10

21-19. Circuit laws for series circuits containing both L and R apply to

(a) R circuits only
(b) individual L and R circuits
(c) LR combination circuits
(d) L circuits only

21-20. Circuit laws for parallel circuits containing both L and R apply to

(a) R circuits only
(b) individual L and R circuits
(c) LR combination circuits
(d) L circuits only

21-21. Explain why the Pythagorean theorem is used to determine impedance values in an *LR* circuit.

21-22. Explain why the signal voltage is actually a variation of the supply voltage in a circuit.

21-23. Why are current and voltage out of phase in an inductive circuit?

21-24. Why does a choke impede circuit current?

21-25. Why is the sawtooth waveform used instead of the sine wave to create high voltages?

21-26. Why is phase analysis necessary when resolving the *LR* circuit for its values of *V*, *I*, and *R*?

21-27. Explain why more than one time-constant period is required to reach 100% of maximum values in the inductive circuit.

21-28. Explain the statement "An inductor will offer a high impedance to a signal on an operating voltage conductor."

21-29. Why do we use the Q of a coil in discussing its applications in a circuit?

21-30. Why does voltage lead current in an inductive circuit?

21–31. A circuit has an inductor and a resistor in series. Current through the resistor is 10 mA. What is current through the inductor? (*Sec. 21-3*)

21–32. A circuit has three parallel inductors. Voltage across the middle one is 10 V ac. What is the voltage across the one closest to the power source? (*Sec. 21-5*)

21–33. Find the impedance for a series circuit having an X_L of 50 and a *R* of 40 Ω. (*Sec. 21-3*)

21–34. What is the impedance when *R* is 60 Ω and X_L is 100 Ω?

21–35. What is the shape of the current waveform when a sine wave is applied to an inductive circuit? (*Sec. 21-2*)

21–36. What is the dc resistance of an open choke? (*Sec. 21-8*)

21–37. What would the dc resistance of a shorted inductor be? (*Sec. 21-8*)

A parallel circuit contains both X_L and R connected to a 70-V source. Calculate the values of I_L, I_T, and I_R using the values given in problems 21–38 through 21–40.

21–38. $R = 200$ Ω, $X_L = 2$ Ω (*Sec. 21-5*)

21–39. $X_L = 4$ Ω, $R = 150$ Ω (*Sec. 21-5*)

21–40. $R = 60$ Ω, $X_L = 60$ Ω (*Sec. 21-5*)

21–41. A 10 mH inductor has a Q of 45 when operated at a frequency of 1.5 MHz. Find its value of R_e. (*Sec. 21-6*)

21–42. A 500-Ω *R* is series connected to an X_L of 400 Ω. Find its impedance. (*Sec. 21-3*)

21–43. A 400-Ω X_L is series connected to a 300-Ω resistance. What is its impedance? (*Sec. 21-4*)

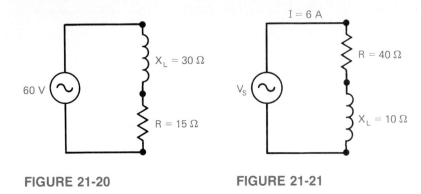

FIGURE 21-20   **FIGURE 21-21**

21–44. A 500-Ω resistance is parallel wired to a 400-Ω X_L. What is the circuit impedance? (*Sec. 21-4*)

21–45. A 400-Ω X_L is connected in parallel with a 300-Ω resistance. What is its Z? (*Sec. 21-4*)

21–46. What current flows in a circuit when 100 V is applied to an impedance of 600 Ω? (*Sec. 21-4*)

21–47. What voltage drop develops across a 200-Ω impedance when 400 mA flows through it? (*Sec. 21-4*)

Use Figure 21–20 to answer problems 21– 48 and 21– 49.

21–48. What is the circuit current, and what are the voltage drops across each component? (*Sec 21-3*)

21–49. What are the values of current and voltage when the resistance is 150 Ω? (*Sec. 21-3*)

21–50. What is the source voltage for the circuit of Figure 21-21? (*Sec. 21-3*)

Use Figure 21–22 to answer problems 21–51 and 21–52.

21–51. What is current through the resistance? (*Sec. 21-5*)

21–52. What is total circuit current? (*Sec. 21-5*)

21–53. Find the impedance for a parallel circuit with a V_s of 200 V and an I of 3 A. (*Sec. 21-5*)

21–54. What is the total circuit current when 600 V is applied to a parallel connected X_L of 40 Ω and an R of 100 Ω? (*Sec. 21-5*)

21–55. What is the Q of a coil having an X_L of 80 Ω and an r_i of 40 Ω? (*Sec. 21-6*)

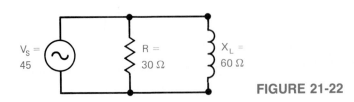

FIGURE 21-22

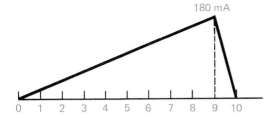

180 mA

0 1 2 3 4 5 6 7 8 9 10   **FIGURE 21-23**

21–56. Find the X_L for a coil with a Q of 500 and an R_i of 10 Ω. (*Sec. 21-6*)

21–57. A choke has a (high, low) impedance to a 10-kHz signal. (*Sec. 21-7*)

21–58. What is the V_L for an inductive circuit of 10 H when there is a rise in current of 5 mA per second? (*Sec. 21-5*)

Use Figure 21–23 to answer problems 21–59 and 21–60.

21–59. Find the V_L during the rise period. (*Sec. 21-10*)

21–60. Find the V_L during the fall period of the wave. (*Sec. 21-10*)

21–1. 90°

21–2. 45°

21–3. V_L leads I_C by 90°.

21–4. V_L and V_A are in phase.

21–5. 18.03 Ω

21–6. 5 Ω

21–7. 18.86 V

21–8. $V_X = 10$ V; $V_R = 16$ V

21–9. 6.71 A

21–10. 5 Ω

21–11. 13.31

21–12. 16.67 Ω

21–13. Inductor

21–14. Low

21–15. High

21–16. 0.0067 s

21–17. 30 μs

Chapter 19-21

Inductive action is presented in these chapters. The use of inductive circuits to move electrical energy from one circuit to another through transformer actions is one principle that is used. Inductive reactance considers the delay in circuit current due to inductive components in the circuits. The final chapter in this section presents methods of solving for circuit values in inductive circuits.

1. The ability to transfer energy from one coil to a second is accomplished with mutual inductance.

2. The unit of mutual inductance is the henry (H).

3. The amount of flux from one coil to a second coil is the coupling coefficient (k).

4. Coupling coefficient is a ratio of the flux from one coil to a second coil.

5. Coupling is classified as being *loose* or *tight*.

6. A zero coupling coefficient occurs when the two coils are at right angles to each other.

7. Mutual inductance includes the coupling coefficient, coil 1 flux, and coil 2 flux.

8. A transformer has an input, or primary winding, an output, or secondary winding, and a core.

9. The value of induced secondary voltage is determined by the turns ratio between primary and secondary windings.

10. Transformers may be classified as voltage step-up, voltage step-down, or voltage-isolation types.

11. When a load device is connected to the secondary winding it is *loaded* by this device and forms a complete electrical circuit.

12. Secondary current depends on secondary voltage and load resistance values.

13. Secondary power and primary power are equal in the ideal transformer. This is known as power transfer.

14. Transformers may have either multiple primary windings or multiple secondary windings; or in some cases, both.

15. An unloaded secondary will produce a higher-than-rated output voltage.

16. Transformer efficiency is the percentage of output to the input $\times$ 100%.

17. The efficiency of a transformer is reduced by eddy currents and their heating effect.

18. A toroid is a highly efficient doughnut-shaped core. It is usually utilized at frequencies above the audio range.

19. Major problems in transformers are opens and shorted turns in internal windings.

20. Opposition to an alternating current is called reactance.

21. Opposition to an alternating current in an inductive circuit is known as inductive reactance.

22. The symbol X_L represents inductive reactance, measured in units of the ohm.

23. An inductive circuit's reactance value affects current flow in the circuit instead of the dc resistance value.

24. Inductive reactance is directly proportional to operational frequency of the circuit.

25. X_L is also directly proportional to the value of inductance; therefore, X_L is proportional to fL.

26. The formula for X_L is $X_L = 2\pi fL$.

27. Total X_L in a series circuit is equal to the sum of the individual X_L values.

28. Total X_L in a parallel X_L circuit is expressed by the formula $1/X_{LT} = 1/X_{L1} + 1/X_{L2}$, and so on.

29. The rules for series resistances and series X_L values are identical.

30. The rules for parallel resistances and parallel X_L values are identical.

31. Ohm's law rules may be used for circuits containing inductive reactance values.

32. Kirchhoff's rules for voltage drops and current division apply to inductive reactance circuits.

33. Voltage and current are out of phase in a circuit containing X_L.

34. Voltage leads current by 90° in the inductive reactance circuit.

35. An inductive circuit has a 90° phase shift between voltage and current.

36. Voltage leads current in an inductive circuit.

37. Rules for current are the same for ac and dc circuits.

38. Circuit laws for series circuits are the same for dc and inductive ac circuits.

39. Circuit laws for parallel circuits are the same for dc and ac inductive circuits.

40. Impedance is opposition to an ac wave in a circuit containing reactance and resistance.

41. Impedance is represented by the letter Z and is measured in units of the ohm.

42. Impedance values may be calculated by use of the Pythagorean theorem: $Z^2 = A^2 + B^2$.

43. The Q of a coil is a measure of its quality.

44. Q is determined as a ratio of X_L to internal resistance.

45. Electron flow in conductors occurs at or near the surface at high frequencies.

46. The effective resistance of an inductor (R_e) is the ratio of X_L to Q.

47. A signal voltage is a varying value of the dc operating voltage.

48. An inductor will choke a signal on an operating voltage conductor. This offers a high impedance to signal variations.

49. Induced voltage can create very high values in an inductor when the fall time of current has a very short time period compared to its use time.

50. A sawtooth waveform is used to induce high voltages.

51. Time required for current or voltage to rise to 63.2% of its full value is called the time constant.

52. Five time-constant periods are required for a rise to 100% of the value.

CHAPTER 22

Capacitance

Another device used to react to changes in a circuit is the capacitor. Capacitors react to changes in voltage as opposed to changes in current for the inductor. This device is a voltage storage unit. It attempts to oppose any changes in the amplitude of voltage in a circuit. Capacitors are used to control voltage levels in units, to transfer signals between stages, and as a part of a resonant circuit.

KEY TERMS

Capacitance
Capacitor
Condenser
Coupling
Dielectric Constant
Dielectric Strength
Electrolytic Capacitor
Energy Field

Farad
Leakage
Microfarad
Nonelectrolytic Capacitor
Picofarad
Voltage Breakdown Rating
Working Voltage Rating

OBJECTIVES

Upon completion of the material in this chapter, you will

1. Understand basic concepts of capacitance.
2. Recognize typical capacitor types.
3. Be able to calculate values for series- and parallel-wired capacitors.
4. Understand the stray capacitance factor.
5. Be able to select basic capacitor types.
6. Recognize capacitor failures.

The study of electrical and electronic theories will identify three basic types of components: resistors, capacitors, and inductors. These three components are considered to be *passive*. This term indicates that these devices do not change their internal values as they operate in an electrical circuit. (As you continue your studies you will find that diodes, tubes, and transistors are considered to be *active* devices because their internal values change as they operate.)

The capacitor utilization is second only to that of the resistor in electrical circuits. The reason for this will become apparent as you read the material in this chapter. In summary, there are three major reasons why we use capacitors.

1. The capacitor has the ability to store an electrical charge.
2. The capacitor will not permit any voltage to exist on its terminals until some of the voltage charge is moved from one of its plates to its other plate.
3. Capacitors are able to react to different frequencies.

The first two characteristics are discussed in this chapter. The third characteristic will be better understood after we have studied capacitor action in alternating-current circuits.

Before we begin the study of the capacitor and its action in an electrical circuit, let us clarify the name of this device. Early in the days of investigations of electrical concepts and applications, this device was thought to be able to reduce the size of the electrical charges. This concept led to the name **condenser**. Electrical charges were thought to be able to be condensed, or compressed, and stored in a small device. As our studies of electrical energy continued, this concept was found to be in error. The current theory related to capacitors is that these devices are able to store an electrical charge. They have the *capacity* to store this charge until required to give it up. The name **capacitor** is now used in almost all electrical work to describe this device. One of the few places where this term has not been readily adopted is in the automotive field. Many of the people working on or with auto electric circuits still refer to the capacitor as a condenser. Keep in mind that capacitors react to *voltage* and inductors react to *current* in the circuit.

Condenser *Obsolete name used for capacitor.*

Capacitor *Device used to store electrical charges in circuit. Opposes changes in voltage in circuit.*

22-1 Capacitor Construction

The physical construction of a capacitor is shown in Figure 22-1. Capacitors consist of two parallel metallic *plates*. The plates are separated by an insulating material called the *dielectric*. In addition to these three components the capacitor has leads, or wires, enabling it to be connected to an electrical circuit.

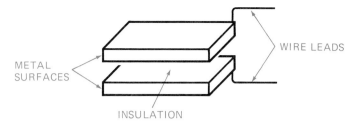

METAL
SURFACES

WIRE LEADS

INSULATION

FIGURE 22-1 The capacitor consists of two metallic surfaces and an insulating material called the dielectric.

The illustration shows only the components making up a basic capacitor. Another section of this chapter shows some of the more common commercial types of capacitors. **Capacitance**, by definition, is the ability of an electrical device to store an electrical charge. This is accomplished when a voltage source is connected to one plate and the negative lead of the source is connected to the plates of the capacitor (Figure 22-2). The positive lead of the voltage source is connected to one plate and the negative lead of the source is connected to the other plate. The plates are electrically neutral before the voltage source is connected. Each plate has an equal quantity of positive and negative charges when it is neutral [Figure 22-2(a)].

Capacitance Ability of device to store an electrical charge.

When the voltage source is connected to the plates of the capacitor, the charges are redistributed. The negative charges on the upper plate [Figure 22-2(b)] are attracted to the positive electrode of the power source. This gives the plate a positive charge. The positive charges on the lower plate are repelled by the positive charges on the upper plate. They are attracted to the negative electrode of the power source. The result of this action is a device containing one positively charged plate and one negatively charged plate.

The plates of the capacitor are charged when a voltage source is connected to them. This establishes a difference in electrical potential between the plates. The capacitor therefore has the ability to accept a voltage charge. The plates of the capacitor are capable of holding an electrical charge.

Energy Field The charges on the plates of the capacitor have another effect. They develop an **energy field** in the dielectric material. This energy field is better known as an electric field. It consists of lines of force between the charged plates of the capacitor. The lines of force are concentrated in the dielectric material. They therefore store energy in the dielectric material.

Energy field Lines of force developed between charged plates of capacitor.

REVIEW PROBLEMS

22-1. Name the physical parts of a capacitor.
22-2. Define the term *capacitor*.
22-3. Where is the electric field in a capacitor?

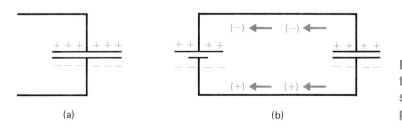

(a)

(b)

FIGURE 22-2 The plates of the capacitor accept the same charges as those of the power source connected to it.

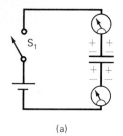
(a)

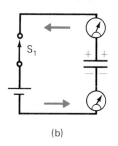
(b)

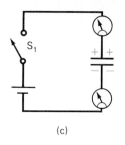
(c)

FIGURE 22-3 An ammeter connected in line between the source and the plates of the capacitor shows electron flow during charge and discharge periods.

22-2 Capacitor Action

A capacitor having electrically neutral plates does not perform much work. When the capacitor's plates are either being charged or being discharged, work is performed. The action involved in charging a capacitor's plates is shown in Figure 22-3. Part (a) shows a capacitor with no electrical charge. Both of its plates have the same quantity of positive and negative charges, and the power source is not connected to the plates. It is electrically neutral. Under these conditions neither of the ammeters in the circuit show any current flow.

When the external power source is connected to the plates of the capacitor, the operational conditions change. Switch S_1 is used to complete the circuit [Figure 22-3(b)]. Charges now flow between the plates of the capacitor and the electrodes of the power source. The ammeters indicate a movement of electrons in the circuit. The upper plate will give up its electrons while the lower plate accepts electrons. The resulting action creates a difference in electrical potential between the plates of the capacitor. The upper plate becomes positive and the lower plate is negatively charged. The quantity of the voltage is equal to the amount of the voltage applied from the voltage source.

The plates of the capacitor are able to store, or hold, the electrical charge [Figure 22-3(c)]. The power source is removed from the circuit by opening switch S_1. The charges on the capacitor's plates remain. When the plates of the capacitor are charged, there is no additional current flow in the circuit. In fact, the only time a current will flow is when the charges on the plates are in motion.

In theory a capacitor is able to hold its charge indefinitely. In practice the charge will slowly drop due to some of the imperfections in the dielectric material. The technical term for this gradual discharge is **leakage**.

Leakage *Term describing effect of neutralization of charges on plates of capacitor.*

The discharge of the capacitor is exactly opposite to the effect of its charge (Figure 22-4). The charged capacitor is shown in part (a). The upper plate has a positive charge and the lower plate has a negative charge. When a short circuit is placed across the plates of the capacitor [part (b)], the extra electrons on the negative plate move to an area that is less negative. In this example the

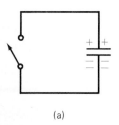
(a)

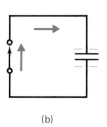
(b)

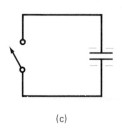
(c)

FIGURE 22-4 During discharge, the capacitor plates both have the same charge.

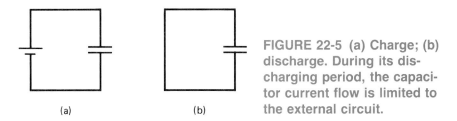

FIGURE 22-5 (a) Charge; (b) discharge. During its discharging period, the capacitor current flow is limited to the external circuit.

(a) (b)

less negative area is the positive plate. The result is the neutralization of the voltages on the two plates, since they are now both electrically equal.

When the electrons are in motion [Figure 22-4(b)], a current flows in the circuit. When the plates are electrically neutral, the current no longer flows. When this occurs the capacitor is considered to be *discharged*. *Current flow* during either the charging or the discharging action of the capacitor is limited to the components in the circuit. There is no current flow through the capacitor. The dielectric material opposes any current flow through the electric field between the plates. All current flow is from one plate, back through the circuit, and then to the other plate. Current flow during charge or discharge of the capacitor is limited to the *external circuit* rather than the capacitor (Figure 22-5).

REVIEW PROBLEMS

22-4. What happens to the voltage on the plates of the capacitor as it charges?

22-5. What method may be used to discharge a capacitor?

22-6. What voltage is measured across the plates of a fully discharged capacitor?

22-7. What is the method used to charge a capacitor?

22-8. How does current flow in a capacitor circuit?

22-3 Units of Capacitance

The basic unit of capacitance has 1 **farad** (F) of capacitance when 1 C of charge is stored when 1 V is applied to its plates.

Farad *Unit of capacitance.*

The formula for this action may be described in terms of voltage and charge as

$$C = \frac{Q}{V} \qquad (22\text{-}1)$$

where C = capacitance (F)

Q = quantity of charge

V = voltage

This formula may be rearranged and shown as

$$Q = CV \qquad \text{or} \qquad V = \frac{Q}{C}$$

EXAMPLE 22-1

A capacitor stores 100 C and has 5 V across its plates. Determine its capacitance in farads.

Solution

$$C = \frac{Q}{V}$$

$$= \frac{100 \text{ C}}{5 \text{ V}}$$

$$= 20 \text{ F}$$

EXAMPLE 22-2

A 5-F capacitor has 150 V across its plates. What charge does it store?

Solution

$$Q = CV$$

$$= 5 \text{ F} \times 150 \text{ V} = 750 \text{ C}$$

EXAMPLE 22-3

How much voltage is developed across a 10-F capacitor with 15 C of charge?

Solution

$$V = \frac{Q}{C}$$

$$= \frac{15 \text{ C}}{10 \text{ F}}$$

$$= 1.5 \text{ V}$$

Microfarad *Term describing value of* $\frac{1}{1,000,000}$ *Farad*

Picofarad *Term describing value of* $\frac{1}{1,000,000,000,000}$ *Farad*

The application of the capacitor in electrical and electronic devices usually requires a unit with values much less than 1 F. Typical units of capacitance are the microfarad and the picofarad. One **microfarad** (μF) is equal to 1×10^{-6} F, or 0.000001 F. One **picofarad** (pF) is equal to 1×10^{-12}, or 0.000000000001 F. An earlier term for picofarad was the micro-microfarad ($\mu\mu$F). This term does not meet SI notation requirements and is no longer in use. One may find it used on older schematic diagrams and should at least be able to recognize its value. Conversion factors for picofarads, microfarads, and farad units are as follows:

$$1 \text{ F} = 1 \times 10^6 \text{ } \mu\text{F}$$

$$1 \text{ F} = 1 \times 10^{12} \text{ pF}$$

$$1 \text{ } \mu\text{F} = 1 \times 10^{-6} \text{ F}$$

$$1 \text{ } \mu\text{F} = 1 \times 10^6 \text{ pF}$$

$$1 \text{ pF} = 1 \times 10^{-12} \text{ F}$$

$$1 \text{ pF} = 1 \times 10^{-6} \text{ } \mu\text{F}$$

EXAMPLE 22-4

Convert the following values into microfarad units.
(a) 0.00005 F
(b) 0.049 F
(c) 760 pF
(d) 10 pF

Solution

(a) 0.00005 F $\times$ 10^6 = 50 μF
(b) 0.049 F $\times$ 10^6 = 49000 μF
(c) 760 pF $\times$ 10^{-6} = 0.00076 μF
(d) 10 pF $\times$ 10^{-6} = 0.00001 μF

EXAMPLE 22-5

Convert the following values into picofarad units.
(a) 0.43 $\times$ 10^{-7} F
(b) 0.05 μF
(c) 0.000085 F
(d) 0.001 μF

Solution

(a) 0.43 F $\times$ 10^{-7} $\times$ 10^{12} = 43,000 pF
(b) 0.05 μF $\times$ 10^6 = 50,000 pF
(c) 0.000085 F $\times$ 10^{12} = 85 $\times$ 10^6 pF
(d) 0.001 μF $\times$ 10^6 = 1000 pF

REVIEW PROBLEMS

22-9. Name the basic unit of capacitance.
22-10. What quantity of farads are in 1 μF?
22-11. What quantity of farads are in 1 pF?
22-12. Convert 15 μF to picofarads.
22-13. Convert 360 pF to farads

22-4 Capacitor Characteristics

Capacitors are produced in a variety of sizes and shapes. In addition, they are available with different types of dielectric materials. Many of these factors are due to the demands of those who use the capacitors in electrical circuitry. In general, the topic of capacitors may be divided into at least these general sections. They are construction, types, and applications for capacitors. With this in mind, let us examine these factors.

Construction of Capacitors A capacitor is made of two metallized surfaces and some type of dielectric material. Both of these affect the specific value of capacitance for the device. The metallized surfaces, or plates, of the capacitor are used to store the electrical charges. It stands to reason, then, that the area of these surfaces is a formative factor in determining the quantity of charge that can be stored. The surface area of each plate is directly proportional to the quantity of charge. A large surface area on the plates of the capacitor will increase its capacitance value. The plate area of some capacitors is deliberately etched to provide a larger total area for a given size of plate. The plate shown in Figure 22-6(a) is smooth and has a specific quantity of surface area. If the plate is etched, or roughened [part (b)], its surface area for the 1-in.² material is increased. This permits a larger quantity of charge for a specific area of plate and provides more capacitance in a given physical size of plate material. The purpose of this type of construction is to build a larger-value capacitor having a small physical size. Another significant factor is the space separating the two plates of the capacitor. The interaction of charges in the two plates is related to the distance between the two plates. The effect of charges on the two plates is inversely proportional to the space between the plates. In other words, as the spacing between the plates of the capacitor is increased, the quantity of capacitance is reduced. We can also state that as the distance between the plates of the capacitor is reduced, the amount of capacitance is increased. This, of course, is effective when all other factors related to capacitors are held constant.

Dielectric constant *Term describing insulating strength of dielectric material.*

A third factor related to the amount of capacitance is the strength of its dielectric material. All materials have a dielectric value. It is called the **dielectric constant** of the material. Some of the more common values are shown below.

Material	Dielectric Constant
Vacuum	1.0
Air	1.006
Paper	2.5
Oil	4.0
Mica	5.0
Aluminum oxide	7
Glass	7.5
Tantalum oxide	25
Ceramic	1200–7500

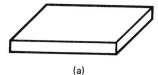
(a)

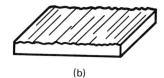
(b)

FIGURE 22-6 One plate of a capacitor. Its surface area is increased when the surface is etched rather than being smooth.

The reference for dielectric material in this instance is air, with a value close to 1.0. The **dielectric strength** determines the amount of voltage and also the amount of capacitance for a given device. The quantity of charge and the value of capacitance increase as the dielectric constant of the material increases. The quantity of capacitance is directly proportional to the dielectric constant of the material. Another way of stating this is to say that when all other factors for the capacitor are equal (plate size, surface area, and spacing) a device using an aluminum oxide dielectric will have more capacitance than a device using paper as its dielectric material. Conversely, a physically smaller capacitor may be produced when the dielectric constant is increased.

A related factor for the dielectric material is its *strength*. The dielectric strength of a material describes its ability to hold an electrical charge without failing. Following are the dielectric strengths for some typical materials.

Dielectric strength *Term describing voltage breakdown rating of dielectric materials.*

Material	Value (V/Mil)
Air	75
Oil	400
Ceramic	75
Paper	1250
Glass	5000
Mica	3000

The rating for these materials is described in units of volts per mil (0.001) of thickness of the materials. The thickness of the dielectric material is one factor in the rating of capacitors. This factor is called the **voltage breakdown rating**. It describes the value of voltage the capacitor can safely store. Voltages higher than this value often cause a failure in the dielectric material and a corresponding failure in the capacitor.

Voltage breakdown rating *Term describing upper limit of voltage for safe capacitor operation.*

The following factors determine the value of capacitance:

1. *Plate size and surface area.* The capacitance value is directly proportional to the area of the plates.
2. *Plate spacing.* The capacitance value is inversely proportional to the distance between the plates of the capacitor.
3. *Dielectric strength of the insulator.* High numeric values for the dielectric material produce larger values for the capacitor.

The formula for determination of the value of capacitance is

$$C = 8.85 \times 10^{-12} K \frac{A}{d} \qquad (22\text{-}2)$$

where C = capacitance (F)
K = dielectric constant (F/m)
A = plate area (m²)
d = distance between plates (m)

REVIEW PROBLEMS

22-14. What effect on total capacitance is there when the plate area is decreased?
22-15. Increasing the distance between the plates of a capacitor has what effect on total capacitance?
22-16. What effect on total capacitance is there when a paper dielectric material is used instead of mica?

Fixed and Variable Capacitors Capacitors are manufactured with both a fixed, or constant, value and as variable devices. The schematic symbols for capacitors used in the United States and considered as standard for the industry are shown in Figure 22-7. Two symbols are shown for fixed-value capacitors. Symbol (a) is generally accepted as a standard. Symbol (b) is used when a polarized, or polarity-sensitive, capacitor is used. The lower curved line of the symbol normally indicates the level connected to circuit common or the more negative level of the capacitor.

Variable capacitors are those whose value can easily be changed. When a variable capacitor is used, its symbol includes an arrowhead [Figure 22-7(c)]. In some applications a dual- or triple-section capacitor is required. These are usually made as a single unit with multiple sections. The schematic symbol for these is shown by adding a dashed line to connect the diagonal arrows of the single unit [Figure 22-7(d)]. A multiple-section variable capacitor is shown in Figure 22-8. This type of capacitor is normally used to tune radio stations. It is also used in applications where two or more capacitance values need to be selected at the same time.

Fixed-value capacitors are usually classified by the type of dielectric material used in their construction. There are two major categories for fixed-value

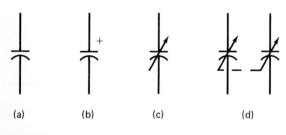

(a) (b) (c) (d)

FIGURE 22-7 Schematic symbols for a variety of capacitors: (a) fixed-value nonelectrolytes; (b) fixed-value electrolytes; (c) single variable; (d) multiple variable.

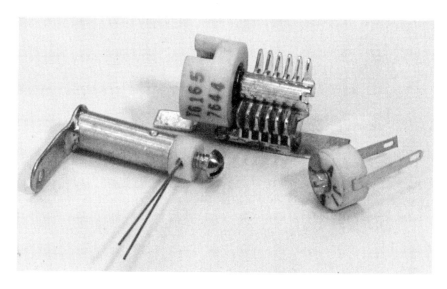

FIGURE 22-8 Variable capacitors are available in a variety of sizes and shapes. (Courtesy E.F. Johnson Company.)

capacitors. These are called electrolytic and nonelectrolytic. This classification is related to the type of dielectric material used in their construction. In general, the largest percentage of capacitors are constructed in a similar manner. The large plate surface area required for these capacitors dictates the need to use a rolled sandwich construction (Figure 22-9). This construction consists of the

FIGURE 22-9 Construction of the fixed-value capacitor. (From Joel Goldberg, *Fundamentals of Electricity*, 1981, p. 182, 183. Reprinted by permission of Prentice-Hall, Englewood Cliffs, New Jersey.)

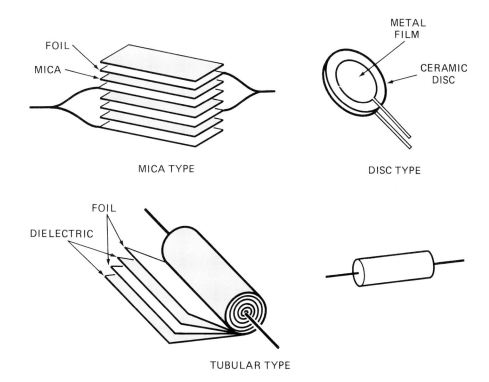

FOIL

MICA

MICA TYPE

METAL FILM

CERAMIC DISC

DISC TYPE

FOIL

DIELECTRIC

TUBULAR TYPE

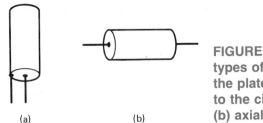

FIGURE 22-10 Two common types of connections from the plates of the capacitor to the circuit: (a) radial; (b) axial.

(a)   (b)

two metallized surfaces separated by the dielectric material. This sandwich is rolled into a compact tubular form, as seen in the illustration. This tubular form is then "packaged" into a plastic or metal container. Leads are connected to each of the plates. The leads become a part of the completed capacitor assembly. Two types of lead packaging is used. One of these is called a *radial assembly* and the other is called an *axial* assembly (Figure 22-10). The radial lead assembly has both leads at the same end of the package. This is so that the component may be mounted vertically on a circuit board. It takes less space on the board when vertically mounted. The axial unit has its leads on its axis, hence its name.

Electrolytic capacitors are constructed using aluminum plates. One of these plates is coated with an aluminum oxide material. This becomes the positive plate of the capacitor. The oxide layer acts as the dielectric material for the capacitor. An electrolyte is impressed on either a paper or a gauze type of material. It and the other plate are the negative electrode of the capacitor. Electrolytic types of capacitors are also produced using a layer of tantalum oxide instead of aluminum oxide for the positive plate. The advantage of using tantalum oxide is the physical size of the capacitor is reduced without any loss in capacitance value. The two types of electrolytic capacitors are shown in Figures 22-11 and 22-12.

One must always be aware that an electrolytic type of capacitor has a very definite polarity. All electrolytic capacitors have polarity markings on their cases. Either the positive lead or the negative lead is marked. The specific

Electrolytic capacitor *Type of capacitor using an electrolyte material for its dielectric.*

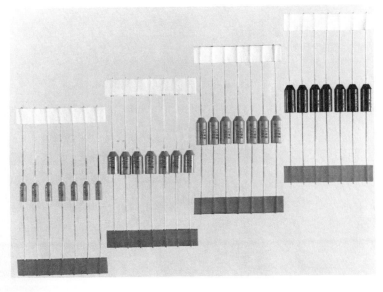

FIGURE 22-11 Aluminum oxide types of capacitors. (Courtesy Mallory Capacitor Company/Emhart Electrical/ Electronic Group.)

marking depends on the manufacturer's specifications. In most of the products the negative lead is connected to the case of the electrolytic capacitor. The positive lead is mounted on some type of insulating material.

When installing an electrolytic type of capacitor the positive lead must *always* be connected to the most positive of the two connections. If one fails to do this, the results may be catastrophic. The electrolytic capacitor is held in a sealed container. When connected in a reverse-polarity manner, the electrolyte in the capacitor emits a gas. This gas expands in the sealed container and something has to give. The result is an explosion and the destruction of the capacitor. The possibility of personal injury also exists under these conditions. Needless to say, one *must* exercise caution and make certain that the polarity of the electrolytic capacitor is observed. It will not explode when its correct polarity is observed during installation or operation.

The only exception to this statement is for a group of capacitors rated as *motor starting* types. These are designed to be used in ac circuits and have no polarity.

Electrolytic capacitors are rated in values of microfarads. In general, those capacitors whose values are 1 μF or larger are classified as electrolytic types. Capacitors having values of less than 1.0 μF are usually nonelectrolytic types.

Nonelectrolytic capacitors may use one of several different types of dielectric material. These types are identified as paper, mica, ceramic, metallized polyester, and polyester film. All of these terms refer to the type of dielectric material used between the plates of the capcitor. The terms are not used to identify the physical casing of the capacitor. Early capacitors were coated with wax to seal them and protect their plates from the effects of moisture and dirt. Today, almost all nonelectrolytic capacitors are sealed in a plastic type of housing. Several different types of nonelectrolytic capacitors are illustrated in Figure 22-13. The sizes and shapes of these nonelectrolytic capacitors are almost limitless.

Construction of the most popular types of nonelectrolyte capacitors varies, depending on the specific type. Physical size requirements play a critical role in the development of these capacitors. Almost all nonelectrolytic capacitors are rated in capacitance values of 1.0 μF or less. Usually, the lower end of readily available capacitors is around 5 pF.

Ceramic capacitors are produced with a ceramic material for their dielectric. The ceramic dielectric is plated with a metal coating. This forms the two plates of the capacitor. Wire leads are attached to this assembly and the unit is covered with a plastic coating. This capacitor's construction is shown in Figure 22-9. The shape of these units has led to their common name of *disk capacitors*.

Mica capacitors are constructed by stacking layers of foil and mica materials. Every second piece of foil is connected to a common lead. This becomes one plate lead. The remaining foils are connected to another common lead. This is the second plate lead. The quantity of foil pieces determines the ca-

FIGURE 22-12 An electrolytic type of capacitor. (Courtesy Mallory Capacitor Company/ Emhart Electrical/Electronic Group.)

FIGURE 22-13 Typical capacitors used in electronics. (Courtesy Mallory Capacitor Company/Emhart Electrical/ Electronic Group.)

pacitance of this unit. The foil and mica sandwich is packaged in a plastic housing. The shape of this housing is rectangular. It is often the size of a small postage stamp.

Paper capacitors have a different type of construction. This capacitor is built of two pieces of foil separated by a paper or plastic insulating material. The requirements for the surface area of the foil necessitate that it to be rolled up into a tubular form. When unrolled the foil is several feet long. Leads are attached to each piece of foil and the assembly is sealed into a plastic package.

Metallized capacitors are constructed using either a polyester or a film dielectric material. Their construction is similar to that used for paper types of capacitors. The major difference is the physical size of the unit. The use of the metallized dielectric permits the production of smaller physical size while maintaining capacitance values.

Marking and Coding Designations Capacitors have two designations. One of these is the value of capacitance. The second is a voltage rating.

The *voltage rating* for a capacitor identifies the *maximum* dc voltage rating of the dielectric material. Often this is described in units called **working voltage rating**. This is the realistic value of dc voltage the capacitor is capable of storing. Any voltage level higher than this rating is capable of destroying the dielectric material. The result is a failure of the capacitor.

Capacitance values are marked on the body of the capacitor in one of two basic ways. One method is to print a numeric value on the body. This value may be in units of either microfarads or picofarads. Normally, electrolytic, paper and the film types of capacitors carry a microfarad rating. Mica and ceramic capacitors are usually produced with values in the picofarad range. The symbols for mica and pica are often omitted since the type of capacitor often dictates its capacitance range.

Working voltage rating
Maximum operating voltage rating for capacitors and other electrical components.

The alternative method of marking values on capacitors is to use a color-coding system. The EIA standard color/number system is used to indicate a value in units of picofarads. Color coding of capacitors is fast becoming a bit of history. In addition, there are several different color-coding systems and little standardization. Since most capacitors produced in the past one or two decades have not used a color-coding system, there is little need to present the variety of marking systems in this book.

There is one marking system used for small capacitors that should be understood. This marking system is used extensively for those film and mica types having short leads and very small bodies. The system uses numbers and a multiplier for this identification. Markings on the body of the capacitor may read "104." This, according to the following chart, indicates a value of 10 × 10,000, or 100,000 units. The small size would indicate a value in picofarads. This capacitor would have a value of 100,000 pF or 0.1 μF.

Multiplier Number	Multiply by:	Tolerance of Capacitor		
		10 pF or less	Letter	Over 10 pF
0	1	0.1	B	—
1	10	0.25	C	—
2	100	0.5	D	—
3	1,000	1.0	E	—
4	10,000	2.0	F	1
5	100,000	—	G	2
6	—	—	H	3
—	—	—	J	5
8	0.01	—	K	10
9	0.1	—	M	20

At times a letter, such as R, may be used in place of the decimal point in the marking. For example, a capacitor marked 3R3 is equivalent to one marked 3.3 pF. A typical capacitor may be marked "131J." This represents a value of 130 pF having a tolerance factor of ±5.

REVIEW PROBLEMS

22-17. Name the two types of fixed-value capacitors.
22-18. What is the difference between radial and axial lead components?
22-19. What materials are used in electrolytic capacitors?
22-20. What materials are used in nonelectrolytic capacitors?
22-21. What can occur when an electrolytic capacitor is installed with reverse polarity?
22-22. What is the voltage rating of a capacitor?

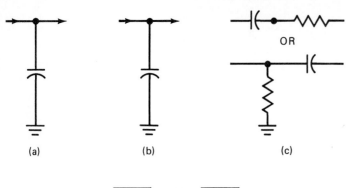

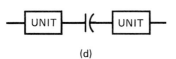

FIGURE 22-14 Typical uses
for capacitors: (a) bypass;
(b) filtering; (c) timing;
(d) coupling.

Capacitor Applications Capacitors have a multitude of functions in electrical circuits. These functions are commonly known as bypass, filter, timing, and coupling (Figure 22-14). Capacitors act in a manner similar to a sponge. The sponge will absorb liquids when it is dryer than the liquid. The sponge will also give up its liquid on demand. Capacitors act in a similar manner when involved in voltage changes in an electrical circuit. Capacitors used in either bypass or filter circuits will attempt to absorb any rise in circuit voltage. If the circuit voltage falls, the capacitor will give its voltage charge back to the circuit. In other words, the capacitor will attempt to maintain a constant level of circuit voltage. This statement is an oversimplification of the total effect on the circuit. Further details are given in Chapter 23.

The rate of charge or of discharge for a capacitor may be utilized in timing circuits. The delay in reaching a predetermined level of charge or discharge is used to operate electrical devices or circuits. This effect is also described in greater detail in Chapter 23.

The final application for capacitors is called **coupling**. The capacitor has the ability to transfer its charge level from one of its plates to the other plate. When one plate becomes more positive, the other plate becomes more negative. Changes in the level of voltage on one plate are transferred, or coupled, to the second plate of the capacitor in this manner. Variable voltage levels, called *signals*, are used to control the operation of electronic devices. The signals are passed from one unit, or section, to the next unit through the action of capacitors. Only the variations are passed to the next section, while the fixed value of operating voltage does not transfer. Details of this action also appear in Chapter 23, where details of capacitor action in circuits are discussed.

Coupling *Ability of device to transfer its signal from one stage to an adjacent stage in the system.*

22-5 Capacitors in Combination

Any of the three basic electrical components may be connected in either series or parallel combinations. The rules related to combinations of capacitors are similar to those for resistances. There are some significant differences. Once one realizes the effect of changing the plate spacing or size, the reasons for these differences become apparent.

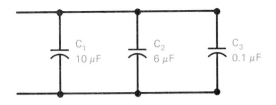

FIGURE 22-15 When capacitors are placed in parallel, the total sum of their plate areas provides an increase in total capacitance values.

Parallel Capacitance When two or more capacitors are connected as a parallel bank, the result is the increase of the total plate area for the combination. When all other factors remain relatively constant, the effect is the increase in the quantity of capacitance. The formula for parallel-connected capacitors is

$$C_T = C_1 + C_2 + C + \ldots + C_n \tag{22-3}$$

Simply stated, total capacitance for a parallel bank of capacitors is the sum of the individual values of the capacitors. Using the circuit in Figure 22-15, the total capacitance, C_T, is equal to $10 + 6 + 0.1$, or 16.1 µF.

When a voltage is applied to this circuit, each capacitor is charged to the same voltage level. The rule for voltages in a parallel bank of capacitors is the same as the rule for voltage in a parallel resistive bank:

$$V_T = V_{c1} = V_{c2} = V_c \ldots = V_{cn} \tag{22-4}$$

EXAMPLE 22-6

What is the total capacitance when four 12-µF capacitors are connected in parallel?

Solution

$$C_T = C_1 + C_2 + C_3 + C_4$$
$$= 12 + 12 + 12 + 12$$
$$= 48 \ \mu F$$

In this example it is possible to use the formula $C_T = 4C$ since all capacitors are equal.

EXAMPLE 22-7

What is the total capacitance when capacitors having values of 0.1, 4, 0.02, 11, and 0.013 µF are parallel connected?

Solution

$$C_T = 0.1 + 4 + 0.02 + 11 + 0.013$$
$$= 15.133 \ \mu F$$

FIGURE 22-16 When capacitors are connected in series, the separation between the outside plates of each creates a smaller quantity of capacitance.

REVIEW PROBLEMS

22-23. A total capacitance value of 1.6 μF is needed. Capacitors having values of 1.0, 0.8, 0.4, and 0.4 μF are available. Which ones will combine for the required value?

22-24. What is the total capacitance when three 2 μF and two 0.25 μF capacitors are parallel connected?

Series Capacitances Capacitors connected in series present a totally different set of circumstances. When capacitors are series connected (Figure 22-16), the *outside* plates of the combination are farther from each other than those of a single capacitor. The result as far as the circuit is concerned is a reduction in the total amount of capacitance. Another way of stating this is that the thickness of the dielectric material between the two outside plates is increased when capacitors are series connected.

Capacitors connected in a series string provide a total capacitance value less than the value of the lowest single capacitor's value. This rule is similar to the reciprocal rule stated earlier for parallel-connected resistors. The formula for series-connected capacitors is also similar. The rule is:

$$\frac{1}{C_T} = \frac{1}{C_1} + \frac{1}{C_2} + \frac{1}{C_3} + \cdots + \frac{1}{C_n} \tag{22-5}$$

This formula may be stated: The reciprocal of capacitance in a series capacitance circuit is equal to the sum of the reciprocal value of each capacitor in the circuit.

EXAMPLE 22-8

Find the total capacitance of a 2-, 8-, and 3-μF capacitor in a series circuit configuration.

Solution

$$\frac{1}{C_T} = \frac{1}{C_1} + \frac{1}{C_2} + \frac{1}{C_3}$$

$$= \frac{1}{2} + \frac{1}{8} + \frac{1}{3}$$

$$= 0.5 + 0.13 + 0.33$$

$$= 0.96$$

$$C_T = \frac{1}{0.96} = 1.04 \ \mu\text{F}$$

Equal capacitor values in a series configuration are handled in the same manner as that used for equal resistances. The quantity of capacitors is divided by the value of one to determine the total capacitive value.

$$C_T = \frac{C}{n} \tag{22-6}$$

where C is the value of one capacitor and n is equal to the number of equal-value capacitors.

EXAMPLE 22-9

Find the total capacitance for three 12-μF capacitors connected in series.

Solution

$$C_T = \frac{C}{n}$$

$$= \frac{12}{3} = 4 \ \mu\text{F}$$

When two capacitors are series connected, we can use a variation of the reciprocal formula. This is stated as

$$C_T = \frac{C_1 \times C_2}{C_1 + C_2}$$

22-6 Voltage Division

Voltage division in capacitive circuits is similar to voltage division in resistive circuits. The parallel bank of capacitors has equal voltage drops across each of its components. This is stated as

$$V_T = V_{C1} = V_{C2} = V_{C3} = \cdots = V_{Cn} \tag{22-7}$$

Voltage drops in a series capacitive circuit are proportional to the reciprocal of the capacitor's value. They are inversely proportional to the individual capacitive values in the circuit. In this type of circuit the smallest capacitor has the largest voltage charge. This fact is explained when we realize that the charge movement in a series capacitive circuit is equal because there is only a path for charge movement. The value of voltage across any capacitor in a series

capacitive circuit is calculated in this manner:

$$V_C = \frac{C_T}{C_X} V_T \qquad (22\text{-}8)$$

where C_T = total circuit capacitance

$\qquad C_X$ = specific capacitive value

$\qquad V_T$ = source voltage

EXAMPLE 22-10

Find individual voltage drop for the circuit in Figure 22-17.

Solution

$$V_{C1} = \frac{C_T}{C_1} V_T$$

$$\frac{1}{C_T} = \frac{1}{C_1} + \frac{1}{C_2}$$

$$= \frac{1}{2\ \mu F} + \frac{1}{8\ \mu F}$$

$$= 0.5\ \mu F + 0.125\ \mu F = 0.625\ \mu F$$

$$C_T = 1.6\ \mu F$$

$$V_{C1} = \frac{1.6\ \mu F}{2\ \mu F} \times 60\ V \qquad V_{C2} = \frac{1.6\ \mu F}{8\ \mu F} \times 60\ V$$

$$= 48\ V \qquad\qquad\qquad = 12\ V$$

FIGURE 22-17

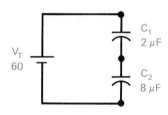

Example 22-10 illustrates the concept of the smallest capacitive value receiving the larger voltage charge in the circuit. The rules developed by Kirchhoff related to voltage drops in a closed loop also apply to this type of circuit. The sum of the individual drops in this type of closed-loop circuit also equal zero. An alternative statement is: The sum of the individual voltage drops are equal to the value of the applied source voltage.

Capacitive effects occur when any two pieces of metallized material are separated by a dielectric material. This could produce some unwanted values of capacitance in an electrical circuit. The quantity of capacitance in a circuit may affect its operation. This is more notable in high-frequency electrical circuits. The effect of any undesirable capacitor value is minimized by proper placement of the wire leads of the capacitor. This is called *lead dress*. Placement of the wire leads of the capacitor to minimize the effect of any undesirable capacitor values may be critical to the correct operation of high-frequency radio-frequency circuits.

A second way of minimizing this effect is aided by the manufacturers of capacitors. Tubular types of capacitors have one of their two plates close to the outside of the unit. This plate is identified by the presence of a printed black ring around one end of the body of the capacitor. (If the body is black, another color is used for this marking.) The outside plate, or foil, is normally connected to circuit common. This will minimize any stray capacitance effects. Ceramic and mica capacitors do not use a wound foil for their plates. They do not have any markings of this type.

22-7 Testing Capacitors

Testing is accomplished by one of two popular methods. These methods include the use of an ohmmeter and or capacitance bridge type of tester. In general, there are three conditions for capacitor failure: shorted, open, and leaky. The shorted unit has a breakdown of its dielectric material. The failure could be caused by higher-than-normal voltage applied to its plates. This punctures the dielectric and creates an undesirable current path between two plates. As a result the capacitor will not hold a voltage charge. Also, if the capacitor is in the circuit in parallel with the source voltage, a low-resistance, high-current short will usually produce a power supply failure.

When the capacitor is open it will not accept a voltage charge. It fails to act in any manner. The result is equipment failure. The symptoms are the same as if the capacitor were removed from the circuit.

A leaky capacitor will not accept a full voltage charge. The charges normally expected to build up on its plates are quickly neutralized due to this leakage condition. A leaky capacitor will provide a lower-than-normal charge to its circuit. This is also indicated by a circuit failure. Less-than-normal operation occurs. There is some circuit operation when a capacitor is leaky. *Ohmmeter tests* are conducted to provide a rapid determination of the capacitor. When using the ohmmeter, keep in mind that this device has its own internal voltage source. This source will place a charge on the capacitor plates. Also be aware that any voltage charge on the plates of the capacitor must be neutralized to make this type of test.

Once the charges on the plates of the capacitor are neutralized, the test results are reasonably valid. Ohmmeter tests of capacitors are more accurate as the capacitance value of the unit increases. Units with picofarad values are not often accurately tested with the ohmmeter. The only accurate test for small capacitors is to test for short circuits. The ohmmeter tests and their evaluation are as follows:

TROUBLE SHOOTING APPLICATIONS

1. *Short tests.* A short-circuited capacitor will provide a low-resistance reading with the ohmmeter. When a steady reading of zero ohms or close to zero ohms is obtained, the capacitor is shorted and needs to be replaced. The term *steady reading* means that the meter indicates a constant value and does not change after the meter's leads are left connected to the capacitor for several seconds. An instantaneous zero ohms reading followed by a steady increase in resistance value does not indicate a short-circuited capacitor.

2. *Leakage test.* A capacitor having a poor dielectric condition is said to be *leaky*. This capacitor will indicate a rise in resistance reading as the meter's source voltage charges the capacitor's plates. The quantity of charge for a leaky capacitor is less than the charge capability of a good capacitor.

3. *Open test.* When the ohmmeter indicates no charging action for the capacitor, it may be open. The reading on the meter should be a high resistance for this condition. Keep in mind the fact that small values of capacitance may not show any low-resistance reading at all. It may be necessary temporarily to bridge the capacitor being tested with a known-good unit to determine if the original unit is defective. This test can be accomplished without removing the original capacitor. If the original unit is open, the test capacitor will restore the function of the circuit.

Another device used to test capacitors is a capacitor bridge (Figure 22-18). It uses a Wheatstone bridge circuit as a basic measuring unit. The leads of the tester are connected to the capacitor to be tested. To make a valid test, it is best to remove one of the capacitor's leads from the circuit. The control

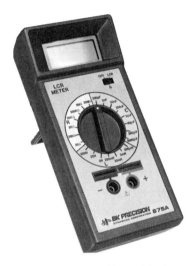

FIGURE 22-18 Portable type of capacitor testing instrument. (Courtesy of B & K-Precision/Dynascan Corporation.)

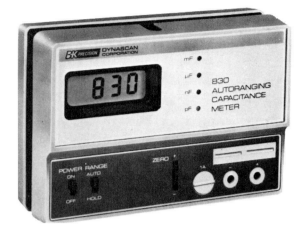

FIGURE 22-19 Autoranging digital type of capacitor tester. (Courtesy of B & K-Precision/Dynascan Corporation.)

knob of the tester is rotated until a reading is obtained. The measured value of the capacitor is compared to its marked value. When these two readings are reasonably close to each other, the capacitor is said to be working properly.

A second type of capacitor tester is shown in Figure 22-19. This is an electronic unit and has a digital readout of the value of capacitance. Some of these are more sophisticated than others. They have automatic ranging and will indicate any leakage while making the measurement. Some models are even able to make accurate measurements of values while the capacitor is still in the circuit.

The technological advances associated with automatic component insertion on circuit boards and automatic wave soldering of the components onto the board often generate some problems. One of these is a failure of the solder connection between the component and the foil on the circuit board. The capacitor may test bad, although in reality the only problem is the failure of the electrical connection between the board and the component. Reheating the connection and applying a small amount of fresh solder often corrects this condition. It is wise to heat the connections to a capacitor shown to be open before removing it from the circuit and replacing it. Test the circuit after doing this, before replacing the capacitor.

TROUBLESHOOTING APPLICATIONS

Capacitor Replacement Replacement is very easy if a stock of exact replacement components is available. If not, the next solution is to order a replacement part from a local source of electrical or electronic parts. Replacement capacitors do not always have to be of the exact value as the original

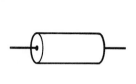

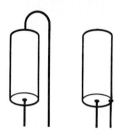

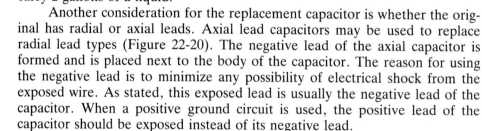

FIGURE 22-20 One method of using a radial capacitor to replace an axial capacitor.

capacitor. Capacitors, as we have stated, carry two markings. One of these is a voltage rating. This rating is the *maximum* permissible voltage. It is possible and permissible to replace a defective capacitor with one having a higher voltage rating. The only limitation is the size of the replacement unit. One must be able to mount the replacement unit in the device.

The reason for this lack of critical value when the replacement's voltage rating exceeds the original capacitor's rating is really simple. The capacitor will charge only to the level of voltage in the circuit where it is being used. Voltage ratings indicate the maximum single value for the capacitor. Increasing the voltage rating for the replacement unit may provide a larger margin for safe operation. It is analogous to using a bucket capable of holding 5 gallons to carry 2 gallons of a liquid.

Another consideration for the replacement capacitor is whether the original has radial or axial leads. Axial lead capacitors may be used to replace radial lead types (Figure 22-20). The negative lead of the axial capacitor is formed and is placed next to the body of the capacitor. The reason for using the negative lead is to minimize any possibility of electrical shock from the exposed wire. As stated, this exposed lead is usually the negative lead of the capacitor. When a positive ground circuit is used, the positive lead of the capacitor should be exposed instead of its negative lead.

The second major consideration when replacing a capacitor is its capacitance value. Like resistors, capacitors have a tolerance rating. This rating is usually low for picofarad values. When looking for an electrolytic type of capacitor, the tolerance rating is much higher. Some of the larger electrolytic capacitors have tolerance limits as high as $+50$ to -20% of the marked value. This permits a great amount of lattitude in the selection of a replacement unit. After all, a 400-μF capacitor having a $+50$ to 20% tolerance can actually be between 600 and 320 μF and still be the proper size of replacement capacitor.

REVIEW PROBLEMS

22-31. An ohmmeter test of a capacitor shows 1 Ω. What is the problem?

22-32. What ohmmeter reading is typical for an open capacitor?

22-33. What is the acceptable capacitance value for a capacitor rated at 200 μF $+20$ to -40%?

1. A capacitor consists of two metallized surfaces, or plates, separated by a dielectric material.

2. Capacitors are rated in units of farads (F). Usually, this rating is in units of microfarads or picofarads.

3. The purpose of a capacitor is to store an electrical charge.

4. Capacitance values are determined by the type of dielectric material, plate surface area, and plate spacing.

5. Current flow in a capacitor occurs only during its charging or discharging period.

6. Dielectric strength of a material is the rating of its voltage breakdown value.

7. Capacitors are produced as either fixed-value or variable-value units.

8. Capacitors are rated as being either electrolytic or nonelectrolytic types. Capacitors higher than 1.0 μF are usually electrolytic types.

9. Electrolytic capacitors have a polarity rating.

10. Installation of electrolytic capacitors in reverse polarity can produce an explosion and failure of the capacitor.

11. Nonelectrolytic capacitors use mica, air, paper, and ceramic as their dielectric material.

12. Capacitor applications include bypass, filter, and coupling and timing circuits.

13. Parallel capacitors act to increase the total value of capacitance (opposite to that of parallel resistances).

14. Series capacitors act to reduce the total value of capacitance (opposite to that of series resistances).

15. Voltage division across series capacitors is inversely proportional to the values of the capacitors.

16. Capacitors may be tested with an ohmmeter or a dedicated capacitor tester.

17. Capacitor failures are either open, shorts, or leakage problems.

18. When replacing a capacitor, both the voltage and capacitance ratings must be considered.

19. Except for timing circuits the value of a replacement capacitor may vary from the original value by the amount of the tolerance.

Answer true or false; change all false statements into true statements.

22–1. Moving capacitor plates closer to each other decreases the amount of capacitance.

22–2. Reducing surface area of one plate of a capacitor will decrease its capacitance.

22–3. Electrolytic capacitors are usually 1.0 μF or higher.

22-4. The voltage rating of a capacitor indicates the strength of its dielectric material.

22-5. Total capacitance for series-connected capacitors is the sum of each unit's value.

22-6. Total capacitance for capacitors in parallel is the sum of each unit's value.

22-7. It is possible to use a 400-V-rated capacitor in a 30-V circuit.

22-8. A shorted capacitor indicates 0 Ω on an ohmmeter.

22-9. Axial lead capacitors may be used to replace radial lead units.

22-10. The positive level of an electrolytic capacitor must be connected to the positive side of the circuit.

MULTIPLE CHOICE

22-11. The material used between the plates of a capacitor is called its

(a) insulation material
(b) dielectric material
(c) separation material
(d) plate-divider material

22-12. One method of increasing the value of capacitance in a unit is to

(a) decrease its plate area
(b) increase its plate area
(c) increase spacing between plates
(d) increase the applied voltage

22-13. Capacitors have

(a) only a voltage rating
(b) only a capacitance rating
(c) only a tolerance rating
(d) all of the above

22-14. Capacitors are used

(a) to bypass, couple, and as a part of a resonant circuit
(b) only to store electrical charges
(c) only as a secondary voltage source
(d) none of the above

22-15. In a capacitive circuit, current flow is limited to

(a) charge periods
(b) discharge periods
(c) both of the above
(d) neither of the above

22-16. One microfared is equal to _____ farads.
(a) 0.001
(b) 0.000,001
(c) 0.000,000,001
(d) 0.000,000,000,001

22-17. The term used to indicate 0.000,000,000,001 F is

(a) 1 μF
(b) 0.001 μF
(c) 1 pF
(d) 0.1 pF

22-18. The material having the highest dielectric strength is

(a) air
(b) paper
(c) mica
(d) oil

22-19. The material having the highest dielectric constant value is

(a) oil
(b) air
(c) ceramic
(d) mica

22-20. A capacitor having small size and a high value of capacitance is usually considered as a(n) _____ type.

(a) electrolytic
(b) ceramic
(c) oiled paper
(d) air

ESSAY QUESTIONS

22-21. Explain the difference between radial and axial types of capacitor construction.

22-22. What effect does the voltage rating have on a capacitor?

22-23. Why are multiple-section capacitors used instead of single-section units?

22-24. Why is a capacitor considered to be a passive device?

22-25. Why is it necessary to consider the tolerance when selecting a replacement capacitor?

22-26. When should only exact values of capacitance be used in replacement action?

22-27. What is the purpose of the notation of the outside foil connection of the capacitor's leads?

22-28. Why is it necessary to maintain the proper polarity of a replacement electrolytic type of capacitor?

22-29. Explain each of the types of markings on a capacitor body.

22-30. Why do series-connected capacitors result in a total value that is less than the lowest single value of capacitance used?

22–31. A 0.005-μF capacitor is charged to 24 V. What is its charge in coulombs? (*Sec. 22-2*)

22–32. What is the capacitance of a unit that stores 140×10^{-6} C of charge when its voltage is 18 V? (*Sec. 22-2*)

22–33. Convert 240 pF to units of microfarads. (*Sec. 22-3*)

22–34. convert 15×10^{-12} farads to picofarads. (*Sec. 22-3*)

22–35. What is the value range for a capacitor rated at 8 μF and a tolerance of $+20$ to -30%? (*Sec. 22-3*)

22–36. One of the plates of a capacitor has a charge of 840 μC. What is the charge on the other plate? (*Sec. 22-2*)

22–37. What is the voltage rating for two 10-μF capacitors connected to a 40-V source? (*Sec. 22-4*)

22–38. What is the total capacitance when three 6-μF capacitors are series connected? (*Sec. 22-5*)

22–39. What is the voltage across the center capacitor in problem 22-28 when a source of 360 V is connected to the string? (*Sec. 22-6*)

22–40. What is total capacitance when four 0.005-μF capacitors are parallel connected? (*Sec. 22-5*)

Use Figure 22-21 to answer problems 22-41 through 22-43.

22–41. What is total capacitance for the circuit? (*Sec. 22-5*)

22–42. What is the voltage on C_2? (*Sec. 22-5*)

22–43. What is total capacitance when C_2 is shorted? (*Sec. 22-7*)

Use Figure 22-22 to answer problems 22-44 through 22-48.

22–44. What is total capacitance? (*Sec. 22-5*)

22–45. What is the voltage across C_3? (*Sec. 22-5*)

22–46. What is total capacitance when C_2 is open? (*Sec. 22-5*)

22–47. What is total capacitance when C_1 is open? (*Sec. 22-5*)

22–48. What is the voltage across C_1 when C_2 is shorted? (*Sec. 22-5*)

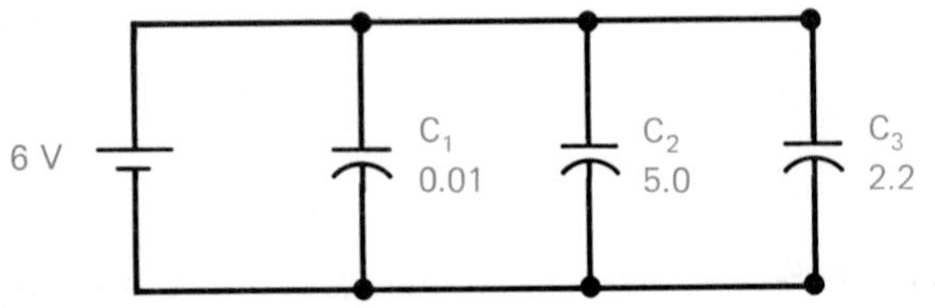

FIGURE 22-22

FIGURE 22-21

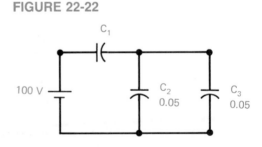

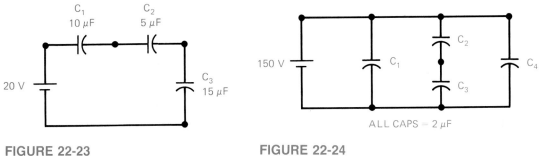

FIGURE 22-23   FIGURE 22-24

Use Figure 22-23 to answer problems 22-49 and 22-50.

22–49. What is total capacitance? (*Sec. 22-5*)

22–50. What is the voltage drop across C_1? (*Sec. 22-6*)

22–51. What is total capacitance when capacitors having values of 0.012 μF, 2500 pF, 0.0082 μF, and 1800 pF are parallel wired? (*Sec. 22-5*)

22–52. What is total capacitance when the capacitors in problem 22-41 are series wired? (*Sec. 22-5*)

Use Figure 22-24 to answer problems 22-53 through 22-55.

22–53. What is voltage across each capacitor? (*Sec. 22-6*)

22–54. What is the voltage across C_2 when the source voltage is doubled? (*Sec. 22-6*)

22–55. What is the voltage across each component when all values are changed to 10 μF? (*Sec. 22-6*)

Use Figure 22-25 to answer problems 22-56 through 22-60.

22–56. What is total capacitance? (*Sec. 22-5*)

22–57. What is the voltage drop across C_2? (*Sec. 22-5*)

22–58. What is total capacitance when C_2 opens? (*Sec. 22-5*)

22–59. What is total capacitance when C_2 is shorted? (*Sec. 22-5*)

22–60. What is the voltage charge across C_2 when C_1 is open? (*Sec. 22-5*)

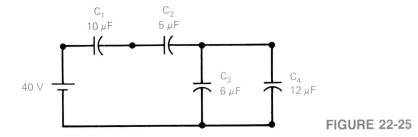

FIGURE 22-25

22–61. The resistance measured across the leads of a capacitor is 0 Ω. What is the problem?

22–62. Resistance measured across the leads of a capacitor is infinite. Is there a problem?

22–63. Two capacitors are series connected across a 100-V dc source. The actual voltage measured across C_1 is 100 V and across C_2 is 0 V. What is the problem?

ANSWERS TO REVIEW PROBLEMS

22–1. Two metalic surfaces and a dielectric

22–2. A capacitor is a device capable of storing an electrical charge.

22–3. In the dielectric between its plates.

22–4. It increases until it is equal to the applied voltage.

22–5. Short circuiting its plates.

22–6. O V.

22–7. Connect it to a power source.

22–8. To the plates and only during either charge or discharge periods.

22–9. The farad.

22–10. One millionth, or 1×10^{-6}.

22–11. One millionth of a millionth, or 1×10^{-12}

22–12. 15,000,000 pF

22–13. 0.00000000036 F.

22–14. Decreases

22–15. Decreases

22–16. Decreases

22–17. Electrolytic and nonelectrolytic

22–18. Axial leads are attached on the axis of the body; radial leads are both in the same end of the capacitor.

22–19. Aluminum oxide and tantalum oxide

22–20. Paper, mica, and ceramic

22–21. It may explode.

22–22. The maximum voltage rating for the dielectric

22–23. Use the 0.8-, 0.4-, and 0.4-μF units.

22–24. 6.50 μF

22–25. 2.0 μF

22–26. 200 V each

22–27. 0.2449 μF

22–28. 2.0 μF

22–29. 48 V

22–30. 24 V

22–31. It is short circuited.

22–32. Infinity

22–33. Between 240 and 120 μF

CHAPTER 23

Capacitive Reactance

The effect of opposition of changes in voltage in the ac circuit is known as capacitive reactance. This effect is similar to that of inductive reactance except that its opposition is to voltage changes in the circuit. Rules applicable to inductive circuits also apply to those circuits having capacitive reactance.

KEY TERMS

Capacitive Reactance X_C
Charge and Discharge
Current

OBJECTIVES

Upon completion of the material in this chapter, you will

1. Understand the effect of ac in a capacitive circuit.
2. Be able to analyze series and parallel capacitive reactance circuits.
3. Understand RC time-constant concepts.
4. Understand the concept of signal coupling.

A voltage applied to a capacitor will charge its plates. When the voltage applied to these plates is an ac voltage, the charge will vary as does the amplitude and polarity of the applied voltage. The charge on the plates of the capacitor may be used to produce a current flow. Current will flow through a load connected to the plates of the capacitor. The quantity of the charges on the plates determines the amplitude and time of current flow through the load.

An electric current will not flow through the dielectric of a capacitor. The charges on the plates will interact and produce the effect of current movement across the dielectric. In Figure 23-1, a capacitor, a lamp, and a 120-V source are connected as a closed-loop series circuit. Current cannot flow through the dielectric of the capacitor. The varying ac voltage will charge one plate with electrons during one half of the ac cycle. This forces electrons away from the other plate of the capacitor. The electrons are attracted toward the positive, or upper, connection of the ac source [Figure 23-1(a)]. When the electrons are in motion through the lamp, the filament will glow and light is produced.

During the second half of the ac cycle the polarity of the power source is reversed [Figure 23-1(b)]. Electron flow through the circuit is now reversed. Electrons formerly attracted to the upper terminal of the power source are now repelled by the electrons at this terminal. They move through the lamp filament to the upper plate of the capacitor. Electrons on the lower plate of the capacitor are now attracted to the lower connection of the power source. As before, the movement of electrons through the filament of the lamp will produce light.

The brightness of the lamp depends on the quantity of electrons the capacitor is able to store on its plates. Larger values of capacitance produce a brighter glow to the lamp. Conversely, a smaller value of capacitance produces a dim light.

Capacitive reactance
Opposition of current flow in an ac circuit containing capacitance.

This reactance to the ac current is one important factor of capacitor action. Reactance is the opposition to an ac current. When reactance is a factor discussed with capacitor action, the term **capacitive reactance** is used to describe it. The symbol used for reactance is the letter X. Values of capacitive reactance

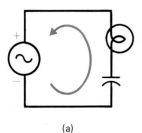

(a)

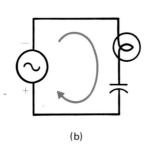

(b)

FIGURE 23-1 The lamp will be lit as long as current flow charges and discharges the plates of the capacitor in the circuit.

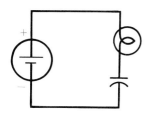

FIGURE 23-2 A direct current will not permit any current flow, the plates of the capacitor do not change their charge level, and the lamp will not glow.

are identified using a combination of X and C (for capacitance). The symbol set X_C is used to identify capacitive reactance values. X_C values are described in units of the ohm.

X_C *Term used to describe capacitive reactance values in circuit.*

The area of the plates of any capacitor is one factor used to determine its amount of charge. Larger plate areas are capable of storing greater charges than are small plate areas. The larger plates will therefore produce a larger current flow through the load as the capacitor discharges into the load. Its reactance is less than that in a smaller capacitor. This is the reason for the brighter lamp when the larger capacitor is used in the circuit.

When a dc source is used (Figure 23-2), the results are different than when an ac source is connected to the capacitor. The lamp will glow for a short period and then be extinguished. The length of glow is determined by the time required to charge the plates to the value of applied voltage. In this circuit the applied dc voltage will charge the capacitor's plates to the value of the applied voltage. Once charged, this voltage on the plates opposes any further current flow through the circuit. All of the applied voltage is across the plates of the capacitor and none develops across the terminals of the lamp. The lamp filament will not glow under these conditions.

The frequency of operation of the voltage has a strong effect on the amount of capacitive reactance in the circuit. A dc voltage has zero frequency and therefore a maximum effect on the amount of X_C in the circuit. As the frequency of operation increases, the reactance is reduced and more current is permitted to flow.

This condition appears to *block* a dc current through the capacitor. The capacitor acts as though it permits an ac current to flow through its plates, but does not allow a dc current to pass across its terminals. These factors can be summarized in this manner:

1. An alternating current will flow in a capacitive circuit when an ac voltage is applied to the circuit.
2. Smaller capacitors will permit less current flow in the circuit. This is a result of a greater amount of X_C, and additional quantities of ohms of reactance opposition.
3. The rate of change of the charges on the plates of the capacitors is a factor that affects current flow and varies X_C.
4. When one has a zero-frequency current, as with dc, the capacitor has infinite resistance, and therefore there is no current flow. The capacitor acts as an open circuit under these conditions.

The effects of capacitive reaction have a great variety of applications in electrical and electronic circuits. When we consider that X_C depends on the

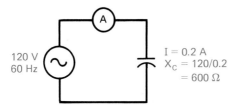

FIGURE 23-3 Current flow in this circuit creates 600 Ω of capacitive reactance.

frequency of operation, it becomes reasonably easy to understand that a capacitor will act to block a dc voltage and act as though it passes an ac voltage.

The alternating charging and discharging of a capacitor's plates makes it seem as though the ac current is flowing in the circuit. The circuit shown in Figure 23-3 will illustrate this point. This circuit consists of an ac source, a capacitor, and an ac ammeter. This ammeter will measure and display the amount of ac current flowing in this circuit. The ac current is actually the quantity of charge and discharge of the plates of the capacitor. Current flow in this circuit is equal throughout the circuit except that there is no current flow through the dielectric of the capacitor.

23-1 Values of X_C

The ratio of V_C/I_C is the quality of opposition to the ac sine-wave current in the circuit. This value is expressed in units of the ohm. Using the circuit shown in Figure 23-3, the values of applied voltage and current flow are 120 and 0.2, respectively. The reactance, therefore, is equal to 120/0.2, or 600 Ω. This is the value of X_C for this circuit. This is the value of current that can be produced by a 120-V sine-wave voltage applied to the plates of this capacitor. When discussing current,

$$X_C = V_C/I_C \qquad (23\text{-}1)$$

When describing frequency and capacitance, the formula for capacitive reactance is

$$X_C = 1/(2\pi f C) \qquad (23\text{-}2)$$

The quantity of X_C is dependent on the amount of capacitance and the applied frequency of the ac voltage to the circuit. When the C in Figure 23-3 is increased, the capacitor will be able to take on a large charge. It will then produce a larger charge current and a larger discharge current. The amount of X_C is smaller as the capacitance is increased.

When the frequency of the ac voltage is increased the rate of charge and discharge of the plates of the capacitor is faster. This will produce a higher rate of current in the circuit. The action, when related to X_C, indicates that the relation of V_C/I_C is less. The result would be a higher current for the same amount of applied voltage. We can state that X_C is less for higher frequencies of operation. In a practical situation, reactance values can be almost any value from near infinity to almost 0 Ω.

REVIEW PROBLEMS

23-1. Rank these capacitor values in the order of reactance, from largest to smallest: 10 μF, 3 μF, 0.5 μF.

23-2. Rank the capacitors in problem 23-1 for their charge current, from largest to smallest.

23-2 Reactance Formula

When attempting to determine the values of X_C we must consider both frequency of operation and the capacitance value. The result of these calculations will produce a value of capacitive reactance, in units of the ohm. The formula for X_C is $X_C = 1/(2\pi f C)$. The frequency values for this formula are in units of hertz and capacitance values are in units of farads. An example of the use of this formula is the determination of X_C when we have a capacitor of 15 μF and an operating frequency of 60 Hz.

$$X_C = \frac{1}{2\pi f C}$$

$$= \frac{1}{2\pi \times 60 \text{ Hz} \times 15 \times 10^{-6} \text{ F}}$$

$$= \frac{1}{0.005655}$$

$$= 176.84 \ \Omega$$

When working with this formula, there are some shortcuts and some items that must be remembered. First, the value of 2π is approximately equal to 6.28. This value represents the circular motion of the generator creating the sine wave. It is possible to simplify the calculations by using a derivation of the formula, 1/6.28. This is equal to approximately 0.159 and thus

$$X_C = \frac{0.159}{f C}$$

Also keep in mind that the value of capacitance, C, is in units of farads in this formula. When using microfarads or picofarads the values of 10 to the powers of 6 or 12 are negative. The remaining factor is that the frequency is in units of hertz.

EXAMPLE 23-1

Find the value of X_C when the capacitor is 2.5 μF and the frequency is 800 Hz.

Solution

$$X_C = \frac{0.159}{fC}$$

$$= \frac{0.159}{800 \text{ H} \times 2.5 \times 10^{-6} \text{ F}}$$

$$= \frac{0.159}{0.002}$$

$$= 79.5 \; \Omega$$

EXAMPLE 23-2

Find the capacitive reactance when a 400-pF capacitor is in a circuit operating at 600 kHz.

Solution

$$X_C = \frac{0.159}{6 \times 10^5 \times 4 \times 10^{-10}}$$

$$= \frac{0.1569}{24 \times 10^{-5}}$$

$$= 662.5 \; \Omega$$

Capacitance and Reactance X_C is inversely proportional to the capacitance in the circuit. This means that the value of X_C increases as the quantity of capacitance in the circuit is reduced.

Frequency and Reactance X_C is inversely proportional to the frequency of the circuit. This statement means that the value of X_C increases as the frequency of operation is reduced in the circuit. It also means that the value of reactance is decreased as the frequency of operation is increased.

The value of capacitance, C, may be found by a modification of the original formula for X_C. Inverting the reactance formula will accomplish this.

$$C = \frac{0.159}{fX_C}$$

EXAMPLE 23-3

Find the value of capacitance when a 250-Ω X_C is required and the operating frequency is 500 kHz.

Solution

$$C = \frac{0.159}{fX_C}$$

$$= \frac{0.159}{0.5 \times 10^{-6} \text{ H} \times 250 \text{ }\Omega}$$

$$= \frac{0.159}{125 \times 10^{-6}}$$

$$= 1272 \text{ pF}$$

EXAMPLE 23-4

Find the frequency of operation for a capacitor of 1.2 μF and an X_C of 1200 Ω.

Solution

$$f = \frac{0.159}{CX_C}$$

$$= \frac{0.159}{1.2 \times 10^{-6} \text{ F} \times 1200 \text{ }\Omega}$$

$$= \frac{0.159}{0.00144}$$

$$= 110.4 \text{ Hz}$$

REVIEW PROBLEMS

23-3. Find the value for a capacitor when the X_C is 800 Ω and the operating frequency is 6 MHz.
23-4. Find the value of capacitance for an X_C of 400 Ω and a frequency of 12 MHz.

23-3 Reactances in Series and Parallel

Capacitive reactance is measured in units of the ohm. It is an opposition to current flow and functions in a manner similar to resistance in the circuit. Because of this, the values of capacitive reactances are in the same manner as resistances in a circuit. In a reactive circuit, three reactive values will add to produce a total ohmic value. Reactance values of 100, 300, and 800 Ω will equal a total value of 1200 Ω of capacitive reactance. The formula for series capacitive reactances is

$$X_{CT} = X_{C1} + X_{C2} + X_{C3}, + \cdots + X_{Cn} \qquad (23\text{-}3)$$

When capacitive reactances are combined into a parallel circuit configuration the total reactance is determined using a formula similar to that used for parallel resistances.

$$\frac{1}{X_{CT}} = \frac{1}{X_{C1}} + \frac{1}{X_{C2}} + \cdots + \frac{1}{X_{Cn}} \qquad (23\text{-}4)$$

Determination of X_C values for equal-value parallel reactances can be found by using the same formula as used for equal-value resistances:

$$X_{CT} = \frac{X_C \text{ (for one unit)}}{n} \qquad (23\text{-}5)$$

where n is the number of equal units. When there are unequal values, the formula is

$$X_{CT} = X_{C1}\frac{X_{C2}}{X_{C1} + X_{C2}} \qquad (23\text{-}6)$$

REVIEW PROBLEMS

23-5. Find the total X_C for two series-connected reactances having values of 600 Ω and 200 Ω.
23-6. Find the total X_C for three 600-Ω reactances parallel connected in a circuit.
23-7. Find the total reactance for two parallel-connected reactances having values of 120 Ω and 300 Ω.

23-4 Capacitive Reactance and Ohm's Law

The formula for the interaction of voltage, current, and resistance in a reactive circuit is similar to that used in a dc circuit. Ohm's law may be used for circuits containing X_C components. In a capacitive reactance circuit the formulas for this are

$$V = IX_C \qquad (23\text{-}7)$$

$$I = \frac{V}{X_C} \qquad (23\text{-}8)$$

$$X_C = \frac{V}{I} \qquad (23\text{-}9)$$

These are illustrated in the examples that follow. Figure 23-4 shows this rela-

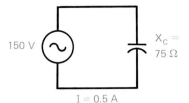

I = 0.5 A   **FIGURE 23-4**

tionship. The value of X_C is found by

$$X_C = \frac{V}{I}$$

$$= \frac{150 \text{ V}}{0.5 \text{ A}}$$

$$= 300 \ \Omega$$

The value of current, I, is determined by

$$I = \frac{V}{X_C}$$

$$= \frac{150 \text{ V}}{300 \ \Omega}$$

$$= 0.5 \text{ A}$$

and the value of voltage, V is:

$$V = IX_C$$

$$= 0.5 \text{ A} \times 300 \ \Omega$$

$$= 150 \text{ V}$$

When two reactances are connected in parallel (Figure 23-5), it is still possible to calculate individual circuit values. Current through each of the capacitors is determined:

$$I_{C1} = \frac{V}{X_{C1}}$$

$$= \frac{150 \text{ V}}{150 \ \Omega}$$

$$= 1.0 \text{ A}$$

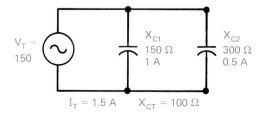

FIGURE 23-5 Determination of X_c for parallel capacitive reactances.

and

$$I_{C2} = \frac{V}{X_{C2}}$$

$$= \frac{150\text{ V}}{300\text{ }\Omega}$$

$$= 0.5\text{ A}$$

Therefore,

$$I_T = I_1 + I_2$$

$$= 1.5\text{ A}$$

The total reactance in this circuit is

$$X_{CT} = \frac{V_T}{I_T}$$

$$= \frac{150\text{ V}}{1.5\text{ A}}$$

$$= 100\text{ }\Omega$$

Current flow in the parallel reactive circuit is similar to current flow in a parallel resistive circuit. The greatest quality of current will always take the path of least reactance. The current-divider rule presented by Kirchhoff may be used in this circuit to quickly determine current through each of the branch circuits. In this example, the value of X_{C1} is one-half of the value of X_{C2}. Since current in Ohm's law is inversely proportional to resistance, it is also inversely proportional to the reactance in the circuit. X_{C1}, being the smaller of the two reactances, will have the greatest amount of current flow. These two reactances have a ratio of 1:2. There are three parts, or components, when the 150-Ω value is divided into the 300-Ω value. Since current is inversely proportional to the ohmic value, X_{C1} will have twice the current flow of X_{C2}. This evaluation provides a rapid method of approximating an answer and aids in the determination if the actual mathematical analysis is correct.

When two reactances are series connected (Figure 23-6), the rules for series resistances apply for the solution of circuit values:

$$X_{CT} = X_{C1} + X_{C2} \tag{23-10}$$

$$= 150\text{ }\Omega + 300\text{ }\Omega$$

$$= 450\text{ }\Omega$$

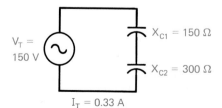

FIGURE 23-6 Determination of X_c for series capacitive reactances.

and

$$I_T = \frac{V_T}{X_{CT}}$$

$$= \frac{150 \text{ V}}{450 \text{ }\Omega}$$

$$= 0.33 \text{ A}$$

In the series circuit current is equal throughout the circuit. Therefore, current through each of the reactances is equal to 0.33 A. It is now possible to determine the voltage drop across each of the reactances:

$$V_{XC1} = IX_{C1}$$

$$= 0.33 \text{ A} \times 150 \text{ }\Omega$$

$$= 50 \text{ V (rounded)}$$

and

$$V_{XC2} = IX_{C2}$$

$$= 0.33 \text{ A} \times 300 \text{ }\Omega$$

$$= 100 \text{ V (rounded)}$$

Kirchhoff's law for voltages in series may be applied in this circuit. The sum of the voltage drops developed across each reactance is equal to the applied voltage. Another way of looking at this circuit is to use an approximation method to determine the voltage drops. The value of X_{C2} is twice as large as X_{C1}. It will therefore develop twice the voltage drop of $_{C1}$. The value of reactance is divisible by 3 in this example. Therefore, X_{C1} is one-third of the total and X_{C2} is two-thirds of the total reactance. In a series resistive circuit the voltage drops are proportional to the values of the resistances. This is also valid for series reactances. One-third of the applied voltage may be measured across X_{C1} and two-thirds of the applied voltage measured across X_{C2}.

23-5 Capacitive Reaction Applications

One of the most common uses for X_C is in systems where it is necessary to block the dc component but permit the passage of the ac component of the voltage. Earlier, it was stated that the electronic signal is actually a variation of the operating voltage in any electronic device. In electronic devices the signal must travel through the system from its input to the output. In addition, a dc voltage is needed to operate the electronic components, such as transistors, integrated circuits, and vacuum tubes.

The basic electronic circuit illustrated in Figure 23-7 may help in the comprehension of the use of X_C in order to pass a signal from one section of a unit to the next section. Each of the triangular symbols represents an electronic amplifier in this figure. A capacitor is used to connect the output of amplifier 1 to the input of amplifier 1. In a system such as this, if the dc operating voltage at the output of the first amplifier were connected directly to the input of

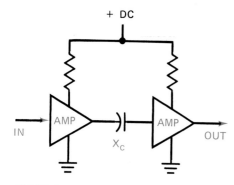

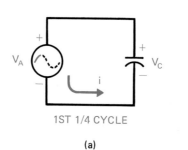

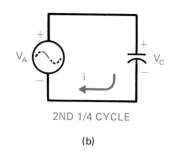

1ST 1/4 CYCLE

2ND 1/4 CYCLE

(a)

(b)

FIGURE 23-7 Use of capacitive reactance to transfer, or couple, signals from one amplifier stage to another amplifier stage.

FIGURE 23-8 Current flow during the first half cycle of the ac wave.

amplifier 2, it would be too high a value for operation of the second amplifier. The design of a system such as this one calls for the transfer of the ac component, or signal, from one stage to the next stage. The variations of operating voltage, or signal, will pass through the X_C component and thus be utilized for proper operation of the second amplifier. This process is called *signal coupling*. It is commonly used in electronic devices such as audio amplifiers, radios, television receivers, video display terminals, and many other consumer-related electronic products. This system is also used in industrial and commercial communications applications.

The value of capacitance is related to the operational frequency of the system (Figure 23-8). Using a value of 150 Ω of X_C, it is possible to determine the value of capacitance required for a variety of operational frequencies:

Frequency	C (approx.)	Typical Application
60 Hz	18 μF	Low audio and power line
1 kHz	1.1 μF	Audio frequencies
10 kHz	0.106 μF	Audio frequencies
1 MHz	0.001 μF	Low-frequency radio (RF)
10 MHz	106 pF	Medium-frequency radio
100 MHz	10.6 pF	High-frequency radio, TV

These data illustrate the reduction in the value of capacitance as the frequency of operating is increased. This is a good time to compare the effects of resistance, X_C, and X_L. In either a dc circuit or an ac circuit the value of resistance, R, remains the same. Both X_C and X_L are frequency dependent. The effects of these two factors are opposite, since X_C will decrease with the frequency of operation while X_L will increase with the frequency of operation in a circuit.

The list above provides some typical values for capacitors in different frequency ranges. These capacitors may be used for signal coupling between stages, signal decoupling, or filter action in electronic circuits. Large values of capacitance are required for low-frequency power supplies and low-frequency audio systems. Low values of capacitance are used for high-frequency operational systems.

REVIEW PROBLEMS

23-8. A 40-μF capacitor has an X_C of 66.3 at 60 Hz. Find its X_C for frequencies of (a) 10 Hz and (b) 120 Hz.

23-9. What size capacitor is required to provide an X_C of 200 Ω at a frequency of 5 kHz?

23-6 Charge and Discharge Current

A sine-wave voltage applied to a capacitor will produce an alternating **charge and discharge current** (Figure 23-8). The two circuits illustrate this action for each of the first two quarters of the total sine wave. Voltage V_C that develops across the plates of the capacitor is constant and the same value as the applied voltage V_A. The reason for this constant value is the fact that the source and the capacitor are connected in parallel with each other. The current flow in the circuit is represented by the lowercase letter i. The specific value of current depends on the charge and discharge of capacitor C. Remember, the sine-wave voltage is constantly changing. It starts at 0 V and rises to its maximum positive value. The voltage then falls back to zero. This action completes the first half-cycle of the ac wave. The action is then repeated for the second half-cycle, only in a negative direction.

Charge and discharge current Current flow in capacitive circuit during charge or discharge of capacitor.

When V_A is increasing, the voltage charge on C is also increasing. The capacitor then builds up, or stores, the charge. Electron current flow during this period is from the source to the negative plate of the capacitor and from the positive plate to the positive terminal of the source. When V_A is decreasing, the voltage charge on C is also decreasing. The capacitor then gives up its charge and attempts to maintain circuit voltage. Electron current flow during this period is reversed. It now flows from the capacitor's negative terminal back to the source. Current values therefore react to these changes. When V_A is not changing, there is no charge or discharge current.

When the polarity of the applied source voltage is reversed (Figure 23-9), electron current flow follows the polarity and voltage amplitude of the source. As the voltage rises during the third quarter of the ac cycle [Figure 23-9(a)], current flows from the negative source terminal to the plate of the ca-

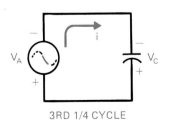

3RD 1/4 CYCLE

(a)

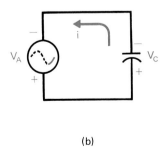

(b)

FIGURE 23-9 Current flow during the last two quarter-cycles of the ac wave.

pacitor. In this illustration the upper plate becomes negative during this half-cycle.

The final quarter-cycle [Figure 23-9(b)] is a repeat of the second quarter-cycle, but with reverse polarity. The source voltage decreases and the capacitor gives up its charge in its attempt to maintain the amplitude of the maximum source voltage. Electron current now flows from the capacitor back to the source.

This illustrates that the sine-wave voltage as applied to the plates of a capacitor will produce a sine-wave current with its alternating charge and discharge cycle. One should note that the electron current flow occurs during the charge and the discharge cycle. It occurs whenever there is a change in voltage applied to the capacitor. During this change both the voltage and the current have the same frequency.

Calculating Capacitive Current The quantity of capacitive current depends on the amplitude of the voltage change in the circuit. Also, the capacitor's size determines the amount of charge current. A capacitor having a high farad rating will produce a larger current than one with a smaller rating. The formula for determining the value of capacitor current is

$$i_C = C \frac{dV}{dt}$$

where i = current flow (A)

C = capacitance (F)

dV = change in voltage

dt = change in time during the same period

For example, suppose that we need to find the capacitive current for a 150-pF capacitor when it charges to 50 V during a period of 10 μs. This can be calculated as follows:

$$i_C = C \frac{dV}{dt}$$

$$= 150 \times 10^{-12} \times \frac{50}{1 \times 10^{-5}}$$

$$= 150 \times 5 \times 10^{-6}$$

$$= 750 \times 10^{-4}$$

$$= 750 \ \mu A$$

This formula for capacitive current is similar to the formula for induced voltage discussed in Chapter 21. In both of the formulas there must be a change in circuit values in order to have the results. This change always occurs when an ac source is utilized. In the inductive circuit the change in voltage occurs when there is a change in current in the circuit. In the capacitive circuit a change in voltage produces the change in circuit current.

The formula used to show the change in capacitive current is used to show

the current for an instant of time during the ac cycle. This will show either the charge or the discharge current across the plates of the capacitor.

EXAMPLE 23-5

Determine the instantaneous value of charge current I_C for a 12-μF capacitor when the voltage difference is 25 V during a period of 10 s.

Solution

$$i_C = C\frac{dV}{dt}$$

$$= 12 \times 10^{-6} \times \frac{25}{10}$$

$$= \frac{0.0003}{10}$$

$$= 30\ \mu A$$

EXAMPLE 23-6

Find the value of *discharge* current for Example 23-5.

Solution

The answer is exactly the same for the value of current flow. This is because the conditions for current flow are identical. Since the current is now flowing in the opposite direction, it is considered to be a negative value. The answer is $-30\ \mu A$.

Phase Relationships of Voltage and Current In an inductive circuit the voltage will lead the current during one cycle. The capacitive circuit has the opposite effect. Current produced by capacitive action will lead any changes due to voltage. Therefore, the current changes lead the voltage changes in the capacitive circuit (Figure 23-10). In a purely capacitive circuit this phase shift is 90°. One memory aid that has proved to be very helpful is the use of the mnemonic phrase. This is a phrase used as an aid in memory. The phrase' ELI the ICEman' is one such phrase used to aid memory for the relationship of voltage and current in circuits containing reactive components. The first portion

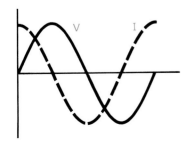

FIGURE 23-10 Phase shift between voltage and current in a capacitive reactive circuit.

of the phrase, ELI, is a reminder that voltage(E) *leads* current (I) in an inductive (L) circuit. In the capacitive (C) circuit the current (I) *leads* the voltage (E). The balance of the phrase is used only as an aid in the memorization of these relationships.

REVIEW PROBLEMS

23-10. What is the phase difference between voltage and current in a capacitive circuit?

23-11. Where during its phase does the voltage have its most positive value?

23-12. Where during its phase does the current have its most positive value?

CHAPTER SUMMARY

1. The voltage charge on the plates of a capacitor may be used to produce a current flow.

2. Capacitor action acts as though an alternating current flows through its dielectric material.

3. Capacitor action effectively blocks a dc current flow.

4. The opposition to current flow in a capacitive circuit is called *capacitive reactance*.

5. Capacitive reactance is measured in units of ohms.

6. Capacitive reactance is described as X_C.

7. Ohm's law is used to determine values of X_C, I, and V in a circuit.

8. The formula for X_C is $1/(2 \pi fC)$ ($\frac{1}{2\pi}$ is equal to 0.159).

9. Capacitive reactance is inversely proportional to the value of capacitance in the circuit.

10. Capacitive reactance in a circuit is determined in the same manner as resistances in either series or parallel circuits.

11. Rules presented by Kirchhoff for voltage drops in a series circuit and current flow at a junction also apply to circuits containing capacitive reactance components.

12. Capacitive reactance principles are used to transfer signals between stages of many electronic devices.

13. An applied voltage will create a current flow in a capacitive circuit.

14. Current leads voltage by 90° in a circuit containing capacitive reactance.

SELF TEST

Answer true or false; change all false statements into true statements.

23–1. Capacitive reactance is measured in units of the ohm.

23–2. Reactance is the opposition to current flow in a dc circuit.

23–3. The phase shift for a circuit containing capacitive reactance is 60°.

23–4. The reciprocal of 2π is equal to 0.159.

23–5. Inductive reactance and capacitive reactance function in the exact way in a circuit.

23–6. Current flow in a capacitive reactance circuit increases with the amount of capacitance.

23–7. Source voltage and capacitor voltage are equal in a parallel capacitive reactance circuit.

23–8. Current flow is constant in a capacitive reactance circuit.

23–9. Capacitive current increases due to increased frequency of the ac source.

23–10. Capacitive current decreases as the quantity of capacitance is increased.

23–11. In the capacitor, electrons are attracted toward the

    (a) negative plate
    (b) positive plate
    (c) negative power source connection
    (d) upper plate

23–12. A discharge capacitor has _____ charges on its plates.
    (a) no
    (b) unequal
    (c) equal
    (d) negative

23–13. The opposition to voltage in a capacitor is called
    (a) capacitor impedance
    (b) capacitor reaction
    (c) impedance reaction
    (d) capacitive reaction

23–14. A voltage having a frequency of 0 Hz offers _____ to a capacitor.
    (a) infinite impedance
    (b) a low impedance
    (c) opposing current activity
    (d) a high current factor

23–15. Capacitance is _____ reactance in a capacitive circuit.
    (a) directly proportional to
    (b) opposite in
    (c) inversely porportional to
    (d) not affected by

23–16. The amount of capacitive reactance in a circuit will _____ as the frequency of operation increases.
    (a) increase
    (b) decrease
    (c) not be affected
    (d) react violently

23–17. Capacitive reactance values connected in series offer
    (a) an increase in total X_C
    (b) a decrease in total X_C
    (c) no opposition of total X_C
    (d) total opposition to voltage in the circuit

23–18. X_C values use the same formula as
    (a) X_L

(b) impedance
(c) series resistances
(d) parallel resistances

23–19. Capacitive current flow in the circuit occurs
(a) continuously
(b) as long as a voltage source is applied
(c) only during charge times
(d) during periods of change in the amplitude of the applied voltage

23–20. Phase relationship of voltage and current in the capacitive reactance circuit is
(a) an in-phase relationship
(b) dependent on the amplitude of applied voltage
(c) dependent on the quantity of current flow
(d) an out-of-phase relationship

ESSAY QUESTIONS

23–21. Why is there a phase shift between voltage and current in the capacitive reactive circuit?

23–22. Explain why a capacitor acts as a sponge in an ac circuit

23–23. Why does a capacitor act as though it permits the passage of an ac voltage across its plates?

23–24. Explain why a capacitive reactance act as though it blocks dc in a circuit?

23–25. How does the frequency of operation of the circuit affect the amount of capacitive reactance?

23–26. Why does X_c permit the passage of a signal and block the operational voltage in a circuit?

23–27. How is X_c used in a filter network?

23–28. Why is the term *reactance* used instead of resistance to describe circuit conditions?

23–29. Explain why Kirchhoff's laws apply in the X_c circuit.

23–30. Explain the relationship between capacitive current and the quantity of capacitance in a circuit.

PROBLEMS

23–31. Find the value of C required when a frequency of 1.5 kHz is applied to 1500 Ω of X_c. (*Sec. 23-2*)

23–32. Find the value of C required for a circuit having an ac input of 120 V at 60 Hz and a X_c of 600 Ω. (*Sec. 23-2*)

23–33. What is the total X_c for three parallel-connected 3600-Ω reactances? (*Sec. 23-3*)

23–34. What is the total X_c for a 500-Ω and a 400-Ω series-connected X_c? (*Sec. 23-3*)

23–35. What is the voltage drop across each component when 120 V ac is applied to parallel-connected reactances having values of 300, 200, and 100 Ω? (*Sec. 23-4*)

23–36. What is the current flow for each of the reactances of problem 23-35? (*Sec. 23-4*)

23–37. What is the voltage drop across each of three series-connected 100-Ω reactances when a source voltage of 300 V ac is applied? (*Sec. 23-4*)

23–38. Two X_c units, having values of 300 and 600 Ω, are series connected across a 400-V ac source. What is the voltage drop across each? (*Sec. 23-4*)

23–39. What current flows through the circuit in problem 23-38? (*Sec. 23-4*)

23–40. What value of capacitance is required for an X_C of 20 Ω when using a frequency of 15 VHz? (*Sec. 23-4*)

23–41. Find the value for X_C using a 750-μF capacitor and a frequency of 1.5 kHz. (*Sec. 23-2*)

23–42. Find the frequency of operation for an X_c of 400 and a capacitance value of 4.0 μF. (*Sec. 23-2*)

23–43. Find the frequencies of operation using an X_c of 500 and capacitances of 10 μF, and 100 pF, and 2 μF. (*Sec. 23-2*)

23–44. A circuit contains a 5-μF capacitor and has 120 V at 60 Hz applied to it. What is the X_C? (*Sec. 23-2*)

23–45. What is the current flow in problem 23-44? (*Sec. 23-4*)

23–46. How much current will flow in problem 23-45 when the capacitance is doubled? (*Sec. 23-4*)

23–47. How much current will flow in problem 23-45 when the frequency is increased to 400 Hz? (*Sec. 23-4*)

23–48. Find the frequency for a capacitance of 5000 pF having a reactance of 4 KΩ. (*Sec. 23-2*)

23–49. A circuit has three capacitive reactances connected in parallel. The values of capacitive reactance are 100, 200, and 400 Ω. Find their equivalent reactance value. (*Sec. 23-4*)

23–50. Find the value of capacitance for each capacitor in problem 23-49 when a signal of 1 kHz is applied to the circuit. (*Sec. 23-4*)

23–1. 0.5 μF, 3 μF, 10 μF

23–2. 10 μF, 3 μF, 0.5 μF

23–3. 33.125 pF

23–4. 33.125 pF

23–5. 800 Ω

23–6. 200 Ω

23–7. 85.7 Ω

23–8. (a) 397.8 Ω; (b) 33.2 Ω

23–9. 0.16 μF

23–10. 90°

23–11. At 90°

23–12. At 0°

CHAPTER 24

Capacitive Circuits

The use of a capacitor in a circuit will react to any changes in the amplitude of the voltage present in the circuit. Almost all electronic power supplies require components that will help maintain a constant level of voltage at the output terminals of the power supply. This is accomplished with the addition of a capacitor in parallel with the output terminals of the supply. The determination of the value of the capacitor is accomplished using circuit analysis methodology. Timing circuits often require one or more capacitors to establish the frequency of the timing unit. In addition, capacitors may be connected in either series or parallel to change the total value of capacitance in the circuit. Persons employed in the repair of electronic equipment may have to create values of capacitance not readily available. Knowledge of capacitive circuit analysis will aid in accomplishing this.

Capacitive Current   Transient Response
RC Time Constant

OBJECTIVES

Upon completion of the material in this chapter, you will

1. Understand the concept of capacitance in the ac circuit.
2. Be able to analyze series and parallel circuits for capacitance reactance.
3. Understand the use of a capacitor as a signal coupling device.
4. Understand the concept of charge and discharge current in a capacitive circuit.

24-1 Need for Circuit Analysis

In Chapter 21 we described the need to analyze circuits containing inductance and resistance. A need also exists for the analysis of circuits containing capacitance and resistance. The need to determine the quantity of reactance in a circuit containing both capacitance and resistance is as important as the analysis of circuits with both inductance and reactance. In this chapter we review the concepts of capacitive reactance and discuss applications.

In Chapter 23 we described the effect of a current in a circuit containing capacitance and an applied voltage. We also compared the phase relationship of the applied voltage and the current in the circuit. This phase relationship is illustrated in Figure 24-1. Note the 90° phase shift in this illustration. In a capacitive circuit the current leads the voltage by 90°. Another way of stating this is to say that the voltage lags the current by 90°.

Figure 24-2 shows the phasor relationship of the two waves. Part (a) shows the phasor relationship when I_C is leading V_C by a clockwise angle of 90° and part (b) shows the phasor relationship when I_C is lagging V_C by an angle of $-90°$. A general statement related to capacitance and reactance is that in any capacitive reactance circuit the current and voltage are out of phase by 90°.

FIGURE 24-2 Phasor relationship between voltage and current showing a 90° phase shift.

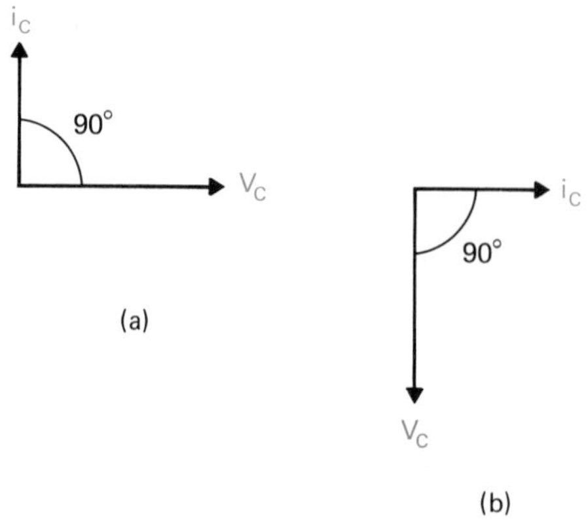

FIGURE 24-1 Relationship of voltage and current in a capacitive reactive circuit.

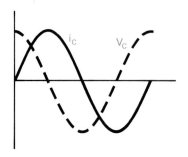

(a)

(b)

Current in the Series Circuit The rules for current in a series resistive circuit apply to circuits containing capacitive reactance as well. In Figure 24-1 there is only one current path. Therefore, the current flow in the ac source, the connecting wires, and to the plates of the capacitor has the same value.

Voltage in the Parallel Circuit Using Figure 24-1 again, we see that the ac source is connected in parallel with the plates of the capacitor. When this occurs, the rules for voltage in a parallel circuit are valid. Voltage applied to the plates of a capacitor has the same value as the source voltage in this circuit. This statement does not change the phase relationship of voltage and current. It is related only to the value of voltage applied to the plates of the capacitor.

Frequency of Voltage and Current The 90° phase shift between voltage and current in the capacitive reactance circuit does not affect the frequency of operation. The voltage and the current have the same frequency in this circuit. When the ac source frequency is 400 Hz, the frequency of the current in the circuit is also 400 Hz. This is true for any ac source frequency.

REVIEW PROBLEMS

24-1. What is the phase relationship between voltage and current in a capacitive reactance circuit?

24-2. What is the phase relationship between V_C and source voltage in a capacitive reactance circuit?

24-3. Does current lead or lag voltage in the capacitive reactance circuit?

24-2 Series Circuits Containing X_C and Resistance

Both resistance and capacitive reactance determine the circuit value for current when they are series connected (Figure 24-3). Each component has its own voltage drop, as produced by current flow in the circuit. The voltage drop for the components is equal to values of IR for the resistance and IX_C for the reactance.

When considering the capacitive reactance by itself, the voltage drop across it lags the current by 90°. The voltage drop across the resistive component

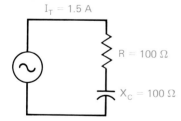

$I_T = 1.5$ A

$R = 100 \ \Omega$

$X_C = 100 \ \Omega$

FIGURE 24-3 Series-connected resistance and capacitive reactance.

has the same phase of the current. This is because there is no phase shift in a purely resistive circuit. The phase difference between the two components requires that their values be added using phasors.

Addition of V_C and V_R Using Phasors Figure 24-4 shows the phasor relationship between current and voltage in the circuit shown in Figure 24-3. Since current is the same in this circuit, the current is shown on the horizontal leg of the diagram. The voltage drop across the resistive component is determined by use of the formula $V = IR$. In this circuit $V = 1.5 \times 100$, or 150 V.

Voltage drop across the capacitive reactance component is determined in a similar manner. In this circuit $V_{XC} = I_C$. Substitution of values in the formula $V_C = 1.5 \times 100$. The value of V_C is 150 V. Since there is a 90° phase shift across the capacitive component, this phasor is drawn as the vertical leg of the triangle.

The triangle shown is a right triangle. Its hypotenuse value is determined by use of the formula

$$
\begin{aligned}
V_T &= \sqrt{V_R^2 + V_C^2} \\
&= \sqrt{150^2 + 150^2} \\
&= \sqrt{22{,}500 + 22{,}500} \\
&= \sqrt{45{,}000} \\
&= 212 \text{ V}
\end{aligned}
\tag{24-1}
$$

The reason for the difference in the sum of the voltage drops in the circuit is due to the phase shift. The peak value of the voltage across the resistance occurs when the peak voltage across the capacitor is at zero. This provides a total voltage in this circuit of 212 V instead of the 300 V we would expect in a purely resistive circuit.

Impedance and Phasors The triangle shown in Figure 24-4 illustrates voltage relationships in the circuit. A similar triangular relationship exists for impedance in this type of circuit (Figure 24-5). When the current is equal through both components, the value of current, I, can be canceled. Impedance values may be found using the impedance formula:

$$
Z = \sqrt{R^2 + X_C^2}
\tag{24-2}
$$

FIGURE 24-4 Phasor relationship for the circuit of Figure 24-3.

FIGURE 24-5 Impedance triangle can be used to show phasor relationship.

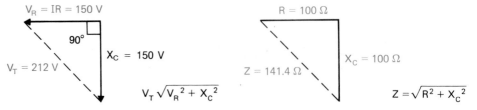

When R and X_C are expressed in units of ohms, the value of impedance is also expressed in units of ohms. In this example the values are substituted in the formula

$$Z = \sqrt{100^2 + 100^2}$$
$$= \sqrt{10,000 + 10,000}$$
$$= \sqrt{20,000}$$
$$= 141 \ \Omega$$

When the value of impedance is inserted into the Ohm's law formula, the current in the circuit may be determined.

$$I = \frac{V}{Z}$$
$$= \frac{212 \text{ V}}{141 \ \Omega} \tag{24-3}$$
$$= 1.5 \text{ A}$$

The total current in this circuit is equal to the applied voltage divided by the impedance. In this circuit the value of voltage is 212 V and the impedance is 141 Ω. The current value is 212/141, or 1.5 A. The phasor sum of the two voltage drops of 150 V is equal to the applied voltage of 212 V. The applied voltage is equal to the value of IZ, or 1.5 × 141, for a value of 212 V.

24-3 Parallel Resistance and Capacitive Reactance

The 90° phase shift between voltage and current applies to branch currents rather than voltage drop in the parallel circuit containing R and X_C. Voltages in the parallel circuit must still be equal.

Each of the branches shown in Figure 24-6(a) has its individual current value. The resistive branch current is determined by the formula $I = V/R$. Current in the capacitive branch is calculated by use of the formula $I = V/X_C$. The currents for each of these branches is 2 A using each of these formulas.

The phasor diagram [Figure 24-6(b)] has the generator voltage V as the reference. This is because the voltage from the generator is equal in this circuit. The phase difference of 90° between current in the resistive circuit and the current in the capacitive circuit is also shown.

FIGURE 24-6 In the parallel circuit, current flows in each branch.

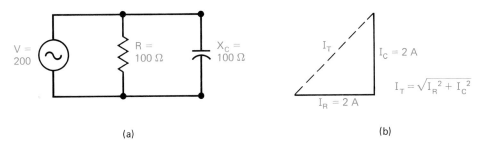

(a)

(b)

The total current in this circuit is determined by use of the formula

$$I_T = \sqrt{I_R^2 + I_{XC}^2}$$
$$= \sqrt{2^2 + 2^2}$$
$$= \sqrt{4 + 4} \qquad \text{(24-4)}$$
$$= \sqrt{8}$$
$$= 2.83 \text{ A}$$

In this circuit the total current is 2.83 A.

Impedance of R and X_C in Parallel The impedance for this circuit is determined by dividing the applied voltage by the total current. The formula $Z = V/I$ is used for this.

$$Z = \frac{V}{I}$$
$$= \frac{200 \text{ V}}{2.83 \text{ A}} \qquad \text{(24-5)}$$
$$= 70.67 \ \Omega$$

This value of 70.67 Ω is the opposition to current flow for the circuit as presented to the leads of the ac source. It is equal to the values of 100 Ω of resistance in parallel with 100 Ω of capacitive reactance.

REVIEW PROBLEMS

24-4. Find the value of I_T for branch currents of 4 A and 8 A.
24-5. Find the value of I_T for branch currents of 10 A and 10 A.

24-4 Voltage-Divider Circuits

Series-connected capacitors will act as a voltage-divider network when they are connected across an ac source. Each of the capacitors will develop a part of the total applied voltage across its terminals. The rule for voltage dividers applies to this type of circuit as well as a dc circuit. The sum of the individual voltage drops in this circuit will equal the value of applied ac voltage.

The value of voltage drop across an individual capacitor is inversely proportional to the capacitance value for the device. This is opposite to the effect of voltage drops in a series resistive circuit. When two capacitors having values of 10 and 50 μF are series connected, the smaller value will develop the largest voltage drop. In this example, the 10-μF capacitor will develop five-sixths of the applied voltage. If a voltage of 100 V is applied to the circuit, then $\frac{5}{6}$, or 83.3 V, will be developed across the plates of this capacitor, and the balance,

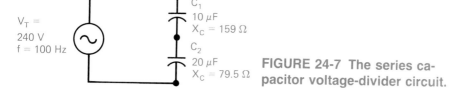

FIGURE 24-7 The series capacitor voltage-divider circuit.

16.7 V, will develop across the 50-μF unit. These two values of 83.3 and 16.7 add to equal the applied 100-V value.

Figure 24-7 illustrates the concept of a series capacitive voltage-divider network. An ac voltage of 240 V is applied to a series circuit containing a 10-μF and a 20-μF capacitor. The total reactance in this circuit is 238.5 Ω. The total current is determined by the formula $I = V/X_C$. This value is 1.006 A. The current is the same through both of the capacitances. The voltage across each reactance is calculated by use of the formula $V = IX_C$. Voltage drop across X_{C1} is 1.006 × 159, or 159.95 V, and the voltage drop across X_{C2} is 1.006 × 79.5, or 79.98 V. The two values of about 160 V and 80 V add to equal the value of applied voltage of 240 V. As stated earlier, the larger value of capacitance has the smaller of the voltage drops in this circuit.

When the series-connected capacitors are connected to a dc source the value of the charges on the plates of each capacitor is similar to the charges on the plates when an ac voltage source is used. In Figure 24-8, two capacitors are series connected to a 240-V dc source. The values of the capacitors are 10 μF and 20 μF. The dc voltage charge on the plates is distributed in inverse proportion to the individual values of capacitance. In this example the total capacitance value is 30 μF. Since the voltage division that occurs is inversely proportional to the individual values, the larger voltage drop will occur across the 10-μF capacitor. The specific value is determined by use of a ratio or proportional formula. The 10-μF capacitance has one-third of the total capacitance in this circuit. The voltage is calculated in this manner:

$$V_1 = \frac{C_T}{C_1} \times 240 \qquad (24\text{-}6)$$

The value of C_T must be determined first if this formula is to be used. This value is 6.67 μF.

$$V_1 = \frac{6.67}{10} \times 240$$

$$= 160 \text{ V}$$

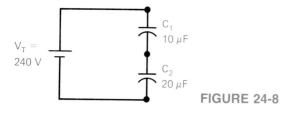

FIGURE 24-8

and

$$V_2 = \frac{6.67}{20} \times 240$$

$$= 80 \text{ V}$$

The values of 160 V and 80 V add to equal the applied source voltage of 240 V.

A second method of determining the voltage drop across series capacitors is to use a ratio method. Since the value of C_1 is one-third of the sum of the two capacitor values, the voltage drop across it is equal to two-thirds of the applied voltage. (Remember that the smaller capacitor has the larger voltage drop.)

$$V_1 = \tfrac{2}{3} \times 240$$

$$= 160 \text{ V}$$

$$V_2 = \tfrac{1}{3} \times 240$$

$$= 80 \text{ V}$$

The method of determining voltage drops for series capacitances applies to circuits using either an ac source or a dc source. The reason is that the plates of the capacitors are charged equally using either of the sources.

REVIEW PROBLEMS

24-6. Find V_1 and V_2 for capacitances of 100 and 150 pF connected in series across a 200-V source.

24-7. Three capacitances of 10, 15, and 30 µF are series connected to a 200-V source. Find the voltage drops.

24-5 Capacitive Current

Capacitive current *Current flow in an ac circuit containing capacitance.*

When a sine-wave voltage is applied to a capacitive circuit, the charge time for a capacitor is equal to the discharge time. Current during the charge and the discharge portion of the cycle is equal. Values of I_C and V_C are calculated using X_C values.

When a waveform having a shape other than a sine wave is used, one is not able to use X_C values for calculation. The use of reactance is limited to sine-wave activity. Current in a non-sine-wave circuit must be calculated using the change in voltage and the change in current, or dV/dt. The formula for this is

$$i_C = C\frac{dV}{dt} \tag{24-7}$$

The use of this formula for determining current values uses Figure 24-9 for its example. The waveform in this illustration is a sawtooth form. Its rise time is linear. The time required for it to rise to its full value is 40 s. The

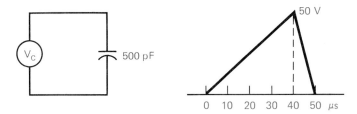

FIGURE 24-9 A sawtooth-shaped waveform provides a linear rise charge time.

linearity of this rise time indicates that there is an equal increase in current for each of the time periods illustrated. The rise time for the current is the period of time when the capacitor is charged. The discharge time for this waveform is much shorter than the rise time. In this illustration the discharge time is 10 μs.

Another factor involved in this circuit is the direction of current flow. During the rise-time period, current flow is from the source to the plates of the capacitor. When the capacitor discharges, the current flow direction is reversed. It now flows from the capacitor back to the source.

The value of charge current for this circuit is determined by use of the formula $i_c = C(dV/dt)$. The value of charge current is

$$i_c = C\frac{dV}{dt}$$

$$= 500 \times 10^{-12}\left(\frac{50}{40 \times 10^{-6}}\right)$$

$$= 5 \times 10^{-10}\left(\frac{50}{4 \times 10^{-5}}\right)$$

$$= 5 \times 10^{-5} \times 12.5$$

$$= 625 \text{ μA}$$

The value of discharge current is greater, since the discharge time is less in this circuit.

$$i_c = C\frac{dV}{dt}$$

$$= 500 \times 10^{-12}\left(\frac{50}{10 \times 10^{-6}}\right)$$

$$= 5 \times 10^{-10}\left(\frac{50}{1 \times 10^{-5}}\right)$$

$$= 5 \times 10^{-5} \times 50$$

$$= 2500 \text{ μA}$$

These wave shapes are similar to those shown in Chapter 20. The difference is in the fact that the voltage and current wave shapes are interchanged. A conclusion from this is that both i_c and v_l depend on the rate of change occurring in the circuit.

24-6 RC Time-Constant Calculations

Transient response Sudden change in circuit current.

A sudden change in current in a circuit is known as its **transient response**. A circuit having nonsinusoidal waveform cannot be measured in the same manner as those having sine waveforms. The value of the change is measured by its **RC time constant**. The formula for this is

RC time constant Time required to charge or discharge capacitor 63.2% of full value.

$$T = RC \qquad \text{(in seconds)} \qquad (24\text{-}8)$$

where T = time-constant value

R = resistance in series with the capacitance

C = value of capacitance (F)

The value of the time constant for a capacitor of 15 μF in series with a 1000-Ω resistor is

$$T = RC$$
$$= 1000 \times 15 \times 10^{-6}$$
$$= 15 \times 10^{-3}$$
$$= 15 \text{ ms}$$

The time constant T represents the time for the voltage across the plates of the capacitor to change by a factor of 63.2%. When the applied voltage is 200 V, the capacitor will charge to 63.2% of 200 for this example. This value is the charge for one time constant. It requires five time constants for the capacitor to be charged to 100% of the applied voltage.

1. There is a phase shift between voltage and current in a circuit containing both resistance and capacitive reactances. Its range is between 0 and 90°.

2. The frequency of voltage and current is the same in a capacitive reactance circuit.

3. Current in a series capacitive reactance circuit is equal throughout the circuit.

4. Addition of voltage drops in the capacitive reactance circuit requires the use of phasors.

5. Impedance and phasor relationships use the same triangulation methods.

6. Branch current in the capacitive reactance circuit is determined by the use of phasor diagram analysis.

7. Series-connected capacitors act as voltage dividers.

8. Voltage drops across series capacitors are inversely proportional to the values of capacitance.

9. Nonsinusoidal waveform analysis cannot use reactance values.

10. Analysis of nonsinusoidal waveform circuits uses the formula $i_C = C(dV/dt)$ for determination of circuit current.

11. RC time constants identify the time for a capacitor to charge to 63.2% of its full charge value.

12. Five time constants are required for a capacitor to charge to the value of the applied voltage.

Answer true or false; change all false statements into true statements.   **SELF TEST**

24–1. Capacitive reactances in series add.

24–2. Source voltage is equal to capacitor voltage in the parallel circuit.

24–3. Individual voltage drops in the series capacitive circuit equal the applied voltage.

24–4. In the series capacitive circuit, the largest capacitor has the biggest voltage drop.

24–5. In the parallel capacitive circuit, the largest capacitor has the smallest voltage drop.

24–6. Phase shift has no effect on the calculation of total current in the capacitive reactive circuit.

24–7. The phase shift in a capacitive circuit has the current leading the voltage.

24–8. A capacitor will charge to the value of the applied voltage during three time constants.

24–9. Current in a nonsinusoidal circuit must be calculated using values of dt and dV.

24–10. In a sawtooth waveform circuit the current is greatest during the shortest period of time (either rise or fall).

24–11. Two electrical waves that start at the same moment in time have a phasor angle of
(a) 0°
(b) 90°
(c) 120°
(d) 180°

24–12. In a parallel capacitive circuit, voltage drops are
(a) equal throughout the circuit
(b) dependent on the values of capacitors
(c) equal to series resistance rules
(d) none of the above

24–13. In a series circuit containing both R and C, the voltage phase relationship between the two components is a(n)
(a) in-phase condition
(b) 45° shift, V lagging
(c) 90° shift, V lagging
(d) 90° shift, I lagging

24–14. A phase shift between voltage and current occurs in the
(a) series X_C circuit
(b) branch circuits of a capacitive circuit
(c) applied voltage to the circuit
(d) none of the above

24–15. Capacitors in series form a
(a) current-dividing circuit
(b) voltage-aiding circuit
(c) current-aiding circuit
(d) voltage-dividing circuit

24–16. Current flow in an X_C circuit during the rise period flows
(a) from the source to the capacitor
(b) from the capacitor to the source
(c) both from the source and the load
(d) both to the source and to the load

24–17. The formula for determination of the RC time constant is
(a) $T = RC$
(b) $R = C/T$
(c) $C = RT$
(d) $T = R/C$

24–18. The number of time constant periods for an almost 100% charge is
(a) 2
(b) 3
(c) 4
(d) 5

24–19. The voltage will fall to _____of its full value during the first time constant.
(a) 63.2%
(b) 70.7%

(c) 29.3%
(d) 36.8%

24–20. In the X_C circuit, the rise and fall relates to the
  (a) current
  (b) resistance
  (c) voltage
  (d) impedance

24–21. Why is there a phase shift between voltage and current in the X_C circuit?
24–22. How are vectors used to describe phase shift?
24–23. What is the relationship of voltage and frequency in the X_C circuit?
24–24. Why is there a voltage drop across series-connected capacitors in the X_C circuit?
24–25. What is meant by the term *transient response*?
24–26. What is the frequency relationship between V and I in the X_C circuit?
24–27. Explain why voltage leads current in the X_C circuit.
24–28. Why does it require five time constants to achieve a full charge on a capacitor's plates?
24–29. Why does it require five time constants to achieve a full discharge on a capacitor's plates?
24–30. What is the relationship between C and R in the X_C circuit?

24–31. Calculate the current for a circuit consisting of 88 Ω of resistance in series with an X_C of 44 Ω connected to a 150-V ac source. (*Sec. 24-1*)
24–32. Calculate the impedance in problem 24-31. (*Sec. 24-2*)
24–33. Find the voltage drops across each component in problem 24-31. (*Sec. 24-1*).
24–34. Calculate the current when the components in problem 24-31 are connected in parallel. (*Sec. 24-1*)
24–35. Find the impedance and current in each branch for the circuit in problem 24-34. (*Sec. 24-1*)
24–36. Two capacitors are series connected across a 6-kV ac source. The values of the capacitors are 500 μF and 2500 μF. Find the voltage drop across each capacitor. (*Sec. 24-2*)
24–37. Find the total circuit current for problem 24-36. (*Sec. 24-2*)
24–38. Draw the schematic diagram for a two-capacitor series circuit containing 20 kΩ of resistance and an unknown value of capacitance connected to a 3-V ac source. (*Sec. 24-2*)
24–39. A 200-Ω resistor is series connected to a 200-Ω X_C. This produces a 2-A current flow in the circuit. Find the value of capacitance required

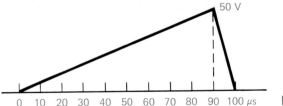

50 V

0 10 20 30 40 50 60 70 80 90 100 μs   **FIGURE 24-10**

to maintain the 200-Ω X_C at frequencies of 120, 400, and 1 kHz. (*Sec. 24-2*)

24–40. Calculate the impedance of the circuit in problem 24-39. (*Sec. 24-2*)

24–41. A circuit contains four series-wired capacitors of 1, 1.5, 10, and 30 pF. An ac voltage of 1.2 kV is connected to the circuit. Find the voltage drops across each capacitor. (*Sec. 24-4*)

24–42. A circuit has a 500-Ω X_C parallel connected to a 300-Ω R and a 72 V source. Find the total current and total impedance of the circuit. (*Sec. 24-3*)

24–43. A 500-pF capacitor is connected to a sawtooth source. The waveform produced is shown in Figure 24-10. Find the charge current and discharge current for this circuit. (*Sec. 24-6*)

24–44. Calculate the circuit currents for the waveform in problem 24-43 when the maximum voltage is 200 V and the time for charge and discharge is reduced by 50%. (*Sec. 24-6*)

24–45. A 50-Ω R and an 80-Ω X_C are series connected to a 120-V source. Find I_T and Z_T for this circuit. (*Sec. 24-2, and Sec. 24-3*)

24–46. Find I_T and Z_T when the components in problem 24-45 are parallel connected. (*Sec. 24-3*)

24–47. A 1.50-kΩ R is series connected to a 500-Ω X_C and a 240-V source. Find I_T and Z_T. (*Sec. 24-2*)

24–48. Find I_T and Z_T when a 500-Ω R series connected to a 1.5-kΩ X_C. (*Sec. 24-2*)

24–49. Find I_T and Z_T when a 500-Ω R is parallel connected to a 1.5-kΩ X_C and a 158 V source. (*Sec. 24-3*)

24–50. A 0.04-μF capacitor is series connected to a 10-k/Ω resistance and a 20-V ac source. Find the values of X_C, I, V_R, and V_C for this circuit at each of the following frequencies. (*Sec. 24-2*)
(a) 0 Hz
(b) 60 Hz
(c) 400 Hz
(d) 15 kHz

ANSWERS TO REVIEW PROBLEMS

24–1. Voltage lags current by 90°.

24–2. They are equal.

24–3. Current leads voltage.

24–4. 8.9 A

24–5. 14.14 A

24–6. 120 V and 80 V

24–7. 100 V, 66.6 V, and 33.33 V

24–8. 1.25 V/μs (rise time) and 5 V/μs (fall time)

24–9. 15.625 μA/μs (rise time)

24–10. 6 s

24–11. 6 ms

Chapters 22–24

The characteristics of capacitance are useful in the control and modification of electrical circuits. This section discusses the physical properties of capacitors; the electrical characteristics of capacitance. In addition, the uses of capacitors in basic circuit application are presented.

1. A capacitor consists of two metallized surfaces, or plates, separated by a dielectric material.

2. Capacitors are rated in units of farads. Usually, this rating is in units of microfarads or picofarads.

3. The purpose of a capacitor is to store an electrical charge.

4. Current flow in a capacitor occurs only during its charging or discharging period.

5. Dielectric strength of a material is the rating of its voltage breakdown value.

6. Capacitors are rated as being either electrolytic or nonelectrolytic types. Capacitors higher than 1.0 μF are usually electrolytic types.

7. Installation of electrolytic capacitors in reverse polarity can produce an explosion and failure of the capacitor.

8. Nonelectrolytic capacitors use mica, air, paper, and ceramic as their dielectric material.

9. Capacitor applications include bypass, filter, and coupling and timing circuits.

10. Parallel capacitors act to increase the total value of capacitance (opposite to that of parallel resistances).

11. Series capacitors act to reduce the total value of capacitance (opposite to that of series resistances).

12. Voltage division across series capacitors is inversely proportional to the values of the capacitors.

13. Capacitors may be tested with an ohmmeter or a dedicated capacitor tester.

14. Capacitor failures are either open, shorts, or leakage problems.

15. When replacing a capacitor both the voltage and capacitance ratings must be considered.

16. The voltage charge on the plates of a capacitor may be used to produce a current flow in a circuit.

17. Capacitor action acts as though an alternating current flows through its dielectric material.

18. Capacitor action effectively blocks a dc current flow.

19. The opposition to current flow in a capacitive circuit is called *capacitive reactance*.

20. Capacitive reactance is measured in units of ohms.

21. Capacive reactance is described as X_C.

22. Ohm's law is used to determine values of X_C, I, and V in a circuit.

23. The formula for X_C is $1/(2\pi fC)$ ($\frac{1}{2\pi}$ is equal to 0.159).

24. Capacitive reactance is inversely proportional to the value of capacitance in the circuit.

25. Capacitive reactance in a circuit is determined in the same manner as resistances in either series or parallel circuits.

26. Rules presented by Kirchhoff for voltage drops in a series circuit and current flow at a junction also apply to circuits containing capacitive reactance components.

27. Capacitive reactance principles are used to transfer signals between stages of many electronic devices.

28. An applied voltage will create a current flow in a capacitive circuit.

29. Current leads voltage by 90° in a circuit containing capacitive reactance.

30. There is a phase shift between voltage and current in a circuit containing both resistance and capacitive reactances. Its range is between 0 and 90°.

31. The frequency of voltage and current is the same in a capacitive reactance circuit.

32. Current in a series capacitive reactance circuit is equal throughout the circuit.

33. Branch current in the capacitive reactance circuit is determined by the use of phasor diagram analysis.

34. Series-connected capacitors act as voltage dividers.

35. Nonsinusoidal waveform analysis cannot use reactance values.

36. Analysis of nonsinusoidal waveform circuits uses the formula $i_C = C(dV/dt)$ for the determination of circuit current.

37. RC time constants identify the time for a capacitor to charge to 63.2% of its full charge value.

38. Five time constants are required for a capacitor to charge to the value of the applied voltage.

CHAPTER 25

RC and *L/R* Time Constants

Effects of inductance and capacitance in a circuit will modify its operational characteristics. These changes are used to the advantage of circuit designers to control current changes or voltage changes. Many electrical circuits used in production welding equipment require short-term electrical power bursts. Time-constant circuits will produce the desired power bursts. Timing circuits used in computers also utilize these concepts.

KEY TERMS

Differentiation   **Transient Time**
Integrators   **Universal Time-Constant Chart**
Time Constant

OBJECTIVES

Upon completion of the material in this chapter, you will

1. Understand the concept of L/R time constants.
2. Recognize wave shapes used in electronic circuits.
3. Understand the short-circuit effect for time-constant circuits.

Electrical circuits may contain resistance, inductance, and capacitance. All of these components may be present in one circuit. Other circuits may be purely resistive, resistive and capacitive, or resistive and inductive. Each of the circuits indicated have their own special set of operating characteristics. These characteristics produce different operating conditions in a circuit. The material in this chapter is specifically related to circuit conditions in those circuits containing inductive reactances or capacitive reactances. As a starting point, let us review the characteristics of a purely resistive circuit. This will be followed by the presentation of X_L and X_C circuit operation.

25-1 Resistive Circuit Operation

A purely resistive circuit is illustrated in Figure 25-1(a). This circuit has a 100-Ω resistor series connected to a 10-V dc source. Switch S is used to connect the source to the 100-Ω resistive load. Figure 25-1(b) illustrates current flow in this circuit when the switch is used to complete the circuit and connect the battery to the 100-Ω resistance. Current immediately rises to the Ohm's law value of 0.1 A. Current is maintained at this 0.1-A level as long as the battery is able to provide a constant voltage.

Figure 25-1(b) an instantaneous rise in current from the zero, or open-circuit value, to the steady-state value of 0.1 A. In reality, this is not a true representation. Current does require a moment in time to rise to its operational level in any circuit. This time period is called *transient time*, or the time required for the current to change from one value to another value. In this circuit, as stated, this transient time is very short as the current rises from zero to 0.1 A.

This same condition occurs when the switch is opened and the battery is disconnected from the load. Current immediately falls to zero as the source is removed from the load. We usually consider the transient time in a purely

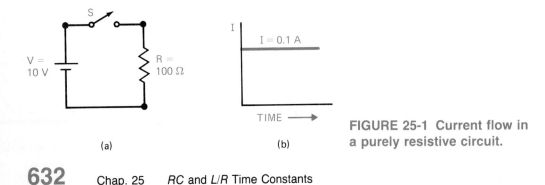

(a)

(b)

FIGURE 25-1 Current flow in a purely resistive circuit.

resistive circuit to be negligible, or close to zero. It is not normally a significant factor in a dc circuit that has only resistance.

25-2 *L/R* Time Constant

Different circuit operational conditions exist when a circuit contains inductance and resistance. The circuit of Figure 25-2 shows a 100-Ω resistance in series with a 1-H inductance connected to a 10-V dc source. When switch S is closed, current will start to flow in this circuit, just as it did in the resistive circuit. The addition of the inductor changes the operating conditions of this circuit. Inductance, you will remember, will attempt to oppose any changes in the level of circuit current. The quantity of EMF applied to the circuit will eventually overcome much of this opposition, but it takes time for this to occur.

The reason for the time factor in an inductive circuit is the fact that the magnetic field surrounding the conductor opposes any changes in the amplitude of the current. A rising current creates a magnetic field that cuts across the conductor and induces a voltage opposite to that of the applied voltages. Eventually, the current in this illustration circuit will rise to its full value of 0.1 A. The rise, however, is not instantaneous.

The same effect occurs when the switch is opened and the current is allowed to drop to 0 A. The magnetic field surrounding the conductor will now collapse. This induces a voltage on the conductor that has the same polarity as the applied voltage. The result is an attempt to maintain the level of voltage and current supplied by the battery source. The magnetic field eventually collapses and the ability to induce a voltage on the conductor no longer exists.

The time required for the rise or the fall of the current in the circuit is identified as the **transient time** for the circuit. Transient time, or response, is measured in terms of a ratio. In an inductive and resistive circuit this ratio is

Transient time *Time required for sudden change in circuit current to occur.*

$$T = \frac{L}{R} \qquad (25\text{-}1)$$

where T = time constant (s)

L = inductance (H)

R = resistance (Ω)

In a practical circuit the resistance is often the value of the resistance of the

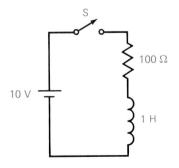

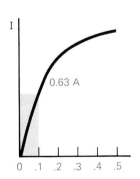

FIGURE 25-2 Circuit having a series-connected resistance and inductance.

wire used to create the windings of the inductor rather than a separate, discrete resistance value.

When the time constant for Figure 25-2 is calculated,

$$T = \frac{L}{R}$$

$$= \frac{1 \text{ H}}{100 \text{ }\Omega}$$

$$= 0.01 \text{ s}$$

Time constant *Time required for current in circuit to rise to 63.2% of full value. Five periods are required for a full charge.*

The **time constant** is defined as the amount of time required for the circuit current to rise to 63.2% of its maximum value. Often this value is rounded to be 63%. Five time constants are required to produce a 100% rise in circuit current.

The same conditions are present when current falls from its maximum value due to the removal of the power source. Current will fall by 63.2% of the maximum value during the first time constant. In the example, current will drop from its maximum of 0.1 A to a value of 0.037 A during the first time constant.

EXAMPLE 25-1

Find the time constant for a 15-H inductor having a resistance of 200 Ω.

Solution

$$T = \frac{L}{R}$$

$$= \frac{15 \text{ H}}{200 \text{ }\Omega}$$

$$= 0.075 \text{ s}$$

EXAMPLE 25-2

Find the time constant for a series circuit containing an inductance of 40 mH and a resistance of 150 Ω.

Solution

$$T = \frac{L}{R}$$

$$= \frac{0.04 \text{ H}}{150 \text{ }\Omega}$$

$$= 267 \text{ }\mu\text{s}$$

EXAMPLE 25-3

Find the time constant when 10 V is applied to an inductance of 5 H and a resistance of 50 Ω.

Solution

$$T = \frac{L}{R}$$

$$= \frac{5 \text{ H}}{50 \text{ Ω}}$$

$$= 0.1 \text{ s}$$

EXAMPLE 25-4

Find the current during the first time constant when the applied voltage produces a maximum current of 250 mA.

Solution

The first time constant produces a current of 63.2% of maximum.

$$I = I_{\text{max}} \times 0.632$$

$$= 250 \text{ mA} \times 0.632$$

$$= 158 \text{ mA}$$

REVIEW PROBLEMS

25-1. Find the time constant for a 10-H series connected with 200 Ω.

25-2. Find the time constant for a 10-H series connected with 2 kΩ.

25-3. Find current for one time constant when a maximum current of 120 mA flows in the circuit.

25-3 Decay of the Magnetic Field Effect

An open circuit has an infinitely high value of resistance. When a circuit containing an inductance and a resistance is disconnected from its power source, the discharge time constant becomes a very short period because the switch acts an an infinitely high resistance. The current fall to zero amperes is much more rapid due to this smaller time-constant period. As a result of this action, a very high voltage is induced across the inductor in the circuit. The amplitude of this induced voltage due to self-induction in the coil will often be much higher than the source voltage of the circuit.

The circuit shown in Figure 25-3 will illustrate the concept of a high induced

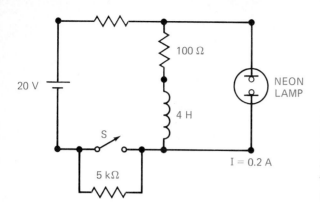

20 V

100 Ω

4 H

S

5 kΩ

NEON LAMP

I = 0.2 A

FIGURE 25-3 A circuit of this type will develop a high voltage during rapid discharge and the lamp will light even though it normally requires a much higher voltage.

voltage during the period immediately after the source power is removed from the circuit. This circuit has a 5-kΩ resistor in series with an inductor. The inductor has a value of 4 H and a resistance of 100 Ω. A switch is connected in parallel with the 5-kΩ resistance. The purpose of the switch is to either add this resistor to the total resistance of the circuit or to remove it from the circuit by short circuiting it. In addition, a neon glow lamp is connected in parallel with the inductor. The operating characteristics of the neon lamp require a value of 90 V in order to ionize the gas in the lamp and make it glow. It appears to be obvious that the 20-V dc source cannot ionize the gas in the lamp under normal operating conditions.

When the switch, S, is closed, the only value of resistance in the circuit is the 100-Ω resistance of the inductor. A current of 0.2 A will flow after a period of five time constants is reached. The time constant for this circuit is

$$T = \frac{L}{R}$$

$$= \frac{4 \text{ H}}{100 \text{ Ω}}$$

$$= 0.04 \text{ s}$$

Five time constants will require 5 × 0.04 s, or 0.2 s. The current in this circuit will be 0.2 A after a period of 0.2 s, or 200 ms.

When switch S is opened, the circuit resistance is increased to 5100 Ω. The relationship of L to R has changed due to the increase in circuit resistance.

$$T = \frac{L}{R}$$

$$= \frac{4}{5100 \text{ Ω}}$$

$$= 0.00078 \text{ s} \quad \text{or} \quad \text{about } 0.0008 \text{ s}$$

Current drop, or decay, becomes close to 0 A after a period of 0.004 s. This is much faster than the time required to charge to inductor's magnetic field.

The very rapid drop in the size of the magnetic field surrounding the inductor will induce a very high voltage across the terminals of the inductor.

The value of this voltage is 1020 V. This value is much higher than the 90 V required to ionize the neon gas in the lamp. It will glow until the ionization voltage drops to less than the 90-V value.

The method of determining the value of the induced voltage during the collapse of the magnetic field is accomplished in the following manner. When switch S is closed, the value of current in this circuit is 0.2 A. Since this is a series circuit, this current is equal throughout the circuit. When switch S is opened, the total resistance rises to 5100 Ω because the 5-kΩ resistance is now in series with the inductor. The magnetic field energy stored in the inductor will flow through the total resistance for the first instant after closing the switch. A current of 0.2 A will flow momentarily through the total resistance of 5100 Ω. The voltage produced at this moment across this resistance is equal to the *IR* value in the circuit.

$$V = IR$$
$$= 0.2 \text{ A} \times 5100 \ \Omega$$
$$= 1020 \text{ V}$$

This value is present only during the first instant that the switch is opened. One can start to comprehend why it is often possible to see a spark across the terminals of a switch when the power to an inductive load is removed. This will often occur when the power to a motor is turned off.

25-4 Applications for Induced Voltage

Two familiar applications for the high value of induced voltage are in devices in common use. One of these is the video display portion of a computer terminal or television receiver. The cathode ray tube used to display the visual images requires a very high operational voltage. This voltage will vary depending on the electrical characteristics of the cathode-ray-tube display. Often, it reaches a value of 24 kV.

A high-frequency power supply is used in the power section of the terminal. This power supply uses a sawtooth waveform as an input voltage. The rapid decline of this waveform when it is connected to an inductor will create a high voltage. This voltage is utilized for proper display tube operation.

A second common use for the high value of induced voltage is the ignition coil in the automobile. Modern automobiles often require voltages around 18 kV for correct spark plug operation. The inductance in the automobile is called the *ignition coil*. This coil operates on the same principle in order to create a high voltage from the 12-V battery in the automobile. The ignition systems uses a switching device in order to have a current flow through the coil. This device is either a transistorized switching circuit or a set of mechanical points. In either case, the circuit is switched off after its magnetic field is established. The time constant for the charge time is much longer than the discharge time. The result is the establishment of a high operating voltage for a short duration. This voltage is on long enough to produce a spark in the engine cylinders and ignite the gas mixture.

25-4. Using Figure 25-3, is the longer time period during charge or discharge of the inductor's field?

25-5. What is the voltage value induced during discharge when 30 V is applied to 6 H having 50 Ω of resistance?

25-5 *RC* Time Constants

When working with capacitive circuits the time constant is measured by using the product of *RC*. This is calculated by use of the formula

$$T = RC$$

The value of *R* is in ohms, the value of capacitance in farads, and the time is in seconds. This is illustrated by using the circuit shown in Figure 25-4.

$$T = RC$$
$$= (2 \times 10^6) \times (4 \times 10^{-6})$$
$$= 2 \times 4$$
$$= 8 \text{ s}$$

Meaning of Time Constant The term time constant refers to the rate of charge or of discharge of the components in the circuits. When the reference is to a *RC* time constant the term indicates the amount of time required to charge a capacitor by 63.2% of its full charge value. It also refers to the time required to discharge the capacitor to 63.2% of its fully discharged voltage value, or 0 V.

The time-constant value in seconds indicates the length of time required for the capacitor to react to the changes in voltage in the circuit. A short time-constant value shows a rapid change. A long time-constant value indicates a slower change in circuit conditions.

After a period of five time constants the capacitor is fully charged (or fully discharged). If the capacitor is still connected to the power source, it will not accept any more charge after the fifth time constant. The voltage applied

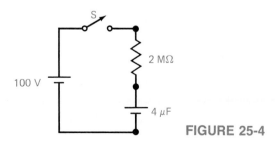

FIGURE 25-4

to the plates of a capacitor does not have to be at 0 V to produce a change in the amplitude of the charge. A capacitor will accept a charge whenever the voltage on its plates is less than the source voltage applied to it. Conversely, the capacitor will discharge any time the voltage on its plates is greater than the source voltage applied to the plates. The charge or discharge voltage is the difference between the voltage on the plates of the capacitor and the voltage from the source.

EXAMPLE 25-5

Find the time constant for a 1.0-MΩ resistor in series with a 0.001-μF capacitor.

Solution

$$T = RC$$
$$= (1 \times 10^6)\Omega \times (1 \times 10^{-9})F$$
$$= 1 \times 10^{-3}$$
$$= 1 \text{ ms}$$

EXAMPLE 25-6

The capacitor in Example 25-5 has 400 V dc applied to it. What is the charge after 1 time constant period?

Solution

During one time constant the voltage charge is 63.2% of the applied voltage. Therefore, 400×0.632 provides the voltage value of 252.8 V.

EXAMPLE 25-7

The capacitor in Example 25-6 is fully charged. What is its voltage after one time constant of discharge?

Solution

The discharge value is 36.8% of the full charge voltage. $400 \times 0.368 = 147.2$ V.

The ability to determine the voltage value for time constants between two and four may be accomplished by use of mathematical formulas. The process requires the computation of 63.2% of the difference between the voltage after the time constant and the full voltage value. This, after five time constants, will show the full charge (or discharge) value.

Another method, probably much easier to use, is shown in Figure 25-5. This is a **universal time-constant chart**. This chart illustrates both the increase and the decay of the voltage due to reactance in a circuit. It is often easier to learn to use this type of chart than it is to attempt to perform a series of

Universal time-constant chart *Chart displaying mathematical equivalents of rise and fall time values.*

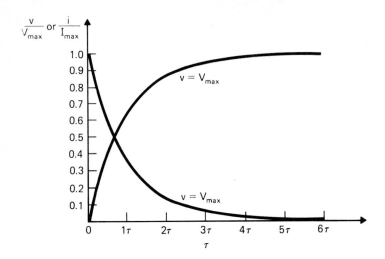

FIGURE 25-5 Universal time chart showing charge and discharge time for an inductive circuit.

mathematical calculations. One significant set of values obtained from this chart is shown below. This is the relation of the time-constant value to the percent of the total values of charge or discharge.

Amplitude (%)	Factor (Time Constants)
63	1
86	2
95	3
98	4
99	5

The fifth time constant does not actually reach a value of 100% of the charge voltage. It is so close, however, that we tend to use this value to represent a full charge or a full discharge in circuit computations.

The curves in Figure 25-5 show both the charge and the discharge, or decay values for the voltages on the plates of the capacitor. Since capacitors react to changes in voltage, the reference here is to the value of the voltage on the plates of the capacitor.

It is possible to use the same set of curves to illustrate the change in current levels in a circuit containing inductance and resistance. When attempting to use the universal time-constant chart for inductances, one must remember that inductance reacts to changes in the current in the circuit. The curves represent changes in circuit current for the time-constant periods when the circuit is inductive instead of capacitive.

The use of the universal time-constant chart will quickly slow the time required to charge a capacitor to a specific percent of its total charge value. For example, using Figure 25-5, it is possible to determine how long it would take to charge a capacitor to 60% of its full voltage. In this example the figure shows that it would take 0.916 time constant to accomplish this. If the time constant is 0.1 s, then 0.1×0.916, or 0.0916 s or 91.6 ms is the time required.

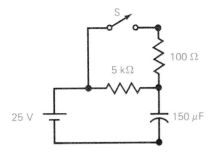

FIGURE 25-6 Circuit demonstrating charge and discharge time for a small and for a relatively large resistance.

25-6 Short Circuiting the *RC* Circuit

A capacitor can be charged over a fairly long period by use of a high value of resistance and thus a long time constant. Once charged, it is possible to discharge the capacitor rapidly through a much smaller value of resistance. The circuit shown in Figure 25-6 illustrates how this will work. This circuit consists of a 25-V power source, or battery, a 5-kΩ resistor, and a large capacitor having a value of 150 μF. The 5-kΩ resistor has a parallel branch. It consists of the open switch (S) and a 100-Ω resistor. The action occurring in this circuit is similar to that when an inductor is used in a circuit like this. The difference between the two circuits is that the inductor will induce a high voltage when discharged and the capacitor will create a momentary high current when it is discharged.

During the time period when the battery is charging, a 5-kΩ resistance is connected in series with the capacitor. Charge current is limited to the Ohm's law value of $I = V/R$, or 25/5000. Charge current is therefore 0.005 A, or 5 mA. The time constant for the charge current is RC, or 5000 × 0.00015. This is calculated as

$$T = RC$$
$$= (5 \times 10^3) \ \Omega \times (15 \times 10^{-5}) \ \text{F}$$
$$= 75 \times 10^{-2}$$
$$= 0.75 \ \text{s}$$

This capacitor is fully charged after a period of five time constants, or 3.75 s. The charge on the capacitor at this time is 25 V.

When the capacitor is discharged as shown in Figure 25-6, the switch S is closed. The total resistance in the circuit under this condition is slightly less than 100 Ω (actually just over 98 Ω). We will use the 100-Ω value just to keep the manipulation of numbers simpler. First, the discharge time constant is determined:

$$T = RC$$
$$= (1 \times 10^2) \ \Omega \times (15 \times 10^{-5}) \ \text{F}$$
$$= 15 \times 10^{-3}$$
$$= 0.015 \ \text{s}$$

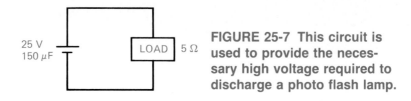

FIGURE 25-7 This circuit is used to provide the necessary high voltage required to discharge a photo flash lamp.

Five time constants will show that the capacitor is fully discharged after a period of 0.075 s. The discharge current is determined by use of Ohm's law:

$$I = \frac{V}{R}$$

$$= \frac{25 \text{ V}}{100 \text{ }\Omega}$$

$$= 0.25 \text{ A}$$

This concept of slow charge and rapid discharge to obtain a momentary high current is useful in industrial or consumer devices. One such device is the photoflash unit on a camera. The photoflash unit contains a capacitor, a charging circuit, and a very low resistance load (Figure 25-7). The low-resistance load is the flashbulb. Its resistance is often less than 5 Ω.

When the capacitor is charged to its full value in a relatively long time period, it is then discharged during a very short period. The higher current flow through the circuit during the discharge period will produce the expected flash. Using our earlier example, the discharge current for a 5-Ω flashbulb is

$$I = \frac{V}{R}$$

$$= \frac{25 \text{ V}}{5 \text{ }\Omega}$$

$$= 5 \text{ A}$$

This value of 5 A is not a normal current for the circuit. It is a momentary high current produced during the initial discharge of the capacitor into a very low resistance load.

REVIEW PROBLEMS

25-6. Find the RC time constant for a 1.5-μF capacitor and a 10-kΩ resistor.

25-7. Find the RC time constant for a 1.5-μF capacitor and a 10-Ω resistor.

25-8. Find the charge and the discharge current for the circuits of (a) problem 25-6 and (b) problem 25-7, when a 15-V dc source is used.

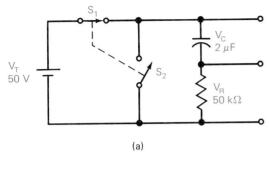

(a)

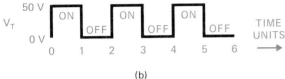

(b)

FIGURE 25-8 Circuit used to develop the waveforms shown in Figure 25-9.

25-7 Wave Shapes for *RC* Circuits

The concept of use of a varying voltage to charge and discharge a capacitor is illustrated with the circuit shown in Figure 25-8(a). This circuit contains a 50-V dc source connected in series with a 2-μF capacitor and a 50-kΩ resistor. Two switches are also included in the circuit. One, S_1, is used to connect the dc source to the series *RC* circuit. The second switch, S_2, will short circuit the *RC* circuit when it is closed. The action of these two switches is on an alternating basis. In other words, S_1 is closed while S_2 is open and then S_1 is opened and S_2 is closed. The use of a dashed line connecting the two switches on the schematic diagram indicates that they are mechanically connected to each other. Since one is shown in the closed, or on, position and the other is shown in the open, or off, position, they will both reverse their electrical connections when they are moved to the opposite position. This is equivalent to applying a square-wave signal source to the circuit [Figure 25-8(b)].

The waveforms produced by this action are shown in Figure 25-9. The wave shapes in Figure 25-9 show three different events related to the *RC* circuit. In part (a) voltage is developed across the capacitor's plates. This is identified as V_C. Four single time-constant units are shown in this drawing. During the first time constant the voltage on the capacitor rises to 63.2% of the 50 V applied to the circuit. In this example, this voltage is 31.6 V. This rise in voltage occurs as the positive portion of the square wave is applied to the circuit.

During the second time constant the input voltage is at zero. The capacitor voltage, therefore, falls to 63.2% of its charged value. In this example the voltage falls to 11.7 V. This occurs during the second *RC* time period.

During the third time constant the voltage is again applied to the plates of the capacitor. The result is an increase in the value of this voltage. The voltage value starts with a value of 11.7 V and rises to 63.2% of the difference between the 50 V applied and the 11.7 V. This value is added to the low voltage of 11.7 V. It is calculated by first finding the difference between 50 and 11.7, or 38.3 V. Once this value is identified, it is multiplied by 63.2% to determine the amount of voltage increase. This value of 24.2 V (38.3 × 0.632) is added to the low-voltage value of 11.7 V. The voltage rise during the third time constant is to a value of 24.2 V + 11.7 V, or 35.9 V.

The fourth time constant indicates a fall in the capacitor plate voltage. It falls to 63.2% of its charged value. This is determined by multiplying the value of 35.9 V by 36.8%. The value of this voltage is 13.25 V at the end of the fourth time constant.

Current in the circuit also rises and falls due to the change in the amplitude of the applied square-wave voltage [Figure 25-9(b)]. We have to assume that the capacitor does not have any charge on its plates when 50 V is initially applied to the RC circuit. Circuit current is then determined by the value of the applied voltage and the amount of resistance. In this circuit the current is $I = V/R$, or 50/50,000. The value of current is 1 mA at this moment.

Current immediately rises to its Ohm's law value of 0.001 A, or 1 mA. The current then falls by 63.2% as the capacitor is charged. This value is 0.37 mA. The capacitor is discharged during the second time interval and the current now falls. This is illustrated by a negative current value because the discharge current is reversed and returns to the source during this part of the cycle. Discharge current is 63.2% of the full current value of 1 mA. It falls by 63.2% of this value. The current level at the end of the second time period is about 0.23 mA. Once again the applied voltage is placed across the RC circuit. Current now rises to a value that is 63.2% of the difference between its low point (−0.23 mA) and the maximum value (1 mA). This value is about 0.77 mA and falls 63.2% of this value. The current value at the end of the third time period is about 0.28 mA.

The final segment of this cycle shows the current again returning to the voltage source. Current starts during this cycle at a level of 0.71 mA and falls to a low value of 0.27 mA at the end of the period.

Figure 25-9(c) shows the voltage drops across the resistor in the circuit. Voltage drops are calculated by using the formula $V_R = IR$. Voltage across the resistor is maximum at the moment the switch S_1 is closed. This value drops to 63.2% of the maximum during the first time constant. The value at the end of this period is 18.4 V. The values of current as determined in the second drawing in this series can now be used to calculate the voltage at the start and end of each time period. For example, the voltage at the end of the first time period is 0.368 × 50, or 18.4 V. Voltage at the start of the second time period is −0.637 × 50, or −31.5 V. This process can be continued to determine the value of voltage for each time constant.

The importance of understanding the process and results of obtaining wave shapes will be clarified as one undertakes additional studies of electrical and electronic concepts. Many timing circuits used in electronic devices are constructed with RC components. The discussion in the following section is related to two of the basic circuits, integrators and differentiators. These circuits are developed by using different time constants. They are illustrated in Figure 25-10 and are developed across either the capacitor or the resistor in the series circuit.

Integrators *Circuit using voltage developed across capacitor in* RC *circuit.*

Integrators This circuit uses the voltage developed across the capacitor in an RC circuit. When this voltage is used as the output of the RC circuit and

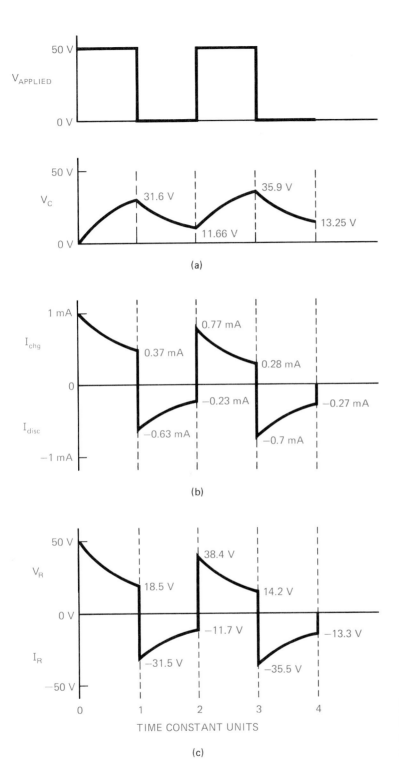

(a)

(b)

(c)

FIGURE 25-9 Waveforms produced by action of the switching circuit shown in Figure 25-8.

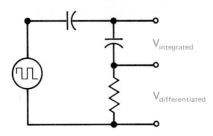

FIGURE 25-10 Circuit used to develop either an integrated or a differentiated wave shape.

connected to a timing circuit, it usually has a relatively long period. The rise or fall in voltage has to develop as it accumulates over this time period.

Differentiation *Voltage developed across resistor in RC circuit.*

Differentiation The voltage that develops across the resistor is an instantaneous value, rising to the value of the applied voltage. The result is a rapid change and a sharp short pulse of voltage.

REVIEW PROBLEMS

25-9. Why is the voltage developed across a capacitor considered as an integrated voltage?

25-10. Why is the voltage across a resistor considered as a differentiated voltage?

CHAPTER SUMMARY

1. Transient time is the time required for a voltage or current to change from one value to another value in a circuit.
2. A circuit containing resistance and inductance has a *L/R* time constant.
3. A circuit containing resistance and capacitance has a *RC* time constant.
4. Five time constants are required to fully charge or fully discharge the reactive component.
5. Current or voltage *rises* in a charging circuit.
6. Current or voltage *decays* in a discharging circuit.
7. Induced voltage is used to create a short-duration high voltage in a circuit.
8. Waveforms for rise and decay during time constants are displayed on a universal time-constant chart.
9. High currents can be created when a capacitor is discharged in a very short period.
10. Output voltages taken across the capacitor in a *RC* circuit are called integrated voltages.
11. Output voltages taken across the resistor in a *RC* circuit are called differentiated voltages.
12. Reactances are applicable only in circuits using sine-wave values.
13. Non-sine-wave circuits use time-constant values instead of reactance values.

Answer true or false; change all false statements into true statements.

25–1. The transient time is the time when a change in circuit values occurs.

25–2. In a dc resistive circuit transient time is almost instantaneous.

25–3. Resistance is inversely proportional to inductance in a circuit containing both values.

25–4. Capacitance is inversely proportional to resistance in a circuit containing both values.

25–5. The time constant is the time in seconds for a change of 63.2% to occur in the circuit.

25–6. Time constants refer only to the rise time in a circuit.

25–7. An induced voltage can be used to develop a very high secondary voltage in an inductive circuit.

25–8. A charged capacitor will create a high voltage when discharged across an inductor.

25–9. Six time constants are required for a full charge or discharge of a capacitor.

25–10. Voltage developed across a capacitor in a *RC* circuit is called the differentiated voltage.

25–11. In a purely resistive circuit, the *V* and *I* are
  (a) out of phase by 90°
  (b) in phase
  (c) out of phase by 120°
  (d) out of phase by 180°

25–12. The relationship between *R* and *L* is
  (a) directly proportional
  (b) inversely proportional
  (c) independent of each other
  (d) none of the above

25–13. The time required for a total change in circuit current is called its
  (a) transient time
  (b) rise time
  (c) fall time
  (d) rise and fall time

25–14. The open circuit has a(n) _____ value of resistance to current flow.
  (a) very low
  (b) medium
  (c) frequency dependent
  (d) infinitely high

25–15. When using high-frequency power supplies, the preferred waveform for operation is a _____ wave.

(a) sine
(b) square
(c) transient
(d) sawtooth

25-16. The formula for resistive and capacitive time constants is

(a) $T = RC$
(b) $T = R/C$
(c) $R = T/C$
(d) $R = TC$

25–17. The number of time constants required for a device to reach 86.46% of its full value is

(a) one
(b) two
(c) three
(d) four

25–18. The voltage developed across a capacitor in an RC circuit is known as the _____voltage.

(a) capacitive
(b) differentiated
(c) integrated
(d) resistive

25-19. The voltage developed across a resistor in an RC circuit is known as the _____voltage.

(a) capacitive
(b) differentiated
(c) integrated
(d) resistive

25–20. The process used to create a voltage higher than the applied voltage in a circuit uses the _____principle.

(a) rapid charge
(b) rapid discharge
(c) normal current change
(d) resistive opposition

ESSAY QUESTIONS

25–21. Explain how the universal time-constant chart is used.

25–22. Explain why the L/R time constant is inversely proportional.

25–23. Explain why the LC time constant is directly proportional.

25–24. Why does the short discharge period create a high voltage in an inductor?

25–25. State some common uses for the rapid discharge rate in an LR circuit.

25–26. Explain what happens when an RC circuit is short circuited.

25–27. Explain why the decay of a *RC* circuit voltage does not fall by 63.2% of its full value.

25–28. Why does the differentiated voltage waveform develop a rapid rise in the voltage across the resistance in the circuit?

25–29. Why does the integrated voltage waveform develop a slow rising or falling wave shape?

25–30. Why do we use time constants when determining rise or fall of *V* or *I* in a circuit?

25–31. Find the time constant for a circuit containing 2.5 kΩ in series with a 0.075-μF capacitor. (*Sec. 25-5*)

25–32. What is the charge on a capacitor after one time constant when a 15-V square wave is applied to it? (*Sec. 25-5*)

25–33. Find the charge on the capacitor in problem 25-32 after time constants of 2, 3, and 4. (*Sec. 25-5*)

25–34. Find the voltage after two time constants of discharge for a capacitor having an initial charge of 25 V. (*Sec. 25-5*)

Find the time constant for the circuit described in problems 25–35 through 25–39.

25–35. $C = 10$ μF, $R = 150$ Ω (*Sec. 25-5*)

25–36. $C = 1$ μF, $R = 1$ kΩ (*Sec. 25-5*)

25–37. $C = 50$ pF, $R = 5$ mΩ (*Sec. 25-5*)

25–38. $C = 0.005$ μF, $R = 6.8$ kΩ (*Sec. 25-5*)

25–39. $C = 0.01$ μF, $R = 1$ MΩ (*Sec. 25-5*)

25–40. Calculate the time required for each *RC* combination in problems 25–35 through 25–39 to reach a full charge. (Assume that they have no charge at the start.) (*Sec. 25-5*)

Use Figure 25-11 to answer problems 25–41 and 25–42.

25–41. Find the capacitor voltage after periods of 10, 20, 30, 40, and 50 μs. (*Sec. 25-5*)

25–42. Find the capacitor voltage after two time constants when the applied voltage is 30 V. (*Sec. 25-5*)

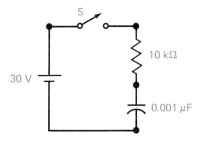

FIGURE 25-11

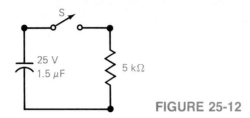

FIGURE 25-12

Use Figure 25-12 to answer problems 25–43 through 25–45.

25–43. What is the capacitor voltage after one time constant? (*Sec. 25-5*)

25–44. What is the capacitor voltage when the switch is closed after 1.5 ms? (*Sec. 25-5*)

25–45. What is the capacitor voltage after the switch is closed for a period of 2.25, 30, and 37.5 ms? (*Sec. 25-5*)

Find the time constant for the circuit combinations given in problems 25–46 through 25–50.

25–46. $L = 10$ H, $R = 10$ kΩ (*Sec. 25-2*)

25–47. $L = 10$ mH, $R = 10$ kΩ (*Sec. 25-2*)

25–48. $L = 50$ mH, $R = 500$ Ω (*Sec. 25-2*)

25–49. $L = 150$ μH, $R = 10$ Ω (*Sec. 25-2*)

25–50. $L = 20$ μH, $R = 100$ Ω (*Sec. 25-2*)

25–51. A 10-μF capacitor is charged to 50 V through a 10-kΩ resistor and then discharged through a 10-Ω load. Find the initial charge and the discharge current. (*Sec. 25-5*)

25–52. A 1.0-μF capacitor is charged through a 1-kΩ resistor to 100 V after five time constants. How long does it take to discharge this capacitor by 85% of its full charge value? (*Sec. 25-5*)

25–53. Find the time constant for a series circuit containing L of 300 μH and R of 75 Ω. (*Sec. 25-2*)

25–54. What is the inductor current after two time constants in problem 25–53? (*Sec. 25-2*)

25–55. How long does it take current in problem 25–53 to reach its maximum steady-state value? (*Sec. 25-2*)

ANSWERS TO REVIEW PROBLEMS

25–1. 0.05 s

25–2. 0.005 s

25–3. 0.07584 A

25–4. Charge time interval

25–5. 11.03 V for first time period; 4.06 V for second time period

25–6. 15 ms

25–7. 15 μs

25–8. (a) 1.5 mA: (b) 1.5A

25–9. It is accumulated over a period of time.

25–10. It is the rapid change occurring across the resistance.

CHAPTER 26

AC Circuits

One of the prime uses of an ac circuit containing both inductive and capacitive reactances is called the resonant circuit. This circuit is related to a frequency or a group of frequencies. It is used to either accept or reject the design frequency or frequencies. A common use is the tuner section of almost all of our radios and television receivers. Resonant, or tuned, circuits are also used in the creation and amplification of radio-frequency signals. The use of the resonant circuit restricts the signal to one specific frequency for proper operation.

KEY TERMS

Apparent Power

Opposite Reactances

Power Factor

Power Factor Correction

Reactive Power

Rectifier

Wattmeter

OBJECTIVES

Upon completion of the material in this chapter, you will

1. Recognize the X_L and X_C components of circuits.
2. Be able to analyze a series ac circuit.
3. Be able to analyze a parallel ac circuit.
4. Understand the concept of real power.
5. Understand the application of ac meters to measure circuit values.

The presentation in this book to this point has involved the understanding of circuits containing only resistance, resistance and inductance, and resistance and capacitance. Many circuits in common use today have an ac voltage applied to them. The factors influencing current and voltage drops still apply to these circuits. Both inductive reactance and capacitive reactance introduce a phase-shift factor. This factor must be considered when one attempts to solve the ac circuit for voltage drops and current flow. The material in this chapter presents methods of solving ac circuit problems.

The dc resistive circuit has no phase shift between voltage and current. They are in phase with each other. In a purely resistive ac circuit there also is a zero phase shift. This is because there are no reactive components in the circuits.

26-1 Purely Resistive Circuits

A series circuit having an ac source will react in a manner similar to when a dc source is applied to it. Figure 26-1 shows two 100-Ω resistors connected in series to a 100-V ac source. The rules established by Ohm and Kirchhoff apply to this circuit. Total resistance is equal to the sum of the individual resistances in the circuit. In this example,

$$R_T = R_1 + R_2$$
$$= 100 \ \Omega + 100 \ \Omega$$
$$= 200 \ \Omega$$

Current flow is determined by use of Ohm's law:

$$I_T = \frac{V_T}{R_T}$$
$$= \frac{100 \ V}{200 \ \Omega}$$
$$= 0.5 \ A$$

Since this is a series circuit the current is equal throughout the circuit, thus:

$$I_{R1} = 0.5 \ A \qquad \text{and} \qquad I_{R2} = 0.5 \ A$$

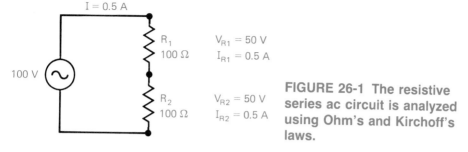

FIGURE 26-1 The resistive series ac circuit is analyzed using Ohm's and Kirchoff's laws.

The voltage drops across the individual resistance are determined as the *IR* drops:

$$V_{R1} = IR_1$$
$$= 0.5 \text{ A} \times 100 \text{ }\Omega$$
$$= 50 \text{ V}$$

and

$$V_{R2} = IR_2$$
$$= 0.5 \text{ A} \times 100 \text{ }\Omega$$
$$= 50 \text{ V}$$

When the components are connected in parallel (Figure 26-2), the rules for parallel circuits having a dc source also apply.

$$R_T = \frac{R_1}{2}$$
$$= \frac{100 \text{ }\Omega}{2}$$
$$= 50 \text{ }\Omega$$

Current in the circuit is determined as follows:

$$I_T = \frac{V_T}{R_T}$$
$$= \frac{100 \text{ V}}{50 \text{ }\Omega}$$
$$= 2 \text{ A}$$

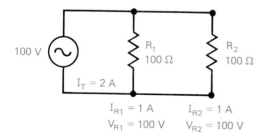

FIGURE 26-2 The parallel resistive ac circuit is also analyzed using the laws of Ohm and Kirchhoff.

and

$$I_{R1} = \frac{V_{R1}}{R_1}$$

$$= \frac{100 \text{ V}}{100 \text{ } \Omega}$$

$$= 1 \text{ A}$$

Since the values of R_1 and R_2 are the same, the current through R_2 is the same as I_{R1}.

26-2 Capacitive Reactive Circuits

The rules for circuits containing capacitive reactances and no resistance are the same as the rules for series resistive circuits. The total reactance for the series circuit shown in Figure 26-3 is the sum of the individual reactances, or 200 Ω. The voltage drops that develop across each reactance add to equal the applied voltage of 100 V. Since these reactances are equal, the voltage drops are also equal. Each reactance has a voltage drop of 50 V in this circuit.

The phase angle of the current when compared to the voltage in this circuit is 90° out of phase. In a capacitive circuit the voltage lags the current by 90°. This may also be expressed as a value of $-90°$.

When the circuit has parallel reactances (Figure 26-4), the rules for a parallel resistive circuit are utilized. In this circuit the total reactance is 50 Ω. Current flow is 2 A for the complete circuit. Current through each branch of the circuit is 1 A. In this circuit the current will lead the voltage in a positive manner since it is a parallel circuit. This may be expressed as a value of $+90°$.

REVIEW PROBLEMS

26-1. Describe the phase angle between voltage and current in Figure 26-3.

26-2. Describe the phase angle between voltage and current in Figure 26-4.

FIGURE 26-3 Reactance values add in the series reactive circuit.

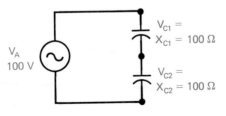

FIGURE 26-4 Rules for parallel resistances apply to parallel reactive circuits.

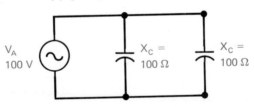

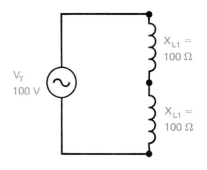

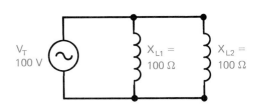

FIGURE 26-5 Rules for series resistances apply for series inductances as well.

FIGURE 26-6 Rules for parallel resistances apply for parallel-connected inductances.

26-3 Inductive Reactive Circuits

The circuit shown in Figure 26-5 is a series inductive circuit. It contains a total of 200 Ω of inductive reactance, since the series-connected reactances add. The values of each reactance are equal; therefore, the voltage drops across the reactances are also equal. In this circuit the voltage drops are 50 V each. The total current flow is determined by use of Ohm's law. The total current in this circuit is 0.5 A. The current is also equal throughout the circuit since it is a series type.

The phase angle between the voltage and the current in an inductive reactive circuit is 90°. In this circuit the voltage leads the current by the 90° factor. The phase angle for this circuit is $+90°$.

When the reactances are connected in parallel (Figure 26-6), the rules for parallel resistances will apply. In this circuit the current flow through each branch is 1.0 A. The total current is the sum of the individual branch currents, or 2.0 A. The branch currents lag the branch voltage by a value of 90°. This is also expressed as $-90°$.

REVIEW PROBLEMS

26-3. Describe the phase angle between voltage and current in Figure 26-5.

26-4. Describe the phase angle between voltage and current in Figure 26-6.

26-4 Cancellation of Opposite Reactances

In any circuit containing components having opposite phase angles the angles will attempt to cancel each other. When two phase angles are out of phase, one being $+90°$ and the second being $-90°$, the cancellation is total. When the phase angles are less than 180° apart, the result is the sum (or difference)

Opposite reactances Circuit where opposite value phase angles are created due to influence of inductor and capacitor.

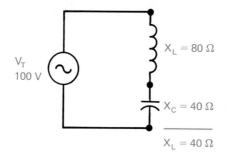

$X_L = 80\ \Omega$

V_T
100 V

$X_C = 40\ \Omega$

$X_L = 40\ \Omega$

FIGURE 26-7 In the series capacitive and inductive circuit, the reactances are 180° out of phase and the total is the sum of the two values.

of the two phase angles. Such is the case for circuits containing both inductive reactance and capacitive reactance. When these two reactance factors are in series, the result becomes the difference in the two reactances. When the two reactances are in parallel, the result is the interaction of the two branch currents. The total current in a parallel circuit containing both inductive and capacitive reactance is a value of current that is the sum of the value of the branch currents.

Series Connection of X_L and X_c In Figure 26-7, two reactances are connected in series. One of these is an inductive reactance. Its value is 80 Ω. The other is a capacitive reactance having a value of 40 Ω. These two reactances are out of phase with each other. The total reactance in this circuit is the difference between the two values, or 40 Ω. Since there is a higher value of inductive reactance than the value of capacitive reactance, the circuit is inductive in nature. The total reactance in the circuit is 40 Ω of inductive reactance, X_L.

Current in this circuit is calculated by using the total reactance value and the value of applied voltage.

$$I_T = \frac{V}{X_L}$$

$$= \frac{100\ \text{V}}{40\ \Omega} \tag{26-1}$$

$$= 2.5\ \text{A}$$

The voltage drops across each of the components is calculated by use of Ohm's law:

$$V_{XL} = IX_L$$

$$= 2.5\ \text{A} \times 80\ \Omega$$

$$= 200\ \text{V} \tag{26-2}$$

and

$$V_{XC} = IX_C$$

$$= 2.5\ \text{A} \times 40\ \Omega \tag{26-3}$$

$$= 100\ \text{V}$$

The individual reactances produce voltages that total a value higher than

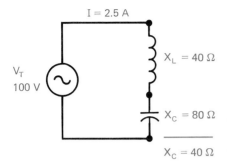

FIGURE 26-8 The larger of the two reactance values dominates the reactance in this circuit.

the applied voltage. The reason for this is that the phase angles of the two voltages are the opposite of each other. In a capacitive circuit the voltage lags the current by a value of 90°. In the inductive circuit the voltage leads the current by a value of 90°. These two factors cancel each other. In this example the voltages $+200$ V and -100 V are added. The result is the value of the applied voltage of 100 V.

When the two values of reactance are reversed (Figure 26-8), the effect is similar. In this circuit the major influence results from the addition of the two reactances. The total circuit reactance is a capacitive value of 40 Ω. Circuit current is the same as in the previous example. It has a value of 2.5 A. The voltage drops across each component are also calculated in the same manner as before. In this circuit the voltage across the X_C component is 200 V and the voltage across the X_L component is -100 V. The net voltage in this circuit is 100 V when these two values are added. This is equal to the value of the applied voltage.

Parallel X_C and X_L Components In Figure 26-9 the two components are connected in parallel. The current through the 80-Ω X_L component is 1.25 A when calculated using Ohm's law. Current through the X_C component is 2.5 A using Ohm's law. Each branch current is calculated without regard for any phase shift.

When the phase angles of the two components are considered, they are 180° out of phase with each other. Therefore, the total current is calculated when one is a positive value and the other is considered a negative value. Total current in this circuit becomes

$$I_T = I_{XC} + (-I_{XL})$$

$$= 2.5 \text{ A} - 1.25 \text{ A} \tag{26-4}$$

$$= 1.25 \text{ A}$$

Since the current in the capacitive circuit leads the voltage and the current in the inductive circuit lags the voltage, the X_C current is considered to be positive.

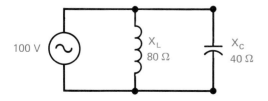

FIGURE 26-9 Current values are determined in parallel reactive circuits.

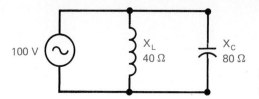

FIGURE 26-10 Current is inductive in this circuit.

When circuit values are reversed (Figure 26-10), the calculations for values are accomplished in the same manner:

$$I_{XC} = \frac{V}{-X_C}$$

$$= \frac{100 \text{ V}}{-80 \text{ }\Omega}$$

$$= -1.25 \text{ A} \qquad (26\text{-}5)$$

and

$$I_{XL} = \frac{V}{X_L}$$

$$= \frac{100 \text{ V}}{40 \text{ }\Omega}$$

$$= 2.5 \text{ A} \qquad (26\text{-}6)$$

In this circuit the major current in inductive. The total current is still determined by adding the two values. The result is a current that is inductive in nature and has a value of the difference between the two currents, or 1.25 A.

REVIEW PROBLEMS

26-5. Using Figure 26-9, find the value of X_L, X_C, and total reactance.
26-6. Using Figure 26-8, find the value of X_L, X_C, and total reactance.

26-5 Parallel Reactive Circuits

Current in a parallel reactive circuit is found by use of phasor addition. The formula for doing this is

$$I_T = \sqrt{I_R^2 + I_X^2} \qquad (26\text{-}7)$$

The total current is calculated for the circuit in Figure 26-11 in the following

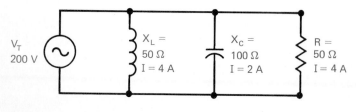

FIGURE 26-11 Phasor addition is used to find total current in this circuit.

manner:

$$I_L = \frac{V}{X_L}$$

$$= \frac{200 \text{ V}}{50 \text{ }\Omega}$$

$$= 4 \text{ A} \qquad (26\text{-}8)$$

$$I_R = \frac{V}{R}$$

$$= \frac{200 \text{ V}}{50 \text{ }\Omega}$$

$$= 4 \text{ A} \qquad (26\text{-}9)$$

$$I_C = \frac{V}{X_C}$$

$$= \frac{200 \text{ V}}{100 \text{ }\Omega}$$

$$= 2 \text{ A} \qquad (26\text{-}10)$$

The next step is to determine to total reactive current. In this circuit it is 2 A, the difference between X_L and X_C. The current is considered to be inductive in nature since the X_L is the dominant factor.

Once the total reactive current is determined, the values of it and the resistance current may be added using phasors.

$$I_T = \sqrt{I_R^2 + I_X^2}$$

$$= \sqrt{4^2 + 2^2}$$

$$= \sqrt{16 + 4} = \sqrt{20}$$

$$= 4.47 \text{ A}$$

The step used to combine the two reactances is illustrated in Figure 26-12. The total reactance in this circuit is inductive in nature, since the current in the X_L branch is 4 A and the current in the X_C branch is 2 A. The total of

FIGURE 26-12 Combination circuit and its phasor diagram.

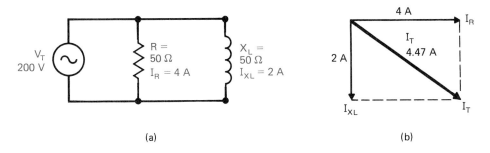

(a)   (b)

these currents is 2 A and its influence is inductive. This may be illustrated by use of a resistance and an inductance in parallel. The phasor diagram is shown in Figure 26-12(b).

Impedance in the circuit is determined by use of the impedance formula:

$$Z = \frac{V}{I_T} \tag{26-11}$$

In this circuit the impedance is 200/4.47, or 44.74 Ω. This is the value of impedance appearing across the source voltage of 200 V.

When multiple resistances and reactances are connected in parallel, the circuit must first be simplified. The easiest method of doing this is to first combine all like component values. In other words, first calculate R_T using all resistance values. Then calculate the combined capacitive reactances and finally the combined inductive reactance values. The procedures for accomplishing this have been discussed in previous chapters. The result is a circuit similar to that illustrated in Figure 26-11. Once this is accomplished the total values can easily be determined. Figure 26-13 illustrates this system.

The order for illustrating the various resistances and reactances has little importance in a parallel branch circuit. This is because they all interact together. Figure 26-13(a) has five discrete components. When they are combined [Figure 26-13(b)], the result is a circuit containing the three components of resistance, capacitive reactance, and inductive reactance. Their respective values are

$$R = 50\ \Omega \qquad X_C = 8\ \Omega \qquad X_L = 18.74\ \Omega$$

FIGURE 26-13 Simplification of a complex circuit by addition of like values. Circuit (a) is equal to circuit (b).

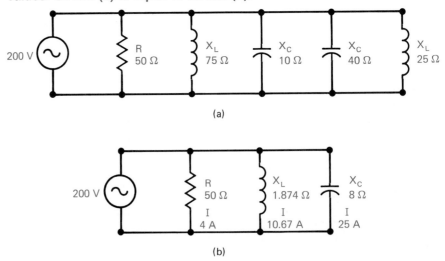

The next step is to determine individual current values, since this is a parallel circuit.

$$I_R = \frac{200}{50} = 4 \text{ A}$$

$$I_{XL} = \frac{200}{18.74} = 10.67 \text{ A}$$

$$I_{XC} = \frac{200}{8} = 25 \text{ A}$$

The dominant current is the X_C current of 25 A. It is subtracted from the X_L current because they are out of phase with each other. The result is an X_C current of 14.33 A and a resistive current of 4 A. Using phasor addition, the current is

$$I_T = \sqrt{I_R^2 + I_X^2}$$
$$= \sqrt{4^2 + 14.33^2}$$
$$= \sqrt{221.349}$$
$$= 14.877 \text{ A}$$

REVIEW PROBLEMS

26-7. Find the total reactance for a parallel circuit containing X_L of 40 Ω and 50 Ω, X_C of 100 Ω and 20 Ω, R of 10 Ω.

26-8. Find the branch currents in problem 26-7 when a source voltage of 50 V is applied to the circuit.

26-6 Series Connection of X_L and X_C Components

Circuits containing both resistances and reactances in series are resolved in the following manner. First, the reactances must be combined by use of phasors. When working with a series circuit such as the one shown in Figure 26-14, the value of reactance in units of the ohm is used with the resistance value to find the impedance of the circuit. In this example, as in any series circuit, the values

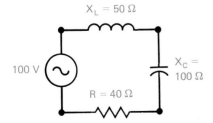

FIGURE 26-14 This circuit has 50 Ω of X_c.

of reactance are 180° out of phase with each other. The result is the sum of the two values, or, in this case, 50 Ω of capacitive reactance.

$$Z = \sqrt{R^2 + X^2}$$
$$= \sqrt{40^2 + 50^2}$$
$$= \sqrt{1600 + 2500} + \sqrt{4100} \qquad (26\text{-}12)$$
$$= 64 \ \Omega$$

Current in this circuit is calculated by use of the formula:

$$I_T = \frac{V}{Z}$$
$$= \frac{100 \ V}{64 \ \Omega}$$
$$= 1.5625 \ A$$

When current is determined and we recognize that in this series circuit the current is equal throughout the circuit, the voltage drops across each component can be determined.

$$V_R = IR$$
$$= 1.5625 \ A \times 40 \ \Omega$$
$$= 62.5 \ V$$
$$V_{XC} = 1.5625 \ A \times 100 \ \Omega$$
$$= 156.25 \ V$$
$$V_{XL} = 1.56 \times 50$$
$$= 78.125 \ V$$

When the values of V_{XC} and V_{XL} are added, their sum is the difference between the two voltages, or 78 V. This sum, when considered as a phasor value, is added to the voltage across the resistance:

$$V_{APP} = \sqrt{V_R^2 + V_X^2}$$
$$= \sqrt{62.5^2 + 78.125^2}$$
$$= \sqrt{3906.25 + 6103.52} \qquad (26\text{-}13)$$
$$= \sqrt{10{,}009.765}$$
$$= 100.05 \ V$$

This value of 100.05 V is a rounded figure and very close to the value of 100 V applied to the circuit.

When multipled components are connected in series (Figure 26-15), the values of like components are combined as a first step in resolving the problem. In this circuit the total value of resistance is 50 Ω, the total X_C is 150 Ω, and the total X_L is 100 Ω. The phase angles of the X_C and X_L components are 180° out of phase; therefore, the total value of the reactive components is the sum

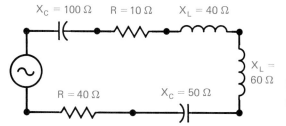

$X_C = 100\,\Omega$ $R = 10\,\Omega$ $X_L = 40\,\Omega$

$R = 40\,\Omega$ $X_C = 50\,\Omega$ $X_L = 60\,\Omega$

FIGURE 26-15 This circuit is simplified by the addition of like values.

of the two values, or 50 Ω. The X_C value is greater than the X_L; therefore, the circuit is capacitive in nature. When reduced to its single values the circuit has 50 Ω of X_C in series with 50 Ω of resistance.

REVIEW PROBLEMS

26-9. Find the total reactance for a series circuit containing 40 Ω R, 120 Ω X_C, and 130 Ω X_L.

26-10. Find the total reactance for a series circuit having 120 Ω of resistance, 40 Ω of X_L, and 80 Ω of X_C.

26-7 Circuits Containing Series and Parallel Components

When a circuit has both a series and a parallel set of components, it must first be simplified. Figure 26-16(a) shows a parallel branch of two 100-Ω inductive reactances. This parallel branch is connected in series with a series string of a 10-Ω resistance and a 60-Ω capacitive reactance. The circuit is considered to be a series circuit with a parallel component. The method of simplifying this is first to combine the parallel components into an equivalent value. The two 100-Ω X_L units exhibit an X_L value of 50 Ω when combined [Figure 26-16(b)]. The combining of values then produces a series circuit containing 10 Ω of resistance and 10 Ω of capacitive net reactance in series [Figure 26-16(c)].

FIGURE 26-16 Simplification of a complex circuit is done using these steps.

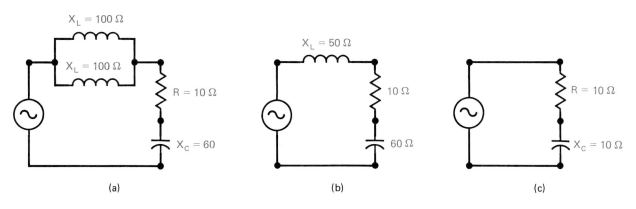

(a)   (b)   (c)

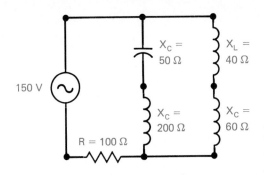

FIGURE 26-17 This circuit also requires simplification before its branch values can be determined.

A more complex circuit is shown in Figure 26-17. There are series components in each of the branch circuits. One contains an X_C of 50 Ω and an X_L of 200 Ω. The second branch has two X_L components in series. The two branches are connected in series with a 100-Ω resistance. The total reactance for the first branch is 150 Ω of X_L. Total reactance for the second branch is 100 Ω of X_L. The total reactance for both branches is found in the same manner as used for parallel resistances:

$$\frac{1}{X_{LT}} = \frac{1}{X_{L1}} + \frac{1}{X_{L2}}$$

$$= \frac{1}{150} + \frac{1}{100} \qquad (26\text{-}14)$$

$$= 0.0067 + 0.01 = 0.0167$$

$$= 60 \ \Omega$$

This value is series connected with the resistance value of 100 Ω. Values for voltage and current can now be determined for this circuit using phasor addition.

$$Z_T = \sqrt{X_L^2 + R^2}$$

$$= \sqrt{60^2 + 100^2}$$

$$= \sqrt{3600 + 10,000} = \sqrt{13,600}$$

$$= 116.62 \ \Omega$$

The current for this circuit is

$$I_T = \frac{V_T}{Z_T}$$

$$= \frac{150 \ V}{116.62 \ \Omega}$$

$$= 1.29 \ A$$

The voltage drops across each of the components are

$$V_{XL} = I_{XL}X_L$$

$$= 1.29 \ A \times 60 \ \Omega$$

$$= 77.4 \ V$$

and

$$V_R = IR$$

$$= 1.29 \text{ A} \times 100 \ \Omega$$

$$= 129 \text{ V}$$

These calculations provide the values for voltage drops across the resistor and the total parallel branch circuit.

26-8 Power in the Reactive Circuit

In a resistive circuit the voltage and the current are in phase with each other. Power in a circuit such as this is the product of voltage and current. When a circuit contains reactive components the voltage and current are not in phase with each other. In one extreme set of circumstances the voltage may be low when the current is high. In another instance the opposite may be true in the circuit. When attempting to calculate power in the resistive circuit the formula to use is $P = VI$. To find power in a reactive circuit, another formula is used:

Reactive power *Electrical power developed in circuit containing inductive or capacitive reactances.*

$$P = I^2R \qquad\qquad (26\text{-}15)$$

This formula produces a value known as the *real power* in the circuit. The circuit shown in Figure 26-18 will show how this is accomplished. In this circuit, real power is found in this manner:

The first step is to determine circuit impedance.

$$Z = \sqrt{R^2 + X_L^2}$$

$$= \sqrt{100^2 + 200^2}$$

$$= \sqrt{50,000}$$

$$= 223.6 \ \Omega$$

The next step is to determine circuit current:

$$I = \frac{V}{Z}$$

$$= \frac{300 \text{ V}}{223.6 \ \Omega}$$

$$= 1.34 \text{ A}$$

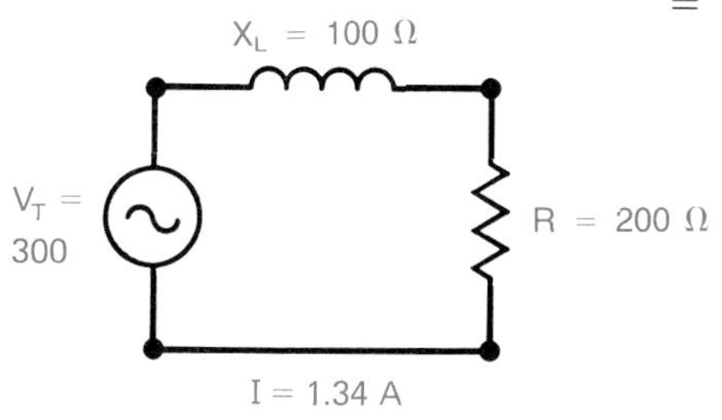

$X_L = 100 \ \Omega$

$V_T = 300$

$R = 200 \ \Omega$

$I = 1.34$ A

FIGURE 26-18 Determination of real power requires the use of the circuit impedance value.

The two values of I and R are used to calculate the real power in this circuit:

$$\text{real power} = I^2R$$

$$= 1.34^2 \text{ A} \times 200 \text{ } \Omega$$

$$= 359.12 \text{ W}$$

Apparent Power The factor of reactance in a circuit produces an out-of-phase condition. If this factor is ignored, the amount of power in the circuit is determined by use of the formula $P = VI$. This value is known as **apparent power** and the term *voltamperes* (VA) is used to describe it. The term *watt* is used to describe units of real power in the circuit.

Power Factor This term is used to indicate the numerical ratio of the factors influencing power in a circuit. In a series circuit, the following formula is used:

$$\text{power factor} = \frac{R}{Z} \qquad (26\text{-}16)$$

In a parallel circuit the following formula is used:

$$\text{power factor} = \frac{I_R}{I_T} \qquad (26\text{-}17)$$

When these formulas are used with the circuit shown in Figure 26-18 the differences are more apparent. For this circuit, the power factor is

$$\text{power factor} = \frac{R}{Z}$$

$$= \frac{100 \text{ } \Omega}{223.6 \text{ } \Omega}$$

$$= 0.447$$

Using the parallel circuit of Figure 26-19, the power factor is

$$\text{power factor} = \frac{I_R}{I_T}$$

$$= \frac{1.5 \text{ A}}{3.35 \text{ A}}$$

$$= 0.45$$

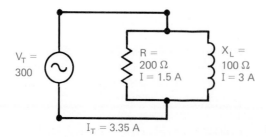

$V_T = 300$ $R = 200 \text{ } \Omega$ $I = 1.5 \text{ A}$ $X_L = 100 \text{ } \Omega$ $I = 3 \text{ A}$

$I_T = 3.35 \text{ A}$

FIGURE 26-19

Correction of Power Factor The values of either of the two power factors are not even close to a factor of unity, or 1. In a practical application the quantity of energy, or power generated by the power company, should be equal to the quantity of power consumed by the user. When the power factor is not close to unity, a power loss occurs. The power-generating company has to increase its operating costs to compensate for the loss of this power. The consumers will be charged a higher rate for power when this condition exists. A remedy for a power factor that is not close to unity is to add some values of opposite reactance to the system. This will offset the phase shift created by the consumer. In a practical situation, the consumer's use of motors and transformers usually makes the circuit inductive. This is offset by adding quantities of capacitance to the circuit.

Power factor correction *Use of additional inductance or capacitance in order to bring V and I values to an in-phase condition in circuit.*

REVIEW PROBLEMS

26-11. Name the units of (a) apparent power and (b) real power.
26-12. What component is added to a capacitive circuit to make the power factor close to unity?

26-9 Measurement of AC Values

The measuring devices for use in alternating-current circuits are not the same as those used in dc circuits. When measuring with a moving needle type of instrument, the ac values will produce an oscillation, or rapid movement left to right, in the meter. Since most meters "rest" on the left side of the scale, this is not practical. Some other means of producing a steady indication must be found. This is accomplished by one of two methods. One of these requires the additions of a component called a **rectifier** in series with the meter movement. The rectifier is a device that limits current flow to one direction in the circuit. When the rectifier is properly series connected to a meter movement, current flow in the circuit will produce an upward movement, or unidirectional movement, to the moving needle in the D'Arsonval meter movement.

Rectifier *Device used to permit flow of current in only one direction in an ac circuit. (Changes ac into pulsating dc).*

A second type of instrument available for the measurement of ac values uses a thermocouple to create an upward movement to the needle of the meter. The thermocouple is a device that produces an electrical charge when it is heated. The thermocouple is made from two dissimilar metals joined together at one end. When this end is heated, a current is produced. The other ends of the thermocouple are connected to a dc meter movement. The flow of electrons created by heat produced a deflection in the meter movement.

The Wattmeter A device called a **wattmeter** is used to measure power in the ac circuit. The wattmeter includes two sets of coils. One of these is connected in series with the ac power line. It measures the ac current in the line. A second coil is connected as a voltmeter in parallel with the ac line. When both connections are complete, the factors of voltage and current influence the deflection of a needle in the movement. Since both current and voltage are necessary to

Wattmeter *Instrument measuring voltage and current and displaying a wattage reading combining both values.*

measure power, both influence the position of the needle. When the two are in phase, the deflection of the needle is high. When the two are out of phase, the deflection of the needle is low. Meter deflection is therefore proportional to the quantity of real power in the circuit.

1. When an ac circuit is purely resistive, the rules presented by Ohm and Kirchhoff may be applied for circuit analysis.
2. Capacitive reactances are added when connected in series.
3. Capacitive reactances are added by the reciprocal formula when connected in parallel.
4. Inductive reactances are added when connected in series.
5. Inductive reactances are added by the reciprocal formula when connected in parallel.
6. X_C and X_L values are opposite-value reactances. Their values cancel each other when they are added together.
7. When series connected, X_C and X_L ohmic values cancel.
8. When parallel connected, X_C and X_L currents cancel.
9. An ac circuit containing X_C, X_L, and R can be reduced to an equivalent circuit containing R and a net X value.
10. Impedance in the reactive circuit uses the formula $Z = V/I_T$.
11. An ac circuit containing reactance uses the formula $P = I^2R$ for the determination of real power.
12. The power factor is used to indicate the relationship of VI power to real power.
13. Ac circuit values may be measured by use of a dc meter and a rectifier.
14. Ac circuit values may also be measured by use of a thermocouple type of meter movement.
15. Ac power is measured by use of a wattmeter.

Answer true or false; change all false statements into true statements.

26–1. The phase shift in a resistive circuit is 25°.
26–2. Voltage lags current in an inductive circuit.
26–3. Voltage lags current in a capacitive circuit.
26–4. Two X_C components in series add to offer a total X_C value.
26–5. The reciprocals of two X_C components in parallel add to offer a total X_C value.
26–6. Two X_L components in series add to offer a total X_L value.
26–7. A dc meter may be used with a rectifier to measure ac circuit values.
26–8. Phasor values are used to calculate impedance in the reactive circuit.
26–9. Phasor values are used to calculate current in the reactive circuit.
26–10. The wattmeter measures both voltage and current in the circuit.

26–11. A capacitive reactive circuit containing no resistance MULTIPLE CHOICE
    (a) uses the rules for series resistances
    (b) uses the rules for parallel resistances
    (c) is equal to a combination series–parallel circuit
    (d) none of the above

26–12. Two equal-value reactances, one X_C and one X_L,
    (a) are in phase with each other
    (b) add to equal zero
    (c) are out of phase by a total of 90°
    (d) has the X_L value lagging the X_C value

26–13. Two equal-value reactances, one X_C and one X_L,
    (a) result in a capacitive circuit
    (b) result in a reactive circuit
    (c) result in an inductive circuit
    (d) result in a resistive circuit

26–14. Current values in the parallel reactive circuit are found by
    (a) use of direct values
    (b) use of phasor values
    (c) use of reactive and resistive values
    (d) use of X_C values only

26–15. The formual $Z = V/I_T$ is used to find
    (a) circuit resistance
    (b) circuit reactance
    (c) circuit impedance
    (d) individual circuit voltage drops

26–16. Why is there a phase shift in a reactive circuit? ESSAY QUESTIONS

26–17. Why do equal values of X_c and X_L result in a resistive circuit?

26–18. Explain the function of a rectifier in the meter circuit when measuring ac values.

26–19. How does a wattmeter measure circuit power?

26–20. Explain the difference between apparent power and real power in a circuit.

26–21. A circuit contains two resistances having values of 1.2 kΩ and 1.8 kΩ.   **PROBLEMS**
What is the total resistance? (*Sec. 26-1*)

26–22. An ac circut contains R of 120 Ω, X_c of 230 Ω, and X_L of 180 Ω. What is the total reactance? (*Sec. 26-2*)

26–23. Find the impedance in problem 26–22. (*Sec. 26-5*)

26–24. A parallel ac circuit has an R of 100 Ω, X_c of 120 Ω, and X_L of 180 Ω. What is the current through each branch when a source voltage of 100 V is applied? (*Sec. 26-4*)

26–25. Find the total circuit current for problem 26-24. (*Sec. 26-4*)

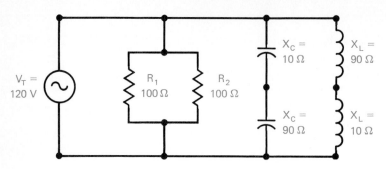

FIGURE 26-20

For problems 26-26 through 26-30, consider a parallel ac circuit that has five branches. The values for each are: R = 400 Ω, X_c = 200 Ω, X_c = 150 Ω, X_L = 450 Ω, and X_L = 50 Ω.

26–26. Find total values of R, X_c, and X_L. (*Sec. 26-4*)

26–27. Determine current through each branch when 150 V is applied to the circuit. (*Sec. 26-4*)

26–28. Calculate the total impedance. (*Sec. 26-4*)

26–29. Calculate the total source current. (*Sec. 26-4*)

26–30. Calculate the real power (*Sec. 26-4*)

26–31. A series circuit contains an X_L of 30 Ω, a X_c of 25 Ω, and a R of 100 Ω. Find the total circuit impedance. (*Sec. 26-4*)

26–32. A series circuit has a R of 150 Ω, a capacitance of 25 μF, and an inductance of 5 H. Find the reactances when 120 V 60 Hz ac is applied to the circuit. (*Sec. 26-4*)

26–33. What is the circuit current for problem 26–32? (*Sec. 26-4*)

26–34. What are the values of reactances when the components in problem 26-32 are connected in parallel to a 120-V 60-Hz source? (*Sec. 26-4*)

26–35. Find the current through each branch for problem 26-34. (*Sec. 26-4*)

Use Figure 26-20 to answer problems 26-37 through 26-40.

26–36. Find the impedance for the circuit in problem 26-32. (*Sec. 26-4*)

26–37. Find the reactance for each branch. (*Sec. 26-4*)

26–38. Find branch currents for each branch. (*Sec. 26-4*)

26–39. Calculate total impedance. (*Sec. 26-4*)

26–40. Calculate voltage drops across each component. (*Sec. 26-4*)

ANSWERS TO REVIEW PROBLEMS

26–1. Voltage lags current by 90°.

26–2. Current leads voltage by 90°.

26–3. Voltage leads current by 90°.

26–4. Current lags voltage by 90°.

26–5. X_L = 80 Ω, X_c = 40 Ω, X_T = 40 Ω

26–6. X_L = 40 Ω, X_c = 80 Ω, X_T = 40 Ω

26–7. X_{LT} = 22.2 Ω, X_{CT} = 16.67 Ω, net X = capacitive

26–8. $I_{XL} = 2.25$ A, $I_{XC} = 2.99$ A, $I_R = 5$ A

 26–9. 10 Ω

26–10. 40 Ω

26–11. (a) VA; (b) watt

26–12. Inductance

CHAPTER 27

The *J* Operator and Complex Numbers

Many of the ac circuit values may be resolved using a system that includes an imaginary number. The values of inductance and capacitance offer a 90° phase shift between voltage and current in the circuit. This effect creates values that are not normally measured, but must be considered in the process of circuit analysis. The complex-number system uses the symbol *j* to represent the reactive quantity and a whole number to represent the resistive values in the circuit.

Admittance Y *j* **Operator**
Complex Number   **Polar Form**
Imaginary Axis   **Rectangular Form**
Imaginary Number   *RLC* **Circuit**
j **Factor**   **Susceptance** B

OBJECTIVES

Upon completion of the material in this chapter, you will

1. Understand positive and negative numbers.
2. Understand the *j*-operator system.
3. Be able to analyze a circuit for its impedance.
4. Be able to use complex numbers in polar and rectangular form.
5. Be able to analyze ac circuits using complex-number form.

27-1 The *j* Operator

Alternating-current circuit analysis requires determination of both the *magnitude* and the *phase angle* of the basic electrical quantities. These two factors of phase angle and magnitude can also be expressed by use of phasors. It is not always possible to show exact phasor values or diagrams in discussing these terms. Often one has to describe them using their mathematical values. The manipulation of these complex values becomes very difficult at times. When the phasor values are shown as complex numbers it is often much easier to combine their values. The system in common use for doing this is directly related to the expression of these values as complex numbers. Once shown in this manner, it becomes relatively easy to manipulate them mathematically. The use of the complex number system permits the addition, subtraction, multiplication, and division of phasor quantities in a very systematic format.

Perhaps the best method of explaining the complex-number system is to start by relating how number systems have developed. People have always required some sort of numbering system in order to express quantities. In the early days of human life, quantities were very simple. People were only required to express quantities in values that were greater than the single unit, or 1. This, of course, was necessary to discuss the number of children in the family, how many days were required to travel between two points, and many other relatively simple things. Numbers were limited to basic units of 1, 2, 3, and so on. As the need developed, it was possible to use a simple subtraction system in order to show a reduction in the previously expressed quantities.

Problems arose when attempting to show parts of the whole system. It was not possible to present fractions of the whole numbers in the early numbering systems. As a result of this limitation, the fractional numbering system was developed. These numbering systems, including the fractions, were very adequate until another need developed. This need was for a method of expressing negative numbers, as required for temperature. This need, of course, did not occur until the development of temperature-measuring devices. In short, a new numbering system was developed as the need for it was created.

As the years progressed, the creation of the square-root numbering system occurred. Consider a series of equations using the square-root system:

$$\sqrt{25} = \pm 5$$

$$\sqrt{2} = \pm 1.414$$

$$\sqrt{1} = \pm 1$$

$$\sqrt{0} = \pm 0$$

These illustrations are not unusual and a system of numbers is available to represent each of the square-root values. Now, consider what happens when one attempts to determine the square root of a negative number:

$$\sqrt{-1} = \text{does not compute}$$

Since it is not possible to show the answer to the problem using the normal numbering system, another system was required. This system uses an **imaginary-number** process. In mathematics, the imaginary number is expressed as *i*. This is expressed as follows:

$$\sqrt{-25} = \sqrt{-1}\,\sqrt{+25} = 5\sqrt{-1} = \pm 5i$$

In electrical and electronic terminology, the letter *i* is used to indicate values of electrical current. Another letter was required to minimize confusion between current values and imaginary number values. The letter *j* is thus used in electrical and electronic mathematics to represent an imaginary number. This is shown as

$$5\sqrt{-1} = \pm 5j \quad \text{instead of} \quad \pm 5i$$

The *j*, or *j* **factor**, has some additional characteristics that are very interesting. These are shown below.

$$j = \sqrt{-1}$$
$$j^2 = \sqrt{-1} \times \sqrt{-1} = -1$$
$$j^2 = j(j^2) = -\sqrt{-1} \tag{27-1}$$
$$j^4 = (j^2)(j^2) = (-1)(-1) = +1$$

The relationship shows that there is no need to use any *j* number having a power greater than 1. This fact is important when attempting to reduce any mathematical expression containing the letter *j*, or *j* factor. The letter *j*, or *j* factor, in electrical problem analysis is presented as the use of the '*j* operator.' This means that the '*j*' portion of the circuit is developed as an imaginary number.

A mathematical expression may contain both real numbers and imaginary numbers. These two numbers can be shown in graphic form. Figure 27-1 shows how real numbers are expressed. The numbers to the right of the vertical dividing line are considered to be positive. The numbers to the left of the vertical line are negative in nature. Any number can be located as a point along the horizontal axis of the graph. In this illustration both a positive $+4$ and a negative -3 are located.

The complex plane shown in Figure 27-2 has both a real and a complex axis. The horizontal axis contains both positive and negative real numbers. The vertical axis represents both positive and negative *j* values. The vertical axis is identified as the **imaginary axis**. All numbers on this imaginary axis are rep-

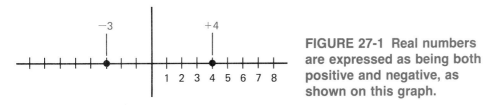

FIGURE 27-1 Real numbers are expressed as being both positive and negative, as shown on this graph.

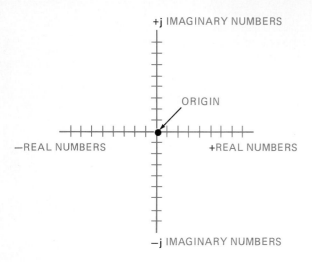

FIGURE 27-2 Use of a complex plane to locate real and imaginary numbers.

resented with the letter j. In this example, the origin of the plane is the zero point in the middle where the two axes cross.

The complex plane is used to represent angular values of numbers (Figure 27-3). Each of the four quadrants of the graph is represented as it relates to the rotation of a vector or phasor through a circular rotation. In this illustration the zero axis is the positive real-number axis. It represents both 0° and 360°. Moving in a counterclockwise direction, the $+j$ axis represents a 90° rotation. As one continues around the circle, the next axis is the negative real axis and is considered as 180° of rotation. The third axis is located at 270° and is the $-j$ axis. Finally, the circular motion is concluded at the zero or 360° point. This plane is divided into four quadrants.

Another method of showing this j-operator system is to relate it to the

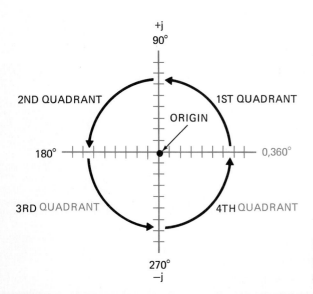

FIGURE 27-3 The complex plane is used to represent angular values.

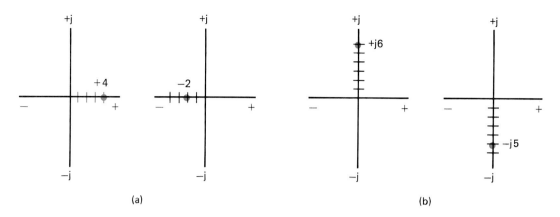

FIGURE 27-4 Location of real (a) and imaginary (b) numbers on the complex plane.

degrees of rotation of a vector or phasor:

$$0° = 1$$
$$90° = j$$
$$180° = j^2 = -1$$
$$270° = j^3 = j^2 \times j = -1 \times j = -j$$
$$360° = \text{same as } 0°$$

Both real and imaginary numbers can be represented on this complex plane Figure 27-4(a) shows how real numbers are located on their respective segments of the plane. Figure 27-4(b) shows the location of imaginary numbers.

When a point does not fall on either axis but falls at a point somewhere in one of the four quadrants, the point is considered to be a *complex number*. This type of number is represented by its coordinates. In Figure 27-5 the point in the first quadrant has a real value of $+2$ and an imaginary value of $+j3$. In

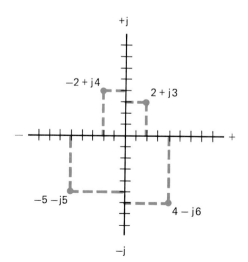

FIGURE 27-5 Relationship and location of complex numbers on the complex plane.

the second quadrant, the point has a real value of -2 and an imaginary value of $+j4$. In the third quadrant, the point is represented by -5 real value and an imaginary value of $-j5$. The final quadrant's point is coordinated at $+4$ and $-j6$.

27-2 The Rectangular Form of Complex Numbers

Complex number *Number having both real and imaginary values.*

A **complex number** is defined as a number that has both a real and an imaginary component. The vector R can be expressed as a complex number. This is expressed as

$$R = A + jB \qquad (27\text{-}2)$$

Rectangular form *Expression of complex number using two values that are 90° apart.*

This expression is the **rectangular form** of the complex number. It is expressed as the sum of the two components that are 90° apart.

When a real and an imaginary number are combined, the result is considered to be a *complex number*. One example of this is the expression $4 + j5$. This is an expression of a complex number. The first number, 4, indicates a value of 4 on the real axis. The second number, $+j5$, shows an imaginary number 5 units away from a common zero reference point on the imaginary axis. This second number is 90° out of phase with the first number. The values of a complex number must be added as phasors.

A series of phasors for complex numbers are shown in Figure 27-6. The use of the $+j$ phasor gives the indication that the value is shown as being up on the 90° line. When the phasor is a $-j$ value, this is shown as being down on the 90° line, or below the zero reference point.

These phasors are shown with the end point of one connected to the start of the second phasor. This indicates that they are in a position to be added. The sum of the two phasors forms the hypotenuse of the right triangle. When the two phasors form a partial rectangle the values represent the fact that the phasors are in *rectangular* form. Another way of stating this is to say that this is the rectangular form of the complex number.

The size of each of the values in the complex-number system has a relation to the position of the phasors. Either the real value or the imaginary value can be the larger. A complex number having a real value larger indicates an angle of less than 45° from the real axis. When the imaginary term is larger than the real term, the angular value is greater than 45° from the real axis. If both values are equal, the angle is exactly 45°.

FIGURE 27-6 Use of a vector to locate the complex number.

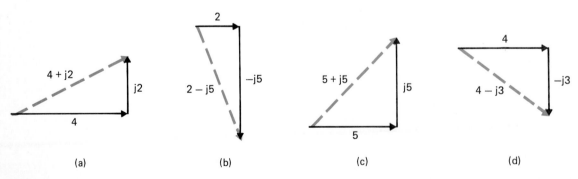

(a)   (b)   (c)   (d)

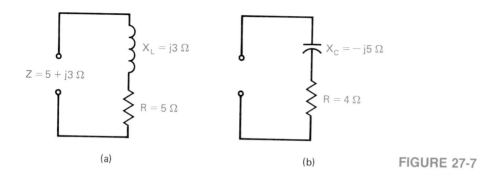

(a)

(b)

FIGURE 27-7

27-3 Complex Numbers in AC Circuits

Phase angles are directly related to the use of a $-j$ to represent $-90°$ values and $+j$ to represent a value of $+90°$. These rules are illustrated by the circuits shown in Figure 27-7. There are some basic rules that apply to all circuits being analyzed using this method.

1. The value of the resistance, R, is represented by a real number. The angle for the real number is $0°$. The resistances are represented by their respective values of $5R$ and $4R$. These may also be stated as $5\ \Omega$ and $4\ \Omega$.
2. The inductive reactance value is stated as a positive value having a $90°$ shift. Values of X_L are considered to be positive. The X_L value for the circuit shown in part (a) has a value of $+j3$ or $j3\ \Omega$. This factor holds for all values of inductive reactance whether they are in series or parallel circuits. Since voltage always lead current in an inductive circuit, the $+j$ value also is used for V_L in circuit analysis.
3. When describing values of capacitive reactance, X_C, a value of $-90°$, or $-j$, is utilized. In the example shown in Figure 27-8(b) the X_C value is expressed as a value of $-j5\ \Omega$. This rule applies for X_C values connected in either series or in parallel with a resistance value. This is true because the value of X_C represents a voltage charge across the terminals of a capacitor. In a X_C circuit the voltage always lags the current by a value of $-90°$. In this example, the $-j$ value is also used to represent the voltage of the capacitor, shown as V_C.

   In summary, values that use either the $+j$ or the $-j$ indicate a phase shift in the circuit. If the circuit is capacitive, the use is a $-j$, due to the lagging of the voltage behind the current. When the circuit is inductive, the voltage leads the current and the term $+j$ is used to show this.

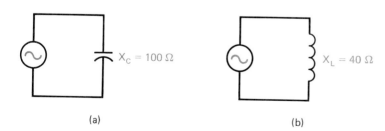

(a)

(b)

FIGURE 27-8 Phasors used to display combinations of complex numbers.

REVIEW PROBLEMS

Express the following in reactangular form.

27-1. $X_L = 24\ \Omega$
27-2. $X_C = 120\ \Omega$
27-3. $X_L = 1000\ \Omega$
27-4. $X_C = 52\ \Omega$
27-5. $R = 1000\ \Omega$

27-4 Total Impedance in Complex Form

The use of the complex number form is a convenient method of stating the impedance of a circuit containing a resistance and a reactance connected in series with each other and to a source. The circuit shown in Figure 27-8 is used to show how this is accomplished. Part (a) has an impedance of $5 + j5\ \Omega$. The impedance formula for this circuit is

$$Z_a = 5 + j3\ \Omega$$

In part (b), the impedance is expressed

$$Z_b = 4 - j5\ \Omega$$

EXAMPLE 27-2

Express the impedance of each of the reactive circuits in Figure 27-9.

Solutions

(a) $Z = R - jX_C$ (27-3)

       $= 100 - j10$

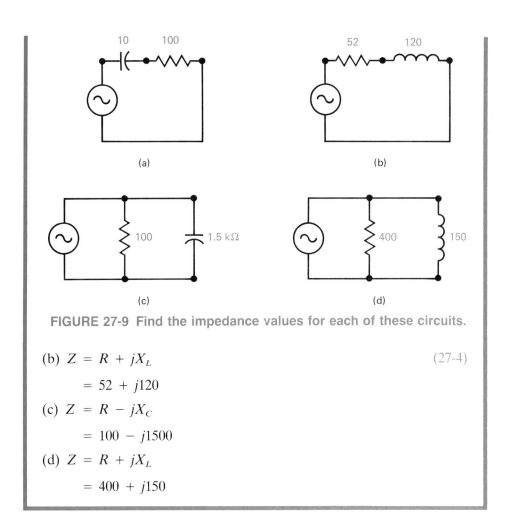

FIGURE 27-9 Find the impedance values for each of these circuits.

(b) $Z = R + jX_L$ (27-4)

      $= 52 + j120$

(c) $Z = R - jX_C$

      $= 100 - j1500$

(d) $Z = R + jX_L$

      $= 400 + j150$

When either the real number or the imaginary number has a value of zero, the preferred method of showing this is to include the zero value. For example, a value of $0\ \Omega$ of R and $5\ \Omega$ of X_L connected in series is written as

$$Z_T = 0 + 5j$$

and a value of $24\ \Omega$ of R and $0\ \Omega$ of X_L in series is written as

$$Z_T = 24 + 0j$$

The purpose of doing this is to maintain the general form of the complex number, $Z = R + jX$. The inclusion of the zero in the general form aids in the reduction of confusion when working with imaginary numbers.

When observing this system, it is possible to exhibit multiple impedances as complex numbers. Either of the following formulas may be used. When a parallel impedance is expressed, the formula is

$$\frac{1}{Z_T} = \frac{1}{Z_1} + \frac{1}{Z_2} + \frac{1}{Z_3} + \cdots \frac{1}{Z_n} \qquad (27\text{-}5)$$

The modification of the formula for two parallel resistances may also be used.

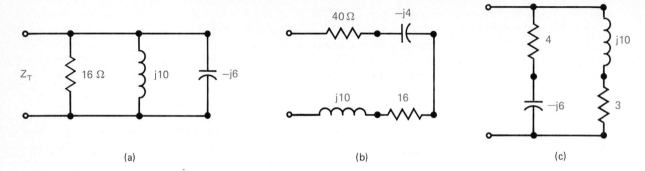

(a)   (b)   (c)

FIGURE 27-10 Utilization of circuit analysis rules for solutions to values of complex numbers.

This is the formula known as "the product over the sum."

$$Z_T = \frac{Z_1 Z_2}{Z_1 + Z_2} \tag{27-6}$$

Similarly, when exhibiting the formula for series-connected impedances, this formula is used:

$$Z_T = Z_1 + Z_2 + Z_3 + \cdots Z_n \tag{27-7}$$

The circuits shown in Figure 27-10 illustrate how these formulas are utilized. The parallel circuit [part (a)] is

$$\frac{1}{Z_T} = \frac{1}{16} + \frac{1}{j10} + \frac{1}{-j6}$$

The series circuit [part (b)] is exhibited as

$$Z_T = (40 + -j4) + (16 + j10)$$

A combination series and parallel circuit [part (c)] is exhibited as

$$Z_T = \frac{(4 - j6)(3 + j10)}{(4 - j6) + (3 + j10)}$$

The series circuit values may be combined by the use of the fundamental rules for combining like terms. In this circuit the values of 40 and 16 Ω are combined for a total of 56 Ω of R. The two j values are also combined. Their values of $+j10$ and $-j4$ are added. The total j value is then $+j6$. In the parallel circuit the same factors will produce a total of both R and j values. The use of the j factor and complex numbers is very important when the combination series and parallel circuit values need to be solved.

REVIEW PROBLEMS

Write the following impedances in complex form.

27-6. $X_C = 10\ \Omega$ and $R = 16\ \Omega$
27-7. $X_L = 23\ \Omega$ and $R = 82\ \Omega$

27-5 Mathematical Operations With Complex Numbers

Imaginary numbers cannot be combined with real numbers because they represent an out-of-phase condition. This out-of-phase factor is 90°. It is possible to combine complex numbers when the following rules are observed:

Addition Two complex numbers are added by combining like terms. Combine the real numbers and then combine the imaginary numbers:

$$(4 - j3) + (-6 + j5)$$
$$= (4 - 6) + (-j3 + j5)$$
$$= -2 + j2$$

Subtraction Two complex numbers are subtracted using the same rule used when the numbers are added:

$$(9 - j7) - (4 - j6)$$
$$= (9 - 4) - (j7 - j6)$$
$$= 5 - j1$$

Multiplication Complex numbers may also be multiplied. The imaginary numbers when multiplied produce real numbers. This is because the unit j^2 is equal to a -1 real number. The -1 value is located on the real axis, not on the imaginary axis. The multiplication of the two imaginary numbers shifts their value by 90° and places it on the real axis. An example of this is

$$(4 - j7) \times (8 + j3)$$
$$= 32 - j56 + j12 - j^2 21$$

Since $j^2 = -1$, this then becomes

$$= 32 - j56 + j12 + (-1)(-21)$$
$$= 32 - j56 + j12 + 21$$
$$= 53 - j44$$

Division It is also possible to divide complex numbers. This is more difficult because it is impossible to divide a real number by an imaginary number. The process requires conversion of the denominator of the complex number into a whole number as an initial step. The process of doing this is known as *rationalizing* the denominator. This is accomplished by multiplying by the *conjugate* of the denominator. The *conjugate* is the numeric value of the complex number when the sign between the numbers is changed. For example, the conjugate of the expression $(6 + j3)$ is $(6 - j3)$. An example of the division process is

$$\frac{10}{1 + j2} = \frac{10(1 - j2)}{(1 + j2)(1 - j2)}$$

$$= \frac{10(1 - j2}{1 - j^2 4}$$

$$= \frac{10(1 - j2)}{5}$$

$$= 2(1 - j2)$$

$$= 2 - j4$$

Multiplication of Two *j* Terms This is accomplished when the two numbers are multiplied together and the two *j* terms are also multiplied together. The answer for the two *j* terms is j^2 and this is equal to a value of -1. The example,

$$j5 \times j3$$

$$= -1 \times 15$$

$$= -15$$

Division of Two *j* Terms This is accomplished by dividing the coefficients and producing a real number. The two *j* factors will cancel as seen in this example:

$$\frac{j16}{j2} = 8$$

REVIEW PROBLEMS

Add the following terms.

27-8. $j8 + j19$
27-9. $j6 + j4$

Subtract the following terms.

27-10. $j28 - j8$
27-11. $j8 - j3$

Multiply the following terms.

27-12. $j6 \times j3$
27-13. $j21 \times -j4$

Divide the following terms.

27-14. $j24$ by $j6$
27-15. $-j6$ by $-j2$

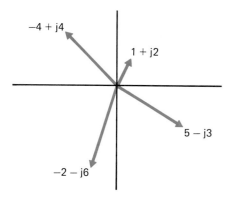

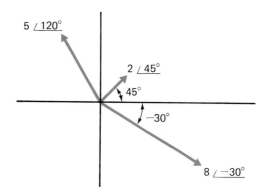

FIGURE 27-11 Complex numbers are also displayed in their rectangular form.

FIGURE 27-12 Complex numbers are also displayed in their polar form.

27-6 Polar Form for Complex Numbers

The material thus far in this chapter explains how a complex number is expressed in *rectangular* form. It is also possible to express the same complex number in a second way. This second way is known as the *polar form* of the complex number. Each, as you will see, has its advantages. Each will also have its disadvantages.

A phasor quantity has both *phase* and *magnitude*. In most electrical and electronic circuits the letters R, V, I, X_L, X_C, and Z are used when the magnitude of the quantities is represented.

Figure 27-11 illustrates a series of phasors in their rectangular form. The lines with arrowheads represent the quantity as the sum of the j coordinate and the real coordinate. The arrow is drawn from the origin to the coordinate point in the complex plane. In this example the phasors are $1 + j2$, $5 - j3$, $-4 + j4$, and $-2 - j6$. Each of these are plotted in the figure. They identify both the direction and the length of the phasor.

These same quantities may be illustrated in **polar form** (Figure 27-12). The polar form has two components. These are the angle of the phasor relative to the positive real axis and the magnitude of the phasor. In this example the values are shown as $2\angle 45°$, $5\angle 120°$, and $8\angle -30°$. Each first number displays the magnitude of the phasor and the $\angle$ symbol is used to represent the angle of the phasor. Each angle is shown as it relates to 0°. The two sets of phasors, as seen in Figures 27-11 and 27-12, both represent the same phasors. They are shown in both the rectangular and the polar form.

Polar form *Expression of complex number using real number and angular form of imaginary number.*

27-7 Conversion between Rectangular and Polar Form

One has to remember that the phasor can be considered to form a right triangle in the complex plane. In Figure 27-13 the vertical axis represents the imaginary value B. The horizontal axis is the real value A. The length of the phasor is the hypotenuse of the triangle and is represented, or equal to,

$$\sqrt{A^2 + B^2} \qquad (27\text{-}8)$$

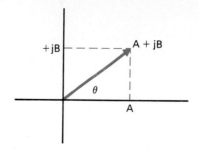

FIGURE 27-13 The phasor will form a right triangle.

This may also be shown as

$$C = \sqrt{A^2 + B^2} \qquad (27\text{-}9)$$

where the letter C is used to represent the magnitude of the phasor.

The angle of the right triangle is represented by the Greek lowercase letter theta (θ). This is equal to

$$\text{angle } \theta = \arctan \frac{B}{A} \qquad (27\text{-}10)$$

Rectangular-to-Polar Conversion In general, the formula for conversion from rectangular to polar form is

$$A \pm jB = \sqrt{A^2 + B^2} \; \underline{/\arctan \frac{B}{A}} = C \; \underline{/ \pm \theta} \qquad (27\text{-}11)$$

EXAMPLE 27-3

Convert this complex number from its rectangular to it polar form:

$$12 - j8.$$

Solution

The magnitude of the phasor $12 - j8$ is

$$C = \sqrt{12^2 + 8^2}$$
$$= \sqrt{208}$$
$$= 14.42$$

The angle is

$$\theta = \arctan\left(\frac{-8}{12}\right)$$
$$= -33.69°$$

Therefore, the complete polar expression for the complex number $12 - j8$ is

$$C = 14.42\underline{/-33.69°}$$

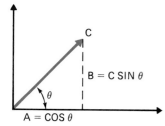

FIGURE 27-14 Polar values may be converted into their rectangular form.

REVIEW PROBLEMS

Convert the following into polar form.

27-16. $8 + j6$

27-17. $4 - j4$

Polar-to-Rectangular Conversion This process is the reverse of the process described previously. The polar form of the complex number provides the angle and the magnitude of the number (Figure 27-14). The C and the θ parts of the triangle must be known in order to use this conversion process. The use of trigonometry will provide this information:

$$A = C \cos \theta$$

$$B = C \sin \theta$$

The general formula for conversion is stated as

$$C \angle \theta = C \cos \theta + jC \sin \theta = A \pm jB \qquad (27\text{-}12)$$

The following example will illustrate how this is accomplished.

EXAMPLE 27-4

Convert the polar value of $\angle 200 - 45°$ into its rectangular value.

Solution

First, separate the real number from the imaginary, or j term.

$$A = C \cos(-45°) = 141.4$$

Then the j component is found:

$$B = 200 \sin(-45°) = -141.4$$

The complex expression in rectangular form is

$$141.4 - j141.4$$

Examples 27-3 and 27-4 should aid in an understanding of how they are accomplished.

REVIEW PROBLEMS

Convert the following into rectangular form.

27-18. $48 \angle -60°$

27-19. $175 \angle -60°$

Summary of Conversion Factors for Complex Numbers In the preceding sections we described how to convert from polar to rectangular form and from rectangular to polar form. There are a few basic rules that will summarize these conversions.

1. Always note whether the angle is smaller than $\pm 45°$ or greater than $\pm 45°$.
2. Always note whether the j term is smaller or larger than the real term.
3. When the angle is between 0 and $\pm 45°$, the j term must be smaller than the real term. (The j term is the opposite side.)
4. When the angle falls between 45 and 90°, and -45 and $-90°$, the j term has to be larger than the real term.
5. When adding or subtracting complex numbers, use the rectangular form.
6. The polar form is usually used for multiplication or division of complex numbers because the results may be obtained more quickly using the polar form.

27-8 Complex Analysis of Alternating-Current Circuits

Reactance and resistance may be solved in the complex form. A circuit containing both resistance and inductive reactance, X_L, may be represented by a phasor that is $+90°$ from R [Figure 27-15(a)]. When the circuit contains R and capacitive reactance, X_C, it is represented by a phasor that is $-90°$ from R [Figure 27-15(b)].

Resistance has no phase angle between current and voltage. It is assigned

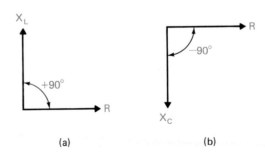

FIGURE 27-15 Use of phasors to represent inductive values (a) and capacitive values (b).

an angle of 0°. In the polar form this is expressed as R $\angle 0°$. When the circuit is purely resistive there is no j term. This provides a rectangular form value of $R + j0$, or R. In the illustration, the value R lies along the real axis. This axis corresponds to 0°.

Reactance produces a 90° phase shift. This shift occurs between current and voltage in the circuit. If the reactance is capacitive, the shift causes the voltage to *lag* the current by 90°. When this is expressed in polar form it is given a $-90°$ angle. This is expressed as X_C $\angle -90°$. When this value is expressed in rectangular form, it is seen as $-jX_C$. In either case the X_C is on the $-j$ axis in the phasor diagram.

When the value is inductive in nature, it is shown as an X_L value. X_L produces a *lag* in the current by 90°. This produces an assignment of an $+90°$ angle. When seen in its polar form, it appears as X_L $\angle 90°$ and in its rectangular form it becomes jX_L. This value lies along the $+j$ axis of the phasor diagram.

Both the capacitive and inductive reactances use the j term to indicate a 90° phase shift. The difference, of course, is the direction of the phase shift.

EXAMPLE 27-5

Using the three circuits shown in Figure 27-16, express each reactance or resistance in both polar and rectangular form.

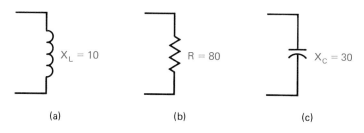

(a)   (b)   (c)

FIGURE 27-16

Solution

(a) Inductive reactance in polar form is $10\angle 90°$ and in rectangular form is $+j10$ Ω.

(b) Resistance in polar form is $80\angle 0°$ Ω and in rectangular form is 80 Ω.

(c) Capacitive reactance in polar form is $30\angle -90°$ and in rectangular form is $-j30$ Ω.

REVIEW PROBLEMS

Write these values in rectangular form.

27-20. $R = 100$ Ω
27-21. $X_C = 85$ Ω
27-22. $X_L = 150$ Ω

27-9 Impedance in Complex Form

In the series RC circuit, impedance may be expressed in either rectangular or polar form. The easiest method of accomplishing this is to determine impedance directly from the circuit. The formula for this is

$$Z = R - jX_C \qquad (27\text{-}13)$$

When using the polar form the formula is

$$Z = \sqrt{R^2 + X_C^2}\ \angle - \arctan\frac{X_C}{R} \qquad (27\text{-}14)$$

This circuit and its phasor diagram are illustrated in Figure 27-17.

In the series RL circuit, impedance may also be expressed in both polar and in rectangular form. Figure 27-18 shows a series X_L circuit and its corresponding phasor diagram. The equation in rectangular form is

$$Z = R + jX_L \qquad (27\text{-}15)$$

and in its polar form

$$Z = \sqrt{R^2 + X_L^2}\ \angle \arctan\frac{X_L}{R} \qquad (27\text{-}16)$$

In the parallel RC circuit (Figure 27-19), the value of impedance may be developed by use of the "product over the sum" method. This is the same formula as that shown in Chapter 7 for two different values of parallel resistances:

$$Z = \frac{(R\angle 0°)(X_C\angle - 90°)}{R - jX_C} = \frac{RX_C\angle 0° - 90°}{\sqrt{R^2 + X_C^2}\ \angle - \arctan(X_C/R)} \qquad (27\text{-}17)$$

$$= \frac{RX_C}{\sqrt{R^2 + X_C^2}\ \angle - 90° + \arctan(X_C/R)}$$

In the parallel RL circuit, the impedance is found in the same manner as

FIGURE 27-17 Display of X_c values in polar form.

FIGURE 27-18 Display of X_L values in polar form.

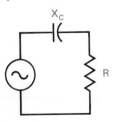
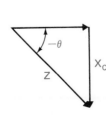
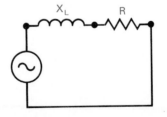
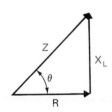

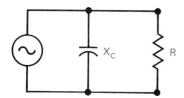

FIGURE 27-19 Parallel *RC* circuit.

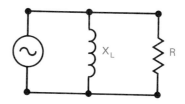

FIGURE 27-20 Parallel *RL* circuit.

used for the parallel *RC* circuit. This, again, uses the "product over the sum" method. This circuit is shown in Figure 27-20.

$$Z = \frac{(R\angle 0°)(X_L\angle 90°)}{R + jX_L} = \frac{RX_L\angle 0° = 90°}{\sqrt{R^2 + X_L^2}\angle 90° - \arctan(X_L/R)} \qquad (27\text{-}18)$$

EXAMPLE 27-6

Use the method above to solve for the circuit impedance in the circuits shown in Figure 27-21.

(a) $Z = R - jX_C$

$$= 50\,\Omega - j100\,\Omega \,\text{in rectangular form}$$

$$= \sqrt{R_P + X_C^2} \,\angle -\arctan(X_C/R)$$

$$= \sqrt{(50^2 + 100^2)} \,\angle \arctan(100/50)$$

$$= 111.8 \,\angle -63.4° \,\Omega \,\text{in polar form}$$

(b) $X = \dfrac{RX_C}{\sqrt{R^2 + X_C^2}\,\angle -90° + \arctan(X_C/R)}$

$$= \frac{(1\,\text{k}\Omega)(2\,\text{k}\Omega)}{\sqrt{1\,\text{k}\Omega^2 + 2\,\text{k}\Omega^2}\,\angle -90° + \arctan(2/1)}$$

$$= 894.4 \,\angle -26.57° \,\Omega \,\text{in polar form}$$

$Z = 894.4\cos(-26.57°) + j894.4\sin(-26.57°)$

$$= 799.9 - j400\,\Omega \,\text{in rectangular form}$$

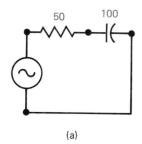
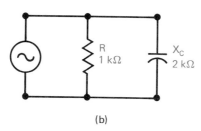

(a) (b)

FIGURE 27-21 Solving for circuit impedances for series or parallel X_c circuits.

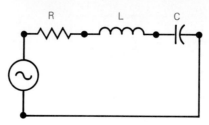

FIGURE 27-22 Series *RLC* circuit.

RLC circuit *Resonant circuit containing resistance, inductance, and capacitance.*

A series ***RLC* circuit** is shown in Figure 27-22. This circuit contains resistance, inductance, and capacitance values. In any circuit containing these three components, a phase shift occurs. The inductive component produces a lag in circuit current and the capacitive component creates a lag in the circuit voltage. These two values will attempt to oppose each other. In fact, when they are equal, their values cancel and the circuit becomes entirely resistive. The total amount of reactance in a series *RLC* circuit is determined by use of the formula

$$X = X_L - X_C \qquad (27\text{-}19)$$

The total amount of impedance for a series *RLC* circuit may be found using either the polar or the rectangular form. The respective formulas are

$$Z_T = R + jX_L - jX_C \qquad (27\text{-}20)$$

and

$$Z_T = \sqrt{R^2 + (X_L - X_C)^2} \; \underline{\big/ \arctan \dfrac{X_L - X_C}{R}} \qquad (27\text{-}21)$$

EXAMPLE 27-7

Find the impedance for the circuit of Figure 27-22, using values of $R = 10\ \Omega$, $C = 500\ \mu F$, and $L = 20\ mH$. The frequency f is 90 Hz.

Solution

The first step is to determine the reactive values of each component.

$$X_C = \frac{1}{2\pi f C}$$

$$= \frac{1}{6.28 \times 90 \times 500 \times 10^{-6}}$$

$$= 3.54\ \Omega$$

$$X_L = 2\pi f L$$

$$= 6.28 \times 90 \times 0.02$$

$$= 11.30\ \Omega$$

In this circuit the inductance reactance is greater than the capacitive

---

reactance value. The total, or net, reactive value is

$$X_T = X_L - X_C$$

$$= 11.3 - 3.54$$

$$= 7.76 \ \Omega$$

In rectangular form the impedance of this circuit is

$$Z = R + jX_L - jX_C$$

$$= 10 + j11.3 - j3.54$$

$$= 10 + j7.76 \ \Omega$$

In polar form the impedance of the circuit is

$$Z = \sqrt{R^2 + X^2} \ \underline{/\text{arctan} \ \dfrac{X}{R}}$$

$$= \sqrt{10^2 + 7.76^2} \ \underline{/\text{arctan} \ \dfrac{7.76}{10}}$$

$$= 12.66 \ \underline{/37.8°} \ \Omega$$

Admittance (Y) and Susceptance (B) When we discuss the values of branch currents in a parallel circuit, it is often easier to add the values of branch currents rather than attempt to combine the reciprocals of the circuit imped- ances. In electrical work, conductance is represented by the letter G. Con- ductance is, of course, the reciprocal of resistance in the circuit. The formula for conductance is $G = 1/R$. This concept is also applied to solution of complex impedance values. The major types of reciprocal values for complex impedances are called **admittance Y** and **susceptance B**. Admittance Y is the reciprocal of impedance. Susceptance B is the reciprocal of reactance. These values may be expressed as seen in the following formulas:

Admittance Y Opposite value of impedance.

Susceptance B Opposite value of reactance.

Conductance: $G = \dfrac{1}{R}$ S   (27-22)

Admittance: $Y = \dfrac{1}{Z}$ S   (27-23)

Susceptance: $B = \dfrac{1}{\pm X}$ S   (27-24)

We describe units of X, Z, and R in units of ohms. The reciprocals of the values of G, Y, and B are described in units of the siemens.

Phase angles for Y or B is the same as current in the circuit. These signs are opposite to those of the angle for Z or X because of their conversion from the reciprocal. The capacitive branch will have a susceptance of $+jB$ and will be the same angle as the branch current. The inductive branch will have a susceptance of $-jB$ and the same angle as the branch current.

In determining the values in a branch circuit, note that the total admittance for the circuit, Y_T, is equal to $G \pm jB$.

EXAMPLE 27-8

Find the values of conductance, capacitive susceptance, inductive susceptance, and admittance for the components in the circuit of Figure 27-23.

Solution

$$G = \frac{1}{R\angle 0°} = \frac{1}{10\angle 0°}$$

$$= 0.1 \angle 0° \text{ S}$$

$$B_C = \frac{1}{X_C\angle -90°}$$

$$= \frac{1}{5\angle -90°}$$

$$= 0.2\angle 90° \text{ S}$$

$$B_L = \frac{1}{X_L \angle 90°}$$

$$= \frac{1}{10 \angle 90°}$$

$$= 0.1 \angle -90° \text{ S}$$

$$Y = G + jB_C - B$$

$$= 0.1 \text{ S} + j0.2 \text{ S} - j0.1 \text{ S}$$

$$= 0.1 \text{ S} + j0.1 \text{ S}$$

$$= 0.1414 \angle -45° \text{ S}$$

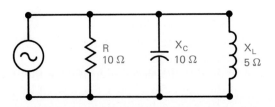

FIGURE 27-23 Circuit used to determine values using recipricals of each individual value.

Once Y is determined, Z can be found:

$$Z = \frac{1}{Y}$$

$$= \frac{1}{0.1414 \, / -45°}$$

$$= 7.07 \, /45° \; \Omega$$

A parallel RLC circuit is shown in Figure 27-24. The fastest and probably the easiest method of solving for circuit impedance in this type of circuit is to use the "sum of the reciprocals" method. The presence and low initial cost of the pocket calculator has made this process very simple and rapid. In the three-branch (or parallel) RLC circuit, the formula for total impedance is

$$\frac{1}{Z_T} = \frac{1}{R/0°} + \frac{1}{X_L/90°} + \frac{1}{X_C/-90°}$$

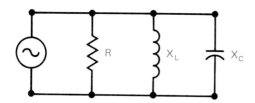

FIGURE 27-24 Parallel RLC circuit.

This formula may be extended to include any quantity of branch paths.

EXAMPLE 27-9

Find the total impedance for the circuit shown in Figure 27-24.

Solution

$$Z = \frac{1}{R/0°} + \frac{1}{X_L/90°} + \frac{1}{X_C/-90°}$$

$$= \frac{1}{100/0°} + \frac{1}{50/90°} + \frac{1}{100/-90°}$$

$$\frac{1}{Z} = 0.01/0° \, S + 0.02/90° \, S + 0.01/-90° \, S$$

$$= 0.01/0° \, S + 0.01 \, /90° \, S$$

$$= 0.1414 \, /45° \, S$$

$$Z = \frac{1}{Y}$$

$$= \frac{1}{0.1414 \angle -45°}$$

$$= 70.7 \angle 45° \ \Omega$$

27-10 Complex Analysis of the Reactive Circuit

The development of methods of solving for circuit values is complete when one has an understanding of how to use them in practical situations. There is a need to determine both voltage drops and current values in the series and in the parallel reactive circuit. The following procedures will aid in the clarification of how this is accomplished.

EXAMPLE 27-10

Find the voltage drops and the total current for the circuit shown in Figure 27-25.

Solution

$$Z = R + jX_L - jX_C$$

$$= 50 + j25 - j60$$

$$= 50 - j35 \ \Omega$$

The next step is to convert the answer into a polar form. This is done because of the ease of using Ohm's law.

$$Z = \sqrt{50^2 + -35^2} \ \angle \ \arctan\left(\frac{-35}{50}\right)$$

$$= 61.44 \ \angle -35°$$

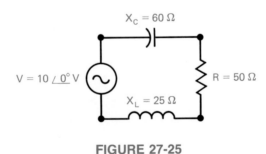

$X_C = 60 \ \Omega$

$V = 10 \angle 0° \ V$

$R = 50 \ \Omega$

$X_L = 25 \ \Omega$

FIGURE 27-25

Use Ohm's law to determine current:

$$I = \frac{V}{Z}$$

$$= \frac{10}{61.44 \angle -35°}$$

$$= 0.163 \angle 35°$$

Use Ohm's law to determine individual voltage drops:

$$V_R = IR$$

$$= (0.163)(50)$$

$$= 8.15 \angle 0°$$

$$I_{XL} = (0.163 \angle 0°)(60 \angle -90°)$$

$$= (0.163)(60) \angle -90°$$

Another example shows how to determine branch currents and total current in the parallel circuit.

EXAMPLE 27-11

Use the circuit and values shown in Figure 27-26 to solve for branch currents and total circuit current.

Solution

First determine each branch current by use of Ohm's law.

$$I_R = \frac{V_S}{R} = \frac{10 \angle 0°}{4 \angle 0°}$$

$$= 2.5 \angle 0° \text{ A}$$

$$I_C = \frac{V_S}{X_C} \text{A}$$

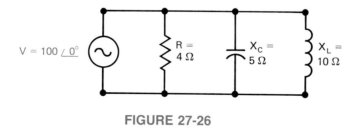

FIGURE 27-26

$$= \frac{10 \angle 0°}{5 \angle -90°}$$

$$= 2 \angle 0 + 90° \text{ A}$$

$$I_L = \frac{V_S}{X_L}$$

$$= \frac{10 \angle 0°}{10 \angle 90°}$$

$$= 1 \angle -90° \text{ A}$$

Total circuit current is determined by use of Kirchhoff's law for current:

$$I_T = I_R + I_C + I_L$$

$$= 2.5 \angle 0° + 2 \angle +90° + 1 \angle -90°$$

$$= 2.5 \text{ A} + j2 \text{ A} - 1j \text{ A}$$

$$= 2.5 \text{ A} + 1j \text{ A}$$

$$= \sqrt{2.5 A^2 + 1A^2} \angle \arctan \frac{1}{2.5}$$

$$= 2.69 \angle 21.8°$$

In this example, the total current is 2.69 A and it leading the source voltage V by 21.88°.

CHAPTER SUMMARY

1. Alternating-current circuit analysis requires determination of both the magnitude and phase of the electrical quantities.
2. In electrical work, an imaginary number is used to identify a 90° phase shift.
3. The term j is used as the imaginary number.
4. A $+j$ indicates a 90° shift and a $-j$ indicates a 270° shift, or $\angle -90°$.
5. The j term is equal to $\sqrt{-1}$.
6. The term j^2 is equal to -1.
7. Real and imaginary numbers can be displayed on a complex plane.
8. A complex number is a value that falls somewhere between the axis of a real number and an imaginary number. This number is represented by its coordinates. An example of this is the complex number $5 + j3$.
9. An inductive reactance circuit will be seen by use of the $+j$ term.
10. A capacitive reactance circuit will be seen by use of the $-j$ term.
11. Impedance of a series reactive circuit may be stated by use of a complex number.

12. Impedance in the parallel reactive circuit uses formulas that are similar to those used for parallel resistive circuits.

13. Impedance in the series reactive circuit uses formulas that are similar to those used for series resistive circuits.

14. Complex numbers can be added, subtracted, multiplied, and divided.

15. Complex numbers may be displayed in either a rectangular or a polar form.

16. The polar form of the complex number is used to multiply and divide.

17. The rectangular form is used for addition and subtraction of the complex number.

18. The polar form of the complex number has two components: the angle of the phasor relative to the real axis and the magnitude of the phasor. An example of the polar form is $2\angle 45°$.

19. Complex numbers can be converted from the polar form to the rectangular form and from the rectangular form to the polar form.

20. The terms *admittance* and *susceptance* are used to represent the reciprocals of impedance and reactance, respectively.

Answer true or false; change all false statements into true statements.   **SELF TEST**

27–1. The i operator is used to represent an imaginary electrical value.

27–2. Ac circuit analysis may be accomplished by using only the magnitude of a value.

27–3. The square root of j is -1.

27–4. In a complex number set, the resistance value is the same as the j value.

27–5. Values of capacitive reactance are shown as $-j$ values.

27–6. In an X_C circuit the voltage is shown as the $-j$ value.

27–7. Complex numbers are added by combining like terms.

27–8. Imaginary numbers, when multiplied, produce other imaginary numbers.

27–9. Complex numbers may be converted between their polar and their rectangular forms.

27–10. A phasor is considered to form a right triangle in the complex plane.

27–11. The letter j used in electrical formulas, represents a(n)   **MULTIPLE CHOICE**
(a) imaginary number
(b) whole number with a 180° phase shift
(c) whole number with no phase shift
(d) none of the above

27–12. The square root of j^3 is equal to
(a) $+1$
(b) -1
(c) 0
(d) none of the above

27–13. The square root of j^4 is equal to
(a) $+1$
(b) -1
(c) 0
(d) none of the above

27–14. Angular values of numbers are shown on
(a) a complex plane
(b) a straight plane
(c) a linear graph
(d) their imaginary axis

27–15. A value of 360° is equal to the j factor of
(a) -1
(b) $-j$
(c) $+1$
(d) $+j$

27–16. The expression $D + j4$ represents a(n)
(a) complex number
(b) real number
(c) imaginary number set
(d) combination number set

27–17. The reciprocal of resistance is
(a) susceptance
(b) conductance
(c) admittance
(d) impedance

27–18. The expression $C - j3$ represents the
(a) polar form of the expression
(b) normal form of the expression
(c) rectangular form of the expression
(d) inverse form of the expression

27–19. When using the j operator form, values of X_C are
(a) $-90°$
(b) $+90°$
(c) in phase, or $0°$
(d) all of the above

27–20. The reciprocal of impedance is
(a) susceptance
(b) conductance
(c) admittance
(d) resistance

ESSAY QUESTIONS

27–21. Why are imaginary numbers used for ac circuit analysis?

27–22. Why are susceptance, admittance, and conductance values used for solution of ac problems?

27–23. Exactly what does the phasor show us?

27–24. What is illustrated by use of imaginary numbers?

27–25. Why don't we use the j operator in dc circuit analysis?

27–26. Why is a triangle used to solve for ac circuit values?

27–27. What are the rotational degrees for the $+j$ values?

27–28. What term is used to represent values of capacitive reactance?

27–29. Explain the difference between the polar form and the rectangular form of a complex number.

27–30. Why is one form used for addition and subtraction of complex numbers and the other form used for multiplication and division of complex numbers?

27–31. Convert the following numbers from their rectangular to their polar values (*Sec. 27-7*).
(a) $6 + j4$
(b) $10 + j5$
(c) $7 - j8$
(d) $70 - j40$

PROBLEMS

27–32. Convert the following numbers from their polar values to their rectangular values (*Sec. 27-7*)
(a) $3\angle 50°$
(b) $5\angle 0°$
(c) $4\angle -60°$
(d) $7\angle 90°$

27–33. Find the impedance of the circuit when 40 Ω of R are connected in series with 25 Ω of X_C. (*Sec. 27–4*)

27–34. Find the impedance of the circuit when 50 Ω of R is connected in series with an X_L of 100 Ω. (*Sec. 27–4*)

27–35. Express the values of problem 27–33 in rectangular form. (*Sec. 27–7*)

27–36. Express the values of problem 27–34 in rectangular form. (*Sec. 27–7*)

27–37. Express the values of problems 27–33 and 27–34 in polar form. (*Sec. 27–7*)

27–38. Find the impedance of the circuit when 40 Ω of R is connected in parallel with 25 Ω of X_C. (*Sec. 27–4*)

27–39. Find the impedance of the circuit when 50 Ω of R is connected in parallel with an X_L of 100 Ω. (*Sec. 27-4*)

27–40. Express the values of problem 27–33 in rectangular form. (*Sec. 27-7*)

27–41. Express the values of problem 27–34 in rectangular form. (*Sec. 27-7*)

27–42. Express the values of problems 27–33 and 27–34 in polar form. (*Sec. 27-7*)

Use Figure 27-27 to answer problems 27–43 through 27–47.

27–43. Find the impedance in polar form. (*Sec. 27-4*)

27–44. Find the total circuit current when the source voltage is $5\angle 0°$. (*Sec. 27-8*)

27–45. Find the impedance value for a series *RLC* circuit containing an R of 75 Ω, an X_C of 80 Ω, and an X_L of 40 Ω. (*Sec. 27-8*)

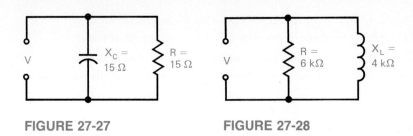

FIGURE 27-27   **FIGURE 27-28**

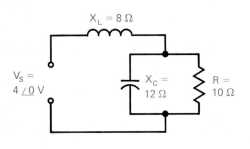

FIGURE 27-29

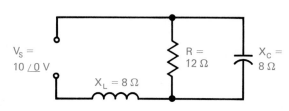

FIGURE 27-30

27–46. Find the impedance value for the *RLC* circuit values in problem 27–45 when they are connected in parallel. (*Sec. 27-8*)

27–47. Find each branch current when the source voltage is 5 $\angle 0°$ (*Sec. 27-8*)

27–48. A series resonant circuit contains the following values: $L = 200$ µH, $C = 50$ pF, and $R = 1.5$ kΩ. It has a source voltage of 6 V. A signal of 2 MHz is applied to the circuit. Find the following. (*Sec. 27-8*)
(a) X_L
(b) X_C
(c) Z_T
(d) I
(e) V_R
(f) V_L
(g) V_C

27–49. Find the total circuit impedance for the circuit in Figure 27-28. (*Sec. 27-9*)

27–50. Find the total circuit impedance for the circuit shown in Figure 27-29. (*Sec. 27-9*)

27–51. Find the total circuit impedance for the circuit shown in Figure 27-30. (*Sec. 27-9*)

ANSWERS TO REVIEW PROBLEMS

27–1. $j24$ Ω
27–2. $-j120$ Ω
27–3. $j1000$ Ω
27–4. $-j52$ Ω
27–5. 1 kΩ

27–6. $16 - j10$

27–7. $82 + j23 \; \Omega$

27–8. $j27$

27–9. $j10$

27–10. $j20$

27–11. $j5$

27–12. -18

27–13. 84

27–14. 4

27–15. 3

27–16. $10 \; \underline{/36.9°}$

27–17. $5.65 \; \underline{/45°}$

27–18. $24 - j41.57$

27–19. $+87.5 - j151.6$

27–20. $100 \; \Omega$

27–21. $-j85$

27–22. $j150$

27–23. $35 \; \underline{/0°}$

27–24. $65 \; \underline{/90°}$

27–25. $93 \; \underline{/-90°}$

CHAPTER 28

AC Network Analysis

The rules used to solve for voltage, current, and resistance in dc networks may also be applied to circuits having ac sources. The process, although similar, is slightly more complicated. This is due to the use of complex numbers for the circuit values. Once the concepts of complex numbers are understood, they may be applied to the ac circuit for analysis purposes.

KEY TERMS

Kirchhoff's Current Law   **Norton's Theorem**
Kirchhoff's Voltage Law   **Thévenin's Theorem**

OBJECTIVES

Upon completion of the material in this chapter, you will

1. Be able to analyze ac reactive circuits using the following formulas:
   (a) Superposition
   (b) Kirchhoff's current law
   (c) Kirchhoff's voltage law
   (d) Thévenin's theorem
   (e) Norton's theorem

There are two basic types of alternating-current circuits. One type is strictly resistive. The second type is both resistive and reactive. Usually, a reference to an ac circuit refers to a circuit having an ac voltage source and containing either capacitive or inductive reactances.

The ac circuits containing reactances are very complex. This is because of the phase shift between voltage and current created by the presence of the reactance values. Values of voltage and current contain both magnitude and a phase angle. These values are expressed in either polar or in rectangular form, as explained in Chapter 27. Material presented in this chapter shows how to use the complex-number system for circuit analysis.

In this chapter each of the basic circuit analysis processes is discussed. The difference between the presentation related to dc circuit analysis and this chapter relating to ac circuit analysis is the use of complex-number systems to solve for circuit values.

28-1 Kirchhoff's Voltage Law

Kirchhoff's voltage law states that the sum of the EMFs and voltage drops in a closed-loop, or series, circuit will equal zero. This is relatively simple when a dc source is used. However, when an ac source is used, the effects of phase shifting due to the addition of reactances to the circuit must be considered. Kirchhoff's law for ac circuits is stated using this statement: In the closed loop the *phasor sum* of the EMFs and voltage drops is equal to zero.

Figure 28-1 illustrates this change from the basic dc source law. The basic change is that one must consider the effects of reactance in this circuit due to the use of the capacitor and inductor. In this figure the voltages are still added:

$$-V_S + V_C + V_R + V_L = 0 \tag{28-1}$$

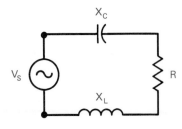

FIGURE 28-1 A series *RLC* circuit requires values in order to be analyzed.

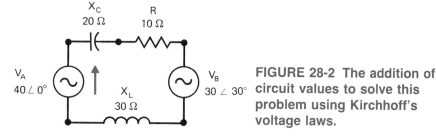

FIGURE 28-2 The addition of circuit values to solve this problem using Kirchhoff's voltage laws.

Actually, one may start at any point in the circuit and add all the voltage drops:

$$V_C - V_S + V_R + V_L = 0$$

$$V_R - V_S + V_C + V_L = 0$$

The net result is equal to zero, as defined by **Kirchhoff's voltage law (ac)**.

When circuit values are included (Figure 28-2), the circuit may be analyzed for its individual voltage drops. This type of circuit analysis is most easily accomplished when the steps described below are followed.

First, start by drawing a schematic diagram of the circuit. This step is followed by marking each component with its ohmic value or its voltage value. Use a $+j$ value for inductive reactances and a $-j$ value for capacitive reactances. Resistive values are shown with their true ohmic value. Once the components are identified, the next step is to identify one voltage source as a reference voltage for the circuit. It is best if you assume current flow in the most convenient direction.

Follow this step by identifying the polarity of the voltage drops that develop due to current flow in the circuit. Each component should have a polarity sign for both of its ends. If you are in doubt as to the polarity of the end of the component, a simple rule will aid in making the proper decision. If you follow the direction of electron current flow, the first sign assigned to the component is negative in nature. If one follows the direction opposite to that of the electron current, the first sign on the component is positive in nature.

All voltage sources must be treated as phasors in the analysis of the circuit. In this type of circuit, the easiest method of adding phasors is to have them in their rectangular form.

Let us see how these steps are applied in a practical circuit.

Kirchhoff's voltage law
(ac) *The phasor sum of the EMFs and voltage drops equal zero in a closed loop.*

EXAMPLE 28-1

Use the schematic diagram shown in Figure 28-2 as the circuit in this analysis. Solve for circuit values using Kirchhoff's voltage law.

Solution

This circuit contains a series resistance of 10 Ω, a value of X_C of 20 Ω, a value of X_L of 30 Ω, and two power sources. The two power sources have values of 40 and 30 V.

The first step, as defined, is to identify the *j* factors of the reactances.

These are shown as $-j20 \, \Omega$ for the capacitor and $+j30 \, \Omega$ for the inductance. The 40-V source is the dominant source and current flow is determined by its influence. The arrows in this circuit indicate the direction *against* the flow of current and are used for the assignment of polarity for each of the components. Kirchhoff's voltage law is applied to this circuit after the polarities have been assigned. Kirchhoff's voltage law in this situation is used to determine circuit current.

$$V_A - IR - (-jX_CI) - V_B - (+jX_LI) = 0 \qquad (28\text{-}2)$$

When values are substituted in the formula, we have

$$(40 - j0) - 10I + j20I - (26 + j15) - j30I = 0$$

After like terms are consolidated or collected, we obtain

$$14 - j15 - 10I - j10I = 0$$

$$I(10 + j10) = 14 - j15$$

$$I = \frac{14 - j15}{10 + j10}$$

$$= \frac{20.5 \angle -46.9°}{14.14 \angle 45°}$$

$$= 1.45 \angle -1.9° \text{ A}$$

Once the circuit current is determined, its value is used to calculate the voltage drops across each of the components in the circuit.

REVIEW PROBLEMS

Use Figure 28-1 to answer problems 28–1 and 28–2.

28-1. Write a formula showing the use of Kirchhoff's voltage law.
28-2. What is the difference between Kirchhoff's voltage law for dc and for ac circuits?

28-2 Kirchhoff's Current Law

Many circuits have multiple current paths. This then presents more than one current in the circuit. One method of solving for the values of circuit currents is to use Kirchhoff's current law. This will reduce the number of currents that must be determined. **Kirchhoff's current law (ac)** states: In an electrical circuit the *phasor sum* of the currents entering a junction is equal to the *phasor sum* of the currents leaving the same junction.

Kirchhoff's current law (ac) *Phasor sum of all currents entering a junction are equal to phasor sum of all currents leaving the junction.*

The circuit shown in Figure 28-3 is used to illustrate how this is accomplished. For the purposes of this presentation, the junction identified as X in the drawing is the reference junction for determining current flow.

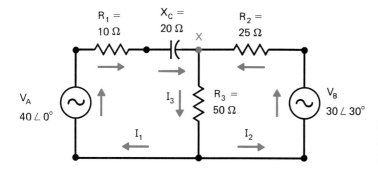

FIGURE 28-3 This problem is solved using Kirchhoff's current laws.

EXAMPLE 28-2

Find the current flow through junction X in Figure 28-3.

Solution

The current identified as I_3 may be replaced by the expression

$$I_1 + I_2$$

Kirchhoff's voltage law for the current path identified as I_1 is

$$V_A - I_1(R_1 - jX_C) - (I_1 + I_2)R_3 = 0 \qquad (28\text{-}3)$$

Substitution of known values:

$$(40 + j0) - 10I_1 + j20I_1 - 50I_1 - 50I_2 = 0$$

Collecting like terms:

$$40 - 60I_1 + j20I_1 - 50I_2 = 0$$

Divide by 20 to simplify the expression:

$$3I_1 - jI_1 + 2.5I_2 = 2$$

The next step is to solve for the current in the path identified as I_2. Applying Kirchhoff's voltage law to this path yields

$$V_B - I_2R_2 - (I_1 - I_2)R_3 = 0$$

$$(26 + j15) - 25I_2 - 50I_1 - 50I_2 = 0$$

$$2.6 + j1.5 - 5I_1 - 7.5I_2 = 0$$

Following this, the first equation is multiplied by 3 to remove the I_2 value. The second equation is then added to the first:

$$9I_1 - j3I_1 + \quad 7.5I_2 = \quad 6$$

$$-5I_1 \qquad \underline{\quad -7.5I_2 = \quad -2.6 - j1.5}$$

$$4I_1 - j3I_1 \qquad\qquad = \quad 3.4 - j1.5$$

and current I_1 is

$$I_1 = \frac{+3.4 - j1.5}{4 - j3}$$

$$= 0.724 + j0.168$$

$$= 0.744\angle 13.1°$$

Next, substitute for I_1 in the first equation:

$$3(0.724 + j0.168) - j(0.724 + j0.168) + 2.5I_2 = 2$$

$$2.34 - j0.22 + 2.5I_2 = 2$$

and

$$I_2 = \frac{-0.340 + j0.22}{2.5}$$

$$= -0.136 + j0.088$$

$$= 0.162\angle 148.1° \text{ A}$$

The current identified as I_2 has an angle of 148°, appears to be in the third quadrant, and is leading the source voltage by the value of 148°. This current direction has to be reversed and is shown as

$$I_2 = 0.162\angle -31.9°$$

The currents through R_3 are added to determine the total current:

$$I_3 = I_1 + I_2$$

$$= 0.724 + j0.168 - 0.1136 + j0.088$$

$$= 0.588 + j0.256$$

$$= 0.641\angle 23.5° \text{ A}$$

The value of 0.744 at an angle of 13.1° leading is supplied by the source V_A and the value of 0.162 at an angle of 32.9 lagging is supplied by V_B.

REVIEW PROBLEMS

28-3. How does Kirchhoff's current law differ from his voltage law?.
28-4. Using Figure 28-3, state the current law.

28-3 Superposition Theorem

When a network contains more than one power source, the process described in the superposition theorem may be utilized. In this theorem, all the power sources except one are replaced by a value equal to their own internal resistance.

If the internal is close to zero, the power source is replaced with a short circuit. Once this is accomplished, the circuit can be solved by use of Ohm's law. This, of course, provides the values of current and voltage drops for the specific power source. The process must be repeated for each of the power sources in the circuit. The circuit used in Example 28-2 is shown again for the explanation of how this process is accomplished.

EXAMPLE 28-3

Using the circuit values shown for Figure 28-3, determine individual circuit values using the superposition theorem.

Solution

First, short out the second power source, V_B (Figure 28-4). The result places R_2 in parallel with R_3. The equivalent resistance for this branch circuit is

$$R_{EQ} = \frac{R_2 R_3}{R_2 + R_3}$$

$$= \frac{50 \times 25}{50 + 25}$$

$$= 16.7 \ \Omega$$

Next, determine total circuit impedance:

$$Z_T = (R_1 + R_{EQ}) - jX_C \qquad (28\text{-}4)$$

$$= 26.7 - j20$$

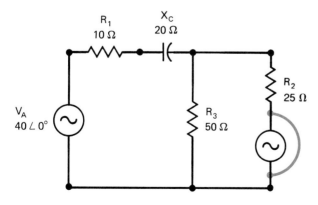

FIGURE 28-4 This dual supply circuit requires use of the superposition theorem for its solution of values.

The next step is to determine circuit current for this source:

$$I_T = \frac{V_A}{Z_T}$$

$$= \frac{40}{26.7 - j20} \qquad (28\text{-}5)$$

$$= 0.958 + j0.718 \text{ A}$$

An analysis of the relationship between R_2 and R_3 shows that they have a ratio of 2:1. Therefore, one-third of the total current will flow through the largest resistance value and two-thirds of the total current flows through the smaller of the two values, or the 25-Ω resistance.

$$I_3 = \frac{0.958 + j0.718}{3}$$

$$= 0.319 + j0.239 \text{ A}$$

Once the values for V_A are established, the process is repeated using the other voltage source. In Figure 28-5, power source V_A is shorted out. First, start by finding the circuit impedance for the parallel branch portion of the circuit:

$$Z_{EQ} = \frac{R_3(R_1 - jX_C)}{R_3 + R_1 - jX_C}$$

$$= \frac{50(10 - j20)}{60 - j20}$$

$$= 12.5 - j12.5$$

The total circuit impedance is found as the next step:

$$Z_T = R_2 + Z_{EQ}$$

$$= 37.5 - j12.5$$

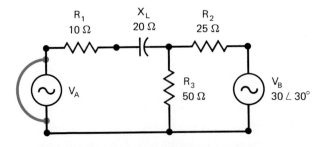

FIGURE 28-5 The second step in solution is the short circuiting of the power sources.

and finally, the total current is determined:

$$I_T = \frac{V_B}{Z_T}$$

$$= \frac{26 - j15}{37.5 - j12.5}$$

$$= 0.504 + j0.562 \text{ A}$$

The final steps in the circuit analysis result in the combination of the two currents through R_3. Let us identify this current as I_{R3}. First, substitute like terms to simplify the solution to this circuit. An analysis of the circuit shows the following to be true:

$$V_{EQ} = I_T Z_{EQ}$$

$$I_2 = \frac{V_{EQ}}{R_3}$$

$$= \frac{I_T Z_{EQ}}{R_3}$$

$$= \frac{(0.504 + j0.562)(12.5 - j12.5)}{50}$$

$$= 0.267 + j0.015 \text{ A}$$

The total current flow through resistance R_3 is equal to the sum of the two currents:

$$I_3 = I_A + I_B$$

$$= 0.319 + j0.239 + 0.267 + j0.15$$

$$= 0.586 + j0.254 \text{ A}$$

$$= 0.640 \angle 23.4°$$

REVIEW PROBLEMS

28-5. How many sources are used for each section of the superposition theorem?

28-6. What is done with the other sources when solving the problem?

28-7. How many current sources are used to find total current through the common resistance?

28-4 Thévenin's Theorem (ac)

This theorem is very convenient when we wish to determine circuit values for a complex circuit under a variety of load conditions. The analysis of this circuit will determine specific values for load voltage, load current, and load power.

Thévenin's theorem (ac) An active network can be replaced with an equivalent circuit consisting of an impedance in series with an ac source.

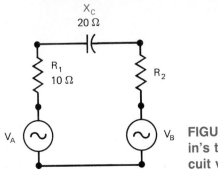

FIGURE 28-6 Use of Thévenin's theorem to solve for circuit values.

There is only one basic principle used for this circuit. It is stated: Any active network (no matter how complex) can be replaced with an equivalent circuit consisting of an impedance that is in series with a constant voltage source.

The two basic conditions required for analysis using Thévenin's theorem are:

1. The voltage source, V_{TH}, is the open-circuit voltage. This is determined by measuring the value of the source voltage with the load disconnected from the circuit.
2. The series impedance, Z_{TH}, is equal to the internal impedance of the original network when it is observed from the terminals of the load. In this analysis, all voltage sources are replaced by their internal impedance values, or short circuits, depending on their original values of impedance.
3. The difference between this ac circuit analysis and the dc circuit analysis described in Chapter 11 is the inclusion of the reactance values and their resulting phase shifts.

Analysis of this circuit may be accomplished using the same circuit as shown for Figure 28-2. The resistance R_3 is considered the load resistance for this circuit. When the load resistance is removed, as required for analysis using Thévenin's theorem, the circuit looks as shown in Figure 28-6. In this Théveninized circuit the two voltage sources oppose each other.

EXAMPLE 28-4

Use the circuit shown for Figure 28-2 to analyze for values using Thévenin's theorem process.

Solution

The Thévenin voltage, V_{TH}, is equal to a value of

$$V_{TH} = V_A - V_B$$

$$= (40 + j0) - (26 + j15) \qquad (28\text{-}6)$$

$$= 14 - j15 \text{ V}$$

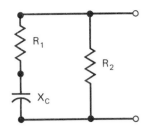

FIGURE 28-7 Equivalent circuit after its reduction.

The total circuit impedance is

$$Z_T = R_1 + X_C + R_2$$
$$= 35 - j20 \ \Omega$$

The current in this equivalent circuit is

$$I_T = \frac{V_{\text{TH}}}{Z_T}$$

$$= \frac{14 - j15}{35 - j20} \tag{28-7}$$

$$= 0.486 - j0.151 \ \text{A}$$

The individual voltage drops are developed using the equivalent circuit
shown in Figure 28-7:

$$V_{R2} = I_T R_2$$
$$= (0.486 - j0.151)25$$
$$= 12.15 - j3.77 \ \text{V}$$
$$V_{\text{load}} = V_{\text{TH}} + V_{R2}$$
$$= (26 + j15) + (12.15 - j3.77)$$
$$= 38.16 + j11.23 \ \text{V}$$

The impedance in series with the circuit is shown as developed using the
equivalent circuit of Figure 28-8. This is accomplished by assuming that

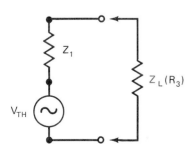

FIGURE 28-8 Thévenin equivalent circuit.

both of the voltage sources are shorted out. The result of the short circuits places R_2 in parallel with the combination of $R_1 + X_C$. This value of series impedance is

$$Z_1 = \frac{(R_1 - jX_C)(R_2)}{(R_1 - jX_C) + R_2}$$

$$= \frac{(10 - j20)(25)}{35 - j20}$$

$$= 11.5 - j7.69 \ \Omega$$

This is the Thévenin equivalent value for this circuit. Current flow through R_3 is determined by the following:

$$I_{R3} = I_T$$

$$= \frac{V_{TH}}{Z_1 + R_3}$$

$$= \frac{38.16 + j11.23}{(11.5 - j7.69) + 50}$$

$$= 0.588 + j0.256 \ \text{A}$$

$$= 0.641 \underline{/23.5°}$$

and

$$V_{R3} = I_3 R_3$$

$$= (0.641 \underline{/23.5°})(50 \underline{/0°})$$

$$= 32.05 \underline{/23.5°}$$

REVIEW PROBLEMS

28-8. A Thévenized network is equivalent to what type of circuit?
28-9. What is done with the load in the process of Thévenizing the circuit?

28-5 Norton's Theorem (ac)

Norton's theorem (ac) *An active network can be replaced with an equivalent circuit consisting of an impedance in parallel with an ac source.*

The final theorem discussed in this chapter is **Norton's theorem**. This theorem, as described in Chapter 11, is used to determine current flow in a complex circuit. It is an adaptation of the Thévenin theorem as it applies to circuit current instead of voltage. In this circuit the Norton values are considered to be in parallel with the voltage source instead of in series, as in the Thévenin analysis.

To be consistent, the same circuit (Figure 28-3) is used for analysis by Norton's theorem. The basic rule for a Norton analysis is:

1. A constant current value identified as I_N is the quantity of current flow in the circuit when the load is short circuited.
2. The parallel impedance, Z_N, is equal to the impedance of the original circuit as observed from the terminals of the load. In this analysis all voltage sources are replaced by their internal impedances or a short circuit.

EXAMPLE 28-5

Determine the current through the load resistance R_3 and the voltage across this load.

Solution

The circuit shown in Figure 28-9 is equal to the original circuit with the load resistance shorted out. In this Nortonized circuit equivalent the currents through the load are equal to the sum of the individual currents. These may be determined by use of the superposition theorem.

The source V_B is shorted and

$$I_A = \frac{V_A}{R_1 - jX_C}$$

$$= \frac{40}{10 - j20}$$

$$= 0.8 + j1.6 \text{ A}$$

When V_A is shorted,

$$I_B = \frac{V_B}{R_2}$$

$$= \frac{26 + j15}{25}$$

$$= 1.04 + j0.6 \text{ A}$$

Thus the total current through the load is

$$I_{R3} = I_A + I_B$$

$$= 0.8 + j11.6 + 1.04 + j0.6$$

$$= 1.84 + j2.2 \text{ A}$$

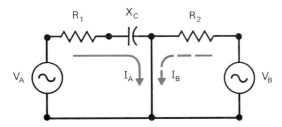

FIGURE 28-9 Nortonized circuit equivalent to Figure 28-3.

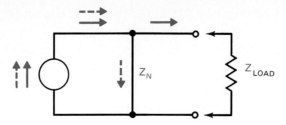

FIGURE 28-10 Norton equivalent circuit.

The next step requires the determination of the parallel impedance for this circuit. The process for doing this is the same as that used to determine the circuit impedance for a Thévenin circuit. Both of the voltage sources are considered to be shorted out. This places R_2 in parallel with $R_1 + X_C$. The series impedance is (borrowing from Example 28-4)

$$Z_N = 11.5 - j7.69 \ \Omega$$

The circuit can now be considered as a Norton equivalent (Figure 28-10). Using a method known as the current ratio method, the value of the load current is determined:

$$I_L = I_N \frac{Z_N}{Z_N + Z_L}$$

$$= (1.84 + j2.2) \frac{11.5 - j7.69}{(11.5 - j7.60) + 50} \tag{28-8}$$

$$= 0.588 + j0.256$$

$$= 0.641 \underline{/23.5°} \ \text{A}$$

This value is equal to the value of I_3 in the original circuit.

The next step is to determine the voltage drop across the load resistance due to current flow in the circuit:

$$V_L = V_{R3}$$

$$= I_L Z_L$$

$$= (0.641 \underline{/23.5°})(50 \underline{/0°})$$

$$= 32.05 \underline{/23.5°} \ \text{V}$$

REVIEW PROBLEMS

28.10. How is a circuit modified to find its Norton equivalent?

28-11. What is done with the internal voltage sources when the circuit is Nortonized?

1. The laws applicable to dc circuit analysis also apply to ac circuit analysis when we consider the phase angles of the values in the circuit.

2. Kirchhoff's voltage law states that the sum of the phasor angles of the voltages in a closed loop equal zero.

3. Kirchhoff's current law states that the phasor sum of the currents entering a junction are equal to the phasor sum of the currents leaving the same junction.

4. Networks containing more than one power source may be analyzed using the superposition theorem.

5. Values for the load under varying conditions in a complex circuit are most easily resolved when using Thévenin's theorem.

6. Current flow analysis in a complex circuit is solved by use of Norton's theorem.

Answer true or false; change all false statements into true statements.

SELF TEST

28–1. Kirchhoff's voltage law may be applied only to dc circuits.

28–2. The phasor information is included when solving an ac circuit.

28–3. Current leaving a junction does not have to equal the current entering the same junction.

28–4. A Thévenized circuit solves for source current.

28–5. The superposition theorem solves for total values for a resistance common to two or more source voltages.

28–6. A Nortonized circuit solves for load current.

28–7. In the Nortonized circuit, the load is short circuited for solution of the problem.

28–8. Both polar and rectangular values are used to solve ac network problems.

28–9. One often has to determine circuit impedance to solve for other circuit values.

28–10. A series circuit is also called a closed loop.

28–11. The use of phasor analysis is required in ac circuit analysis because of
    (a) phase shifting between V and I values
    (b) voltage differences in the circuit
    (c) current-value changes in the circuit
    (d) variable R values in the circuit

MULTIPLE CHOICE

28–12. Kirchhoff's law for voltage drops is related to
    (a) open-loop circuits
    (b) open-current circuits
    (c) closed-loop circuits
    (d) multiple-current-path circuits

28–13. One must consider _____ when applying Kirchhoff's voltage law.
 (a) phasor sum of the currents
 (b) phase relation of current and load
 (c) phasor sum of the EMF and voltage drops
 (d) phasor sum of the applied EMF

28–14. One initial step in the use of Kirchhoff's voltage law is to identify
 (a) current polarities
 (b) source EMF polarity
 (c) source and load polarities
 (d) none of the above

28–15. Kirchhoff's law for current division in the ac circuit is related to
 (a) open-loop circuits
 (b) open-current circuits
 (c) closed-loop circuits
 (d) multiple-current-path circuits

28–16. One must consider _____ when applying Kirchhoff's current law.
 (a) phasor sum of the currents
 (b) phase relation of current and load
 (c) phasor sum of the EMF and voltage drops
 (d) phasor sum of the applied EMF

28–17. One initial step in the use of Kirchhoff's current laws is to identify
 (a) current polarities
 (b) source EMF polarity
 (c) source and load polarities
 (d) none of the above

28–18. When using the superposition theorem to solve for circuit values in an ac circuit:
 (a) power sources must be all considered at the same time
 (b) all power sources except one are short circuited for each major step of the solution
 (c) only one power source is considered
 (d) none of the above

28–19. The system that reduces the complex circuit to one constant EMF source and a series resistance is _____ theorem.
 (a) Norton's
 (b) the superposition
 (c) Millman's
 (d) Thévenin's

28–20. The system that reduces the complex circuit to one constant value and a parallel resistance is _____ theorem.
 (a) Norton's
 (b) the superposition
 (c) Millman's
 (d) Thévenin's

ESSAY QUESTIONS

28–21. Explain the difference in the use of circuit analysis rules for dc and ac circuits.

28–22. Which circuit analysis system is used for multiple-source systems?

28–23. Why is Thévenin's theorem used for systems having varying load conditions?

28–24. How does Kirchhoff's current law apply to an ac circuit?

28–25. Why is a series circuit known as a closed loop?

28–26. Express each of the following in rectangular form. (*Sec. 28-1*)
    (a) $R = 7.5\ \Omega$
    (b) $X_L = 10\ \Omega$
    (c) $X_C = 3\ \Omega$

28–27. Express each of the following in polar form. (*Sec. 28-1*)
    (a) $R = 7.5\ \Omega$
    (b) $X_L = 10\ \Omega$
    (c) $X_C = 3\ \Omega$

PROBLEMS

Use Figure 28-11 to answer problems 28-28 through 28-30.

28–28. Find total circuit impedance in both polar and rectangular form. (*Sec. 28-1*)

28–29. Find the circuit current. (*Sec. 28-2*)

28–30. Find the voltage drops across each of the components. (*Sec. 28-1*)

Use Figure 28-12 to answer problems 28-31 through 28-33.

28–31. Find total circuit impedance in both polar and rectangular form. (*Sec. 28-1*)

28–32. Find the voltage drops for each component. (*Sec. 28-1*)

28–33. Write the formula for Kirchhoff's voltage law for the circuit. (*Sec. 28-1*)

28–34. Find the voltage drops for each component in the circuit of Figure 28-13. (*Sec. 28-1*)

FIGURE 28-11

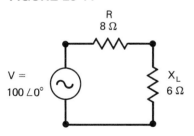

FIGURE 28-12

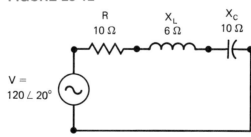

FIGURE 28-13

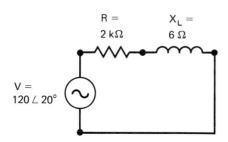

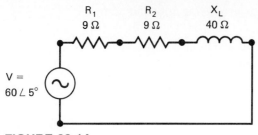

FIGURE 28-14

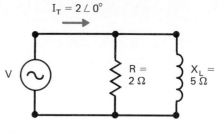

FIGURE 28-15

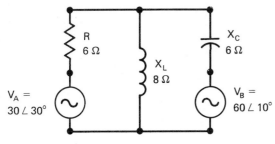

FIGURE 28-16

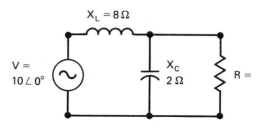

FIGURE 28-17

28–35. Find the voltage drops across R_2 and X_L for the circuit in Figure 28-14. (*Sec. 28-1*)

Use Figure 28-15 to answer problems 28-36 through 28-38.

28–36. Find the current through both components. (*Sec. 28-2*)

28–37. Is the current through the X_L component leading or lagging the voltage? (*Sec. 28-2*)

28–38. Change the values to $V = 60\underline{/0°}$ V, $R = 12\ \Omega$, and $X_L = 10\ \Omega$. Solve for the voltage drops across the two components. (*Sec. 28-1*)

28–39. Use Figure 28-16 and solve for I_{XL} using the superposition theorem. (*Sec. 28-3*)

28–40. Use the circuit in Figure 28-17 and solve for the Thévenin equivalent circuit. (*Sec. 28-4*)

ANSWERS TO REVIEW PROBLEMS

28–1. $-V_S + V_C + V_R + V_L = 0$

28–2. The ac law includes the phase angle of the value.

28–3. It is related to the current at any one junction of the circuit instead of the voltage drops around the circuit.

28–4. Using the junction of X_C and R_1, the formula is stated as $I_{XC} = I_{R1}$.

28–5. One

28–6. They are short circuited.

28–7. All of them are added.

28–8. A series circuit

28–9. It is disconnected.

28–10. It is considered as a parallel circuit with the load short circuited.

28–11. They are replaced with their impedance value, or a short circuit.

Chapters 25–28

The material in these chapters covers RC and L/R circuits and their applications in alternating-current circuits.

1. Transient time is the time required for a voltage or current to change from one value to another value in a circuit.

2. A circuit containing resistance and inductance has a L/R time constant.

3. A circuit containing resistance and capacitance has a RC time constant.

4. Five time constants are required to fully charge or fully discharge the reactive component.

5. Current or voltage *rises* in a charging circuit.

6. Current or voltage *decays* in a discharging circuit.

7. Induced voltage is used to create a short-duration high voltage in a circuit.

8. Waveforms for rise and decay during time constants are displayed on a universal time-constant chart.

9. High currents can be created when a capacitor is discharged in a very short period.

10. Output voltages taken across the capacitor in a RC circuit are called integrated voltages.

11. Output voltages taken across the resistor in a RC circuit are called differentiated voltages.

12. Reactances are applicable only in circuits using sine-wave values.

13. Non-sine-wave circuits are time-constant values instead of reactance values.

14. When an ac circuit is purely resistive, the rules presented by Ohm and Kirchhoff may be applied for circuit analysis.

15. Capacitive reactances are added when connected in series.

16. Capacitive reactances are added by the reciprocal formula when connected in parallel.

17. Inductive reactances are added when connected in series.

18. Inductive reactances are added by the reciprocal formula when connected in parallel.

19. X_C and X_L values are opposite-value reactances. Their values cancel each other when they are added together.

20. When series connected, X_C and X_L ohmic values cancel.

21. When parallel connected, X_C and X_L currents cancel.

22. An ac circuit containing X_C, X_L, and R can be reduced to an equivalent circuit containing R and a net X value.

23. Impedance in the reactive circuit uses the formula $Z = V/I_T$.

24. An ac circuit containing reactance uses the formula $P = I^2R$ for the determination of real power.

25. The power factor is used to indicate the relationship of VI power to real power.

26. Ac circuit values may be measured by use of a dc meter and a rectifier.

27. Ac circuit values may also be measured by use of a thermocouple type of meter movement.

28. Ac power is measured by use of a wattmeter.

29. Alternating-current circuit analysis requires determination of both the magnitude and phase of the electrical quantities.

30. In electrical work, an imaginary number is used to identify a 90° phase shift.

31. The term j is used as the imaginary number.

32. A $+j$ indicates a 90° shift and a $-j$ indicates a 270° shift, or $\angle -90°$.

33. The term j is equal to $\sqrt{-1}$.

34. The term j^2 is equal to -1.

35. Real and imaginary numbers can be displayed on a complex plane.

36. A complex number is a value that falls somewhere between the axis of a real number and an imaginary number. This number is represented by its coordinates. An example of this is the complex number $5 + j3$.

37. An inductive reactance circuit will be seen by use of the $+j$ term.

38. A capacitive reactance circuit will be seen by use of the $-j$ term.

39. Impedance of a series reactive circuit may be stated by use of a complex number.

40. Impedance in the parallel reactive circuit uses formulas that are similar to those used for parallel resistive circuits.

41. Impedance in the series reactive circuit uses formulas that are similar to those used for series resistive circuits.

42. Complex numbers can be added, subtracted, multiplied, and divided.

43. Complex numbers may be displayed in either a rectangular or a polar form.

44. The polar form of the complex number is used to multiply and divide.

45. The rectangular form is used for addition and subtraction of the complex number.

46. The polar form of the complex number has two components: the angle of the phasor relative to the real axis and the magnitude of the phasor. An example of the polar form is $2\angle 45°$.

47. Complex numbers can be converted from the polar form to the rectangular form and from the rectangular form to the polar form.

48. The terms *admittance* and *susceptance* are used to represent the reciprocals of impedance and reactance, respectively.

49. The laws applicable to dc circuit analysis also apply to ac circuit analysis when we consider the phase angles of the values in the circuit.

50. Kirchhoff's voltage law states that the sum of the phasor angles of the voltages in a closed loop equal zero.

51. Kirchhoff's current law states that the phasor sum of the currents entering a junction are equal to the phasor sum of the currents leaving the same junction.

52. Networks containing more than one power source may be analyzed using the superposition theorem.

53. Values for the load under varying conditions in a complex circuit are most easily resolved when using Thévenin's theorem.

54. Current flow analysis in a complex circuit is solved by use of Norton's theorem.

CHAPTER 29

Resonance

Resonance is the term used to describe a condition in a circuit containing inductive reactance, capacitive reactance, and resistance. In this condition the two reactances are equal and opposite in phase; thus they cancel each other out. The remaining value is purely resistive. The circuit reacts to the value of resistance and either accepts or rejects a specific frequency. Resonant circuits are used to "tune" to radio stations, accept specific ranges of frequencies or ranges of frequencies in order to minimize interference.

KEY TERMS

Bandwidth
Detuning
Half-Power Frequencies
Lower Cutoff Frequency
Parallel Resonant
Resonance

Resonant Circuit
Selectivity
Series Resonant Circuit
Tank Circuit
Upper Cutoff Frequency
Varactor Capacitor

OBJECTIVES

Upon completion of the material in this chapter, you will

1. Understand the concept of resonance.
2. Be able to analyze series resonant circuits.
3. Be able to analyze parallel resonant circuits.
4. Be able to determine resonant frequency.
5. Be able to determine the bandwidth of a resonant circuit.

The concept called *resonance* is an important factor in electronic circuits. This factor is used in communications equipment to tune to a specific frequency. Almost all communications devices, including radios, radio-telephones, video display terminals, and television receivers use resonant circuits. The use of the resonant circuit permits the selection of a single frequency for transmission and reception of the broadcast signal.

Resonance occurs when a specific set of electrical conditions exist. These conditions are directly related to the factors of capacitive reactance and inductive reactance as well as resistance. The effect of the interaction of these factors as it relates to a specific frequency is presented in this chapter under the name of *resonance*.

29-1 The Effect of Resonance

In Chapter 23, the relationship between the reactances and frequency was presented. The inductive reactance in a circuit increases as the frequency of the ac voltage presented to it is increased. Conversely, the amount of capacitive reactance in a circuit decreases as the frequency of operation is increased. Since one of these factors is increasing and the other is decreasing, there is one point in a circuit containing both where the two will cancel each other. When both reactances are canceled, only resistance remains in the circuit at the specific frequency. When the two reactances cancel each other, the circuit is said to be at **resonance**. This is known as a **resonant circuit.**

Resonance *Circuit containing resistance, capacitance, and inductance where reactances cancel each other at a specific frequency.*

Resonant circuit *RLC circuit used to 'tune' to specific frequency.*

All circuits containing inductive and capacitive components are resonant at some specific frequency. These types of circuits are known as *LC* circuits. The fact that the circuit also contains resistance is assumed when discussing this type of circuit. In many instances this circuit is known as an *RLC* circuit because of the resistance factor. The combination of the *LC* factors make the circuit resonant. If the *LC* values are held constant and the frequency of the ac voltage applied to the circuit is changed, the circuit is no longer resonant. It then acts as any other circuit containing inductance and capacitance.

As mentioned earlier, the major use of a resonant circuit is found in its ability to *tune* to a specific frequency. This, of course, depends on the exact values of the capacitance and the inductance in the circuit. A classic example of this type of circuit is shown in Figure 29-1. In this illustration the input circuit of a FM broadcast band radio is used. The antenna of any radio picks up all the signals being broadcast in the area. These all induce a voltage onto the antenna wiring of the radio. A resonant circuit is used to select one specific

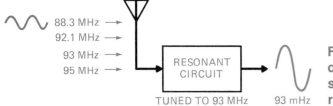

FIGURE 29-1 A resonant circuit is used to "tune" to a specific frequency in the FM radio.

88.3 MHz →
92.1 MHz →
93 MHz →
95 MHz →

RESONANT CIRCUIT

TUNED TO 93 MHz

93 mHz

frequency and to ignore all others. The specific frequency in this example is 93 MHz. This frequency is accepted by the resonant circuit and the others are rejected. The result is shown at the output of the block as a larger sine wave. At the point of resonance the signal at the output is large and the nonresonant signals are rejected.

29-2 Series Resonance

The series circuit shown in Figure 29-2 will illustrate the concept of resonance. This circuit contains a resistance of 10 Ω, a capacitor of 2.65 μF, and an inductance of 2.65 H. The values of reactance are 1000 Ω for both the capacitor and the inductor. In this circuit the reactances are 180° out of phase with each other. Therefore, they cancel. The result is a circuit containing 10 Ω of resistance. This circuit is resonant at 60 Hz, the frequency of the ac signal at the input of the circcuit.

Current in the Series Resonant Circuit In this circuit the total current is determined by use of Ohm's law. Current equals 50 V/10 Ω, or 5 A. In any series resonant circuit the current is limited by the total resistance. Often, the total resistance is the resistance of the wires in the inductor. This is usually a very low value since most inductors are wound using a copper wire having a very low resistance factor.

Series resonant circuit R, L and C components in series to form a resonant circuit

A response curve for this circuit is shown in Figure 29-3. The vertical portion of the graph shows the current and the horizontal portion displays the frequency of operation. In this example the resonant frequency is 60 Hz. At

FIGURE 29-3 Response curve for the circuit in Figure 29-2.

FIGURE 29-2 A series *RLC* circuit will resonate at a specific frequency.

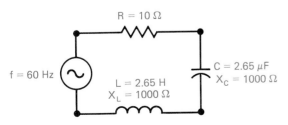

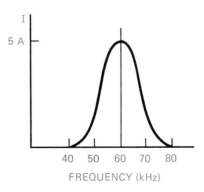

the point of resonance the current is highest. When the voltage applied to the circuit is lower in frequency than the resonant point, a lesser amount of current flows in the circuit. This is also true when the applied voltage's frequency is higher than the point of resonance. This is shown in the illustration. Current is much greater at the 60-Hz frequency that it is at any other frequency.

The reason for the difference in current values is due to the reactance in the circuit. When the operating frequency is higher than resonance, X_L is higher than X_C. The net result is a greater amount of ohmic value to the circuit and a reduction of current flow. When the operating frequency is lower than resonance, X_C is higher than X_L and the total ohmic value is greater than when X_C and X_L are equal and cancel each other.

In summary, when the X_C and X_L values in a circuit are equal and 180° out of phase, they cancel each other. The net result is the resistance in the circuit. This is usually a low value. The low resistance permits a high value of current flow. When the operating frequency is above the resonant frequency, the circuit is inductive in nature. When the operating frequency is lower than the resonant frequency, the circuit is capacitive. In either of the nonresonant conditions the ohmic value of reactance adds to the value of resistance. The result of the addition of these two ohmic values is a reduction in current in the circuit.

Impedance in the Series Resonant Circuit When operating at the point of resonance the circuit reactance values of capacitance and inductance cancel each other. As a result the impedance of the resonant circuit is at a minimum value. This value is actually 0 Ω. When the reactances cancel the phase relationship between voltage and current is also zero and they are considered to be in phase with each other. This places the voltage from the source in phase with the current flow in the resonant circuit.

Voltage and Current in the Series Resonant Circuit A resonant circuit is unique because it has the ability to create a high voltage across either the inductor or the capacitor at the point of resonance. The circuit shown in Figure 29-4 will illustrate this point. This circuit has 2.65 H of inductance, 10 Ω of resistance, and 2.65 µF of capacitance connected in series to a 50-V ac source. The series resonance calculations below illustrate the various factors of reactance, current, and voltage as they apply to the circuit components for a variety of frequencies. All values of voltage and reactance are rounded to the nearest whole number.

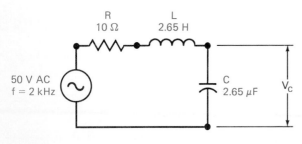

FIGURE 29-4 Characteristics of the series resonant circuit.

f (Hz)	X_L (Ω)	X_C (Ω)	Net Reactance		Impedance Z	IX_L (A)	V_{XL}	V_{XC}
			$X_C - X_L$	$X_L - X_C$				
40	666	1502	836	—	836	0.0598	40	90
50	832	1202	370	—	370	0.135	112	162
60	1000	1000	0	0	10	5.0	5000	5000
70	1165	858	—	307	307	0.163	189	140
80	1331	751	—	580	580	0.086	114	65

These calculations illustrate the ability of a series resonant circuit to produce a higher-than-applied voltage across the reactive components at the point of resonance. When the applied frequency is not at resonance, the voltage drops across the components are very small. Since the voltages created by the high value of current flow at resonance are 180° out of phase with each other it is necessary to connect the output of this circuit across either the inductor or the capacitor, but not across both.

REVIEW PROBLEMS

Use the series resonance calculations above to answer the following questions.

29-1. What frequency is the resonant frequency?
29-2. At what frequency is the highest capacitor voltage drop?
29-3. At what frequency is the highest inductor voltage drop?

29-3 Parallel Resonance

A **parallel resonant** circuit has the capacitor and the inductor in parallel with each other. A resistance is included in the circuit. This resistance represents the winding resistance of the inductor and it is usually shown as being in series with the inductor (Figure 29-4). A parallel resonant circuit is often called a *tank circuit*. This is because the parallel resonant circuit is able to store electrical energy. The energy is stored until it is required for circuit operation. Under practical conditions, this storage time is often in units of microseconds.

Parallel resonant *RLC* circuit where components are connected in parallel with each other.

Conditions for Parallel Resonance. Parallel resonance occurs when X_C is equal to X_L. This is the same as stated for a series resonant circuit. It is based on the same assumption. This assumption is that the two reactances are 180° out of phase and therefore cancel each other. In the parallel resonant circuit the current flow through the inductor and capacitor are equal. They flow in the opposite directions and therefore cancel each other. In the circuit of Figure 29-5, the voltage applied to both components is equal. In an inductive circuit, the voltage leads the current by 90°. In the capacitive circuit the voltage lags the current by 90°.

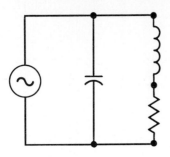

FIGURE 29-5 A parallel resonant circuit has resistance in series with its inductance.

The fact that the two currents are out of phase by 180° results in a total current flow in the circuit of 0 A. In the parallel resonant circuit very little, if any, current is provided by the source. In a practical circuit there is a little current flow due to the resistance in series with the inductor. This small current does flow to compensate for the energy consumed by the resistor.

Line Current In the parallel resonant circuit the line current is minimal. This is because of the very high circuit impedance at resonance. The circuit shown in Figure 29-6 is used to illustrate the conditions existing for a parallel resonant circuit. This information is expressed in the following parallel resonance calculations (all values of voltage rounded to the nearest whole number).

f (Hz)	X_L (Ω)	X_C (Ω)	I_L (A)	I_C (A)	Net Current V/X_C	Net Current V/X_L	I_T (A)	Z_T (Ω)
40	666	1502	0.075	0.033	0.042	—	0.042	1,190
50	832	1202	0.06	0.042	0.018	—	0.018	2,778
60	1000	1000	0.05	0.05	0	0	0.000005	10,000
70	1165	858	0.043	0.059	—	0.016	0.016	3,125
80	1331	751	0.037	0.067	—	0.030	0.030	1,667

Note the values of current and impedance for the various frequencies. This points out the different values of I and Z for the frequencies below, at, and above resonance.

The reason for minimum line current is seen when one considers the opposing current direction due to opposite phases in the inductive and capaci-

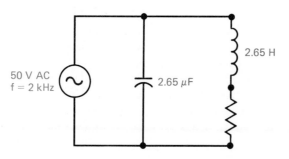

50 V AC
f = 2 kHz

2.65 µF

2.65 H

FIGURE 29-6 Characteristics of the parallel resonant circuit.

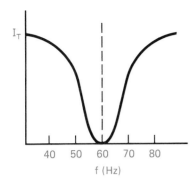 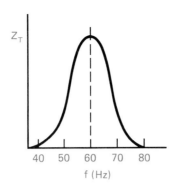 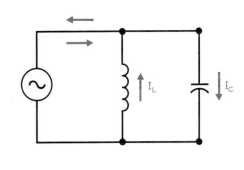

FIGURE 29-7 Graphic representation of current and impedance in the parallel resonant circuit.

tive circuits. When the two phases are 180° apart in phase, the currents will cancel each other. The net result is a current value of close to 0 A. The fact that there is close to 0 A of current leads one to accept that the impedance of the circuit must be very high if this is to occur. The opposing current flow are shown in Figure 29-7. This, too, will help to illustrate how the currents cancel each other since they flow in opposite directions due to the 180° phase difference between them.

Line Voltage   In the parallel resonant circuit the line voltage is equal to the source voltage. Since both the inductive and capacitive components are connected in parallel with the source, their voltage is equal to the applied source voltage.

The Tank Circuit   Earlier in this section the term **tank circuit** was used. This description applied to a parallel-connected resonance circuit. When we consider the interaction between I_L and I_C, they appear to cancel each other. This is true when they are compared to the applied source. However, when we eliminate the source from consideration, the currents remain in the parallel branch. When the source is removed, the two currents will circulate inside the branch circuit. The energy stored in the L and C components is able to develop a complete sine wave to an output circuit even when the input signal to the tank circuit is less than a complete sine wave. This creates a flywheel effect in the circuit since the two currents are following each other in the branch.

Tank circuit   Term used to describe parallel resonant circuit.

29-4 Bandwidth of a Resonant Circuit

The tables used earlier in this chapter showed that the maximum current in a series resonant circuit occurred at the frequency of resonance and that the current dropped rapidly on either side of this frequency. **Bandwidth** is described as the range of frequencies between which the current is equal to or greater than 70.7% of the peak value at resonance. This factor of bandwidth is often shown as BW. It is a very important characteristic of any resonant circuit.

Bandwidth   Range of frequencies between two points greater than 70.7% of maximum signal value.

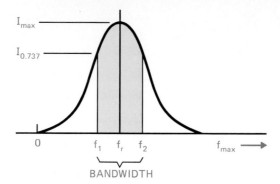

FIGURE 29-8 Current response curve for a parallel resonant circuit.

Series Resonant Circuit Figure 29-8 shows the response curve for a series resonant circuit. The term f_r is used to describe the center frequency, or frequency of resonance. The frequency identified as f_1 shown to the left of f_r is called the **lower cutoff frequency**. Its value is 70.7% of the maximum current in this circuit. The frequency identified as f_2 is shown above the resonant frequency point. It is called the **upper cutoff frequency**. It, too, is a value of 70.7% of the maximum current in this type of circuit.

Lower cutoff frequency
Lower frequency of bandwidth value.

Upper cutoff frequency *Upper frequency of bandwidth value.*

EXAMPLE 29-1

A series resonant circuit has a maximum current of 250 mA at resonance. Find the value of current at the cutoff frequencies.

Solution

The current at both cutoff frequencies is 70.75% of maximum current.

$$I_{f1} = I_{f2} = 0.707 I_{max}$$
$$= 0.707 \times 0.25 \text{ A} \qquad (29\text{-}1)$$
$$= 0.17675 \text{ A}$$

Parallel Resonant Circuit In the parallel resonant circuit the *impedance* is maximum at the resonant frequency. The bandwidth for a parallel resonant circuit is defined as related to the impedance curve rather than to the current curve. Bandwidth is defined in the same manner as that used for the series resonant circuit. It is still related to a value of 70.7% of maximum. In this type of circuit the maximum value is related to impedance rather than current. Bandwidth parameters are described by the formula:

$$Z_1 = 0.707 Z_{max}$$
$$Z_2 = 0.707 Z_{max} \qquad (29\text{-}2)$$

This is illustrated in Figure 29-9. In this illustration the values have been changed to represent impedance rather than current. The bandwidth values are described as those between f_1 and f_2 in the illustration.

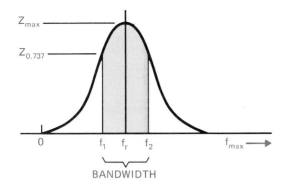

FIGURE 29-9 Impedance values for the parallel resonant circuit.

Zmax —

Z0.737 —

0   f₁ f_r f₂   f_max ⟶

BANDWIDTH

Bandwidth Formula Bandwidth for both the series and the parallel resonant circuit is the range of frequencies between the two cutoff frequencies. These are shown in Figures 29-8 and 29-9. The formula for determination of this is

$$BW = f_2 - f_1 \qquad (29\text{-}3)$$

The value of the center frequency can be found by using the value of the two frequencies, f_1 and f_2. This is accomplished by use of this formula:

$$f_1 = f_r + \frac{BW}{2} \qquad (29\text{-}4)$$

or

$$f_2 = f_r - \frac{BW}{2} \qquad (29\text{-}5)$$

EXAMPLE 29-2

Find the bandwidth of a resonant circuit having an upper cutoff frequency of 15 kHz and a lower cutoff frequency of 5 kHz.

Solution

$$BW = f_2 - f_1$$
$$= 15 \text{ kHz} - 5 \text{ kHz}$$
$$= 10 \text{ kHz}$$

EXAMPLE 29-3

Find the resonant frequency for a circuit having a lower cutoff frequency of 88 MHz and a bandwidth of 150 kHz.

Solution

$$f_r = f_1 + \frac{BW}{2}$$
$$= 88 \text{ MHz} + \frac{0.15 \text{ MHz}}{2}$$
$$= 88.075 \text{ MHz}$$

29-5 Half-Power Frequencies

Half-power frequencies
Another name for the two 70.7% points of the bandwidth determination.

Often the frequencies at the 70.7% points are called the **half-power frequencies**. The reason for this is because the power of the circuit at the 70.7% points is one-half of the maximum power delivered at the resonant frequency of the circuit. This is proven in the following manner. When the circuit is at resonance.

$$P_{max} = I_{max}^2 R \qquad (29\text{-}6)$$

and the power at f_1 and f_2 is

$$P_{f1} = I_{f1}^2 R$$

$$= (0.707 I_{max})^2 R$$

$$= (0.707)^2 I_{max}^2 R$$

$$= 0.5 I_{max}^2 R = 0.5 P_{max}$$

REVIEW PROBLEMS

29-4. What percent of total current is the current at cutoff?

29-5. At what point between the cutoff values does the center frequency occupy?

29-6. What is the power level at the half-power points?

29-7. Find the half-power values for a resonant circuit having a maximum power of 1.6 W.

29-8. Find the bandwidth of a resonant circuit having an upper cutoff frequency of 15 kHz and a lower cutoff frequency of 7.0 kHz.

29-6 Selectivity

Selectivity *Ability of resonant circuit to accept a specific frequency and reject all adjacent frequencies.*

The curves shown in Figures 29-8 and 29-9 are also known as *selectivity curves*. **Selectivity** is defined as the description of how well the resonant circuit responds to a specific frequency and rejects all other frequencies. A greater value of selectivity is identified by a narrow, or smaller, bandwidth.

It is very easy to assume that a resonant circuit accepts only those fre-

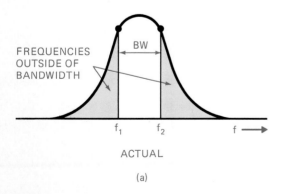

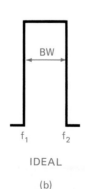

FREQUENCIES
OUTSIDE OF
BANDWIDTH

BW

f_1 f_2 $f \longrightarrow$

ACTUAL

(a)

BW

f_1 f_2

IDEAL

(b)

FIGURE 29-10 Bandwidth determination of the circuit frequency response curve.

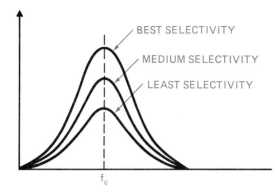

BEST SELECTIVITY

MEDIUM SELECTIVITY

LEAST SELECTIVITY

f_c

FIGURE 29-11 Selectivity curves are rated and seen as values of voltage.

quencies that are within its bandwidth. This is not a true statement. In reality, the resonant circuit accepts all frequencies, including those outside its bandwidth. Those signals that fall outside the bandwidth parameters are attenuated to a great degree. In general, it can be stated that the farther the frequencies are from the center frequency, the greater the amount of attenuation of the signal. The selectivity curve shown in Figure 29-10(a) is considered as an ideal response curve. Notice that the actual response curve does not have vertical components but exhibits a gradual slope from the peak out to the limits of the frequency. The ideal bandwidth is shown in Figure 27-10(b).

The relationship between the amplitude and sharpness of the response curve and the amount of selectivity is shown in Figure 29-11. Each of the curves in this drawing has a 70.7% cutoff point. The shortest, or lower, curve has a much wider response curve than that shown for the upper curve. The upper selectivity curve has the best selectivity and the smallest bandwidth.

When one considers that the radio frequencies used for all types of broadcasting services have different stations adjacent to each other, the need for a sharp response curve and narrow bandwidth becomes important. If the receiving device has a wide bandwidth, it is possible to tune to more than one station at the same time. On the other hand, if the bandwidth of the receiving device is too narrow, it may be possible to eliminate a portion of the received signal.

An example of this condition is the ability to receive an FM stereo radio broadcast signal. The bandwidth for this signal is 150 kHz. If the receiver has a bandwidth of less than 150 kHz, a portion of the received signal will be attenuated. This could result in the stereo component not being processed properly. The stereo component could even be eliminated under a very narrow bandwidth design in the receiver.

29-7 The Quality Factor (Q) of a Circuit

The figure of merit, or quality factor, of a circuit is defined as the ratio of reactive power to resistive power. It is the ratio of the power in the inductor to the power in the resistor in series with the inductor. The Q of a circuit is determined at the resonant frequency for the circuit. The formula for this is

$$Q = \frac{I^2 X_L}{I^2 R} \qquad (29\text{-}7)$$

This formula can be simplified by the elimination of like terms:

$$Q = \frac{X_L}{R} \qquad (29\text{-}8)$$

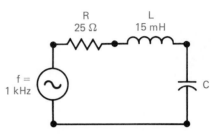

FIGURE 29-12

EXAMPLE 29-4

Find the Q at resonance of the circuit shown in Figure 29-12.

Solution

$$
\begin{aligned}
X_L &= 2\pi f L \\
&= 2 \times 3.14 \times 1000 \text{ Hz} \times 0.015 \text{ H} \\
&= 94.2 \ \Omega \\
Q &= \frac{X_L}{R} \\
&= \frac{94.2 \ \Omega}{25 \ \Omega} \\
&= 3.76
\end{aligned}
$$

EXAMPLE 29-5

Find the Q at resonance for the circuit shown in Figure 29-13.

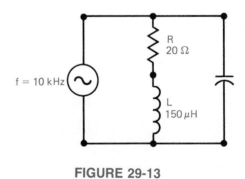

FIGURE 29-13

Solution

$$X_L = 2\pi fL$$
$$= 2 \times 3.14 \times 10000 \text{ Hz} \times 0.15 \text{ H}$$
$$= 9420 \ \Omega$$

$$Q = \frac{R}{X_L}$$
$$= \frac{9420 \ \Omega}{20 \ \Omega}$$
$$= 471$$

Note that in both of the example circuits, the amount of capacitance has no relation to the determination of circuit Q.

Circuit Q as a Multiplier In the series RCL circuit the voltages developed across L and across the C components depend on the Q of the circuit. At the point of resonance the full amount of the source voltage develops across the resistance because the combined voltages of X and C are zero. This, of course, is due to their opposite polarities. An example of this is shown in Figure 29-14. Using Ohm's law, $V_L = IX_L$ and $V_C = IX_C$. The voltages developed across L and C may be identified using the following steps:

$$V_1 = IX_L \tag{29-9}$$

$$Q = \frac{R}{X_L}$$
$$X_L = QR \tag{29-10}$$

QR may be substituted for X_L in any equation, since they are equal:

$$V_L = IQR = IRQ$$

The values of V_S and IR are equal at resonance; therefore,

$$V_L = QV_S \tag{29-11}$$
$$V_C = QV_S$$

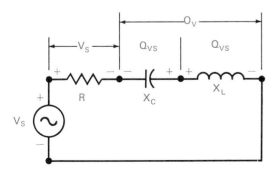

FIGURE 29-14 Cancellation of X_C and X_L values in the series RLC circuit.

The equations above show that circuit Q multiplied by the source voltage provides the value of voltage that develops across both L and C at resonance. When the value of Q is high, the values of V_L and V_C may also be much greater than the source voltage. Do not forget that the polarities of V_L and V_C are opposite and will cancel each other. The result is a net reactive voltage of zero for the circuit.

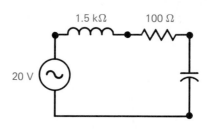

FIGURE 29-15

EXAMPLE 29-6

Find the values of V_C and V_L for the circuit shown in Figure 29-15.

Solution

$$Q = \frac{X_L}{R}$$

$$= \frac{1500\ \Omega}{100\ \Omega}$$

$$= 15$$

$$V_L = V_C Q V_S$$

$$= 15 \times 20$$

$$= 300\ \text{V}$$

Bandwidth and Circuit Q In a practical circuit the value of the resonant circuit Q affects the bandwidth of the circuit. A high value of circuit Q will create a narrow bandwidth. Low values of circuit Q will offer a wide bandwidth. The high value of circuit Q usually indicates a better selectivity value for the circuit. The relationship of circuit Q and bandwidth is expressed as

$$\text{BW} = \frac{f_r}{Q} \tag{29-12}$$

EXAMPLE 29-7

Find the bandwidth of the circuit shown in Figure 29-16.

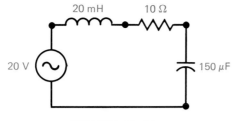

FIGURE 29-16

Solution

$$f_r = \frac{1}{2\pi \sqrt{LC}}$$

$$= \frac{1}{2 \times 3.14 \times \sqrt{0.02 \times 0.00015}}$$

$$= \frac{1}{6.28 \sqrt{0.000003}}$$

$$= \frac{1}{0.01088} = 91.88 \text{ Hz}$$

$$X_L = 2\pi f L$$

$$= 2 \times 3.14 \times 9.19 \times 0.02$$

$$= 11.55 \ \Omega$$

$$Q = \frac{X_L}{R}$$

$$= \frac{11.55}{10}$$

$$= 1.155$$

therefore

$$\text{BW} = \frac{f_r}{Q}$$

$$= \frac{91.88}{1.155} \text{ Hz}$$

$$= 79.9 \text{ Hz}$$

REVIEW PROBLEMS

29-9. What is the circuit Q factor?
29-10. What is the formula for circuit Q?
29-11. Is the circuit Q high or low for a narrow-bandwidth signal?

29-8 Resonant Circuit Applications

Resonant circuits are used in several different ways. These include tuning and filtering of electronic signals. The methods of tuning, or accepting, an electronic signal are discussed in this section. The topic called filtering is presented in Chapter 30.

Tuning of an electronic signal is accomplished by use of a resonant circuit. The circuit may be either "fixed tuned" or "variable tuned." Both types of circuits are in common use in all types of communications equipment. One example of a fixed-tuned circuit is the type of tuning used for citizens' band and marine radios. The operator of the radio is only able to select certain specific frequencies. All tuning is accomplished by selection of a specific channel from the front panel of the radio. This is true for both the transmitter and receiver frequencies. Under normal operating conditions it is impossible for the operator to change the frequency of the specific channel.

The other type of tuning is called variable tuning. Under this scheme the operator is able to adjust both the transmitting and receiving frequencies as required. This type of transmitter adjustment requires a license from the federal government in the United States. Receiver adjustments do not require any license. In fact, almost all AM and FM receivers use an adjustable tuning process. The operator is able to tune to any portion of the broadcast band simply by adjusting the tuning control knob.

Both the fixed-tuned and the variable-tuned circuits use the same type of components. Usually, the tuned circuit is nothing more than an *LC* circuit operating at its resonant frequency. When a fixed-tuned circuit is used, both the *L* and the *C* components have a fixed value. The resonant frequency is predetermined by use of a chart or by calculation. A circuit of this type is shown in Figure 29-17(a).

A selectable type of tuning is accomplished using this circuit and a switching network. In a switch selection system either the inductor or the capacitor is changed by use of a switch. The switch usually has several sets of contacts. Each set of contacts is connected to a different-value component. Rotation of the switch selects the specific value required for resonance. This type of switching arrangement is shown in Figure 29-17(b). The switch may have as many sets of contacts as required for the number of frequencies being selected.

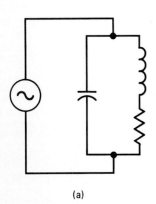

(a)

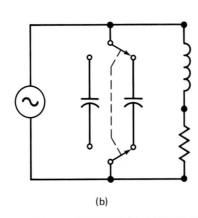

(b)

FIGURE 29-17 Use of a switching circuit to select values for resonance.

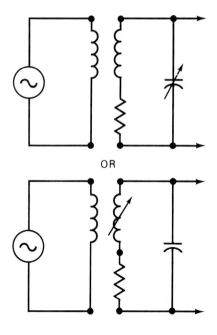

OR

FIGURE 29-18 Either the value of capacitance or inductance is changed in order to "tune" to a different frequency.

Two methods of adjusting the resonant frequency of the tuned circuit is shown in Figure 29-18. One of these uses a fixed-value inductor and a variable-value capacitor. This type of tuning is often used in AM and FM receivers. The output signal is developed across the capacitor in this circuit. The second method shown uses a fixed value of capacitance and a variable inductor. Usually, the inductor's value is varied by adjustment of a core inside its windings. This procedure has been used very commonly in automobile radios. The reason for this is that the variable capacitor would become dusty and, when in this condition, become intermittent in operation. Use of a variable inductor core eliminates this problem. The result was a much more reliable system.

The technology of today has done away with both types of tuning systems. The receivers manufactured today use a solid-state device called a **varactor capacitor**. This device has no moving components. It could be described as a voltage-controlled capacitor. Application of a dc voltage sets the value of capacitance. Use of an adjustable voltage source produces the required amounts of capacitance for the resonant circuit.

Varactor capacitor
Semiconductor diode used as a capacitor by varying the applied voltage to its terminals.

29-9 Detuning

Detuning describes a resonant circuit that is not resonant at the desired frequency. If we are attempting to tune to a frequency of 1.6 MHz, for example, we find that the circuit is resonant at that frequency by design. The impedance of the circuit is very high at resonance and the signal also has a very high amplitude. When the resonant circuit is improperly tuned and we are still attempting to receive a signal on 1.6 MHz, the amplitude of the received signal is very low. This concept is called detuning, or mistuning. The result is a less-than-desired amplitude of the signal.

Detuning *A resonant circuit not resonant at the desired frequency.*

29-12. State three methods used to vary the resonance of a circuit.

29-13. What is another name for detuning?

CHAPTER SUMMARY

1. A resonant circuit permits the selection of one specific frequency.
2. Factors affecting resonance are inductance, capacitance, and resistance.
3. When inductive reactance and capacitive reactance cancel, the resistance in the circuit is the only remaining factor.
4. Resonant circuits are known as *LC* or *RCL* circuits.
5. Resistance is very low in a series resonance circuit.
6. Current flow is high in a series resonant circuit.
7. Impedance is low in a series resonant circuit.
8. High voltages may be created across the inductive and capacitive components in a series resonant circuit.
9. Voltages created in a series resonant circuit oppose each other by 180°.
10. A parallel resonant circuit is called a *tank circuit*.
11. In a parallel resonant circuit X_C and X_L cancel because they are 180° out of phase.
12. The bandwidth of a resonant circuit is defined as the points above and below center frequency where the voltage, current, or impedance is 70.7% of the peak value.
13. Bandwidth of the resonant circuit must be adequate to process the signal.
14. Q is the quality factor, or figure of merit, for a circuit.
15. Selectivity is the ability of a circuit to respond to a specific frequency and reject all other frequencies.
16. Tuning a resonant circuit is accomplished by adjusting either the capacitance or the inductance of the circuit.
17. Small or narrow bandwidth is indicative of better selectivity.

SELF TEST

Answer true or false; change all false statements into true statements.

29-1. A resonant circuit is a tuned circuit.

29-2. Inductive reactance is 90° out of phase with capacitive reactance.

29-3. A parallel resonant circuit has a high impedance at resonance.

29-4. Half-power points are equal to 70.7% points.

29-5. A detuned series resonant circuit has a high impedance at resonance.

29-6. A high-Q resonant circuit has poor selectivity.

29-7. A parallel resonant circuit has a high impedance at resonance.

29-8. A series resonant circuit has a high impedance at resonance.

29-9. A series resonant circuit has a high impedance when not at resonance.

29-10. Bandwidth of a circuit is the range of frequencies between the upper and lower cutoff frequences.

29–11. A circuit is said to be at resonance when
    (a) resistance and reactance cancel
    (b) resistance and X_L cancel
    (c) X_L and X_C cancel
    (d) resistance and X_C cancel

29–12. In a resonant circuit, decreasing the value of X_L will
    (a) decrease the resonant frequency
    (b) modify the circuit resistance
    (c) increase the resonant frequency
    (d) change the wavelength of the voltage

29–13. One of the major functions of a resonant circuit is to react to
    (a) current changes
    (b) voltage changes
    (c) a specific frequency
    (d) none of the above

29–14. In the series resonant circuit:
    (a) voltage is high at the point of resonance
    (b) current is high at the point of resonance
    (c) impedance is low at the point of resonance
    (d) V, I, and Z have no relation to resonance

29–15. In the parallel resonant circuit:
    (a) voltage is high at the point of resonance
    (b) current is high at the point of resonance
    (c) impedance is low at the point of resonance
    (d) V, I, and Z have no relation to resonance

29–16. A parallel resonant circuit is also know as a(n)
    (a) tank circuit
    (b) receptive circuit
    (c) inductive loop
    (d) voltage-developing circuit

29–17. the range of frequencies at which the current is equal to or greater than 70.7% of the peak value at resonance
    (a) defines the 70.7% points of the circuit
    (b) defines the lower frequency point of the circuit
    (c) defines the bandwidth of the resonant circuit
    (d) has little relation to resonance

29–18. The points of the circuit where the power is at 70.7% of maximum are known as the _____ points.
    (a) 70% power
    (b) 30% power
    (c) half-power
    (d) resonant

29–19. The term used to describe the ability of a circuit to respond to one frequency and reject all others is
    (a) sensitivity
    (b) selectivity
    (c) band rejection
    (d) bandwidth value

29–20. The highest value of resonant voltage develops in a circuit having a
 (a) low Q factor
 (b) medium Q factor
 (c) high Q factor
 (d) high RLC factor

ESSAY QUESTIONS

29–21. Explain the term *detuning*.

29–22. How does bandwidth affect circuit tuning?

29–23. Explain the difference between selectivity and sensitivity.

29–24. Explain why X_L and X_C cancel at resonance.

29–25. Why is a resonant circuit also known as a RLC circuit?

29–26. What is the effect of attempting to tune a specific frequency when the components are not able to resonate at that frequency?

29–27. Explain the relationships among current, impedance, and voltge in a series resonant circuit.

29–28. Explain the relationships among current, impedance, and voltage in a parallel resonant circuit.

29–29. Explain how the three methods of tuning a resonant circuit function.

29–30. How does bandwidth of a resonant circuit affect the ability to select one radio or TV station and reject all others?

PROBLEMS

29–31. Find the value of X_C at resonance when X_L is 1.5 KΩ. (*Sec. 29-2*)

29–32. Find the impedance of a series resonant circuit having an L of 25 mH, a C of 12 μF, and a R of 25 Ω. (*Sec. 29-2*)

29–33. Determine the f_r for a parallel resonant circuit with a C of 28 μF, an L of 5 H, and a R of 15 Ω (*Sec. 29-3*)

29–34. Determine the f_r for problem 29-33 when the R is decreased to 5 Ω. (*Sec. 29-3*)

29–35. Determine the f_r for a parallel resonant circuit using the values of problem 29-34. (*Sec. 29-3*)

29–36. Determine the C for a resonant circuit with a f_r of 0.80 MHz and an L of 10 mH. (*Sec. 29-3*)

29–37. A resonant circuit has an X_L of 3 kΩ and an R of 50 Ω. The resonant frequency is 6 Hz. What is the bandwidth? (*Sec. 29-3*)

29–38. A series resonant circuit has a Q of 40. The source voltage is 20 V. Find the values of V_L and V_C at resonance. (*Sec. 29-2*)

29–39. What is the bandwidth of a circuit when the lower cutoff frequency is 1.8 KHZ and the upper cutoff frequency is 2.4 KHz. (*Sec. 29-4*)

29–40. The bandwidth of a resonant circuit is 850 Hz. Its resonant frequency is 8 KHz. What are its cutoff frequencies? (*Sec. 29-4*)

29–41. A resonant circuit has an X_L of 4 kΩ and an R of 45 Ω. Determine its BW when the resonant frequency is 4 KHz. (*Sec. 29-4*)

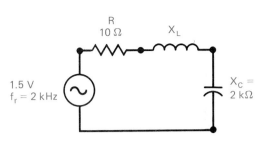

FIGURE 29-19

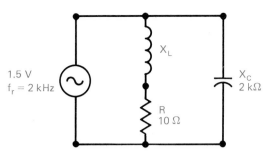

FIGURE 29-20

29–42. Determine the value of C for a circuit with a f_r of 180 KHz and an L of 150 μH. (*Sec. 29-4*)

29–43. Determine the value of X_L at resonance for Figure 29-19. (*Sec. 29-2*)

Use Figure 29-20 to answer problems 29-44 through 29-49.

29–44. Determine the Q. (*Sec. 29-3*)

29–45. Determine the bandwidth. (*Sec. 29-3*)

29–46. Determine the impedance. (*Sec. 29-3*)

29–47. Find the value of Q. (*Sec. 29-3*)

29–48. Find the value Q for the circuit in problem 29-42 when the R is increased to 100 Ω. (*Sec. 29-3*)

29–49. Find the values of current. (*Sec. 29-3*)

29–50. A series resonant circuit has an X_L of 2.5 KΩ and and R of 5 Ω. Determine the following values. (*Sec. 79-7*)
  (a) Circuit Q
  (b) X_C
  (c) Current when 75 mV is applied to the circuit
  (d) V_{XC}

29–1. 60 Hz

29–2. 60 Hz

29–3. 60 Hz

29–4. 70.7%

29–5. Midpoint

29–6. $0.5\ P_{MAX}$

29–7. 0.8 W

29–8. 8 KHz

29–9. Figure of quality

29–10. $Q = X_L/R$ (series); $Q = R/X_L$ (parallel)

29–11. High

29–12. Varying the capacitance, varying the inductance, or using a varactor diode

29–13. Off-resonance

CHAPTER 30

Filters

The electric or electonic filter is a circuit used to accept or reject a certain frequency or group of frequencies. A filter device is utilized in the dc power supply to remove undesired changes in voltage or current. A filter is used on a communications device to remove frequencies outside the range of operation. Filters are also used on the power lines of computers to attenuate undesired momentary high-voltage conditions known as "spikes." In consumer products the filter is also used to accept a specific frequency. An example of this is the color information transmitted from a television transmitter. This signal information is accepted by one circuit in the receiver and used to turn on the color-processing section of the receiver.

Active Filter
Amplifier
Bandpass Filter
Bypass
Capacitor Coupling
Crystal Filters
Cutoff Frequency
DC Component
Ferrite Beads

Filter System
High-Pass Filter
Low-Pass Filter
Parallel Resonant
Pass Band
Passive Filter
Series Resonant
Stop Band
Transistor

OBJECTIVES

Upon completion of the material in this chapter, you will

1. Understand the concept of combination ac and dcs circuit values.
2. Understand the concept of signal coupling.
3. Recognize filter circuits.
4. Be able to calculate values for electronic filter circuits.

A filter is a device for separating one item from another. In the automobile, an oil filter is used to remove unwanted particles of metal and other contaminants from the lubricating oil in the engine. In the world of electrical and electronic applications a filter is a device used to either accept or reject a specific range of frequencies. Electrical filters are either *active* or *passive*. The **active filter** depends on the action of a transistor, a tube, or an integrated circuit. The **passive filter** depends on the action of resistors, capacitors, and inductors.

The material in this chapter is related to the passive filter. Often the name of the filter circuit will identify its type of action in the circuit. For example, a low-pass filter will permit the passage of a specific range of low frequencies and attenuate higher frequencies. The specific range of frequencies may also be named, or identified, in the title of the filter. Electrical filters are often called low-pass, high-pass, bandpass, and even band-stop filters. In addition, one might find a 19-kHz bandpass filter in the FM stereo decoder section of a radio.

One example of a **filter system** is shown in Figure 20-1. In this example there are several frequencies applied to the filter section of the system. This particular filter is designed to pass only one set of frequencies. This frequency set is centered around 15 kHz. Any other frequency not very close to 15 kHz is attenuated, or rejected, by the filter. The result is seen at the output of the filter block. Only the frequency of 15 kHz is available at the output of this unit.

A second example of a filter system is shown in Figure 30-2. Here there are two filter systems. One of these is called a **low-pass filter** and the other is called a high-pass filter. The low-pass filter [Figure 30-2(a)] operates as its name implies. It permits the low frequencies to pass and attenuates the higher frequencies. In this example the two frequencies are 100 and 20 kHz. The low-pass filter action permits the low frequency of 20 kHz to pass and blocks, or attenuates, the 100-kHz high frequency. The **high-pass filter** [Figure 30-2(b)] works in the opposite manner. It permits the passage of the higher frequency and attenuates the lower frequency. The output of this high-pass filter is the 100-kHz frequency and the 20-kHz frequency is attenuated. Filters such as

Active filter *Filter circuit using vacuum tube or transistor circuitry.*

Passive filter *Filter circuit using capacitors, resistors, and inductors.*

Filter system *Electrical circuit used to attenuate or enhance specific frequencies.*

Low-pass filter *Circuit designed to permit passage of low frequencies and attenuation of higher frequencies.*

High-pass filter *Circuit designed to attenuate low frequencies and pass high frequencies.*

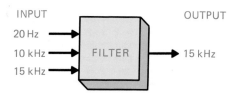

INPUT

20 Hz
10 kHz   FILTER
15 kHz

OUTPUT

15 kHz

FIGURE 30-1 A filter is designed to eliminate a specific range of frequencies in an electrical circuit.

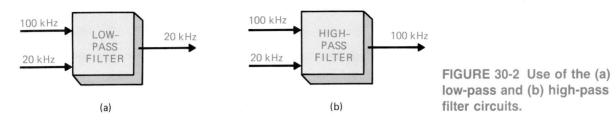

FIGURE 30-2 Use of the (a) low-pass and (b) high-pass filter circuits.

these are commonly used in communictions equipment as well as consumer entertainment devices.

30-1 AC and DC Combination Circuits

An electrical circuit may include a varying component. When the dc voltage changes at some consistent rate, it is said to fluctuate or pulsate. In fact, many of the voltages; available from a system that changes ac into dc are, for this reason, known as pulsating dc voltages. Figure 30-3 illustrates this point. The voltage shown has only one polarity. It is always positive in this example. This voltage does vary from a high of $+4$ V to a low of $+2$ V. It cannot be identified as a steady or constant value, due to its variations. This type of voltage is known as a *pulsating* dc voltage. The values may be separated into a dc component and an ac component. The ac component is the voltage that varies from $+2$ to $+4$ V. The **dc component** would be a steady voltage at the midpoint of the variations. In this example the dc component has a value of $+3$. The reason for this analysis is because the two voltages are said to add together when both are applied to the same circuit.

Dc component *Average value of signal voltage in system.*

The waveform shown in the upper portion of Figure 30-4(a) represents a varying ac voltage having a value of 2 V from peak to peak. The difference between the highest value and the lower value is 2 V when one subtracts -1 from $+1$. These values are added to the steady dc voltage value of $+3$ V shown in the lower portion of part (a). The two voltages are combined in one circuit in Figure 30-4(b). The result of this addition is a pulsating voltage ranging between $+4$ and $+2$ V.

FIGURE 30-4 A varying dc voltage does not have to start at the zero voltage level.

FIGURE 30-3 Varying, or pulsating, dc voltage waveform.

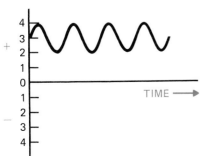

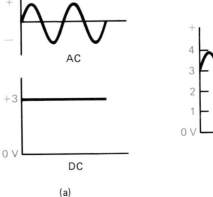

(a)

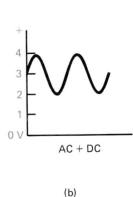

AC + DC

(b)

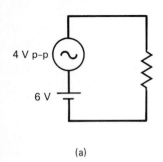

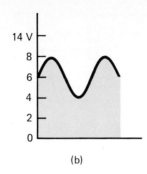

4 V p-p

6 V

(a)

14 V
8
6
4
2
0

(b)

FIGURE 30-5 Combination of dc and ac voltages creates the varying dc voltage shown in part (b).

Another circuit showing this action is seen in Figure 30-5(a). Here a dc source of + V is combined with an ac source of 4 V peak to peak. The result of this combination is shown as the varying dc voltage in Figure 30-5(b). The voltage observed across the resistor has both an ac component and a dc component. It varies from a high of +8 V to a low of +4 V when the dc and the ac components are added together

The requirements of the circuit in which the filter is used will determine the operaing frequency as well as the type of filter circuit. In a signal processing system the ac component is desired. A filter in this system is used to remove the dc component in order to process the signal, or ac component. The dc component in a system of this type is actually the level of dc voltage required for proper circuit operation. Another type of circuit may require removal of the ac component and leave only the dc portion of the voltage (Figure 30-6). The output of a rectifier system is called a pulsating dc voltage. This voltage is one-half of the input ac voltage. The pulsating dc voltage requires more conditioning before it can be used to operate transistors or integrated circuits. A filter network is used to remove the pulsations and leave only a "pure" dc voltage.

REVIEW PROBLEMS

Use the Figure 30-5 to find the following values

30-1. Peak-to-peak value of the ac component
30-2. Highest peak value of the voltage
30-3. Rms value of the ac component

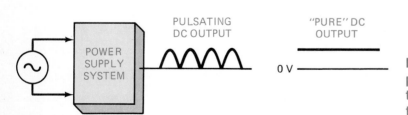

PULSATING
DC OUTPUT

"PURE" DC
OUTPUT

0 V

POWER
SUPPLY
SYSTEM

FIGURE 30-6 The output of a power supply system is filtered to remove any variations from its voltage level.

30-2 Types of Coupling

Almost all electrical and electronic signal processing systems require both an operating voltage and a means of transferring the signal from one stage to the next stage through the system. This method of transferring the desired component is an example of filtering. Two of the more comon methods of coupling use either a transformer or a capacitor. The basic principles of transformer action and capacitor action are used for this purpose. As stated earlier, the principles do not change, only the applications. One must be able to apply the basic principles in order to comprehend how the circuit functions.

Transformer Coupling The concept of transformer action occurs only when there is a change in the value of the current flow in the primary of the transformer. The circuit shown in Figure 30-7 illustrates how this applies. The primary circuit has both an ac component and a dc component. The value of the combined voltages produce a varying voltage having a 6-V p-p value and a dc value of 20 V. The combined values of these voltages produce a voltage that varies between $+23$ and $+17$ V.

Transformer action occurs when there is a change in the current flow in the primary winding of the transformer. The current change in the primary occurs when the voltage changes from $+17$ to $+23$ V. The value of the dc component has no effect on the induced voltage process in the transformer. Transformer action also inverts the voltage at the secondary of the device. This is illustrated by the inverted waveform. Voltages induced in the secondary are the result of the varying current in the primary. In this example assume a turns ratio of 1:1. The secondary voltage is 6 V p-p. Note that only the variations of the input voltage appear at the secondary winding of the transformer. The dc component in the primary is blocked and does not transfer to the secondary.

The frequency of the voltage waveform at the secondary winding is totally dependent on the frequency of the ac component of the primary voltage. Any frequency that can be processed by a transformer may be used for this purpose.

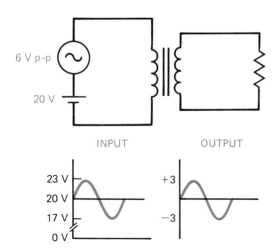

FIGURE 30-7 Transformer coupling of the variations of the voltage create the ac voltage at the secondary winding of the transformer.

This is an example of signal processing without the inclusion of the dc component in the secondary, or output, portion of the circuit. In addition, there is a 180° phase shift in the voltages of the primary and the secondary. One could call this an example of a high-pass filter system since the frequency of the dc component is 0 Hz and any frequency used has a higher order than that of the dc voltage.

Capacitor Coupling One of the more common types of signal coupling is known as capacitor coupling. **Capacitor coupling** is used in audio amplifier circuits to process signals from the input to the output of the system. Capacitor coupling circuits are used in amplifiers because the capacitor occupies less space than required by a transformer. In addition, on a practical side, capacitors usually are less costly than transformers and certainly weigh less. All of these factors have influenced the application of capacitors are signal coupling devices in electronic systems.

The circuit shown in Figure 30-8 is an example of capacitor coupling. This circuit is an *RC* coupling circuit. It consists of a series capacitor and a resistance connected from the output side of the capacitor to common. In a circuit such as this the combination dc and ac voltage is applied between the capacitor and circuit common. The capacitor will charge to the level of the dc component. This is the same value as the average dc voltage in the circuit. The dc voltage cannot appear across the resistance in the circuit because it is blocked by the capacitor. Only the changes in voltage that are transferred to the right-hand plates of the capacitor. This is another example of a high-pass filter system.

In this circuit the voltage on the plates of the capacitor will be a constant 40 V. The action of the ac component causes a rise in this value to 45 V and a fall to a low of +35 V. The average voltage across the capacitor is 40 V when one considers that the peaks and valleys of the voltage cancel each other out.

The variations in the ac component in this circuit produce a current flow in the resistance, *R*, in the circuit. A voltage drop is created across the resistance when current flow occurs. The value of this voltage is equal to the ac component

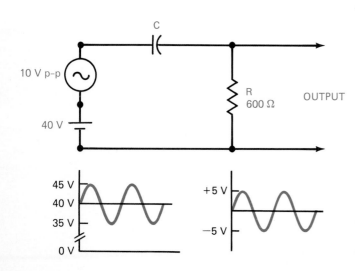

FIGURE 30-8 Capacitors are also used to couple, or transfer, signal voltages from one stage to another stage in electronic devices.

applied to the capacitor. This is true since the variations of the ac component occur at different moments in time.

There is no phase shift in this illustration. The reason for this is found in the practicality of the circuit. Almost all *RC* coupling circuits require a value of *R* that is about 10 times the value of X_C. The amount of reactance is almost nil under these conditions. The phase angle for a ratio of 10:1 is less than 6° and is considered to be zero for practical purposes.

EXAMPLE 30-1

Find the value of X_C so that X_C is $\frac{1}{10}$ the value of *R* (use Figure 30-8).

Solution

$$X_C = 0.1R$$
$$= 0.1 \times 600 \ \Omega$$
$$= 60 \ \Omega$$

EXAMPLE 30-2

Find the value of the sine-wave output for Figure 30-8 when a 10-V–5-kHz signal is applied to the circuit.

Solution

$$C = \frac{1}{2\pi f X_C}$$
$$= \frac{0.159}{5000 \times 60}$$
$$= 0.53 \ \mu F$$

$$V_{out} = \frac{R}{\sqrt{R^2 + X_C^2}} V_{in} \tag{30-1}$$
$$= \frac{600}{\sqrt{600^2 + 60^2}} \times 10$$
$$= \frac{600}{\sqrt{360,000 + 3600}} \times 10$$
$$= \frac{600}{603} \times 10$$
$$= 0.995 \times 10 = 9.95 \ V$$

The output voltage is almost equal to the input voltage without the dc component of 40 V.

Bypass Coupling When traveling on a highway a bypass is considered as a roadway around something, usually an urban area. In electrical work the term **bypass** is used to indicate a path around a component. Often a capacitor is used in an application where the signal is bypassed around a resistance or inductance. The result of this action is a dc voltage without any ac component. Such a circuit is illustrated in Figure 30-9 where the capacitor is connected in parallel with the load. Any variations that are presented to the parallel connection of R and C are *bypassed* to circuit common. The output voltage appears as a constant dc value. This is an example of allow-pass filter, since the zero frequency dc is passed on to the load and the higher-frequency ac component is shunted, or bypassed, to circuit common.

> **Bypass** *Use of capacitor to maintain constant voltage value at specific point in circuit.*

An example of how the bypass system functions is found in almost all ac power supply systems. These systems have an input of an ac voltage. Their output is *rectified*, or converted, into a pulsating dc voltage. The waveform for this pulsating dc voltage is shown in Figure 30-10. In most applications, the pulsating dc voltage requires further processing before it can be utilized by an electronic circuit. This processing is known as *filtering*. The result of adding a filter section to the output of the rectifier system will produce an almost steady dc voltage. Electronic filter capacitors are installed in parallel with the load and output of the rectifier block of the power supply.

The value of any filter capacitor may be determined by use of the capacitive reactance formula. Often a value of reactance of around 10 Ω is desired for this type of circuit.

FIGURE 30-9 A bypass type of coupling circuit used to eliminate voltage variations in the circuit.

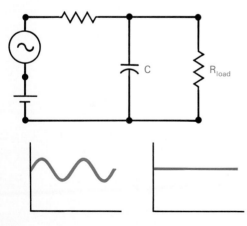

FIGURE 30-10 The output of the rectification system requires filtering or bypassing of the voltage variations.

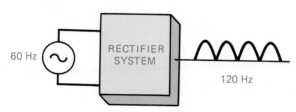

In Example 30-3, the value of 133 μF is not standard for the industry. When selecting such a capacitor, it is common practice to use the next highest readily available standard commercial value.

Another application for the bypass capacitor is shown in Figure 30-11. The device shown in this illustration is a **transistor**. It is connected in a configuration called an **amplifier**. One element of the transistor is known as the *emitter*. In a practical amplifier circuit the emitter element should not have any signal action on its emitter element. (Additional discussion about transistor action is presented in Chapter 31.) The signal present at the emitter element is *bypassed* to circuit common by the addition of the capacitor, C. This action appears as a short circuit for the signal while permitting normal current flow through the elements of the transistor. The voltage at the emitter is maintained at a constant dc value without any variations as a result of the application of the emitter bypass capacitor.

Transistor *Solid state device used as controlled variable resistance in circuit.*

Amplifier *Electronic circuit using a small amount of power to control a larger amount of power.*

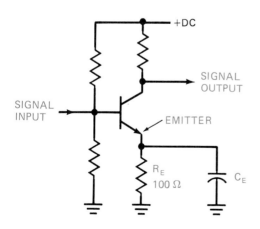

FIGURE 30-11 Bypassing of an element of an active device will "lock" the voltage at a constant value.

EXAMPLE 30-4

Find the value of emitter bypass capacitor keeping its reactance about $0.1R_{emitter}$ when a frequency of 8 kHz is applied to the circuit.

Solution

$$X_C = 0.1R_{emitter}$$

$$= 0.1 \times 100$$

$$= 10 \ \Omega$$

and

$$C = \frac{1}{2\pi f X_C} = \frac{0.159}{f X_C}$$

$$= \frac{0.159}{8000 \times 10}$$

$$= \frac{0.159}{80,000}$$

$$= 1.98 \ \mu F$$

(Use of a 2.0-μF capacitor is recommended.)

REVIEW PROBLEMS

30-6. What is the value of X_C in Example 30-4 when the value of $R_{emitter}$ is increased to 1.5 kΩ?

30-7. Find the value of emitter bypass capacitor when $R_{emitter}$ is 1.5 kΩ and a 10-kHz signal is applied.

30-3 Low-Pass Filters

Discussion previously in this chapter identified the term low-pass filter. It is an electronic device that permitted only low frequencies to pass through it while attenuating higher frequencies. There are some specific characteristics related to the low-pass filter. These are explained in conjunction with Figure 30-12.

Pass band *Range of frequencies not attenuated by filter.*

The term **pass band** is used to define the range of frequencies permitted to pass through the filter. This range depends on the design of the filter components. The upper end of the pass band is identified as the **cutoff frequency**, f_C. In Figure 30-12 the cutoff frequency is 70.7% of the maximum output of the filter.

Cutoff frequency *Upper frequency limit of filter operation.*

Stop band *Set of frequencies attenuated by filter action.*

Stop band is another term applied to the description of filter action. The **stop band** is that set of frequencies where the filter attenuates the signals. An

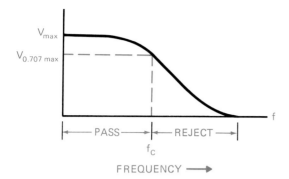

FIGURE 30-12 Frequency response curve for a low-pass filter.

example of this is a filter acting as a low-pass filter. The filter is designed to cut off frequencies above 20 kHz. The *pass band* describes those frequencies below the cutoff point of 20 kHz. The *stop band* describes those frequencies above the cutoff frequency point of 20 kHz.

REVIEW PROBLEMS

30-8. A low-pass filter has a cutoff frequency of 15 kHz. What is its stop band?

30-9. What is the pass band of the filter in problem 30-8?

30-4 Choke-Type Filters

An inductor is known as a *choke* in the world of electricity and electronics. This is an old term and probably a carryover from the days when the inductor was thought to choke off the flow of current. The term *choke* is still used to describe circuits using inductors to remove any variations in circuit current and voltage.

The circuit shown in Figure 30-13 is used to illustrate how the choke filter circuit functions. In this circuit the inductor, L, is in series between the input and the output of the circuit. A resistor, R, is connected in parallel across the output terminals of the filter. The value of reactance of the inductor is very high at the frequency of the input signal when it is compared to the value of the resistance, R. Under these conditions the signal is attenuated and only the dc component of the circuit appears at the output terminals of the filter. Another

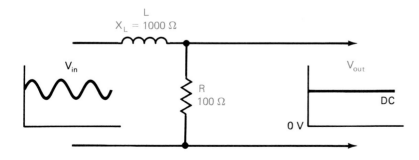

FIGURE 30-13 Circuit used to illustrate a choke filter network.

way of describing this is to state that the signal voltage is dropped across the inductor and none appears across the resistor.

EXAMPLE 30-5

Find the output voltage for the choke filter shown in Figure 30-13 when the input is a 3 $V_{\text{peak to peak}}$ signal riding on 6 V dc. The X_L of the choke is 1000 Ω at the signal frequency.

Solution

First find the value of the sine wave.

$$V_{\text{out}} = \frac{R}{\sqrt{R2 + X_L^2}} V_{\text{in}}$$

$$= \frac{100}{\sqrt{100^2 + 1000^2}} \times 3$$

$$= \frac{100}{\sqrt{1,010,000}} \times 3$$

$$= \frac{100}{1004.98} \times 3$$

$$= 0.2985 \text{ V}_{\text{p-p}}$$

This value represents the value of the sine wave at the output of the filter. It is obvious that the sine-wave voltage is attenuated and almost eliminated.

Ferrite beads *Compressed powdered ferrite devices used as filter devices in radio frequency circuits.*

The chokes described to this point operate at low frequencies. Most of them are either iron-cored or ferrite-cored coils of wire. Another type of choke is used at high RF frequencies. This choke consists of a hollow piece of ferrite material, either doughnut shaped or tubular. The conductor is passed through the center of the ferrite material. Any radio-frequency signals on the conductor are choked at the point of the ferrite material. The specific composition of the ferrite material determines the frequency range of operation. These chokes are commonly known as **ferrite beads** in the electronics field.

30-5 Low-Pass Filter Circuits

The types of low-pass filters shown in Figure 30-14 are typical of those used in electronic circuits. Part (a) shows an "L"-type circuit. It actually has the shape of an inverted capital letter L. This type of filter could also be called a "choke input" filter since the choke is first in line and is the input connection to the filter circuit. Figure 30-14(b) shows a "T" type of filter. Its configuration is similar to that of the capital letter T. Figure 30-14(c) shows a configuration that is similar to the symbol used to represent the Greek capital letter pi.

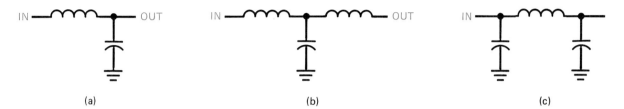

FIGURE 30-14 Low-pass filters used to couple between the source and the load.

The interesting part of all of these filter networks is that in each the capacitor is connected in parallel with the load, input, and output, and the inductors are connected in series between the input and output circuits. The circuits shown are actually only a portion of the system requirements. A filter circuit is connected between two other circuits in the system. Electron current flow is still from the negative of the power source, through the load, and then back to the positive terminal of the power source. Electron current flow through the filter portion of the system occurs in the same manner (Figure 30-15).

Since the filter circuit is connected in series between the power source and the load, it is possible to place the series components in either the positive or the negative side of the wiring. The usual practice is to place these components in the positive side of the circuit. When very high voltages are encountered in circuits, it may be more practical to install these series components in the negative, or common, side of the wiring. This may produce some extra wiring in the circuit. It also raises the common connection at the load above the value of circuit common. In defense of this type of circuit, the components do not require high-voltage breakdown values. This reduces both the cost and the physical size for those components required for this type of circuit. In many practical circuits a series resistor is used instead of the inductor. The size and cost factors involved in production often dictate this substitution. The circuit will respond in a similar manner with this substitution.

30-6 High-Pass Filters

The high-pass filter permits those frequencies *above* the cutoff frequency to pass and attenuates frequencies below the cutoff frequency point. An example of this is shown in Figure 30-16. This drawing is the reverse of that shown for

FIGURE 30-15 Electron current flow between source and load.

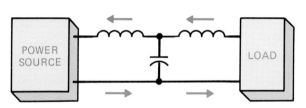

FIGURE 30-16 Frequency response curve for a high-pass filter.

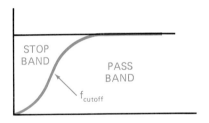

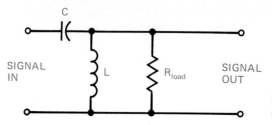

FIGURE 30-17 Circuit diagram for a high-pass filter.

the low-pass filter (Figure 30-12). The stop band is lower in frequency than the pass band. The curve shown in this illustration is opposite from that used for the low-pass filter. It does perform in the same manner as the low-pass filter.

The circuit diagram shown in Figure 30-17 is that of a high-pass filter network. Note that the placement of the capacitor and the inductor are reversed from that used for the low-pass filter circuit. The high-pass filter network may use all of the same circuit configurations used by the low-pass filter. The only difference is the reversal of the positions of the L and C components. The choke in this type of circuit has a high reactance above the cutoff frequency. As a result of this, the voltage in the circuit is developed across the choke and little voltage develops across the capacitor. Below the cutoff frequency the resistance acts as a short circuit and no voltage will develop across its terminals.

These terms also apply to other types of filters. A filter could have both an upper cutoff frequency and a lower cutoff frequency. This depends on its specific use. Since the bottom end of the low-pass filter is a frequency of zero, it is not possible to have a lower cutoff frequency. When the filter is a high-pass type, it will have a cutoff frequency that is lower than its pass band. When the filter is a bandpass type, it will have both an upper and a lower cutoff frequency.

30-7 Bandpass Filters

Bandpass filter *Circuit designed to pass a mid-range of frequencies and attenuate both above and below the desired range.*

It is possible to process a group of frequencies other than those passed or attenuated by low-pass filter or high-pass filter action. Such a filter network is called a **bandpass filter**. The bandpass filter is a combination of a low-pass and a high-pass filter. The curves for such a filter are shown in Figure 30-18. This filter is used as a part of an FM stereo decoding system in an FM stereo receiver.

In this type of receiver the stereo components of the broadcast signal occupy a group of frequencies that fall between 23 and 53 kHz. The stereo decoder section of the receiver must attenuate signals of all other frequencies.

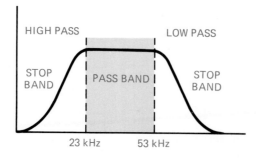

FIGURE 30-18 The bandpass filter permits a specific band of frequencies to pass and attenuates all other frequencies.

The bandpass filter is used to pass this group of frequencies. The low-pass filter section is used to cut off all frequencies above 530 kHz. The high-pass filter section cuts off all frequencies below 23 kHz. This is but one application of the bandpass type of filter network. The passband of this filter is the difference between the lower cutoff frequency and the upper cutoff frequency. In this example the passband is 30 kHz.

REVIEW PROBLEMS

30-10. What is the passband of a filter with cutoff frequencies of 20 Hz and 20 kHz?

30-11. What is the passband of a bandpass filter with a low cutoff frequency of 40 MHz and an upper cutoff frequency of 50 MHz?

30-8 The Resonant Filter

A filter may be designed as either a series resonant circuit or a parallel resonant circuit. The characteristics of the specific type of resonant circuit apply to these circuits when they are included as part of a total circuit configuration. The specific characteristics related to circuit Q and bandwidth also apply. These types of filters may be either a bandpass or a band-stop filter.

Series Resonant Filters This type of filter is shown in Figure 30-19. It consists of a series-connected inductance and capacitance. The **series resonant** circuit has minimum impedance and maximum current at the resonant frequency. This characteristic develops a minimal voltage drop across the filter network and a maximum voltage drop across the load resistor. This, of course, occurs only at the resonant frequency.

> Series resonant *Series RLC circuit having components resonant at one frequency.*

Two series resonant circuits are illustrated in Figure 30-19. Both exhibit the same characteristics. The series resonant circuit always has a minimum impedance at resonance. When the *LC* network is in series with the input [Figure 30-19(a)], the majority of the voltage drop in the circuit occurs across

FIGURE 30-19 A resonant filter network may be either (a) series or (b) parallel with the load.

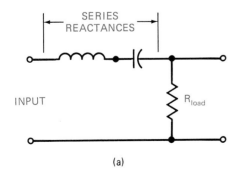
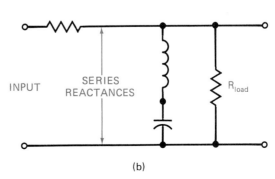

(a)   (b)

the load resistance. This is known as a *bandpass* type of filter. When the *LC* components are connected in parallel with the load, there is a minimum voltage drop across the load and a maximum amount of current flows through the filter. In Figure 30-19(b), the maximum voltage drop is developed across the series resistance. This is one example of a *band-stop* filter network.

REVIEW PROBLEMS

30-12. What is characteristic of the impedance of a series resonant filter?
30-13. What is characteristic of the voltage drop across a series resonant filter?

Parallel resonant *Parallel RLC circuit resonant at one specific frequency.*

Parallel Resonant Filters Two **parallel resonant** circuits are shown in Figure 30-20. One of these has the *LC* circuit in series with the load resistance. The other circuit has the *LC* components in parallel with the load resistance. A parallel resonant *LC* circuit has a high impedance at resonance. This means that the voltage drop developed across the *LC* circuit at resonance is high and the current is low.

The circuit shown in Figure 30-20(a) is a voltage-divider network. Voltage drops are developed across the series *LC* circuit and the load resistance. Since the impedance of the *LC* circuit is much higher than that of the load resistance, the majority of the voltage drop develops across the *LC* circuit. The voltage in this type of circuit is the signal voltage. Since most of this signal voltage develops across the series *LC* circuit, there is very little, if any, developed across the load. This type of circuit will impede the signal. It is known as a *band-stop* type of circuit.

The circuit shown in Figure 30-20(b) has the *LC* components connected in parallel with the load resistance. In this type of circuit the voltage drop developed across the *LC* circuit is also developed across the load resistance. The load resistance has to be a high resistance value. This type of circuit develops maximum signal voltage at resonance. This maximum voltage develops across the *LC* circuit and the load. The voltage, again, is the input signal voltage for the circuit. This type of *LC* filter circuit is known as a *bandpass* type because of this action.

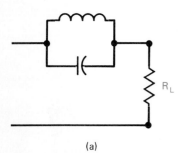

(a)

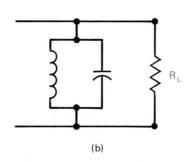

(b)

FIGURE 30-20 Parallel resonant filter networks are also connected in (a) series or (b) parallel with the load.

Combination Filters When the circuit requirements demand it, the series resonant filter may be combined with a parallel resonant filter. The combination of these filters creates a filter having a very narrow frequency characteristic.

Crystal Filters The *piezoelectric* effect found in quartz often is used in electronic circuits. A thin piece of quartz will create a mechanical vibration at one frequency. The frequency is dependent on the physical characteristics of the quartz. The schematic symbol for a quartz crystal is shown in Figure 30-21. The crystal, as it is known, will produce a voltage when it is twisted, compressed, or expanded. A voltage at the same frequency as the point of resonance of the crystal will create one of these physical actions. When used as a filter, the crystal will have a very high Q factor and provide a very narrow bandwidth for any signals.

Crystal filters Quartz material having ability to attenuate specific frequencies.

30-9 Filter Design

A majority of the better types of electronic filters are constructed of capacitors, inductors, and resistors. Their purpose is to either enhance or attenuate a range of frequencies. Usually, the name associated with the filter accurately describes the type of frequency response for the filter network. As stated earlier, there are certain basic terms used to describe the characteristics of all electronic filter networks. These terms include *pass band*, *stop band*, and *roll-off point*. The use of these terms is shown in Figure 30-22.

Low-Pass Filters Figure 30-22(a) shows the circuit for a low-pass filter. Its purpose is to attenuate the higher frequencies and to pass the lower frequencies that are presented at its input. The filter consists of a series inductance, L, and a parallel capacitance, C. These are shown with a resistance representing the load for the circuit. In this low-pass filter network, the capacitor offers a low-impedance path for all of the higher frequencies present in the circuit. The

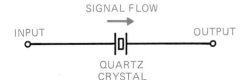

FIGURE 30-21 A crystal material will also act as a filter for a signal frequency.

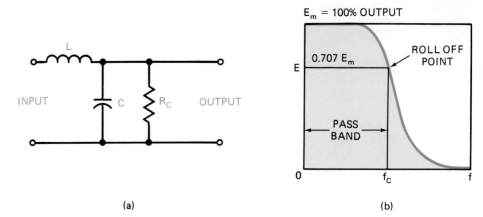

FIGURE 30-22 Design of the low-pass filter network.

inductance offers a low impedance to the lower frequencies. Since the inductance is in series, the low frequencies pass. The parallel-connected capacitance offers a low impedance path to circuit common for the higher frequencies. Since the X_L components are low compared to the values of R_C, the higher frequencies are bypassed to circuit common.

Figure 30-22(b) shows the curve used to indicate how the frequency terms are utilized. The shaded portion of the illustration shows the portion of the frequencies that are passed. The term E_m represents the maximum voltage presented to the filter circuit. The 0.707 point (or 70.7%) of the maximum voltage value is identified as the rolloff point for the circuit. This is the value of the signal level that is passed by the circuit. The signal frequencies that are higher than this roll-off point are attenuated by the circuit. They fall into the stop-band section of the graph. They are attenuated and are bypassed to circuit common at this frequency.

The method of calculation for a low-pass filter uses three formulas to obtain the proper circuit values for the filter network. These formulas are

$$L = \frac{R_L}{2\pi f_C} \tag{30-2}$$

$$C = \frac{1}{2\pi f_C R_L} \tag{30-3}$$

$$f_C = \frac{1}{2\pi\sqrt{LC}} \tag{30-4}$$

where L = inductance (H)

C = capacitance (F)

R_L = load resistance (Ω)

f_C = cutoff frequencies (Hz)

High-Pass Filters The design for a high-pass filter circuit is shown in Figure 30-23(a). This filter uses the same components as does the low-pass filter. In

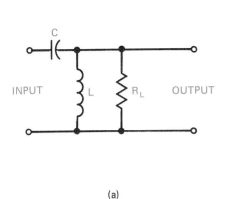

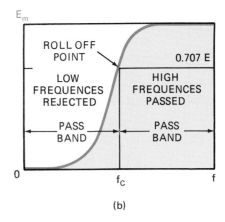

(a)

(b)

FIGURE 30-23 Design of the high-pass filter network.

the high-pass filter the position of the inductance and capacitance are reversed. The capacitor is connected in series with the signal path and provides a low-impedance path for the higher frequencies. The parallel-connected inductor offers a low-impedance path to low frequencies. Figure 30-23(b) illustrates this concept. It is a reversal of the form shown for a low-pass filter network.

The formulas used to determine the values for the L and C components used in the high-pass filter are

$$L = \frac{R_L}{2\pi f_C} \qquad (30\text{-}5)$$

$$C = \frac{1}{2\pi f_C R_L} \qquad (30\text{-}6)$$

Bandpass Filters Use the circuits shown above as well as a series resonant LC circuit. This circuit [Figure 30-24(a)] has two parallel sections. These consist of L_2 and C_2 as one filter network and L_3 and C_3 as the second parallel resonant network. Both of these resonant circuit offer a high impedance to the specific range of frequencies we desire to attenuate. The series resonant circuit consists of L_1 and C_1. This circuit provides a low-impedance path for the desired fre-

FIGURE 30-24 Design of the pass-band filter network.

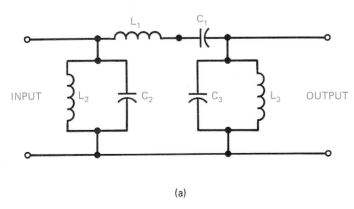

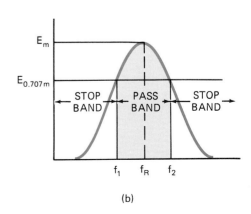

(a)

(b)

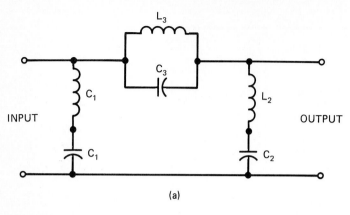

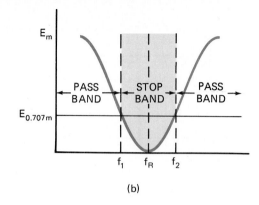

(a)

(b)

FIGURE 30-25 Design of the stop-band filter network.

quencies of the signal. The response curve for the bandpass filter network is shown in Figure 30-24(b). Note that there are two stop-band sections to this curve. One is higher in frequency than the pass band and the other is lower in frequency. Filters of this type are often used as tunable resonant circuits in communications equipment. It is possible to substitute variable capacitors for the fixed values shown in the circuit. The use of the variable capacitor permits one to "tune" in the desired frequencies.

Band-Stop Filters Band-stop filters often use the same components as those of the bandpass filter networks. The difference is in the manner in which they are constructed. The circuit for a band-stop filter network is shown in Figure 30-25(a). This filter utilizes two series resonant circuits connected across the input and output of the network. These two series resonant circuits provide a low-impedance path to the specific frequencies one wishes to attenuate, or stop. They consist of L_1, L_2, C_1, and C_2. The parallel resonant circuit made up of L_3 and C_3 are series connected between the other two networks. This parallel section provides a high-impedance path for the undesired signal frequency. The response curve for this type of filter network is shown in Figure 30-25(b).

30-10 Filter Types

In addition to the basic series resonant or parallel resonant filters described previously, the filter may also be identified as having a specific commercial application. Electrical and electronic filters often found as discrete devices include low-pass radio-frequency filters, high-pass radio-frequency filters, power-line interference filters, and terrestrial interference filters. The low-pass radio-frequency filter is used to eliminate unwanted RF single frequencies from being transmitted by any type of broadcast transmitter. A low-pass RF filter is built into the output section of a citizens' band transmitter-receiver unit. This filter attenuates any second harmonics of the transmitter. The second harmonic of the broadcast frequency of a CB radio just happens to be the same frequency as that used by television channel 2. Another method of attenuating unwanted RF signals is to use a high-pass filter on the receiver. The high-pass filter for

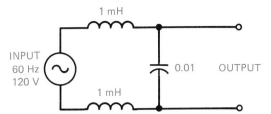

FIGURE 30-26 A filter circuit for the 60-Hz power line.

a television receiver will attenuate those frequencies below 50 MHz. This is another method of preventing unwanted RF signals from interfering with television reception.

Many motors, RF circuits, and fluorescent lamps have the ability to create unwanted radio-frequency signals. These signals are often carried by the 60-Hz power lines. This type of signal is unwanted. It will interfere with the normal signal processing of computer, radios, and television receivers. A device called a power-line filter is inserted in the ac line between the ac source and the unit requiring the ac power. The schematic diagram for a typical ac power-line filter is shown in Figure 30-26. It consists of a small inductance in series with the power line and a small value of capacitance placed in parallel with the power line. This type of filter will attenuate much of the interference created by the motor, lamp, or RF device.

Another filter becoming better known to us is called a *terrestrial interference* filter. Current trends in technology have given us the satellite television receiver system. The signals from the satellites centered 22,300 miles over the earth's equator have very little amplitude by the time they reach the earth. In the process of transmission to the earth, these signals may pick up some additional signals. These are unwanted types of signals that create "sparklies" in the signal. The sparkly is a type of interference. A filter designed to attenuate the frequencies of the sparklies is used. The result is a better quality of picture on the receiver.

REVIEW PROBLEMS

30-16. What is the name of the electrical effect for a quartz crystal?
30-17. What type of electronic equipment uses a low-pass filter?
30-18. What type of electronic equipment uses a high-pass filter?

1. A passive filter consists of inductive and capacitive components.
2. An electronic filter may be used to remove unwanted signals.
3. An electronic filter may be used to pass specific frequencies.
4. A low-pass filter attenuates higher frequencies.
5. A high-pass filter attenuates low frequencies.
6. A low-pass filter is used to remove variations of the dc voltage from a power supply.

CHAPTER SUMMARY

7. A bypass coupling capacitor is used to remove unwanted voltage variations from one section of an electronic device.

8. The pass band is the range of frequencies permitted to pass through the filter.

9. The cutoff frequency is that frequency where the filter attenuates the signal.

10. A filter is connected in series between two specific points in a circuit.

11. The stop band is a set of frequencies where filter action attenuates the signal.

12. A bandpass filter is used to permit a specific range of frequencies to pass and it rejects all other frequencies.

13. A series resonant filter is used as an attenuator for a signal frequency.

14. A parallel resonant filter is used to accept a signal frequency.

15. Series and parallel resonant circuit are often combined in filter networks.

16. A quartz crystal may be used as a filter device.

SELF TEST

Answer true or false; change all false statements into true statements.

30-1. The low-pass filter attenuates low frequencies.

30-2. The high-pass filter attenuates low frequencies.

30-3. The cutoff frequency is determined by the drop is signal voltage to 70.7% of maximum value.

30-4. A filter is used to remove any variations from a dc voltage.

30-5. Transformers will pass the dc component and attenuate the ac component of a voltage.

30-6. Capacitors will pass the dc component and attenuate the ac component of a voltage.

30-7. A bypass capacitor eliminates signal voltage variations at its point of insertion.

30-8. A resistor is often used in place of an inductor in a filter network.

30-9. A series resonant filter develops a high voltage across its LC components.

30-10. The crystal filter has a very high Q at resonance.

MULTIPLE CHOICE

30-11. A high-pass filter is used to
  (a) attenuate high frequencies
  (b) attenuate low frequencies
  (c) pass a midrange of low frequencies
  (d) none of the above

30-12. A bandpass filter is used to
  (a) reject a specific group of frequencies
  (b) attenuate all frequencies except the desired group

(c) attenuate all low frequencies

(d) attenuate all high frequencies

30-13. A filter network used to reject one specific group of frequencies in one part of the spectrum being processed is called a _____ filter.
  (a) high-pass
  (b) bandpass
  (c) low-pass
  (d) band-reject

30-14. The process of passing an electrical signal from one stage in a device to the next stage is called
  (a) signal transfer
  (b) transfer voltage
  (c) signal coupling
  (d) information transfer

30-15. Almost all electrical and electronic devices that use signal processing actually
  (a) vary the operating voltage in the device for processing
  (b) vary the type of transistors to process the signal
  (c) cause a voltage to change at the power source

30-16. Transformer coupling occurs when the
  (a) current in the primary induces a voltage on the secondary
  (b) voltage in the primary induces a voltage on the secondary
  (c) magnetic field in the primary induces a voltage on the secondary
  (d) magnetic field in the secondary induces a voltage on the primary

30-17. A bypass coupling capacitor is used to
  (a) pass signals on to the following stages of the device
  (b) maintain operating current in the power source
  (c) couple signals between stages
  (d) "lock" the operating voltage, or pass any signal variations to circuit common

30-18. The process of removing unwanted circuit voltage or current variations in the power supply is known as
  (a) filtering
  (b) voltage decoupling
  (c) current decoupling
  (d) bypassing

30-19. A filter choke required to operate in the radio-frequency range uses a core material made of
  (a) ferrite
  (b) iron
  (c) laminated steel
  (d) brass

30-20. A crystal type of filter uses
  (a) ceramic capacitors
  (b) crystal capacitors
  (c) quartz materials
  (d) a clear plastic substance

30–21. Explain the relationship between function and the name of filter networks.

30–22. Where is a low-pass filter used in electronic equipment?

30–23. Explain the term *stop-band filter*.

30–24. Why is a series resonant filter network used as an attenuator for a signal frequency?

30–25. Why is a parallel resonant filter network used to accept a specific frequency?

30–26. Explain the term *cutoff frequency*.

30–27. Why are series and parallel resonant filter networks combined in some systems?

30–28. Why is it necessary to remove the voltage variations from the output of a dc power supply?

30–29. Explain how a crystal functions as a filter.

30–30. Explain how the electrical characteristics of the components used in a filter networkk affect the frequencies of operations.

PROBLEMS

30–31. A signal has a peak voltage of 1.5 V. What is its voltage amplitude at its cutoff frequency? (*Sec. 30-9*).

30–32. A bandpass filter has a low-frequency cutoff of 25 kHz and a high-frequency cutoff of 49 kHz. What is its bandpass? (*Sec. 30-9*)

30–33. What is the value of a sine-wave voltage of 15 V p-p riding on a dc voltage of 40 V? (*Sec. 30-1*)

30–34. What is the value of f for an X_C of 120 Ω (Use a value of 0.1 for the C) (*Sec. 30-9*)

30–35. Find the value of the sine-wave output using Figure 30-8 and values of 25 V and 10 kHz. (*Sec. 30-9*)

30–36. Find the values of the sine-wave in problem 30-35 using values of 100 V and 100 Hz. (*Sec. 30-9*)

30–37. A sine-wave signal of 5 V p-p is riding on a 20-V dc voltage. What is the approximate signal voltage output from a high-pass filter? (*Sec. 30-1*)

30–38. What would you expect to find for the signal output in problem 30-37 if a low-pass filter is used? (*Sec. 30-5*)

30–39. Find the value of emitter bypass capacitor for the circuit in Figure 30-11 when a 45-kHz frequency is applied. (*Sec. 30-2*)

30–40. Find the value of C in problem 30-39 when a frequency of 45 MHz is used. (*Sec. 30-2*)

30–41. A high-pass filter has a cutoff frequency of 120 Hz. Which of the following frequencies are attenuated? (*Sec. 30-6*)
(a) 60 Hz
(b) 80 Hz
(c) 100 Hz

(d) 130 Hz

(e) 180 Hz

30–42. A bandpass filter accepts a 10-kHz band between 20 and 30 kHz. Which of the following frequencies are attenuated? (*Sec. 30-7*)

(a) 8 kHz

(b) 15 kHz

(c) 25 kHz

(d) 35 kHz

(e) 40 kHz

30–43. Explain the difference between the coupling capacitor and the bypass capacitor. (*Sec. 30-2*)

30–44. Find the value of X_C for the circuit in Figure 30-8 when $R = 3\ k\Omega$ and $f = 100$ Hz. (*Sec. 30-2*)

30–45. Find the value of the sine-wave output for Figure 30-8 when a 5-V, 10-MHz signal is applied. (*Sec. 30-4*)

30–46. Using Figure 30-9, find the value of C when the X_C is 20 Ω and the frequency of operation is 60 Hz. (*Sec. 30-2*)

30–47. Find the value of C for problem 30-46 when the frequency of operation is 16 kHz. (*Sec. 30-2*)

30–48. Using Figure 30-11, find the value of C for an operational frequency of 30 MHz. (*Sec. 30-2*)

30–49. Find the value of C for problem 30-48 when the frequency of operation is 1.6 MHz. (*Sec. 30-2*)

30–50. Write a general statement for the value of capacitance in a bypass circuit as it relates to an increase in the frequency of operation of the circuit. (*Sec. 30-2*)

30–1. 4 Vp-p

30–2. 8 V

30–3. 1.4144 V

30–4. 28.28 V p-p

30–5. 35 V p-p

30–6. 150 Ω

30–7. 0.106 μF

30–8. Frequencies above 15 kHz

30–9. Frequencies below 15 kHz

30–10. 19.98 kHz

30–11. 10 MHz

30–12. It is low.

30–13. It is low.

30–14. It is high.

30–15. It is high.

30–16. Piezoelectric

30–17. Radio broadcast transmitters

30–18. All types of receivers

ANSWERS TO REVIEW PROBLEMS

Chapters 29–30

A combination of resistance, capacitance, and inductance will resonate at one specific frequency. This concept is used to either pass or reject specific frequencies. The material in these chapters discusses how the concept of resonance is used for these purposes.

1. A resonant circuit permits the selection of a specific band of frequencies.

2. Factors affecting resonance are inductance, capacitance, and resistance.

3. When inductive reactance and capacitive reactance cancel, the resistance in the circuit is the only remaining factor.

4. Resonant circuits are known as *LC* or *RCL* circuits.

5. Resistance is very low in a series resonance circuit.

6. Current flow is low in a parallel resonant circuit.

7. Impedance is high in a parallel resonant circuit.

8. High voltages may be created across the inductive and capacitive components in a series resonant circuit.

9. Voltages created in a series resonant circuit oppose each other by 180°.

10. A parallel resonant circuit is called a tank circuit.

11. A passive filter consists of inductive and capacitive components.

12. An electronic filter may be used to remove unwanted signals.

13. An electronic filter may be used to pass specific frequencies.

14. A low-pass filter attenuates higher frequencies.

15. A high-pass filter attenuates low frequencies.

16. A low-pass filter is used to remove variations of the dc voltage from a power supply.

17. A bypass coupling capacitor is used to remove unwanted voltage variations from one section of an electronic device.

18. The pass band is the range of frequencies permitted to pass through the filter.

19. The cutoff frequency is that frequency where the filter attenuates the signal.

20. A filter is connected in series between two specific points in a circuit.

21. The stop band is a set of frequencies where filter action attenuates the signal.

22. A bandpass filter is used to permit a specific range of frequencies to pass and it rejects all other frequencies.

23. A series resonant filter is used as an attenuator for a signal frequency.

24. A parallel resonant filter is used to accept a signal frequency.

25. Series and parallel resonant circuit are often combined in filter networks.

26. A quartz crystal may be used as a filter device.

CHAPTER 31

Semiconductor Diodes and Transistors

Modern technology has produced solid-state devices known as semiconductor diodes and transistors. These miniature devices replaced the vacuum tube and require a lot less power for their operation, as well as being much smaller in size. Diodes and transistors are combined in even smaller units known as integrated circuits. The basic concept applying to the operation of these devices is that of a variable resistance in the circuit. Variations in their resistance when combined with a fixed-value series resistor will produce variations in the level of operating voltage. These variations are called the "signal" in any device able to process the signal.

When the fundamental circuit concepts are understood as they apply to these solid-state devices, one is able to both design and service equipment having solid-state components. Keep in mind that the concepts do not change; only their applications change.

OBJECTIVES

Upon completion of the material in this chapter, you will

1. Recognize types of semiconductor devices.
2. Understand the concept of the PN junction.
3. Recognize typical applications for semiconductors.
4. Understand the concept of amplification.
5. Be able to analyze an amplifier using load-line analysis procedures.
6. Troubleshoot transistor and diode circuits.

The presentation to this point is related to discussions about passive devices. The passive devices are but three: the resistor, the inductor, and the capacitor. These devices are static in nature. In other words, their values do not change as they operate. In the field of electronics there is another set of devices. These are known as *active* devices. The reason for this is that their characteristics change as they operate. In most cases the change is actually a variation in the amount of internal resistance in the device. The reason for any change is the influence of some outside force. This force includes voltage, current, light, or pressure.

The common devices included in the family of active devices are the *diode,* the *transistor,* and the *integrated circuit.* Each of these devices is a part of a larger family. In other words, the diode is a generic name that includes the signal diode, the power diode, the zener diode, the photodiode, and the varactor diode. These are only some of the many discrete devices identified as diodes. A similar group of devices falls under the general category of transistor. An even larger family of devices is known as the integrated circuit.

It is far beyond the scope of this book to explain each of these many devices in great detail. The basic concepts of how they function are explained. Further details of their operation and design are explained in many other excellent books related to electronics.

31-1 Semiconductor Material

Semiconductors *Electronic devices consisting of a resistive material that is able to conduct electrons when proper polarity voltage is applied.*

Doping *Addition of conductive material into an insulator material.*

The devices named at the beginning of this chapter all are classified as **semiconductors**. This is due to their construction. They are all made of a semiconductor material. The majority of the semiconductors use a base material that is silicon. A small quantity of semiconductors are constructed using a germanium semiconductor material. Both of these materials are insulators in their pure states. When they are used in the production of diodes, transistors, or integrated circuits, the pure silicon is *doped* with an impurity. The **doping** process changes the pure insulator into a semiconductor.

When the silicon or germanium is used as a semiconductor it has the qualities of a resistor. The amount of resistance in the semiconductor material is directly related to the amount of doping that occurs during manufacture of the material. An insulator normally has an infinite quantity of resistance. The doping changes the insulator into a lesser value resistance. In fact, the doping changes the insulator–semiconductor material into a voltage-sensitive resistance.

The explanation of the voltage-sensitive resistance is quite simple. The semiconductor's internal resistance value will change as a voltage is applied to it. The amount of internal resistance will decrease as the voltage applied to the material is increased. This permits a greater amount of current flow through the material.

Semiconductor material is doped in one of two ways. One way adds an inpurity containing extra electrons. The addition of extra electrons into the material changes it from a pure insulator into a material containing negative charges. This type of semiconductor material is known as *N*-type semiconductor material. It is considered to have a negative charge. It gives up its electrons easily.

The second type of doping adds an impurity having a lack of electrons. The lack of electrons makes this material positive in nature. In a purely technical discussion the lack of electrons in semiconductor material is known as **holes**. It is known as *P*-type semiconductor material. This material will accept electrons very easily.

Holes *Positive charges in semiconductor material.*

The PN Junction A piece of P-type or N-type semiconductor material standing alone has little practical purpose as an active device. Only when two or more of these materials are combined can semiconductor action occur. When a piece of P-type and a piece of N-type semiconductor material are combined (Figure 31-1), the junction of the two materials is known as a **PN junction**. The PN junction has an added characteristic. This characteristic is a sensitivity to the polarity of the applied voltage. The PN junction in reality does not consist of the joining of two discrete pieces of semiconductor material, but it actually consists of the infusion of one type of material into the surface of the other type of material. This is accomplished during the manufacturing process by use of heat.

PN junction *Connection between positive and negative types of semiconductor material.*

The PN junction forms a voltage-polarity-sensitive resistance device. Figure 31-2(a) has the positive element of the power source connected to the P-type material and the negative element of the power source connected to the N-type material. When the power source connections are made in this manner, the electrons in the N-type material are forced toward the PN junction. The positive charges, known as *holes*, are forced toward the PN junction in the P-type material. The meeting of these two oppositive types of charges effectively

FIGURE 31-2 A forward-biased PN junction (a) exhibits low resistance values and the reverse-biased PN junction (b) has a high resistance.

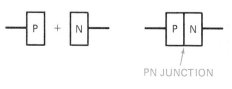

FIGURE 31-1 The semiconductor PN junction consists of both P- and N-type semiconductor materials.

PN JUNCTION

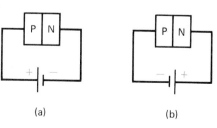

(a)　　　　(b)

neutralizes the junction. The result is a very low resistance across the PN junction. This type of junction with its voltage polarity is called a **forward-biased** junction.

The forward-biased PN junction exhibits a very low resistance. When a voltage is applied to this junction, it forces a current flow through the semiconductor device. Since the junction has a relatively low resistance, a low value of voltage will permit a relatively large current flow. When a silicon semiconductor junction is forward biased the voltage drop occurring across it amounts to about 0.7 V. Similarly, a germanium PN junction has a voltage drop of about 0.2 V when it is forward biased.

When the polarity of the applied voltage is reversed [Figure 31-2(b)], the charges in both the P-type material and the N-type material move away from the PN junction. They move toward the connections to the power source. This forms an area at the PN junction that has a lack of any charges. The PN junction becomes an insulator under these conditions. This is known as a **reverse-biased** junction. The internal resistance of this junction is extremely high. There is little, if any, current flow through the PN junction when it is reverse biased. The voltage drop that develops across the junction is very high under these conditions

31-2 Basic Applications

There are three basic circuits for active devices. These are known as *rectifiers*, *amplifiers*, and *oscillators*. Each of these uses semiconductors in its applications. Prior to the introduction of the semiconductor, vacuum tubes were used for these purposes. Since vacuum tubes are used in only a very small percentage of applications, they are not included here. A brief description of the three basic circuits follows.

Rectifiers are devices or circuits used to permit a one-way direction of current flow in a circuit. The rectifier circuit is used primarily to convert an ac voltage into a dc voltage. The dc voltage is required for correct circuit operation. The rectifier is similar to a voltage-polarity-controlled switch in its action.

Amplifiers are devices or circuits used to enlarge a signal. The enlargement may be either the signal voltage or the signal current, or in some applications, both the voltage and the current.

Oscillators are devices or circuits used to create an electronic signal. The signal is created at a specific frequency. The frequency of the signal is usually dependent on the L and C components in the circuit.

31-3 Diode Characteristics

The semiconductor diode, or **diode**, has certain basic characteristics. These are used in a variety of circuits. The fundamental characteristics are (1) forward current and (2) reverse voltage. Each of these are a part of the information needed when one is designing, replacing, or ordering replacement components.

Forward current describes the maximum quantity of current that can safely be allowed to flow through the diode and its junction. The flow of electron current through any resistance results in the creation of heat. Too great an amount of heat in a semiconductor will destroy the device.

Forward biased Voltage applied to junction that develops a low resistance at that point.

Reverse biased Voltage applied to PN junction that increases junction resistance.

Rectifiers Devices used to permit single-direction current flow in an ac circuit.

Amplifiers Devices using a small amount of power to control a larger amount of power.

Oscillators Devices used to create an electronic signal voltage.

Diode Two-element semiconductor device.

Forward current Maximum current flow when diode is forward biased and has low junction resistance.

ANODE CATHODE

ELECTRON FLOW

FIGURE 31-3 Schematic symbol for a semiconductor diode and its elements.

Reverse voltage is a term applied to the conditions existing when the diode is reverse biased. The diode exhibits a very high resistance when it is reverse biased. The voltage drop developed across its terminals is also very high. The diode has a **breakdown voltage rating**. This rating is the maximum voltage that can be applied to the diode when it is reverse biased without its insulation breaking down. When the insulation of the diode breaks down, it becomes a conductor. The reverse voltage rating is the maximum voltage limit for the diode prior to it breaking down and becoming a conductor.

Reverse voltage *Voltage drop across high resistance PN junction when reverse biased.*

Breakdown voltage rating *Maximum voltage PN junction can withstand without failing when reverse biased.*

31-4 Diode Elements and Symbols

The diode has two elements. The derivation of the term *diode* is from *di*, or two electrodes, or elements. The two elements of the diode are known as the **anode** and the **cathode**. These and the schematic symbol for the solid-state diode are shown in Figure 31-3. The straight-line portion of the symbol is the cathode element. The arrow formation end of the diode is the anode. Electron current flow through the diode is from its cathode to its anode element. This direction of current flow makes the cathode element *negative with respect to its anode* element.

Anode *Positive element of diode.*

Cathode *Negative element of diode.*

31-5 Diode Action

When the diode cathode has a negative voltage charge on it and the anode has a positive voltage charge, the diode is forward biased. The internal resistance of the diode is very low. It is on the order of 10 Ω to about 1000 Ω for low-current-rated diodes and less for diodes rated 1 A or higher. When a voltage is applied to the diode's elements (Figure 31-4), the polarity of the applied voltage determines the internal resistance of the diode. When an ac voltage is applied to a circuit containing the diode and a resistance, a series circuit is formed. The rules for any series circuit apply to voltage drops, current flow, and total resistance in this type of circuit.

When the applied ac voltage places a positive charge in the anode of the diode, it is forward biased. the internal resistance of the diode becomes very low. Actually, the majority of the resistance in the circuit at this time is the value of the fixed resistor. Current flow is limited by this resistor's value and the amplitude of the applied voltage. The voltage drop across the diode is either

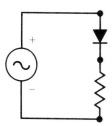

FIGURE 31-4 When the diode is forward biased, most of the applied voltage develops across the series resistance.

0.7 or 0.2 V, depending on the materials used in the construction of the diode. The majority of the voltage drop occurs across the fixed resistance as long as the value of the applied voltage is greater than either 0.7 or 0.2 V. Again, the basic rules for a series circuit apply in this situation.

When the polarity of the ac voltage is reversed, the negative terminal of the source is applied to the anode of the diode. The diode is reverse biased under this condition. The internal resistnce of the diode becomes infinitely high. This resistance is much higher than the value of the fixed resistance in the series circuit. The majority of the voltage drop occurring now appears across the elements of the diode. Current flow is limited to a very low value, due to the high total circuit resistance.

In summary, the diode is actually a variable resistance whose value is controlled by the applied voltage. This is why the semiconductor diode is classified as an *active* device. The actual value of the internal resistance of the diode is controlled by the amplitude of the applied voltage.

REVIEW PROBLEMS

31-1. Which element of the diode is positive when the diode is forward biased?

31-2. What is the voltage drop across the silicon diode when it is forward biased?

31-3. In the circuit of Figure 31-4, where is the greatest voltage drop when the diode is forward biased?

31-6 Diode Use

The majority of the use of the diode is as a *rectifier*. The rectifier is a one-way switch. In a rectifier circuit the diode switches between a forward-biased mode and a reverse-biased mode. A schematic diagram for a basic rectifier circuit is shown in Figure 31-5. This system permits a voltage drop to develop across the load when the diode is forward biased. The forward-bias condition occurs each one-half of the cycle of the ac wave. The result of this action is that one-half of the input ac wave appears across the load resistance during one full cycle of the input ac wave. This type of rectifier is known as a **half-wave rectifier** because of the output voltage waveform shape. In a practical rectifier system the output voltage requires filter action in order to change it into a pure dc voltage. This is not shown in the figure.

Half-wave rectifier *Circuit converting half of ac wave into pulsating dc voltage.*

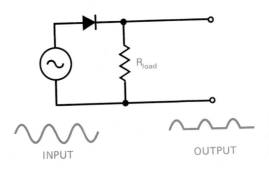

FIGURE 31-5 Use of the diode to change ac into pulsating dc.

INPUT   OUTPUT

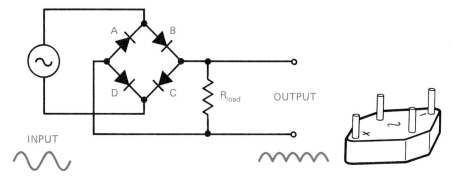
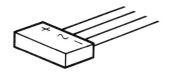

FIGURE 31-6 Four diodes are connected as a bridge rectifier. Typical bridge rectifier packages are also shown.

A second type of rectifier system is called the **full-wave bridge rectifier**. Its schematic diagram is shown in Figure 31-6. The full-wave bridge rectifier system uses four diodes. These are connected as shown, with two anodes connected on one pair of diodes and two cathodes connected on the other pair of diodes. Diodes *b* and *c* are forward biased during the half-cycle of the ac input voltage when the voltage waveform is positive. Diodes *a* and *d* conduct during the half-cycle when the input voltage is negative. Both sets of diodes are connected in the same manner to the resistance representing the load. Current flow through the load is in the same direction during both half-cycles of the input ac waveform. The waveform appearing across the load is also shown in this drawing. Its frequency is twice that of the input ac wave.

Bridge rectifiers may be constructed using four discrete diodes. The bridge rectifier diode set is available in a single package. These types of packages are shown in the bottom of the same figure. The major difference between the packages shown is the amount of current each is able to handle without failure. Some of the larger bridge rectifier packages have a metal heat-conductive surface on the bottom of the package. This type of bridge rectifier is mounted onto a metal surface of the electronic device. The metal surface of the unit acts as a *heat sink* and helps dissipate the heat generated during operation away from the bridge rectifier package. This method permits cooler operation of the unit and prolongs its operating life.

Full-wave bridge rectifier *Circuit converting both halves of ac wave input into same polarity pulsating dc voltage.*

31-7 Diode Identification

Semiconductor diodes are marked in two ways. One of these is the identification of the cathode end of the diode package. The cathode end of the package is usually marked using a band of color. When the diode is packaged in black plastic, the band is white or silver. When the diode package is a light color, the band is dark. In either case the band indicates the cathode of the diode. Some manufacturers use the diode symbol instead of the band of color.

The second method of marking a diode is to use a number. The Institute of Electrical and Electronic Engineers (IEEE) formed a Joint Electronic Device Engineering Council (JEDEC). This council developed a numbering system for American electronic devices. The JEDEC system for semiconductor diodes uses a "1N——" system. The first number, 1, indicates a diode. The letter N is used as the specific letter for semiconductor diodes. A diode using this system would be identified using the number and letter system starting with 1N and followed by a two-, three-, or four-number identifier. The last set of digits represents the specific information about the device. A device having the identifier of 1N4001, for example, has the same electrical characteristics regardless of which of the many semiconductor manufacturers produces it. The best method of determining the specific values of any semiconductor device is to look them up in a specifications manual.

Another method of identifying diodes is their use. Generally speaking, diodes fall into two classes. One of these is known as the *signal*, or *low-power class*, and the other is called the *power class*. Signal diodes usually operate at low current levels of under 100 mA. Their reverse voltage rating is often less than 50 V. Signal diodes are used as demodulators, signal switches, and for signal detection.

The power diode has a much higher rating for both its reverse voltage and its current capacity. Power diodes are able to handle voltages of over 1000 V and currents in excess of 200 A. These values may be considered as close to the maximum for a single diode. Some power diodes are rated at values less than these maximum values.

It is almost always possible to replace a defective diode with one having a higher rating. The only major limitations to this practice are the physical size of the replacement diode and/or critical circuit requirements.

31-8 Testing Semiconductor Diodes

TROUBLESHOOTING APPLICATIONS

Since the semiconductor diode is a voltage-dependent resistance it is possible to test its resistance to determine if it is in working condition. The ohmmeter is a very dependable device for this purpose. The ohmmeter has a built-in power source. This is a dc source and has a constant polarity. The process for using the ohmmeter to test a diode is very simple. A circuit for conducting this test is shown in Figure 31-7. The positive, or red, lead of the ohmmeter is usually connected to the positive lead of its internal battery. When the two leads of the ohmmeter are connected across the terminals of the diode, a bias voltage is applied to the diode. If the diode is forward biased the resistance reading will be under 5 kΩ. The specific value depends on the range setting of the ohmmeter. Usually, a range of $R \times 1$ kΩ is used for this test. If the diode is forward biased, the voltage at the test leads of the ohmmeter will produce a current flow and the meter will read a resistance value.

When the test leads, or the diode leads, are reversed, the diode is reverse biased. Its internal resistance becomes very high. Little, if any, current flow occurs under this condition. The ohmmeter will indicate an infinitely high value of resistance. In many tests the ohmmeter will not indicate any reading of resistance.

The diode test is therefore a test of its resistance under forward- and reverse-biased conditions. If the diode tests indicate the same resistance values in both directions, the diode has failed. It must be replaced if its internal resistance is the same when it is forward biased and when it is reverse biased. This is true when the resistance is low in both directions and also when it is high in both directions.

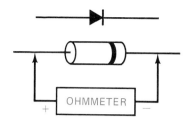

FIGURE 31-7 Use of an ohmmeter to measure the values of forward- and reverse-biased junction resistances.

31-9 Transistors

The development of the device known as the **transistor** was an attempt to produce a device that would emulate the triode, or three-element, vacuum tube. The reason for this is elementary when one considers some of the problems inherent in the use of the vacuum tube. First, the tube requires large quantities of energy in order to heat it to operating temperature and conditions. The transistor does not need this energy. Second, the energy used to heat the tube must be dissipated to maintain proper operating conditions. Third, the spatial requirements for a tube, its socket, and associated wiring are much greater than those required for the transistor. The first electronic computer required a room of considerable size. The heat from its tubes could be used to heat a house. The tubes required massive quantities of air conditioning in order to keep them cool enough for proper operation. A computer capable of performing faster and more complex operations and having greater capacity is available at the present time. It can be held in one's hand! This reduction in size, power, and temperature is the result of the development and application of the transistor.

The term *transistor* is from the words *transfer* and *resistance*. This is the process used in the control of the transistor. There are two basic types of transistors in current use. One of these is called the *bipolar* transistor and the other is known as the *field-effect transistor*, or FET. The FET most closely resembles the operation of the vacuum-tube triode. (The triode is a three-element vacuum tube.) For reasons not known to this author, the FET was not the transistor originally developed in great quantities. The bipolar transistor is the most popular type of transistor in use today for discrete applications. The schematic symbols for the bipolar transistor are shown in Figure 31-8. The identification of the elements of this transistor are the same, but their polarities are opposite. The NPN type is shown in part (a). The arrow on its emitter element points away from the other two elements. The other elements are known as the base and the collector. The PNP type is shown in part (b). The difference between it and the NPN type is the direction of its arrow. This arrow points toward the other elements of the transistor.

Transistor *Three-element semiconductor device.*

(a)   (b)

FIGURE 31-8 Schematic symbols for the junction transistor: (a) NPN; (b) PNP. The arrowhead indicates the polarity of the transistor.

Bipolar transistor *Transistor having both positive and negative polarity elements.*

The **bipolar transistor** consists of two semiconductor diodes connected back to back. The emitter is considered to be the source of conduction in the device. The collector is the element where the conduction ends. The base element is the control element for the transistor. The base element is very thin compared to the other two elements. Note the word "conductor" instead of electrons in this explanation. The reason for this is the difference in conduction between the NPN and PNP types of transistors. In the NPN type electron flow starts at the emitter element. The majority of electron flow then moves past the base and to the collector element. Only about 2 to 5% of electron flow is found in the base element.

The PNP transistor's electron flow is in the opposite direction. Electron flow starts at the collector, flows past the base, and then flows to the emitter element. Charge conduction in the PNP transistor is in the form of positive charges known as *holes*. Hole charge movement in the PNP transistor is the same as electron flow in the NPN type. Hole movement is the opposite direction from that of electron charges. In either case, the results are the same.

31-10 Transistor Markings

Transistors are identified by the same JEDEC system used for diodes. The JEDEC identifier for a transistor is a 2N—— system. Almost all transistors using this system will start with the identifier 2N followed by two, three, or four numbers. Specific data about the individual transistor type are given in the specifications manual. It is impossible to attempt to memorize data about each transistor. Often, people remember the specifications for types they commonly use. The lead identification is also a bit difficult. Not all transistors have the same basing arrangement for their leads. Here, too, the best suggestion is to look up the specific information in a semiconductor reference manual.

31-11 Transistor Types

Transistors are generally classified as being either low-power or high-power types. The low-power transistors are also classified as being either small-signal types or switching types. The specific classification depends on the application for the transistor. Small-signal types are used in amplifier systems. Switching types are used in digital computer systems.

High-power transistors are commonly known as power transistors. Their use is usually limited to output stages of most types of solid-state equipment. The power transistor is usually used where larger quantities of current handling is required.

A diode junction exists between the emitter element and the base element of the transistor. This junction is normally forward biased when the transistor is installed in a circuit. Most of the transistors in use today are made from a silicon semiconductor material. The emitter-to-base junction will have a voltage drop of 0.7 V when it is forward biased. This small quantity of voltage will *control* electron flow through the transistor. The current flow through the emitter-to-base circuit amounts to between 2 and 5% of the total current flow

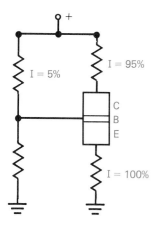

FIGURE 31-9 Current flow in the junction transistor.

in the transistor. Since the emitter-to-base junction is forward biased, a small change in the voltage between these elements is used to *control* a much greater current flow between the emiter and the collector elements.

The base-to-collector junction of the transistor is reverse biased. Due to the method of constructing the bipolar transistor, current flow occurs in an unusual manner. Figure 31-9 shows a modified block diagram for this device. In a NPN type of transistor, electrons flow from the emitter to the base. This represents 2 to 5% of the total current flow. They also flow from the emitter, past the base, and to the collector. This path represents the majority of the current flow, or from 95 to 98% of the total current. Since the base-to-collector junction of the transistor is reverse biased, a much larger voltage drop occurs across this junction.

The voltages applied to the elements of the transistor greatly influence the quantities of current flow. The emitter-to-base voltage difference is less than 1 V. The base-to-collector voltage difference is much greater than 1 V. When current starts to flow through the transistor, the greatest attraction for it is toward the collector element. The majority of current flow through the transistor occurs from the emitter, past the base, and to the collector, due to the higher voltage present at the collector.

Consider the emitter-to-base circuit as a low resistance in the circuit. The actual amount of voltage between these two elements controls the current flow through the transistor. When the voltage drop between these two elements is zero, there is no current flow. The transistor is in a condition known as **cutoff**. The diode junction called the emitter-to-base junction is not forward biased unless there is a difference in voltage between its elements. When the voltage on these two elements is equal, the diode does not conduct. All current flow through the transistor is shut off. On the other hand, when the emitter-to-base voltage rises above the 0.7 V level, it is easier for current to flow through the base and on to the collector. When the transistor is exhibiting its maximum currentflow, an increase in the emitter-to-base voltage will not permit the flow of any additional current. This condition is known as **saturation**.

The schematic drawing in Figure 31-10 is that of a typical bipolar transistor circuit. The transistor shown in this circuit is a NPN type. Resistances R_{B1} and R_{B2} are connected as a voltage-divider circuit. These two resistances establish the voltage at the base element. Resistance R_e is used to establish the voltage

Cutoff *Operational point of semiconductor device where it is not able to conduct.*

Saturation *Operational point of semiconductor where it cannot be turned on further.*

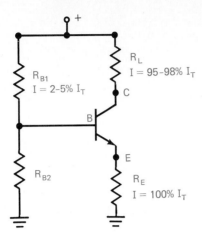

FIGURE 31-10 Current flow using a schematic diagram for the junction transistor.

at the emitter element at an operational value. When this transistor is in a working circuit, the difference between the emitter and the base voltages will be 0.7 V (assuming that this is a silicon transistor).

Resistance R_L is called the load resistor. It is a fixed-value resistance connected in series between the + terminal of the supply voltage and the collector ellement of the transistor. In this application, the emitter-to-collector terminals of the transistor act as a variable resistance in the circuit. The addition of a variable resistance symbol in Figure 31-11 shows this.

The variable resistance symbol represents the internal resistance of the emitter-to-collector connections. Current flow in this circuit does not recognize the emitter-to-collector connection as a transistor. As far as the current flow is concerned, these connections are equal to a variable resistance. The circuit "sees" two resistances connected in series. One of these is the fixed resistance R_L and the other is the variable resistance between emitter and collector. This variable resistance is identified as R_{E-C}. The circuit represents two resistances connected as a series string.

Current flow through the two resistances develops a voltage drop across each resistance. The amplitudes of the voltage drop across each component are directly related to their respective values. When both R_i and R_{E-C} have equal values of resistance, the voltage drops across them are also equal. When R_L has a higher value of resistance then R_{E-C}, the voltage drop across R_L is

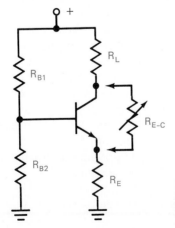

FIGURE 31-11 The junction transistor acts as a variable resistance in the circuit.

larger. This is the same format as applied to two resistances in series and their respective voltage drops. Current flow through the two resistances is totally dependent on the value of the applied voltage and the value of the total resistance in the circuit. This is related directly to the rules established by Ohm and Kirchhoff.

31-12 Field-Effect Transistors

The FET functions in a manner that is similar to the bipolar transistor. The major difference between these two devices is initially found in their construction. The basic construction and the schematic symbol for an N-channel FET are shown in Figure 31-12. The FET is constructed from a solid piece of semiconductor material. This material has a ring of material fused on it. This ring of material is known as the *gate*. The gate functions in a manner that is similar to that of the base of the bipolar transistor. The FET shown is called an N-channel FET. Electron flow occurs from the source to the drain. The amount of electron flow is regulated by the voltage charge placed on the gate element. When the voltage on the gate is increased, the area occupied by the gate increases. This effectively increases the resistance between source and drain. The increased resistance restricts current flow. When the voltage charge on the gate is reduced, the area of the gate is also reduced. This effectively lowers the internal resistance of the FET. More current flows through it in this state. FET construction is either as an N-channel or as a P-channel type. The difference is in the type of material used for the source to drain of the FET.

Field-effect transistor *Transistor type using electrostatic field to control electron flow.*

The FET has a very high input impedance or resistance. Because of this, it is often used as an input semiconductor in FM radios or other devices that normally have a low signal amplitude input. The construction of the FET requires that certain precautions be taken. Some types of FETs are very susceptible to static electrical charges. These must be handled with extreme care and under the conditions described by the manufacturer. If this is not done, the device may be accidentally destroyed.

REVIEW PROBLEMS

31-4. Name the three elements of the bipolar transistor.
31-5. Name the three elements of the FET.
31-6. Which elements of the bipolar transistor act as the control?
31-7. Which elements of the bipolar transistor act as a variable resistance?

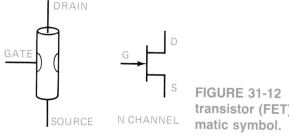

FIGURE 31-12 Field-effect transistor (FET) and its schematic symbol.

31-13 Testing Transistors

TROUBLESHOOTING APPLICATIONS

The bipolar transistor may be checked by use of an ohmmeter. Since this device is constructed of two diodes, it can be tested using the same techniques as those used for the diode. The ohmmeter connected between the emitter and base will test for the validity of this junction. Resistance values should be low in one direction and high when the ohmmeter leads are reversed. This is also true for the base-to-collector junction. Readings that are the same ohmic value when the leads are reversed at one junction indicate a defective transistor.

When the ohmic value is low in both directions across one junction, the junction is short circuited. When the ohmic value is high in both directions across the same junction, the junction is open. In either of the tests, the transistor must be replaced with a functionally good unit.

Another test one may make on the bipolar transistor is one using a voltmeter. This test is conducted with the operating voltage turned on for the system. The difference in voltage measured between the emitter and the base of the transistor whould be about 0.7 V if the transistor is a silicon type (and most of them are). This is not an exact value. It may vary by as much as 0.3 V in either direction from the ideal voltage value. If this voltage is zero, the transistor's emitter-to-base junction is shorted. If the voltage is much higher, the transistor has an internal condition known as *leakage*. Leakage creates an abnormally high resistance between emitter and base. This can be identified by the abnormally high voltage measured between these two elements.

There are two ways of measuring the voltage drop between emitter and base (Figure 31-13). One of these places the negative lead of the voltmeter directly on the emitter led or its connection to the circuit board. The other meter lead is placed on the base lead or its connection to the circuit board. This method provides a direct reading of the emitter-to-base voltage drop [see Figure 31-13(a)]. The second method requires two readings and a slight amount of mathematics. One reading is made between the emitter element and circuit common. This reading is noted and the second reading is then made. The second reading measures the voltage drop between the base and circuit common. The difference between these two voltage values is the emitter-to-base voltage drop [see Figure 31-13(b)]. The selection of the specific method is left to the person conducting the test.

A second type of test requires knowledge of the type of circuit. If the circuit is similar to that used in Figure 31-10 or 31-11, one can measure the

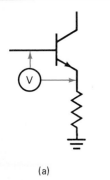
(a)

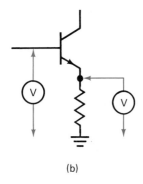
(b)

FIGURE 31-13 Measurment of the voltage difference between base and emitter is accomplished using either of these methods.

voltage drop existing between collector and circuit common. This voltage should be roughly one-half of the source voltage in the popular common-emitter amplifier circuit configurations. If this voltage measures the same value as the source voltage, the transistor is in the state of cutoff. If the voltage drop at this point in the circuit is close to zero, the transistor is in a state of saturation. Both are not normal for a transistor in this circuit configuration. Further investigation into why these conditions exist is required.

There are some excellent transistor testers currently available from manufacturers of electronic test equipment. Some of these are types one uses without the removal of the semiconductor from the circuit. Others require the removal of the device prior to a realistic test. The best type of tester is one that is comfortble for the user. This decision is only made after using both types in practical applications.

REVIEW PROBLEMS

31-8. What is the normal voltage drop between emitter and base for a silicon transistor?
31-9. What is the voltage drop across a shorted emitter-to-base junction?
31-10. Using the circuit shown in Figure 31-11, what condition exists when the collector-to-common voltage is equal to the source voltage?

31-14 Other Semiconductor Devices

There are a great many different semiconductor devices. In fact, there are too many to cover in this book. A few of these are described in this section. Descriptions of other types and greater detail about these are presented in books specifically related to the study of semiconductors. Some of the more common semiconductor devices are briefly described here.

Light Emitting Diodes Typical light-emitting diode (LED) assemblies are shown in Figure 31-14(a). The schematic symbol for the LED is a diode symbol with the addition of arrows [Figure 31-14(b)]. The arrows indicate an emission from the diode. **LED** construction uses the principle stating that the PN junction will emit light when a current passes through it. The standard diode is packaged in a housing that does not show this light. The LED uses a colored plastic housing and usually some sort of magnifying unit inside the housing. The color of the plastic and the type of material from which the LED is produced determine the color of the LED. The magnifier is used to increase the size and glow of the light.

LED construction uses a gallium arsenide material for its glow effect. This is one of a family of gallium materials used in the production of the LED family. Typical commercially available packages for LEDs are also shown in this figure. LEDs are used in place of incandescent lamp bulbs as indicators. They have a much longer life than the lamp bulb. The LED also consumes less current during operation. They are smaller than lamps and do not require a base for

LED *Light emitting diode.*

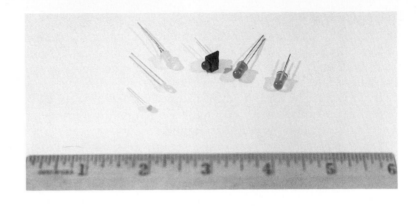

(b)

FIGURE 31-14 (a) Light-emitting diodes; (b) schematic symbol.

their mounting. All of these factors have influenced their use in electrical and electronic equipment.

LED Bar Units The bar unit is an outgrowth of the basic LED. A plastic prism is placed over the LED [Figure 31-15(a)]. Light is transmitted equally through the prism. When the LED glows the entire surface of the prism glows in the same color. This principle is used as an enlargement of the basic LED. It has applications in home entertainment units as the signal-strength indicator in a receiver or an audio unit.

LED Readouts Another application for the LED bar unit is shown in Figure 31-15(b). Seven bar units are combined into one package in this application. The device is known as a *seven-segment readout*. the seven segments are controlled by other circuits in the system. The specific elements lighted by the circuit provides a numeric indication. Numbers between zero and nine can be displayed by this device. Several seven-segment readouts are often combined into one larger unit when numbers of two or more digits are required in one display.

Zener Diodes When one requires a constant-voltage output from a dc power source, a device known as a **zener diode** may be used. This device looks the same as any other diode. It does have a special schematic symbol, shown within the basic circuit diagram of Figure 31-16. The zener diode is installed in reverse of other diodes. Its anode is connected to circuit common and its cathode is connected through a series resistance to the positive terminal of the power

Zener diode Voltage-regulating diode.

PRISM

LED

(a)

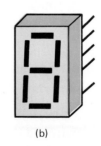

(b)

FIGURE 31-15 Use of the LED in a seven-segment readout.

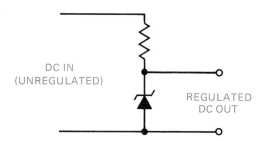

FIGURE 31-16 A zener diode is used as a voltage regulator or reference in the circuit.

DC IN (UNREGULATED)

REGULATED DC OUT

source. The resistance of the diode will vary under circuit operation. It attempts to maintain a constant voltage across its terminals under varying current conditions in the circuit. In this manner it acts as a voltage regulator. The zener diode is placed in parallel with the load. When the load resistance decreases, the zener diode's internal resistance will increase. This will restore the voltage to its original value. When the load resistance increases, the zener resistance will decrease. This also restores the output voltage to its original value.

Photodiodes The **photodiode** is a light-sensitive device. (Figure 31-17). The resistance of the photodiode will change as light strikes its surface. Ideally, the photodiode is a high resistance in a dark environment. When light strikes its surface its resistance decreases. When it is connected as shown in this figure, the voltage across it is very high in the absence of light. The voltage across the diode will decrease as light strikes its surface. The output of this circuit controls another circuit. The other circuit is controlled by the variable voltage of the photodiode circuit. Another type of control occurs when the fixed resistance in the circuit is replaced by another type of device. Current flow through the circuit due to light changing the resistance of the photodiode operates the other device.

Photodiode *Light sensitive diode that is controlled by the quantity of light striking its surface.*

Phototransistors The base of the **phototransistor** is light sensitive. This method if used to control emitter-to-collector current. The light striking the base element replaces the voltage described earlier in the section of this chapter describing basic transistor action. One of the more practical applications for the phototransistor is shown in Figure 31-18. This device is known as an *optocou-*

Phototransistor *Transistor using light to control electron flow through its junctions.*

FIGURE 31-17 The photodiode changes its internal resistance when exposed to light.

FIGURE 31-18 Use of a photo transistor and a LED as an optical coupler.

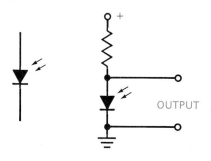

OUTPUT

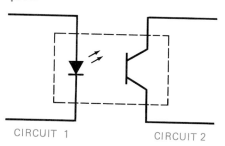

CIRCUIT 1　　　CIRCUIT 2

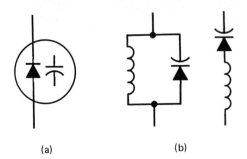

(a)   (b)

FIGURE 31-19 The varactor diode is used as a variable capacitor in a circuit.

pler. It is also known as an *optoisolator*. This device is used to couple signals between two circuits when the two circuits cannot be connected together. Circuits such as these require total isolation and the opto device is a convenient method of transferring the signal between stages under these conditions.

Varactor Diode A device briefly described in an earlier chapter is known as the **varactor diode**. Its graphic symbol is shown in Figure 31-19(a). Either of the arrangements shown in Figure 31-19(b) is appropriate for this diode. The varactor diode exhibits the qualities of a capacitor in the circuit. A capacitor consists of two surfaces on which charges can collect. The two surfaces are separated by an insulator called the dielectric. The PN junction of a diode has similar characteristics. It is because of this that it may be used as a variable diode. The variations in the junction spacing are controlled by the voltage applied to the diode. This is normally used to control current flow through the diode. It can also be used to control the amount of capacitance between the two parts of the diode. Varactor diodes are used as tuning diodes in electronic circuits. A voltage placed across the diode's terminals will change its capacitance. When the varactor diode is connected as a part of a resonant circuit, the voltage variations on it will change the circuit's point of resonance. Sets of varator diodes are used in the tuner sections of television receivers.

Varactor diode *Diode using varying applied voltage to change its capacitive characteristics.*

REVIEW PROBLEMS

31-11. What type of diode emits light?
31-12. What type of diode is affected by light?
31-13. What semiconductor devices make up an optocoupler?
31-14. What is the basic principle for operation of the varactor diode?

CHAPTER SUMMARY

1. Semiconductor devices are considered as active devices because they change their state during operation.
2. Common semiconductor materials are silicon and germanium.
3. Two common types of semiconductor materials are known as P type and N type.
4. N-type semiconductor material has an excess of electrons.

5. P-type semiconductor material has an excess of positive charges (or a lack of electrons).

6. A semiconductor diode is a two-element device made up of both P- and N-type material.

7. The joining of the P- and N-type material is called the PN junction of the diode.

8. The junction may be either forward biased or reverse biased by the addition of a voltage charge on its elements.

9. A forward-biased junction has low resistance, a low voltage drop, and permits high current flow.

10. A reverse-biased junction has high resistance, high voltage drop, and permits little, if any, current flow.

11. The voltage drop across a forward-biased silicon junction is 0.7 V. Across a forward-biased germanium junction the voltage drop is 0.2 V.

12. Basic applications for active devices include rectification, amplification, and oscillation circuits.

13. The diode acts as a voltage-dependent electronic switch.

14. Diodes are identified by a number system starting with 1N. The cathode end of the diode is marked on its body.

15. Diodes may be tested using an ohmmeter.

16. Transistors are three-element devices that emulate vacuum-tube operation.

17. The emitter-to-base junction of a transistor is forward biased. The base-to-collector junction is reverse biased.

18. Transistors are identified by a number system starting with "2N." Identification of the elements requires use of a manual of specification.

19. The transistor acts as a variable resistor in a circuit. Its emitter-to-collector resistance is controlled by the emitter-to-base conditions.

20. The diode junctions of a transistor may be tested in a manner that is similar to the one used for the diode.

21. Other semiconductor devices include the LED, photodiode, phototransistor, and the optocoupler.

Answer true or false; change all false statements into true statements.   SELF TEST

31–1. The emitter-to-base junction of a transistor is forward biased.

31–2. The base-to-collector junction is always forward biased.

31–3. The majority of current flow in a transistor occurs between base and collector.

31–4. A photodiode will increase its resistance when exposed to light.

31–5. A phototransistor will change its emitter-to-collector resistance when exposed to light.

31–6. A diode may be tested by use of an ohmmeter.

31–7. A properly functioning silicon diode will have a 0.7-V drop across its junction when reverse biased.

31–8. A zener diode is used as a voltage-regulating diode.

31-9. Current flow through the NPN transistor is opposite in direction to that of the PNP transistor.

31-10. The arrowhead points toward the other elements in the schematic symbol for a NPN transistor.

MULTIPLE CHOICE

31-11. When the voltage placed across the PN junction results in a low resistance, the junction is said to be
(a) turned on
(b) forward biased
(c) reverse biased
(d) voltage biased

31-12. The term used to describe the maximum quantity of current that a diode is capable of handling is called
(a) maximum current rating
(b) current rating (maximum)
(c) forward current rating
(d) reverse current rating

31-13. When a diode is reverse biased, its internal resistance is
(a) high
(b) low
(c) medium
(d) current dependent

31-14. The circuit used to permit current flow in only one direction is called a(n)
(a) oscillator
(b) amplifier
(c) rectifier
(d) logical conditioner

31-15. The circuit used to process both halves of the ac wave and change it into a pulsating dc voltage is called a _____ system.
(a) half-wave rectifier
(b) voltage doubler
(c) pulse changing
(d) full-wave rectifier

31-16. The transistor uses _____ to control its output elements.
(a) a small change in base-to-emitter voltage
(b) a small change in emitter-to-collector voltage
(c) a small emitter-to-collector current
(d) a large base-to-emitter voltage

31-17. The conditioner of a tube or transistor stating that it is "on" as much as possible is known as
(a) saturation
(b) cutoff
(c) full on
(d) state on

31–18. The diode type that changes its capacitance value is known as the _____diode.
    (a) capacitor
    (b) voltage-controlled
    (c) variable
    (d) varactor

31–19. The voltage drop developed across a properly working silicon diode is.
    (a) 0.7 V
    (b) 1.7 V
    (c) 0.2 V
    (d) 1.2 V

31–20. The voltage drop developed across a properly working base-to-emitter junction of a germanium transistor is.
    (a) 0.7 V
    (b) 1.7 V
    (c) 0.2 V
    (d) 1.2 V

ESSAY QUESTIONS

31–21. Explain why the PN junction is able to change its resistance as a voltage is applied to it.

31–22. Explain why the solid-state devices act as variable resistances in a circuit.

31–23. Explain why the reverse-biased junction has its resistance characteristics.

31–24. Draw the basic circuit for a half-wave rectifier. Explain how it produces the pulsating dc output voltage.

31–25. Draw a basic circuit for a full-wave bridge rectifier system. Explain how it creates its pulsating dc voltage output.

31–26. Explain how the transistor amplifier circuit is able to "amplify."

31–27. Explain this statement: "A signal is used at the input of a transistor circuit to create a larger signal having the same shape at its output terminals."

31–28. Explain how a diode acts as a voltage-dependent electronic switch.

31–29. Explain how the basic diode can be used as a photodiode.

31–30. Explain how the basic diode can be used as a voltage-controlling device.

PROBLEMS

31–31. What current flows through the collector of a silicon transistor when 100 mA flows in the emitter? (*Sec. 31-9*).

31–32. What is the voltage drop across a 1.5-kΩ resistor connected in series with the collector of a transistor when the voltage drop across the resistor is 10 V? (*Sec. 31-9*)

31–33. A LED has a normal current flow of 10 mA. How much current flows in a seven-segment LED display when the number 4 is displayed? (*Sec. 31-14*).

31–34. A diode has a 4.0 V on its cathode and 3.3 V on its anode. Is it forward or reverse biased? (*Sec. 31-3*).

31–35. The voltage on the emitter of a NPN transistor is 2.8 V. The base of the same transistor has 3.5 V. Is this correct for proper operation? (*Sec. 31-13*).

31–36. A diode has -6 V on its cathode and -5.3 V on its anode. Is it forward or reverse biased? (*Sec. 31-3*).

31–37. A diode's resistance measures 400 Ω when the positive lead of the ohmmeter is on its anode. The same diode measures 400 Ω when the negative lead of the ohmmeter is on its anode. Is this device a diode? (*Sec. 31-8*).

31–38. The base of a PNP transistor measures 5.2 V. The voltage on its emitter is 5.9 V. Is the transistor functional with these voltages? (*Sec. 31-13*).

31–39. What is the internal resistance of a silicon diode that measures 0.7 V across its junction and has 250 mA flowing through it? (*Sec. 31-8*)

31–40. How much current will flow across the PN junction of a silicon diode when its junction voltage is 250 V and its internal resistance is 150 MΩ? (*Sec. 31-8*)

TROUBLESHOOTING PROBLEMS

31–41. A semiconductor diode has a voltage of 3.8 V on its anode and a voltage of 4.2 V on its cathode. Will a current flow through it?

31–42. In a series circuit consisting of a 10-kΩ resistor and a semiconductor diode, the diode has a voltage of 45 V measured from its anode to circuit common and the resistor has a voltage of 65 V measured from the lead connected to the diode to circuit common. There is no current flow in the circuit. What is wrong?

31–43. In the above circuit, the load resistance is 5000 Ω and the voltage from the rectifier is 24-V dc. A current meter in the circuit measures 1.5 mA. Assuming the load resistance is correct, what else could be wrong in this circuit?

31–44. An ohmmeter measures 5 kΩ when its $(+)$ lead is connected to the anode of a diode and its $(-)$ lead is connected to the cathode of the diode. When the leads are reversed, the ohmmeter shows the same 5-kΩ reading. What is wrong?

31–45. A voltmeter's leads are placed on the base and the emitter of a transistor. The voltage measured is 2.4 V. What is wrong?

31–46. A NPN-junction transistor's elements are measured. The values are: $E = 2.3$ V, $B = 1.6$ V, $C = 10$ V. Is this correct?

31–47. The resistance between the emitter and base junction of a transistor is measured as infinity. When the ohmmeter leads are reversed, the resistance reading is 4 kΩ. Is this proper?

31–48. Using the circuit shown in Figure 31-17, a light is shined on the photo diode. A meter is used to monitor the current through the circuit and shows no difference in the quantity of current when the light is removed. What is wrong?

31–49. Using the circuit shown in Figure 31-18, the voltage across the secondary, or output, of the optocoupler shows a change when the current flow across the diode connections is varied. Is this correct?

31–50. The seven segment LED assembly shown in Figure 31-15 has one segment that does not glow during its normal operation. When a voltage source is connected to this segment it does glow. Is the problem in the LED assembly or the circuit operating it?

31–1. The anode

31–2. 0.7 V

31–3. Across the load resistance

31–4. Emitter, base, and collector

31–5. Source, gate, and drain

31–6. Emitter and base

31–7. Emitter and collector

31–8. 0.7 V

31–9. 0 V

31–10. Cutoff

31–11. Light-emitting diode (LED)

31–12. Photodiode

31–13. LED and phototransistor

31–14. Change of capacitance

**ANSWERS TO
REVIEW
PROBLEMS**

CHAPTER 32

Semiconductor Applications

The applications for semiconductors seem almost endless. Each day we read about additional uses for diodes, transistors, and integrated circuits. The requirements of the manufacturing industries for minimal failure rates mandated the use of integrated circuits instead of discrete devices in circuit applications. The IC is also easier and less expensive to install in production activities. Applications of complex circuits in small spaces have dictated the development of integrated circuitry. Regardless of the device being used, the basic rules related to current flow and voltage drops apply equally to all types of semiconductor circuits.

Block Diagram
Common-Base Amplifier
Common-Collector Amplifier
Common-Emitter Amplifier
Diode Switching
Emitter Follower
Feedback

Gain
Integrated Circuit
Oscillator
Power Supply
Rectifier Systems
Signal Processing
Switching Transistors

OBJECTIVES

Upon completion of the material in this chapter, you will

1. Recognize typical applications of semiconductors.
2. Understand the concept of rectification.
3. Recognize typical rectifier circuits.
4. Understand the concept of amplification.
5. Recognize typical amplifier circuits.
6. Understand the concept of transistor switching.
7. Understand the basic concepts of operational amplifiers.

The applications of the active devices discussed in Chapter 31 have probably done more to influence our lives than any other recent invention. If one would take the time to list on a piece of paper all the electronic devices used during one week, the list would be quite long. This is particularly true when one considers the electronics now in the home, office, automobile, and other common places. Perhaps one major influence we tend to overlook is the computer used to monitor the gasoline we purchase for use in our automobiles. Almost all do-it-yourself gasoline service stations use a computer to tell the operator how much gas purchased as well as the total price of the sale. This, of course, is but one of the many uses of electronics we accept as an everyday occurrence.

The applications of electronics are based on the rules and concepts studied earlier in this book. As stated previously, the rules and concepts do not change. It is the applications that are modified to offer new or improved operation. The fundamental rules presented by Ohm, Kirchhoff, and Watt have yet to change. Circuits used in today's technology still use basic series, parallel, or combination series–parallel configurations. Components such as resistors, capacitors, and inductors are still used in the same manner. Only the applications for these components have been updated to meet the needs of today's world. With these thoughts in mind, let us look at some of the basic applications for active devices. This will provide the necessary knowledge for success in the field of electronics technology.

32-1 Diode Applications

Diode devices have several applications. Two of the most basic ones are in power supply systems and switching circuits. There are many others, of course, but these applications will provide some fundamental background for understanding.

Rectifier systems *Electronic circuit used to convert ac into dc for operation of other circuits requiring dc voltages.*

Power supply *System using rectifier and other components to provide proper levels of operational power for electronic circuits.*

Rectifier systems are used to change an ac voltage and current into a dc voltage and current. The principal use of this type of circuit is called the **power supply** section of an electronic unit. Almost all electronic devices require a dc voltage for their operation. Due to the influence of some of the early inventors and discovers of electricity, the power supplied by our electric companies is in the form of alternating current. This has to be altered and converted into a dc form for use in electronic units. The process used to accomplish this is called *rectification*.

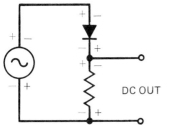

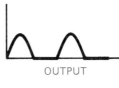

DC OUT

OUTPUT

FIGURE 32-1 The half-wave rectifier system produces one-half of the input ac wave.

32-2 Half-Wave Rectification

The circuit shown in Figure 32-1 is known as a *half-wave rectifier*. It is a series circuit consisting of an ac source, a diode, and a resistance. The resistance is used to represent the load. The load, in electrical and electronic terms, is the unit where some work is accomplished. This load may be any sort of electrically operated device, from a simple lamp to a complex computer used to control manufacturing functions. Since this is a series circuit the basic rules relating to current, resistance, and voltage drop are applicable.

The ac voltage applied to this circuit has two peak values. One of these occurs when the waveform is at its highest positive point. This happens at about 90° of the rotation of the wave. The second occurs when the wave is at its maximum negative point of 270°. The first point is shown on the drawing using colored letters. The upper part of the sine wave is positive at this moment in time. Using Kirchhoff's rules, the polarities of each of the components are identified in the circuit.

The polarities assigned during the first half-cycle of the ac wave provide a forward bias to the diode. Its resistance is very low when compared to the resistance of the load. It is almost impossible to measure this value, but it is possible to use the fixed 10-kΩ resistance to determine circuit values. The first thing to do requires the conversion of the ac sine-wave value into a peak-to-peak value. This is required to understand why the value of dc voltage is as high as it is.

The rms value of the sine wave does not appear at the output of the rectifier system. The waveform appearing in the figure and at the output is a peak voltage form. The rms value must be converted into a peak value in order to determine its voltage:

$$V_{peak} = V_{rms} \times 1.414$$
$$= 100 \times 1.414 \qquad (32\text{-}1)$$
$$= 141.4 \text{ V}$$

This is the value of voltage that develops across the load during the first half-cycle of the input ac wave. The reason for this is due to the diode action in the circuit. The application of the ac wave provides a forward bias during the first half-cycle of the wave. When the diode is forward biased the voltage across its terminals (assuming that it is a silicon diode) is 0.7 V. The balance of the peak voltage drop occurs across the load. Since the 0.7 V is such a small value, it is usually ignored during calculations. The peak voltage developed across the load during this half-cycle is 141.4 V. For the purists, one must subtract 0.7 V from this value in order to determine the actual peak voltage:

$$V_{\text{load}} = V_{\text{peak}} - V_{\text{diode}}$$

$$= 141.4 - 0.7 \qquad (32\text{-}2)$$

$$= 140.7 \text{ V}$$

The waveform for the voltage appearing across the load is also shown in Figure 32-1.

When the input ac wave has its second half-cycle the polarities of the voltage applied to the circuit are reversed. This is also shown in the drawing. This half-cycle is indicated by the polarity signs in the circles. All polarities are reversed during the second half-cycle. The bias on the diode is shown as being reversed as well. The resistance of the diode goes infinitely high when it is reverse biased. The voltage drop occurring across the components is also reversed at this time due to the change in resistance values in the circuit. The largest resistance develops the largest voltage drop in any series circuit. This rule applies to this series string as well as any other series string.

When the diode is reverse biased the majority of the applied peak voltage drop occurs across the diode. Very little, if any, voltage drop occurs across the load. This is shown in the waveform portion of the figure as the zero line voltage value. Since the load represents that portion of the circuit where work is being performed, there is no work performed during the second half-cycle of the input ac wave. The efficiency of this system is 50% since it is on one-half of the time and off during the other half of the input wave time. This system is known as a *half-wave rectifier system* because of this action.

Current flow during each half-cycle is determined by use of Ohm's law.

$$\text{First half-cycle: } I_{\text{load}} = \frac{V}{R}$$

$$= \frac{140.7 \text{ V}}{10,000 \text{ }\Omega}$$

$$= 14.07 \text{ mA} \qquad (32\text{-}3)$$

$$\text{Second half-cycle: } I_{\text{load}} = \frac{V}{R}$$

$$= \frac{140.7 \text{ V}}{100,000,000 \text{ }\Omega} \text{ (approx.)}$$

$$= 1.47 \text{ }\mu\text{A}$$

Current flow, as indicated, during the second half-cycle is almost zero.

32-3 Full-Wave Rectification

The efficiency of the rectifier system is doubled by use of either of two alternative systems. Both of these systems utilize both halves of the input ac wave in order to provide an output voltage. These two systems are known as the *full-wave center-tapped* rectifier and the *full-wave bridge* rectifier systems. Both systems require more components than those used in the half-wave system.

Full-wave center-tapped rectifier systems require two diodes and a special transformer. The transformer used in this system must have a connection at the center of its secondary winding. The secondary voltage from this type of transformer is divided into two equal parts. The action occurring in the center-tapped secondary of the transformer is shown in Figure 32-2. A total of 12 V rms is developed across the full secondary winding of the transformer. One-half of this volume is developed across each of the halves of the winding. Six volts is available from both the upper half-winding and lower half-winding. Transformer action develops the polarities shown on the drawing for one-half of the cycle of the input ac wave. During the second half of the cycle all of the polarities are reversed. This schematic diagram for a full-wave center-tapped rectifier system is shown in Figure 32-3. This system uses two diodes, a center-tapped secondary transformer, and a single resistance to represent the load. During the first half-cycle current flow utilized one of the diodes. This is due to the polarities of the two half-winding imposes a forward bias on diode (A). The lower half-winding imposes a reverse bias on diode (B). Electron current flow is from the negative, or center tap of the transformer, through the load,

FIGURE 32-3 Use of two diodes and current flow through a common load produces full-wave rectification.

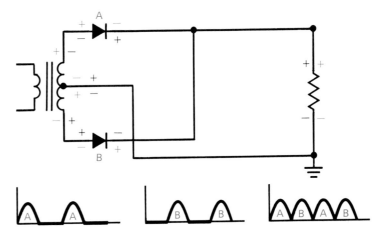

FIGURE 32-2 A center-tapped transformer produces two out-of-phase output voltages.

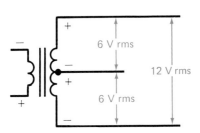

through diode (A), and to the upper connection of the upper half-winding of the transformer. There is no current flow through the lower half-winding during this period. The waveforms shown below the circuit indicate the output voltage during the (A) portion of the cycle.

When the polarities of the input ac voltage are reversed, the upper half-winding no longer is able to forward bias diode (A). The lower half-winding now imposes a forward bias on diode (B). Electron flow now occurs from the center tap of the transformer, through the load, through diode (B) and to the lower connection of the lower half of the transformer. The waveforms shown under the circuit indicate the output voltage for the B portion of the input ac wave.

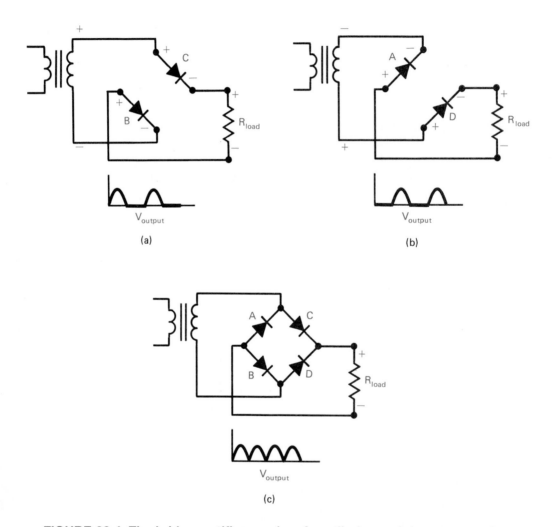

FIGURE 32-4 The bridge rectifier requires four diodes and does not need a center-tapped type of transformer.

The direction of electron current flow during both half-cycles is in the same direction through the load. The result of this is also shown below the circuit. The output waveform for the full-wave center-tapped rectifier system is twice that of the half-wave rectifier system. Therefore, the efficiency of this system is twice that of the half-wave system.

The output voltage from the full-wave center-tapped system is dependent on the peak value of one-half of the full secondary voltage of the tranformer. This is true because only one-half of the transformer is operating at a time. The current flow through the secondary winding of the transformer alternates between each half of the secondary winding.

Full-wave bridge rectifier systems do not require the use of center-tapped transformer secondary winding. The circuit for a full-wave bridge rectifier system is shown in Figure 32-4. This system requires four diodes connected as shown in the circuit. Advanced applications of the bridge rectifier circuit utilize commercially packaged diode sets. These have the four diodes encapsulated in one package. Both the encapsulated and the individual diodes function in the same manner. The advantage of the packaged unit is the ease of installation.

The process used for the full-wave bridge circuit is shown in Figure 32-4 (a) and (b). Part (a) shows the polarities applied to the dioides during the first half cycle of the ac wave. During this half-cycle diode *b* and *c* are forward biased and the other two diodes are reverse biased. Conduction is through diode (B), the load, and finally diode (C). During the second half-cycle [Figure 32-41(b)], the other two diodes conduct while the first two are reverse biased. The diodes that are reverse biased are considered to be disconnected from the circuit during their respective half-cycles.

The complete circuit for the full-wave bridge rectifier system is shown in Figure 32-4 (c). This is the combination of the two other half-circuits shown in parts (a) and (b). Current flow during each of the half-cycles occurs in the same direction through the resistance representing the load. The output voltage waveform is shown with the same part of the drawing. The output voltage is equal to the peak value of the input ac voltage. The frequency of the output voltage is twice that of the input voltage. The quantity of current flow still depends on the value of the peak ac input voltage and the value of the resistance of the load.

REVIEW PROBLEMS

These questions relate to a full-wave bridge rectifier system.

32-5. How many diodes conduct during the first half-cycle of the ac input wave?

32-6. How many diodes conduct during the second half-cycle of the ac input wave?

32-7. How does the output frequency compare to the input frequency?

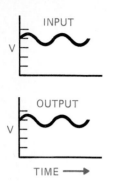

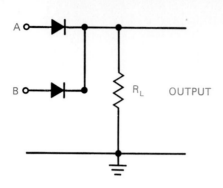

FIGURE 32-5 When a positive voltage is applied to either of the diodes, it is forward biased and current flow will occur through it.

32-4 Diode Switching

Diode switching *Forward biasing of diode in order to control signal flow in circuit*

The second major function for a diode is to act as a switch. The switching action is dependent on the dc bias placed across the terminals of the diode. Earlier in this book the concept of a signal riding on a dc voltage was discussed. This same concept is used when diodes are used as switches. The circuit shown in Figure 32-5 will illustrate this concept. The circuit shows two diodes. The cathode ends of both of the diodes are connected to the same load resistance. In this circuit neither diode is forward biased until a signal of some sort is connected to its anode. If the signal is only an ac wave it has both positive and negative voltages. The only time the diode will be forward biased is when the ac signal is positive. The output across the load will appear as a half-wave rectified voltage. This is not desirable for signal processing.

When the signal is a part of a dc voltage, as shown in the waveform drawing, the dc component of the voltage will forward bias the diode. The ac component of the voltage will then be impressed on the load along with the value of the dc voltage (minus the 0.7 v drop across the diode).

32-5 Transistor Circuits

One of the major functions of the transistor is to amplify a signal. The second major function is to act as a switch. Both of these functions work on the principle of the transistor being able to change its internal resistance between emitter and collecter when the emitter-to-base voltage is changed. Applications in audio amplifiers, all types of radios, television recorders, and receivers, as well the principles of transistor operation described in Chapter 31. The material in this section presents specific circuits for transistors. The presentation is limited to bipolar transistors since they are the most commonly used types.

Switching transistors *Transistors having ability to change their operational conditions very rapidly and used to control electronic circuits.*

Switching Transistors Computers and the machine control units use many transistor switching circuits. The basic transistor switching circuit is shown in Figure 32-6. This circuit consists of two base-bias resistors R_{B1} and R_{B2}. The voltage on the base of the transistor is adjusted by the ratio of these two resistors. Normally, the base bias establishes the condition of cutoff for the transistor. When the transistor is in cutoff, the internal resistance between emitter and collector is very high. A series circuit formed by the load resistance,

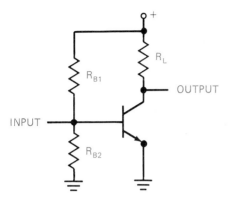

FIGURE 32-6 Basic switching transistor circuit.

R_1, and the emitter-to-collector connections of the transistor. The internal resistance of the transistor is much higher than that of the load resistance. The largest voltage drop in this series circuit develops across the terminals of the transistor. The output terminal of the circuit is connected between circuit common and the collector element of the transistor. The output voltage of this circuit is close to the value of the source voltage.

The collector-to-common voltage of a switching transistor changes between two values. These values are (1) close to the value of the source voltage and (2) close to the value of circuit common voltage. In the circuit shown in Figure 32-6, these specific values will fall between 4.5 and 5.0 V if the output voltage is close to source. When the output of the transistor is close to circuit common the voltage will be between 0.5 and 0 V. In the computer these values are known as high, on, or 1 when the voltage is close to the value of the source voltage. When the output is near 0 V the output is said to be low, off, or 0. These terms are used to describe digital circuit values.

The signal input for this circuit is between circuit common and the base of the transistor. When the signal is a square wave (Figure 32-7), the transistor will act as a switch. The signal voltage changes the emitter-to-base voltage difference from close to zero to close to 1.0 V. When the emitter-to-base voltage is near zero, the transistor is cut off. Its emitter-to-collector resistance is high and the output voltage between common and the collector is also high. This is illustrated by the square waves in the figure. During the first time unit the input voltage is low and the output voltage is high.

During the second time period the input voltage rises to close to 1 V. This changes the internal conditions of the transistor. Its internal resistance drops

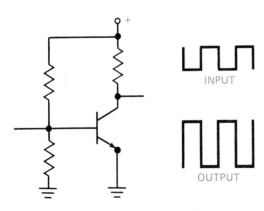

FIGURE 32-7 Use of an input signal to create an output square wave in the switching transistor circuit.

to a very low value. The transistor is now in saturation due to the change in the difference between emitter and base voltages. When the transistor is in saturation, the voltage at its output, from collector to circuit common, falls to a value close to the common or ground voltage value. The output voltage during the second time period is low.

The relationship between input and output voltage is an inverse one in this type of circuit. When the input is "high," the output is "low." This relationship is reversed when the input is low. When a constant-frequency square wave is used as the input signal, the output is also a square wave. When the input does not change at a constant rate but remains either high or low, the transistor is performing as a switch.

The circuit shown in Figure 32-8 is that of a transistor switch. The collector element of the transistor is connected in series to one lead or relay coil. The other end of the relay coil is connected to source voltage. When the input signal places the transistor in saturation, the majority of the voltage drop in the circuit develops across the terminals of the relay coil. It is energized and performs its normal operation at this time. When the input voltage creates a condition of cutoff in the transistor, its internal resistance goes high and the voltage in the circuit shifts from the relay coil to the elements of the transistor. The coil is deenergized and acts as if it is turned off at this time.

The type of circuit action, known as switching, is one of the functions of the transistor. The other major function is that of amplification. The amplifier is considered an analog type of function. That is, the function is not limited to either on or off, as described for the switching circuit. The analog transistor circuit will provide a change in the level of the output voltage between the extremes of cutoff and saturation. In fact, a good many amplifier circuits will not permit the operating conditions of the transistor to reach cutoff or saturation.

REVIEW PROBLEMS

32-8. What is the transistor condition when its output voltage is high?
32-9. What is the transistor condition when its output voltage is low?
32-10. What is the shape of the input wave for a switching transistor?

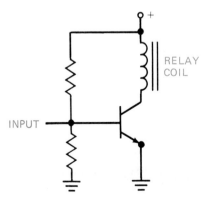

FIGURE 32-8 A switching transistor may be used to control a secondary circuit, such as the relay coil shown here.

Transistor Amplifiers There are three basic transistor amplifier circuits. The names of these circuits are related to **signal processing** and not to electron current flow through the transistor. Electron current flow does not change its direction or function in these three ciircuits. Each of the circuits is based on the same theory: that a small change between emitter and base voltage will create a larger change between emitter and the collector of the transistor. Keep this in mind as these three circuits are explained. The three circuits are called the *common-emitter*, the *common-collector*, and the *common-base* amplifier circuits.

    Common-emitter amplifier circuits are named because the signal input is between emitter and base and the signal output is between emitter and collector. The emitter element is common to both the input and the output circuits of the amplifier. The circuit configuration for the amplifier is shown in Figure 32-9. The bias on the base of the transistor places its operating point halfway between the extremes of cutoff and saturation. When the operation of the transistor is at this point, an input signal will vary the input voltage between emitter and base. The input voltage in this example has a value of 0.1 V peak to peak. This voltage, when added to the fixed base voltage, will produce a small change in the total base voltage. The 0.1 V p-p voltage changes the base voltage to extremes of 0.2 V to 0.1 V (0.7 + 0.1 = 0.8 V and 0.7 + −0.1 = 0.6 V).

    The small change in base voltage produces a larger change in the voltage between emitter and collector. The change in emitter-to-collector voltage is about 10 V peak to peak in this example. Ten volts peak-to-peak is equal to +5 V and −5 V. This is added to the value of dc voltage at the collector element. The designal of an amplifier of this type sets the collector-to-common voltage at about one-half of the source voltage when no signal is applied to the circuit. In this example the value of the collector voltage is +6 V with no signal present.

    The output signal voltage swing is between the value of the source voltage and the voltage at circuit common. In this example this swing is between +11 and +1 V. This is the process called *amplification*. By definition, amplification is the use of a small amount of power to *control* a larger amount of power. When one considers the values of voltage and current in the input circuit as being very small, this statement will make more sense. Under normal operation,

Signal processing *Action in electronic device permitting movement of information from one stage to next stage in system.*

Common-emitter amplifier *Transistor circuit having signal input at base, signal output at collector, and emitter common to both input and output.*

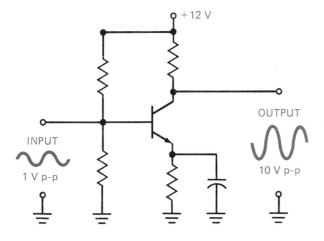

FIGURE 32-9 When the input conditions permit operation between cutoff and satura-tion, the transistor is used as an amplifier. This circuit is the common-emitter (CE) am-plifier.

the base circuit current is in values of microamperes and the base voltage change is less than 1 V. Power for the input circuit is on the microwatt level with values like these. The output circuit of a transistor amplifier probably has current flow in the milliampere range and voltage of 8 to 10 V. The power for this output circuit will be in the range of hundreds of milliwatts. The factor of microwatts controlling milliwatts is the concept called *amplification*. Other practical circuits may have output values in the watt range. Values are not limited just to those used as examples.

Also note that the emitter element has a bypass, or decoupling, capacitor connected from its element to circuit common. This is used to remove any signal from the emitter. A signal on the emitter of this transistor amplifier will be out of phase with the signal at the collector. The two oppose each other and the net result is a smaller-than-desired output signal. Signal processing in this amplifier circuit accomplishes two major items. These are a voltage gain and a signal phase inversion. Both are typical of the common-emitter amplifier circuit.

REVIEW PROBLEMS

32-11. Where is the input connection to the common-emitter amplifier?
32-12. Where is the output connection to the common-emitter amplifier?
32-13. What is the phase relationship between input and output signals?
32-14. Is there voltage gain with this amplifier?

Transistor Gain The amount of amplification in any amplifier circuit is dependent on several factors. These include the design of the transistor, the amount of signal applied, and the components in the circuit. Transistors are designed to have a current gain factor. This factor is known as the "beta" of the transistor. The beta is the relationship between output and input, or the current **gain** of the device. Beta is a number. It refers to the amount of amplification for a transistor regardless of the amplitude of the input signal. This factor is true only as long as the transistor does not enter cutoff or saturation. The beta factor tells us how much amplification we can expect from the specific transistor.

Gain *Ratio of output V, I, or W to input values of V, I or W.*

EXAMPLE 32-1

Find the signal gain for a transistor having a voltage gain of 100 and an input signal of 2 mV

Solution

$$\text{Output signal gain} = \text{voltage gain} \times V_{\text{input}}$$

$$= 100 \times 0.002 \qquad (32\text{-}4)$$

$$= 0.2 \text{ V}$$

In this example, the output signal will be 0.2 V when the input signal is 0.002 V.

The values shown in Example 32-1 and in the test have only considered maximum conditions. The amplifier used in the example has a gain factor of 100. This indicates that the output signal will be 100 times greater than the input signal. This is true for any value of input signal used. If the maximum value of input signal is 0.002 A before any distortion of the output signal can occur, a signal of 0.0014 A will also have a gain of 100 and the output voltage will be 0.14 A. This maintains the beta factor of 100. The reduced output voltage is created due to the reduced value of input voltage to the circuit.

Common-base amplifier circuits are also named because of the manner in which the signal in processed. A circuit diagram for the common-base amplifier is shown in Figure 32-10. This amplifier has a voltage-divider network connected between source and circuit common. This network consists of R_{B1} and R_{B2}. Its center connection is to the base of the transistor. A bypass capacitor, C_b, is used to decouple any signal present at the base of the transistor. The emitter-to-collector circuit has two series resistances. One of these, R_L, is connected between the collector and source. The other, R_E, is connected between the emitter and circuit common. The input connection to the circuit is between circuit common and the emitter element. The output connection of the circuit is between common and the collector of the transistor.

The basic concept for the transistor action is the relationship between the voltage developed between the base and the emitter. The common emitter and switching circuits will hold, or lock, the emitter voltage at some specific level. The voltage to the base is then varied in some manner to create transistor action. The concept of holding the base at a fixed value and changing the emitter voltage will accomplish the same results. After all, the purpose is to change the relationship between the two voltages. As long as it is accomplished, the manner by which it is done is not the greatest concern.

The common-base amplifier circuit input is at the emitter element of the transistor. The input signal will create a change of voltage on the emitter. When the input signal rises, the *difference* between base and emitter voltage *decreases*.

Common-base amplifier *Transistor amplifier using emitter for input elements and collector for output elements.*

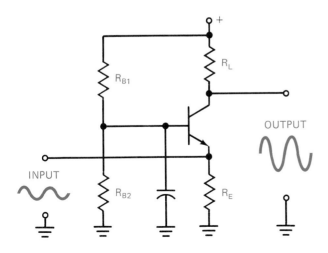

INPUT

OUTPUT

FIGURE 32-10 An input signal connected across the emitter resistor and an output connected between collector and circuit common creates this common-base (CB) amplifier.

This effectively increases the internal resistance between emitter and collector. The voltage at the collector element also rises.

When the input signal voltage drops, the difference between the emitter-to-base voltage becomes greater. The internal resistance of the emitter-to-collector voltage of the transistor also decreases. The voltage measured from common to collector falls. The difference between the changes between emitter and base and the changes between emitter and collector will create a voltage gain in this circuit. The voltages between emitter and collector are much greater than those of the input circuit. This amplifier will produce a voltage gain. The phase relationship of the input and the output signals is different from that of the common-emitter circuit. In the common-base amplifier circuit the signals are in phase with each other.

REVIEW PROBLEMS

32-15. Where is the input connection to the common-base amplifier?
32-16. Where is the output connection to the common-base amplifier?
32-17. What is the phase relationship between input and output signals?
32-18. Is there voltage gain with this amplifier?

Common-collector amplifier *Transistor amplifier using base for input and emitter for output elements.*

Common-collector amplifiers (Figure 32-11) are the third circuit configuration. The major difference in this circuit is that the load resistance, R_L, is connected between the emitter and circuit common and there is no collector resistance. In fact, the collector element has a bypass, or decoupling, capacitor connected to it.

Signal processing is from the input at the base to the output at the emitter element of the transistor. Any voltage drop occurring in the output circuit must occur across either the emitter resistor or the emitter to collector connection of the transistor. The output of this circuit is taken across the emitter resistor. Since the voltage measured from base to circuit common is always greater than the voltage measured from the emitter to circuit common, this circuit never exhibits a voltage gain.

Circuits of this type usually exhibit a current gain. This is related to the fact that the emitter element has 100% of the circuit current and the base element has only 2 to 5% of the circuit current. When the output voltage is

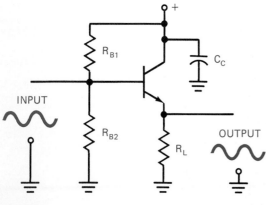

FIGURE 32-11 An input connected between base and common and an output connected across the emitter resistor creates this common-collector (CC) amplifier.

taken across the emitter resistance, the larger amount of current in this circuit will be used with the voltage drop across it to create a power gain.

The phase relationship of the signal voltages for the common-collector circuit show that they are in phase with each other. The voltage gain for this circuit is less than 1. This is because the voltage at the emitter element is always less than the voltage at the base of the transistor. The common-collector circuit is often called an **emitter-follower** circuit because the signal follows the emitter circuit.

Emitter-follower *Another name for the common collector amplifier configuration.*

REVIEW PROBLEMS

32-19. Where is the input connection to the common-collector amplifier?
32-20. Where is the output connection to the common-collector amplifier?
32-21. What is the phase relationship between input and output signals?
32-22. Is there voltage gain with this amplifier?

32-6 Oscillators

The third basic application for the transistor is as an oscillator. The **oscillator** circuit is capable of creating an electronic signal. It does this at a consistent rate, or frequency. Electronic oscillators can be designed to create signals at almost all frequencies in the spectrum. The specific frequency of operation depends on the values of the components used in the construction of the circuit. A block diagram for a basic oscillator system is shown in Figure 32-12. The oscillator uses a circuit similar to that of an amplifier. The difference between the amplifier circuit and the oscillator circuit is the **feedback** path. Something must be added to an amplifier circuit to sustain the oscillations that occur. This addition is the return of a controlled amount of the output signal back to the input of the oscillator. The oscillator is also unique in that it requires no input signal for it to operate. The oscillator requires a positive feedback type of signal. It also requires some form of frequency-determining components. The basic requirements for any electronic oscillator circuit include the active device, such as a transistor, and a resonant circuit.

The action of an oscillator may be described in the following manner. A resonant LC circuit is used to create an oscillation. Only a very small voltage is normally required to start the oscillations. The signal created by the oscillations of the LC circuit is then amplified by use of an active device. A portion of the amplified signal is fed back to the LC circuit used to initiate the oscil-

Oscillator *Electronic circuit used to create a repetitive waveform*

Feedback *System where a portion of the output signal is returned, or fed back, to the input for control purposes.*

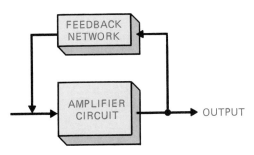

FIGURE 32-12 An amplifier circuit having a positive feedback path is used as an oscillator.

lations. This is normally done by use of either inductive or capacitive coupling circuits.

The energy from the output that is fed back to the *LC* circuit must be in phase with the oscillations and strong enough to overcome any losses of signal occurring when the circuit starts its oscillation. The signal that is fed back maintains the oscillations because it is at the same frequency as the original oscillations. The operating power required for the oscillator circuit is obtained from the power source of the electronic unit. The final statement related to oscillators is that the output from the oscillator is normally coupled to an amplifier circuit by the use of either an inductive or a capacitive coupling network.

The large variety of oscillator circuits for both transistors and vacuum tubes precludes a full discussion in this book. One basic type of oscillator circuit is shown in Figure 32-13. The transformer, T_1, establishes the feedback path for this circuit. In any electronic device, there will be some current flow when power is applied to the circuit. The current, of course, starts at zero when the power is not applied. Initially, there will be a rise in the level of current flow in the emitter-to-collector circuit. This change in the level of current flow also passes through one winding of the transformer. The change in current flow in this winding will produce a moving magnetic field around the winding. The moving magnetic field will induce a voltage on the other winding of the transformer. The induced voltage will change the bias on the base of the transistor. The changing bias will then change the current flow through the emitter-to-collector circuit.

Oscillation is sustained by the varying current flow through the transistor and the effect of the feedback voltage on the base circuit. A rise in collector circuit current will produce a reduction in base voltage because of the voltage induced on the transformer feedback winding. This in turn reduces the collector current. When the collector current drops, the voltage induced on the base circuit rises. This happens because the magnetic field around the transformer winding is now moving in the opposite direction. A rise in the base voltage will turn on the transistor an additional amount and start the process all over again.

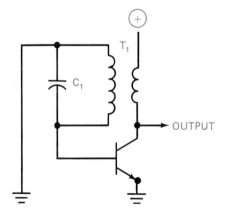

FIGURE 32-13 The energy formed in the inductances provides a feedback path for this oscillator circuit.

REVIEW PROBLEMS

32-23. What is necessary for oscillation to occur?
32-24. What components are used for frequency determination in the oscillator circuit?

32-7 Integrated Circuits

The **integrated circuit** is a development of transistor technology. The integrated circuit is a generic term used to describe the hundreds of different microcircuits. Most integrated circuits are made up of many miniature transistors, diodes, resistors, and capacitors. All of these types of components are produced on one small silicon chip. The size of the basic integrated circuit (IC) is about $\frac{1}{16}$ of an inch square. Several typical IC packages are shown in Figure 32-14.

Integrated circuit Microcircuit containing many components mounted on a common base material.

Integrated circuits are generally classified in one of several different manners. Among these classifications are such titles as analog ICs, linear ICs, and digital ICs. Most of these are available in a typical package having 8 to 16 pins on it. These are shown in the figure. Another method of classifying the IC has to do with the complexity of the circuits on the chip. Generally speaking, this type of classification includes LSI, VLSI, and MSI types. These initials represent large-scale integration, very-large-scale integration, and medium-scale integration. The specific identification is dependent on the quantity of components inside the IC. These terms are rather broad, so no attempt is made to identify the exact quantity of components in each of them.

The schematic diagram for a stereo decoder IC is shown in Figure 32-15. The quantity of components is fairly large in this circuit. It is considered as a SSI (small-scale integration) construction IC. It is almost impossible for the person servicing a unit containing ICs to fully comprehend all the circuitry in the chip. Almost all the IC manufacturers have reverted to a **block diagram** of the circuits on the IC. A typical block diagram for the same circuit is shown in Figure 32-16. It is much easier for persons working with the IC to use. The repair of any IC amounts to its replacement. There is little need to be able to

Block diagram Presentation of unit as functional blocks rather than in schematic diagram form.

FIGURE 32-14 Typical integrated circuit assemblies.

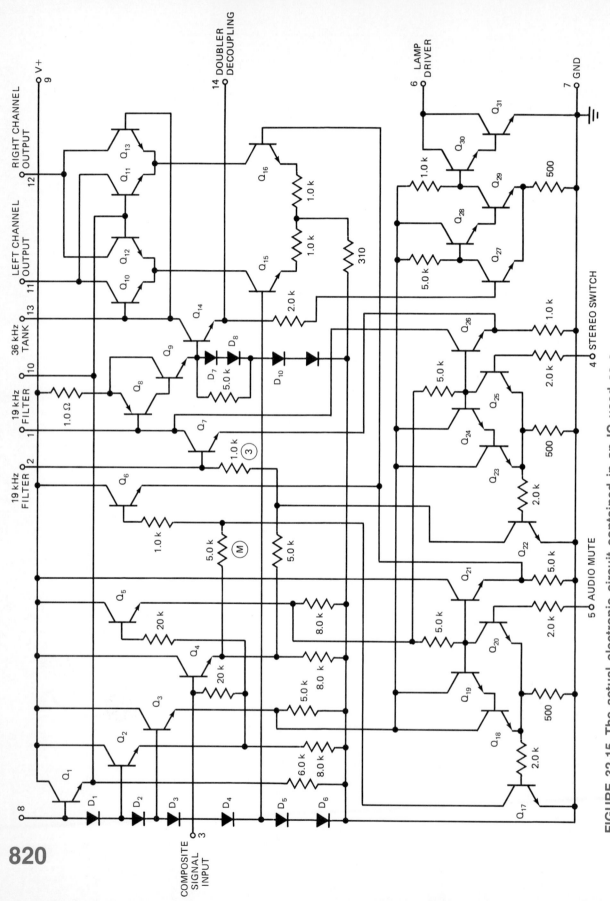

FIGURE 32-15 The actual electronic circuit contained in an IC used as a stereo decoder in an FM radio.

820

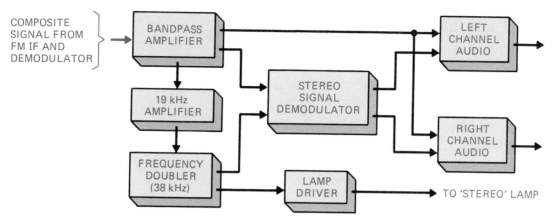

FIGURE 32-16 The block diagram for the stereo decoder IC shown in Figure 32-15.

see the specific circuit on the chip. The ability to follow signal flow or current flow paths into and out of the IC is the best manner in which to service and comprehend the system.

The terms *digital* and *linear* refer to the types of circuits on the IC chip. A digital circuit is one using transistors as switches. The voltage levels on a digital chip are either close to the value of the source voltage or close to the value of the circuit common voltage. Switching occurs between these two values. The *linear* IC uses analog circuits. One is apt to find an amplifier or an oscillator circuit on a linear IC.

REVIEW PROBLEMS

32-25. Name the classification of IC chips.
32-26. What types of signals are processed by a digital IC?
32-27. What types of signals are processed by a linear IC?

---

1. The diode is used as a rectifier and as a switch in circuits.
2. A single-diode circuit is called a half-wave rectifier circuit.
3. Half-wave rectifier circuits produce a voltage drop across the load during one-half of the input ac cycle.
4. The output voltage from a rectifier system is equal to the peak ac input voltage.
5. Full-wave rectifiers use both halves of the input ac wave.
6. The output waveform frequency of the full-wave system is twice that of the input ac frequency.
7. The full-wave center-tapped system requires a center-tapped transformer and uses two diodes.

CHAPTER SUMMARY

8. The full-wave bridge system uses four diodes and does not require a center-tapped transformer.

9. Any frequency ac voltage may be used with a rectifier system.

10. The diode is used as a switch when it is forward biased and a voltage with both dc and ac components is used.

11. Transistors are used as switches when they are operated beyond, or into, their cutoff or saturation modes.

12. Transistors are used as amplifiers when they are operated between cutoff and saturation limits.

13. Three basic transistor amplifier circuits are: common emitter, common collector, and common base.

14. Common-emitter circuits have an input at the base and an output at the collector.

15. Common-collector circuits have an input at the base and an output at the emitter.

16. Common-base circuits have an input at the emitter and an output at the collector.

17. Signal phase inversion occurs only in the common-emitter configuration.

18. Voltage gain occurs in the common-emitter and common-base circuits.

19. An oscillator is a device used to create an electronic signal at a specific frequency.

20. The frequency of oscillation depends on the values of the L and C components.

21. An oscillator is basically an amplifier circuit with positive feedback added to it.

22. Integrated circuits are microminiature circuits containing many transistors, diodes, resistors, and capacitors in a single IC chip.

SELF TEST

Answer true or false; change all false statements into true statements.

32–1. The common-emitter amplifier has an output signal that is 180° out of phase with its input signal.

32–2. The common-base circuit output signal is in phase with its input signal.

32–3. The term "gamma" is used to describe transistor gain.

32–4. A half-wave rectifier requires a series circuit and a load in addition to the input ac.

32–5. The output frequency of a full-wave bridge rectifier system is the same frequency as its input ac.

32–6. The output voltage of a rectifier system is equal to the peak value of the ac input voltage.

32–7. A switching diode is reverse biased when it is turned on.

32–8. A switching transistor is either in cutoff or in saturation.

32–9. When a switching transistor is in cutoff, its collector voltage is low.

32–10. Current through the switching transistor is high when it is in saturation.

MULTIPLE CHOICE

32–11. The semiconductor acts as a _____ in the circuit.
  (a) variable diode
  (b) variable resistance
  (c) fixed-value resistance
  (d) none of the above

32–12. The dc output voltage of a half-wave rectifier system is equal to _____ times the rms input voltage value.
  (a) 0.636
  (b) 0.707
  (c) 1.414
  (d) 2.828

32–13. The dc output voltage of a full-wave center-tapped rectifier system is equal to _____ times the rms input voltage value.
  (a) 0.636
  (b) 0.707
  (c) 1.414
  (d) 2.828

32–14. In the full-wave center-tapped rectifier system, each diode conducts
  (a) during alternate half-cycles of the input ac
  (b) during both half-cycles of the input ac
  (c) only during the negative half-cycle of the input ac
  (d) none of the above

32–15. The number of diodes required for the full-wave bridge rectifier system is
  (a) 1
  (b) 2
  (c) 3
  (d) 4

32–16. In the junction transistor, when the base-to-emitter voltage is changed:
  (a) the base-to-collector resistance is increased
  (b) the collector-to-emitter resistance is increased

32–17. A switching transistor operates

    (a) between saturation and cutoff
    (b) at saturation and cutoff
    (c) between the operating point and cutoff
    (d) between saturation and the operating point

32–18. The input signal for a common-emitter amplifier is

    (a) between emitter and collector
    (b) between emitter and base
    (c) at the emitter
    (d) between collector and base

32–19. Gain for a transistor amplifier is calculated by

    (a) dividing output by input
    (b) dividing input by output
    (c) multiplying input times output
    (d) multiplying stage 1 by stage 2 gain

32–20. The basic difference between an amplifier and an oscillator is that the

    (a) amplifier requires feedback
    (b) amplifier has gain
    (c) oscillator does not need a power source
    (d) oscillator has a feedback circuit

ESSAY QUESTIONS

32–21. What happens to the resonant frequency of an oscillator when the value of its feedback capacitance is decreased?

32–22. What type of circuit is left when the feedback system of an oscillator is removed?

32–23. What types of signals are processed by a linear integrated circuit?

32–24. A transistor common-collector amplifier has a relay coil connected between its emitter and circuit common. Is the transistor in cutoff or in saturation when the coil is energized?

32–25. A transistor amplifer has a relay coil connected between its collector and the source voltage. Is the transistor in saturation or in cutoff when the relay coil is energized?

32–26. Explain why the transistor acts as a variable resistance in the circuit.

32–27. Explain why the applied voltage appears across the terminals of the rectifier diode when it is reverse biased.

32–28. Why is a center-tapped transformer required for full-wave rectification when two diodes are used?

32–29. Draw the three basic transistor amplifier circuits and explain how each functions.

32–30. Explain the difference between an analog and a digital integrated circuit.

32–31. The voltage across a 100-Ω emitter resistor of a transistor amplifier is 2.0 V. What is the current flow through the transistor? (*Sec. 32-5*)

32–32. The voltage supplied to a common-collector amplifier is 12.0 V. Voltage measured at the collector element is also 12.0 V. Is this correct? (*Sec. 32-5*)

32–33. The gain of a transistor is 150. What is its output when the input signal is 20 mV p-p? (*Sec. 32-5*)

32–34. An amplifier circuit has an input current of 20 μA flowing through 10 kΩ and an output current of 150 mA flowing through 5 kΩ. What is its voltage gain? (*Sec. 32-5*)

32–35. What is its poweer gain of problem 32–34? (*Sec. 32-5*)

32–36. A 10-MHz oscillator is required for a system. What is the frequency of its resonant circuit? (*Sec. 32-6*)

32–37. What would you expect the voltage drop across the junction of a silicon diode to be when it is forward biased? (*Sec. 32-5*)

32–38. What value of voltage would you expect to measure between emitter and the base of a properly operating silicon transistor? (*Sec. 32-5*)

32–39. What is the current gain for the circuit in problem 32–33? (*Sec. 32-5*)

32–40. What is the power gain for the circuit in problem 32–33? (*Sec. 32-5*)

32–41. A diode is connected in series with the secondary winding of a transformer and a load. The cathode of the diode is connected to the transformer winding. An oscilloscope monitoring the waveform across the load indicates a positive-going waveform. Is this correct?

32–42. A center-tapped transformer secondary is connected to a pair of diodes, forming a full-wave center-tapped rectifier circuit. An oscilloscope monitoring the output of the rectifier system shows half-wave rectification. What is wrong?

32–43. A rectifier diode is connected in series with the secondary winding of a transfromer and a load, forming a half-wave rectifier system. A voltmeter placed across the load indicates a lack of any voltage. What is wrong?

32–44. Using the circuit shown in Figure 32-8, the problem is that the relay coil cannot be activated. Measurement of the transistor voltages shows $E = 0.0$ V, $B = 0.0$ V, and $C = +12$ V. What is the problem?

32–45. Using the circuit of Figure 32-9, the signal at the collector has the same phase as the signal at the base of the transistor. Is this correct?

32–46. In the same circuit, the conditions now show that the output signal has an inverted phase, but it is the same amplitude as the input signal. A measurement of the emitter of the transistor shows an in-phase signal. What is the problem?

32–47. The circuit shown in Figure 32-11 has an in-phase signal slightly smaller in amplitude when compared with the signal at the base of the transistor. Is this correct?

32–48. Using the circuit shown in Figure 32-8, the voltage at the collector of the transistor measures 0.0 V and you know that the source voltage is 24.0 V. What would you expect to be wrong in this circuit?

32–49. Using the circuit shown in Figure 32-8, the voltage at the collector of the transistor measures 24.0 V and you know that the source voltage is 24.0 V. What would you expect to be wrong in this circuit?

32–50. In the diagram of Figure 32-12, it is possible to inject a signal at the base of the transistor and it will be amplified at the collector terminal. The circuit does not oscillate. Where would you expect the problem area to be?

ANSWERS TO REVIEW PROBLEMS

32–1. 0.7 V is silicon and 0.2 V is germanium.
32–2. Across the load
32–3. 50 or less
32–4. 60 Hz
32–5. Two
32–6. Two
32–7. Twice the input frequency
32–8. Cutoff
32–9. Saturation
32–10. Square
32–11. Emitter and base
32–12. Emitter and collector
32–13. 180° shift
32–14. Yes
32–15. Common and emitter
32–16. Common and collector
32–17. In phase
32–18. Yes
32–19. Common and base
32–20. Common and emitter
32–21. In phase
32–22. No

32–23. Feedback, a resonant circuit, and an amplifier circuit

32–24. L and C

32–25. LSI, VLSI, and MSI

32–26. Square waves

32–27. Signal waves

CHAPTER 33

Operational Amplifiers

One form of the integrated circuit is the operational amplifier. This device is used to compare differences in voltage levels for system control purposes. It is also used as an impedance-matching device as well as the performance of mathematical functions. The operational amplifier has found uses in almost every type of electronic device because of its versatility.

KEY TERMS

Adder

Differential Amplifier

Inverting

Negative Feedback

Noninverting

Op Amp

Open-Loop Gain

Parameters

Positive Feedback

Saturation Voltage

Voltage Comparator

OBJECTIVES

Upon completion of the material in this chapter, you will

1. Understand the characteristics of the operational amplifier.
2. Be able to describe basic uses of the operational amplifier.
3. Be able to explain the use of the terms *differentiate, summing, inverting,* and *noninverting* as used with the op amp.
4. Be able to calculate basic circuit gain and other parameters for the op amp.

The basic concepts relating to the integrated circuit were presented in Chapter 32. One of the most prominent applications for the integrated circuit is when it is used as an *operational amplifier*. This device is found in a great many different applications. The operational amplifier, or *op amp* as it is usually called, is often used as the sole active device in an active filter circuit. It is also utilized as a differential amplifier, instrumentation amplifier, voltage amplifier, and impedance-matching device as well as in many other applications.

A block diagram for an op-amp circuit is shown in Figure 33-1. The basic circuit configuration consists of two power sources, two inputs, and a single output connection. All operational amplifiers have this type of circuit configuration.

The op amp may be defined as a direct coupled semiconductor device (using the latest technologies) having a very high gain factor. It is a voltage amplifier and is fabricated on a single piece of silicon material. The basic circuit for an op amp may be constructed using discrete transistors or, in older circuits, the vacuum tube, but this is not economically feasible at present. A manufacturer's data sheet for a typical op amp is shown in Figure 33-2. This particular device is identified as a number LM741 operational amplifier. As seen in the data sheet, the specific circuit uses several transistors, resistors, and even a capacitor. All of this is produced on a single small piece of semiconductor material. The integrated-circuit op amp is then "packaged" in one of several industry standard packages for integrated circuits. Three of the more common packages are illustrated at the bottom of the data sheet.

Power supply requirements for operational amplifier devices demand a dual-polarity voltage source. The typical power supply connection for an op amp is shown in Figure 33-3. One must use either a dual voltage supply or, as an alternative, design a voltage-divider network with circuit common at the midpoint of the voltage-divider circuit, as shown in the illustration. The typical

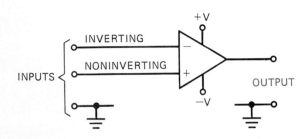

FIGURE 33-1 Block diagram for the basic operational amplifier.

Absolute Maximum Ratings

	LM741A	LM741E	LM741	LM741C
Supply Voltage	±22V	±22V	±22V	±18V
Power Dissipation (Note 1)	500 mW	500 mW	500 mW	500 mW
Differential Input Voltage	±30V	±30V	±30V	±30V
Input Voltage (Note 2)	±15V	±15V	±15V	±15V
Output Short Circuit Duration	Indefinite	Indefinite	Indefinite	Indefinite
Operating Temperature Range	$-55°$C to $+125°$C	$0°$C to $+70°$C	$-55°$C to $+125°$C	$0°$C to $+70°$C
Storage Temperature Range	$-65°$C to $+150°$C	$-65°$C to $+150°$C	$-65°$C to $+150°$C	$-65°$C to $+150°$C
Lead Temperature (Soldering, 10 seconds)	$300°$C	$300°$C	$300°$C	$300°$C

Electrical Characteristics (Note 3)

PARAMETER	CONDITIONS	LM741A/LM741E MIN	TYP	MAX	LM741 MIN	TYP	MAX	LM741C MIN	TYP	MAX	UNITS
Input Offset Voltage	$T_A = 25°$C										
	$R_S \leq 10\ k\Omega$					1.0	5.0		2.0	6.0	mV
	$R_S \leq 50\Omega$		0.8	3.0							mV
	$T_{AMIN} \leq T_A \leq T_{AMAX}$										
	$R_S \leq 50\Omega$			4.0							mV
	$R_S \leq 10\ k\Omega$						6.0			7.5	mV
Average Input Offset Voltage Drift				15							$\mu V/°$C
Input Offset Voltage Adjustment Range	$T_A = 25°$C, $V_S = ±20V$	±10				±15			±15		mV
Input Offset Current	$T_A = 25°$C		3.0	30		20	200		20	200	nA
	$T_{AMIN} \leq T_A < T_{AMAX}$			70		85	500			300	nA
Average Input Offset Current Drift				0.5							nA/$°$C
Input Bias Current	$T_A = 25°$C		30	80		80	500		80	500	nA
	$T_{AMIN} \leq T_A \leq T_{AMAX}$			0.210			1.5			0.8	μA
Input Resistance	$T_A = 25°$C, $V_S = ±20V$	1.0	6.0		0.3	2.0		0.3	2.0		$M\Omega$
	$T_{AMIN} \leq T_A \leq T_{AMAX}$, $V_S = ±20V$	0.5									$M\Omega$
Input Voltage Range	$T_A = 25°$C							±12	±13		V
	$T_{AMIN} \leq T_A \leq T_{AMAX}$				±12	±13					V
Large Signal Voltage Gain	$T_A = 25°$C, $R_L > 2\ k\Omega$										
	$V_S = ±20V$, $V_O = ±15V$	50									V/mV
	$V_S = ±15V$, $V_O = ±10V$				50	200		20	200		V/mV
	$T_{AMIN} \leq T_A \leq T_{AMAX}$, $R_L \geq 2\ k\Omega$,										
	$V_S = ±20V$, $V_O = ±15V$	32									V/mV
	$V_S = ±15V$, $V_O = ±10V$				25			15			V/mV
	$V_S = ±5V$, $V_O = ±2V$	10									V/mV
Output Voltage Swing	$V_S = ±20V$										
	$R_L \geq 10\ k\Omega$	±16									V
	$R_L \geq 2\ k\Omega$	±15									V
	$V_S = ±15V$										
	$R_L \geq 10\ k\Omega$				±12	±14		±12	±14		V
	$R_L \geq 2\ k\Omega$				±10	±13		±10	±13		V
Output Short Circuit Current	$T_A = 25°$C	10	25	35		25			25		mA
	$T_{AMIN} < T_A < T_{AMAX}$	10		40							mA
Common-Mode Rejection Ratio	$T_{AMIN} \leq T_A \leq T_{AMAX}$										
	$R_S \leq 10\ k\Omega$, $V_{CM} = ±12V$				70	90		70	90		dB
	$R_S \leq 50\ k\Omega$, $V_{CM} = ±12V$	80	95								dB

FIGURE 33-2 Data sheet for the 741 op amp as provided by the manufacturer. (Courtesy of National Semiconductor.)

PARAMETER	CONDITIONS	LM741A/LM741E			LM741			LM741C			UNITS
		MIN	TYP	MAX	MIN	TYP	MAX	MIN	TYP	MAX	
Supply Voltage Rejection Ratio	$T_{AMIN} \leq T_A \leq T_{AMAX}$, $V_S = \pm20V$ to $V_S = \pm5V$										
	$R_S \leq 50\Omega$	86	96								dB
	$R_S \leq 10\,k\Omega$				77	96		77	96		dB
Transient Response	$T_A = 25°C$, Unity Gain										
Rise Time			0.25	0.8		0.3			0.3		µs
Overshoot			6.0	20		5			5		%
Bandwidth (Note 4)	$T_A = 25°C$	0.437	1.5								MHz
Slew Rate	$T_A = 25°C$, Unity Gain	0.3	0.7			0.5			0.5		V/µs
Supply Current	$T_A = 25°C$					1.7	2.8		1.7	2.8	mA
Power Consumption	$T_A = 25°C$										
	$V_S = \pm20V$		80	150							mW
	$V_S = \pm15V$					50	85		50	85	mW
LM741A	$V_S = \pm20V$										
	$T_A = T_{AMIN}$			165							mW
	$T_A = T_{AMAX}$			135							mW
LM741E	$V_S = \pm20V$			150							mW
	$T_A = T_{AMIN}$			150							mW
	$T_A = T_{AMAX}$			150							mW
LM741	$V_S = \pm15V$										
	$T_A = T_{AMIN}$					60	100				mW
	$T_A = T_{AMAX}$					45	75				mW

Note 1: The maximum junction temperature of the LM741/LM741A is 150°C, while that of the LM741C/LM741E is 100°C. For operation at elevated temperatures, devices in the TO-5 package must be derated based on a thermal resistance of 150°C/W junction to ambient, or 45°C/W junction to case. The thermal resistance of the dual-in-line package is 100°C/W junction to ambient.

Note 2: For supply voltages less than ±15V, the absolute maximum input voltage is equal to the supply voltage.

Note 3: Unless otherwise specified, these specifications apply for $V_S = \pm15V$, $-55°C \leq T_A \leq +125°C$ (LM741/LM741A). For the LM741C/LM741E, these specifications are limited to $0°C \leq T_A \leq +70°C$.

Note 4: Calculated value from: BW (MHz) = 0.35/Rise Time(µs).

FIGURE 33-2 *Continued*

Op amp *Shortened name for operational amplifier.*

Noninverting *Op amp having same phase input and output signal.*

Inverting *Op amp having phase reversal between input and output signals.*

op amp requires both a positive and a negative power source. Normally, these both have the same voltage value.

Input and output connections for the **op amp** are shown in Figure 33-4. The op amp requires two input connections and has but one output connection. One of the input terminals is designated as a **noninverting** input and the other is identified as an **inverting** input. The noninverting input terminal is also shown as the positive (+) terminal, while the inverting terminal is shown as the negative (−) connection. A circuit common, or ground, is also required. The op amp has a single output terminal in addition to circuit common.

The use of the + and − signs for the two inputs has a direct relation to

FIGURE 33-3 Power supply requirements for the op amp demand a dual-voltage dual-polarity source.

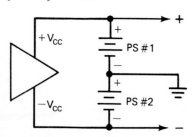

FIGURE 33-4 Most operational amplifiers have both inverting and noninverting input terminals.

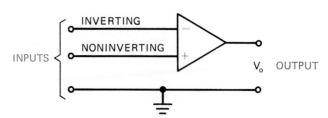

the polarity of the output signal. The voltage, or signal, applied to the positive, or noninverting, input terminal will appear at the output with the same phase as the input signal. A signal or voltage applied to the negative, or inverting, input terminal will appear at the output in an inverted, or 180° phase shifted, signal.

There are two additional input connections shown on the schematic diagram and on the terminal outline for the amplifier. These are identified as "offset null" terminals. The purpose of these two terminals is to balance the input voltage values for each of the two input connections. A look at the schematic diagram for the device will show the input null connections in series with each of the input transistors. These, then, are used to balance the two input circuits so that both have exactly the same amount of amplification.

REVIEW PROBLEMS

33-1. How many input terminals are used on a single op amp?
33-2. How many output terminals are used with an op amp?
33-3. What is the phase angle of the output signal when an input signal having a −45° phase angle is applied to the inverting input?
33-4. What is the phase angle of the output signal when the same signal is applied to the noninverting input terminal?

33-1 Operational Characteristics

The manufacturer of an operational amplifier provides data sheets describing the operating **parameters** of the device. These are usually provided as a part of the total data sheet information for a specific op amp. When we have limited knowledge about the terms describing this information, it is difficult to understand what all these data show. We summarize the parameters for operational amplifiers as follows:

Parameters Performance values of devices.

Term	Typical	Ideal
Input resistance	500 kΩ to high MΩ	Infinite
Voltage gain	50 k to 10 M	Infinite
Output resistance	About 100 Ω	0 Ω
Output voltage	About 90% of $+ V_{cc}$ and 90% of $- V_{cc}$	Infinite
Frequency response	Dc to 100 MHz (depending upon operational conditions)	Dc to infinity

Output voltage of the operational amplifier is also known by another name. This name is the **saturation voltage**, V_{sat}, of the op amp. This value is determined by utilizing the values provided in the data sheet. For the LM741 the saturation

Saturation voltage Value of voltage required to produce saturated condition in op amp.

value is 90% of the source voltage V_{CC}. In the case of an op amp having a source voltage of $+15$ and -15 V, the value of V_{sat} is determined:

$$+V_{sat} = +V_{source} \times 0.9 \tag{33-1}$$

$$-V_{sat} = -V_{source} \times 0.9 \tag{33-2}$$

Therefore, the output voltage of this op amp will be between the 90% values of both voltages. Calculate this:

$$+V_{sat} = +15 \text{ V} \times 0.9 = +13.5 \text{ V}$$

$$-V_{sat} = -15 \text{ V} \times 0.9 = -13.5 \text{ V}$$

Therefore, the maximum signal value for the output voltage will be

$$V_{out(max)} = +V_{sat} - (-V_{sat})$$

$$= +13.5 \text{ V} - (-13.5 \text{ V}) \tag{33-3}$$

$$= +13.5 \text{ V} + 13.5 \text{ V} = 27.0 \text{ V}$$

This value is the highest value of output voltage obtainable from this op-amp circuit with the indicated supply voltages. The actual value of output voltage can be less than this amount, but never greater than 27.0 V.

REVIEW PROBLEMS

33-5. What is the maximum signal value for an op amp when the operating voltages are ± 25 V?

33-6. What is the maximum signal voltage for an op amp when the operating voltages are ± 12 V?

Open-loop gain *Amount of amplification for op amp without use of a feedback network.*

    Open-loop gain is another term used to describe the quantity of the amplification in the op-amp circuit. The definition of this term is: the amount of gain for the circuit when a feedback circuit is not utilized. A circuit having feedback is known as a *feedback* type of circuit. The term A_{OL} has been identified as used to describe open-loop gain. The output voltage of an op-amp circuit is found using the general formula

$$V_{out} = A_{OL}V_d \tag{33-4}$$

where A_{OL} is the open-loop output voltage and V_d is the differentiation of voltages at the input terminals of the device. This value is calculated when one knows the value of A_{OL} and the values of V_{CC}, or supply voltage.

EXAMPLE 33-1

Find the value of V_{CC} when the V_{supply} is ± 20 V and the A_{OL} is 25,000.

Solution

First, find the maximum output voltage values when the op amp is at saturation.

$$+V_{sat} = +V_{CC} \times 0.9$$
$$= +20 \text{ V} \times 0.9 = +18 \text{ V}$$
$$-V_{sat} = -V_{CC} \times 0.9$$
$$= -20 \text{ V} \times 0.9 = -18 \text{ V}$$

Therefore,

$$A_{OL} = 36 \text{ V}$$

Next, find V_d:

$$V_d = \frac{36 \text{ V}}{25,000}$$
$$= 0.00144 \text{ V} \quad \text{or} \quad 1.44 \text{ mV}$$

When the open-loop gain of the op-amp circuit is higher, the same process is used to determine V_d.

EXAMPLE 33-2

Find the value of V_d when the V_{supply} is ± 20 V and the A_{OL} is 60,000.

Solution

The same value of V_{sat} is used. In this circuit it is 36 V.

$$V_d = \frac{36 \text{ V}}{60,000}$$
$$= 0.0006 \text{ V} \quad \text{or} \quad 600 \text{ } \mu\text{V}$$

REVIEW PROBLEMS

33-7. What is the output voltage for an op amp when it has a gain of 20,000 and its V_d is 24?

33-8. What is the output voltage for problem 33-7 when the op-amp gain is 75,000?

Differential Amplifier In addition to its ability to invert a signal, the op amp is able to *differentiate* between the values of its two input voltages. What this means is that the op amp's output will be the result of the *difference* between the two values of voltage at its input terminals (Figure 33-5). If, for example, the two input voltages have the same amplitude and are applied to both inputs, they will appear at the output terminal 180° out of phase with each other and

Differential amplifier Ability to differentiate between two signal voltages and created sum or difference of the signals at output.

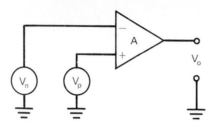

FIGURE 33-5 The two inputs are identified as shown in this diagram.

the two values will cancel. The net result is an output voltage (V_o) value of zero.

The symbol A is used to represent the open-loop gain of the op amp. This is the amount of gain for the op amp without any signal or voltage feedback from its output terminal to one of its input terminals. The term A_{OL} is used to identify the open-loop gain of the amplifier circuit. Gain values for operational amplifiers typically are between 50,000 and 10,000,000, as indicated in the earlier summary table. This does not mean that the gain cannot be less than 50,000, for there are some devices having gains on the order of 20,000. The term V_n represents the voltage applied to the negative, or inverting, terminal. The term V_p is used to represent the voltage applied to the positive, or non-inverting, terminal of the device. Both of these are measured from their respective input terminal to circuit common.

The differential input voltage is calculated by the formula

$$V_o = A(V_p - V_n) \tag{33-5}$$

The operational amplifier amplifier's output is the difference between the two input signals, hence the label of being a *differential amplifier*. The complete block diagram circuit for this differentiational amplifier is shown in Figure 33-6. Included in the diagram are a load resistance, Z_L, and the identification of V_n, V_p, A, and V_o.

REVIEW PROBLEMS

33-9. Find the differential input voltage for an op amp when V_p is $+5$ V and V_n is -5 V.

33-10. Find the output voltage for an op amp having an A of 60,000, a V_p of $+3$ V, and a V_n of -3 V.

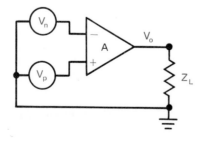

FIGURE 33-6 Block diagram for the differentiating operational amplifier.

33-2 Circuit Configurations

There are some basic circuit configurations for the operational amplifier. These include the single input, a dual input, and those circuits with feedback networks. The basics of these circuits are described in this section. *Single-input* amplifier configurations are similar to the one shown in Figure 33-7. In this circuit one of the input terminals is connected to circuit common. Either of the two input connections may be connected to circuit common, depending on the specific requirements of the circuit. If one requires an inverted output signal, the input is connected to the inverting input terminal. If an in-phase output signal is required, the input is connected to the noninverting input terminal and the inverting input terminal is connected to circuit common.

The output voltage or signal values may be described in either of the following terms. When the inverting input terminal is used, this formula is used:

$$V_o = -AV_n \qquad (33\text{-}6)$$

and when the noninverting terminal is utilized:

$$V_o = +AV_p \qquad (33\text{-}7)$$

is the formula describing this.

The voltage difference for an operational-amplifier circuit having one of its input terminals connected to circuit common is simple to determine. All one has to do is to measure the actual value of the input voltage. This is the value of V_d under these conditions.

Feedback The process of returning a signal or voltage obtained from the output of a circuit to its input circuitry is known as *feedback*. Feedback occurs in operational-amplifier circuits when a portion of the output is returned to one of the two major inputs of the amplifier. Actually, feedback can occur when any value up to 100% of the output signal is returned to the input. Feedback may be either *positive* or *negative* in nature. The difference in the terminology depends on the action occurring due to the feedback conditions.

Positive feedback occurs when the effects of feedback increase the action of the input signal or voltage. An example of positive feedback is the reaction of your body when attempting to stop running. If you were to attempt to come to an immediate stop, the momentum of your body will try to keep you moving forward. This action is considered to be the output of the action you are attempting. If you were to use a fixed object, such as a wall of a building, in

Positive feedback *Use of an in-phase signal to control amplification of device.*

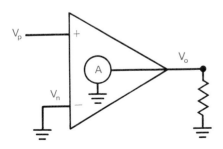

FIGURE 33-7 **A typical circuit configuration for the op amp has one of its inputs connected to circuit common.**

an effort to try to stop, your pushing against the building is an example of positive feedback. The increased effort due to your momentum will aid in stopping your movement. The pushing effect is an example of input activity. The total effect may be considered as the result of your output increasing your input efforts. Since most operational amplifiers have almost infinite gain, there is little need for a positive feedback network.

Negative feedback *Use of an out-of-phase feedback signal to control amplification of device.*

Negative feedback occurs when the effects of feedback decrease the action of the input signal or voltage. An example of this is the action required to hit a tennis ball. The tennis ball has no horizontal motion when it is thrown straight up into the air in front of you and hit. There is only vertical motion at this time. A specific amount of effort is required to hit the ball away from you in a horizontal direction. If you hit the ball toward a wall and it bounces off the wall and back toward you, it has a certain quantity of horizontal energy. If you were to hit the ball with the same amount of effort as you did when it had only vertical motion, it would not travel as far. This is because of the energy in the flight of the ball.

The energy in the tennis ball as it returns to you is similar to the effects of negative feedback in an amplifier circuit. The energy you used to hit the ball when it was in its vertical flight path is similar to an output signal. The energy in the ball as it flies toward you is considered to be a feedback type of system. If you were to hit the ball with the same amount of energy as you initially used, the ball will not travel very far. This feedback opposes your efforts to hit the ball. It has reduced your effort to hit the ball back to your opponent. The result of the horizontal motion of the ball has decreased your input effort to hit it.

In general, the effects of negative feedback tend to reduce the amount of input energy in an electrical or electronic circuit. In the case of an op amp, positive feedback will place the amplifier in a saturated mode and negative feedback will reduce the total gain in the circuit to less than that of saturation. In other words, positive feedback defeats the purposes of using feedback in this type of circuit.

A circuit utilizing the concept of negative feedback is shown in Figure 33-8. First, let us establish that this circuit is not in the saturated mode. The next step is to determine the value of V_o, the output voltage. The first step in doing this is to determine the value of V_{sat}. In this circuit the noninverting input is connected to circuit common. The difference between the voltage at the inverting input and common is equal to $+20\ V \times 0.9$, or $+18\ V$. Assuming a

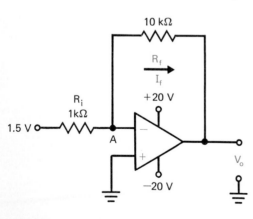

FIGURE 33-8 Feedback circuit utilizing negative feedback.

gain factor of 25,000, the value of V_d is 18 V/25,000, or 720 µA. Since this voltage is almost equal to zero and very close to the value of common, or ground, we can state that it is equal to a virtual common, or ground, value. Let us call V_A equal to 0 V under these conditions.

Most op amps have a very high input impedance. Because of this, very little, if any, input current flows through the op amp and almost all of the input current will flow through the input resistance R_i. Almost all the input current will also flow through the feedback resistance, R_f. The current flow through the input resistance, therefore, is

$$I_{in} = \frac{V_{in} - V_A}{R_i} \tag{33-8}$$

$$I_f = \frac{V_A - V_{out}}{R_f} \tag{33-9}$$

Since these currents are equal, as previously stated,

$$\frac{V_{in}}{R_i} = \frac{V_{out}}{R_f} \tag{33-10}$$

Using the values provided in the circuit, the formula becomes

$$\frac{1.5 \text{ V}}{1000} = \frac{-V_{out}}{10,000}$$

$$V_{out} = -15 \text{ V}$$

If this circuit was in the saturation mode, the output voltage would be equal to 90% of the V_{CC} value or 18 V. Since V_{out} is less than this value, the circuit is not in saturation and can be considered to be operating properly.

EXAMPLE 33-3

Using the op-amp circuit in Figure 33-9, calculate the output voltage of the amplifier.

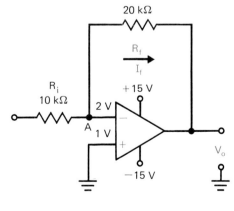

FIGURE 33-9

Solution

In this circuit the value of V_A is equal to 2 V. The next step involves finding the values of current.

$$I_{in} = \frac{V_A - V_{in}}{10,000}$$

$$I_{out} = \frac{V_{out} - V_A}{20,000}$$

We have determined that in the type of circuit $I_{in} = I_{out}$, therefore,

$$\frac{V_A - V_{in}}{10,000} = \frac{V_{out} - V_A}{20,000}$$

$$\frac{2 - (-1)}{10,000} = \frac{V_{out} - 2}{20,000}$$

$$V_{out} = 8 \text{ V}$$

REVIEW PROBLEMS

Use Figure 33-9 to find the following values.

33-11. Output voltage when V_A is 3 V
33-12. Output voltage when R_+ is 15 Ω

33-3 Inverting Amplifier

Example 33-3 is actually an inverting amplifier type of circuit. This is one type of practical circuit used with operational amplifiers. The circuit shown in Figure 33-10 is typical for an inverting op-amp arrangement. Notice that the input is connected at the inverting terminal and the noninverting terminal is connected to circuit common. The current flow through R_{in} is equal to the current flow through R_f under these conditions. The voltage at point A is equal to virtual ground, or 0 V, for practical purposes. The voltage drop developed across R_i is equal to the value of voltage presented at V_{in}. Current flow through R_{in} is

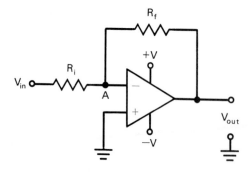

FIGURE 33-10 Inverting op-amp circuit configuration.

equal to V_{in}/R_i. In addition, the voltage drop developed across R_f is equal to V_{out}. Current through R_f is equal to $-V_{out}/R_f$. Since these two currents are equal, this equation may be stated:

$$\frac{V_{in}}{R_i} = \frac{-V_{out}}{R_f} \qquad (33\text{-}11)$$

This may also be shown as

$$\frac{V_{out}}{V_{in}} = \frac{-R_f}{R_i} \qquad (33\text{-}12)$$

These two expressions are used to determine the gain of the op-amp circuit. In this arrangement, the polarity of the input signal is not important. The input signal can be either dc or any form of ac signal. The following examples will aid in an understanding of these circuits.

EXAMPLE 33-4

Find the gain for the op-amp circuit shown in Figure 33-11.

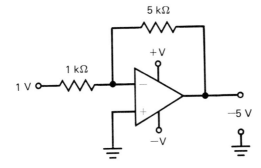

FIGURE 33-11

Solution

$$\text{gain} = \frac{V_{out}}{V_{in}}$$

$$= \frac{-5\ \text{V}}{1\ \text{V}} \qquad (33\text{-}13)$$

$$= -5$$

This may also be determined by

$$\text{gain} = \frac{-R_f}{R_i}$$

$$= \frac{-5000}{1000} \qquad (33\text{-}14)$$

$$= -5$$

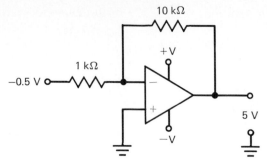

FIGURE 33-12

EXAMPLE 33-5

Find the gain for the op-amp circuit shown in Figure 33-12.

Solution

$$\text{Gain} = \frac{V_{out}}{V_{in}}$$

$$= \frac{5 \text{ V}}{-0.5 \text{ V}}$$

$$= -10$$

Examples 33-4 and 33-5 used a dc input voltage. When the input voltage is a signal, the rules described previously are still applicable.

EXAMPLE 33-6

Find the output voltage for the circuit shown in Figure 33-13.

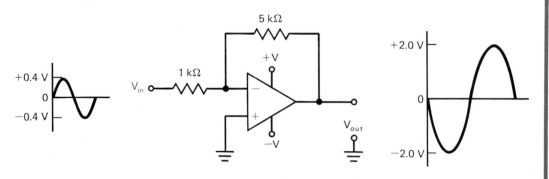

FIGURE 33-13

Solution

The first step in the analysis of this circuit is the determination of the value of the input signal. In this example, the signal swings between +0.4 and −0.4 V. This gives a value of 0.8 V p-p for the input signal.

Using the values of R_f and R_i, the gain of this circuit is

$$\text{gain} = \frac{-R_f}{R_i}$$

$$= \frac{-5000}{1000}$$

$$= -5$$

The output voltage value therefore is

$$\frac{V_{out}}{V_{in}} = \frac{-R_f}{R_i}$$

$$\frac{V_{out}}{0.8} = \frac{-5000}{1000}$$

$$= -5$$

$$V_{out} = -5 \times 0.8$$

$$= -4.0 \text{ V}$$

Since the input signal is connected to the inverting terminal of the op amp, the output signal is 180° out of phase with the input signal.

REVIEW PROBLEMS

33-13. What is the phase relationship between output and input of the inverting amplifier circuit?

33-14. What terminal is connected to circuit common in the inverting amplifier circuit?

33-15. Where is the feedback resistance connected in the inverting amplifier circuit?

33-4 Noninverting Amplifier

Some electronic systems require an output signal that is in phase with the input signal to the circuit. When this is a requirement, a noninverting op-amp circuit is used. The circuit shown in Figure 33-14 is one example of a noninverting type of op-amp circuit. In this circuit the input to the op amp is connected to the noninverting terminal. The inverting terminal has the resistance R_i connected between it and circuit common. The feedback resistance R_f is also connected to the inverting input terminal, as it is in the inverting amplifier configuration. The methods used to determine gain for the inverting amplifier may also be used to calculate the gain for a noninverting amplifier circuit.

There are certain assumptions that one must make when attempting to determine circuit gain.

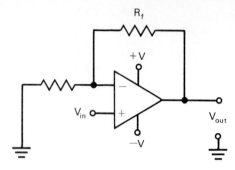

FIGURE 33-14 Noninverting op-amp circuit configuration.

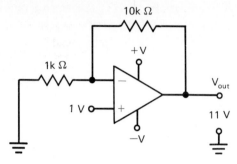

FIGURE 33-15 Circuit used to calculate op-amp gain.

1. The system is not at the point of saturation.
2. The voltage at point A is the same value as V_{in}.
3. The voltage drop across R_f is equal to the difference between V_{out} and V_{in}.
4. The voltage drop across R_i is equal to $V_{in} - 0$ V.
5. The current through R_i is equal to current through R_f.

These factors may be used to calculate the gain for the noninverting op amp circuit shown in Figure 33-15.

EXAMPLE 33-7

Determine the gain for the circuit in Figure 33-15.

Solution

When solving this problem, one must use the value of the voltage at the noninverting terminal as a part of the calculations.

$$\text{Gain} = \frac{V_{out}}{V_{in}} = 1 + \frac{R_f}{R_i}$$

$$= 1 + \frac{R_f}{R_i}$$

$$= 1 + \left(\frac{10,000}{1000}\right)$$

$$= 11$$

The polarity of the input signal has a direct effect on the polarity of the output signal. When the input signal at the noninverting input terminal of the op amp is negative in nature, the output signal is also negative. In the case of Example 33-7, the output for an input of -1 V would be -11 V. This, of course, does not affect the gain of the circuit.

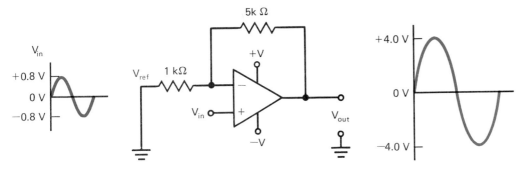

FIGURE 33-16 Gain calculations are possible when operating conditions are known.

When the input signal is in the form of a sine wave, the same principles apply. Using the circuit shown in Figure 33-16, the gain may be determined. First, the value of the input voltage is equal to 1.6 V p-p. The gain for this circuit is equal to $1 + (R_f/R_i)$, $1 + (5000/1000)$, or 6. The output voltage gain for this circuit is six times the value of the input signal voltage. When the input signal voltage is 1.6 V p-p, the output signal voltage is 1.6 V × 6, or 9.6 V p-p.

REVIEW PROBLEMS

33-16. What is the phase relationship between output and input of the noninverting amplifier circuit?
33-17. What terminal is connected to circuit common in the noninverting amplifier circuit?
33-18. Where is the feedback resistance connected in the noninverting amplifier circuit?

33-5 Voltage Comparator

Another major function for the operational amplifier is that of a **voltage comparator**. A basic circuit configuration for the voltage comparator is shown in Figure 33-17. One of the two inputs to the op amp is identified as a reference input. It is shown as V_{ref} in the figure. The other terminal is identified as the terminal and is shown as V_{in}. The reference voltage is obtained from some predetermined point in the electronic circuit. The input voltage is obtained from a different point in the circuit. The circuit does not utilize any feedback network and because of this the output signal voltage swings from the positive saturation voltage value to the negative saturation voltage value. In this type of circuit when V_{in} is greater in value than V_{ref}, the differential voltage is positive and the output waveform goes to the positive saturation voltage level. On the other hand, when V_{ref} is greater than V_{in}, the output voltage is negative and goes to the negative saturation voltage level. The waveform representing this activity is also shown in the figure.

Voltage comparator *Op-amp circuit used to compare two input signals and output the difference between them.*

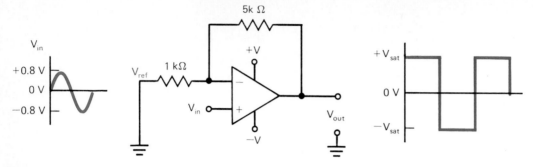

FIGURE 33-17 Voltage comparator circuit using an op amp.

When the input waveform is in the form of a sine wave, the same action occurs. The circuit shown in Figure 33-18 will illustrate this point. In this circuit the noninverting input terminal is connected to circuit common and the inverting input terminal is used as V_{in}. A sine-wave signal is injected at the input and designated as V_{in}. The output waveform is shown as an inverted square wave. This is due to the action of the op-amp circuit as it swings between its negative saturation voltage value and its positive saturation voltage value.

An illustration of a practical type of circuit utilizing the voltage sensor type of circuit is shown in Figure 33-19. This circuit includes two light-emitting diodes (LEDs). The two LEDs are used to indicate the polarity of the output voltage from V_{out}. The LEDs will light when the voltage drop across their circuit is high. Each LED circuit consists of the LED and its series resistance. One LED, identified as "low," will be turned on when a major voltage drop occurs across its circuit. The other LED will light when a major voltage drop appears across its circuit. Both LEDs cannot be on at the same time in this circuit.

The input for this circuit is at the inverting terminal of the op amp. The input sine-wave voltage is high during the first half of its cycle. The output voltage is low during this period. When this condition exists, the LED identified as "low" will have a large voltage drop across its circuit. The resulting current flow will produce a glow to the "low" LED. The "high" LED does not have any significant voltage drop during this half-cycle and it is not turned on. When the input waveform reverses, as it does during its second half-cycle, the low LED goes out and the high LED is now turned on. This, again, is due to the voltage drop developing across its terminals under this set of conditions. This

FIGURE 33-18 Use of a sine-wave input to produce a square-wave output from the op amp.

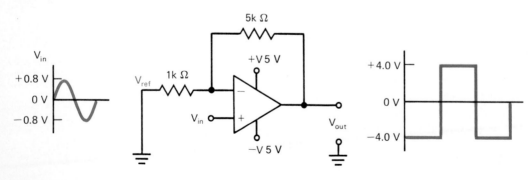

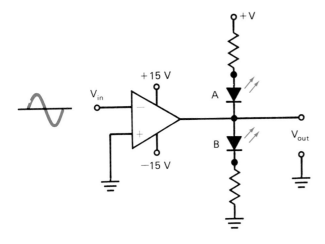

FIGURE 33-19 Voltage-sen-
sing circuit using LED diodes
to indicate highs and lows.

is one example of the use of the operational amplifier as a voltage comparator
and output indicating circuit.

REVIEW PROBLEMS

Use Figure 33-19 to answer problems 33–19 and 33–20.

33-19. What is the value of voltage applied to the low LED circuit?
33-20. What is the voltage drop across R_2 and the high LED circuit when
the input signal is low?

33-6 Op-Amp Adder

The final application for the operational amplifier presented in this chapter
shows how it can be used for mathematical operations. The circuit shown in
Figure 33-20 is called an **adder** circuit. In this circuit there are two input circuits
connected to point A and the inverting input terminal of the op amp. The two
inputs are identified as R_{i1} and R_{i2}. R_{i1} has an input voltage of 3.0 V and R_{i2}
has an input voltage of 1.5 V. Both of the input circuits are connected to their
respective 1-kΩ resistor to point A. Current will flow through both of the
resistors to point A. Current flow through R_{i1} is 3 mA and current flow through

Adder *Op amp circuit used to
sum the voltage at each of its
inputs.*

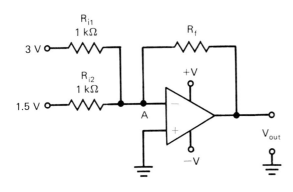

V_{out}

FIGURE 33-20 The op amp is
used as an adder in this circuit.

R_{i2} is 1.5 mA. The rule concerning current entering a junction as identified by Kirchhoff is applied to point A. The total current at point A is 4.5 mA. Earlier discussions identified the fact that the current at the input of the op amp is equal to the current flow through the feedback resistance R_f.

Since the voltage at the noninverting input terminal is zero, the voltage at point A with respect to circuit common is equal to the sum of the two input voltages. In this example this value is 4.5 V. This is also the value of the output voltage since the current flow in the circuit is equal. A circuit of this type is used to add input voltage values. This will be true unless any of the voltage values force the op amp into saturation.

EXAMPLE 33-8

Find the value of output voltage for the circuit shown in Figure 33-21.

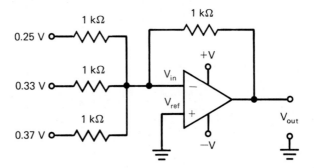

FIGURE 33-21

Solution

First, find the total current for the input circuit. I_{in} is equal to I_f, and this value is used to calculate V_{out}.

$$I_f = \frac{0.25 \text{ V}}{1 \text{ k}\Omega} + \frac{0.33 \text{ V}}{1 \text{ k}\Omega} + \frac{0.37 \text{ V}}{1 \text{ k}\Omega}$$

$$= 0.00025 + 0.00033 + 0.00037$$

$$= 0.00095 \text{ A}$$

Next, use the value of I_f and the value of R_f in Ohms law to find V_{out}.

$$V_{out} = 0.00095 \text{ V} \times -1000$$

$$= -0.95 \text{ V}$$

This type of summing circuit may also be used in a op-amp circuit having gain if the total gain of the circuit is the product of the gain and the voltage calculated as V_{out}.

1. An operational amplifier is an electronic device having two inputs and one output.
2. Most op amps requires a dual-voltage and dual-polarity power supply.
3. Op amp inputs are identified as inverting and noninverting.
4. A signal connected to the inverting input will produce a 180° phase-shifted output signal.
5. The output of the op amp when a noninverting signal terminal is used is in phase with the input signal.
6. Gain in the op amp is very high, often approaching infinity.
7. Input resistance of the op amp is very high, often approaching infinity.
8. Open-loop gain is the quantity of gain in the op amp without any feedback network.
9. Op amps are used as differential amplifiers when they differentiate between the two input voltage values.
10. When all or any portion of the output signal is returned to the input, a feedback network is established.
11. Positive feedback aids the signal output.
12. Negative feedback opposes the signal output.
13. Most op-amp feedback systems utilize negative feedback.
14. Op amps may be used as voltage comparators.
15. Op amps are also used to provide mathematical functions for circuits.

CHAPTER SUMMARY

Answer true or false; change all false statements into true statements.

SELF TEST

33–1. The operational amplifer has three inputs terminals and one output terminal.

33–2. Most op amps require a dual-voltage, dual-polarity power source.

33–3. Positive feedback is required in op-amp circuits.

33–4. Gain for the op amp can reach an infinite level.

33–5. The ideal op amp does not draw any current from its input circuit.

33–6. Op amps can be used for mathematical procedures.

33–7. Differentiation occurs in the output circuit of the op amp.

33–8. Open-loop op-amp circuits require a feedback network.

33–9. Signals connected to the inverting input of the op amp will appear as a 90° phase-shifted output signal.

33–10. The op amp will operate successfully with signals ranging from dc to high-frequency sine waves.

MULTIPLE CHOICE

33–11. The minimum number of input terminals for an op amp is _____ and the maximum number of output terminals is _____.
(a) 5, 2
(b) 2, 2
(c) 4, 1
(d) 2, 1

33–12. Most operational amplifiers require
(a) single-voltage-polarity power supply systems
(b) single-current-polarity power supply systems
(c) dual-voltage-polarity power supply systems
(d) triple-voltage-polarity power supply systems

33–13. The term used to describe the amount of amplification in the op-amp circuit is
(a) open-loop gain
(b) loop voltage gain
(c) closed-loop gain
(d) amplified gain

33–14. Gain in the op-amp circuit is controlled by
(a) an input network
(b) a feedback network
(c) a voltage control network
(d) forward back

33–15. The op amp using the sum of its two input voltages is called a
(a) summing amplifier
(b) differential amplifier
(c) voltage difference amplifier
(d) none of the above

33–16. A type of voltage return system that reduces the quantity of gain is known as
(a) negative feedback
(b) positive feedback
(c) voltage return
(d) signal feedback

33–17. The inverting amplifier configuration will provide an output signal that is _____ with the input signal.
(a) 45° out of phase
(b) 90° out of phase
(c) 180° out of phase
(d) in phase

33–18. The noninverting amplifier configuration will provide an output signal that is _____ with the input signal.
(a) 45° out of phase
(b) 90° out of phase
(c) 180° out of phase
(d) in phase

33–19. When one of the op-amp input circuits is held at a constant voltage level, or at circuit common, the circuit is called a
(a) reference amplifier
(b) current comparator
(c) voltage comparator
(d) current reference

33–20. The amount of open-circuit gain of an op amp is
(a) close to infinity
(b) not possible to calculate
(c) close to zero
(d) dependent on the applied voltage in the circuit

33–21. Why is the op amp built on one IC body?

33–22. Explain one method of obtaining a dual-voltage dual-polarity power source from a single voltage source.

33–23. What is meant by the term *saturation*?

33–24. How does the term *open-loop gain* apply to the op amp circuit?

33–25. Why is negative feedback used in the op-amp circuit?

33–26. Why is an inverting amplifier used in electronic systems?

33–27. When is a noninverting amplifier used in electronic systems?

33–28. When is a voltage comparator circuit used?

33–29. When is a summing amplifier circuit used?

33–30. Describe the expected output waveforms for the inverting and the noninverting op amp.

33–31. An op amp requires a V_{CC} of ± 18 V. What is its saturation voltage with a 90% saturation level? (*Sec. 33-1*)

33–32. What is the saturation voltage when the V_{CC} is 24 V? (*Sec. 33-1*)

Use Figure 33-22 to answer problems 33–33 through 33–35.

33–33. Calculate the output voltage. (*Sec. 33-1*)

33–34. Change the value of R_f to 20 kΩ and calculate V_o.

33–35. Change the value of V_{CC} to ± 30 V and calculate V_o. (*Sec. 33-1*)

Use Figure 33-23 to answer problems 33–36 through 33–38.

33–36. Calculate the output voltage. (*Sec. 33-2*)

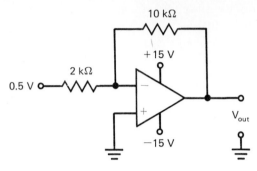

FIGURE 33-22

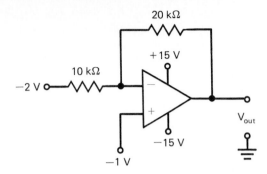

FIGURE 33-23

33–37. Change the value of R_f to 20 kΩ and calculate V_o. (*Sec,. 33-2*)

33–38. Change the value of V_{CC} to ±30 V and calculate V_o. (*Sec. 33-2*)

Use Figure 33-24 to answer problems 33–39 and 33–41.

33–39. Calculate the output voltage. (*Sec. 33-2*)

33–40. Change the value of R_f to 20 kΩ and calculate V_o. (*Sec. 33-2*)

33–41. Change the value of V_{CC} to ±30 V and calculate V_o. (*Sec. 33-2*)

Use Figure 33-25 to answer problems 33–42 through 33–44.

33–42. Calculate the output voltage. (*Sec. 33-2*)

33–43. Change the value of R_f to 20 kΩ and calculate V_o. (*Sec. 33-2*)

33–44. Change the value of V_{CC} to ±30 V and calculate V_o. (*Sec. 33-2*)

Use Figure 33-26 to answer problems 33–45 through 33–47.

33–45. What is the output phase angle when the sine wave shown in Figure 33-26(a) is injected at the input of the op amp? (*Sec. 33-3*)

33–46. What is the output phase angle when the wave shown in Figure 33-26(b) is injected at the input of the op amp? (*Sec. 33-3*)

33–47. What is the output phase angle when the wave shown in Figure 33-26(c) is injected at the input of the op amp? (*Sec. 33-3*)

33–48. What is the gain for the circuit shown in Figure 33-22? (*Sec. 33-3*)

33–49. Show the value of V_o and the phase of the signal at the output for the circuit shown in Figure 33-27. (*Sec. 33-3*)

FIGURE 33-24

FIGURE 33-25

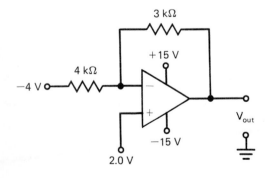

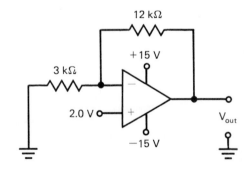

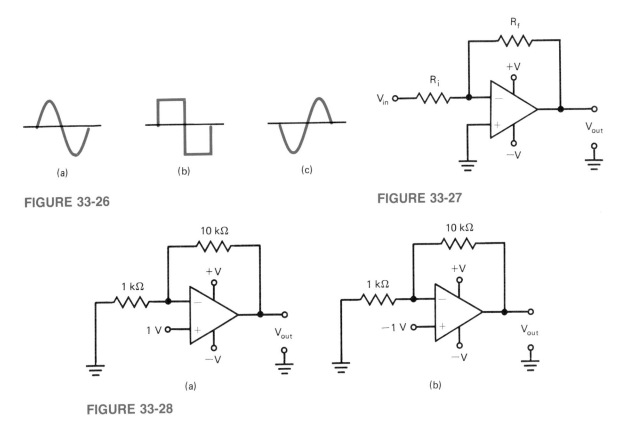

FIGURE 33-26

FIGURE 33-27

FIGURE 33-28

33–50. Show the phase of the output signal for the circuits shown in Figure 33-28. (*Sec. 33-3*)

33–51. Is the statement 'It is not possible to conveniently repair an integrated circuit' a correct one? If so, how are problems corrected in the circuit utilizing integrated-circuit technology?

33–52. Using the circuit of Figure 33-16, the output signal is inverted and smaller than the input signal. Is this correct?

33–53. Using the same figure, the expected output waveform is a sine wave. When measured, the output waveform is seen as a square wave. Is this correct? If not, what is the major problem item?

33–54. Using Figure 33-19, which LED should be on during the second half cycle of the input signal?

33–55. The output gain of an op amp is 2.0 when measured. The input signal has a value of 2.4 V. Is this correct operation?

33–36. The circuit shown in Figure 33-20 has a noninverted output signal. Is this correct?

33–1. Two

33–2. One

33–3. $+135°$

33–4. $-45°$

33–5. ± 22.5 V

33–6. ± 10.8 V

33–7. 0.0012 V, or 1.2 mV

33–8. 0.00032 V, or 320 μV

33–9. 10 V

33–10. 10,000

33–11. 9 V

33–12. 4.5 V

33–13. 180° phase shift

33–14. Noninverting input terminal

33–15. From the output to inverting input terminal

33–16. In phase

33–17. Neither terminal

33–18. From the output to the inverting input terminal

33–19. $+15$ V

33–20. 113.5 V

33–21. 2.58 V

33–22. -2.58 V

Chapters 31–33

The applications of the basic concepts of resistance, inductance, and capacitance constantly change as new ideas are developed. The concepts of these do not change. One must be aware of how these concepts apply to basic circuitry. The material in this section is an attempt to explain how some of the fundamental ideas are used in practical circuits.

1. Semiconductor devices are considered as active devices because they change their state during operation.

2. Common semiconductor materials are silicon and germanium.

3. Two common types of semiconductor materials are known as P type and N type.

4. N-type semiconductor material has an excess of electrons.

5. P-type semiconductor material has an excess of positive charges (or lack of electrons).

6. A semiconductor diode is a two-element device made up of both P- and N-type material.

7. The joining of the P- and N-type material is called the PN junction of the diode.

8. The junction may be either forward biased or reverse biased by the addition of a voltage charge on its elements.

9. A forward-biased junction has low resistance, a low voltage drop, and permits high current flow.

10. A reverse-biased junction has high resistance, high voltage drop, and permits little, if any, current flow.

11. The voltage drop across a forward-biased silicon junction is 0.7 V. Across a forward-biased germanium junction the voltage drop is 0.2 V.

12. Basic applications for active devices include rectification, amplification, and oscillation circuits.

13. The diode acts as a voltage-dependent electronic switch.

14. Diodes are identified by a number system starting with 1N. The cathode end of the diode is marked on its body.

15. Diodes may be tested using an ohmmeter.

16. Transistors are three-element devices that emulate vacuum-tube operation.

17. The emitter-to-base junction of a transistor is forward biased. The base-to-collector junction is reverse biased.

18. Transistors are identified by a number system starting with 2N. Identification of the elements requires use of a manual of specifications.

19. The transistor acts as a variable resistor in a circuit. Its emitter-to-collector resistance is controlled by the emitter-to-base conditions.

20. The diode junctions of a transistor may be tested in a manner that is similar to the one used for the diode.

21. Other semiconductor devices include the LED, photodiode, phototransistor, and the optocoupler.

22. The diode is used as a rectifier and as a switch in circuits.

23. A single-diode circuit is called a half-wave rectifier circuit.

24. Half-wave rectifier circuits produce a voltage drop across the load during one-half of the input ac cycle.

25. The output voltage from a rectifier system is equal to the peak ac input voltage.

26. Full-wave rectifiers use both halves of the input ac wave.

27. The output waveform frequency of the full-wave system is twice that of the input ac frequency.

28. The full-wave center-tapped system requires a center-tapped transformer and uses two diodes.

29. The full-wave bridge system uses four diodes and does not require a center-tapped transformer.

30. Any frequency ac voltage may be used with a rectifier system.

31. The diode is used as a switch when it is forward biased and a voltage with both dc and ac components is used.

32. Transistors are used as switches when they are operated beyond, or into, their cutoff or saturation modes.

33. Transistors are used as amplifiers when they are operated between cutoff and saturation limits.

34. Three basic transistor amplifier circuits are: common emitter, common collector, and common base.

35. Common-emitter circuits have an input at the base and an output at the collector.

36. Common-collector circuits have an input at the base and an output at the emitter.

37. Common-base circuits have an input at the emiter and an output at the collector.

38. Signal phase inversion occurs only in the common-emitter configuration.

39. Voltage gain occurs in the common emitter and the common-base circuits.

40. An oscillator is a device used to create an electronic signal at a specific frequency.

41. The frequency of oscillation depends on the values of the *L* and *C* components.

42. An oscillator is basically an amplifier circuit with positive feedback added to it.

43. Integrated circuits are microminiature circuits containing many transistors, diodes, resistors, and capacitors in a single IC chip.

44. An operational amplifier is an electronic device having two inputs and one output.

45. Most op amps require a dual-voltage and dual-polarity power source.

46. Positive feedback is required in op-amp circuits.

47. Gain for the op amp can reach an infinite level.

48. The ideal op amp does not draw any current from its input circuit.

49. Op amps can be used for mathematical procedures.

50. Differentiation occurs in the output circuit of the op amp.

51. Open-loop op-amp circuits require a feedback network.

52. Signals connected to the inverting input of the op amp will appear as a 90° phase-shifted output signal.

53. The op amp will operate successfully with signals ranging from dc to high-frequency sine waves.

Appendices

APPENDIX A
Glossary

AC abbreviation for "alternating current".

ACTIVE DEVICE device, such as a transistor or diode, capable of controlling voltage or current and producing gain or switch action in a circuit.

ACTIVE FILTER filter network utilizing an active device, such as a transistor or an integrated circuit, as one of the components.

ADDER operational-amplifier circuit having multiple inputs at one of its terminals; the sum of the voltages at this terminal controls the amplifier.

ADMITTANCE (Y) ease at which an ac current flows in a circuit; the reciprocal of impedance (Z) $(Y = 1/Z)$.

AIR GAP nonmagnetic break in a magnetic circuit; usually the opening between the poles of a magnet.

ALKALINE CELL chemical cell using a potassium hydroxide electrolyte to create electrical energy.

ALKALINE CELL OR BATTERY cell using an electrolyte of an alkaline material to produce a voltage.

ALTERNATING CURRENT (AC) current that reverses its direction on a regular basis.

ALTERNATOR mechanical device used to convert rotary motion or energy into electrical energy; associated with the production of ac voltage and current.

AMMETER instrument used to measure flow of electrons in a circuit.

AMPERE (A) unit of electrical current.

AMPERE-TURN unit of magnetizing force.

AMPLIFICATION process of using a small amount of power to control a larger amount of power in a circuit or system.

AMPLIFIER device or circuit used for amplification.

AMPLITUDE height of electrical wave measured from one point of reference to its maximum value.

ANALOG METER type of measuring device where output is continuously proportional to the forces providing the deflection of the meter needle.

ANALOG METER MOVEMENT dial scale of meter providing proportional movement of dial in response to quantity being measured.

ANODE positive element of an electrical device or battery.

APPARENT POWER product of voltage and current in an ac circuit when the two are out of phase with each other; measured in units of the voltampere.

ARMATURE rotating member of a motor or a generator.

AUDIO FREQUENCY (AF) range of frequencies starting about 20 Hz and extending up to 20 KHz; frequencies able to be heard by human beings.

AUTOPOLARITY type of measuring device having ability to display the polarity of the measured value.

AUTO RANGING type of electrical measuring device having circuitry that automatically selects the proper range for display of circuit values.

AUTOTRANSFORMER transformer having one winding used for both primary and secondary functions.

AVERAGE VALUE used in ac circuits to describe voltage or current; has a value of 63.7% of peak value.

AXIAL ASSEMBLY construction where the wire leads of the device are located on its axis.

BALANCED BRIDGE CIRCUIT (*see* BRIDGE CIRCUIT) in a bridge circuit, where the voltage drops at the center connections of the two series strings are equal.

BAND group or range of frequencies.

BANDPASS circuit used to permit a specific range of frequencies to pass and to attenuate all others.

BANDPASS AMPLIFIER circuit permitting a specific range of frequencies to be amplified and passed on to another circuit.

BANDPASS FILTER electronic filter network designed to permit the passage of a range of frequencies while attenuating all other frequencies presented to it.

BAND STOP circuit used to attenuate a specific band of frequencies.

BANDWIDTH frequencies falling between upper and lower limits of specific band of frequencies.

BATTERY device used to change chemical energy into electrical energy.

B–H MAGNETISM CURVE curve showing stages of magnetization of ferrous material; also shows demagnetization of the material.

BIPOLAR TRANSISTOR semiconductor device having ability to conduct both negative and positive charges.

BLEEDER RESISTOR resistance connected across output of power source filter network used to discharge capacitors in the circuit.

BLOCK DIAGRAM display of functional sections or blocks of a device rather than the specific circuit components.

BRANCH one leg of a parallel circuit.

BRANCH CIRCUIT (*see* **BRANCH**)

BRIDGE (a) circuit used to achieve a balance between two parallel paths; (b) undesired conductive path between two conductors on a printed circuit board.

BRIDGE CIRCUIT electrical circuit consisting of two parallel branches, each containing a two-resistance string; voltage or current values are compared between the center connections of the branch resistances for control of a circuit connected between the two center connections.

BRUSHES used in motors and generators as connectors between rotary and non-rotary parts.

BUS BAR heavy conductor used to carry large amounts of electrical current.

BYPASS CAPACITOR used as a low-impedance path around a circuit component or element.

C (a) symbol for capacitance; (b) abbreviation for unit of electrical charge, the coulomb.

CAPACITANCE ability of a component to store an electrical charge.

CAPACITIVE CURRENT quantity of current flow in circuit flowing in a capacitor.

CAPACITIVE REACTANCE opposition to ac voltage provided by capacitor action.

CAPACITOR device used to store electrical charges.

CARBON-ZINC CELL chemical voltage source using carbon and zinc for the composition of its elements.

CATHODE negative electrode of a semiconductor diode.

CB CIRCUIT (*see* common-base amplifier)

CC CIRCUIT (*see* common-collector amplifier)

CE CIRCUIT (*see* common-emitter amplifier)

CELL term used to describe a single chemical energy source.

CERAMIC material used for capacitors having a high dielectric constant.

CHARGE quantity of electrical energy stored by a capacitor or battery.

CHARGE CURRENT current flow during the period of charge of a capacitor's plates.

CHIP integrated circuit device.

CHOKE inductor having a high level of inductive reactance.

CIRCUIT BREAKER resettable circuit protection device.

CIRCUIT HIGH SIDE term used to describe the part of the circuit connected to the positive terminal of the power source.

CIRCUIT LOADING effect of the load resistance value on a circuit; usually results in a lower-than-normal resistance value, additional current flow, and a lower-than-normal output voltage.

CIRCUIT LOW SIDE term used to describe the portion of the circuit connected to the common or negative power source.

CIRCULAR MIL cross-sectional area of round conductor having a diameter of 0.001 in.

CLOSED LOOP term used to describe a series circuit, one having only one path for current flow.

COAXIAL CABLE cable used to carry radio-frequency signals having two concentric conductors.

COIL inductor consisting of mutiple turns of wire.

COIL RESISTANCE ohmic value of wire used to wind coil.

COLOR CODE industry standard value set using colors to represent a numbering system.

COMMON point of reference in an electric or electronic circuit, often the negative terminal of the power source; also called "ground".

COMMON-BASE AMPLIFIER transistor amplifier circuit with input between emitter and circuit common and the output connected between its collector and circuit common.

COMMON-COLLECTOR AMPLIFIER transistor amplifier circuit with its input connected between base and circuit common and its output connected between its emitter and circuit common.

COMMON-EMITTER AMPLIFIER transistor amplifier circuit having its input connected between base and circuit common and its output connected between collector and circuit common.

COMMUTATOR used in rotary machinery to maintain polarity of voltage.

COMPLEX NUMBER one number having both real and imaginary terms; in electrical usage having both j terms and real terms.

CONDENSER archaic term used to describe a capacitor.

CONDUCTANCE (G) ability to carry an electrical current; reciprocal of resistance; $G = 1/R$.

CONDUCTOR wire or other material permitting ease of flow of electrons.

CONTINUITY very low resistance path for electrical current, often close to $0\ \Omega$; used to indicate a conductive path.

CONVENTIONAL CURRENT movement of electrical charges from the positive terminal of the power souce, through the load, to the negative terminal of the source.

COSINE trigonometric function describing the ratio of the adjacent side to the hypotenuse of a right triangle.

COULOMB (C) unit of electrical charge equal to 6.25×10^{18} electrons.

COUPLING transfer of signal from one circuit to another circuit.

COUPLING CAPACITOR device used to transfer signal or voltage from one circuit to another circuit.

COUPLING COEFFICIENT quantity of coupling or signal transfer, between two circuits.

CPS cycles per second; replaced by term "Hertz."

CRT cathode ray tube; used to display electrical waveforms.

CRYSTAL FILTER electronic filter device made up of a piece of crystal material; it acts as a bandpass filter when a range of signals is presented to its terminals.

CURRENT (I) movement of electrons past a given point during 1 second.

CURRENT BLOCKING describes ability of capacitor to block current flow in circuit.

CURRENT DIVIDER parallel electrial circuit providing current to more than one circuit.

CURRENT SOURCE power source used to supply electrical current to load.

CUTOFF circuit condition where the active device is not able to conduct any electrons.

CUTOFF FREQUENCY specific frequency for a circuit identified as point of attenuation of signal.

CYCLE one complete set of events describing a repetitive waveform.

D'ARSONVAL METER type of electrical meter movement; uses moving-coil principle to indicate quantity of current flow in armature of coil.

DC direct current; current flow in only one direction in the circuit.

DC COMPONENT that portion of the operating voltage and signal having an average or constant value.

DEFLECTION SENSITIVITY (a) change in direction of movement in cathode ray tube due to the application of a force field; (b) value of current flow in meter movement producing a full-scale deflection.

DEGAUSSING neutralization of magnetic field.

DELTA NETWORK three electrical components connected in the form of a triangle.

DETUNING changing one of the components in a resonant circuit, resulting in a change in the frequency of resonance.

DIELECTRIC insulating material used to store electrical charges.

DIELECTRIC CONSTANT (*k*) concentration of electric field in a dielectric material.

DIELECTRIC STRENGTH maximum voltage a dielectric material can stand without breaking down and becoming a conductor.

DIFFERENTIAL AMPLIFIER operational amplifier circuit using the difference between the voltages at its two input terminals.

DIFFERENTIATION voltage obtained across a resistance in an *RC* circuit.

DIGITAL CIRCUIT electrical circuit using square-wave type of signal for system control.

DIGITAL METER type of measuring instrumentation having a numerical, or digital, readout rather than a moving needle meter.

DIODE two-element electrical device used to control direction of current flow.

DIODE SWITCHING circuit action where signal directions are controlled by changing the voltage on a diode in the circuit.

DIRECT CURRENT (DC) electrical current flow in one continuous direction; has a constant polarity.

DISCHARGE CURRENT current flow during period when plates of a capacitor are discharged.

DISCRETE COMPONENT term used to describe an individually packaged electronic device rather than a packaged composite circuit.

DOPING introduction of a second type of material into a pure material, such as silicon or germanium; permits the flow of electrons or positive charges through the material.

DPDT arrangement of contacts of switch or relay; represents double pole–double throw arrangement.

DPST terminology for switch contacts or relay contacts; represents double pole–single throw arrangement of contacts.

EDDY CURRENT current circulating around iron core of a inductor.

EFFECTIVE RESISTANCE value of ac equivalent to an equal amount of dc required to heat a device 1°C.

EFFECTIVE VALUE used with sine-wave ac and is 70.7% of peak value of wave; also known as rms value and has same heating effect as comparable amount of dc voltage.

EFFICIENCY ratio of the power output to the power input of a circuit or system.

EIA COLOR CODE color code to number system used by the Electronic Industries Association as a standard for marking values on components.

ELECTRIC CHARGE (*see* CHARGE)

ELECTRIC FIELD (*see* FIELD)

ELECTRODE terminal used to emit, collect, or control electron flow.

ELECTROLYTE solution used to form ion charges.

ELECTROLYTIC CAPACITOR capacitor using an electrolyte material for its dielectric.

ELECTROMAGNET device consisting of coil of wire and iron core that forms a magnet when an electrical current flows through it.

ELECTROMAGNETIC INDUCTION process of creating voltage on secondary wire due to movement of magnetic field and wire.

ELECTRON negative component of atom structure of material.

ELECTRON FLOW movement of electrons past a given point from the negative terminal of the power source, through the load, to the positive terminal fo the power source.

EMF electromotive force; measured in units of the volt.

EMITTER element of bipolar-junction transistor.

ENERGY FIELD charges on the plates of a capacitor.

EQUIVALENT RESISTANCE effective value of resistance equal to the values of all resistances in the circuit or subcircuit.

EXCITATION describes the value of voltage used to create a magnetic field in the coils of a motor or generator.

FARAD (F) unit of capacitance; stores 1 coulomb of charge when 1 volt is applied.

FARADAY'S LAW describes the effects of magnetic induction; states that the generation voltage increases with additional flux or a faster rate of change of the flux field.

FEEDBACK circuit taking a portion or sample of the output and returning it to the input of the circuit for control purposes.

FERRITE nonmetallic magnetic material.

FERRITE BEAD circular piece of ferrite material having a hole in its center; it is placed around a wire or conductor and acts as a choke to high-frequency signals.

FERROMAGNETIC magnetic property of iron or similar metals that are able to be strongly magnetized by an outside magnetic field.

FIELD either a magnetic or electrical group of lines of force.

FIELD-EFFECT TRANSITOR (FET) semiconductor device using a varying electrical field to control current flow through it.

FIELD WINDING portion of a motor or generator providing the magnetic field.

FIGURE OF MERIT numeric value assigned to a component rating its ability to be used in a circuit.

FILTER general term used to describe circuit used to react to certain frequencies.

FILTER CAPACITOR capacitor used in circuits to attenuate changes in circuit voltage.

FILTER CHOKE inductor used to attenuate variations in the current.

FLUCTUATING DC variations to the dc voltage value with a constant polarity.

FLUX magnetic lines of force.

FLUX DENSITY quantity of magnetic flux in a given area.

FLYWHEEL EFFECT term used to describe the continued oscillations of a circuit as the energy in it is released.

FORWARD CURRENT current flow in a semiconductor when it is forward biased.

FREQUENCY (*f*) quantity of repetitions per second for a periodic wave; measured in units of the Hertz (Hz).

FULL-SCALE DEFLECTION term used to describe amount of current required to produce full movement of meter needle across face or scale of instrument.

FULL-WAVE RECTIFIER circuit used to convert both halves of the ac wave into pulsating dc voltage.

FUSE protection device having meltable link.

FUSIBLE RESISTOR resistor having the qualities of a fuse; designed to fail under an overcurrent condition.

GAIN (A) ratio of output to input; also known as "amplification."

GALVANIC CELL voltage source using electrochemical principles.

GALVANOMETER sensitive measuring instrument used to measure electrical charges.

GAUSS (G) unit of flux density in magnetic circuit; equal to one magnetic line of force per square centimeter.

GENERATOR mechanical device used to convert rotary motion into electrical energy.

GERMANIUM (Ge) material used to make semiconductors.

GIGA (G) measurement prefix representing 1,000,000,000, or 1×10^9.

GROUND reference point for electrical measurement; also used as the common return point in electrical circuits.

HALF-POWER FREQUENCIES 70.7% of maximum signal amplitude used to define limits of frequency response for a resonant circuit.

HALF-WAVE RECTIFIER circuit used to convert half of the ac wave into pulsating dc voltage.

HARMONIC multiple of fundamental frequency.

HEAT SINK mechanical device used to move heat away from the device creating it; often used for mounting of semiconductors or power resistors.

HENRY (H) unit of inductance.

HERTZ (Hz) number of cycles in 1 second.

HIGH-PASS FILTER circuit used to pass high frequencies and to attenuate low frequencies.

HOLE positive electrical charge existing in semiconductor material.

HOLE CURRENT movement of positive electrical charges in a semiconductor material.

HOT RESISTANCE resistance of a device at its normal operating temperature.

HYPOTENUSE part of a right triangle; side opposite the 90° angle.

HYSTERESIS effect of magnetic induction lagging behind the applied magnetizing force in a circuit.

Hz hertz; means cycles per second (formerly cps).

IC integrated-circuit device.

IMAGINARY NUMBER number out of phase by 90°; in electrical terms, the letter *j* is used to represent this value.

IMPEDANCE MATCH process of making output impednace of one circuit equal to input impedance of a second circuit to achieve maximum transfer of power.

INDUCTANCE (*L*) process in inducing voltage from movement of magnetic field across a conductor.

INDUCTION process of magnetic field of one conductor creating a voltage in a second adjacent conductor using magnetic fields motion for this effort.

INDUCTIVE COUPLING use of a transformer or other inductive device to transfer signals from one circuit to another circuit.

INDUCTIVE REACTANCE opposition of an inductor to an ac voltage or current.

INDUCTOR wire formed in shape of coil showing properties in induction when current is passed through the wires in coil.

IN PHASE relationship of two electrical waves, both starting at the same moment in time.

INSTANTANEOUS VALUES amplitude of voltage or current at one instant of time.

INSULATOR nonconducting material having extremely high resistance.

INTEGRATED CIRCUIT (IC) miniatured set of circuit

components mounted on a standard-size body or holder; a microminiature circuit.

INTERNAL RESISTANCE effective resistance in a voltage source; identified as R_i.

INVERSE RELATION reciprocal relation; an increase in one variable will produce a decrease in the other variable.

INVERTING operational amplifier circuit where the output is inverted from, or out of phase with, the input signal.

ION atom having an electrical charge.

IR DROP difference of voltage measured across the terminals of an electrical or electronic device.

ISOLATION TRANSFORMER device having equal voltage values of primary and secondary; main purpose is to isolate two circuits electrically.

j **FACTOR** influence of the imaginary number *j* on a circuit's voltage and current.

j **OPERATOR** imaginary number used in analysis of ac circuits; represents phase angle of 90°.

JOULE (J) unit of work or energy.

JUNCTION (a) meeting and connection of two or more conductors; (b) physical connection of two semiconductor materials.

k term used to represent coefficient of coupling between two coils.

KILO (k) prefix representing 1000, or 1×10^3.

KIRCHOFF'S CURRENT LAW (KCL) sum of all currents entering and leaving a junction. Used in both ac and dc circuit analysis.

KIRCHOFF'S VOLTAGE LAW (KVL) sum of all voltages in a closed loop; used for both ac and dc circuits.

LEAKAGE condition producing a reduction in quantity of electrons on plates of a capacitor, or its ability to store charges. Also describes a trickle of charge carriers into PN junction under reverse bias.

LEAKAGE FLUX amount of magnetic flux not contained in the air gap, or useful part, of a magnetic circuit.

LECLANCHE CELL battery or cell constructed of carbon and zinc.

LED light-emitting diode.

LENZ'S LAW describes quantity of current induced in circuit due to variations of magnetic field.

LINEAR IC integrated-circuit unit containing transistors, capacitors, and resistors in the form of amplifiers, oscillators, and rectifiers.

LINEAR RELATION straight-line relation between two variables; an increase in one is directly proportional to an increase in the other.

LINEAR SCALE meter dial face having equal divisions between each marking.

LINE CURRENT current flow in main source of an electrical system supply.

LINE VOLTAGE voltage of the main supply source of an electrical system.

LOAD term used to describe a device using electrical power in the performance of some type of work; often used to describe radios, computers, etc.

LOADED VOLTAGE DIVIDER (*see* VOLTAGE DIVIDER) condition of voltage-divider network when a load is placed across one of its components.

LOADING EFFECT decrease in source voltage as load current increases.

LOOP closed path or series circuit.

LOW-PASS FILTER circuit designed to pass low frequencies and to attenuate high frequencies.

L/R **RISE TIME** relation of inductance and resistance required to increase circuit current to 63.2% of its full value.

LSI large-scale integration of integrated-circuit devices.

MAGNETICALLY HARD term describing magnetic material having high retention ability for magnetic properties.

MAGNETICALLY SOFT term describing type of magnetic material having a low rate of retention of magnetic properties.

MAGNETIC DENSITY quantity of magnetic lines in a specific area.

MAGNETIC FIELD area in which magnetic lines of force exist.

MAGNETIC FLUX magnetic field.

MAGNETIC INDUCTION process of creating a voltage on a secondary metallic surface as magnetic lines of force move across the surface.

MAGNETIC POLE one portion of magnet where lines of force concentrate; may be either the N or the S poles.

MAGNETIC RETENTION describes ability of material to retain its magnetism once magnetizing force is removed.

MAGNETIC STRENGTH magnitude of the magnetic field.

MAGNETISM property of certain materials allowing them to exert forces on other magnetic materials; can also produce voltages in conducting bodies.

MAGNETOMOTIVE FORCE (MMF) ability to create or produce magnetic lines of force; measured in ampere-turn units.

MAGNITUDE size; quantity assigned to one unit for comparison purposes.

MAXWELL (Mx) unit of magnetic flux representing one line of force in a magnetic field.

MEGA (M) prefix used to represent 1,000,000 or 1×10^6.

MESH CURRENT currents in closed circuit or path.

METER MOVEMENT mechanical portion of electrical meter.

METER MULTIPLIER resistance added in series with meter movement used to extend the range of voltage measurements.

METER SHUNT resistance value added to basic meter movement used to increase range of current measurement; normally has value of resistance less than that of the movement.

MICRO measurement prefix equal to 1/1,000,000, or 1×10^{-6}.

MICROFARAD 1/1,000,000 of a farad (unit of capacitance).

MICROPROCESSOR integrated-circuit assembly capable of controlling a computer or other automated equipment.

MILLI (m) measurement prefix equal to 1/1000, or 1×10^{-3}.

MILLIAMMETER measuring instrument capable or measuring current values in units of the milliampere.

MILLMAN'S THEOREM analysis of circuit where two or more power sources are reduced to one for mathematical analysis purposes.

MOTOR device using electrical energy to create rotary motion.

MSI medium scale integration for integrated-circuit devices.

MULTIMETER electrical measuring instrument capable of measuring voltage, current, and resistance; uses range and function switches for selection of proper range and function.

MULTIPLIER series resistance used in voltmeter circuits to extend basic range of measurement.

MUTUAL INDUCTION (Lm) process of magnetic field from one coil of wire inducing a voltage on another coil of wire.

NANO (n) metric prefix for 1/1,000,000,000, or 1×10^{-9}.

NC term for "no connection" used in connector terminology, or "normally closed" in switch terminology.

NEGATIVE FEEDBACK system of returning an out-of-phase signal from the output to the input of a device or circuit.

NODE common connection point for two or more branch circuits.

NONCONDUCTOR material which will not permit the flow of electrons.

NONELECTROLYTIC CAPACITOR capacitor having a dielectric material composed of a nonelectrolyte material.

NONINVERTING operational amplifier circuit where the input and output are in phase.

NONLINEAR SCALE meter dial marking having unequal spcing between markings.

NORTH POLE positive magnetic pole, having north-seeking properties.

NORTON CURRENT current flow in circuit determined by application of Norton's thereorem.

NORTON EQUIVALENT RESISTANCE calculated value of resistance in circuit during application of Norton's theorem.

NORTON'S THEOREM method of reducing complicated electrical circuit to one current source and shunt or parallel resistance.

OERSTED (Oe) unit of magnetic field intensity.

OHM (Ω) unit of electrical opposition or resistance.

OHMMETER measuring instrument used to measure resistance in a circuit.

OHM'S LAW relationship of voltage, current, and resistance in circuit; $V = I \times R$.

OHMS PER VOLT term used to describe sensitivity rating of voltmeter.

OPEN BRANCH CIRCUIT in a parallel circuit, where one of the branches is open.

OPEN CIRCUIT (a) a break in the normal conducting path for electron flow; (b) an incomplete electric circuit; (c) term used to describe a break in normal current flow path, has infinitely high resistance.

OPEN LOOP GAIN amount of amplification in an operational amplifier circuit without any feedback.

OPERATING VOLTAGE voltage value required for proper operation of circuits.

OPERATIONAL AMPLIFIER (op amp) integrated circuit amplifier used to perform mathematic functions, voltage comparison and differentiation.

OSCILLATOR electrical circuit used to create a periodic wave.

OSCILLOSCOPE measuring device used to display voltage waveforms.

OUT OF PHASE reference to two electrical waves starting at different times.

OUTPUT VOLTAGE term used to describe the signal or operating voltage present at the output terminals of a block or stage.

PARALLEL CIRCUIT electrical circuit having two or more branch paths for current flow.

PARALLEL CONNECTION (*see* PARALLEL CIRCUIT)

PARALLEL RESONANT CIRCUIT *RLC* circuit having *L* and *C* components connected in parallel.

PARTIAL SHORT CIRCUIT condition in a parallel circuit where only one of the branches has a short, or in an inductor, where some of the turns of the coil short out.

PASS BAND describes the band of frequencies permitted to pass through a filter network.

PASSIVE DEVICE electrical devices, such as resistors, capacitors, or inductors whose values do not change due to application of voltage or current.

PASSIVE FILTER filter network consisting of passive devices, such as resistors, capacitors, and inductors.

PC BOARD term for printed circuit or wiring board.

PEAK-TO-PEAK VALUE (p-p) amplitude of wave from its positive peak to its negative peak; equal to 2.828 times its rms value.

PEAK VALUE amplitude, either positive or negative, of waveform equal to 1.414 times its rms value.

PERMANENT MAGNET (PM) piece of steel or other magnetic material so strongly magnetized that it will not give up its magnetism.

PERMEABILITY ability of material to concentrate magnetic lines of force.

PERMEANCE ability of material to retain magnetic properties; opposite of magnetic reluctance.

PHASE angular relation of voltage and current in an ac circuit.

PHASE ANGLE angle between phasor and reference line.

PHASOR line used to represent direction and magnitude of voltage or current with reference to time.

PHASOR IMPEDANCE impedance of an ac circuit determined by use of phasors.

PHASOR SUM sum of the phasors used to represent circuit values in an ac circuit.

PHOTODIODE semiconductor diode device having the ability to change its internal resistance due to light striking it.

PHOTOELECTRIC material whose internal resistance changes when exposed to light rays.

PHOTORESISTOR semiconductor device having the ability to change its total resistance value when exposed to a light source.

PHOTOTRANSISTOR light sensitive transistor; its internal resistance changes as light strikes its junctions.

PICO (p) metric unit representing value of 1/1,000,000,000,000, or 1×10^{-12}.

PICOFARAD 1/1,000,000,000,000 of a farad.

PN JUNCTION junction of P-type semiconductor material with N-type material.

POLAR FORM complex number system providing phase angle and magnitude of the number.

POLARITY property of electrical charges; may be either positive (+) or negative (−) in form.

POSITIVE FEEDBACK return of an in-phase signal from the output to the input of a system or device.

POTENTIAL difference in voltage between two points in a circuit.

POTENTIOMETER variable resistance used in circuit to control voltage values.

POWER (P) time rate of performing work in electrical terminology, $P = I \times V$; the unit of power is the Watt.

POWER FACTOR ratio of actual power to apparent power in an electrical system.

POWER FACTOR CORRECTION use of capacitors or inductors to reduce any phase difference in an ac circuit.

POWER RESISTOR resistor having a power rating of 3 watts or more.

POWER SUPPLY device used to create operating power for circuits.

POWER TRANSFER process of moving power from one circuit to a second circuit.

POWER TRANSFORMER device used as a source of proper operating voltage, or power, for an electrical or electronic unit.

PREFERRED VALUES set of values identified by Electrical Industries Association used to establish a standard replacement component value set.

PRIMARY CELL OR BATTERY nonrechargeable chemical device used to create electrical energy.

PRIMARY WINDING input winding of transformer; connected to voltage source.

PRINTED CIRCUIT electric circuit etched on an insulating base material; each path on the etched board is a conductive path.

PRINTED WIRING copper conductive paths on plastic-base material.

PULSATING DC VALUE the ac component on a dc wave or voltage.

PULSE brief but sharp rise or fall of electrical voltage or current.

Q used to describe figure of quality or merit of resonant circuit.

QUALITY (*see* Q)

R symbol for resistance.

RADIAL ASSEMBLY construction where both leads of a resistor or capacitor are located on same end of the device.

RADIO FREQUENCIES (RF) general classification of range of electromagnetic waves having ability to be radiated; generally, those frequencies ranging above 30 kHz.

RCL CIRCUIT resonant circuit containing resistance, capacitance, and inductance.

RC **TIME CONSTANT** time for circuit to charge to 63.2% of its full charge value in a circuit having resistance and capacitance.

REACTANCE opposition of an ac current offered by an inductor or capacitor.

REACTIVE POWER total power available in a circuit containing reactances.

REAL NUMBER in electrical formulas, any number not containing the imaginary *j* value.

REAL POWER unit of electrical power showing net amount of power consumed by resistance in circuit; measured in units of the watt.

RECIPROCAL RELATION condition where as one variable decreases the other variable will increase in value.

RECTANGULAR FORM complex number representation using "$A + jB$."

RECTIFIER device used to control current flow; used to convert ac into pulsating dc for operation of electronic circuits.

RELAY electromagnetic device having a coil and set of contacts used to control circuit connections.

RELUCTANCE opposition to magnetic flux in circuit; corresponds to effects of resistance in electrical circuits.

RESISTANCE (R) opposition to flow of electrical current in units of the ohm.

RESISTANCE WIRE conductor having a value of high resistance.

RESISTOR device made of carbon or resistance wire used to control current flow or develop voltage drops in a circuit.

RESISTOR STRING group of resistors connected in series.

RESONANCE condition of circuit where $X_C = X_L$ in an *RLC* circuit producing a maximum value of *I*, *V*, or *Z*.

RESONANT CIRCUIT circuit containing R, L, and C values that will either accept or reject a specific frequency; frequency depends on specific values of the components.

RESONANT FREQUENCY specific frequency created by combination of L and C in a circuit.

REVERSE VOLTAGE maximum voltage permitted at a PN junction when it is reverse biased.

RHEOSTAT variable resistance used to control circuit current.

RMS VALUE used in ac wave form; represents 0.707 of peak value of wave; also known as the "effective value" of the ac wave.

ROBOTICS term applied to computer control of machinery or machine processing.

ROTATING FIELD describes a motor or generator where the field coils rotate instead of the armature windings.

ROTOR rotary component of motor or generator.

SATURATION condition where changes of input in circuit no longer have an effect on the output.

SAWTOOTH WAVE periodic wave having a linear amplitude of rise and fall time; often one of two times is longer than other.

SCHEMATIC DIAGRAM electrical drawing of the circuit; uses standard symbols to represent components.

SECONDARY CELL OR BATTERY rechargeable chemical voltage source.

SECONDARY LOADING effect of the load resistance value on the output voltage of a transformer's winding.

SECONDARY WINDING coil of wire in transformer connected to load.

SELECTIVITY ability of resonant circuit to accept one specific frequency and reject all other frequencies.

SELF-INDUCTANCE (L) effect of inductance on the coil or wire that produces it.

SEMICONDUCTOR pure chemical material, normally an insulator, having some other type of material added to its chemical composition, changing it into a type of material permitting a controlled amount of electron flow.

SERIES AIDING electrical circuit where two or more sources are connected in manner that results in the sum of their individual values.

SERIES CIRCUIT electrical circuit having one path for current flow.

SERIES CONNECTION (*see* SERIES CIRCUIT)

SERIES OPPOSING circuit connection where two or more power sources are connected so that their polarities oppose each other; the net result is the difference of the two voltages.

SERIES–PARALLEL circuit containing both a series and a parallel section.

SERIES RESISTANCES two or more resistances connected end to end, to form a portion of a series circuit.

SERIES RESONANT RLC circuit where components are connected as a series string.

SHIELD metallic form placed around electrical compo-

nents or wires in order to prevent electrical interference from outside source.

SHORT CIRCUIT undesired very low resistance path in circuit; resistance values of close to zero ohms.

SHORT CIRCUIT BRANCH condition in a parallel circuit where a short circuit exists in one of the branch paths.

SHORT CIRCUIT CURRENT current flow in circuit when load is short circuited and equal to zero ohms.

SHUNT electrical bypass; used in current meters to provide a path around meter movement.

SI abbreviation for Système International used to describe units of measurement.

SIEMENS (S) unit of conductance (formerly called "mho") reciprocal of resistance; $S = 1/R$.

SIGNAL (a) variations in the amplitude of the operating voltage of a device created by the information being processed in the device; (b) information in the form of electrical energy.

SIGNAL VOLTAGE variations in amplitude of operating voltage in unit produced by some input signal.

SILICON (Si) element used in manufacture of semiconductor devices.

SINE function of angle equal to the ratio of the opposite side of the right triangle to its hypotenuse.

SINE WAVE electrical wave created when amplitude varies in proportion to the sine function of the angles of rotation.

SINGLE-PHASE VOLTAGE describes type of voltage obtained from one phase of a generator's rotation.

SINUSOIDAL having the form of a sine wave.

SKIN EFFECT in radio-frequency circuits, the current flow is on the outside of the conductor, or skin of the conductor.

SLIP RINGS used in alternator to provide connection for brushes and permits generated energy to be removed from unit.

SOLENOID electromagnet having hollow core used to convert electrical energy into physical motion.

SOUTH POLE polar region of magnet seeking south; normally considered to be equal to a ($-$) pole of an electromagnet.

SPDT electrical switch contact arrangement; single-pole, double-throw arrangement.

SPECIFIC RESISTANCE value of resistance for a defined length, volume, or area.

SPST electrical contacts; represents single-pole, single-throw set of contacts.

SQUARE WAVE square or rectangular periodic wave.

STATIC ELECTRICITY electrical charges at rest, or not in motion.

STATOR stationary portion of rotary motor or generator.

STEP-DOWN TRANSFORMER device in which the secondary voltage is less than that of the primary voltage value.

STEP-UP TRANSFORMER device where secondary winding voltage is greater than the value of the primary voltage.

STOP BAND FILTER electronic filter network designed to attenuate specific frequencies.

STORAGE CELL OR BATTERY chemical device for creating electrical energy; can be recharged and reused.

SUBSYSTEM functional portion of a total system.

SUPERPOSITION THEOREM used to analyze a network containing multiple power sources by isolating sources, solving for individual currents and then adding results.

SUSCEPTANCE (B) reciprocal of reactance in an ac circuit; $B = 1/X$.

SWITCH mechanical device used to control power or to direct signals in an electric or electronic device.

SWITCHING TRANSISTORS circuit using transistors operating at both cutoff and saturation; acts as a switch for voltages in the circuit.

TANK CIRCUIT electrical circuit consisting of inductance, resistance, and capacitance; also called resonant circuit.

TANTALUM component used in the construction of some types of electrolytic capacitors.

TAPPED WINDINGS connections to winding of transformer located between start and end of a winding used to provide a portion of the total output value of the winding.

TEMPERATURE COEFICIENT description of how value of component varies with changes in its temperature.

TESLA (T) unit of flux density in a magnetic circuit.

THERMISTOR resistance able to change its value due to changes in its temperature.

THÉVENIN EQUIVALENT CIRCUIT reduction of circuit to its Thévenin values.

THÉVENIN RESISTANCE resistance value of circuit after being Théveninized.

THÉVENIN'S THEOREM method of reducing a complex electrical circuit to one having a single voltage source and a series resistance.

THÉVENIN VOLTAGE quantity of voltage present in Théveninized circuit.

THREE-PHASE POWER electrical power generated by three windings on the generator spaced 120° apart.

TIME CONSTANT unit used to describe rise of voltage or current to 63.2% of maximum or fall of V or I by 63.2% of its maximum value.

TOLERANCE allowable deviation from the marked value of a component.

TOROID doughnut-shaped electromagnetic form.

TOTAL RESISTANCE (R_T) term used to describe the total quantity of resistance in a circuit.

TRANSDUCER device used to convert one form of energy into another form; in electrical work, used to create a voltage for the input to the circuit or to change a voltage into another form for output purposes.

TRANSFORMER device having a mininum of two windings used to isolate, step up, or step down voltage values in an ac circuit.

TRANSIENT value of V or I in circuit caused by a momentary abrupt change.

TRANSISTOR semiconductor device used to replace vacuum tubes; has a minimum of three active elements.

TRICKLE CHARGER device used to recharge cells; uses a slow current rate, or trickle, of electrons to recharge the cells.

TRIGONOMETRY mathematical analysis of angles and triangles.

TUNED CIRCUIT resonant circuit.

TURNS RATIO relationship of windings of wire on transformer primary and secondary coils.

UHF ultrahigh frequency; Ranges from 300 MHz to over 1000 MHz.

UNIVERSAL TIME-CONSTANT CHART chart showing percentages required during five time constants for voltage or current to reach 100% value; used in reactive circuits.

VARACTOR CAPACITOR solid-state capacitor whose capacitance value is changed as a voltage applied to its plates is varied.

VARIABLE RESISTOR resistance unit whose total resistance may be reduced from a maximum value by adjustment.

VECTOR line used to represent direction and magnitude in space.

VHF description of range of radio frequencies; very high frequencies, ranging between 30 and 300 MHz.

VLSI very-large-scale integrated circuit.

VOLT (V) unit of electrical pressure; describes the difference in electrical potential between two points in a circuit.

VOLTAIC CELL term used to describe a basic chemical cell, often used as a standard for voltage references.

VOLTAGE (V) difference in electrical pressure measured between two points in a circuit.

VOLTAGE BREAKDOWN RATING rated safe value of operating voltage for a device.

VOLTAGE COMPARATOR operational amplifier circuit where the difference in voltage at the two input terminals determines the phase and amplitude of the output signal.

VOLTAGE DIVIDER circuit consisting of two or more series-connected components used to establish voltage levels less than the input value of voltage to the circuit.

VOLTAGE DROP difference in electrical potential between two points in a circuit caused by a current flow through a resistance or load.

VOLTAGE DROP POLARITIES assignment of either a (+) or (−) sign to a component in a circuit; specific sign depends on relationship of the end of the component to the power source.

VOLTAGE REGULATOR device used to maintain a constant voltage at some point in the circuit; will do this as input voltage or output load conditions vary.

VOLTAGE SOURCE unit used to provide electrical energy to load; establishes potential difference across its two output terminals.

VOLTAMPERE (VA) unit of apparent electrical power; equal to $I \times V$.

VOLTMETER measuring instrument used to indicate amplitude of voltage in circuit.

VOLTMETER LOADING effect on circuit being measured by addition of voltmeter; usually provides a less-than-normal reading.

VOM instrument used to measure voltage, current, and resistance; has multiple ranges and functions.

WATT (W) unit of electrical power; $P = I \times V$.

WATTMETER instrument used to measure electrical power.

WATT'S LAW relationship of voltage, current, and power as defined by Watt.

WAVEFORM shape of electrical wave plotted against time using rectangular coordinates.

WAVELENGTH used to describe the distance between two identical points in a repeating electrical wave.

WEBER (Wb) unit of magnetic flux.

WET CELL term used to describe a rechargeable chemical cell using a liquid electrolyte.

WIRE GAGE standard system for measurement and identification of wires and capacities.

WORK in electrical terms, the result of a current flowing through a device, the result being a functional activity.

WORKING VOLTAGE RATING one term used to describe the characteristics of a capacitor; the maximum safe value for voltage applied to its plates.

WYE NETWORK three components connected in shape of the letter Y, where all three have one common connection in the center of the configuration.

X_C capacacitive reactance; $X_C = 1/(2\pi f C)$.

X_L inductive reactance; $X_L = 2\pi f L)$.

Y symbol for admittance in the ac circuit; also reciprocal of impedance; $Y = 1/Z$.

Y NETWORK term also used to describe the wye-circuit configuration.

Z symbol for impedance; opposition to current in an ac circuit.

ZENER DIODE semiconductor diode used as a voltage-regulating device in a circuit.

ZERO-OHMS RESISTOR a jumper wire contained in the body of a resistor; required for machine insertion of components in circuit boards.

ZERO-OHMS VALUE quantity of resistance having almost no measurable amount of resistance.

ZINC CHLORIDE CELL chemical cell using zinc and zinc chloride as a basis for creating electrical energy.

APPENDIX B
IEEE Symbols (selected values)

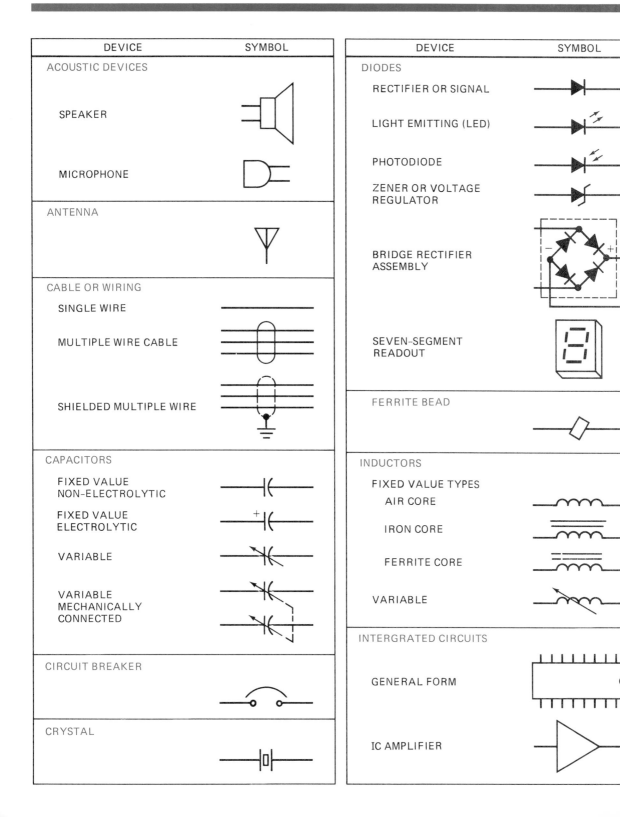

DEVICE	SYMBOL	DEVICE	SYMBOL
ACOUSTIC DEVICES		DIODES	
SPEAKER		RECTIFIER OR SIGNAL	
MICROPHONE		LIGHT EMITTING (LED)	
ANTENNA		PHOTODIODE	
		ZENER OR VOLTAGE REGULATOR	
CABLE OR WIRING		BRIDGE RECTIFIER ASSEMBLY	
SINGLE WIRE			
MULTIPLE WIRE CABLE		SEVEN-SEGMENT READOUT	
SHIELDED MULTIPLE WIRE		FERRITE BEAD	
CAPACITORS		INDUCTORS	
FIXED VALUE NON-ELECTROLYTIC		FIXED VALUE TYPES AIR CORE	
FIXED VALUE ELECTROLYTIC		IRON CORE	
VARIABLE		FERRITE CORE	
VARIABLE MECHANICALLY CONNECTED		VARIABLE	
CIRCUIT BREAKER		INTERGRATED CIRCUITS	
		GENERAL FORM	
CRYSTAL		IC AMPLIFIER	

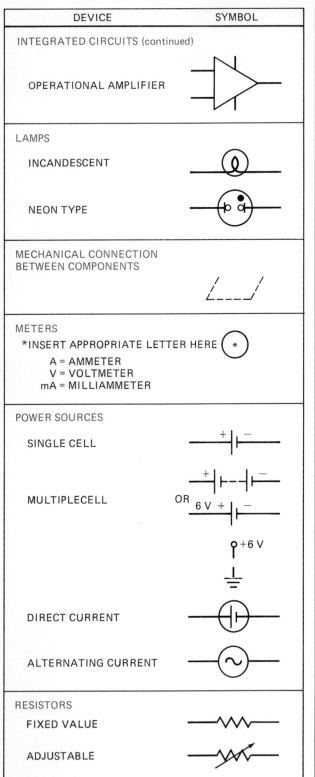

DEVICE	SYMBOL
INTEGRATED CIRCUITS (continued)	
OPERATIONAL AMPLIFIER	
LAMPS	
INCANDESCENT	
NEON TYPE	
MECHANICAL CONNECTION BETWEEN COMPONENTS	
METERS	
*INSERT APPROPRIATE LETTER HERE	
A = AMMETER	
V = VOLTMETER	
mA = MILLIAMMETER	
POWER SOURCES	
SINGLE CELL	
MULTIPLECELL	OR
DIRECT CURRENT	
ALTERNATING CURRENT	
RESISTORS	
FIXED VALUE	
ADJUSTABLE	

DEVICE	SYMBOL
RESISTORS (continued)	
VARIABLE	
THERMISTOR	
VARISTOR	
TAPPED	
PHOTO RESISTOR	
SWITCHES, CONTACTS	
SINGLE POLE SINGLE THROW (SPST)	
DOUBLE POLE SINGLE THROW (DPST)	
SINGLE POLE DOUBLE THROW (SPDT)	
DOUBLE POLE DOUBLE THROW (DPD)	
MULTIPLE CONTACT ROTARY	
TRANSFORMERS	PRIMARY SECONDARY
GENERAL SYMBOL	
IRON CORE	
FERRITE CORE	
TAPPED SECONDARY	

DEVICE	SYMBOL
TRANSFORMERS (continued) SHIELDED	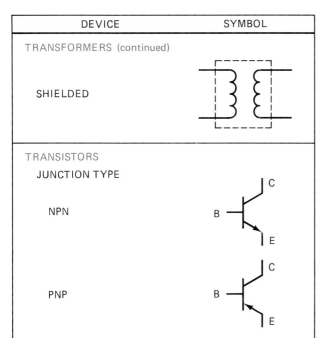
TRANSISTORS JUNCTION TYPE NPN PNP	

DEVICE	SYMBOL
TRANSISTORS (continued) FIELD EFFECT TYPE N CHANNEL P CHANNEL	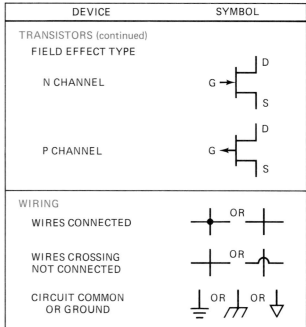
WIRING WIRES CONNECTED WIRES CROSSING NOT CONNECTED CIRCUIT COMMON OR GROUND	

APPENDIX C
Trigonometric Functions

Angle	Sin	Cos	Tan	Angle	Sin	Cos	Tan
0°	0.0000	1.000	0.0000	45°	0.7071	0.7071	1.0000
1	0175	.9998	.0175	46	.7193	.6947	1.0355
2	0349	.9994	.0349	47	.7314	.6820	1.0724
3	0523	.9986	.0524	48	.7431	.6691	1.1106
4	0698	.9976	.0699	49	.7547	.6561	1.1504
5	0872	.9962	.0875	50	.7660	.6428	1.1918
6	.1045	.9945	.1051	51	.7771	.6293	1.2349
7	.1219	.9925	.1228	52	.7880	.6157	1.2799
8	.1392	.9903	.1405	53	.7986	.6018	1.3270
9	.1564	.9877	.1584	54	.8090	.5878	1.3764
10	.1736	.9848	.1763	55	.8192	.5736	1.4281
11	.1908	.9816	.1944	56	.8290	.5592	1.4826
12	.2079	.9781	.2126	57	.8387	.5446	1.5399
13	.2250	.9744	.2309	58	.8480	.5299	1.6003
14	.2419	.9703	.2493	59	.8572	.5150	1.6643
15	.2588	.9659	.2679	60	.8660	.5000	1.7321
16	.2756	.9613	.2867	61	.8746	.4848	1.8040
17	.2924	.9563	.3057	62	.8829	.4695	1.8807
18	.3090	.9511	.3249	63	.8910	.4540	1.9626
19	.3256	.9455	.3443	64	.8988	.4384	2.0503
20	.3420	.9397	.3640	65	.9063	.4226	2.1445
21	.3584	.9336	.3839	66	.9135	.4067	2.2460
22	.3746	.9272	.4040	67	.9205	.3907	2.3559
23	.3907	.9205	.4245	68	.9272	.3746	2.4751
24	.4067	.9135	.4452	69	.9336	.3584	2.6051
25	.4226	.9063	.4663	70	.9397	.3420	2.7475
26	.4384	.8988	.4877	71	.9455	.3256	2.9042
27	.4540	.8910	.5095	72	.9511	.3090	3.0777
28	.4695	.8829	.5317	73	.9563	.2924	3.2709
29	.4848	.8746	.5543	74	.9613	.2756	3.4874
30	.5000	.8660	.5774	75	.9659	.2588	3.7321
31	.5150	.8572	.6009	76	.9703	.2419	4.0108
32	.5299	.8480	.6249	77	.9744	.2250	4.3315
33	.5446	.8387	.6494	78	.9781	.2079	4.7046
34	.5592	.8290	.6745	79	.9816	.1908	5.1446
35	.5736	.8192	.7002	80	.9848	.1736	5.6713
36	.5878	.8090	.7265	81	.9877	.1564	6.3138
37	.6018	.7986	.7536	82	.9903	.1392	7.1154
38	.6157	.7880	.7813	83	.9925	.1219	8.1443
39	.6293	.7771	.8098	84	.9945	.1045	9.5144
40	.6428	.7660	.8391	85	.9962	.3872	11.43

Angle	Sin	Cos	Tan	Angle	Sin	Cos	Tan
41	.6561	.7547	.8693	86	.9976	.0698	14.30
42	.6691	.7431	.9004	87	.9986	.0523	19.08
43	.6820	.7314	.9325	88	.9994	.0349	28.64
44	.6947	.7193	.9657	89	.9998	.0175	57.29
				90	1.0000	.0000	∞

APPENDIX D
Copper Wire Table

Standard Annealed Copper Wire Solid American wire gage

1	2	3	4	5	6	7	8
						Allowable Current Capacity (A)	
Gage Number	Diameter (mils)	Area (cir mils)	Resistance [Ω/1000 ft, 25°C (77°F)]	Weight (lb/1000 ft)	Rubber Insulation	Varnished Cambric Insulation	Other Insulations
0000	460.0	211,600.0	0.0500	641.0	225	270	325
000	410.0	167,800.0	0.0630	508.0	175	210	275
00	365.0	133,100.0	0.0795	403.0	150	180	225
0	325.0	105,500.0	0.100	319.0	125	150	200
1	289.0	83,690.0	0.126	253.0	100	120	150
2	258.0	66,370.0	0.159	201.0	90	110	125
3	229.0	52,640.0	0.201	159.0	80	95	100
4	204.0	41,740.0	0.253	126.0	70	85	90
5	182.0	33,100.0	0.319	100.0	55	65	80
6	162.0	26,250.0	0.403	79.5	50	60	70
7	144.0	20,820.0	0.508	63.0			
8	128.0	16,510.0	0.641	50.0	35	40	50
9	114.0	13,090.0	0.808	39.6			
10	102.0	10,380.0	1.02	31.4	25	30	30
11	91.0	8,234.0	1.28	24.9			
12	81.0	6,530.0	1.62	19.8	20	25	25
13	72.0	5,178.0	2.04	15.7			
14	64.0	4,107.0	2.58	12.4	15	18	20
15	57.0	3,257.0	3.25	9.86			
16	51.0	2,583.0	4.09	7.82	6		
17	45.0	2,048.0	5.16	6.20			
18	40.0	1,624.0	6.51	4.92	3		
19	36.0	1,288.0	8.21	3.90			
20	32.0	1,022.0	10.4	3.09			
21	28.5	810.0	13.1	2.45			
22	25.3	642.0	16.5	1.95			
23	22.6	509.0	20.8	1.54			
24	20.1	404.0	26.2	1.22			
25	17.9	320.0	33.0	0.970			
26	15.9	254.0	41.6	0.769			
27	14.2	202.0	52.5	0.610			
28	12.6	160.0	66.2	0.484			
29	11.3	127.0	83.4	0.384			
30	10.0	100.0	105.0	0.304			

Standard Annealed Copper Wire Solid American wire gage (*continued*)

1	2	3	4	5	6	7	8
						Allowable Current Capacity (A)	
						Varnished	
Gage Number	Diameter (mils)	Area (cir mils)	Resistance [Ω/1000 ft, 25°C (77°F)]	Weight (lb/1000 ft)	Rubber Insulation	Cambric Insulation	Other Insulations
31	8.9	79.7	133.0	0.241			
32	8.0	63.2	167.0	0.191			
33	7.1	50.1	211.0	0.152			
34	6.3	39.8	266.0	0.120			
35	5.6	31.5	335.0	0.0954			
36	5.0	25.0	423.0	0.0757			
37	4.5	19.8	533.0	0.0600			
38	4.0	15.7	673.0	0.0476			
39	3.5	12.5	848.0	0.0377			
40	3.1	9.9	1070.0	0.0299			

APPENDIX E
Electronic Industries Coding Systems

E-1 Basic Color Coding System

Color	Significant Figure	Decimal Multiplier		Tolerance	Voltage Rating
Black	0	1		±20%	—
Brown	1	10	1×10^2	±1%	100
Red	2	100	1×10^3	±2%	200
Orange	3	1,000	1×10^4	±3%	300
Yellow	4	10,000	1×10^5	GMV[a]	400
Green	5	100,000	1×10^6	±5%	500
Blue	6	1,000,000	1×10^7	±6%	600
Violet	7	10,000,000	1×10^8	±12.5%	700
Gray	8	0.01†	1×10^{-2}	±30%	800
White	9	0.1†	1×10^{-1}	±10%	900
Gold	—	0.1	1×10^{-1}	±5%	1000
Silver	—	0.01	1×10^{-2}	±10%	2000
No color	—	—	—	±20%	500

[a]GMV, Guaranteed minimum value.
†Optional coding.

E-2 Solid-State Component Color Coding

Many solid-state components are so small that it is almost impossible to print their part numbers on the body of the device. The standard color coding system uses the JEDEC numbering system to assume that the first number and letter are 1N for a diode. In this marking system, the color bands are grouped near the marking indicating the cathode end of the diode. Each band represents a significant digit, or number, in this system. There is no multiplier band for diode markings. The system uses a number, number, letter designation. When more than three color bands are used, the band farthest away from the cathode-marked end of the diode is always the letter designator. All other bands represent numbers, since diodes may have two, three, or four numbers.

Color	Digit	Suffix Letter
Black	0	—
Brown	1	A
Red	2	B
Orange	3	C
Yellow	4	D
Green	5	E
Blue	6	F
Violet	7	G
Gray	8	H
White	9	J

E-3 Standard Color Coding Systems for Capacitors and Inductors

There are several methods in use for marking the values of ceramic, mica, and tubular capacitors using a color-coding system. Many capacitors have a temperature rating in addition to their value and tolerance. The illustrations included in this appendix show some of the more common methods of marking on these devices. The information in this section will aid in an understanding of how to recognize the marking system in order to "read" the values on the capacitors and inductors.

Color Code for Ceramic Capacitors

Color	Significant Figure	Decimal Multiplier	Capacitance Tolerance More Than 10 pF (%)	Less Than 10 pF (%)	Temp. Coeff. (ppm/°C)
Black	0	1	±20	2.0	0
Brown	1	10	±1		−30
Red	2	100	±2		−80
Orange	3	1000			−150
Yellow	4				−220
Green	5				−330
Blue	6		±5	0.05	−470
Violet	7				−750
Gray	8	0.01		0.25	30
White	9	0.1	±10	1.0	500

Capacitor Characteristic Code

Color Sixth Dot	Temperature Coefficient (ppm/°C)	Capacitance Drift
Black	±1000	±5% + 1 pF
Brown	±500	±3% + 1pF
Red	±200	±0.5%
Orange	±100	±0.3%
Yellow	−20 to +100	±0.1% + 0.1 pF
Green	− to +70	±0.05% + 0.1 pF

EIA Temperature Characteristic Code for Disk Ceramic Capacitors

Minimum Temperature (°C)		Maximum Temperature (°C)		Maximum Cap. Change Over Temp. Range	
X	−55	2	+45	A	±1.0%
Y	−30	4	+65	B	±1.5%
Z	+10	5	+85	C	±2.2%
		6	+105	D	±3.3%
		7	+125	E	±4.7%
				F	±7.5%
				P	±10%
				R	±15%
				S	±22%
				T	−33%, +22%
				U	−56%, +22%
				V	−82%, +22%

EIA Designations for Capacitor Temperature Coefficient

Industry	EIA
NP0	COG
N033	S1G
N075	U1G
N150	P2G
N220	R2G
N330	S2H
N470	T2H
N750	U2J
N1000	P3K
N2200	R3L

EIA Capacitance Tolerance Codes

Letter	Tolerance
C	$\pm\frac{1}{4}$ pF
D	$\pm\frac{1}{2}$ pF
F	± 1 pF or $\pm 1\%$
G	± 2 pF or $\pm 2\%$
J	$\pm 5\%$
K	$\pm 10\%$
L	$\pm 15\%$
M	$\pm 20\%$
N	$\pm 30\%$
P	$-0\%, +100\%$
W	$-20\%, +40\%$
Y	$-20\%, +50\%$
Z	$-20\%, +80\%$

E-4 Power Transformer Lead Color Coding

Transformers used to provide operating power to electronic devices have an EIA standard color coding for their leads. This coding system aids in the identification of the proper pair, or set, of leads connected to a specific winding on the transformer.

1. *Primary winding*: black (if tapped connections are used, then black is common)
2. *Tap*: black-and-yellow-striped
3. *Finish*: black-and-red-striped
4. *High-voltage secondary winding*: red with a red-and-yellow-striped lead for the center tap, if used
5. *Rectifier filament (if used)*: yellow with a yellow-and-blue-striped lead for the center tap, if used
6. *Filament winding 1*: green with a green-and-yellow-striped lead for the center tap
7. *Filament winding 2*: brown with a brown-and-yellow-striped lead for the center tap
8. *Filament winding 3*: slate with a slate-and-yellow-striped lead for the center tap

APPENDIX F
Electrical/Electronic Safety Rules

Perhaps one of the most common occurring accidents in the field of electrical work is related to electrical shock. Many of the people who have worked in this occupation have a tendency to overlook the possibility of permanent physical damage or death because of accidental contact with electricity. The technician has to have a healthy respect for safety and must always work with safe work habits. Some of the more common safety-related rules are presented in this section. Learn to use them and to respect them. You may not have a second chance if you make a mistake because you were not careful!

1. Never work alone. Always have someone else near or working with you in case of an accident.
2. Use electrical safety devices, such as ground-fault interrupters and three-wire electrical power systems.
3. Use an isolation transformer when working on ac-dc-operated devices.
4. Learn to short circuit the operating supply to circuit common before working on a circuit that is turned off.
5. Be careful when using tools that may cause a short circuit.
6. Always work with one hand. The other hand should **not** be touching a metallic surface.
7. Wear safety glasses when working on an environment where flying objects are present. This includes soldering and cutting the ends of wires after they are connected to the circuits.
8. Remove the victim of an electrical shock from the power source without becoming a victim yourself.
9. Do not engage in horseplay or any game type of activity when someone is working on an electrical circuit.

APPENDIX G
Conversion Units Used in Electrical Systems

Element	Unit	Milli Unit	Micro Unit	Pico Unit	Kilo Unit	Mega Unit
Inductance	1 henry	10^{-3} mH	10^{-6} μH	10^{-12} pH		
Capacitance	1 farad	10^{-3} mF	10^{-6} μF	10^{-12} pF		
Conductance	1 siemens	10^{-3} S	10^{-6} μS			
Resistance					10^3 kΩ	10^6 MΩ
Voltage	1 volt	10^{-3} mV	10^{-6} uV		10^3 kV	10^6 MV
Current	1 ampere	10^{-3} mA	10^{-6} μA		10^3 kA	
Power	1 watt	10^{-3} mW	10^{-6} μW		10^3 kW	10^6 MW
Frequency	1 hertz	10^{-3} mHz	10^{-6} μHz		10^3 kHz	10^6 MHz
Time	1 second	10^{-3} ms	10^{-6} μs			

APPENDIX H
U.S. Customary-to-Metric Conversions (SI Metric Units)

Prefix	Symbol	Multiplication Factor	
exa	E	10^{18}	= 1,000,000,000,000,000,000
peta	P	10^{15}	= 1,000,000,000,000,000
tera	T	10^{12}	= 1,000,000,000,000
giga	G	10^{9}	= 1,000,000,000
mega	M	10^{6}	= 1,000,000
kilo	k	10^{3}	= 1,000
hecto	h	10^{2}	= 100
deca	da	10^{1}	= 10
unit		10^{0}	= 1
deci	d	10^{-1}	= 0.1
centi	c	10^{-2}	= 0.01
milli	m	10^{-3}	= 0.001
micro	μ	10^{-6}	= 0.000,001
nano	n	10^{-9}	= 0.000,000,001
pico	p	10^{-12}	= 0.000,000,000,001
femto	f	10^{-15}	= 0.000,000,000,000,001
atto	a	10^{-18}	= 0.000,000,000,000,000,001

APPENDIX I
Basic Formulas for Series and Parallel Component Values

The following formula is used by adding the individual component values:

$$n_t = n_1 + n_2 + n_3 + \cdots + n_n$$

Resistance of resistors in series
Voltage of sources or cells in series
Inductances of coils in series
Inductive reactances of coils in series
Capacitance values of capacitors in parallel
Capacitance reactances of capacitors in series
Current of resistances in parallel

The following formula is used by adding the reciprocal values of the components:

$$\frac{1}{n_t} = \frac{1}{n_1} + \frac{1}{n_2} + \frac{1}{n_3} + \cdots + \frac{1}{n_n}$$

Resistance of resistors in parallel
Capacitance of capacitors in series
Capacitive reactance of capacitors in parallel
Inductive reactance of coils in parallel
Inductance of coils in parallel

APPENDIX J
Chip Component Identification and Replacement

Lead length of components used in VHF and UHF as well as microwave circuits is very critical. The length of wire used to connect a component to a circuit board may affect its frequency of operation. Very small rectangular devices, known as "chip" components, are used to overcome this undesirable affect. This type of component is soldered directly to the surface of the circuit board. Components manufactured as chip components include resistors, capacitors, and transistors. Replacement components other than transistors use a general type of marking system in order to indicate their values. These are indicated below.

J-1 Chip Capacitors

The capacitive value of replacement types of chip capacitors is indicated on the bottom surface of the capacitor. The system uses a two-digit set of numbers to indicate the value of the capacitor. When the value is less than 100 pF, the value is indicated by a one- or two-digit number expressing capacity in units of picofarads. For example:

0.5	0.5 pF
0.75	0.75 pF
1.0	1.0 pF
2.5	2.5 pF
33.0	33.0 pF
82.0	82.0 pF

When the capacitive value is 100 pF or greater, the value is indicated by an alphanumeric coding. The letter precedes the number and is used to express a numerical value to be multiplied by the following number.

Letter	Value	Letter	Value
A	10	N	33
B	11	P	36
C	12	Q	39
D	13	R	43
E	15	S	47
F	16	T	51
G	18	U	56
H	20	V	62
J	22	W	68
K	24	X	75
L	27	Y	82
M	30	Z	91

Note: I and O are not used.

For example:

$$A1 = 10 \times 10^1 = 100 \text{ pF}$$
$$N2 = 33 \times 10^2 = 3300 \text{ pF}$$
$$S3 = 47 \times 10^3 = 47,000 \text{ pF}$$

J-2 Chip Resistors

The value of the chip resistor is marked on the bottom surface of the resistor. A three-digit number system is used. This system also uses the standard EIA color coding number system. The value is read using a digit, digit, multiplier number set. For example:

$$330 = 33 \times 10^0 \ \Omega$$
$$561 = 56 \times 10^1 \ \Omega$$
$$123 = 12 \times 10^3 \ \Omega \text{ or } 12 \text{ k}\Omega$$

Note: A 0-Ω resistor (jumper wire) has a solid color body. It is either red or green in color.

J-3 Chip Replacement Procedures

Use the following procedure to remove a chip component.

1. Heat and add solder to all connections of the chip to be removed. (The solder holds the heat and makes chip removal easier.)
2. Grasp the chip with a pair of tweezers and alternately heat all of the chip connections until all solder is melted.
3. Carefully "twist" the chip to remove it from the board. It may be held in place with an epoxy glue. Do not attempt to remove the chip from the board until after it has been loosened by the twisting action. The board foil may be damaged if this is not followed (see Figure J-1).

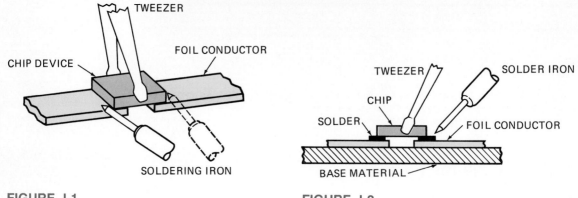

FIGURE J-1

FIGURE J-2

J-4 Chip Installation

Use the following procedure to replace a chip component.

1. Hold the chip in place using a pair of tweezers. Tack-solder one contact to the PC foil. Be sure to tin the copper foil first (see Figure J-2).
2. Solder the remaining contact points to the foil using a small-diameter solder of 0.5 mm or smaller (see Figure J-3).

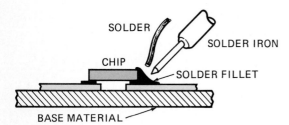

FIGURE J-3

APPENDIX K
The Troubleshooting Process

The process used to diagnose and repair any electrical or electronic device is similar to the process required to service any other device. It involves some basic steps that will apply to almost every type of service-related problem. This approach includes the requirements for "debugging" a new circuit or the repair of a simple electrical switching circuit. The basic steps are described as follows:

1. Always consider safe work habits as a primary step.
2. Use the senses you have for an initial approach to troubleshooting and ultimate repair. These senses include sight, hearing, smell, feel, and touch.
   (a) Use your eyes to look for basic signs of problems. Are the connections required for an input and an output actually in place? Is the connection to the power source connected? (Many consumer repairs have been accomplished simply by connecting the power cord to the wall outlet!) Look for charred or discolored circuit boards or components as areas of trouble. There is a humorous story being told stating that all electrical and electronic parts are manufactured with a quantity of "smoke" inside them. When these parts are operated correctly, the smoke remains inside. If they are malfunctioning some of the smoke comes out. If all the smoke is let out, the part is no longer functional. Although, this is not a true story, it does highlight the relationship of smoke to a component that has failed.
   (b) Listen for snapping noises, indicating an arcing condition due to a broken or disconnected wire. A sizzling sound is usually indicative of a burning component.
   (c) Smell for trouble. The odor of hot electronic components is unique. Once you have had the privilege of smelling this, you will never forget it. The nose is also able to detect heat. When a component that is hot is placed under the nose, the difference in temperature is detected.
   (d) Touch parts to find out if they are hotter than normal. This of course, must be done with care. Electrical components often do become warm as current flows through them. Some heat is normal, but an overheated component may be discovered in this manner.
   (e) The most important sense you have is in your brain. You have the ability to analyze each of the senses and relate the analysis to a determination of either correct or incorrect activity.

The troubleshooting process should be accomplished using a very specific procedure. Actually, in the beginning the procedure seems to be very slow. As you learn to follow its steps, it becomes second nature and very easy to use. It will also permit you to be very successful in the area of diagnostics and repair.

The first statement to be recognized is that any electrical or electronic device produced commercially did function properly at one time. This device does not have any human qualities and cannot think. You, the service person, are the only one able to use a brain to rationally evaluate measurements and functions.

Once you have reminded yourself of the statement in the preceding paragraph, the next step is to develop an understanding of how the device is supposed to work. Analyze the operation of the device in its present condition to determine which of its normal functions are not operating correctly.

Use the literature provided by the manufacturer (or other sources) to identify sections or blocks of the unit. Each block is related to a function. The block diagram becomes a road map for operation of the unit. Signal paths are shown in this diagram. An intimate knowledge of all electronic circuits is not required to use the block diagram and see how it related to functions. Eliminate all sections of the device that seem to be functioning correctly. Any problems are not related to these sections.

Use an instinctive approach. Use the indicators on the front panel or an operating control board as an aid in relating problems to blocks. The indicators become one end of the function section of a block diagram. The block diagram will show the blocks used to control the indicators. This information aids in recognition of the problem area.

Develop the concept of the specific types of signal flow and current flow path systems. There are only a few of these. Use this knowledge to develop a systematic approach to troubleshooting. As an example, the linear (or in line) flow path has only a single row of blocks. If this is the type of flow path, the suspected area may be identified as being from the start to the finish of this path. A set of brackets may be considered to be around this section. If there is a valid input and no output, the first major test to make is in the *middle* of the system. This test splits the path into two parts. A valid signal at the midpoint indicates that the first half of the system is OK. The problem is now narrowed down to the second half of the system. Again, a test is made at the midpoint of the second half. The results of this test determine where the next test is to be made.

Each test becomes a very efficient method of locating the problem area. When the test is made at the midpoint of the system, one-half is immediately eliminated. This is much more efficient than a process requiring testing of each section. The truth of this statement is apparent when you consider a system having 20 or more blocks in a linear flow path. The first tests eliminates 10 of these blocks. The second test eliminates another 5 blocks. Two test have eliminated 15 of the 20 blocks from the problem area. This method is much faster that a block-by-block testing system.

After the specific problem block is identified, the system required is dependent on the type of equipement. If the block is related to a plug-in module or board, the simplest method of immediate repair is replacement of the module or board. This process returns the equipment to operational level as quickly as possible. At a later time the service department will probably want to inspect and possibly repair the defective item.

When the components are not on a plug-in board, a schematic diagram is required for proper circuit analysis. As stated earlier in the book, there are but a few basic circuits in use. The difference comes in their applications. The schematic diagram is used to identify the current paths in the device. Current flow diagnosis is similar to that used for signal processing. Determine the path for current flow and measure voltage drops against the values provided by the manufacturer in the service literature. Any major differences indicate an abnormal current path. Diagnosis of this path with knowledge of voltage-drop development and open-circuit conditions will usually locate the defective components.

Learn to use your knowledge of component failure as an aid in the repair of any circuit. For example, resistors seldom fail in an operational mode. This is not to say that they *never* fail, but normally, some other component failure will cause a resistor failure. Learn to look for these major causes of failure that result in the secondary component failure. Replacing both devices is required for a proper repair. When only the resistor is replaced, the unit is certain to fail again very soon. Recalls are expensive and nonproductive. Learn to minimize them.

Finally test the unit for a reasonable period of time before it is returned to operation. This will indicate that it does not have additional problems. It also reduces the probability of a callback to repair the unit properly.

APPENDIX L
Answer to Selected Problems

CHAPTER 2

2–11. b
2–13. c
2–15. a
2–17. d
2–19. a
2–39. Amount of current flow when 1 V is applied across 1 Ω resistance
2–41. 1/1,000,000 or 1×10^{-6} A
2–43. 46.67 A
2–45. 6.08C
2–47. Resistance
2–49. Ac changes in magnitude and polarity, dc does neither
2–53. (a) 5c; (b) 2c; (c) 6c; (d) 30c; (3) 220c
2–55. 2 A
2–57. 40 mA
2–59. (a) 6.67 mA; (b) 66.67 μA

CHAPTER 3

3–11. d
3–13. c
3–15. c
3–17. a
3–19. a
3–31. Three pole, double throw
3–33. 24 V = 0.24 A; 100 V = 1 A; 400 V = 4 A; 9 V = 0.009 A
3–37. Open
3–39. #12 wire
3–41. 9.76 Ω
3–43. 33.08 Ω
3–45. 48.8 kV to 182.07 kV
3–47. No, fuse voltage should be zero.

CHAPTER 4

4–11. c
4–13. a

4–15. b
4–17. d
4–19. a
4–25. $R_T = 20$ kΩ, $P = 10$ W
4–27. 2.5 W
4–28. Rheostat used to vary current, potentiometer used to vary voltage
4–31. 33 k $\pm$ 10%; 100 k $\pm$ 20%; 27 $\pm$ 5%; 9 k $\pm$ 5%; 510 $\pm$ 5%
4–33. 250 mW; 14.4 kW; 1.2 kW; 211.764 mW; 4.7 mW
4–35. $R = 1$ kΩ; $P = 250$ W
4–37. One R is shorted
4–39. No, it is out of tolerance
4–41. No, out of tolerance
4–43. No, maximum value is zero ohms

CHAPTER 5

5–11. b
5–13. d
5–15. a
5–17. b
5–19. b
5–31. 60,000 V
5–33. 768 W
5–35. 6 A; 2.4 A; 1 A; 333 mA
5–37. 5 Ω; 50 Ω; 25 Ω; 62.5 kΩ
5–39. 900 V
5–41. (a) 150 kW; (b) 500 kW; (c) 8.7 kW; (d) 0.18 kW
5–43. (a) 6300 V; (b) 0.49 V; (c) 0.001683 V; (d) 180 V
5–45. 90 mA
5–47. 500 mA
5–49. 960 W
5–51. 3 Ω
5–53. 17.14 Ω
5–55. 31.2 V
5–57. 33.3 μA

5–61. To stop any current flow through the chest area of your body

5–63. 0.010 A

5–65. Remove source of shock, use artificial respiration, call for help

5–67. Open

5–69. Open across the second resistor or a short across the first one

CHAPTER 6

6–11. c

6–13. d

6–15. d

6–17. c

6–19. a

6–31. 60 V

6–33. 44 V

6–35. 66 W

6–37. 10 V

6–39. 50 V

6–41. 15 kΩ

6–43. 4.5 V

6–45. 50 mA

6–47. 12.5 V

6–49. 3 V

6–51. 16.67%

6–53. 4 mA

6–55. 54 V

6–57. 240 mW

6–59. 4 mA

6–61. 25 kΩ

6–63. 250 V

6–65. $V_{R1} = 75$; $V_{R2} = 30$; $V_{R3} = 60$; $V_{R4} = 60$; $V_{R5} = 150$

6–67. 8 mA

6–69. 19 kΩ

6–71. $V_{R1} = 26.3$; $V_{R2} = 0$; $V_{R3} = 0$; $V_{R4} = 21$; $V_{R5} = 52.7$

6–73. 526.3 mW

6–75. $V_{R1} = 100$ V, all others = 0 V

6–77. R_3 is shorted

6–79. R_4 is open or R_1, R_2, and R_3 are shorted

CHAPTER 7

7–11. c

7–13. c

7–15. d

7–17. c

7–19. a

7–27. 20 Ω

7–29. $R_T = 133.33$ Ω, $I_T = 0.3$ A

7–31. $I_T = 1.55$ A; $P_T = 46.57$ W

7–33. 28 W

7–35. $R_T = 20$ Ω; $I_T = 4$ A

7–37. $R_T = 480$ Ω; $I_T = 62.5$ mA

7–39. 2.637 A

7–41. 37.5 Ω

7–43. 1.5 A

7–45. 162 Ω

7–47. 500 Ω

7–49. 260 mA

7–51. 110 mA

7–53. $I_1 = 150$ mA; $I_2 = 650$ mA; $I_3 = 300$ mA; $I_4 = 500$ mA

7–55. $R_1 = 1$ k; $R_2 = 231$; $R_3 = 500$; $R_4 = 300$ k

7–57. 50 Ω

7–59. 1 A; 1 A

7–61. 100 V

7–63. 33.33 Ω

7–65. 1 A

7–67. 600 Ω each

7–69. $I_{R2} = 6$ A; $R_T = 10$ Ω

7–71. 101 Ω

7–73. Series connected: $R = 6$ Ω; $I = 10$ A; parallel connected: $R = 24$ Ω; $I = 5$ A

7–75. $R_2 = 100$ Ω, 100 V, 1 A; $R_3 = 100$ Ω, 100 V, 1 A

7–77. 100 V; 25 mA

7–79. 0 V

7–81. Current is reduced due to the increase in R_T

7–83. R_2 has opened.

CHAPTER 8

8–11. c

8–13. b

8–15. a

8–17. a

8–19. a

8–31. 1 kΩ

8–33. $V_{R4} = V_B - V_A$

8–35. 800 μA

8–37. V_{R1}, $V_{R2} = 177.2$; V_{R3}, $V_{R4} = 272.8$ V

8–39. 6.23 mA; 7.22 kΩ

8–41. $V_1 = 12.46$; $V_2 = 13.85$; $V_3 = 18.69$; $V_4 = 11.67$; $V_5 = 7$ V

8–43. $I_1 = 28.57$ mA; $I_2 = 4.67$ mA; $I_3 = 9.52$ mA; $I_4 = 14.29$ mA

8–45. $R_1 = 15$ kΩ, 150 V, 10 mA
$R_2 = 200$ Ω, 30 V, 150 mA
$R_3 = 4$ kΩ, 120 V, 30 mA
$R_4 = 1$ kΩ, 120 V, 120 mA
$R_5 = 937.5$ Ω, 150 V, 160 mA

8–47. $R_1 = 9.4$ V; $R_2 = 118.8$ V, $R_3 = 18.8$ V, $R_4 = 11.8$ V, $R_5 = 7.1$ V, $R_6 = 4.7$ V

8–49. 123.1 mA

8–51. $P_1 = 337.5$ W; $P_2 = 253.125$ W; $P_3 = 84.375$ W; $P_5 = 450$ W; $P_T = 1350$ W

8–53. $R_1 = 400$; $V_{R3} = 200$; $I_{R3} = 200$ mA; $I_{R4} = 500$ mA; $R_4 = 500$; $I_{R5} = 500$ mA

8–55. All voltages = 5 V

8–57. 22.22 mA

8–59. 20 mA

8–61. 12 Ω

8–63.

	V_{R1}	V_{R2}	V_{R3}	V_{R4}	V_{R5}
X-grounded	−20	−40	−60	−80	−100
Y-grounded	+40	+20	−20	−40	−60
Z-grounded	+100	+80	+60	+40	+20

8–65. R_1 is reduced to 1/2 of its normal resistance value

8–67. 16 V

8–69. 22.4 V

8–71. Open

8–73. Yes

CHAPTER 9

9–11. c

9–13. a

9–15. d

9–17. d

9–19. d

9–29. 300 V; 50 mA

9–31. 20 mA

9–33. 10 mA

9–35. $V_{R4} = 9.97$; $V_{R5} = 19.93$

9–37. $V_{R1} = 10.13$; V_{R2}, $V_{R3} = 9.935$; $V_{R5} = 19.87$

9–39. $R_1 = 61.5$; $R_2 = 272.7$; $R_3 = 40$; $R_4 = 400$

9–41. $R_1 = 840$; $R_2 = 120$; $R_3 = 80$; $R_4 = 160$

9–43. $I_{R1} = 112$ mA; $I_{R2} = 28$ mA

9–45. $I_1 = 333$ mA; $I_2 = 667$ mA; $I_3 = 1$A

9–47. 0 Ω

9–49. R_2 open

9–51. R_3 open

9–53. 0 V, source is shorted

9–55. R_1 has increased or R_2 has decreased.

CHAPTER 10

10–7. c

10–9. a

10–17. $I_{R1} = 47.3$ mA
$I_{R2} = 118.2$ mA; $I_{R2} = 65.5$ mA

10–21. 1.27 A

10–23. $I_{R1} = 2.82$ A; $I_{R2} = 2.82$ A; $I_{R3} = 2.68$ A; $I_{R4} = 141$ mA

CHAPTER 11

11–11. a

11–13. a

11–15. d

11–23. 318 mA

11–25. 357.8 mA

11–27. 181.8 mA

11–29. $R_{TH} = 166.67$ Ω; $V_{TH} = 26.67$

11–35. 26 V

CHAPTER 12

12–11. b

12–13. b

12–15. c

12–17. b

12–29. 26.46 to 27.54 V

12–31. (a) 0 μA; (b) 900 μA; (c) 99.9 mA; (d) 999.9 mA; (e) 9.9 mA

12–33. 16.67 mA

12–35. One half scale

12–37. Parallel to R_1 at points A and B

12–39. 1 mA

12–41. 20 V

12–43. Voltmeters are connected in parallel

12–45. Voltmeter loading

CHAPTER 13

13–11. c

13–13. b

13–15. b

13–17. d

13–27. 333.3 Ω

13–29. 100 W

13–31. $I_{SC} = 24$ A; $I_{OC} = 0$ A

13–33. Nothing is wrong.

13–35. No

CHAPTER 14

14–11. b
14–13. c
14–15. b
14–21. 26.32 mT
14–23. 40 mT

CHAPTER 15

15–11. d
15–13. b
15–15. d
15–21. 375 At
15–23. 120 At
15–25. 1083.3
15–27. 0.667
15–29. Increased speed will increase voltage and increasing field strength will also increase voltage

CHAPTER 16

16–11. b
16–13. d
16–15. b
16–21. 16 H
16–23. 4 H
16–25. 397 mH
16–27. 0 H
16–29. 500.067 mH
16–31. 20 V
16–33. 1.33 μH
16–35. 5.67 H
16–37. 44.5 mH
16–39. 2.9 H
16–41. 40 mH
16–43. 433 Ω
16–45. Source voltage
16–47. Yes, connect the two in series
16–49. Inductance decreases as the core is removed.

CHAPTER 17

17–11. c
17–13. b
17–15. d
17–17. b
17–19. a
17–31. 50 kHz
17–33. 16.97 V_p

17–35. 250 Hz
17–37. 90.03 V
17–39. 60 Hz
17–41. 200 mA
17–43. 300 Ω
17–45. $V_p = 33.94$; $V_{p-p} = 67.88$; $V_{AVG} = 21.61$ V
17–47. 90°
17–49. 282.8 μA
17–51. 120 Hz
17–53. 5.45 mA
17–55. 1.38 W
17–57. 3.93 mA
17–59. 6.36 V; 0.1667 Hz

CHAPTER 18

18–11. c
18–13. a
18–15. d
18–27. 120 Hz
18–29. 62.12 V
18–37. 2.9 A
18–39. 2.5 A
18–41. 52.08 A
18–43. Rotating field; rotating armature; reverse field and armature
18–45. Use of an electromagnet

CHAPTER 19

19–11. c
19–13. c
19–15. b
19–17. b
19–19. c
19–31. 1:5
19–33. 720 V
19–35. $I_p = 750$ mA; $P_p = 90$ W
19–37. 200 W
19–39. 1:0.577
19–41. 2:3
19–43. 0.83
19–45. 600 V
19–47. 12 V
19–49. 0.9 A
19–51. 7.5:1; 5:1; 1:1; 5:4; 8:25
19–53. 45 V
19–55. 60 W
19–57. 108.42 W
19–61. Open winding (either primary or secondary)

19–63. Shorted turns
19–65. 0 V

CHAPTER 20

20–11. c
20–13. b
20–15. d
20–21. 427 Ω
20–23. 6.786 kΩ
20–25. 754 Ω
20–27. 80 Ω
20–29. 33.33 Ω
20–31. 49 Ω
20–33. 130 mH
20–35. 2.04 A
20–37. $V_{L1} = 76.93$; $V_{L2} = 15.38$; $V_{L3} = 23.07$; $V_{L4} = 7.69$
20–39. Increase
20–41. $L_T = 8.57$ mH; $X_{LT} = 808$ Ω; $X_{L1} = 1414$ Ω; $X_{L2} = 1885$ Ω
20–43. (a) 20; (b) 66.67; (c) 166.67; (d) 1666.67; (e) 250 kΩ
20–45. (a) 15.9 A; (b) 4.77 A; (c) 1.91 A; (d) 191 mA; (e) 1.27 mA
20–47. 15.9 mH

CHAPTER 21

21–11. a
21–13. c
21–15. d
21–17. c
21–19. c
21–31. 10 mA
21–33. 64 Ω
21–35. Sine wave
21–37. 0 Ω
21–39. $I_L = 17.5$ A; $I_R = 466.67$ mA; $I_T = 17.506$ A
21–41. 2093
21–45. 240 Ω
21–47. 80 V
21–49. $I_T = 392$ mA; $V_L = 11.77$; $V_R = 58.83$
21–51. 1.5 A
21–53. 66.67 Ω
21–55. 2
21–57. High
21–59. 0.02 L

CHAPTER 22

22–11. b
22–13. d
22–15. c
22–17. c
22–19. c
22–21. 120 nC
22–23. 0.00024 μF
22–25. 5.6-9.6 μF
22–27. Minimum 20 V each in series, 40 V each in parallel
22–29. 120 V
22–31. 7.21 μF
33–33. 0 F
22–35. 33.33 V
22–37. 0 F
22–39. 2.73 μF
22–41. 0.0245 μF
22–43. C_1 and C_4 = 150 V each; C_2 and C_3 = 75 V each
22–45. Same answer as 22–53
22–47. 21.8 V
22–49. Probably not

CHAPTER 23

23–11. b
23–13. d
23–15. c
23–17. a
23–19. d
23–31. 0.0707 μF
23–33. 1200 Ω
23–35. All voltages are 120 V
23–37. All voltages are 100 V
23–39. 444.44 mA
23–41. 0.14 Ω
23–43. 31.83 Hz; 3.183 Hz; 159.15 Hz
23–45. 226.2 mA
23–47. 1.51 A
23–49. 57.14 Ω

CHAPTER 24

24–11. a
24–13. c
24–15. d
24–17. a

24–19. d
24–31. 1.525 A
24–33. $V_R = 134.16$; $V_C = 67.08$
24–35. $I_R = 1.7$ A; $I_C = 3.41$ A; $Z_T = 39.35$ V
24–39. 120 Hz = 6.63 µF; 400 Hz = 1.99 µF; 1 kHz = 0.796 µF
24–41. 666.67 V; 444.44 V; 66.67 V; 22.22 V
24–43. $I_{CHARGE} = 277.78$ µA; $I_{FALL} = 2.5$ mA
24–45. $Z_T = 94.34$ Ω; $I_T = 1.27$ A
24–47. $Z_T = 1.58$ kΩ; $I_T = 151.79$ mA
24–49. $I_T = 333.1$ mA; $Z_T = 474.34$ Ω

CHAPTER 25

25–11. b
25–13. a
25–15. c
25–17. b
25–19. b
25–31. 187.5 µs
25–33. 12.97 V; 14.25 V; 14.72 V
25–35. 1.5 ms
25–37. 0.25 psec
25–39. 10 msec
25–41. 18.96 V; 25.94 V; 28.5 V; 29.45 V; 29.8 V
25–43. 9.2 V
25–45. 4.27 V; 24.5 V; 25 V
25–47. 1 µs
25–49. 15 µs
25–51. 5 mA; 5 A
25–53. 4 µs
25–55. 20 µs

CHAPTER 26

26–11. d
26–13. d
26–15. c
26–21. 720 Ω
26–23. 130 Ω
26–25. 1.038 A
26–27. $I_R = 375$ mA; $I_{C1} = 750$ mA; $I_{C2} = 1$ A; $I_{L1} = 333$ mA; $I_{L2} = 3$ A
26–29. 1.63 A
26–31. 100.1125 Ω
26–33. 67.22 mA
26–35. $I_A = 800$ mA; $I_C = 1.3$ A; $I_L = 63.66$ mA

CHAPTER 27

27–11. a
27–13. b
27–15. c
27–17. b
27–19. a
27–31. (a) $7.2\,\angle 33.69°$
(b) $11.2\,\angle 26.6°$
(c) $10.63\,\angle -48.8°$
(d) $80.6\,\angle -29.74°$
27–33. 47.17 Ω
27–35. $40 - j25$
27–37. $50 + j25$
27–39. 44.72 Ω
27–41. $50 + j100$
27–43. $7.07\,\angle 45°$
27–45. 85 Ω

CHAPTER 28

28–11. a
28–13. c
28–15. d
28–17. a
28–19. d
28–27. (a) $7.5\,\angle 0°$; (b) $10\,\angle 90°$; (c) $3\,\angle -90°$
28–29. 10 A
28–31. $10 - j4$; $10.77\,\angle -21.8°$
28–33. $V + V_R + V_L + V_C = 0$
28–35. $V_L = 54.72$; $V_R = 12.31$
28–37. Lagging

CHAPTER 29

29–11. c
29–13. c
29–15. a
29–17. c
29–19. b
29–31. 1.5 kΩ
29–33. 13.46 Hz
29–35. 13.46 Hz
29–37. 5.9 to 6.1 Hz
29–39. 600 Hz
29–41. $3911 - 4089$ Hz
29–43. 2 kHz
29–45. $1990 - 2010$ Hz
29–47. 200
29–49. 0.15 A

CHAPTER 30

30–11. b
30–13. d
30–15. a
30–17. d
30–19. b
30–31. 1.06 V
30–33. +47.5 V to + 32.5 V

CHAPTER 31

31–11. b
31–13. a
31–15. d
31–17. a
31–19. a
31–31. 95 to 98 mA
31–33. 40 mA
31–35. Yes
31–37. No
31–39. 2.8 Ω
31–41. No
31–43. Bad diode
31–45. Bad (leaking) transistor
31–47. Yes
31–49. Yes

CHAPTER 32

32–11. a
32–13. c
32–15. d
32–17. b
32–19. a
32–31. 0.02 A
32–33. 3 V_{p-p}
32–35. 28,125,000
32–37. 0.7 V
32–39. 150
32–41. No
32–43. Open diode or open transformer
32–45. No
32–47. Yes
32–49. Open transistor emitter to collector

CHAPTER 33

33–11. d
33–13. a
33–15. a
33–17. c
33–19. a
33–31. 32.4 V
33–33. +2.5 V
33–35. +5 V
33–37. +4 V
33–39. +3 V
33–41. −4 V
33–43. 26
33–45. −180°
33–47. In phase

Index

R

Radial leads, component, 93
Radio frequency choke, 428
Range extension, meter, 336, 343
Ratings, voltmeter, 342
RC:
    circuits, 643
    time constants, 638, 643
    waveshapes, 643
Reactance, 518
    formula, 597
    frequency, 598
    parallel, 599
    series, 599
Reactive circuit power, 667
Rectangular form, complex number, 680
Rectangular-to-polar conversion, 688
Rectification:
    half-wave, 806
    full-wave, 805
Rectifier, 52, 782, 784, 699
    full-wave bridge, 809
    full-wave center-tapped, 807
Relative permeability, 394
Relay, 59
Resistance, 22, 76, 123
    coil, 435
    instrument, 349
    parallel wired, 148, 270
    series wired, 142
    unknown branch, 187
    voltmeter, 340
Resistor, 81
    adjustable, 95
    axial lead, 93
    bank, 214
    carbon composition, 81
    combinations, 102
    current sensitive, 99
    failures, 93, 99
    film type, 82, 89
    fusible, 101
    low power, 81
    markings, 86
    multiplier rating, 88
    photo, 99
    power, 82
    radial lead, 93
    rated, 76
    thermistor, 99
    tolerance, 85
    variable, 76, 95
    voltage dropping, 160

voltage-dependent, 99
    zero-ohms, 91
Resonance, 730
    parallel, 733
    series, 734
Resonant:
    circuit, 434, 730
        bandwidth, 735
    filter, 765
Reverse biased, 782
Reverse voltage, 783
RF choke, 428
Rheostat circuit, 97
RLC circuit, 694
Robotics, 6
Root-mean-square value, 453
Rotating field generator, 472

S

Safety, 345
Safety devices, 270
Saturation, 397, 789
Saturation voltage, 833
Sawtooth wave, 459
Schematic diagram, 51
Secondary:
    cell, 363
    loading, 501
    winding, 497
Segmented commutator, 30
Selectivity, 738
Semiconductor, 20, 22, 40, 780
Sensitivity rating, meter, 328
Series, 207
    capacitances, 578
    circuit, 66, 130, 142, 264, 271
        Ohm's law, 150
    power, 157
    reactances, 615, 658
    resistances, 145, 615
    resistance banks, 218
    resonance, 736
        current, 731
        filters, 765
        impedance, 732
        voltage and current, 732
    string, 214
    voltage drops, 147
Series-parallel, 206, 214, 237
    analysis, 210
    current, 213
    identification, 207